World Regional Geography Book Series

Series Editor

E. F. J. De Mulder, DANS, NARCIS, Utrecht, Noord-Holland, The Netherlands

What does Finland mean to a Finn, Sichuan to a Sichuanian, and California to a Californian? How are physical and human geographical factors reflected in their present-day inhabitants? And how are these factors interrelated? How does history, culture, socio-economy, language and demography impact and characterize and identify an average person in such regions today? How does that determine her or his well-being, behaviour, ambitions and perspectives for the future? These are the type of questions that are central to The World Regional Geography Book Series, where physically and socially coherent regions are being characterized by their roots and future perspectives described through a wide variety of scientific disciplines. The Book Series presents a dynamic overall and in-depth picture of specific regions and their people. In times of globalization renewed interest emerges for the region as an entity, its people, its landscapes and their roots. Books in this Series will also provide insight in how people from different regions in the world will anticipate on and adapt to global challenges as climate change and to supra-regional mitigation measures. This, in turn, will contribute to the ambitions of the International Year of Global Understanding to link the local with the global, to be proclaimed by the United Nations as a UN-Year for 2016, as initiated by the International Geographical Union. Submissions to the Book Series are also invited on the theme 'The Geography of…', with a relevant subtitle of the authors/editors choice. Proposals for the series will be considered by the Series Editor and International Editorial Board. An author/editor questionnaire and instructions for authors can be obtained from the Publisher.

This book series is published in cooperation with the International Geographical Union (IGU). The IGU is an international, non-governmental, professional organization devoted to the development of the discipline of Geography. The purposes of the IGU are primarily to promote Geography through initiating and coordinating geographical research and teaching in all countries of the world.

More information about this series at https://link.springer.com/bookseries/13179

Haris Gekić • Aida Bidžan-Gekić
Nusret Drešković • Ranko Mirić • Péter Reményi

The Geography of Bosnia and Herzegovina

Between East and West

Springer

Haris Gekić
Department of Geography
University of Sarajevo
Sarajevo, Bosnia and Herzegovina

Aida Bidžan-Gekić
Department of Geography
University of Sarajevo
Sarajevo, Bosnia and Herzegovina

Nusret Drešković
Department of Geography
University of Sarajevo
Sarajevo, Bosnia and Herzegovina

Ranko Mirić
Department of Geography
University of Sarajevo
Sarajevo, Bosnia and Herzegovina

Péter Reményi
Department of Political Geography,
Development and Regional Studies
University of Pécs
Pécs, Hungary

ISSN 2363-9083 ISSN 2363-9091 (electronic)
World Regional Geography Book Series
ISBN 978-3-030-98522-6 ISBN 978-3-030-98523-3 (eBook)
https://doi.org/10.1007/978-3-030-98523-3

Cover illustration: Anton Balazh-Fotolia.com

This Springer imprint is published by the registered company Springer Nature Switzerland AG
The registered company address is: Gewerbestrasse 11, 6330 Cham, Switzerland

Preface

In front of you is the first complete book on the geography of Bosnia and Herzegovina, which in a systematic way presents its most important geographical characteristics, starting from the geographical position all the way to the geographical regions. Bosnia and Herzegovina is unfortunately best known for the war in the first half of the 1990s. We hope that this book will contribute to raising the profile of Bosnia and Herzegovina and show readers the numerous geographical advantages and interesting potentials it has.

Geography as a fundamental and applied science, general education, and professional subject in its essence and position in the system of sciences has played a very prominent role in shaping our view of the world. Determining cause-and-effect relationships and the complexity of processes in geographical space and the human environment is a fundamental task of geography. Observing and meaningfully explaining the conflicts between nature and humans, caution in exploitation, and careful management of space require fundamental preliminary work, which is nurtured in its research by geography. In this book, we have presented the geographical characteristics of Bosnia and Herzegovina and the key problems in the solution of which geographical science is of great importance.

This book, *The Geography of Bosnia and Herzegovina*, is the result of many years of work on researching the geographical features of this very interesting country. Numerous field observations were made, which were stationary and route type in different seasons.

By its character, the book is also a scientific documentary, because due to its scientific basis and high informativeness and connection, it certainly represents the interest of economic, social, and political subjects. It was done with the purpose of contributing to finding the best solutions in further economic and social, and otherwise sustainable development. The book can be of great use in the development of various spatial and regional plans and projects, and significantly contribute to the improvement of social development in Bosnia and Herzegovina.

We especially emphasize the importance of the book for geographers, teachers, and students of geography and other sciences who, for the first time, have the opportunity to get acquainted with Bosnia and Herzegovina in a systematic geographical way; therefore, the book can serve as a university textbook as well. We hope that readers around the world will get better acquainted with the geographical reality of Bosnia and Herzegovina and especially its natural and anthropogenic beauties through the book. Bosnia and Herzegovina, in terms of its socio-geographical characteristics, is certainly in the true sense a geographical space between West and East.

The book comprises a total of 15 chapters grouped into 5 parts according to content. The first part consists of two chapters dealing with geographical position, size, shape, spatial coverage, and boundaries. The second part is devoted to physical-geographical characteristics. It consists of six chapters dealing with geology, geotectonics, geomorphology, climate, water, soil, flora, fauna, and protected areas. The third part deals with human geography and consists of four chapters, starting from historical geography through political geography and population geography to urban and rural geography. The fourth part explains the main economic characteristics of Bosnia and Herzegovina in two chapters, natural resources and economic development. The final part deals with regionalization and regions of Bosnia and Herzegovina.

Four of the five authors of this book are employed by the Department of Geography, Faculty of Science, University of Sarajevo. The study of geography in Bosnia and Herzegovina began with the establishment of the Cathedra of Geography in the Faculty of Philosophy, University of Sarajevo, in 1950. With the establishment of the Faculty of Science in Sarajevo, as a separate teaching and scientific institution in 1960, the Cathedra of Geography grew into the Department of Geography, the most revered higher education institution in Bosnia and Herzegovina in the field of geographical science.

In the modern period, the teaching-scientific process at the level of the Department of Geography is organized through the following sub-organizational units: Cathedra of Physical Geography, Cathedra of Human Geography, Cathedra of Regional Geography, Cathedra of Geology, GIS Center, Institute of Regional and Spatial Planning, and Cabinet for Geography in Education. The teaching process takes place in three study directions: geography in education, regional and spatial planning, and tourism and environmental protection, through all three cycles of study according to the Bologna principles.

The geographers of the Department also worked through the Geographical Society of Bosnia and Herzegovina. The scientific journal *Geographical Review* has been published since 1957. After the 1992, the Geographical Society of the Federation of Bosnia and Herzegovina, in co-organization with the Department of Geography, Faculty of Science, University of Sarajevo, have organized five congresses of geographers of Bosnia and Herzegovina with international participation.

Sarajevo, Bosnia and Herzegovina — Haris Gekić
Sarajevo, Bosnia and Herzegovina — Aida Bidžan-Gekić
Sarajevo, Bosnia and Herzegovina — Ranko Mirić
Sarajevo, Bosnia and Herzegovina — Nusret Drešković
Pécs, Hungary — Péter Reményi

Acknowledgments

The photographs in this book are the result of numerous field research conducted by us over the past few years. We tried to bring the most important geographical objects, phenomena, and processes in Bosnia and Herzegovina as close as possible to the photographs, thus making the book as interesting as possible to read.

We would like to thank all the reviewers for their time and effort in giving valuable comments and suggestions as well as professional assistance, which in methodological and substantive terms have improved the quality of the book.

We sincerely thank Naida Softić and Slaviša Vračar for the translation.

Special thanks to the Agency for Statistics of Bosnia and Herzegovina, the Institute for Statistics of the Federation of Bosnia and Herzegovina, and the Institute of Statistics of Republika Srpska, as well as numerous ministries and professional institutes in Bosnia and Herzegovina and their departments, which enriched the content of the book with their statistical and other data.

On a personal front, the extremely great efforts in writing the book and the very completion of the work were possible thanks to the constant inspiration and boundless support of our families, whom we sincerely thank. The research conducted required a lot of effort, patience, self-discipline, and sacrifice.

Haris Gekić
Aida Bidžan-Gekić
Ranko Mirić
Nusret Drešković
Péter Reményi

Contents

About the Authors

Haris Gekić was born in 1984 in Zenica, Bosnia and Herzegovina. He completed his primary and secondary schooling in Bugojno. He studied geography at the University of Sarajevo. Haris graduated in 2007, completed his master's thesis in 2011 and doctoral dissertation in 2016 in the field of human geography. He has been working at the Department of Geography at the Faculty of Science, University of Sarajevo, since 2007, currently as associate professor of human geography. He is chief of Cathedra for Human Geography in the Department of Geography. Haris is deputy chief editor of *Geografski pregled*, the oldest geography scientific journal in Bosnia and Herzegovina. His research interests include population geography, urban geography, rural geography, and political geography of Bosnia and Herzegovina. He is one of the chief organizers of the Congress of Geographers of Bosnia and Herzegovina. Haris is the author of 5 books and more than 50 papers. He is a member of the Geographical Society of the Federation of Bosnia and Herzegovina. Haris has participated in six COST EU projects.

Aida Bidžan-Gekić was born in 1982 in Sarajevo, Bosnia and Herzegovina. In 2005, she completed an undergraduate program in the Department of Geography, Faculty of Science, University of Sarajevo. She received her master's degree (2011) and doctorate (2016) from the Department of Geography. She has been working at the Department of Geography at the Faculty of Science, University of Sarajevo, since 2008, currently as associate professor of regional geography. Aida is technical editor of *Geografski pregled*, the oldest geography scientific journal in Bosnia and Herzegovina. The spheres of her scientific research are world regional geography and tourism geography of Bosnia and Herzegovina. She is member of the Geographical Society of the Federation of Bosnia and Herzegovina. Aida is one of the organizers of the Congress of Geographers of Bosnia and Herzegovina. She is the author of more than 50 papers and 4 books. Aida has participated in four scientific projects.

Ranko Mirić was born in 1965 in Visoko, Bosnia and Herzegovina. He studied geography at the University of Sarajevo. Ranko graduated in 1994 and started his doctoral thesis in the field of regional geography. He has been working at the Department of Geography at the Faculty of Science, University of Sarajevo, since 1996, currently as associate professor of regional geography. He is president of the Geographical Society of the Federation of Bosnia and Herzegovina and chief editor of *Geografski pregled*, the oldest geography scientific journal in Bosnia and Herzegovina. He was vice dean of the Faculty of Science, University of Sarajevo, from 2016 to 2020. His field of research covers regional geography of Bosnia and Herzegovina, tourism regions, geographical regionalization, and regional geography of the European Union. He is one of the chief organizers of the Congress of Geographers of Bosnia and Herzegovina, and the International Tourism and Hospitality Management Congress. Ranko is the author of more than 100 papers.

Nusret Drešković was born in 1966 in Plav, Montenegro. He graduated in geographical studies from the Department of Geography, Faculty of Science, University of Sarajevo, in 1994. He has been working at the Department of Geography at the Faculty of Science, University of Sarajevo, since 1995, currently as full professor of physical geography and GIS. He has participated in more than 25 scientific projects. He was president of the Geographical Society of the Federation of Bosnia and Herzegovina for 8 years. Currently, he is dean of the Faculty of Science, University of Sarajevo. His field of research covers climatology, meteorology, and GIS. Nusret is one of the chief organizers of the Congress of Geographers of Bosnia and Herzegovina, and the International Tourism and Hospitality Management Congress. He is the author of more than 100 papers and several books.

Péter Reményi was born in 1978 in Pécs, Hungary. He received his master's degree in geography and history in 2002 and 2017, respectively, from the University of Pécs. Péter received his doctorate in geography from the Doctorate School of Earth Sciences at the University of Pécs in 2009. He has been working at the Institute of Geography at the University of Pécs since 2005, currently as associate professor. His research interest covers political geography, Balkan studies, state and nation building and border studies. He is one of the chief organizers of the Hungarian Conference on Political Geography, deputy director of the Center for Eastern Mediterranean and Balkan Studies, editor of the Mediterranean and Balkan Forum, member of the Hungarian Geographical Society. Péter is the secretary of the Political Geography Sub-committee of the Committee of Social Geography of the Hungarian Academy of Sciences. He is the author of more than 100 papers.

Part I

Geographical Position, Borders and Size of Bosnia and Herzegovina

1 Geographical Position, Spatial Coverage, and Size of Bosnia and Herzegovina

Abstract

This chapter deals with the issue of the geographical position of Bosnia and Herzegovina. The geographical position is a complex category that shows the spatial relationship of one phenomenon, object, or locality with other phenomena, objects, or localities. Depending on the research objective, this category is determined, inter alia by an analytical study of the features of mathematical geography, natural, geographical, socio-economic-geographical, and political geographical elements, but also the regional geographical position as a synthetic result. This chapter elaborates not only the main elements of the geographical position of Bosnia and Herzegovina, but also the size and shape of the territory and spatial coverage, which are presented from the point of view of their specific nature with regard to the position of Bosnia and Herzegovina in the European and global contexts. Bosnia and Herzegovina has a very significant geopolitical position in contemporary times where the influences of the most powerful global forces intertwine.

Keywords

Territory of Bosnia and Herzegovina · Extreme points · Landscape · Shape · Tourist-geographical position

1.1 Geographical Position, Size, and Extreme Points

Bosnia and Herzegovina is a European country that existed as a sovereign state way back in the early Middle Ages. Its international sovereignty was reaffirmed in modern times in 1992, when it was admitted to membership of the United Nations as a full-fledged member. The 1995 Dayton Peace Agreement administratively divided Bosnia and Herzegovina into two entities: the Federation of Bosnia and Herzegovina and the Republika Srpska, while in 2000 the territory of Brčko in the northeast became a separate district.

The name Bosnia and Herzegovina is singular. It refers to a single territory. However, sometimes in the everyday life of the population, the southern part of the country is called Herzegovina, but there is no clearly defined and established border. Today's name of the state of Bosnia and Herzegovina was created at the time when the then Austro-Hungarian Monarchy was given a mandate for its occupation at the Berlin Congress in 1878. As part of the occupation mandate, the new name of the country was actually coined and used in the official version in German—*Bosnien und Herzegowina*. The real name of the current territory of Bosnia and Herzegovina until the Berlin Congress existed under the name of Bosnia. The origin of the words "Bosnia" and "Herzegovina" will be discussed in more detail in the chapter about historical geography.

Bosnia and Herzegovina has a very important geographical position on a European scale, since it includes the contact area with the wider area of the Pannonian Basin (area close to the river Sava in the north) and the Central Mediterranean (Adriatic Sea in the south). The dominance of the Dinarids among these regional units is the third important determinant of its geographical position. As part of the establishment of the European network of corridors, the territory of Bosnia and Herzegovina represents one of the links of Central and Eastern Europe with the Mediterranean. Bosnia and Herzegovina is situated in the central part of the northern temperate climate zone, i.e., between 42° N and 45° N and 15° E and 19° E. The maximum meridian (longitudinal) extent of Bosnia and Herzegovina is about three arc degrees, while in terms of its latitudinal extent that is featured by parallels, it comes close to four arc degrees.

Bosnia and Herzegovina's regional-geographical position itself is highly complex. Its affiliation with the European Geo-space implies its inclusion as part of traditional classification in the region of Southern Europe, while the modern

H. Gekić et al., *The Geography of Bosnia and Herzegovina*, World Regional Geography Book Series,
https://doi.org/10.1007/978-3-030-98523-3_1

political, economic, and social elements place this country in the lower hierarchical regional framework of Southeast Europe. Also, the articulation of its position in terms of the physiognomic (and often political) aspects gives it the strong regional attributes of a Balkan country, which is more specific and quite commonly defined and identified in the literature by the term of the so-called Western Balkans.

With its total size of its territory of 51,209.2 km^2, Bosnia and Herzegovina is a predominantly continental country. The mainland accounts for 51,197 km^2, which is approximately 99.98% of the country's size. Only 12.2 km^2 (about 0.2% of the total size of Bosnia and Herzegovina's territory) in the area of the Neum-Klek Bay accounts for the maritime waters, through which Bosnia and Herzegovina exercise the right of access to international waters. By the criteria of its location, it represents for the most part an area affiliated with the mid-mountain relief, where there is a clear domination of altitudes over 500 m above sea level. The lowland relief (up to 200 m above sea level) accounts for only 13.53% of the territory. By the criteria of the total size of its territory, it ranks as the 125th country in the world, seen from the point of view of ranking the sovereign states and dependent territories. In terms of the European continent, Bosnia and Herzegovina occupy the 27th place, according to the size of the sovereign country's territory. Considering that the 45th parallel runs through our extreme northern areas (north of Cazin, Prijedor and Derventa), Bosnia and Herzegovina is situated closer to the equator compared to its distance from the northern geographic pole (Fig. 1.1).

Bosnia and Herzegovina is a southern European country situated in the western part of the Balkan Peninsula, more precisely within the block of geographic coordinates (φ, λ) 42°33′00″ and 45°16′30″ N and 15°44′00″ and 19°37′41″ E. Its shape resembles a right-angled triangle, which is also the main feature of the state emblem of Bosnia and Herzegovina (its state flag and coat of arms). Its approximate position is equidistant from the equator and the north geographic pole, but most of its territory (over 90% of the area) is situated south of 45° north latitude (Table 1.1). The distance from the endpoints in the North-South direction is about 314 km and in the East-West direction about 309 km (Fig. 1.2).

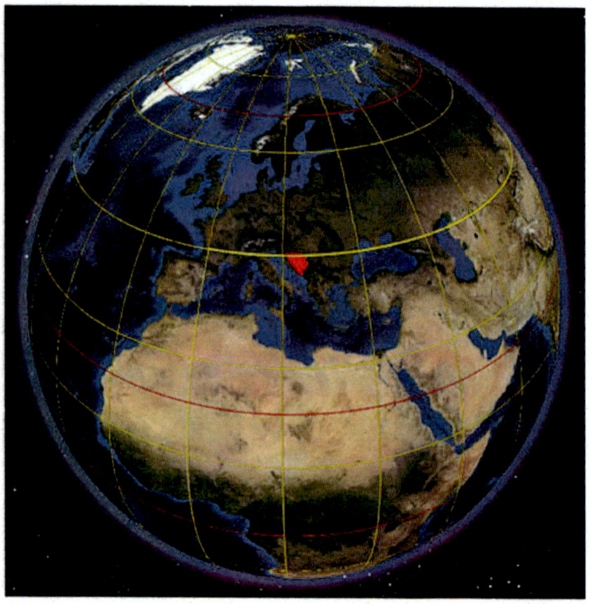

Fig. 1.1 Mathematical geographic position of Bosnia and Herzegovina. (Drešković and Mirić 2017)

1.2 Shape

The shape of the country, except for the external morphological appearance, at first glance does not reflect some deeper political-geographical features. However, this aspect of the country does not remain without its influence on the internal administrative, transportation, social, and economic organization. A compact shape, similar to the appearance of a regular polygon, is favorable because it creates potentially good preconditions for the optimal development of the internal spatial organization.

In terms of its shape, Bosnia and Herzegovina belongs to the category of compact countries where the distance between their geometric center and peripheral or bordering zones is approximately equal in all directions. For Bosnia and Herzegovina, this is an average of 110–130 km, bearing in mind that the geographical center of Bosnia and Herzegovina is located in the area of Krčevine, Municipality of Vitez (Jahić et al. 2014). Although in terms of internal organization, compact countries are the most favorable, in reality, it often happens that the areas with the highest population density or social and economic dynamics are on the very edge of the territory, causing the formation of different social and cultural spaces within the country. This is not the case with Bosnia and Herzegovina, but unfortunately, this advantage was disrupted mainly by establishing an inter-entity boundary line in late 1995. To some extent, the shape of Bosnia and Herzegovina resembles a triangle, with a vertex at the border with the Republic of Serbia in the town of Bijeljina, and the base in Velika Kladuša on the border with the Republic of Croatia, and in Trebinje on the border with the Republic of Montenegro.

If we compare the lengths of the border with the area of the territory, the result is 3 km of the border per 100 km^2 of the territory. The elongation of the territory is approximately equal in directions. The elongation index of the territory is 1.37 and, therefore, Bosnia and Herzegovina has a compact shape—i.e., when all boundaries are approximately equal (elongation index 1–1.5). The coefficient of compactness of the country, which we define as the ratio of the area of the country to the area of the circle, having the same perimeter, for Bosnia and Herzegovina is almost equal to 100, indicating

Table 1.1 Geographical coordinates of extreme points in Bosnia and Herzegovina

Direction	North latitude	East longitude[a]	Municipality	Settlement
North	45° 16′ 30″	16° 55′ 56″	Kozarska Dubica	Gradina Donja
South	42° 33′ 00″	18° 32′ 24″	Trebinje	Podštirovnik
East	44° 03′ 00″	19° 37′ 41″	Bratunac	Žljebac
West	44° 49′ 30″	15° 44′ 00″	Bihać	Bugar

According to FZS (2020)
[a]By Greenwich

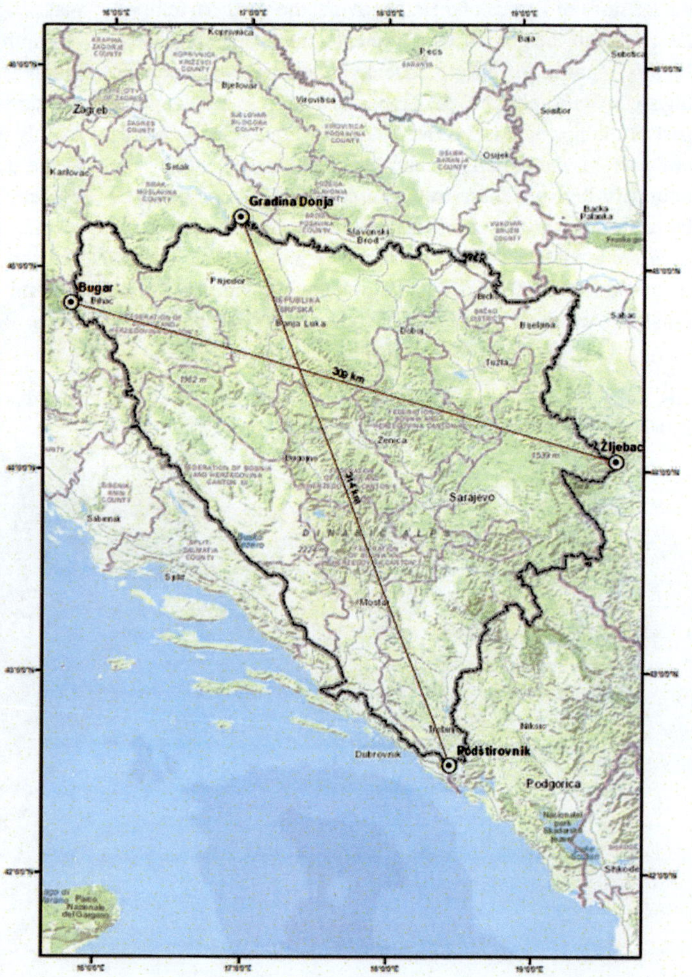

Fig. 1.2 Extreme geographical points and their air distance in Bosnia and Herzegovina

the precise compactness of the territory. Unfortunately, due to the lack of internal cohesion, the compactness of the territory loses its significance.

1.3 Landscape Zones and Spatial Coverage

The territory of Bosnia and Herzegovina is situated in the transition area from the northern subtropical to the northern temperate landscape. The territory of Bosnia and Herzegovina lies approximately in the middle of the northern hemisphere on the Eurasian continental bloc. These basic elements are underlying the essence of natural-geographical features and regional-geographical structure of Bosnia and Herzegovina. In terms of geotectonic-geomorphological determinants, the territory of Bosnia and Herzegovina belongs to the mountain system of the Dinarids as the dominant geotectonic unit of the Balkan Peninsula, i.e. to the western part of the Mediterranean zone of a geologically younger mountain range (a zone within the Alpine-

Himalayan orogen). The mountain system of the Dinarids covers the largest central part of Bosnia and Herzegovina; this system gradually descends towards the north where it passes over to the peripheral parts of the Pannonian Plain (the so-called Peripanonian Bosnia), while in the southwest and south it descends abruptly (and sometimes steeply) towards the Adriatic Bay.

Due to the determinants of the astronomical-geographical position, the clearly pronounced influences of two geographical zones of the northern hemisphere alternate regularly in the territory of Bosnia and Herzegovina: the northern temperate zone and the northern subtropical zone change regularly. In this regard, we differentiate two landscape zones with highly specific physical-geographical features: the southern part of the northern temperate landscape zone and the northern part of the northern subtropical landscape zone. The boundary between these two landscape zones is represented by a wide transition zone that roughly coincides with the orographic watershed of the Black Sea and Adriatic basins. In the territory of Bosnia and Herzegovina, we can distinguish two physical geographical landscape regions within the northern temperate landscape zone: Bosnian-Pannonian Landscape Region with subatlantic-western area (Kozara-Prosara, Lijevče-Motajica, Šamac-Brčko) and continental-east area (Semberija); and the Inner Dinaric Landscape Region with subatlantic-western area (Bosnian Krajina, Central Bosnia) and continental-east area (Eastern Bosnia) (Fig. 1.3).

The southern border of the Bosnian-Pannonian region has a predominantly comparative direction of extension and its shape mainly follows the morphology of the major mountain massifs along the southern edge of the Pannonian Basin (Konjuh, Ozren, Borja, Uzlomac, Manjača, and Majdan planina). The border separating it from the northern subtropical landscape zone is, in fact, a zone of variable widths that mainly follows the direction of the dominant mountain peaks at the transition between the Inner and Outer Dinarids. The climatic influences of both zones are also the basis used to identify this border line. The Northern Subtropical Landscape Zone in Bosnia and Herzegovina is represented by the Outer Dinaric Landscape Region, which includes the following areas: Southwestern Bosnia and Herzegovina, Highland

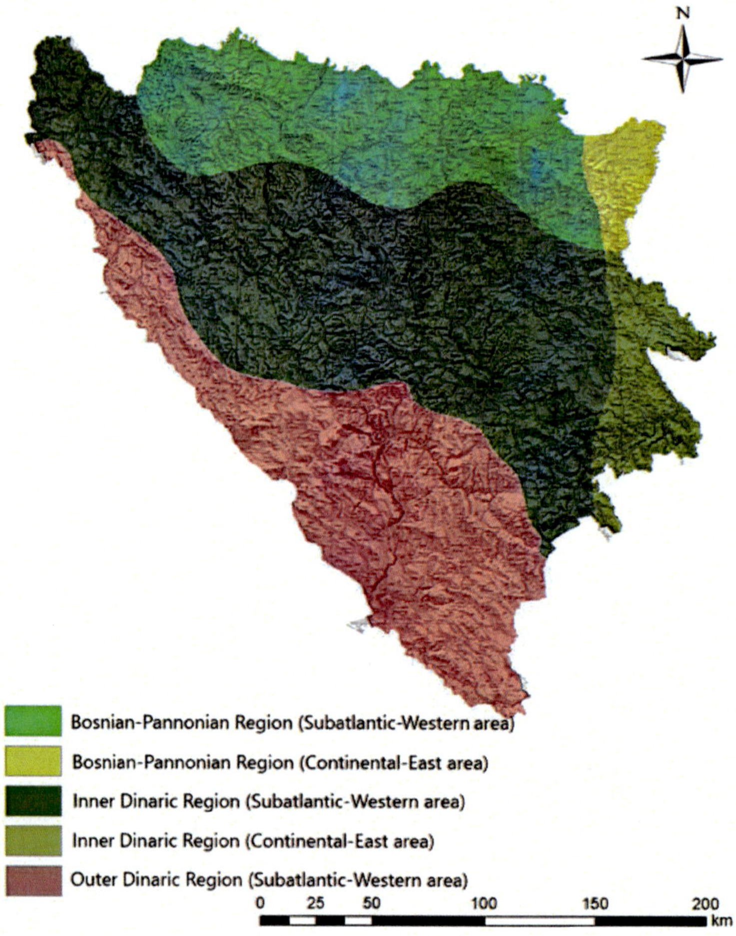

Fig. 1.3 Landscape regions of Bosnia and Herzegovina

Bosnia and Herzegovina, and Lowland Litoral Herzegovina (Drešković and Mirić 2017).

Bosnia and Herzegovina is not a landlocked state per definition; however, its short corridor to the Adriatic does not provide it an economically viable exit to the international waters. The situation seems even less favorable by the construction of the Croatian bridge to the Pelješac peninsula over the entrance of the bay. Bosnia and Herzegovina also has one exclave. It is the village of Međurečje, which is entirely surrounded by the territory of the municipality of Priboj in the Republic of Serbia. During the former SFR Yugoslavia, the Municipality of Priboj built the complete infrastructure in Međurečje (buildings, school, ambulance, police station, and other facilities) to institute the headquarters of its local community Sastavci, which surrounds Međurečje. Međurečje is officially a populated place in the municipality of Rudo in eastern Bosnia and Herzegovina. The village area is about 396 ha, and according to the 2013 Census in Bosnia and Herzegovina, it had 176 inhabitants. Traffic connections, communal infrastructure, electricity, and telecommunication network of the village are oriented towards the municipality of Priboj. Residents of Međurečje pay taxes in Bosnia and Herzegovina. Austro-Hungarian Empire separated this place from the Ottoman Empire as early as 1878. According to locals, a Bosnian fugitive gave one of his wives 400 hectares of land in Sastavci, and since the annexation of Bosnia and Herzegovina, the property has followed the fate of Bosnia and Herzegovina (Fig. 1.4).

The main climate properties of the regional-geographical position of Bosnia and Herzegovina are conditioned by its astronomical-geographical location in the central parts of the European continent, the distribution of air masses, and the features of relief as the most significant modifier. It belongs to an area that is situated in the contact zone between the extreme southern parts of the northern temperate and the extreme northern parts of the northern subtropical zone. The

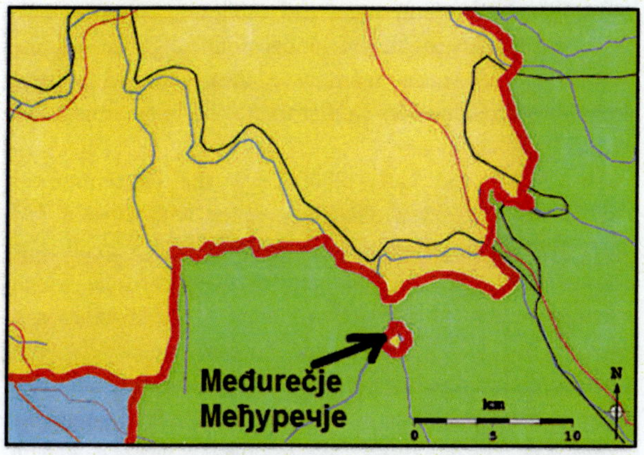

Fig. 1.4 Međurečje, the exclave of Bosnia and Herzegovina in the Republic of Serbia. (Licence Creative Commons 1.0 Map of Međurečje)

effect of its locational distribution is that the dominant climate type in the territory of Bosnia and Herzegovina (over 75% of its territory) is represented by moderately warm rainy climates (climate type C).

In terms of hydrography, most of the territory of Bosnia and Herzegovina belongs to the Black Sea basin, and to a lesser extent to the Adriatic basin. Such a hydrographic position of Bosnia and Herzegovina determines its geographical affiliation with the Posavina-Danube region in the north and the Adriatic-Mediterranean region in the south. The specificity of the relief structure has created a disproportion of the spatial coverage between these basins, as a result of which about three quarters (or 74.7%) of Bosnia and Herzegovina's territory belongs to the Black Sea basin and about one quarter (or 25.3%) to the Adriatic Sea basin. The basic biogeographical characteristics of the geographical position are determined by the location of its territory in the Holarctic biogeographical realm. There are 3 biogeographical regions that are typical for the territory of Bosnia and Herzegovina: Mediterranean region, Euro-Siberian-Boreo-American region, and Alpine-High Dinaric region (Fig. 1.5). Within these regions, there are also six biogeographical areas of Southeast Europe (Pannonian, pre-Pannonian, Illyrian-Moesian, Dinaric, pre-Adriatic, and Adriatic) that stretch through Bosnia and Herzegovina.

1.4 Geopolitical Position

Viewed in the light of Bosnia and Herzegovina's geopolitical position, its position is clearly determined by the factors that can in a certain sense be regarded as unfavorable, due to the possibility of radicalization of geopolitical and geostrategic relations in the region and a number of pronounced secessionist processes ongoing in the country, which has a destructive effect on its political stability and makes difficult the country's incorporation into Euro-Atlantic integration. The main historical-geographical properties of Bosnia and Herzegovina are a reflection of the geographical and historical-cultural habitus of Bosnia and Herzegovina's society that exists in a territory where there has been a centuries-long strong process of merging and permeation of civilizational and cultural influences of the European and Asian nations and their leading monotheistic religions. Based on these elements, at the current moment of development it is possible to identify Bosnia and Herzegovina's affiliation with the European, Mediterranean, and Turkish-Oriental civilization-cultural circle, which results in a very complex multicultural, multi-denominational, and multiethnic state union.

Bosnia and Herzegovina is a small country in terms of population. According to the results of the 2013 Population Census, Bosnia and Herzegovina had 3,531,159 inhabitants, which results in an average population density of about

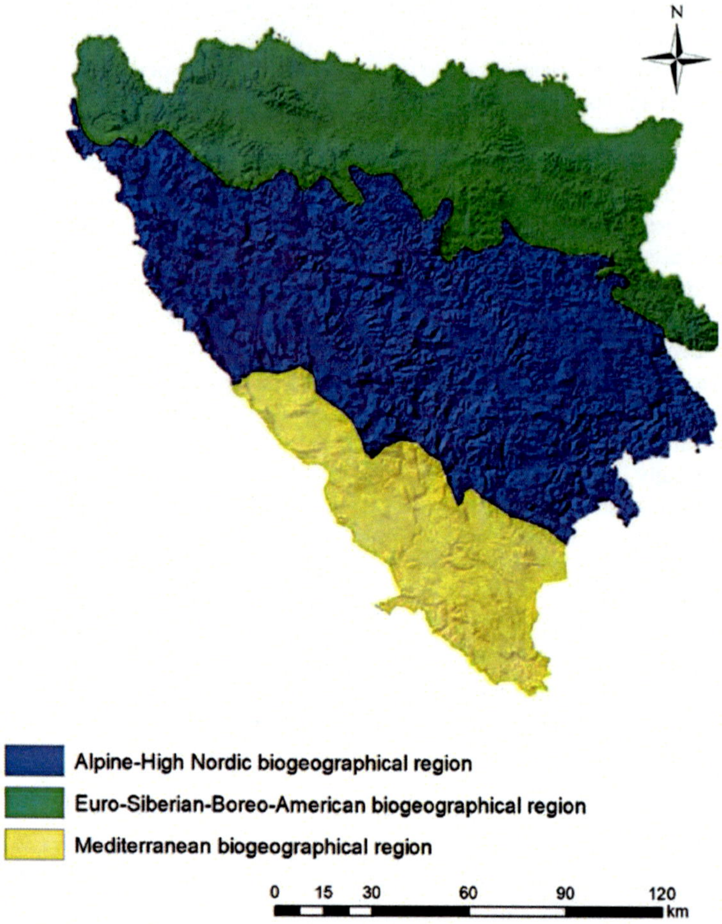

Fig. 1.5 Biogeographical regions in Bosnia and Herzegovina

69 inhabitants/km^2. About 43% of the total population reside in urban settlements of Bosnia and Herzegovina today (BHAS 2016). It is located on the very border of the European Union. The area it occupies serves as a land bridge between East and West. From the earliest times to the most recent times, strong economic and ethnic currents have been circulating throughout this area. Throughout the last few centuries, Bosnia and Herzegovina has maintained the continuity of regional political individuality (Klemenčić 1997).

Bosnia and Herzegovina is a country in South Eastern Europe on the Balkan Peninsula. In a historical perspective, the Balkans used to be a central region in development terms, but at least since the eighteenth century, it became a periphery/semi-periphery of ascending Western Europe. The Balkans, from a geopolitical perspective, is a buffer zone where spheres of various power centers intersect. The geographical basis of this is the location of the peninsula, at the crossroads of civilizations, continents, major transcontinental maritime, and land routes as well as lying in between traditional power centers. Controlling the Straits has been a power struggle between the maritime great power of the times (Britain or USA) and Russia/Soviet Union for centuries. The political struggles around the trans-Balkan pipelines resemble these struggles. Currently the most influential powers in the region includes the Germany led EU, Russia, the United States, Turkey, and China, besides whom smaller states may have local/regional influence in different aspects (like Austria in the banking sector, Hungary in Vojvodina, or the Gulf states in Bosnia and Herzegovina). As a consequence of recent global political processes, the Balkans is seen once again as a theater of geopolitical struggles of great powers similar to the one in the nineteenth century.

The 1990s and early 2000s saw the dissolution of Yugoslavia into seven successor states with Bosnia and Herzegovina being one of them. This created a small-state environment in the western part of the peninsula, which constitutes the direct environment for Bosnia and Herzegovina, which shares land borders exclusively with former member republics of Yugoslavia. The internationalization of former internal borders created a unique neighborhood and cross-border environment where several infrastructure systems, economic, personal, or institutional relations have been established over the now international border. More important is that the people of Bosnia and Herzegovina all have kin groups in the neighboring countries, resulting in a very sensitive neighborhood relationship. Furthermore, these countries share decades of joint history, which was put to an end by the armed conflict in Bosnia and Herzegovina in which all of the current neighbors or their citizens were somehow involved. The peace which ended the conflict decentralized the country, partly along ethnic lines which further complicates Bosnia and Herzegovina's relationship with its neighbors. Now all of the former Yugoslav republics, including Bosnia and Herzegovina, are heading towards the EU, some of them already joined, while the others being at different stages of the enlargement process. The region can be also seen as an enclave within the EU.

This is the geopolitical environment Bosnia and Herzegovina cannot escape and has to live with. Local consequences include the extremely strong influence of Russia in the entity of Republika Srpska, as well as Chinese investments in the infrastructure sector or the sometimes complicated but still overwhelmingly important European financial support.

The current geopolitical position of Bosnia and Herzegovina is the result of a long historical-geographical development, which is why the problems also are similar to those mentioned in the chapter about historical geography. However, with the independence, numerous weaknesses of the geopolitical position of Bosnia and Herzegovina came to the fore. Due to the shape of the Republic of Croatia, which forms the semicircle around Bosnia and Herzegovina (the so-called Croatian geopolitical horseshoe) and the direct border contact with the Republic of Serbia and the Republic of Montenegro, Bosnia and Herzegovina is in a kind of encirclement (Klemenčić 1997).

Therefore, Bosnia and Herzegovina has been a subject of great interest mainly to its two neighbors: Serbia and Croatia, for many years now. Even during the Ottoman rule in Bosnia, Serbia, together with Montenegro, had territorial and national pretensions towards Bosnia, with the desire that, after the Ottoman withdrawal from Bosnia, it would join Serbia.

In 1844, Garašanin wrote about this in his work Načertanije. He begins with the historical fact that the people of Bosnia are Bosniaks of all three faiths: Muslims, Orthodox, and Catholics. Garašanin writes: "... the main thing for the idea of unification of Bosnia with Serbia should be to win over the (Bosnian friars) ... in addition, as a third instance, a short and general folk history of Bosnia could be printed, in which the fame and names of those Bosniaks who converted to Mohammedan faith should not be left out... It will not be a difficult task for Serbia to have a greater influence on the eastern religion of (Orthodox) Bosniaks. More precaution and attention to this requires that Catholic Bosniaks be won over. At the head of these are the Franciscan friars" (Garašanin 1844).

The international community representatives, primarily the European Union and the United States, are helping Bosnia and Herzegovina maintain its territorial integrity and sovereignty. In view of all the foregoing, Bosnia and Herzegovina does not have a favorable political-geographical position in the modern sense of the term, especially having in mind the constant destabilizing activities at the global level between the great world powers of the United States and the Russian Federation, where both countries see this territory as their area of interest.

Unfortunately, BiH has such a geostrategic position that it represents a border state between two cultures: Eastern and Western. Both are integrated through the permeation of three cultural and civilizational circles: Greco-Roman, Jewish-Christian, and Arab-Islamic. Moreover, instead of being an advantage, it is taken as a disadvantage due to the imperial-chauvinist territorial pretensions of its neighbors, Serbia, Montenegro, and Croatia. This permeation of all three civilizational and cultural circles makes Bosnian culture-specific; it makes what we today know as the Bosnian spirit, today, the main pillar of the defense of the Bosnian state and its people.

In terms of its economic progress, the EU integration is a crucial, strategic, political, and economic priority for Bosnia and Herzegovina. Inclusion into the Central European Free Trade Agreement (CEFTA) is one of the mechanisms running towards the faster socio-economic development and stability. Also, as part of the EU accession process, Bosnia and Herzegovina cooperates with the countries of the Western Balkans within the so-called Berlin Process, which is a complementary long-term aspect of integration into European economic and political associations.

1.5 Traffic and Tourist-Geographical Position

The modern traffic-geographical position of Bosnia and Herzegovina is represented by its potential to serve as a point of connection of the Central and Eastern Europe with the Adriatic forefront. A very favorable position with regard to the main European traffic routes pertains primarily to the system within the European Agreement of Main International Arteries-AGR, where this route is marked as E-73 and is part of the Pan-European Corridor Vc as one of the most important branches of the Trans European Motorways (TEM) and the Trans European Rail (TER) project. The planned route stretches over in the direction of Budapest, Osijek, Sarajevo, and Ploče, and represents the most important infrastructure project in Bosnia and Herzegovina. From the point of view of its tourist-geographical position, according to the tourist regionalization designed by the World Tourism Organization

(UN WTO), Bosnia and Herzegovina occupies a place in the European tourist region and makes part of the Southern (Mediterranean) subregion. Bosnia and Herzegovina is involved mainly in European tourist flows or routes, and as such it can be included in three European tourist routes: the Western European tourist route, in which the final destination of tourists is the Iberian Peninsula and the Maghreb countries; the Central European tourist route leading through the Alps to the Apennine Peninsula and the surrounding islands, but also to Tunisia; the Eastern European tourist route, in which the final destination of tourists is the Balkan Peninsula and the Middle East (Bidžan and Gekić 2015).

The territory of Bosnia and Herzegovina is located on the eastern tourist corridor, i.e., the direction of tourist movements from the home areas of Europe to the eastern Mediterranean, which indicates that it has a continental tourist-geographical position. Since the previously mentioned tourist routes have intercontinental importance, i.e., they connect Europe and Asia, but Europe and Africa also, it can be considered that Bosnia and Herzegovina has an intercontinental tourist-geographical position according to the size of its influence.

In addition to this transit function, Bosnia and Herzegovina also has a contact function of its geographical position, which is reflected, among other things, in the fact that it borders three countries: Serbia in the east, Montenegro in the south, and north and west with Croatia. This, in good neighborly relations, has a positive effect on foreign tourism, primarily excursion and transit movements. In the bay of Neum-Klek, it exits to the Adriatic Sea. Its central position in relation to its neighboring countries increases its tourist contact zone. As it is a mountainous, lowland, and Mediterranean country, and due to its cultural and historical development, there are various landscape motifs in Bosnia and Herzegovina. We find these complex tourist motives in different combinations of natural and anthropogenic elements. The polyvalence of the tourist-geographical position of Bosnia and Herzegovina is reflected in the possibility for tourists to experience the Peripannonian Bosnia, the Dinaric mountain massif, as well as the Mediterranean area in which various anthropogenic motifs are incorporated (Bidžan and Gekić 2013).

References

BHAS 2016, Agency for Statistics of Bosnia and Herzegovina (2016) Census of population, households and dwellings in Bosnia and Herzegovina 2013, Sarajevo

Bidžan A, Gekić H (2013) Influence of natural disasters on the development of touristic regions in Bosnia and Herzegovina and the World. J Int Environ Appl Sci 8(3):446–454

Bidžan A, Gekić H (2015) Regional aspect of tourism development in Bosnia and Herzegovina. Int J Innov Appl Stud 11(4):886–893

Drešković N, Mirić R (2017) Regionalna geografija Bosne i Hercegovine I [Regional geography of Bosnia and Herzegovina I]. Faculty of Science, University of Sarajevo, Sarajevo

FZS (2020) Statistical yearbook of the Federation of Bosnia and Herzegovina 2020. Institute for Statistics of FBiH, Sarajevo

Garašanin I (1844) Program nacionalne i spoljašnje politike Srbije na koncu 1844. godine [The program of national and foreign policy of Serbia at the end of 1844]. In: Čović B (ed) Izvori velikosrpske agresije, Zagreb, p 75

Jahić H, Spahić M, Mezetović A (2014) Geografski centar Bosne i Hercegovine [Geographical Center of Bosnia and Herzegovina]. Acta Geogr Bosn Herzeg 1:73–82

Klemenčić M (1997) Bosna i Hercegovina—Geopolitika [Bosnia and Herzegovina—geopolitics]. In: Klemenčić M (ed) Atlas Evrope. Leksikografski zavod Miroslav Krleža, Zagreb, p 308

Map of Međurečje available at with Licence Creative Commons 1.0. https://sr.wikipedia.org/wiki/%D0%94%D0%B0%D1%82%D0%BE%D1%82%D0%B5%D0%BA%D0%B0:Sastavci.png. Accessed on 16 June 2021

Check for updates

2 Borders of Bosnia and Herzegovina

Abstract

The development and characteristics of the borders of Bosnia and Herzegovina are presented through a description of the most important attributes related to their character, historical-geographical evolution, but also their spatial coverage and general physical geographical features. In the past centuries, Bosnia and Herzegovina has survived many significant changes in its status, territorial integrity, geopolitical, and economic-geographical position. It was independent and territorially complete, occupied, and divided on the geopolitical and economic-geographical periphery of the great empires (Ottoman and Austro-Hungarian), but also in the center of the common Slavic unitary (Kingdom of Yugoslavia) and federal state (SFRY). In modern times, Bosnia and Herzegovina has borders with three countries: the Republic of Croatia, the Republic of Serbia, and the Republic of Montenegro. The total length of Bosnia and Herzegovina's borders with neighboring countries is 1538 km. There are still border disputes with the Republic of Croatia and the Republic of Serbia.

Keywords

Length of borders · Demarcation · Border Disputes · Territorial sea · Hydrographic borders

2.1 Origin of the Borders

Throughout its long history, Bosnia and Herzegovina has always had some aspects of statehood manifested in various forms. In this regard, its borders have changed throughout history (Prelog 1912). The current borders of Bosnia and Herzegovina have been confirmed several times by a series of international treaties, conventions during two and a half centuries (from 1699 to 1943), and the decisions of AVNOJ (Anti-Fascist Council for the National Liberation of Yugoslavia) from 1943 and 1945.

The most important elements of the international legal definition of the modern borders of Bosnia and Herzegovina are derived from the decisions of AVNOJ from 1945, when its borders were based on those determined by the Treaty of Berlin in 1878, with some later corrections at Sutorina, Bosansko Grahovo, Bihać, and East Herzegovina area (Najetović 2010). Within such established borders, Bosnia and Herzegovina acquired the international subjectivity of an independent state on April 6, 1992. She became a full member of the United Nations on May 21, 1992, as an independent, sovereign, free, unified, integral, and indivisible state within the borders it has inherited since 1918 (Ministry of Foreign Affairs of BiH 2009).

The demarcation of the successor states of the former Yugoslav community was decided within the International Conference on Yugoslavia, which included the Arbitration Commission for Borders, better known as the Badinter Commission, named after its president, French lawyer Robert Badinter, formed by the Council of Ministers of the European Economic Community. On August 27, 1991, this Commission dealt primarily with the legal framework of the borders of the former Yugoslav republics and on that occasion made several concrete conclusions, of which, for this title, the most important are:

- The issue of succession of states should be resolved based on international law and based on equity, bearing in mind that all interested countries are bound by convincing norms of general international law, especially those concerning respect for fundamental human rights, peoples' rights, and national minorities.
- SFRY is disintegrating and no longer exists as a state; no successor state has the right to continue the membership of the SFRY in international organizations.

H. Gekić et al., *The Geography of Bosnia and Herzegovina*, World Regional Geography Book Series,
https://doi.org/10.1007/978-3-030-98523-3_2

- The borders between the former federal units are inherited and cannot be changed by force, but only by agreement—by the principle of *uti possidetis*.
- The succession of Bosnia and Herzegovina happened on March 6, 1992, after a referendum held on February 29 and March 1, 1992, where the majority of the population voted for a democratic and independent state of Bosnia and Herzegovina. With the Dayton Peace Agreement, as well as those preceding, Bosnia and Herzegovina inherited the external borders, which the Republic of Bosnia and Herzegovina, with some minor border changes inherited by agreement of the then republics concerning the border segments near Bihać, Kulen Vakuf, Drvar, Bosansko Grahovo, Duvno, Gacko, Foča and Drina river islands downstream from Zvornik, had until 1992 (Andrassy et al. 1998.)

2.2 Current Borders and Disputes

The borders of Bosnia and Herzegovina are mainly natural-hydrographic and orographic. The total length of country's borders with neighboring countries is 1538 km. The land border is 774 km long, while the water border is 764 km (Agency for Statistics of FBiH 2020). The total length of the border with the Republic of Croatia is 932 km, of which the land border is 494 km, and the water border is 438 km (maritime boundary in the area of Neum and Klek in the length of 13.4 km). The border in the north and north-west towards the Republic of Croatia is mainly hydrographic. It consists of the river Sava (from the estuary of Una to the estuary of Drina and Sava rivers), then the river Una (from its estuary with Sava upstream to the settlement of Donji Rakani, south of Novi Grad) (Fig. 2.1), the right orographic watershed of the rivers Žirovnica, Glina and Korana rivers to the mountain of Plješivica. The western and southwestern border of our country towards the Republic of Croatia is mainly orographic and leads to the mountain ranges of Plješivica, Dinara, and Kamešnica, then the right orographic watershed of the river Trebižat—a tributary of Neretva river.

In the area of Neum and Klek, Bosnia and Herzegovina has a 24.4 km long coastal facade overlooking the Adriatic Sea. The southern border with the Republic of Croatia, in the area of the Bosnian coast, follows the coastline in Neum, in the middle of the bay of Mali Ston in the length of 13.4 km of maritime boundary, and then the border is orographic and goes to the mountains of Treskavac, Neprobić, Ravni vrh, Oštra glava, Stražišće, Sniježnica, and Orjen, which separate the South Croatian coast from the Mediterranean East Herzegovina Mesoregion of Bosnia and Herzegovina (Lepirica 2009).

The agreement on the state border between the Republic of Croatia and Bosnia and Herzegovina was signed in Sarajevo on July 30, 1999 by Presidents Franjo Tuđman and Alija Izetbegović. It was compiled with the help of foreign experts. After the signing, no party ratified it according to its constitutional regulations. But according to its Article 22, that Agreement is temporarily applied from the day of its signing, and it is concluded for an indefinite period of time. Clause 3 provides that: "Either Contracting Party may terminate this Agreement at any time, by written notice, through diplomatic channels. In such event, the Agreement shall terminate 6 months after the date of receipt by the other Party of the notice of termination." Notwithstanding this provision, if either party canceled the provisional application of that Agreement, and it has not so far, the border between Croatia and Bosnia and Herzegovina would remain the same until both countries have concluded an agreement on their

Fig. 2.1 The Una River near Novi Grad. (Photo by Ranko Mirić)

new border (Degan 2015). Bosnia and Herzegovina has the following border disputes with the Republic of Croatia: on the Adriatic Sea in the area of Neum, on the river Una between Kostajnica and Hrvatska Kostajnica, and in the lower reaches of the river Una.

The maritime demarcation between Bosnia and Herzegovina and Croatia is specific, because the territorial sea of Bosnia and Herzegovina is surrounded by the internal waters of Croatia, making the situation quite unique. The authorities of Bosnia and Herzegovina are of the opinion that Croatia is violating the UN Convention, because it unilaterally applies a direct line of separation (drawn between the island of Vodenjak-Hvar to Cape Proizda-Korčula), as it was in the time of the former Yugoslavia. In doing so, Croatia has constituted its territorial sovereignty and internal waters, closing the sea area of Bosnia and Herzegovina (Mihajlović 2012).

According to the Convention on the Law of the Sea and the applicable customs in international relations, the narrow access to the sea near Neum is a sufficient reason for Bosnia and Herzegovina to have a territorial sea. The maritime border defined within the border agreement reached in 1999 was defined as "the middle line between the mainland of the Republic of Croatia and Bosnia and Herzegovina, in accordance with the 1982 Convention on the Law of the Sea" (Klemenčić and Topalović 2009).

The total length of the border of Bosnia and Herzegovina with the Republic of Serbia is 357 km, of which the land boundary is 84 km, and the river boundary is 273 km. Part of the northern border Bosnia and Herzegovina shares with Serbia (and Vojvodina) is 27 km (the course of Sava river from Brezovo polje to the estuary of Drina river). The northeastern border with Serbia is a river. It stretches along the lower course of the Drina river and part of its middle course upstream from Zvornik, where there are artificial reservoirs of HPP (Hydro Power Plant) "Zvornik" (Fig. 2.2) and HPP "Perućac." The eastern border of Bosnia and Herzegovina starts from Zemljica on the right valley side of Drina river downstream from the estuary of Žepa river, from where it emerges on the highest ridge of the Zvijezda mountain and cuts the valleys of Bijelo and Crni Rzav and the terrain of Starovlaška visija. Then, the short state river border towards Serbia is represented by the rivers Uvac (about 5 km) and Lim (about 4 km) over Guvnin, and through the right orographic watershed of Jablanica, a tributary of Uvac and the slopes of Zvijezda exists to the Drina river. Specificity of the mentioned part of the border is the exclave Međurečje. It is located in the border zone surrounded on all sides by the territory of Serbia. This part of Bosnia and Herzegovina will probably face state delimitation commissions in the near future.

Bosnia and Herzegovina has several open border disputes with neighboring Republic of Serbia. First of all, it refers to the exclave of Međurječje, which administratively and territorially belongs to the municipality of Rudo in Bosnia and Herzegovina, and is surrounded on all sides by the territory of the municipality of Priboj in Serbia. Another problem is the part of the Belgrade-Bar railway that leaves the territory of Serbia for a length of 12 km and passes through the village of Štrpce in the municipality of Rudo. In addition to the above, two hydroelectric power plants, "Zvornik" and "Bajina Bašta," are also in dispute. Both are included in Serbia's power system, although they were built on the Drina River, which is the border river between the two countries. Also, the river Drina in the lower part of its course, thanks to the strong meandering, significantly changed its course in relation to the period when the middle of the riverbed was determined as the border (Halilović and Suljić 2016).

The border between Bosnia and Herzegovina and Montenegro is 249 km long, of which 196 km represents the land boundary, and 53 km is the water boundary. The southern and southeastern border goes from Orjen (the southernmost point of Bosnia and Herzegovina, Podštirovnik in the municipality of Trebinje, is located on the border of Bosnia and Herzegovina, Croatia and Montenegro) to the northwestern slopes of Zvijezda mountain (where the border

Fig. 2.2 Border between Bosnia and Herzegovina (area A) and the Republic of Serbia (area B) on the River Drina. (Photo by Ranko Mirić)

of our country with the Republic of Montenegro exists to the Drina river), the border with Montenegro goes through the mountain range of Bijela and Skorča gora, the eastern shore of Lake Bileća, Somina, the southeastern rim of Gatačko polje and the mountain ranges of Lebršnik, Volujak, and Maglić, then the rivers Piva and Tara, the slopes of Obzir, Ljubišnja, Okladnik, Kovač-planina, and Gradina. This mountain, Gradina, represents the state border between Bosnia and Herzegovina, and Montenegro (Lepirica 2009; Agency for Statistics of FBiH 2020). Currently, there are no border issues between two countries.

In administrative and political terms, the state border of Bosnia and Herzegovina consists of the external borders of 35 peripheral Bosnian municipalities, then the external borders of the five cantons of the entity of Federation of Bosnia and Herzegovina and the external borders of another Bosnian entity of Republika Srpska. Bosnia and Herzegovina has 83 border crossings, of which 55 are international, and 28 are local-border crossings. At the border with the Republic of Croatia, there are 30 international border crossings with different traffic regimes and 20 local-border crossings for border traffic. At the border with the Republic of Serbia, there are 12 international border crossings with different traffic regimes and 4 local-border crossings for border traffic. There are 6 international border crossings and 4 local border crossings for border traffic on the border with Montenegro. Bosnia and Herzegovina has 7 common locations at border crossings with the Republic of Croatia (3 on the territory of Bosnia and Herzegovina, and 4 on the territory of the Republic of Croatia), i.e., locations where officials of both countries perform their competencies at the same border crossing. In the coming period, it is necessary to seek the possibility of establishing additional common locations, especially with neighboring countries with which Bosnia and Herzegovina does not have this practice (Council of Ministers of Bosnia and Herzegovina 2020; Border Police of Bosnia and Herzegovina 2020). There are not any ongoing international arbitrage processes currently between Bosnia and Herzegovina and neighboring countries.

References

Agency for Statistics of FBiH (2020) Statistical yearbook. Available at https://docs.google.com/gview?url=http://fzs.ba/wp-content/uploads/2020/12/Godisnjak2020.pdf. Accessed on 20 Aug 2021

Andrassy J, Bakotić B, Vukas B (1998) Međunarodno pravo 1 [International law 1]. Školska knjiga, Zagreb, pp 276–277

Border Police of Bosnia and Herzegovina (2020). Available at http://www.granpol.gov.ba/Content/Read/10?title=Onama. Accessed on 20 Aug 2020

Council of Ministers of Bosnia and Herzegovina (2020). Available at http://www.vijeceministara.gov.ba/akti/prijedlozi_zakona/default.aspx?id=21135&langTag=hr-HR. Accessed on 16 Aug 2020

Degan V-Đ (2015) Delimitacija i demarkacija vanjskih granica Republike Hrvatske. Poredbeno pomorsko pravo 54(169):57–71

Halilović A, Suljić A (2016) Border/territorial disputes of Bosnia and Herzegovina with neighboring countries. Zbornik radova PMF 12: 49–64

Klemenčić M, Topalović D (2009) Morske granice u Jadranskome moru [Maritime borders in the Adriatic Sea]. Geoadria 14(2):311–324

Lepirica A (2009) Historijsko-geografski razvoj granica Bosne i Hercegovine [Historical and geographical development of the borders of Bosnia and Herzegovina]. In: Terzić R (ed) Proceedings of the Faculty of Science, vol 6. University of Tuzla, pp 53–65

Mihajlović S (2012) Crtanje granica na Zapadnom Balkanu: Ljudska perspektiva- Izvještaj za Bosnu i Hercegovinu. Centar za sigurnosne studije, Sarajevo

Ministry of Foreign Affairs of BiH (2009). Available at http://www.mvp.gov.ba/vanjska_politika_bih/multilateralni_odnosi/ujedinjene_nacije_un/bih_i_un/?id=16. Accessed on 12 Feb 2021

Najetović DŽ (2010) Kontinuitet razvoja državne i političke vlasti u BiH od Austrougarskog perioda do prve sjednice zajedničkog Parlamenta BiH [Continuity of development of state and political power in BiH from the Austro-Hungarian period to the first session of the joint Parliament of BiH]. In: Hasanović I (ed) Proceedings of the scientific conference constitutional and legal development of BiH (1910–2010). University of Tuzla, Faculty of Law, pp 183–206

Prelog M (1912) Povijest Bosne od najstarijih vremena do propasti Kraljevstva [History of Bosnia from the earliest times to the collapse of the Kingdom]. J. Studničke, Sarajevo, pp 15–43

Part II

Physical Geography of Bosnia and Herzegovina

3 Geology and Geotectonics of Bosnia and Herzegovina

Abstract

The geological composition of the area of Bosnia and Herzegovina is very complex and was formed due to very intricate endodynamic processes, which resulted in the appearance of a complex petrographic and lithostratigraphic structure represented by igneous, sedimentary, and metamorphic rocks. This chapter elaborates the contents from the aspect of their geochronological occurrence in Bosnia and Herzegovina at the level of eras and periods, from the oldest to recent geological formations. A detailed account of the distribution of the most significant formations with petrographic and lithological characteristics and adequate cartographic representations (maps of Paleozoic, Mesozoic, and Cenozoic deposits) at the level of the era is given. The second part of the chapter presents the general geotectonic features of Bosnia and Herzegovina with interpretations of the most important tectonic elements (faults, overthrusts, etc.), and, in this regard, the seismological features of the area. Also, there is a tabular data on the share of the main geological formations in the area of Bosnia and Herzegovina at the level of periods and epochs presented, determining the spatial dominance of Mesozoic deposits (almost 2/3 of the area of Bosnia and Herzegovina).

Keywords

Rocks · Deposits · Tectonics · Mesozoic · Physical geography

3.1 Geological Formations and Petrographic Composition

The geological composition of the territory of Bosnia and Herzegovina is very complex. It was formed due to very complex endodynamic processes through the long geological history of this area, during which very diverse formations of igneous, sedimentary, and metamorphic rocks were formed (Herak 1987). In such an aquatic and terrestrial environment, very diverse flora and fauna have developed.

As the oldest geological formations in Bosnia and Herzegovina, complexes of shale rocks have been identified: gneisses, mikashists, and phyllites, which resemble real archaic rocks only in their petrographic composition (Scotese 1989). However, their exact age has not been determined yet, although they probably came to be through a metamorphosis of Old Paleozoic and possibly Eozoic sediments (based on the Geological Map of SFRY 1974). The complexes of these rock masses spread in Motajica, Prosara, and a small part of the central Bosnian shale mountains. However, recent research results have shown that the rock complexes at these sites are not the same age. The oldest are the metamorphites of Motajica, in the core of which a large granite laccolith of sharply concentric structure has been found (Fig. 3.1). The center is typical granite, surrounded by a ring of contact metamorphic rocks, ranging from gneiss through phyllites and argiloshists onwards to even less altered clays, siltstone, and sandstones.

Prosara is entirely a shale mountain, with several breakthroughs of microgranite porphyry on which lean crystalline shales are similar to gneiss only in contact zones. The rest, most of this mountain is made up of phyllites of the Old Paleozoic age. In both mountains, the rock complexes more or less go deep into the Algonkian. The youngest are metamorphosed older Paleozoic strata in part of the central Bosnian shale mountains (Nikolić 1990).

Here the "archaic" differs significantly from the "archaic" of Motajica and Prosara mountains, primarily due to the lack of granite, but also because crystalline shales similar to gneiss and mikashist are relatively subordinate and do not form distinct zones as in the previous two mountains. In fact, they appear as inserts in other crystalline shales. These rock complexes are located in the central part of these mountains;

H. Gekić et al., *The Geography of Bosnia and Herzegovina*, World Regional Geography Book Series,
https://doi.org/10.1007/978-3-030-98523-3_3

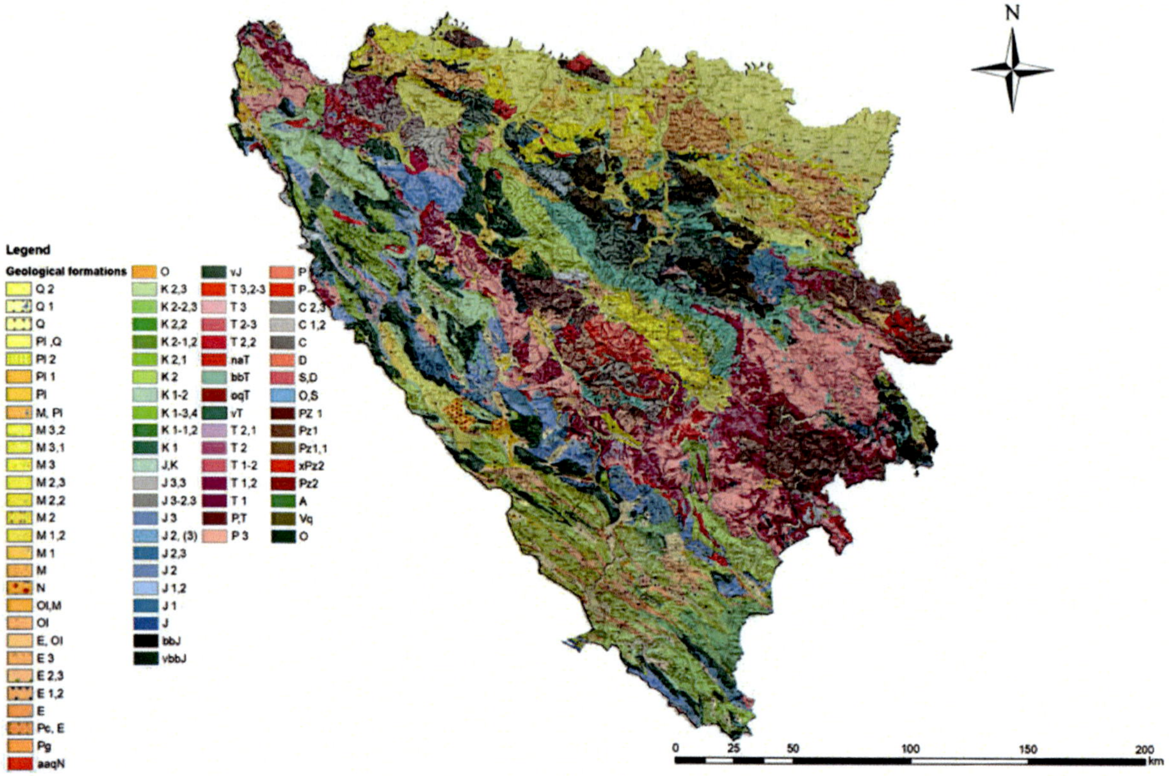

Fig. 3.1 Geological formations of Bosnia and Herzegovina

Table 3.1 Geochronological and surface relations of the main geological formations in Bosnia and Herzegovina by periods

Period	% of total area
Quaternary	11.5
Tertiary-Quaternary	1.1
Tertiary	16.4
Cretaceous	24.7
Jurassic-Cretaceous	3.3
Jurassic	11.7
Triassic	19.2
Permian-Triassic	0.8
Permian	0.6
Carboniferous	1.6
Upper Paleozoic	4.1
Devonian	0.4
Silurian-Devonian	0.5
Ordovician-Silurian	0.3
Lower Paleozoic	0.4
Rocks of unknown age	3.4

Based on Geological map of SFRY 1974 and field research of authors

more precisely, they extend from Busovača in the south towards the Vranica mountain (the most expansive area around Fojnica and Busovača) (Table 3.1).

In Bosnia and Herzegovina, Cenozoic deposits are represented by 29%, Mesozoic 59.7%, Paleozoic 7.9%, and rocks of unknown age with 3.4% of the total area. The geological formations of the following periods are the most represented in terms of share: Cretaceous 24.7%, Triassic 19.2%, Tertiary 16.4%, Jurassic 11.7%, and Quaternary 11.5%.

3.1.1 Paleozoic Deposits

Paleozoic deposits in Bosnia and Herzegovina belong to the Bosnian-Internal Albanian zone of limestone and schists (Čičić 1979). These deposits are located in terrains that are more or less Dinaric direction and differ from each other, except for their presence in isolated geographical areas, in lithological composition, age, and degree of wretchedness (Fig. 3.2). Observing from northwest to southeast, these are: Sana-Una Paleozoic, Ključ Paleozoic, Central Bosnia Paleozoic, Eastern Bosnia Paleozoic (Drina Paleozoic), and Southeastern Bosnia Paleozoic (Paleozoic of Prača).

The Sana-Una Paleozoic includes terrain built of Paleozoic deposits, located in the Una river valley in the area of Bosanski Novi, Piskavice, Bronzani Majdan, Kozica Valley, Sanski Most, Budimlić-Japare, and Rudica (Čičić 1979). The boundaries of this area in the north and southwest were determined by the geotectonic method (based on abnormal relationships with younger deposits). Where tectonic elements are not emphasized, the boundaries were determined based on topographic criteria. There was a lot of ambiguity in determining the age of certain formations, especially the older Paleozoic. This primarily refers to the determination of graptolites in some parts of the southern Ljubija ore mine, based on which a significant portion of the Paleozoic deposits was included in the Silurian or Ordovician.

At the same time, most authors of such attitudes quickly overcame the fact that the layers with graptolites cannot be logically separated from the layers with carbon fossils. After numerous discussions and repeated paleontological analyses, it was established that the rock complexes recognized as Silurian are, in fact, of the Lower Carboniferous age (Anđelković 1978). Of the Precarboniferous sediments, the existence of Upper Devonian sediments was determined at three locations: Paljevine (about 8 km south of the Ljubija mine), Blagaj (near Bosanski Novi), and Rasavci.

Upper Devonian sediment deposits in Blagaj and Rasavci are currently the only sites with documented pre-carboniferous sediments. Limestones with primitive goniatites from Paljevina were excavated at a depth of 60 m, so they cannot be included with certainty in the Precarboniferous formations in the surface projection. Middle and Lower Carboniferous deposits occupy the largest part of the Sana-Una Paleozoic. Taken as a whole, they look quite monotonous in the lithological sense, although a large number of different petrographic members can be singled out in them. The primary criteria for such determination were lithological and facial differences in deposits and tectonic movements that disrupt regular stratification in many places. Lower Carboniferous age was determined based on coral findings (Anđelković 1988). These are, in fact, gray-to-dark gray dolomites, often anchorite limestones, and partly sandy-shale layers located northeast of Sanski Most (Sasina, Klašnica, Kladovita, and Razboj valleys).

Clastic, Lower Carboniferous deposits from the Namurian-Moscowian phase have been identified south of the Retka site (area of southern Ljubija ore mine) and at several locations in the eastern part of the Una-Sana Paleozoic. Upper Carboniferous strata with dark gray limestones, schists, and sandstones are also considerably distributed in the area south of the Ljubija ore mines. Based on the mutual difference and the fact that they occur in larger masses, three elements have been singled out within the carbon deposits: clastic formations with volcanic elements, carbonate rocks, and marine mineral deposits. Ključ Paleozoic includes an area of very irregular shape, elongated in the direction of

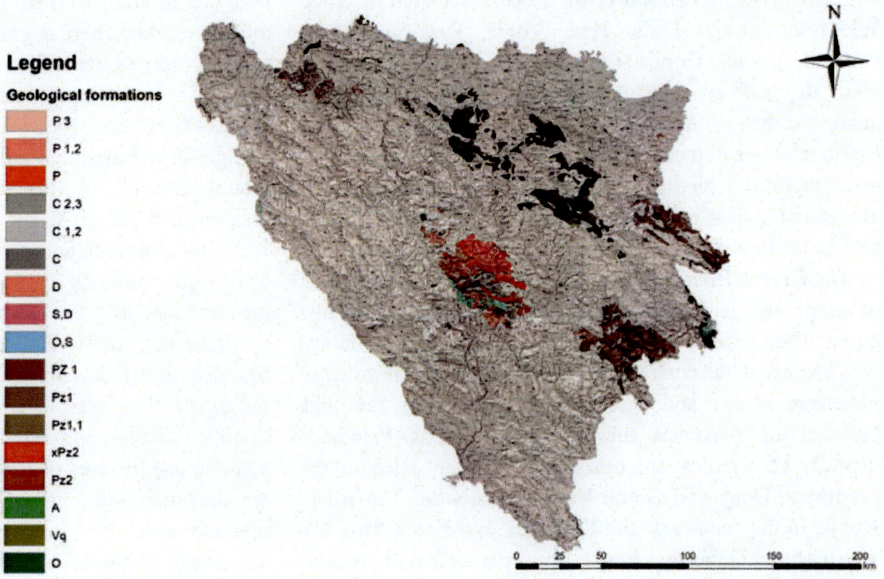

Fig. 3.2 Paleozoic geological formations of Bosnia and Herzegovina

Fig. 3.3 Tectonic breccias. (Photo by Nusret Drešković)

Gornja Sanica. The area is located northwest of Ključ, in the area around Pudin Han, Vojići, Ramići, Kopljenica, Muhamed-begova Prisjeka, Biljani, Pištanica, Budelj, and Gornja Sanica. Paleozoic deposits are represented here by Carboniferous and Permian-Triassic formations. Carbon deposits are composed of clasts (clayey-feldspar-sericite-quartz shales) and fine-grained to coarse-grained conglomerate quartz sandstones. As a product of the intense tectonic activity, tectonic breccias bound by clay material are very common (Fig. 3.3).

The carbon deposits are partly built of breccia limestone. Carbon has so far been determined in three mutually separated areas: the lower course of the Jarčevac and Panjin streams, around Muhamed-begova Prisjeka, and in the village of Vrelo.

Permo-Triassic deposits cover most of the area of Ključ Paleozoic: around Pudin Han, Vojići, Ramići, Biljani, Pištalina, Budelj, Gornja, and Donja Sanica. The clasts of these deposits are made up of hematite-sercite-feldspar-quartz sandstones and light gray to dark purple siltstones. Carbonates are represented by porous limestone, gypsum, and anhydrite. Isolation of special lithostratigraphic units is almost impossible because carbonate and clastic facies alternate in depth and laterally.

The Central Bosnian Paleozoic, from the aspect of spatial coverage and geological diversity, can be divided into two areas: Vranica area, and Jezero and Sinjakovo area. Vranica area belongs to the central Dinarides; i.e., it lies in the zone of Paleozoic shales and Mesozoic limestones, in the area between the Busovača and Vrbas dislocations. Paleozoic deposits of Vranica are distributed between Jajce in the northwest, Donji and Gornji Vakuf in the west, Tarčin and Konjic in the southeast, and Busovača in the east. This is a group of mountains that form one morphological unit whose uniqueness stems from their geological structure. In the lower parts of the slope, they are built of rhyolite and shale, while, in their highest parts, there are limestones, dolomites, and marbles. The uniqueness of the geological structure is reflected in the identical geomorphological processes, resulting in a similar morphology of this group of mountains. Paleozoic deposits are represented from insufficiently determined Silurian to Permo-Triassic.

Silurian deposits were determined by the superposition method, according to which they lie below the Devonian. Within the Vranica area, three subareas can be distinguished where assumed Silurian deposits are represented: the wider area of Busovača, the area of Gornji Vakuf and central Vranica, and the area of the extreme southwestern part of Vranica. From the aspect of primary petrographic composition, pelitic and less often psammitic rocks are represented, metamorphosed (high degree of metamorphosis), mainly into sercite-chlorite-quartz shales or only in quartzites. Devonian deposits are fully represented. Their potential is estimated at about 400 m. The Lower Devonian sediments are represented by light gray limestones and dolomites that form the highest mountainous part of Vranica. There can also be located in the area around Gornji Vakuf where gray-blue limestone slab with little dolomite and marble predominates, and southeast of Vitruša, towards Dusina and Tarčin, where dolomites predominate. Middle Devonian sediments are represented by massive, rarely stratified limestones and dolomites. Marbles are less common. Middle Devonian limestone sediments dominate in three separate areas: the area of Ločika, Krstac, and Smiljevača; the area of Dobruška planina; and the area of Rog, Goletica, and Brezovača. Middle Devonian dolomite sediments are also present in several separate areas: the Vitruša area, Otunjski vis, the Pogorelica mountain, the Deževica area (especially in Gradac) as well as the dolomite zone in the Kreševo-Tarčin series. Upper Devonian deposits have been found only in the Dusina area. The carbon formations of the Vranje area have also not been determined with certainty. The Lower Carboniferous deposits, represented by gravelly meta-sandstones, are located on Medenik and Brezovača near Rostovo and the complex of metamorphic rocks in the extreme southwest and southeast the area (southwestern slopes of Vranica from Mračaj near Gornji Vakuf to Konjic and Bradina in the southeast). The Permian sedimentary complex is represented by red conglomerates, sandstones, and clays, and gypsum occurrences are common, and unlike other petrographic-lithological units, it is well stratified and not metamorphosed. Permian deposits were identified in two mutually separate subregions. The first subregion is the area of Bugojno—Travnik—Jajce, where bojnička, oparska, and travnička deposits have been identified. The second sub-area is the area of Gornji Vakuf—Konjic. Upper Permian deposits represented by red clasts and anhydrite were found here (Drešković and Mirić 2017).

Jezero and Sinjakovo area is located west-northwest of Jajce and is a smaller, limited area. The largest area with Paleozoic deposits is the sub-area Jezero, within which several separate areas gravitate to the Pliva river. Devonian deposits are fully developed in the Pliva river valley, from the Perućica stream to the village of Jezero, where their capacity is estimated at around 250 m. Devonian chronostratigraphy is complete; i.e., all three floors are represented: lower, middle, and upper. The deposits are represented by different types of limestone (from reddish-pink and lumpy in the Perućica valley, dark plate to bank gray) (less often by marble and sercite-chlorite-quartz shales). Of the carbon deposits, only the Lower Carboniferous ones were found in the area of Debela Kosa and Rostovača, where sercite-chlorite-quartz shales predominate. Of the Permian, only Upper Permian deposits are present, represented by red quartz breccias and red quartz sandstones. They occur in the area downstream from Volar, southwest of Đumezlija, on the northern slopes of Sinjakovo, and the ridge of Krmarica.

The Drina Paleozoic covers the area west and southwest of the Drina river, covering an area of about 1200 km^2. The northern and northeastern borders are the Drina river, while in the south and southeast, the border is represented by a broad zone of Dinaric-oriented younger deposits. Slightly wavy morphosculptures of the low and middle mountain type make this area a particular geomorphological unity. The relatively wide valleys of the rivers: Drina, Drinjača, Tišća, Križevica, and Glogovska rijeka; also most often, the gorges of their tributaries deeply dissect the existing relief structure. Old Paleozoic deposits in this area have not been determined with certainty to date. However, a large number of authors believe that the breakthroughs of highly harvested phyllites and chlorite-quartz-muscovite meta-sandstones with a pronounced granular structure can be counted in these sediments. Considering that the mentioned deposits were found in the borane structure, in the core of the anticline, under the sediments of the Lower Carboniferous, and their habitus is quite similar to gneiss-Mikashist, we can talk about the Old Paleozoic age of deposits at the sites of Tegar (southeast of Bratunac), and Kolarska rijeka (east of Srebrenica). In addition to these, in the area of Drinjača near Zvornik, shoots of highly crystallized serpentine and amphibolite were found. The Lower Carboniferous deposits are represented by clayey-coal shales, phyllite shales, phyllites, meta-sandstones, and black layered lenticular limestones. They mainly occur in the vertices of anticlines, although they can also occur in the morphologically lowest parts of valleys of the Drina, Jadar, Bukovica rivers, etc. They are also widespread in the profiles of Šutorina, Tegarska rijeka, Mlječvanska rijeka, Zapoljska rijeka, Brezovačka rijeka, and Biljača. A total of five lithological members have been singled out in the Middle Devonian deposits: heart-clay shales, meta-sandstones, conglomerates, iron and manganese ores, and orthometamorphic rocks and tuffs (Fig. 3.4). In the lowest horizons, black limestones with

Fig. 3.4 Tuffs rock near the Duga waterfall. (Photo by Aida Bidžan-Gekić)

clayey and phyllitic shales are represented, replaced by sandstones with rare carbonate inserts. In the highest horizons are orthometamorphites and tuffs, and at the very top of the pillars are sandstones with rare lenses of conglomerates. Middle Carboniferous deposits were found in Šutorinska rijeka near Milići, Mlječavska rijeka, Pašino brdo near Srebrenica as well as in Ljeskovik-Karno near Srebrenica.

Prača Paleozoic occupies an area of about 1300 km^2, stretching from the southern and southeastern slopes of Jahorina and Romanija mountains, through the valley of the Prača, Ćehotina, and Drina rivers to Foča, from where it crosses into Montenegro. The boundary of Paleozoic deposits is represented by Triassic limestones, which surround the entire area. The area of the Prača Paleozoic is a distinctly mountainous area with several high mountains: Jahorina (1909 m), Zelengora (2015 m), Volujak (2242 m), Ljubišnja (2338 m), etc. The geomorphological formation of the terrain is directly related to the lithological-petrographic composition of the substrate, in which sandstones and shales predominate, which is why these terrains are characterized by a large number of deep stream valleys, slopes, and surfaces. Silurian was found in the stratigraphy of the Prača Paleozoic. These deposits have been identified at two sites: Veliki potok (south of Prača) and Vlaška stijena, and are represented by reddish-shale-fold and crystalline-shale and fibrous limestones. Devonian deposits are represented mainly by precipitous limestones and have so far been paleontologically confirmed only along the prominent peaks of the mountains of southeastern Bosnia (example of Klek and Borovac) and in the valleys of Osanica, Drežnica, and Biturajska rijeka. In the younger Paleozoic deposits, the prominent lithological members are shales and sandstones with limestone deposits. The Lower Carboniferous was determined in the vicinity of Prača, Podgrab, Obrovci, Podkoran, in the area of Foča, in the upstream part of the Bistrica riverbed Osanica and the Odseka valley. The Middle Carboniferous is almost nowhere wholly separated and is represented mainly by sandstones and shale sandstones (Fig. 3.5). The Upper Carboniferous deposits are of continental development and were established in the vicinity of Foča, the Drina valley, and near Zubovići. Perm complexes are represented around Prača, near Han-Orahovica, in the southern foothills of Gola Jahorina, near Boljoradina, Vihar, Lunj, Razbojište, in the canyon valley of Radojna river. The Permo-Triassic sediments, occupying a large part of the area of southeastern Bosnia from Pale and Prača in the north to Šćepan-polje in the south and from Trnovo in the west to Dobro polje in the east, are much more widespread.

3.1.2 Mesozoic Deposits

The Mesozoic in Bosnia and Herzegovina is represented by sedimentary, igneous, and volcanogenic-sedimentary complexes, covering an area of about 35000 km^2. In some parts of the range, sandy marls, clays, sandy limestones, and metamorphosed quartz-mica sandstones lie concordantly in relation to the bedrock. In contrast, in many places (especially in the inner Dinarides), they lie transgressively and discordantly over older or younger Paleozoic deposits Čičić 1984). This transitional complex is usually referred to as the Permo-Triassic.

During the Mesozoic, most of the territory of Bosnia and Herzegovina belonged to the Mediterranean geosynclinals (Josipović et al. 1980). In the conditions of continuous sedimentation of marine material, over a long period, sediments have been deposited whose capacity, together with the

Fig. 3.5 Sandstone at the Drina valley. (Photo by Aida Bidžan-Gekić)

Table 3.2 Geochronological and surface relations of the main geological formations in Bosnia and Herzegovina by epoch (largest by share)

Era	Period	Epoch (age)	Lithology	P (%)
Cenozoic	Quaternary	Holocene in general	Unconsolidated sediments	8.87
Mesozoic	Cretaceous	Upper Cretaceous (santonian)	Limestones, dolomites, clastites: a-flysch	3.02
	Cretaceous	Upper Cretaceous (turonian-santonian)	Limestones, limestones with horns, marls, dolomites	4.43
	Cretaceous	Lower Cretaceous in general	Limestones, dolomites, clastites	5.61
	Jurassic-Cretaceous	Jurassic-Cretaceous	Clastites, limestones	3.32
	Jurassic	Upper Jurassic (malm)	Limestones, dolomites, breccias	3.31
	Jurassic	Middle-upper Jurassic (dogger-malm)	Limestones, dolomites, clastites: a-flysch	3.11
	Triassic	Middle-upper Triassic	Dolomites, limestones, clastites	10.17
	Triassic	Lower Triassic (werfen)	Limestones, clastites	3.99
Paleozoic	Upper paleozoic	Upper paleozoic in general	Clastites, limestones, argiloshists, phyllites	3.53
Unknown	Unknown	Unknown	Peridotites	3.09

Based on Geological map of SFRY 1974, Geological map interpreter SFRY 1974 and field research of authors

products of submarine volcanism, is in some places over several kilometers (especially in the high karst zone) (Nikolić 1989). At the end of the Mesozoic, especially during the Tertiary, their strong uptake and uplift occurred due to strong tectonic movements, and a very complex geotectonic and geomorphological structure was created.

The Triassic is a period in Bosnia and Herzegovina during which continuous sedimentation took place, although there were short periods in which there were changes in the regime of deposition and fluctuations in sea levels. Such paleogeographic conditions enabled the spread of Triassic formations in all parts of Bosnia and Herzegovina (Čičić 1984). They are less represented in the high karst zone, where they are located below the powerful Jurassic-Cretaceous complexes (Table 3.2), and where they build the nuclei of anticlines such as Ilica, Vijenac, Staretina, Krug, Čabulja, Velež, Lastva, and others. They have also been identified in the wider Mostar Basin area.

A much wider distribution of Triassic deposits is found in central Bosnia, where, with minor interruptions, they extend from Bihać in the northwest to Foča and Tjentište in the southeast. These deposits were discovered on the northeastern and eastern slopes of Majevica mountain and in the east of Bosnia, in the area of Javor mountain. The Lower Triassic sediments are represented by red and white quartz-mica sandstones, marls, marly-sandy limestones, sandy marls, sandstones, black bituminous limestones, and less frequently pure limestones and dolomites (Čičić 1978). Developments of these sediments have been determined in the peripheral area of the Drina, Prača, Sana, and Central Bosnian Paleozoic and the central Dinarides. The Middle Triassic is also represented with several developments, the most important of which are: limestones and dolomites with zones of alpine development, precipice limestones, and dolomites, as well as volcanogenic-sedimentary rocks (Fig. 3.6). They are widespread in the vicinity of Sarajevo, Vlasenica, and the outer Dinarides. Upper Triassic deposits (limestones with ammonites, limestones, and dolomites with megalodons and dolomites without fauna) were developed in the area of Romanija, Bjelašnica, Visočica, and the outer Dinarides, as well as in the vicinity of Gacko, Kupres, and the like.

The Jurassic is represented by sedimentary, igneous, and eruptive rocks that occupy an important place in the lithostratigraphic column of the Mesozoic. The presence of all superposition members of the Jurassic period has been determined, although the Dogger creations are quite difficult to determine due to the very scarce fauna they contain. In the outer Dinarides, Jurassic deposits are represented mainly by carbonate sediments. In the zone of Paleozoic shales and Mesozoic limestones zone, Jurassic sediments are represented by clasts and in the ophiolite zone by volcanogenic-sedimentary formations and limestones. Lias-Dogger deposits are widespread in the outer Dinarides and parts of the inner Dinarides and are composed of sandy-marly limestones, marls, hornblende, limestones, and dolomites. Upper Jurassic sediments are most often represented by massive limestones (outer Dinarides, Vlašić, Jajce, Mrkonjić-Grad, etc.) and flysch (south and southwest of the ophiolite zone) (Fig. 3.7).

Chalk is characterized by sediments of many different properties. The first group consists of carbonates in the so-called outer Dinaric belt, represented by mighty deposits of limestone and dolomite, up to about 2 km thick. They extend in a relatively wide belt from Bihać in the northwest to the border with Montenegro in the southeast. The area of the middle course of the Vrbas river, in which Cretaceous deposits of this type are also present, belongs to the Dinarides transition zone. A very distinct paleogeographic unit consists of sediments, the so-called Bosnian zones, represented by flysch deposits (sandstones, marls, marly and sandy limestones) (Fig. 3.8). The potential of these deposits in the average value ranges from 300 to 400 m.

The mentioned zone stretches from Banja Luka in the northwest and passes by Sarajevo and Zenica, from where it turns south to Ulog, and further along the Dinaric route, it extends to the border with Montenegro. The central ophiolite

Fig. 3.6 Middle Triassic limestones in the downstream part of the canyon valley of the Moščanica river. (Photo by Aida Bidžan-Gekić)

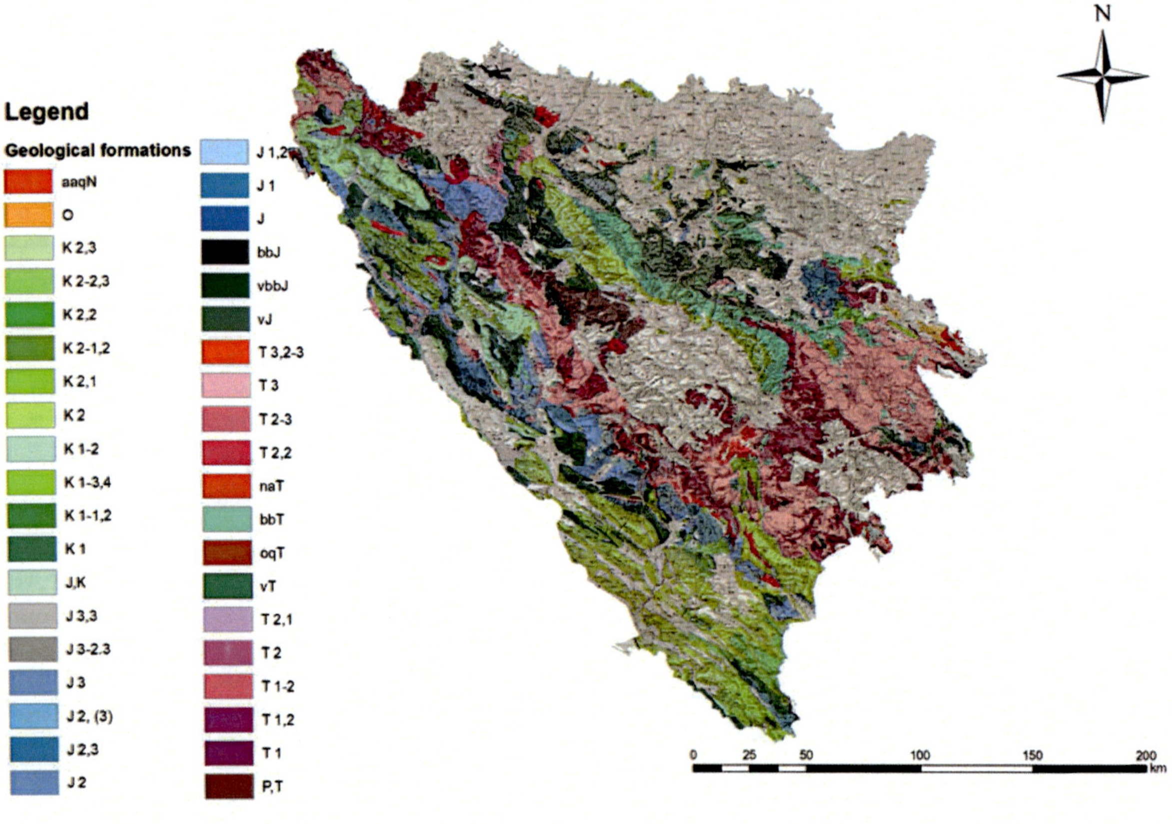

Fig. 3.7 Mesozoic geological formations of Bosnia and Herzegovina

zone is also characterized by carbonate-clastic and partly flysch sedimentation. Younger Upper Cretaceous carbonate and flysch deposits are present in northern and northeastern Bosnia.

Based on the presented facts, it can be concluded that the Mesozoic period in Bosnia and Herzegovina is characterized by exceptional tectonic activity, which resulted in the formation of striking fault zones that mark the boundaries of basic

Fig. 3.8 Flysch deposits near Sutjeska Rijeka. (Photo by Ranko Mirić)

geotectonic regions (Petković and Nikolić 1979). Within the mentioned units, numerous detected and masked faults and fault zones were formed. Numerous earthquakes of tectonic character occurred during the entire geological history of this area and in the recent period. During the last few centuries, the recorded earthquakes have been characterized by very high magnitudes and had catastrophic consequences (such as Banja Luka, the Montenegrin coast, Sarajevo, etc.).

3.1.3 Cenozoic Deposits

The Cenozoic is a period during which, on the territory of Bosnia and Herzegovina, intensive tectonic movements, intense folding, and interruptions in the deposition regime took place. Strong volcanic activity is also present, which was reflected in sedimentation conditions and general sediment types and the construction of very complex structural relationships within the lithostratigraphic column and the complex morphosculptural complex of the terrain. Numerous faults are also present in the whole area, especially in Herzegovina and western Bosnia, where older, mainly Upper Cretaceous sediments have been deposited along the reverse faults over different sections of the Paleogene (Fig. 3.9).

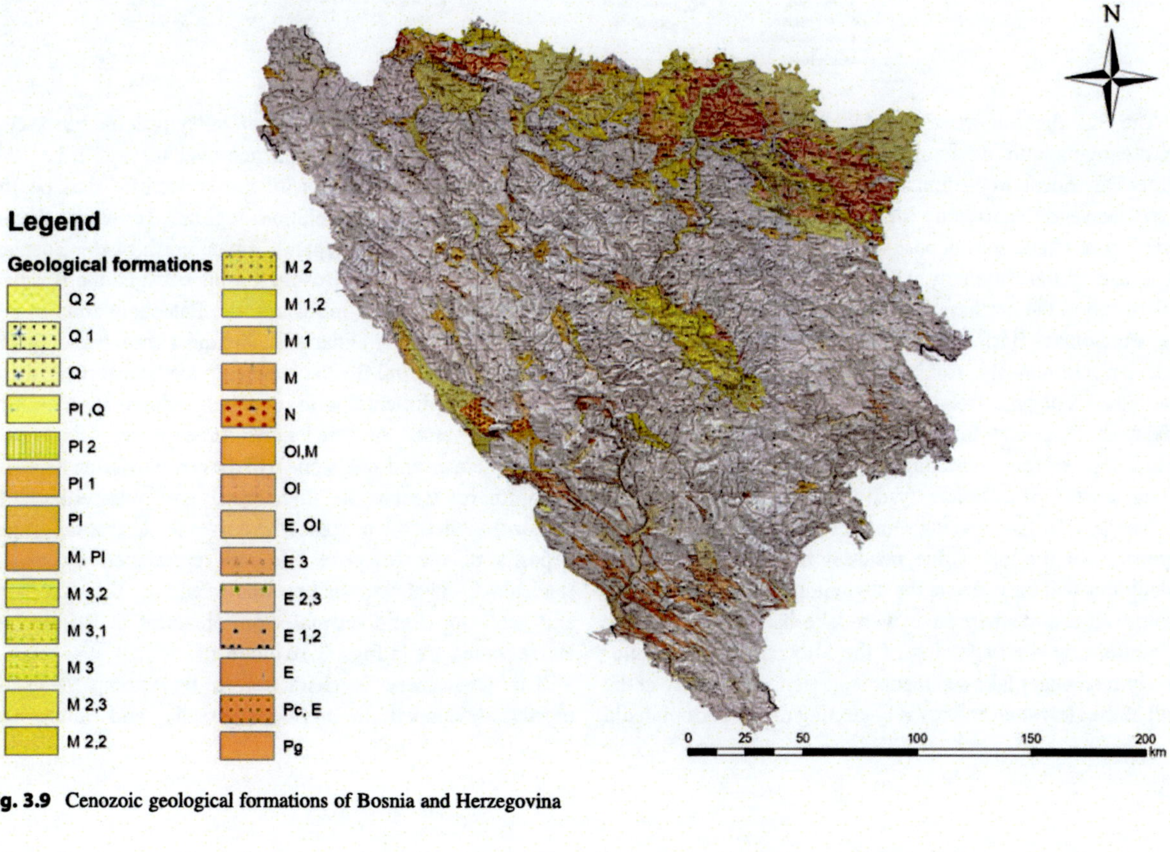

Fig. 3.9 Cenozoic geological formations of Bosnia and Herzegovina

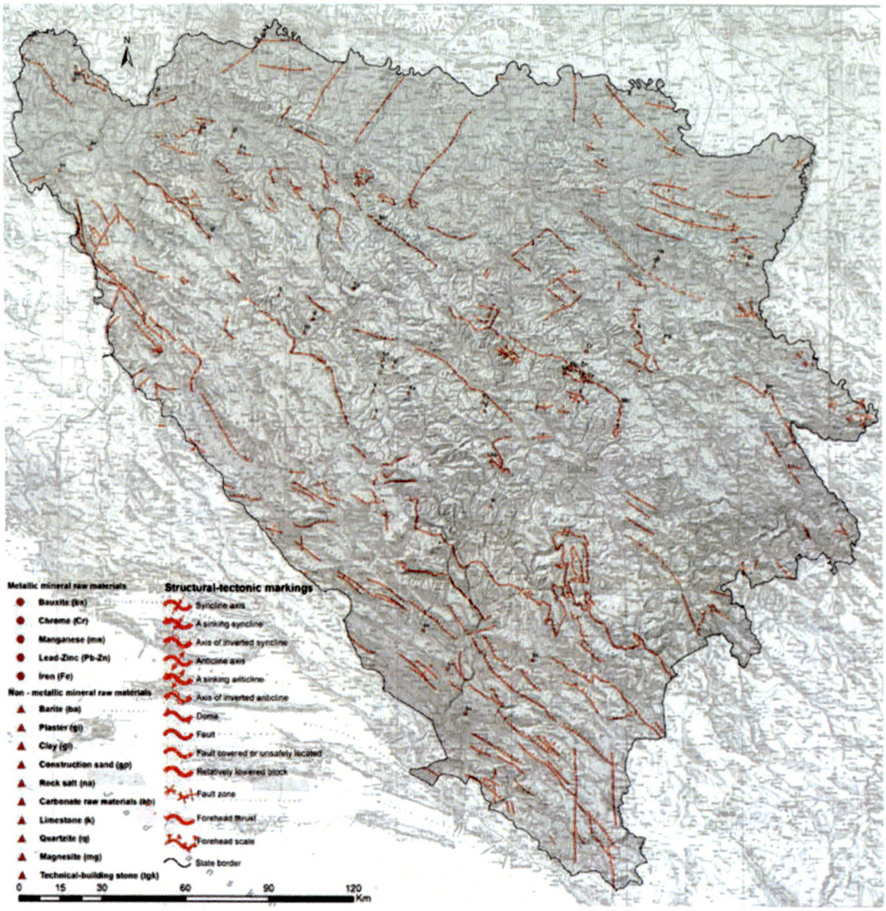

Fig. 3.10 General geotectonic features of Bosnia and Herzegovina

The Paleogene is most abundant in the outer Dinarides (Herzegovina and western Bosnia) and northern Bosnia (inner Dinarides). In the first area, it is represented by Eocene marly or sandy calcareous limestones that lie discordantly over Upper Cretaceous deposits and are not characterized by complete spatial continuity. Unlike the external dynasty, the Paleogene in the northern Bosnian area (the southern rim of the Pannonian Basin—the zone of trenches and horsts) extends continuously, from the Una river through Kozara, Motajica, Vučjak, Trebovac to the eastern branches of Majevica. Paleocene-Eocene limestones and clasts represent it in a total of three horizons, the third of which has all the characteristics of complete flysch (Čičić 1977).

Neogene in the Bosnian Dinarides occurs in two forms: terrestrial-limnic and marine, resulting from paleogeographic conditions that prevailed at the beginning of this period when almost all sedimentary areas were lake basins. Under these conditions, at the beginning of the Miocene, almost exclusively freshwater lake sediments were deposited. Only at the end of the Helvetian and at the beginning of the Tortonian did the southern part of the Pannonian valley become block-cut, with the descent taking precedence over the uplift, i.e., the ingression and transgression of the Paratethys, thus beginning the marine sedimentation regime. Terrestrial-limnic deposits occur in three separate areas: north of the Bosnian serpentine zone, central Bosnia (basins between the Bosnian serpentine zone in the north and the Paleozoic zone in the south), and the outer Dinarides. The most complete deposits are present in central Bosnia, where, in the lowest parts, there are pelitic sediments with coal on which marly-sandy formations rest. At the same time, there are clasts (conglomerates and sandstones) in the cover. Marine deposits in northern Bosnia lie transgressively over either the Mesozoic-Paleozoic or somewhere over the terrestrial-limnic deposits of the Miocene and are represented by marls, sandstones, clays, various limestones, and the like. The general geotectonic and seismotectonic features of Bosnia and Herzegovina are in Figs. 3.10 and 3.11.

The Quaternary is characterized by various deposits, mostly developed in northern Bosnia and along the

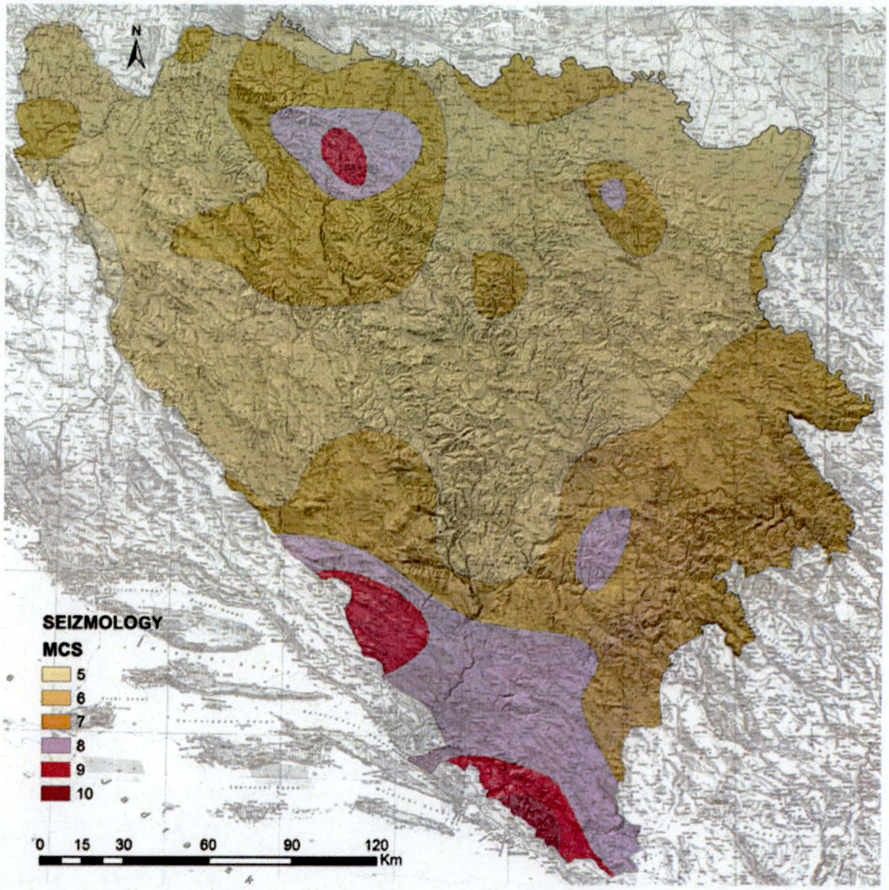

Fig. 3.11 General seismotectonic features of Bosnia and Herzegovina (MCS—Mercalli–Cancani–Sieberg scale)

southwestern edge of the Pannonian Basin, and they are present in intermountain Quaternary depressions such as Spreča, Prijedor, Sarajevsko polje, and fields in southwestern Bosnia and Herzegovina. The sinking of the terrain was compensated by the deposition of river and land sediments. The melting of the Dinaric and Alpine glaciers at the end of the last glacial in the relief of Bosnia and Herzegovina is observed as a section of a large Pleistocene terrace known as another terrace that occupies a large area.

In the valleys and karst fields of Bosnia and Herzegovina, Quaternary deposits are well preserved on the bottom and sides and on the slopes of the surrounding mountains. They are much thicker on the lower and flatter terrains and are represented by alluvial detritus of different lithological-petrographic compositions, which reflect the geological composition of the landscape where the erosion took place. They are mainly of the Young Pleistocene and Holocene ages and often fill the bottoms of caves as well. Holocene sediments are represented by deposits of river and lake origin and tufa deposits, peat bogs, etc. (terrigenous deposits and molasses). They are of great importance in determining current geological processes.

References

Anđelković M (1978) Tektonska rejonizacija Jugoslavije [Tectonic regionalization of Yugoslavia]. Zbornik radova IX kongresa Geologa Jugoslavije, Sarajevo

Anđelković M (1988) Geologija Jugoslavije – Tektonika [Geology of Yugoslavia – Tectonics]. IRO Građevinska knjiga, Beograd

Čičić S (1977) Geologija Bosne i Hercegovine – Kenozoik [Geology of Bosnia and Herzegovina – Cenozoic]. Geoinžinjering, Sarajevo

Čičić S (1978) Geologija Bosne i Hercegovine – Magmatizam i metalogenija [Geology of Bosnia and Herzegovina – magmatism and metallogeny]. Geoinžinjering, Sarajevo

Čičić S (1979) Geologija Bosne i Hercegovine – Paleozoik [Geology of Bosnia and Herzegovina – Paleozoic]. Geoinžinjering, Sarajevo

Čičić S (1984) Geologija Bosne i Hercegovine – Mezozoik [Geology of Bosnia and Herzegovina – Mesozoic]. Geoinžinjering, Sarajevo

Drešković N, Mirić R (2017) Regionalna geografija Bosne i Hercegovine I [Regional geography of Bosnia and Hercegovina I]. Prirodno-matematički fakultet Univerziteta u Sarajevu, Sarajevo

Geological map interpreter SFRY 1974, 1:500,000, Savezni geološki zavod Jugoslavije, Beograd

Geological map of SFRY 1974, Sarajevo, Zagreb, Belgrade, Savezni geološki zavod Jugoslavije, Beograd

Herak M (1987) Geologija – Tektonika i dinamika Zemlje [Geology – tectonics and dynamics of earth]. Školska knjiga, Zagreb

Josipović J, Mojičević M, Ostojić Đ, Jahić M (1980) Paleohidrogeološke karakteristike Bosne i Hercegovine [Paleohydrogeological

characteristics of Bosnia and Herzegovina]. Geoinžinjering, Sarajevo
Nikolić P (1989) Geotektonika [Geotectonics]. Naučna knjiga, Beograd
Nikolić P (1990) Osnovi geologije i Opšta geologija [Fundamentals of geology and general geology]. Naučna knjiga, Beograd
Petković K, Nikolić P (1979) Osnovi geologije [Fundamentals of geology]. Naučna knjiga, Beograd
Scotese CR (1989) Mezozoic and cenozoic plate reconstructions. Elsevier, Oxford

4 Geomorphology of Bosnia and Herzegovina

Abstract

Geomorphological characteristics of the territory of Bosnia and Herzegovina in this chapter are observed from the aspect of morphometric, morphological, and morphogenetic features of the terrain. A detailed presentation of hypsometric characteristics is presented and elaborated from the spatial aspect through differentiation at the regional level, explaining the most important morphostructural elements and natural-geographical features (climatic-hydrographic and pedological-vegetation elements). The second part of the chapter presents the morphogenetic characteristics of the relief of Bosnia and Herzegovina from the aspect of dominant fluvial-denudation, fluvial-accumulative, fluvial-karst, karst, glacial-nivation, and abrasion types. The spatial distribution of defined morphogenetic types, with accompanying geomorphological forms followed by pictorial and tabular representations (area share of main morphogenetic types with lithological composition and relief classes by altitudes), is presented in detail in the text.

Keywords

Morphology · Karts · Hypsometry · Mountains · Physical geography

4.1 Morphological Characteristics of the Relief

From the geomorphological aspect, Bosnia and Herzegovina belongs to the western part of the Mediterranean zone of the younger mountain ranges. The skeleton of the modern relief was built during the Alpine tectonic cycle, i.e., during the Old Alpine tectogenesis when the outer, central, and inner Bosnian Dinarides were separated. During the Mesoalpine tectogenesis, basic tectonic overthrusts were formed, and in the Young Alpine one, a modern morphostructural basis of the relief was formed with a period of intensive folding, faulting, and overthrusting (Paleogene tectonics) and a period of radial tectonics and fracturing of already formed structures which lasts today (neotectonic period) (Drešković and Mirić 2017). These processes are manifested as differential movements of the earth's crust in the form of relative vertical displacements; i.e., some blocks rise and others sink, with the entire Dinaric massif having a rising trend. These modern tectonic relief lines are partially modeled by the subsequent action of a total of four Quaternary glaciations with periods of interglaciation and eustatic sea-level fluctuations. Recent exogenous processes in humid climate conditions have formed a very complex morphosculptural complex of the terrain of Bosnia and Herzegovina. From the aspect of morphometry, the relief of Bosnia and Herzegovina mostly belongs to the middle mountain relief (about 35.6%), while at altitudes of over 1500 m, it accounts for about 3% (Table 4.1).

Some parts of the terrain have a deep vertical and horizontal breakdown, like the mountainous area in the central Dinarides, with high, almost vertical, angles of inclination and, in general, a very pronounced total relief energy (Fig. 4.1). In contrast, low-lying areas in northern Bosnia are characterized by a low degree of overall fragmentation and larger complexes of low-level land within the mountainous area (Bušatlija 1972). In combination with climatic elements and geological background, these general morphological relations directly reflected on the biogeographical, especially phytogeographical specifics of Bosnia and Herzegovina. According to the above indicators, the area of Bosnia and Herzegovina can be divided into three natural geographical regions (Bušatlija 1983).

The northern parts of Bosnia and Herzegovina-Peripannonian Bosnia belong to the southern edge of the Pannonian Basin. From the geomorphological aspect, it is the area of a slightly undulating plain divided by numerous river flows. The relief highlights positive morphostructures in the form of ridges and massifs (antiforms), often of the

H. Gekić et al., *The Geography of Bosnia and Herzegovina*, World Regional Geography Book Series,
https://doi.org/10.1007/978-3-030-98523-3_4

Table 4.1 Hypsometric characteristics of the relief of Bosnia and Herzegovina

Altitude (m)	Area (%)
0–200	13.53
201–500	26.16
501–1000	35.64
1001–1500	21.64
1501–2000	2.99
Above 2000	0.04
Total	**100.00**

Dresković and Mirić (2017)

Fig. 4.1 General morphological characteristics of the relief of Bosnia and Herzegovina

monoclinic type, and negative morphostructures—basins and depressions (synforms). In the western part of the area, Kozara mounting stands out, and, in the eastern part, Ozren and Majevica mountains with an average height of about 950 m. Basins and Quaternary depressions represent the lowest parts of the terrain. The largest basins are: Prijedor-Omarska-Dubica, Podgraci-Vrbaška, Prnjavor, Srednjo-Posavski, and Tuzla. Quaternary depressions are: Ivanjsko-omarska, Lijevče polje, Bosanskobrodska, Srednjoposavska, Semberija, and Spreča. Prosara, Motajica, Vučjak, Trebovac, and others rise from these lowest terrains with heights from 300 m to 700 m.

In the internal structure of most of the Posavina mountains, very rugged thrust faults have been found. Their geological composition consists mainly of shales, flysch, volcanic deposits, fewer carbonates, etc. On the slopes of the mountains and in the river valleys, fluvial, defluxion, diluvial, and disruption-dispersion morphosculpture dominate. To the south, the terrain slowly rises to 1000 m, which is also the border of this area with central Bosnia. The backbone of the hydrographic network of this area consists of larger Posavina rivers, right tributaries of the Sava River: Una, Sana, Vrbas, Bosna, and Drina, with their tributaries. In addition to hydrographic, these watercourses also have a very important climatological function because their valleys draw air masses from the north, northwest, and northeast deep into the interior of the mountain-valley area (Fig. 4.2).

More specifically, the temperate-continental climate prevails in this area, with quite pronounced thermal contrasts: harsh winters and warm summers. The thermal regime at the average annual temperatures shows zonal characteristics: the warmest areas are in the northeast, from where the isotherm

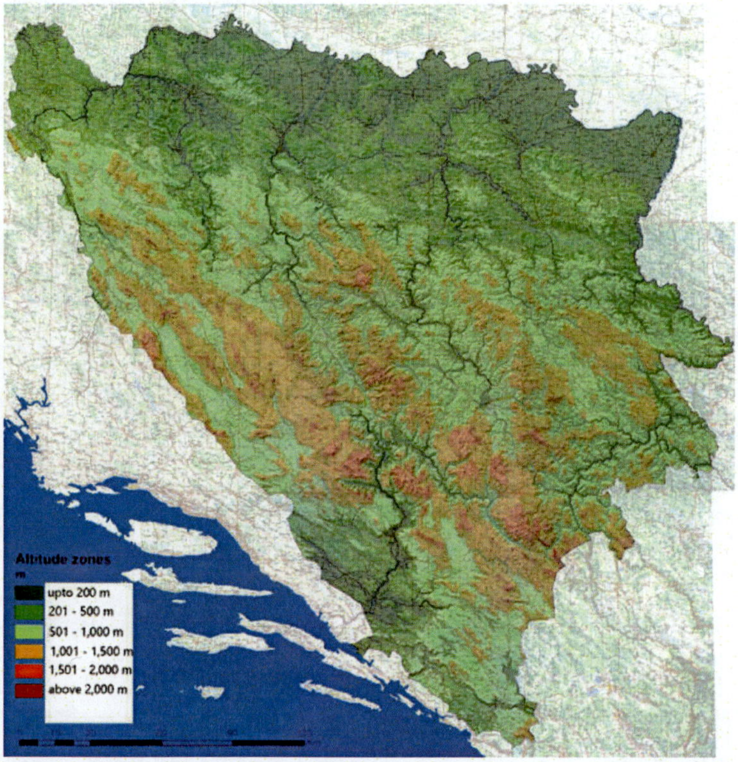

Fig. 4.2 Morphological-morphometric characteristics of the relief of Bosnia and Herzegovina

values decrease towards the mountain-valley zone of Bosnia and Herzegovina. Annual precipitation heights range from 700 mm to 1200 mm. The structure of precipitation is dominated by rain, with snow being a characteristic of all winter months.

Most of the territory of Bosnia and Herzegovina is a mountain-valley area with inter-mountain, inner mountain, and foothill relief, which stretches from Peripannonian Bosnia in the north to the fold zone between the outer and central Dinarides, which consists of high Dinaric mountains: Grmeč, Srnetica, Vitorog, Vranica (Fig. 4.3), Bitovinja, Bjelašnica (Fig. 4.4), Treskavica, Zelengora, and Maglić to the south (Drešković and Mirić 2017).

This is a distinctly mountainous area dominated by a sharia structure with wide overthrusts. Mountains are of tectonic antiforms (horst anticlines and anticlines) with heights often over 2000 m and epigenetic forms (Bušatlija 1977). They have a very inconsistent and stepped longitudinal profile. Their provision's direction depends on their angle in relation to the directrixes of the basic morphostructures, transverse, diagonal, or transverse. This type of valley has, in the upper part of the flow, mostly all right tributaries of the Sava: Una with Sana and Unac, Bosna with Usora, Spreča, Krivaja, and Lašva, and Drina with Prača and Drinjača. Slightly larger areas with low terrain are found in the central part of the area, and among them, the most famous is the Sarajevsko-Zeničko polje, as well as some terrains in the composite extensions of Drina river (Višegrad, Foča, and Goražde). They are tectonically predisposed, filled with Neogene carbon deposits, and subsequently formed mainly by a fluvial process. The overall morphosculpture is dominated by fluvial forms represented, on the one hand, by the riverbed and valley sides and, on the other hand, by proluvial alluvium, alluvial accumulative terraces, alluvial field plains, and river islands. On the sloping sides, there is a sloping morphosculpture formed by the processes of sheet washing, gullying, earth flowing, and sliding. In addition to them, glacial, paleoabrasion, and paleoeol forms are found in fragments, most often as relics. Within this area, several tectonic units can be distinguished: central ophiolite zone, anticlines of Drina and Prača Paleozoic, Jurassic and Cretaceous flysch zone, Central Bosnian shale mountains, Manjača-Grmeč zone, Sana-Una Paleozoic, and several freshwater Neogene basins.

"Climatic characteristics in these areas show more pronounced thermo-pluviometric contrasts compared to the previous area. More specifically, the basic characteristics of this climate are harsh winters with absolute temperature minimums up to −35.0 °C, while summers are warm and with absolute maximum temperatures up to 36.0 °C" (Drešković and Mirić 2017). The average annual rainfall is from 1000 mm to 1200 mm. The structure of precipitation in the lower hypsometric levels is dominated by rain, while in the highest hypsometric levels, it is dominated by snowfall (Drešković and Mirić 2014).

Fig. 4.3 Vranica Mountain—Prokoško Lake. (Photo by Nusret Drešković)

Fig. 4.4 Bjelašnica—Olympic Mountain. (Photo by Aida Bidžan-Gekić)

South of mountain-valley Bosnia to the state border with Croatia and Montenegro, in the southwest, south, and southeast, there is an area of Bosnian high karst. It is an area of very powerful karst deposits, whose depth is often over 3000 m. Mountain systems have a predominantly Dinaric direction of extension (northwest-southeast), dominated by mountains with altitudes of over 2000 m: Prenj, Čvrsnica (Fig. 4.5), Vran (Fig. 4.6) and Cincar. There are mainly covers and spines in their internal structure that are subsequently very fractured and broken by radial processes. Between the mountains or over them, vast plateaus and karst-corrosive surfaces and fields have developed, gradually descending towards the Adriatic basin. A small number of fields have been cut into Paleogene flysch deposits (Dabar and Fatnica), while most of them have been developed on limestones and dolomites. Among the most important are: Glamočko (Fig. 4.7), Livanjsko, Duvanjsko, Kupreško, Nevesinjsko, Popovo, Gatačko, and other smaller fields. Most of these fields have fissure channels that have a dual hydrographic function: during floods, they become springs, so the bottoms of the fields flood and become lakes, while during drought, lake water flows into the interior of the limestone section.

The dominant morphosculpture was developed on a limestone-dolomite substrate and is represented by a karst-corrosion relief of the Holokarst type, with all possible forms:

Fig. 4.5 Massif of Čvrsnica. (Photo by Ranko Mirić)

Fig. 4.6 Massif of Vran. (Photo by Aida Bidžan-Gekić)

Fig. 4.7 Glamočko polje—view from Cincar. (Photo by Nusret Drešković)

karst fields, karst bays, various sinkholes, ravines, shells, muzzles, oysters, bogs, etc. Macro- and most mesoforms are tectonic predisposed and then shaped by karst-corrosion, partly fluvial and gravitational, process (polyphase). The Holokarst is also typical of numerous karst underground forms, which are also characterized by characteristic morphosculpture.

Of the watercourses of this area, Neretva river is significant with its tributaries, which is deeply cut into the mountain morphostructure, which, among other things, has a climatic significance (Fig. 4.8). In addition, valleys are particularly pronounced. According to their morphology, they can be completely closed (steephead, blind) or open, and according to their hydrological function, they can be dry or periodically supplied with water. Some of them have streams where water sinks along the bottom or perimeter of the field (blind streams), such as Trebišnjica, Zalomka, Šujica, etc. This region is under the dominant influence of the Adriatic Sea, which significantly increases the thermal-pluviometric values. Average annual temperatures range from 12 °C to 15 °C. At the level of values of annual and monthly isotherms, it can be observed that in the thermal regime of the whole region, positive temperatures prevail; i.e., there are positive assumptions for the development of grass vegetation. Average January temperatures are quite high and range from 3 °C to 5 °C, while summers are dry and hot (absolute maximum temperatures from 40 °C to 45 °C). The average annual rainfall ranges from 1200 mm to above 2000 mm in mountainous areas. Snow is rare, though not impossible, occurrence. This orographic structure of the terrain of Bosnia and Herzegovina has a sizeable modifying effect on its natural basis. The area of Peripannonian Bosnia is wide open to the influence of cold continental air coming from the north and northeast. In the valleys of larger rivers (Una with the Sana, Vrbas, Bosna, and Drina), which have a predominantly meridional flow direction, these influences extend deep into the interior of the mountain-valley area. But, despite the vast openness of the Pannonian Basin, it can be stated that the penetrations of these air masses are significantly reduced by the provision of the Alpine, Carpathian, and Balkan massifs

Fig. 4.8 Neretva river in Konjic. (Photo by Aida Bidžan-Gekić)

that enclose this basin from the north, northeast, and east. Cold air penetrations are possible along with the lower passes and the Danube valley from the north and the Đerdap gorge from the east.

The southern and southwestern parts of Bosnia and Herzegovina stretch directly along the Adriatic coast. The westernmost point is located on Plješevica, only about 60 km as the crow flies from the sea. Going to the southeast, it is getting closer to the coast, and in large part, that distance is only 1 km, and in the bay of Neum – Klek, the border breaks out on the seashore.

Given the above situation, it would be logical to expect that maritime impacts are felt deep in the interior of Bosnia and Herzegovina, which in reality is not the case. Namely, the coastal façade of the Dinara and Velebit mountain systems limits these impacts to the narrow coastal area. Deeper into the continental interior, the Adriatic climate penetrates the Neretva valley and its tributaries, although this influence is inversely related to the distance from the coast. In fact, the actual Adriatic climate is felt approximately as far as Mostar, and in a significantly modified variant, these influences spread in the upper part of the stream and the valleys of its tributaries, and the lateral karst fields. It is also important to note that this modified variant of the Adriatic climate does not change the continental character of these areas but somewhat mitigates it. Its presence can be determined based on certain types of vegetation characteristic only for the Mediterranean.

4.2 Morphogenetic Characteristics of the Relief

In the conditions of the mentioned geological-tectonic relations and under the direct influences of the climate and the river network, a very diverse morphosculpture has been developed, genetically conditioned (Petrović 1982). More specifically, in the area of the Bosna River Basin, various accumulative-erosive forms of fluvial, pluvial, nival, permafrost, karst, and gravitational morphosculpture have been developed. The overall morphosculpture is dominated by fluvial forms. Fluvial-erosive forms are primarily represented by riverbeds and valley sides and numerous other denudation products that take place on clastic sediments. There is a sloping morphosculpture formed by fluviodenudation processes of sheet washing, abrasion, shedding, earth flowing, and sliding on the sloping sides. These processes deliver the productive pedological layer and unbound rock material to the immediate alluvial plateau (especially on the steep valley sides of those watercourses whose spring catchment area is located in the area of the most prominent mountain peaks). The erosive operation of watercourses is evident in the processes of regressive erosion, under the influence of which watersheds are lowered and destroyed, so watercourses often take on stream capture characteristics. The intensity of fluvial destruction is visible mainly along the entire length of the left and right tributaries of the Bosna river, except in the immediate zone of the mouth. Therefore, according to the geological structure of the terrain of these flows, the longitudinal profile is highly inconsistent and is characterized by the presence of numerous small rapids and waterfalls whose dimensions are often decameter. The morphological development of this type of relief is represented by all energy classes—from 1 to 8 (Zeremski and Bušatlija 1992).

Fluvial-accumulative forms are present on low soil complexes up to about 500 m high, i.e., along the alluvial terrace of Una river with Sana, Vrbas with larger tributaries, Bosnia with larger tributaries, Drina with larger tributaries, Sava with larger tributaries (in the immediate catchment area), and Neretva with larger tributaries. They are represented by proluvial alluvium, alluvial accumulative terraces, alluvial flat plains, and river islands (Fig. 4.9).

Fig. 4.9 Plains near Velika Kladuša. (Photo by Ranko Mirić)

The development of this type of relief can generally be classified into all eight energy classes. Glacial-nivation-cryogenic morphosculptures have recently been encountered only as relict forms of relief, which only with a detailed reconstruction indicate their origin. They are developed only on the ridges and massifs of the mountain-valley area at altitudes above 1800 m. The exogenous type of morphosculpture is represented by the remains of the exacerbation process performed by the snow-ice mass during the Pleistocene glaciation. These are, first of all, relics of cirques and waves and microforms such as striae, sheepback, etc. Accumulative forms are represented by walls of moraine material or accumulations of sheet-glacial or fluvial-glacial material on plateaus at the bottom of slopes at altitudes between 1200 m and 1700 m. A special place in the formation of glacial relief belongs to the phenomena in diluvium, which is manifested in the highest hypsometric regions with remnants of glacier and moraine movements, leaving very often bare rock or large blocks of stones. Large slopes and sharp ridges in zones with this type of morphosculpture are expected, conditioning particular soil formation and maintenance dynamics. In addition, it is possible to determine the presence of smaller morphological bays whose genesis is more related to the combined work with constant running water than exclusively to the karst process.

The fluvial-karst morphosculpture is widespread on the limestone-dolomite substrate and is represented by numerous macro-, meso-, and micro-surface and underground relief forms. The most significant and seemingly most frequent representatives of surface forms of karst relief are karst fields, karstic depressions, sinkholes, and different types of karrens. Karst fields are the most striking surface forms in karst and are characterized by the presence of all other forms of karst morphosculpture, which is why they have the characteristics of the Holokarst (Musa and Drešković 2005). The most typical and most developed karst fields are formed in the region of southwestern Bosnia and Herzegovina. The largest karst fields are: Livanjsko polje (405 km^2), Nevesinjsko polje, Popovo polje, Gatačko polje, Glamočko polje, Duvanjsko polje, and others (Milanović 2006). In addition to the above, there are also examples of mero-karst (in which karstic depressions are present), doline karst (with different types of sinkholes), etc. In this region, there are numerous examples of limestone-dolomite rock material exposed karst. The processes of peneplanation and terrain leveling combined with the non-tectonic uplift of the structural base are especially marked by the formation of karst plateaus. Underground karst morphosculpture also developed on the dominant limestone-dolomite substrate. The prominent erosive representatives are numerous caves (over 30), among which the most famous are: Vjetrenica, Orlovača, Bijambare (Upper, Lower (Fig. 4.10) and Đuričina) caves, Megara, Pašina, Klokočevica, and many others. Numerous highly developed karst accumulative forms such as stalactites and stalagmites, stalactones, cave draperies, etc., have been identified in them.

Fig. 4.10 Bijambare Lower cave. (Photo by Aida Bidžan-Gekić)

One of the unique products of the fluvial-karst process of limestone substrate dissolution is visible through scarce pedological accumulations in the form of red (or terra rosse). The intensity of the entire fluvial-karst process is visible, especially in the climatic region of Mediterranean Bosnia and Herzegovina (or Herzegovina). In addition to the above, physical (temperature and frost) destruction and biological decomposition of the parent rock material can be singled out as a significant geomorphological agent (Červanjov and Bušatlija 1986). This process significantly contributes to the overall surface erosion. The most common way of transporting material from the place of origin at mountain peaks are grooves of decimeters to meter dimensions called screes. They end in direct contact with the alluvial plains, and in these places, smaller or larger rock accumulations of varying degrees of fragmentation and shape are formed, called the talus. By merging several different talus cones into one, scree slopes are formed. These are the sites of the main accumulation of destroyed rock material and pedological substrate on the sloping sides, which, furthermore, represent source areas for additional geomorphological processes. The intensity of these processes is directly dependent, in addition to climatic specifics, on the overall (horizontal and vertical) relief and the type of geological base.

Very illustrative examples of the diversity of this type of morphosculpture are slumps and creeps. These are pieces of rock material of very high capacity, which were transferred to the bottom slope parts after being torn from the parent

Table 4.2 Morphogenetic relief types of Bosnia and Herzegovina

Morphogenetic type	Lithology	Class of relief (m)	Area (%)
Fluvial-denudation	Alternation of clays and sandstones, sand and gravel, limestones outside karst and fluvio karst, clays and metamorphites, metarnophytes, alternation of fine clasts with diabases, magmatites, fine clasts (clays), magmatites	5–30	0.71
		30–100	4.65
		100–300	14.68
		300–800	23.71
		>800	3.12
Fluvial-accumulation	Fine clasts (clays), sand and gravel, alternation of clays and sandstones, metamorphites, coarse detritus	0–5	6.98
		5–30	10.09
		30–100	0.37
		100–300	0.39
		300–800	0.10
		>800	0.06
Fluviokarst	Alternation of clays and sandstones	30–100	0.39
		100–300	1.68
		300–800	0.22
		>800	0.04
Karst	Fine clasts (clays), metamorphites, coarse clasts (sandstones), alternation of clays and sandstones, magmatites	0–5	0.48
		5–30	2.16
		30–100	3.09
		100–300	10.30
		300–800	11.69
		>800	0.69
Glacio-nivation	Metamorfites	5–30	0.10
		30–100	0.23
		100–300	1.32
		300–800	0.71
		>800	0.30
Abrasion	Alternation of clays and sandstones	0–5	0.34
		5–30	0.91
		30–100	0.08
		100–300	0.35
		300–800	0.04

Based on Geomorphological map of SFRY 1992 and field research of authors

carbonate-dolomite complexes. At the places of separation, traces are visible in the form of slump niches and drips. Their deposition is mainly chaotic, given that the length of transport is determined by their capacity, intensity of agents, etc. The main source areas are higher hypsometric levels of most morphostructures, especially in the mountain-valley macroregion of Bosnia and Herzegovina, as indicated by their identical mineralogical-petrographic composition.

The high morphological fragmentation of the relief was reflected in the dynamics of the formation of a large number of slopes of very different degrees of development. In relation to the five-point scale, all categories of slopes are visible, although areas with a specific predominant range can be singled out. Small angles of inclination (from 0° to 10°) are present in the areas of alluvial plateaus of all main rivers and their larger tributaries in both catchment areas, in downstream flow segments, and plateau areas on higher hypsometric floors.

In terms of morphological function, these are areas of recent accumulation. Larger angles of inclination (over 45°) have been developed in the zones of highly dissected relief belonging to the source ridges and middle parts of the basin, especially in the mountain-valley macroregion of Bosnia and Herzegovina. Intensive destructive processes of sheet washing, sliding, gullying, earth flowing, shedding, landsliding, and dissolving the parent rock substrate are visible on these slopes. The high dynamics of the sloping relief also resulted in a vast scale of its exposure. In this regard, the presence of all slope exposures in relation to the scale of eight sides of the world can be stated, except in the zones of flat terrain. To a large extent, their greatest intensity overlaps with the more sloping relief in the zones of the source headwaters of the watercourses and the highest mountain peaks (Table 4.2 and Fig. 4.11).

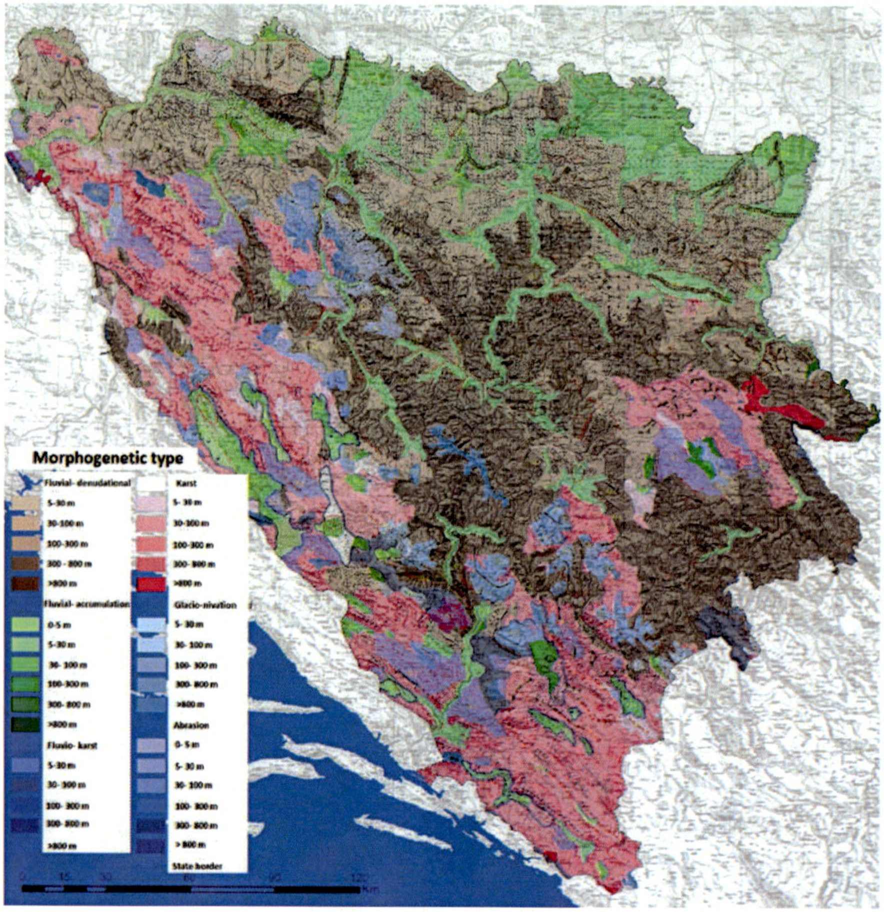

Fig. 4.11 Morphogenetic types of relief of Bosnia and Herzegovina. (Based on Geomorphological map of SFRY 1992 and field research of authors)

References

Bušatlija I (1972) Morfostrukturne i morfoskulpturne karakteristike reljefa Bosne i Hercegovine [Morphostructural and morphosculptural characteristics of the relief of Bosnia and Herzegovina]. IX kongres geografa Jugoslavije. Geografsko društvo Bosne i Hercegovine, Sarajevo

Bušatlija I (1977) Geologija Bosne i Hercegovine, knjiga III- Kenozojske periode – regionalni dio – visoke planine [Geology of Bosnia and Herzegovina, book III- Cenozoic periods – regional part – high mountains]. Geoinžinjering, Sarajevo

Bušatlija I (1983) Principi geomorfološke rejonizacije tektonike i reljefa Jugoslavije [Principles of geomorphological regionalization of tectonics and relief of Yugoslavia]. Geografski pregled 26–27:111–120

Červanjov I, Bušatlija I (1986) Novi aspekti morfološke analize reljefa [New aspects of morphological analysis of relief]. Geografski pregled 28–29:5–14

Drešković N, Mirić R (2014) Klimatska regionalizacija Bosne i Hercegovine [Climate regionalization of Bosnia and Herzegovina]. Zbornik radova III kongres geografa Bosne i Hercegovine. Posebna izdanja Geografskog društva u F BiH, Sarajevo, pp 280–294

Drešković N, Mirić R (2017) Regionalna geografija Bosne i Hercegovine I [Regional Geography of Bosnia and Hercegovina I]. Faculty of Science, University of Sarajevo, Sarajevo

Milanović PT (2006) Karst Istočne Hercegovine i Dubrovačkog primorja (Karst of Eastern Herzegovina and Dubrovnik litoral). Asocijacija speleoloških organizacija Srbije, Beograd

Musa S, Drešković N (2005) Mokro polje u zapadnoj Hercegovini – pull apart struktura [Mokro field in western Herzegovina – pull apart structure]. Zbornik radova I kongresa geografa Bosne i Hercegovine. Geografsko društvo u Federaciji BiH, Sarajevo

Petrović D (1982) Geomorfologija [Geomorphology]. Građevinska knjiga, Beograd

Zeremski M, Bušatlija I et al (1992) Geomorfološka karta Jugoslavije [Geomorphological map of Yugoslavia], 1:500 000. Savez geomorfoloških društava SFR Jugoslavije, Beograd

5 Climatology of Bosnia and Herzegovina

Abstract

The general climatic characteristics of Bosnia and Herzegovina are observed from the aspect of the most important climatic factors and elements. The emphasis is on the influence of the relief structure through the exposure of the northern parts to the impact of cold continental air coming from the north and northeast and the limitation of the influence of the Adriatic Sea to the mountains of the Dinaric direction in the south. Also, an important aspect is the barometric depressions, i.e., the distribution and directions of air mass movement through the territory of Bosnia and Herzegovina. These most important elements influenced the characteristics of basic meteorological elements; i.e., temperature and precipitation, for which spatial models for Bosnia and Herzegovina (with longitudinal and transversal profiles) were developed using qualitative methodological kriging analyzers and presented with a large number of analytical thematic maps and graphs. Based on the previous contents, in the following part, the climatic regionalization of Bosnia and Herzegovina, according to Köppen's classification, is elaborated. A detailed division into climate classes, types, subtypes, and variants was made, which were interpreted in great detail from the aspect of spatial coverage and general climatic characteristics and their share in the area of Bosnia and Herzegovina. Thematic graphic and cartographic representations accompany the description of hierarchical units classified in the climate distribution.

Keywords

Climate · Weather · Temperature · Precipitation · Physical geography

5.1 General Climatic Characteristics

The general climatic characteristics of Bosnia and Herzegovina are determined by the relationship between the interaction of modifier influences of climatic factors on the one hand and the quantitative and qualitative values of climatic elements on the other. The most important climate modifier is the orographic structure of the terrain of Bosnia and Herzegovina. As already pointed out in the previous chapter, the area of Peripannonian Bosnia is wide open to the influence of cold continental air coming from the north and northeast. In the valleys of larger rivers (Una with the Sana, Vrbas, Bosnia, and Drina), which have a predominantly meridional flow direction, these influences extend deep into the interior of the mountain-valley area (Đug and Drešković 2008). But, despite the vast openness of the Pannonian Basin, it can be stated that the penetrations of these air masses are significantly reduced by the provision of the Alpine, Carpathian, and Balkan massifs that enclose this basin from the north, northeast, and east. Cold air penetrations are possible along with the lower passes and the Danube valley from the north and the Đerdap gorge from the east (Drešković 2011). Climate analysis is based on meteorological data from 1980 to 2010 (Meteorological data 1980–2010).

The southern and southwestern parts of Bosnia and Herzegovina stretch directly along the Adriatic coast. The westernmost point is located on Plješevica, only 60 km as the crow flies from the sea. Going to the southeast, it is getting closer and closer to the coast, and in large part, that distance is only 1 km, and in the bay of Neum-Klek, the border breaks out on the seashore. Given the above situation, it would be logical to expect that maritime impacts are felt deep in the

H. Gekić et al., *The Geography of Bosnia and Herzegovina*, World Regional Geography Book Series,
https://doi.org/10.1007/978-3-030-98523-3_5

interior of Bosnia and Herzegovina, which in reality is not the case given that the mountains of the macroregion of high karst and mountain-valley macroregion of Bosnia and Herzegovina orographically limit the impacts of the changed Adriatic climate to macroregions of low Herzegovina.

5.2 Barometric Depressions

Weather conditions and general climatic characteristics of Bosnia and Herzegovina on an annual basis are predominantly determined by the influence exerted by two different types of air masses. The first is air masses of low air pressure—barometric depressions that form over the northwestern part of the Atlantic Ocean and the Mediterranean Sea. The second type consists of air masses of high air pressure—anticyclones that form over the area of the large islands of the North Atlantic, i.e., above the Eurasian and northern parts of the African mainland. Their climatic influence is directly dependent on the physicochemical characteristics and the size and frequency of air masses that operate throughout the year or in part of it in the wider region of Bosnia and Herzegovina. The more pronounced influence of one or other type of air mass is directly dependent, above all, on the period of the year in which they operate and the dominant relief units of Bosnia and Herzegovina.

The influence of barometric depressions on weather conditions is subject to the basic regularity related to their movement. More specifically, numerous synoptic measurements and observations have concluded that the main barometric depressions, as the basic trajectories along which they move, in addition to sea surfaces, prefer large lowland relief complexes, i.e., bypassing pronounced orographic barriers. In connection with that, a scheme of the movement of the main barometric depressions was made, among which the most famous is Van Bebber, who, after many years of analysis of synoptic maps, came to data on their established directions and occurrence frequency. This scheme was later supplemented (given that synoptic observations were significantly improved) so that certain (minor) corrections were made both in terms of preferred trajectories and in terms of the frequency of barometric depressions (Dreškovič 2011). The mean value of the frequency of barometric depressions relative to the annual average can be observed (Fig. 5.1).

By analyzing the above scheme, it can be determined that the weather conditions and climatic specifics of Bosnia and Herzegovina are affected by certain (different) paths that are, according to the modified Van Bebber scheme of barometric depressions, marked as V-b, V-c, and V-d. All three trajectories are part of the Mediterranean arm of the trajectories coming from the Bay of Biscay area, schematically marked V. In the Gulf of Genoa or Rijeka, these depressions fork into the three branches. During the spring and autumn climatological seasons, the V-b depressions move toward Bratislava and Krakow, partly influencing the weather conditions in the extreme northwestern parts of country.

Along the V-c trajectory, barometric depressions are most frequent during the period May–June and in the second half of autumn. These depressions directly impact the weather conditions in the region of northern Bosnia, causing the formation of the main precipitation maximum in late spring and the first half of summer (May–June maximum), and the secondary precipitation minimum in the second half of autumn (October–November maximum). Barometric depressions moving along the V-d path are of particular importance for the weather conditions in the Mediterranean

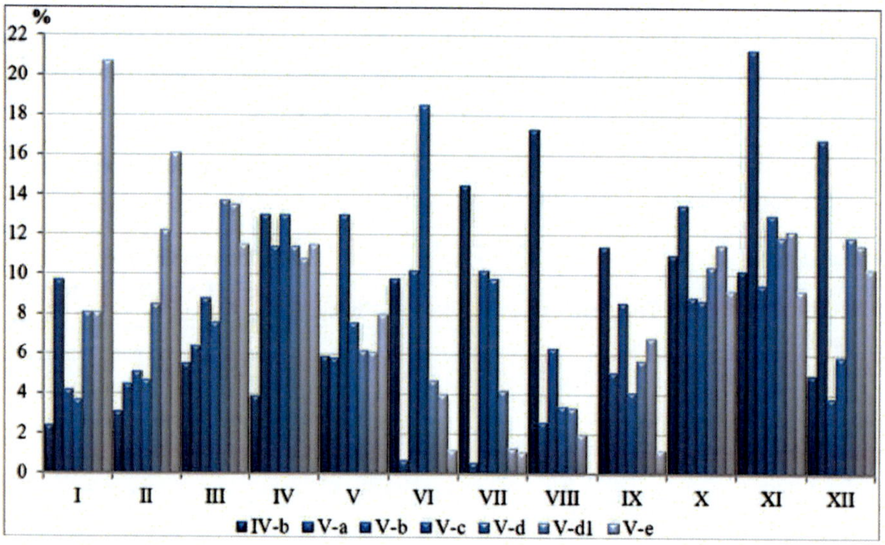

Fig. 5.1 Frequency of barometric depressions in % (Y axis) according to established trajectories (modified Van Beber scheme) operating in the Balkan Peninsula region (X axis). (Dreškovič 2011)

region of Bosnia and Herzegovina and are most frequent during the colder period of the year, especially during March, November, and December, thus causing winter precipitation maximums in this region.

In addition to the above scheme, the weather and climate conditions in Bosnia and Herzegovina are mainly influenced by extra-schematic “penetrations” of cold and humid air from the polar regions of the North Atlantic. These air masses, which are formed and move in accordance with the mechanisms of the general circulation of the atmosphere, are especially frequent in the period May–June, during which rainy and cool weather conditions form over Bosnia and Herzegovina, followed by increased wind, which can last longer than 1 week.

During the summer season in Bosnia and Herzegovina, there are very favorable conditions for the formation of local depressions, usually in a more energetically developed relief. Their result is very intense, and short-term showers of thermal convection formed and excreted from cumulonimbus clouds of very pronounced vertical development. A common form of precipitation emitted from these clouds is hail, which can cause severe damage to crops or technical infrastructure.

In general, it can be stated that the effects of (extratropical) barometric depressions are present throughout the year, but they are many times more pronounced during the transition and winter seasons, while their impact during the summer period is weak. Their impact on the weather situation in Bosnia and Herzegovina is expressed in the form of pronounced regional disturbances, i.e., the penetration of humid and colder air masses that cause several days of worsening weather: increased cloudiness, windier weather, and lower air temperatures. The temporal continuity in these weather conditions can be disrupted by the penetration of dry and cold air masses originating from the continental interior of the European continent or the Arctic regions.

Unlike the previous ones, anticyclonic air masses do not have such “stable” path and frequency patterns. They are usually of smaller spatial coverage and smaller occurrence frequencies. In relation to the physical characteristics of the air masses from which they are formed, anticyclones can generally be divided into summer and winter and are primarily seasonal in terms of the time of year in which they operate. For the general weather and climatic characteristics of Bosnia and Herzegovina, the most important centers of high air pressure are the Azores High, the Siberian High, and the Arctic anticyclone. The Azores High has a dominant effect on summer, i.e., the Siberian and Arctic High on winter weather conditions over the European continent.

Climatic elements in the wider Balkan region during the summer half of the year are directly influenced by the northeastern branch of the Azores High, which penetrates from the subtropical Atlantic Ocean into the Mediterranean and deeper into the continental interior. On the other hand, the field of high air pressure—the southern branch of the Arctic anticyclone is established during the summer over the northeastern and central part of the European continent, under the influence of which is part central-northern Bosnia. There are very frequent periods when the northeast branch of the Azores and the south-southwestern branch of the Central European summer anticyclone merge, so a very stable field of high air pressure is established over the entire territory of Bosnia and Herzegovina. The weather is then dominated by clear, sunny, and windless days with temperatures that often take on tropical characteristics. These barometric relationships can last up to ten days on average. They are usually disturbed by (already mentioned) local depressions of thermal convection, so the summer anticyclone is characterized as less stable than the winter one (Drešković 2004).

Winter anticyclone is usually characterized by two prevailing weather conditions: sunny and cold or cloudy and cold weather. The first type of weather condition in the winter anticyclone usually occurs in the first period of anticyclone development, i.e., when intensive air descent causes adiabatic heating and evaporation of existing clouds. In this type of anticyclone, the weather is clear and windless or with slight advection, which continuously intensifies radiating the substrate during the long night period. This process will be further intensified if the topographic substrate is covered with snow, resulting in the establishment of inverse thermal stratification of air, in which dry and heavier air is “nestled” in the valley or valley bottom, while the warmer is above it. Suppose the mentioned process of intensive cooling of the contact layer of air has several days. In that case, it is usual to establish the dew point temperature at which all water vapor in the air will condense, i.e., the formation of stratus type of clouds, which occurs at the contact mountain peaks. Then a winter anticyclone is established over these areas, characterized by cloudy, cold, and windless weather. A significant feature of winter anticyclones is temperature inversions, characterized by inverse thermal stratification of the air, causing vegetation turns in the height of vegetation zonation. A typical area for such synoptic situations is the mountain-valley area of Bosnia and Herzegovina, where winter anticyclones are the most represented in terms of intensity and duration. In the transitional climatological seasons, the effects of barometric depressions and anticyclones on general climatic conditions are mixed with each other without any particular spatio-temporal pattern in the time of occurrence, frequency of occurrence, and intensity of action.

5.3 Air Temperatures

The average annual air temperature for Bosnia and Herzegovina is about 10.9 °C, with significant differences at the level of the two existing climate zones. More specifically, in the northern temperate climate zone, the value of the average annual temperature is around 9.7 °C. In comparison, the stated value in the Mediterranean climate zone is 12.1 °C (Drešković and Mirić 2016). It was also determined that thermal contrasts in Bosnia and Herzegovina are very pronounced, given that the average annual temperature in the zones of the highest mountain peaks in the southeastern Bosnian highlands is negative and amounts to −1.4 °C (the highest peak of Maglić), while in the zone coastal belt of Neum around 15.9 °C. The stated thermal difference of about 18.0 °C was established at a horizontal distance of only 85.5 km as the crow flies (the peak of highest Maglić-Župa bay), i.e., about 0.21 °C/km, which is about 50 times more pronounced than the average horizontal temperature changes in the Northern Hemisphere. This thermal specificity is a consequence of morphological-morphometric relations because the height differences change by about 30 m/km on average on the mentioned profile. Compared to the Mediterranean, the determined thermal contrasts in the continental climate zone are somewhat less pronounced, considering that the temperature difference at the level of average annual temperatures is slightly lower—around 16.0 °C. The horizontal temperature drop is about 0.07 °C/km, while the altitude change is less pronounced—12.9 m/km.

A very illustrative indicator of the horizontal change in temperature in Bosnia and Herzegovina is its changes in the overall transverse and longitudinal profile. For this purpose, lines for two temperature profiles were constructed on the thermal model for mean annual air temperatures: Channel of Mali Ston—the confluence of the Drina and Sava rivers—transverse profile (Fig. 5.2), Maglić mountain—Glina river (the village of Ponikva)—longitudinal profile" (Drešković and Mirić 2017).

The transverse thermal profile has been withdrawn from the middle of the Channel of Mali Ston, across the Neum Bay and hinterland (Žaba mountain), Hutovo blato, Dubravska plateau, Velež mountain, northwestern slopes of Crvanj mountain, Visočica mountain, western slopes of Bjelašnica and Igman mountains, Trebević, Sarajevo, Ozren and Bukovik mountains, the western slopes of Sljemenska and Javor and Javornik mountains, Spreča valleys, Majevica mountains, Sapna valleys, Visoka glava, Bijeljina, and Brodac hills, to the confluence of the Drina and Sava rivers. The total length of the airline of profile is about 262.5 km, with the length of the topographic profile being many times longer. The average temperature variation along the transverse profile zone in the range of 3.0–15.5 °C can be determined, which is quite representative considering that the profile line is located west of the southeastern Bosnian highlands. One of the primary conclusions deduced from the analyzed graph is that the profile line generally has the opposite shape from the morphological-morphometric changes in the relief of the space through which the profile zone passes.

More specifically, the profile can fully confirm the already stated theoretical views on the decrease in mean annual temperature with an increase in altitude in the whole of Bosnia and Herzegovina by an average of about 0.55 °C/100 m, so that higher hypsometric levels correspond to lower temperatures and vice versa.

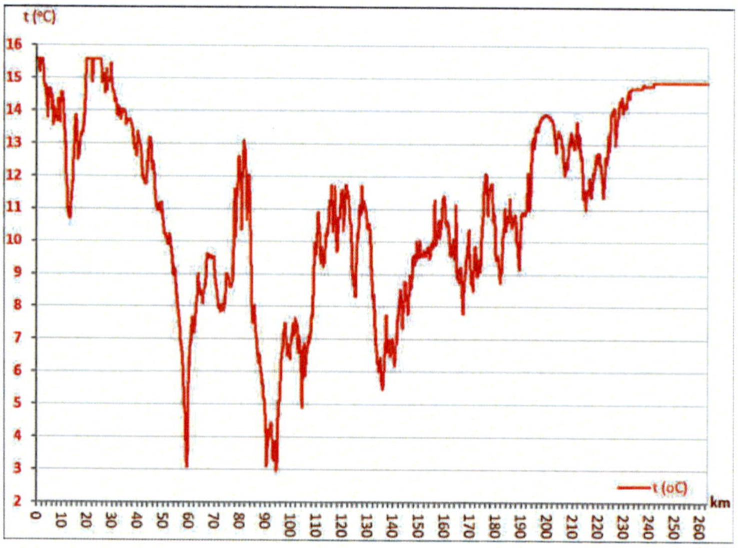

Fig. 5.2 Transverse thermal profile Mali Ston channel—confluence of the Drina and Sava from 1980 to 2010. (Drešković and Mirić 2017)

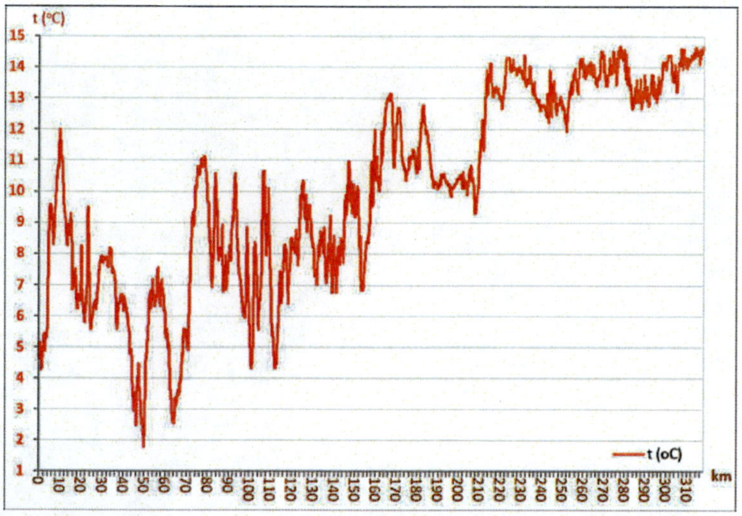

Fig. 5.3 Longitudinal thermal profile: Maglić-Ponikve (river Glina) from 1980 to 2010. (Drešković and Mirić 2017)

The longitudinal thermal profile covers the area northwest of the top of Maglić mountain over Trebova mountain, Lelija mountain, Treskavica mountain, Bjelašnica mountain, Zujevina river valley, northwestern slopes of Bitovnja, Pogorelica, Zec and Vranica, Radovan, Komar, southwestern slopes of Radalj mountain, then Jajce, Plivsko jezero, Mrkonjić-Grad, Manjača mountain, Sana river valley (near Donje Sokolovo) and further southwest of Sanski Most, Majdanska planina, Una river valley (near Ivanjska village), Mašena glavica, Bužimica river valley (the village of Čaglice), Gladno brdo to the valley of the river Glina (near the village of Ponikve). Considering that this profile line passes in the Dinaric direction through Bosnia and Herzegovina, its air length is higher and amounts to about 317 km (Fig. 5.3). The general characteristics of the temperature profile are conceptually entirely in line with the views expressed for the previous profile. The only difference is in the range of temperature variation, considering that it is somewhat more pronounced in the longitudinal profile zone: from 1.5 to 14.5 °C. It should certainly be noted that this is partly because the profile line of the longitudinal profile is not related to the very top of Maglić mountain but is positioned northwest of it, as a result of which the variation is mitigated by about 1.0 °C. It is also important to point out that there are specific differences in the trend of spatial change of air temperatures from one to other parts of Bosnia and Herzegovina (Figs. 5.4, 5.5, and 5.6). More specifically, on the profile from Neum to Brčko, the air temperature first has a general downward trend of 15.6 °C to about 3.0 °C (the first 95 km), from where the general temperature trend tends to rise to about 14.0 °C. The profile can also identify a continuous general trend of increasing air temperatures from 2.0 °C (in the area of the Maglić mountain) to about 14.5 °C in the wider area of Velika Kladuša.

The basic thermal feature observed in the annual flow of air temperatures in the entire area of Bosnia and Herzegovina is the existence of only one distinct temperature maximum and minimum, which on average occur one month after the solstice positions of the Sun. More specifically, maximum air temperatures are present during the climatological summer period (June–August), with a maximum in July. Exceptions to this rule are the zones of the highest mountain peaks, given that, at altitudes above 1900 m above sea level, temperature maxima are time-positioned to another month–August (Fig. 5.6). Such temperature relations are a consequence of the reduced heating surface in the highest hypsometric levels, as a result of which the time required for heating the contact surface and accumulation of heat in the soil is somewhat longer. It should certainly be noted that the determined monthly temperature differences in these zones are very small and range from 0.2 to 0.5 °C. A similar situation exists for the minimum temperature extremes, given that they are timed to the winter climatological period, i.e., December–February. The minimum air temperatures occur in January, with the proviso that they move to February in the highest mountain zones (above the mentioned altitude limit). The month-on-month temperature differences between the extreme months in these hypsometric ranges are also quite small, from 0.1 to 0.3 °C.

Among the mentioned extreme periods in the annual air temperature regime in Bosnia and Herzegovina, there are also transitional climatological seasons, spring and autumn, with similar thermal characteristics (Drešković 2011). The only specificity that can be especially emphasized is that during the spring season (March–May), in accordance with the astronomical factor, temperatures continuously increase towards the summer period—on average by about 4.5–5.0 °C, while in autumn during the air temperature is reversed; i.e., temperatures decrease from the beginning of

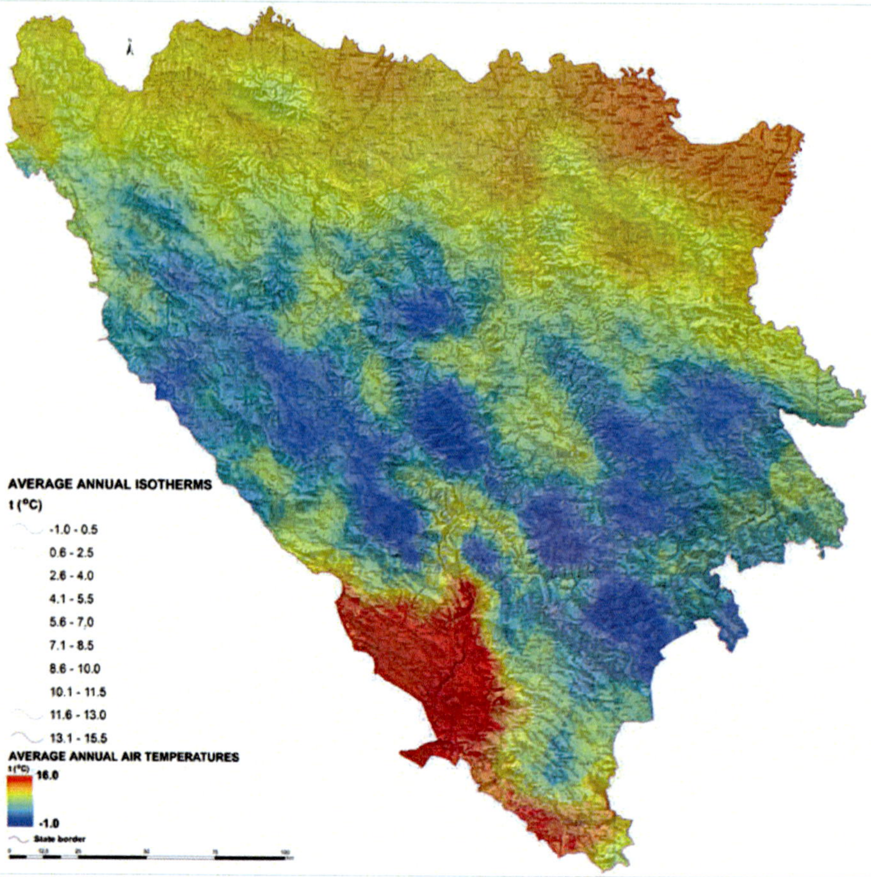

Fig. 5.4 Geographical distribution of annual isotherms in Bosnia and Herzegovina from 1980 to 2010

autumn to the beginning of winter, with an average month-on-month variation of about 5.0 °C. Also, numerous examples found that a slightly lower seasonal average characterizes the spring season than autumn; that is, at the entire Bosnia and Herzegovina, autumn is warmer by about 1.0 °C compared to spring. Based on the above distribution of extremes in the annual flow of air temperature, it can be concluded that, in Bosnia and Herzegovina and the wider region, the climate type of temperate latitudes is predominantly represented. Furthermore, as it is situated practically in the central part of the northern temperate climate zone, with four clearly separated climate periods, said climate type can be identified as a true type of temperate latitudes. In accordance with the macro-relief elements, the mentioned climate type in Bosnia and Herzegovina occurs in two variants: continental and maritime, which spread in the areas of dominant continental or maritime influence. However, it is necessary to state that, concerning the said general scheme of the annual thermal regime in Bosnia and Herzegovina, a very frequent occurrence of the time shift of the maximum, i.e., minimum air temperature is very common.

5.4 Precipitation

The mean annual precipitation height for the territory of Bosnia and Herzegovina is about 1255 mm, obtained by the analysis of the annual precipitation GRID spatial model and its accompanying geodata, but the spatial distribution variations are very distinctive with the values between 706 mm (direct Sava river confluence—wider area around the settlement Domaljevac) and 3259 mm (mt. Orjen—wider area of the peak Velika Jastrebica—1856 m above sea level) (Drešković 2015). Based on the given data, it can be concluded that almost five times less precipitation is emitted annually in the driest areas of Bosnia and Herzegovina. Differences in the spatial distribution of precipitation are also pronounced at the level of the two primary climate regions. More specifically, in the continental climate region of Bosnia and Herzegovina, the average annual precipitation is about 1080 mm and varies between 706 mm to about 2335 mm (in the wider zone of the highest mountain peaks of Zelengora). The value of this analog in the Mediterranean region of Bosnia and Herzegovina is higher and amounts to about 1645 mm, with limit values of variation from 920 mm

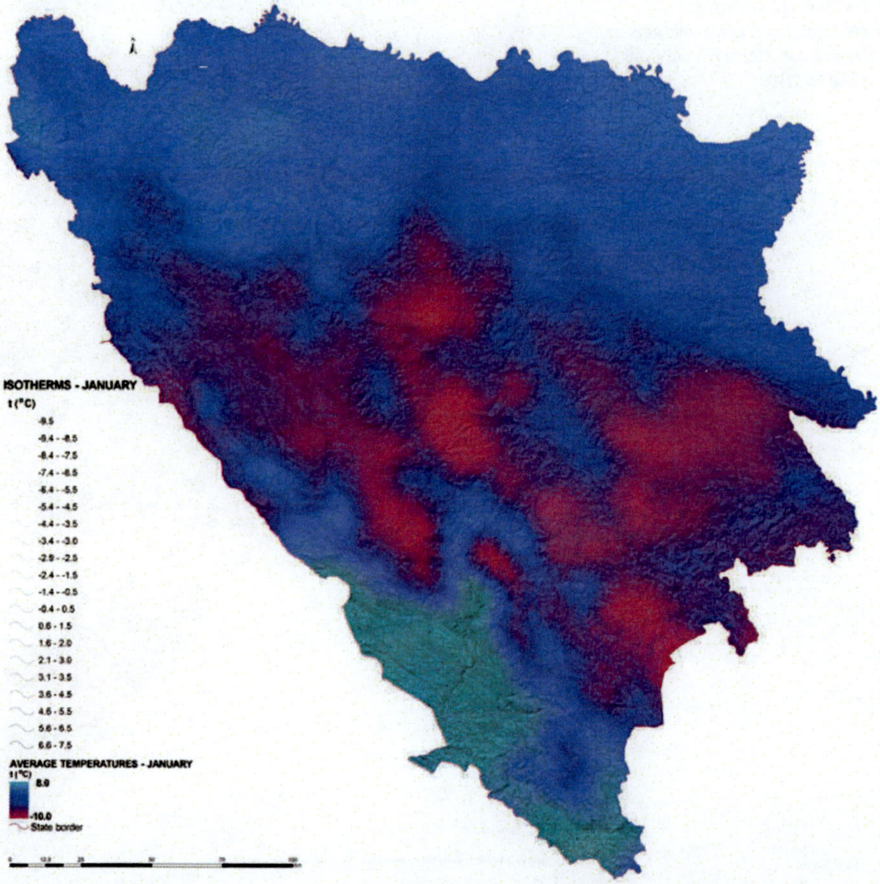

Fig. 5.5 Geographical distribution of January isotherms in Bosnia and Herzegovina from 1980 to 2010

(wider zone around the Rama hydro accumulation) to 3259 mm. The horizontal change in the average annual precipitation in the two mentioned climate zones is also different, with the general conclusion that it decreases from the coastal zone in the south to the Peripannonian area in the north of Bosnia and Herzegovina.

For a more detailed presentation of the horizontal change in the average annual precipitation, one transverse and a longitudinal precipitation profile Drešković (2011) constructed two following graphs: Župa bay—Sava river (village of Batković)—transverse profile (Fig. 5.7), Volujak mountain—Glina river (village of Ponikva)—longitudinal profile (Fig. 5.8) (Drešković 2011).

Quantitative values of isohyets also completely fit to the real values, obtained through monitoring of the meteorological network. The main contribution of this work is that after the implementation of the Ordinary Kriging of spatial interpolation, each grid cell, with 20×20 m resolution, has its associated values for the amount of precipitation (Drešković and Đug 2012).

The transverse precipitation profile was drawn from the village of Zavrelje in Bosnia and Herzegovina (in the direction from the central part of the Župa Bay), from where it descends over part of the Dubrava karst plateau to Popovo polje valley (around the village of Gomiljani) and further over Ljubomirsko polje to mountain Viduša, then eastern slopes of Baba and Gatačka Bjelašnica mountains, part of Vučevo plateau, Zelengora mountain, eastern foothills of Lelija mountain and Treskavica mountain, Bistrica river valley, east Jahorina mountain slope, Prača river valley, Glasinac plateau, Sljemenska planina, western Jahorina mountain valley, Drnjača river valley, the extreme eastern part of the Sprečko polje, the southeastern slopes of the Majevica mountain, the wider area of Semberija to the river Sava (wider zone of the village of Batković). The total length of the airline of profile is 264.8 km, with the length of the topographic profile being many times longer. As it can be seen from the given graph for the transverse precipitation profile, it can be observed that the average variation of annual precipitation along the profile zone ranges from 790 to 2230 mm, which suggests that it is positioned slightly west of the zone with the maximum amount of precipitation, located in the area of Orjen mountain. It can also be stated that the shape of the precipitation profile line significantly deviates from the shape of the topographic profile, which indicates the dominant importance of the geographical

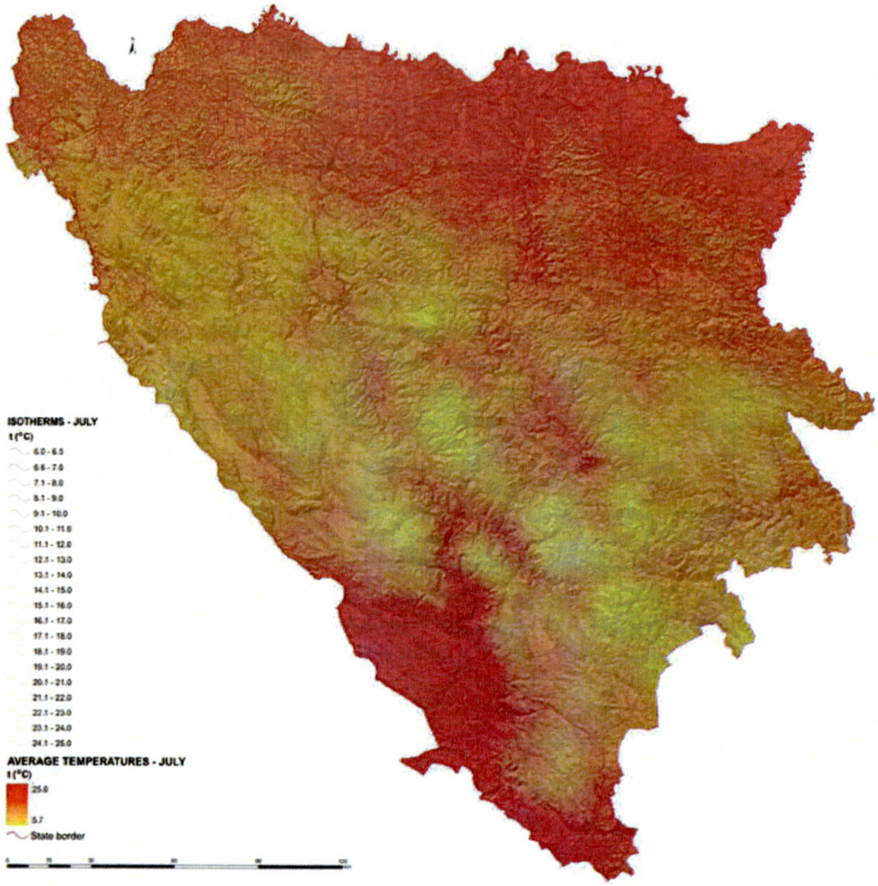

Fig. 5.6 Geographical distribution of July isotherms in Bosnia and Herzegovina from 1980 to 2010

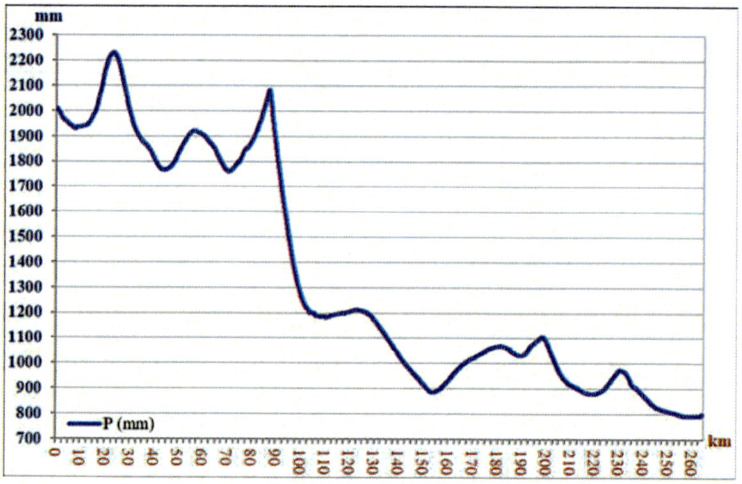

Fig. 5.7 Transverse pluviometric profile: Župa bay—Sava river (Batković settlement zone) from 1980 to 2010. (Drešković 2011)

position and degree of continentality of an area to its annual precipitation height. In other words, in the zone of the transverse profile, the area with the maximum amount of precipitation is located directly in the hinterland of the Adriatic Sea. In contrast, with increasing distance of the precipitation station (deeper inland), the amount of precipitation, as a rule, decreases. In this particular case, the annual precipitation from the Adriatic coast to the Sava river in Bosnia and Herzegovina decreases by an average of about 5.5 mm/km as the crow flies.

Certainly, there are spatial deviations from the mentioned average decrease, where the precipitation after the increase

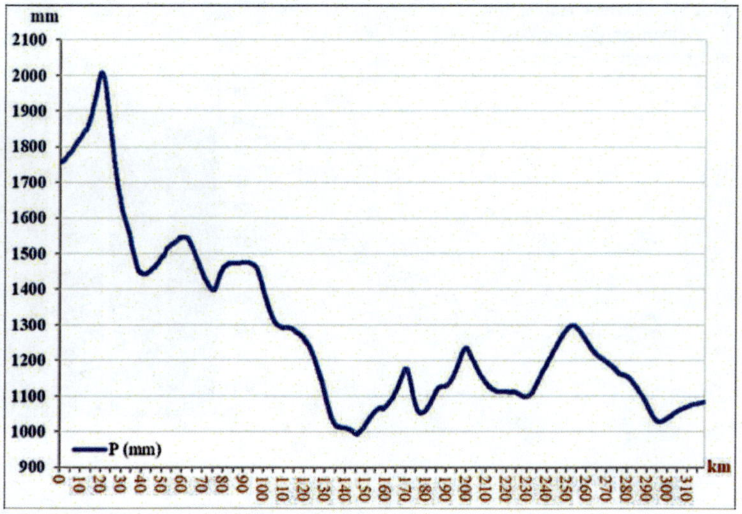

Fig. 5.8 Longitudinal pluviometric profile: mountain peaks of Volujak—river Glina (settlement Poljana) from 1980 to 2010. (Drešković 2011)

after decreasing to a certain point and then decreases again after reaching a certain point. Such is the case in the analyzed profile in the wider zone of mountain morphostructures of the southeastern Bosnian highlands, where precipitation increases from about 1600 mm in the mountain foothills to about 2000 mm in the zone of the highest mountain peaks. This is a consequence of the increased intensity of condensation-sublimation processes with the increase of altitude on the windward sides in the conditions of a constant supply of moist and warm air from the south. In the northern foothills of these mountain morphostructures, the intensity of the mentioned processes weakens, considering that an immense amount of moisture has already been excreted, so that the precipitation height drops quite abruptly to below 1300 mm. With fewer quantitative-spatial "disturbances," the trend of decreasing annual precipitation continues all the way to the Sava river (Drešković 2015).

The longitudinal precipitation profile includes the area starting from the southwestern slopes of the highest mountain peaks of Volujak, from where it extends over the wider zone of mountain peaks of Zelengora mountain to southwestern foothills of Lelija mountain, Visočica mountain peaks, southwestern slopes of Bjelašnica mountain, Ivan mountain, Bitovnja mountain, the highest mountain peaks of Vranica mountain on the southwest, Radovan mountain, Vinačka gorge, northeastern slopes of Lisina mountain, Podrašnjičko polje, Sana river valley (near the village of Ljubine), Mrežnica mountain, Majdanska planina, Una river valley, and Bužimica river valley to Glina river (near the village of Poljana). The total length of the profile line is about 318 km. Similar to the previous precipitation profile, the degree of variation in precipitation height is pronounced and ranges within the limit values from 960 mm (in the wider zone of the Uskopaljska valley) to 2300 mm (wider area of the highest mountain peaks of Zelengora).

In relation to the position of both profile lines, it can be generally stated that in the southern and southeastern and partly in the southwestern part of Bosnia and Herzegovina, the average annual precipitation is many times above the country average, while in the central and northwestern part it decreases below the average. These facts suggest that the southwestern, southern, and southeastern parts of Bosnia and Herzegovina, which are "richer" in precipitation, are under a stronger climatic influence of the Adriatic Sea than the precipitation "poorer" part located north of the orographic watershed, which consists of the highest mountains of the central Dinaric belt (from Plješevica mountain in the northwest to Volujak and Maglić mountains in the southeast). These facts represent a practical confirmation of the previously stated theoretical conclusion that with increasing altitude, the annual precipitation in the whole of Bosnia and Herzegovina increases by an average of about 23 mm/100 m. However, in regional geographic climate analyses, it should be borne in mind that the degree of continentality predominantly influences the stated precipitation relations and must undoubtedly be analyzed to eliminate significant errors in determining the amount of precipitation.

The pluviometric regime in Bosnia and Herzegovina has pronounced spatio-temporal variations as a result of the geographical position of Bosnia and Herzegovina at the transition from the extreme northern parts of the northern temperate climate zone to the northern parts of the northern subtropical climate zone. In this regard, the genesis of precipitation in the subtropical part of Bosnia and Herzegovina is related to the cyclonic activity that develops on the polar front. Since the polar front with cyclones (in the Northern Hemisphere) moves deeper towards the Equator during the winter period, the zone with intense precipitation (among other areas) across the Mediterranean Sea penetrates the Eurasian and African mainland. In this way, the primary

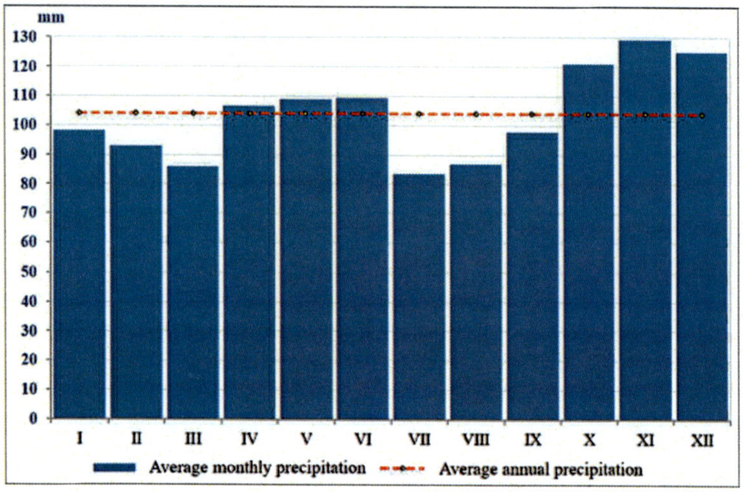

Fig. 5.9 Average annual flow of pluviometric regime in Bosnia and Herzegovina from 1980 to 2010

precipitation season is formed during the climatological winter season.

In the temperate climate zone, due to the intense summer warming of landmasses, the process of thermal convection is also expressed, which creates the necessary assumption that precipitation is excreted during the period of the primary precipitation minimum. With this in mind, it can be stated that, due to the orographic influences of mountain systems, precipitation does not have a regular geographical distribution, i.e., the shape and quantitative values of isohyets are primarily influenced by relief dynamics, primarily through the formation of specific precipitation conditions at the local level. In general, it can be concluded that the intensity and spatio-temporal dynamics of the pluviometric regime in Bosnia and Herzegovina, apart from the mentioned dominant regional western circulation and related frontal and orographic types of precipitation, are significantly influenced by thermal convection processes affecting the formation of convective precipitation types. For these reasons, at the level of the entire Bosnia and Herzegovina, it is possible to talk only about the general features of the pluviometric regime. The average monthly precipitation is 104.5 mm (Figs. 5.9 and 5.10).

In this regard, the basic feature that can be observed in the annual flow of precipitation in the entire territory of Bosnia and Herzegovina is the existence of two distinct precipitation periods. The first of them, characterized by a markedly elevated periodic average, is positioned in time for the second half of autumn and the beginning of the climatological winter season, i.e., October–December. Then, on average, about 125 mm or approximately 120% of the average annual precipitation is excreted per one periodic month, which is why this period has the characteristics of the main precipitation maximum in Bosnia and Herzegovina. A significantly reduced periodic average characterizes the second characteristic precipitation period compared to the annual value; i.e., during this period of the precipitation minimum, an average of about 85 mm or 82% of the average monthly value is excreted. The period of the main precipitation minimum in Bosnia and Herzegovina occurs during the middle and second half of summer, i.e., in the period July–August. Between the above-mentioned characteristic precipitation periods, the pluviometric regime has transitional features; i.e., in its annual course, periods of somewhat less pronounced increased and decreased precipitation heights, secondary precipitation periods can be further identified (Figs. 5.11 and 5.12). The secondary precipitation maximum is time-positioned for the second half of the spring and the beginning of the climatological summer season, i.e., in April-June, when the average monthly precipitation is 108.5 mm or about 101% of the average monthly precipitation.

The secondary precipitation minimum coincides with the period of the end of the winter and the beginning of the climatological spring season, i.e., during the time interval February–March when, on average per period month, about 90 mm or 87% is excreted.

The secondary precipitation minimum coincides with the period of the end of the winter and the beginning of the climatological spring season, i.e., during the time interval February–March when, on average per period month, about 90 mm or 87% is excreted. However, it is necessary to emphasize again that in relation to the given general pluviometric scheme in Bosnia and Herzegovina, there are very pronounced regional differences, mainly observed from the aspect of the main climatic regions. In this regard, in Bosnia and Herzegovina, two basic pluviometric types can be distinguished: continental and maritime (mediterranean). Their main feature is that in the continental pluviometric type, most of the precipitation is excreted in the warmer period of the year (April–September), while in the maritime pluviometric type, the period with maximum precipitation is positioned in the colder part of the year (October–March).

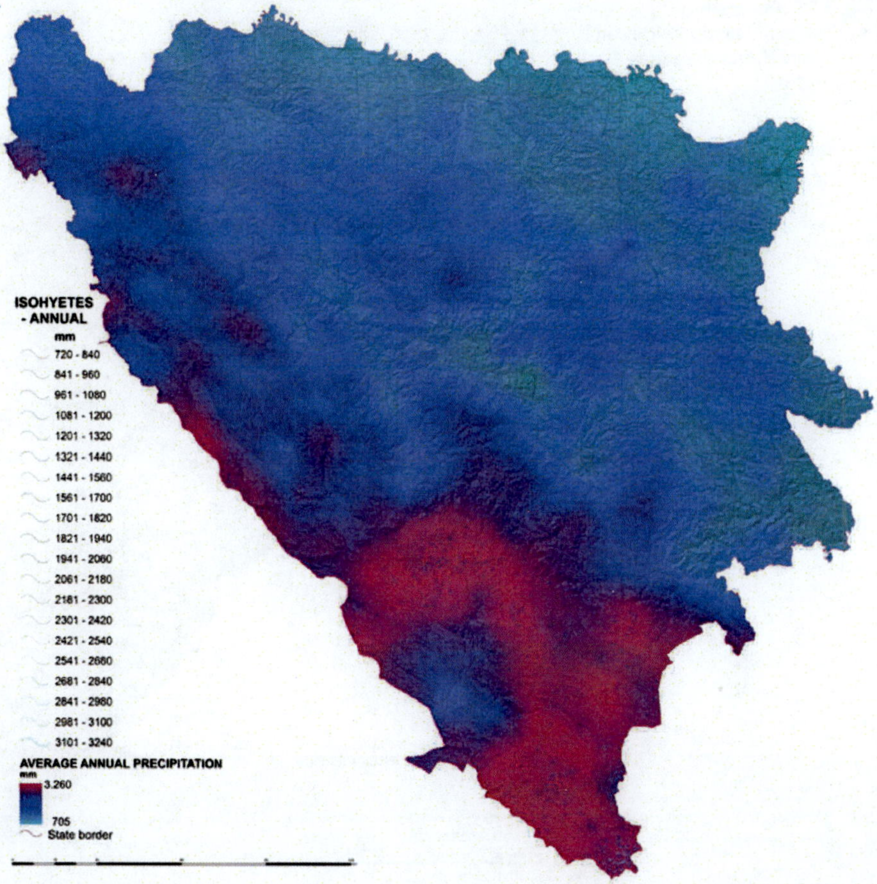

Fig. 5.10 Geographical distribution of annual isohyets in Bosnia and Herzegovina from 1980 to 2010

The continental pluviometric type is under the dominant influence of the climatic characteristics of the European mainland, and in Bosnia and Herzegovina, it is not so clearly differentiated into the stated distinct precipitation periods (Fig. 5.13).

The reason for this pluviometric scheme lies in the fact that the specific analysis included data from all precipitation stations of the continental climate zone, thus neglecting the influence of altitude. The average monthly precipitation in this climate zone is about 90 mm, with the proviso that, despite the above facts, in the annual course of the pluviometric regime at the general level can be distinguished precipitation periods with increased or decreased precipitation.

The main precipitation period is positioned at the end of the spring and the first half of the climatological summer season, i.e., May–July. The value of the periodic average is about 101.5 mm or approximately 113% of the average of all months. The main precipitation minimum is positioned in time at the end of the winter and the beginning of the climatological spring season, i.e., February–March, during which only 69 mm or 77% is excreted on average. Secondary precipitation periods are less pronounced compared to the average of the entire Bosnia and Herzegovina. The secondary precipitation maximum falls in October–November with an average of about 99 mm, which is identical to the monthly average on an annual basis. Also, the weakly expressed secondary precipitation minimum is positioned in time for the second half of the summer and the beginning of the autumn climatological season, i.e., July–September, during which about 95 mm or about 96% are excreted on average.

The presented data concludes that the annual flow of the pluviometric regime in this climate region has a markedly inconsistent relationship with the annual thermal regime. Regional pluviometric specifics, in relation to the given average, are a very pronounced need to be analyzed separately. The main feature of the maritime pluviometric type is that the main precipitation season is time-positioned for the colder period of the year when it is excreted on average per month of the season about 167 mm or approximately 122% compared to the annual average of the entire climate region of about 137 mm (Fig. 5.14).

The mentioned pluviometric dynamics result from the regional geographical distribution and intensity of relative humidity and cloud cover. More specifically, most of this region (as already pointed out) has karst features. Unlike the

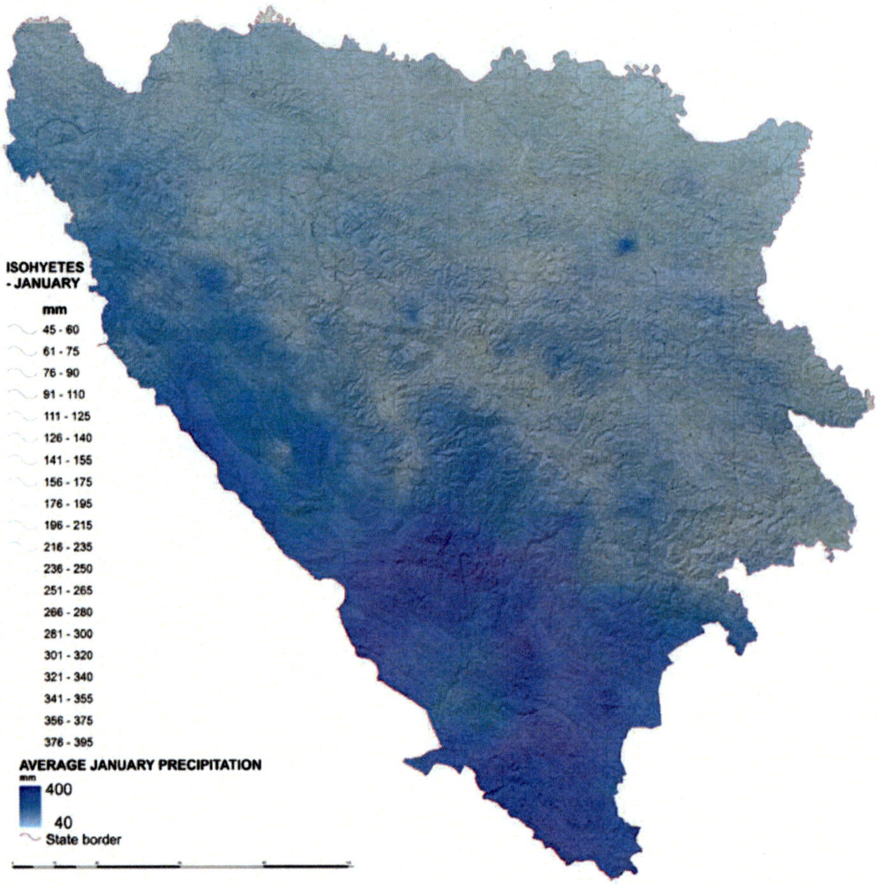

Fig. 5.11 Geographical distribution of January isohyets in Bosnia and Herzegovina from 1980 to 2010

previous one, in the annual course of the pluviometric regime, it is possible to distinguish the distinct precipitation periods more clearly. The main precipitation maximum is positioned in time for the second half of autumn, and the first half of the climatological winter season; i.e., it falls in the period October–December. The average monthly value of the periodic average is about 186 mm, which is about 135% compared to the annual, monthly average. The main precipitation minimum occurs during the climatological summer season, i.e., in June–August, when on average only 83.5 mm or about 30% of the annual, monthly average is excreted per month. The secondary precipitation maximum occurs on average in February, with an average monthly precipitation of about 163 mm. The secondary precipitation minimum is not so clearly separated, although it can be stated that it occurs on average during the end of the winter and the first half of the climatological spring season, i.e., in the period March–April, with a periodic average of 133 mm.

Based on the presented facts, it can be concluded that in the Mediterranean climate region of Bosnia and Herzegovina, the annual flow of the pluviometric regime has a highly correlated (almost inverted) flow in relation to the thermal one. The two basic types of pluviometric regime are listed on the spatial-temporal dynamics as the basic climatological factor with the primary influence. In connection with this are the features of the thermal regime, relative humidity, cloudiness, and windiness, which directly determine the precipitation dynamics in the area.

In relation to the mentioned pluviometric scheme in Bosnia and Herzegovina, transitional types, i.e., areas in which the influences of continental and maritime pluviometric variants intertwine, must be singled out due to the very pronounced dynamics of the relief. Therefore, it is necessary to define the border at the first level—the zone that separates the maritime from the continental type in Bosnia and Herzegovina, i.e., to determine the degree of continentality of the area. Most pluviometric criteria for assessing the degree of continentality are based on the division of the year into its warmer and colder parts. One of the more successful criteria in climate analysis is more precisely defined changes in the pluviometric regime with distance from the coastal zone (Drešković 2011).

This criterion is based on the distribution of precipitation in the warmer and colder part of the year, so that if in the period from October to February (colder half of the year), more than 50% of the annual precipitation is excreted, the

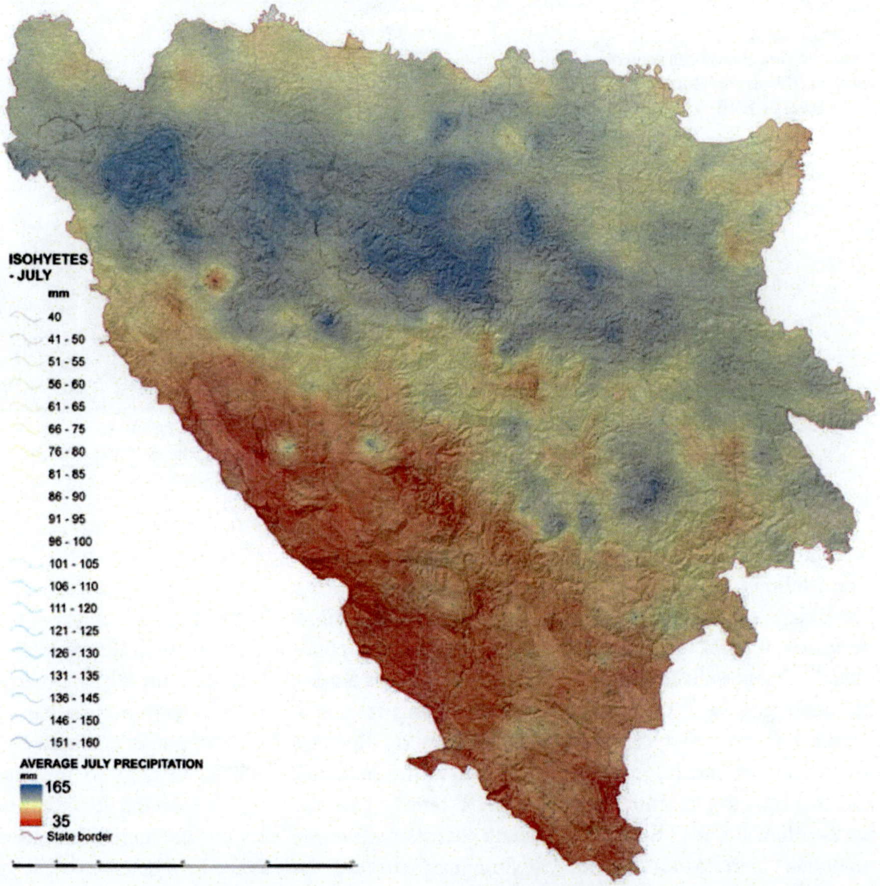

Fig. 5.12 Geographical distribution of July isohyets in Bosnia and Herzegovina from 1980 to 2010

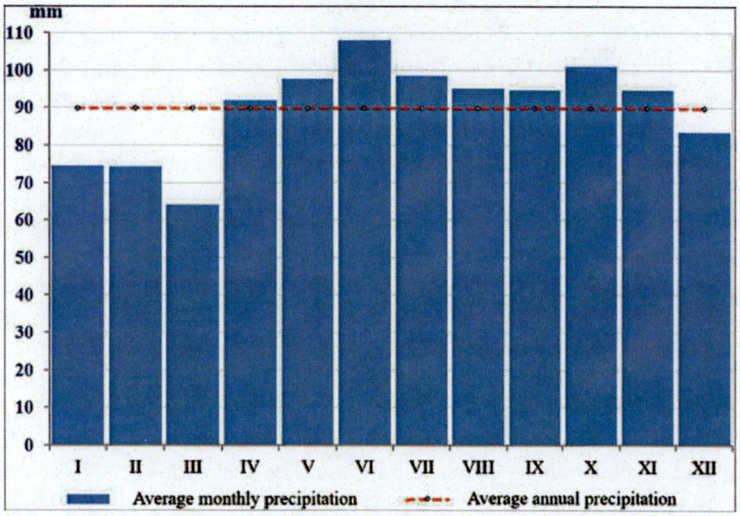

Fig. 5.13 Average annual flow of the pluviometric regime of the continental climate zone in Bosnia and Herzegovina from 1980 to 2010

pluviometric regime has maritime characteristics. Subsequently, if the period March–September (warmer half of the year) has less than 50% of the total annual precipitation, then we are talking about the continental pluviometric regime. The boundary isocontinental line has the value: q = 50%, meaning that all areas in Bosnia and Herzegovina located north of the said boundary have continental and vice versa, areas located south of it have a maritime pluviometric regime.

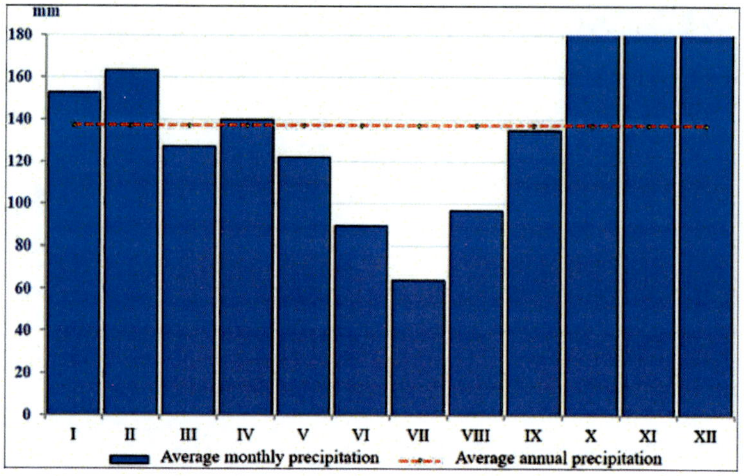

Fig. 5.14 Average annual flow of pluviometric regime of maritime (Mediterranean) climate zone in Bosnia and Herzegovina from 1980 to 2010

The general characteristic of the spatial position of the isocontinental line in Bosnia and Herzegovina is that their shape follows the largest relief forms quite well, which is a practical confirmation of the stated position on the direct influence of relief on the penetration of maritime and continental influences. Isocontinental line $q = 50\%$ enters Bosnia and Herzegovina in the wider zone of the Dinara mountain (between Orlovac and Veliki Bat) in a northeasterly direction to the Šator mountain, from where it bends to the southeast and descends to the central part of Glamočko polje over the northeastern slopes of Staretina mountain. Across parts of the mountain morphostructures of Slovinje and Hrbina, the boundary isotalanthosis reaches the southwestern and western rims of Kupreško polje, from where it bends westward to the mountain Ljubuša and south of the Rama hydro accumulation further to the southwestern slopes of the mountains Pogorelica and Bitovnja. Isotalanthosis $q = 50\%$, north of the Jablanica hydro accumulation, further reaches the canyon valley of the Neretva river (between the mountain morphostructures of Bjelašnica and Treskavica, and Prenj and Visočica), and following the canyon valley further extends to northwestern slopes of Lelija mountain, the northwest slope of Zelengora, the wider area of Perućica from where, across the Vučevo plateau, it descends into the canyon of the river Tara (in the wider area of the mountain Obzir) (Drešković 2015). As observed on the plan, the analyzed border isocontinental line has the Dinaric, i.e., northwest-southeast, direction.

Areas south of the mentioned boundary line-zones have a predominant maritime pluviometric regime. The total variation of the isocontinental line in this band is 11.5% (i.e., between 39.5% and 50%), indicating lower precipitation amplitudes throughout the year. Pronounced Adriatic maritime influences are most felt in the coastal zone and the immediate hinterland. More specifically, the isocontinental line of $40\% \leq q < 45\%$, which is a representative of pronounced maritime influences, covers the marine waters of Neum-Klek and the immediate hinterland, which is closed by the mountain Žaba in the north. Smaller areas with this type of continentality also exist in the subregion of eastern Herzegovina: Hercegovačka Bjelašnica, Popovo polje valley, Fatničko polje, Ljubomirsko polje, the wider area of the Bileća hydro accumulation, etc.

The space, enclosed by the isocontinental line $45\% \leq q < 50\%$, belongs to the altered maritime pluviometric type, given that the quantitative indicators of the pluviometric regime are somewhat more altered compared to the previous type. In other words, in the annual course of precipitation, apart from the winter precipitation maximum and the summer precipitation minimum, more pronounced secondary characteristic precipitation periods can be seen, positioned between them in time. This pluviometric type includes a continuous area that extends in the form of a wide belt from Gnjat mountain and the southern slopes of Šator mountain in the northwest, from where through Livanjsko polje and Buško hydro accumulation further extends to Duvanjsko polje, Midena mountain, northern slopes Zaveli, Lib-mountain, Vran-mountain, Baćina mountain, Jablanica hydro-accumulation, northeastern slopes of Prenj mountain, Visočica mountain, Crvanj mountain, Lelija mountain, Zelengora mountain, Maglić mountain, Volujak mountain, Lebršnik mountain, Gatačko polje, Baba mountain, Gatačka Bjelašnica mountain, Trusina mountain, western and southwestern slopes of Velež mountain, Nevesinjsko polje and parts of Dabar polje in the southwest.

This pluviometric type was also found in the Neretva river valley (downstream of the village of Grabovica), which, in addition to the Mostar valley, includes Mostarsko blato and the downstream segments of the Neretva river tributaries up to the previous boundary of isocontinental in the south. Similar to the previous one, for this type of pluviometric continentality, there is a general, Dinaric direction of extension, which disturbs the dominant mountain morphostructures, in which the isocontinental bend along their foothills.

North of the boundary of isocontinental ($q = 50\%$) in Bosnia and Herzegovina, there is an area where the maritime influences first weaken and then completely disappear in the annual course of the pluviometric regime, i.e., where the effects of the hinterland grow with distance from the border isocontinental. The first such class of pluviometric continentality is within the following values: $50\% \leq q < 55\%$, and mainly covers the area of the region of central Bosnia and Herzegovina. More specifically, the mentioned continuous zone of the area distribution of pluviometric continentality extends from the mountain Osječenica in the northwest through the mountain Bobar, the mountain Jadovnik, the mountain Dinara, Grahovsko polje, the spring of the river Unac, Korita plateau, Bobija mountain, Čemernica mountain, most of Hrbina mountain, Kuprečko polje, Cincar mountain, Krug mountain, northern slopes of Raduša mountain, Ravašnica mountain, Vranica mountain, Zec mountain, Pogorelica mountain, Bitovnja mountain, Ivan mountain, southern and southwestern slopes of Bjelašnica mountain, southwestern slopes of Treskavica mountain, northern and northeastern slopes of Lelija, Ruda mountain, Trebova mountain to Ljubišnja mountain in the extreme northwest. The basic pluviometric characteristic in the annual course of precipitation is that maritime influences are still present, reflected in the presence of winter precipitation maximum, which is partly shifted to the second half of the autumn climatological season.

The presence of maritime influences can also be seen through the time positioning of the periods of precipitation minima that slowly move towards the end of the summer and the beginning of the autumn season. Secondary precipitation extremes are also quantitatively somewhat more pronounced compared to the altered marine variant.

The following type of pluviometric continentality, transitional continental, is defined by the following limit values: $56\% \leq q < 60\%$. Regionally and geographically, this type mainly spreads within central Bosnia and Herzegovina, with the extreme northern parts partially entering the area of the region of northern Bosnia and Herzegovina. In the northwest, the zone of continuous surface distribution covers the mountain Plješevica, from where it further extends southwest to mountain Grmeč, Petrovačko polje, mountain Osječenica, mountain Klekovača, southwestern slopes Šiša gora, Podovi plateau, Pliva river spring catchment area, Vrbas river valley (Uskoplje and Vinac gorge), mountain Radalj, mountain Vilenica, mountain pass Komar, mountain Kruščica, the area between the villages of Kreševo and Visoko, the valley of the river Bosna from Kakanj to the spring of Bosnia river (Sarajevo, Vogošća, Semizovac, Ilijaš, and Visoko valley—valley morphological extensions with gorge valleys between them), northeastern slopes of Bjelašnica mountain, Igman mountain, Jahorina mountain and Drina river valley from Foča to Goražde valley-basin expansion. In the pluviometric regime of these areas, the influence of the Euro-Asian mainland dominates the position and quantitative values of extreme precipitation periods.

More specifically, summer precipitation has the characteristics of the main precipitation maximum, while the secondary maximum is time-positioned in the climatological winter season. A similar analogy applies to periodic minima, which are time-positioned at the end of the winter and beginning of the spring climatological season—the primary precipitation minimum, and at the end of the summer and the first half of the autumn climatological season—the secondary precipitation minimum.

The most pronounced continental influences in Bosnia and Herzegovina are present in areas where the continentality index has $q \geq 61\%$. These areas represent a surface continuity of the transitional continental type and cover practically the entire region of northern Bosnia and Herzegovina. More specifically, this type of pluviometric continentality covers this region's entire lowland and mid-mountain relief between the Sava river in the north and the transitional mid-mountain and high-mountain zones of the previous type in the south. In relation to the main pluviometric characteristics in the quantitative-temporal dynamics of the annual course of precipitation height, the mentioned space has a continental type of pluviometric continentality. It should certainly be noted that, with the increase of latitude, the value of the continentality index increases from 56% (in the mentioned transition zone in the south of the area) towards the highest degree. More specifically, the areas that extend along the alluvial plateau of the Drina river downstream from Zvornik to the confluence with the Sava river and the area of the immediate Sava river basin downstream of Orašje have the highest pluviometric continentality ($q \geq 65\%$). The main feature in the annual course of precipitation refers to the time positioning of the main precipitation maximum for June–August.

The main precipitation minimum falls during the first half of the autumn climatological season, with a minimum in October. The secondary precipitation maximum is relatively evenly distributed over two months, November and December, while the secondary precipitation minimum is time-positioned in the second half of the winter and the beginning of the climatological spring season. These facts represent a practical confirmation of the position stated in the previous chapter: at the average annual precipitation level, VPG is positive, with an average value of about 23 mm/100 m. Due to the very pronounced influence of climatic factors (such as relief, distance from the Adriatic Sea, the degree of continentality, etc.), the complexity in the spatial distribution of precipitation in Bosnia and Herzegovina is very pronounced. The precipitation index quantitatively indicates that the difference in complexity of spatial distribution of precipitation may be attributed to influence of Adriatic Sea, relief and Pannonian Basin (Mihailović et al. 2015).

5.5 Main Climate Types

Taking into account the presented temperature and precipitation indicators on monthly and annual flows of thermal and pluviometric regime, as well as the spatial distribution of vegetation cover, the climate regionalization of Bosnia and Herzegovina into climate types according to the criteria of Köppen climate classification was performed. Exact research of thermal and pluviometric models has established that three climatic classes are represented in Bosnia and Herzegovina: C (moderately warm and humid climates), D (forest-snow or boreal climates), and E (snow climates). Climate class C has a dominant surface share in Bosnia and Herzegovina—about 69%, which is quite expected given that its distribution is associated with areas of low mountain and mid-mountain relief, which occupy about 75% of the state territory. Climate class E has the lowest surface share—about 1% and at the same time occurs in only five isolated areas of surface distribution (in the zone of the highest mountain peaks of the region of Central Bosnia and Herzegovina). According to the pluviometric regime, all the mentioned climatic classes (C, D, and E) are further differentiated into the main climatic types. In this regard, in Bosnia and Herzegovina, climate class C occurs at the level of two main climate types: Cf—moderately warm and humid climate and Cs—Mediterranean climate. Climate class D occurs at the level of only one main climatic type Df—humid boreal climate. Within the snowy climates in Bosnia and Herzegovina, only one main climate type ET can be singled out—Tundra climate (Drešković & Mirić 2014) (Fig. 5.15).

5.5.1 Cf—Moderately Warm and Humid Climates

Cf main climate type continuously occupies the area of the regions of northern and partly (in the form of a kind of "precipitation" islands) of central Bosnia and Herzegovina, up to about 1200 m above sea level (Drešković 2004). Its fundamental climate characteristic is that, in the average annual pluviometric regime, it is not possible to single out a pronounced dry period; i.e., precipitation is relatively evenly distributed throughout the year so that the driest month has an average precipitation height: $p > 40$ mm and more than 1/3 of precipitation compared to the month with the average maximum precipitation. Having in mind these pluviometric determinants and the conducted analyzes, it can be concluded that this climate type, in addition to being present in the regions of northern and central Bosnia and Herzegovina, has the most extensive spatial distribution in Bosnia and Herzegovina—27,469.13 km^2 and the largest number; it occurs in over 100 isolated areas, with an area ranging from several to several hundred thousand hectares. In the Cf climate area, an average of 700 mm to about 1200 mm of precipitation is excreted annually, relatively evenly distributed throughout the year.

5.5.2 Cs—Mediterranean Climates

Cs main climate type is spread only in the climatic region of Mediterranean Bosnia and Herzegovina. Its fundamental climate characteristic is that the annual course of the pluviometric regime can be distinguished by an arid period, which is positioned in time for the climatological summer season. As a rule, the driest month is July with the following pluviometric characteristics: $p \leq 40.0$ mm and $p < 1/3$ of precipitation in relation to the month with the highest average precipitation. According to the presented indicators for the July thermal regime, its dominant spatial representation in the mentioned climatological region is within the average altitude range from 0 to 900 m above sea level. According to D climates, its northern boundary is formed by the mean July isotherm: $t \geq 18$ °C. The average annual amount of precipitation in the Cs climate type averages between 950 mm to about 3260 mm, which are mostly (over 80%) concentrated in the colder period of the year, while the warmer ones are characterized by high weather conditions dominated by clear, very warm and windless weather.

5.5.3 Df—Humid Boreal Climates

According to the established indicators for the pluviometric regime, the main climate type Df appears in Bosnia and Herzegovina as the only precipitation variant of the climate class. This practically means that the distribution for D climate class has already been described simultaneously as the distribution zone Df of climate type. Consequently, in this climate type, the basic climate characteristic in the pluviometric regime is the predominant, even distribution of precipitation; i.e., in the annual course of precipitation, it is not possible to single out a pronounced dry period. In this regard, the driest month has the following characteristics: $p > 40$ mm and $p >$ from 1/3 of precipitation compared to the month with the average maximum amount of precipitation. The average annual precipitation in this climate type in Bosnia and Herzegovina, in relation to the climatological region, is characterized by pronounced quantitative

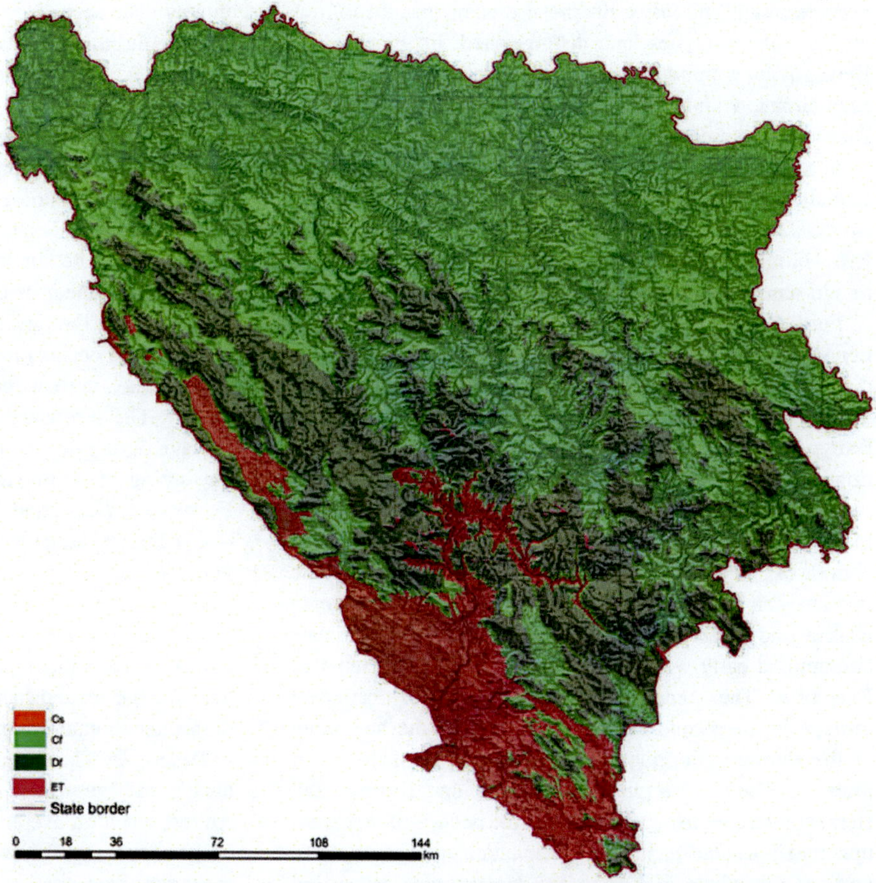

Fig. 5.15 Geographical distribution of climate types in Bosnia and Herzegovina from 1980 to 2010. (Drešković and Mirić 2014)

variability, which on average ranges between 750 and 2950 mm.

Based on the highlighted attitudes, it can be concluded that only the C climate class is further spatially differentiated into two precipitation variants. A review of spatial statistics at the level of summary surface representation of all five identified main climate types in Bosnia and Herzegovina was also made (Fig. 5.16).

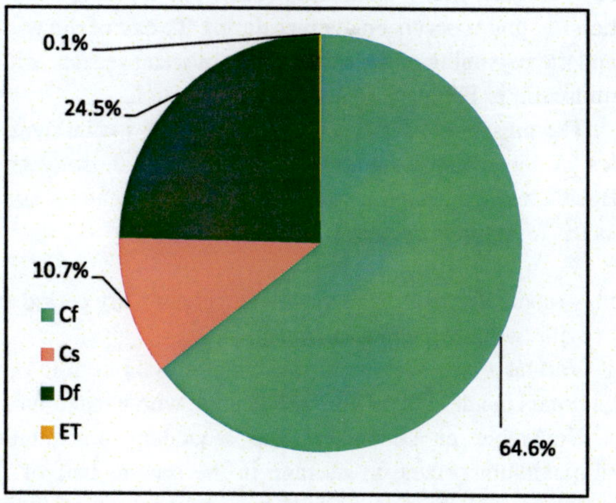

Fig. 5.16 Area share of climates in Bosnia and Hrzegovina from 1980 to 2010

5.6 Main Climate Subtypes

Based on the thermal regime of characteristic months, i.e., average monthly temperatures in January and July, the main climate types in Bosnia and Herzegovina are further differentiated to the level of climate subtypes by adding a third letter in the label (Drešković 2011). In Cf main climate type, according to the presented results on quantitative-spatial dynamics of average annual and July isotherms (given in the previous chapter), the following climate subtypes can be distinguished: Cfa—moderately warm and humid climate with hot summer, Cfb—moderately warm and humid climate with warm summer, Cfc—moderately warm and humid climate with fresh summers.

In the Cs main climate type, similarly to the previous Cf climate type, further climatic differentiation can be performed based on the quantitative-spatial dynamics of average annual and July isotherms. Based on that, the following climate subtypes are distinguished in Bosnia and Herzegovina within this main climate type: Csa—Mediterranean climate with hot summer, Csb—Mediterranean climate with warm summers.

In the Df climate type, also on the basis of quantitative-spatial dynamics of average annual, January and July isotherms, two climate subtypes can be distinguished: Dfb—humid boreal climate with warm summers, Dfc—humid boreal climate with fresh summers.

Further, more detailed climatic differentiation of the mentioned climate subtypes is performed based on the time distribution of precipitation maximums and minimums during the year, based on which a number of variants of climate subtypes can be singled out. More specifically, in the annual flows of pluviometric regimes in Bosnia and Herzegovina, two more or less pronounced periods that have precipitation maxima characteristics, and two more or less pronounced periods of precipitation minima, distributed between periodic maxima, can be distinguished. One precipitation maximum in Bosnia and Herzegovina is on average positioned in late autumn and early winter with a maximum in November or December. The second precipitation maximum occurs on average in the second half of the spring and the beginning of the summer climatological season, with a maximum in May or June. Precipitation minima in Bosnia and Herzegovina are distributed between the periods of precipitation maxima. One period of precipitation minimum at the level of the whole of Bosnia and Herzegovina occurs on average during the second half of summer, i.e., in the period July–August, with a minimum in July. The second precipitation minimum occurs on average during the end of the winter and the beginning of the spring climatological season, with a minimum in February or (less often) in March.

The presented facts about the quantitative-spatial dynamics of the average January, July, and annual isohyets in Bosnia and Herzegovina, the following letters can be added as the fourth index symbol:

x"—moderately warm rainy climate without a dry period and with two precipitation maxima:

1. Variant x"w—The main precipitation maximum is positioned in late autumn and early winter, with a maximum in November or December. The secondary precipitation maximum occurs on average in the second half of the spring and the first half of the summer climatological season, with a maximum in May or June. Precipitation minima are time-positioned between them. The main precipitation minimum at the level of the whole of Bosnia and Herzegovina occurs on average during the second half of summer, i.e., in the period July–August, with a minimum in July. The secondary precipitation minimum occurs on average during the end of the winter and the beginning of the spring climatological season, with a minimum in February or (less often) in March.
2. Variant: x"s—The main precipitation maximum occurs on average in the second half of the spring and the first half of the summer climatological season, with a maximum in May or June. The secondary precipitation maximum is positioned in late autumn and early winter, with a maximum in November or December. The main precipitation minimum at the level of the whole of Bosnia and Herzegovina occurs on average during the second half of summer, i.e., in the period July–August, with a minimum in July. The secondary precipitation minimum occurs on average during the end of the winter and the beginning of the spring climatological season, with a minimum in February or (less often) in March.

sx"—Mediterranean climate with summer dry period and with two precipitation maxima:

In this climatic subvariant's annual course of the pluviometric regime, distinct precipitation periods can also be singled out. The main precipitation maximum is positioned in time for the second half of autumn, and the first half of the winter climatological season; i.e., it falls from October to December. The primary precipitation minimum occurs during the summer climatological season, i.e., June–August when on average, only 83.5 mm or about 30.0% of the annual monthly average is excreted per month. The secondary precipitation maximum occurs on average in February, while the secondary precipitation minimum is not so clearly separated, although it can be estimated that it appears on average during the end of the winter and the first half of the spring climatological season, i.e., in the period March–April.

According to the above climatic indices, the already mentioned climatic subtypes with the following subvariants can be singled out in Bosnia and Herzegovina: Cfax"s—moderately warm and humid climate with hot summer and no dry period; Cfbx"s—moderately warm and humid climate with warm summers and no dry periods; Cfcx"w—moderately warm and wet climate with fresh summer and no dry period; Csasx"—Mediterranean climate with hot summers and summer droughts and with two precipitation maxima; Csbsx"—Mediterranean climate with warm summers and summer dry periods and with two precipitation maxima; Dfbx"w—humid boreal climate with warm summers and without dry period; Dfcx"w—humid boreal climate with fresh summer and without dry period; ET—tundra climate.

References

Drešković N (2004) Klima Sarajeva [Climate of Sarajevo]. Master thesis, Faculty of Science University of Sarajevo, Sarajevo

Drešković N (2011) Klimatski tipovi Bosne i Hercegovine prema W. Köppenovoj klimatskoj klasifikaciji [Climatic types of Bosnia and Herzegovina according to W. Köppen's climate classification]. Doctoral dissertation, Faculty of Science University of Sarajevo, Sarajevo

Drešković N (2015) Spatial distribution of annual rainfall in Bosnia and Herzegovina using Spatial Kriging Interpolator. Geografski pregled 36:95–107

Drešković N, Đug S (2012) Applying the Inverse Distance Weighting and Kriging methods of The spatial interpolation on the Mapping the Annual Precipitation in Bosnia and Herzegovina. In: Seppelt R (eds) 2012 International congress on environmental modelling and software managing resources of a limited planet, Leipzig

Drešković N, Mirić R (2014) Klimatska regionalizacija Bosne i Hercegovine [Climate regionalization of Bosnia and Herzegovina]. Zbornik radova III kongres geografa Bosne i Hercegovine, pp 280–294

Drešković N, Mirić R (2016) Horizontal air temperature changes as a basis for the regional climate differentiation of Bosnia and Herzegovina. Geografski pregled 37:51–65

Drešković N, Mirić R (2017) Thermal regime of annual air temperature in Bosnia and Herzegovina. Revija za Geografijo 12(1):109–120

Đug S, Drešković N (2008) Prirodno naslijeđe Kantona Sarajevo [Natural heritage of Canton Sarajevo]. Kantonalni zavod za zaštitu kulturno-historijskog i prirodnog naslijeđa, ArchDesign, Sarajevo

Meteorological data 1980–2010, Meteorological yearbooks of Bosnia and Herzegovina SHMZ Beograd and FHMZ Sarajevo, Sarajevo

Mihailović DT, Drešković N, Mimić G (2015) Complexity analysis of spatial distribution of precipitation: an application to Bosnia and Herzegovina. Atmos Sci Letters. Royal Meteorological Society 16: 324–330

6 Hydrography of Bosnia and Herzegovina

Abstract

The general characteristics of the natural hydrographic system in Bosnia and Herzegovina is its presence in the form of surface and underground type of watercourses and its consistency of predominantly freshwater and, to a lesser extent, saltwater bodies. In most parts of Bosnia and Herzegovina (clastic rocks), a developed surface river network with a customarily developed catchment area is formed. The areas of Bosnia and Herzegovina with dominant limestone and limestone-dolomite deposits are dominated by underground watercourses (highly developed underground hydrographic systems), specific to the Dinaric karst area. The marine aquatic area, located in Neum-Klek bay, is the only saltwater area that belongs to Bosnia and Herzegovina. Through the general hydrological characteristics of the catchment areas of Bosnia and Herzegovina, differentiation was performed according to the large basins of the Black and Adriatic Seas. The fundamental morphometric and hydrological indicators of the mentioned sea basins are presented, and six major river flows of the Black Sea (Bosna, Una, Vrbas, Drina, Glina with Korana, and the immediate Sava basin) and Adriatic (Neretva, Trebišnjica, and Cetina with Krk) basins. Given this distribution, the chapter follows a detailed analysis of morphometric (river basin area, river network density, coefficients, longitudinal profile characteristics, etc.) and hydrological characteristics (water level elements) for all nine isolated river basins. Particular emphasis is placed on the complexity and very detailed inventory of the river network of these flows.

Keywords

Waters · Rivers · Lakes · Sea · Physical geography

6.1 General Characteristics of the Hydrographic System

The hydrographic system of Bosnia and Herzegovina was formed under the intensive modifying influence of the geological structure of the terrain, as well as climatic and geomorphological characteristics (especially at the local level). The natural hydrographic system in Bosnia and Herzegovina is developed and appears in the form of surface and underground type of runoff; i.e., it includes both freshwater and freshwater bodies. More specifically, most of the hydrographic system consists of freshwater that occurs in surface and groundwater runoff, directly dependent on the hydrogeological characteristics of rock masses in some parts of Bosnia and Herzegovina. Accordingly, it can be said that, in most of Bosnia and Herzegovina, where clastic rocks are formed on the surface, a surface river network has been developed with a small share of groundwater in the form of phreatic outcrops (Drešković and Mirić 2017).

Due to the low permeability coefficient of clastic deposits, most precipitation water flows off the surface through a watercourse system that usually has a standard hydrographic catchment area. On the other hand, underground runoff dominates in parts of Bosnia and Herzegovina, whose surface is predominantly built of limestone and limestone-dolomite rock masses, given that, under the influence of intense neotectonic movements, these rock masses are primarily cracked and have the function of hydrological collectors. The rainwater's subsequent chemical action within the rock masses formed a significantly developed underground hydrographic system (or systems) consisting of cracks, fissures, channels, caverns, extensions, etc., through which groundwater circulates to the lower aquifer. In accordance with the mutual relations within these systems, part of the

H. Gekić et al., *The Geography of Bosnia and Herzegovina*, World Regional Geography Book Series,
https://doi.org/10.1007/978-3-030-98523-3_6

groundwater erupts on the terrestrial topographic surface in the form of various hydrological objects such as springs, ponds, estavelle, etc., which form a particular hydrographic system of the Dinaric karst. Most of the waters of the underground karst system erupt on the seabed in the form of submarine springs, sea mills, etc. A smaller part of the karst area of Bosnia and Herzegovina is characterized by surface runoff which is limited to Neogene flysch deposits or deposits that have a clastic structure and hydrological insulators. In this way, larger watercourses were formed, forming the backbone of these areas' surface river network.

Of the relatively high total annual precipitation of about 1250 mm which is discharged to the mainland of Bosnia and Herzegovina, the surface runoff accounts for a little more than half. More specifically, the surface runoff coefficient is about 0.57; i.e., the average annual runoff is about 1200 m^3/s. Of this value, the surface runoff in the Black Sea basin (Danube basin) accounts for about 722 m^3/s and about 433 m^3/s in the Adriatic basin (Drešković and Mirić 2017).

Following the predominant geological-geotectonic characteristics of the terrain and the dominant types of geomorphological processes, it can be stated that the dendroid type of the river network predominates. Only exceptionally, in parts of the terrain with dominant radial tectonics, riverbeds are tectonic predisposed, and runoffs on the plan acquire certain rectilinear forms. Such a more pronounced case in Bosnia and Herzegovina is with the tectonic trench of the Sava River, which is also the boundary of separation between two first-order geostructural units: the Dinarides and the Panonids.

The total length of the permanent surface river network in Bosnia and Herzegovina (obtained by planimetry of Topographic map 1: 200,000) is about 20,919 km or, on average, for the entire country about 0.41 km/km^2 (Table 6.1).

The estimated values of reserves in aquifers of intergranular porosity for the Black Sea basin are about 3.93 m^3/s and about 1.05 m^3/s for the Adriatic basin. Water reserves in the karst-fissure porosity type zones in the Black Sea basin amount to about 28.26 m^3/s, while in the Adriatic basin, this value is approximately 35.94 m^3/s. Significant amounts of water that participate in the freshwater balance are lakes in Bosnia and Herzegovina. According to the existing data, the total area of the lake waters is about 181.6 km^2, with the most significant part being covered by artificial reservoirs—about 150.4 km^2. The marine reservoir located in the Neum-Klek bay is the only saltwater area that belongs to Bosnia and Herzegovina. This area covers part of the Neum Bay and half of the Channel of Mali Ston, with about 12.2 km^2 representing the territorial waters of Bosnia and Herzegovina. As part of the Adriatic Sea, the marine waters of this bay play a crucial role in the climatic conditions of the southern part of Bosnia and Herzegovina (Fig. 6.1).

Analogous to the geological composition of the terrain and the general physical-geographical conditions in Bosnia and Herzegovina, according to the structure of porosity, several different types of aquifers can be distinguished: terrain with aquifers of intergranular porosity, terrain with aquifers of intergranular and crack porosity, terrain with cavernous-crack porosity aquifers, terrain with crack porosity aquifers, terrains practically without aquifers (Consortium for drafting the water management strategy of the Federation of BiH 2010).

6.2 General Characteristics of the Basin Areas

In accordance with the general physical-geographical characteristics, especially under the influence of geological-tectonic and orographic specifics, in Bosnia and Herzegovina, two main catchment areas belonging to the Black and Adriatic Seas basins can be distinguished. More specifically, the Black Sea basin in Bosnia and Herzegovina includes the Peripannonian and most of the mountain-valley Bosnia and Herzegovina, where the river systems of rivers are entirely or partially formed: Sava (immediate basin),

Table 6.1 Basic morphometric indicators of surface river network in BiH

Catchment area (basin)	Total length (km)
Direct basin of Sava	3197
Glina and Korana	500
Una	2748
Vrbas	2654
Bosna	5911
Drina	3168
Basin of Black Sea in BiH	18178
Neretva	1962
Trebišnjica	299
Cetina and Krka	480
Basin of Adriatic Sea in BiH	2741
Total in BiH	20919

Based on the Hydrological data 1980–2010 and field research of authors

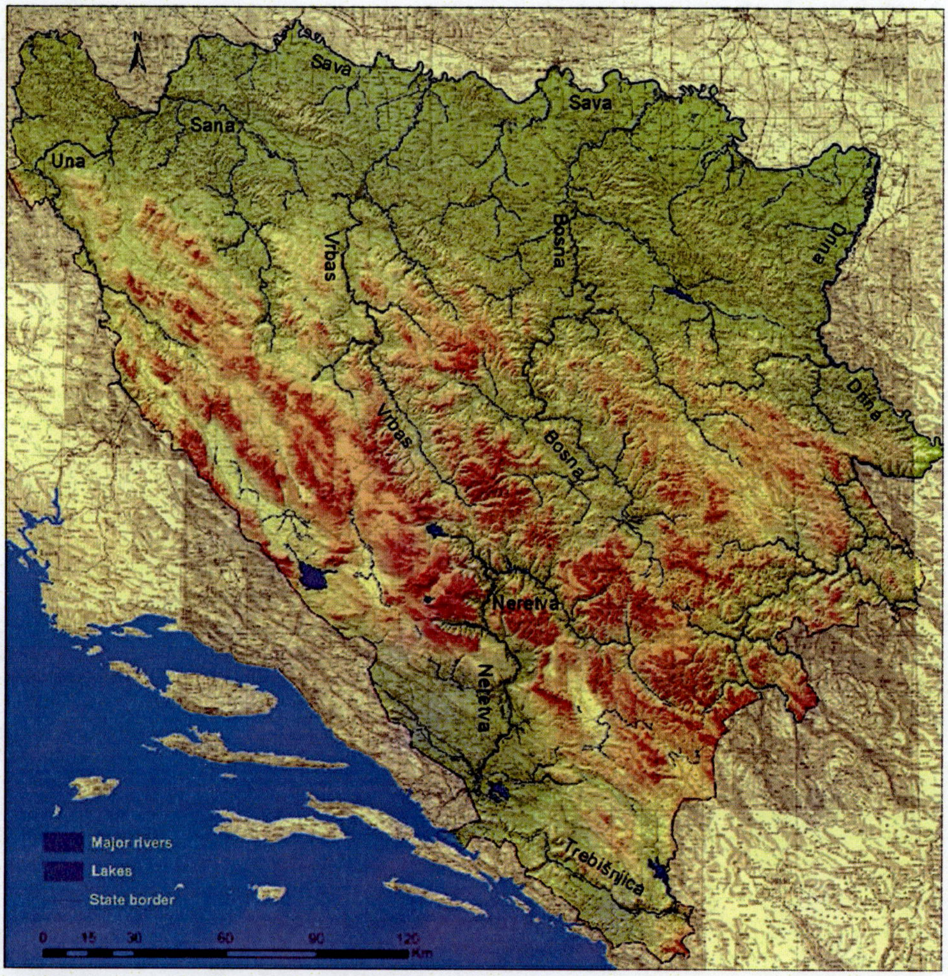

Fig. 6.1 Basic hydrographic system of Bosnia and Herzegovina

Glina, Korana, Una, Vrbas, Ukrina, Bosna, and Drina and their tributaries. The spatial coverage of the Black Sea basin in Bosnia and Herzegovina is 38223.5 km^2 (or 74.7%).

The Adriatic basin covers the area of the Mediterranean and the macro-region of high karst, i.e., the area of the southern part of Bosnia and Herzegovina. The area of the Black Sea basin includes the basins of the rivers: Neretva, Trebišnjica, Cetina, and Krka and their tributaries. The area of the Adriatic Sea basin is 2955.5 km^2, which represents about 25.3% of the total area of Bosnia and Herzegovina (Table 6.2).

In relation to the dominant morphological-geological-geotectonic features, these main catchment areas are further divided into nine sub-catchment areas, six of which belong to the Black and three to the Adriatic Sea Basin in Bosnia and Herzegovina (Fig. 6.2).

The Black Sea basin in Bosnia and Herzegovina includes the following sub-basins (Fig. 6.3): Immediate basin of Sava river, Glina and Korana, Una, Vrbas, Bosna, Drina. The following three sub-basins have been singled out in the Adriatic Sea basin area: Neretva, Trebišnjica, Cetina, and Krka.

Following the stated hydrological differentiation of the hydrological system of Bosnia and Herzegovina, a summary at the level of the catchment areas of the main rivers is given in the following sub-chapters.

6.2.1 Glina and Korana River

Considering the dominant clastic structure of the terrain of the region of northwestern Bosnia, it can be said that the surface river network is well developed, with a significant share of surface runoff, amounting to approximately 0.6. In accordance with the orographic characteristics of the basin, it can be said that the longitudinal profiles of larger watercourses in the basin are mostly mild; i.e., they are characterized by minor falls amounting to a maximum of

Table 6.2 Basic hydromorphological indicators of the surface river network in Bosnia and Herzegovina

Catchment area (river basin)	Area (%)	Inflow (l/s/km^2)
Direct basin of Sava	10.4	11.4
Glina and Korana	1.4	24.6
Una	15.6	26.1
Vrbas	12.3	19.9
Bosna	21.0	16.1
Drina	14.0	21.2
Basin of Black Sea in BiH	74.7	119.3
Neretva	15.5	38.1
Trebišnjica	4.4	49.4
Cetina and Krka	5.4	34.6
Basin of Adriatic Sea in BiH	25.3	122.1
Total in BiH	100.0	241.4

Based on the Hydrological data 1980–2010 and field research of authors

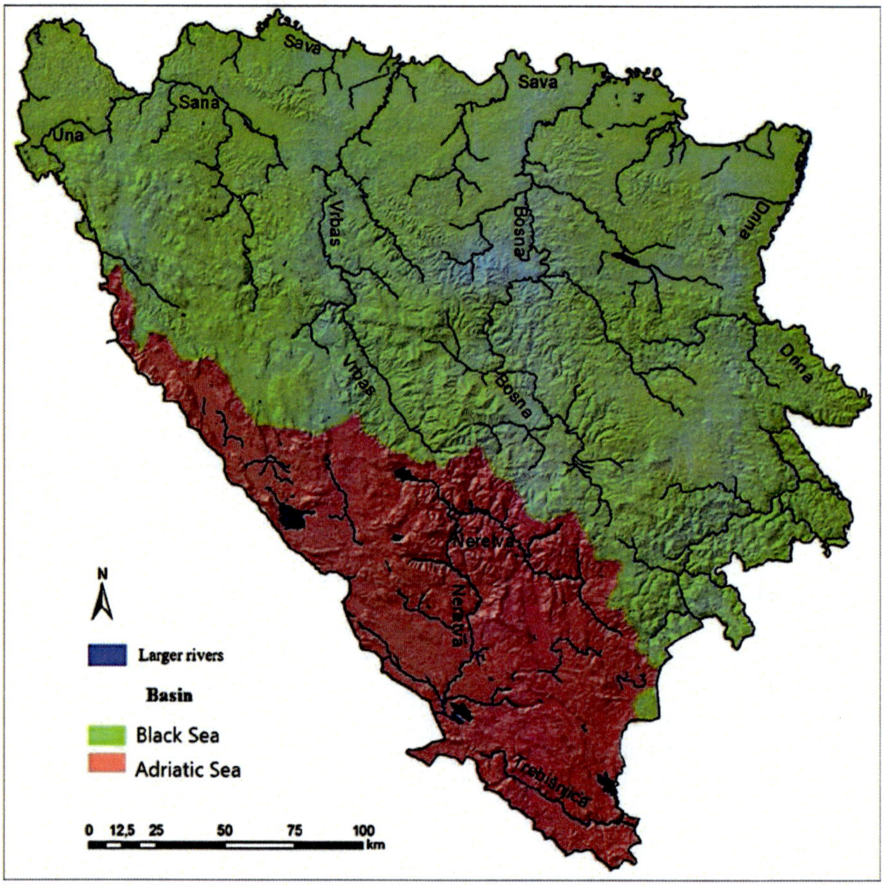

Fig. 6.2 Main basins in Bosnia and Herzegovina

5‰. For this reason, watercourses have more flat features characterized by a continuous and even inflow of water mass. Exceptions are the areas of the spring head formed in the zones of mountainous areas, where watercourses have a less pronounced torrential character.

The hydrographic backbone of the source headwaters of the Korana River in the area of this sub-basin in Bosnia and Herzegovina consists of the rivers Toplica and Mutnica, while the rivers Kladušnica and Glinica with their tributary system form the Glina watercourse. The sub-basin area of these watercourses in Bosnia and Herzegovina is about 705.6 km^2.

In the sub-basin areas of these watercourses and their tributaries, hydrological stations have been formed to observe the basic elements of the river regime. Still, the series of hydrological observations are quite short and partially

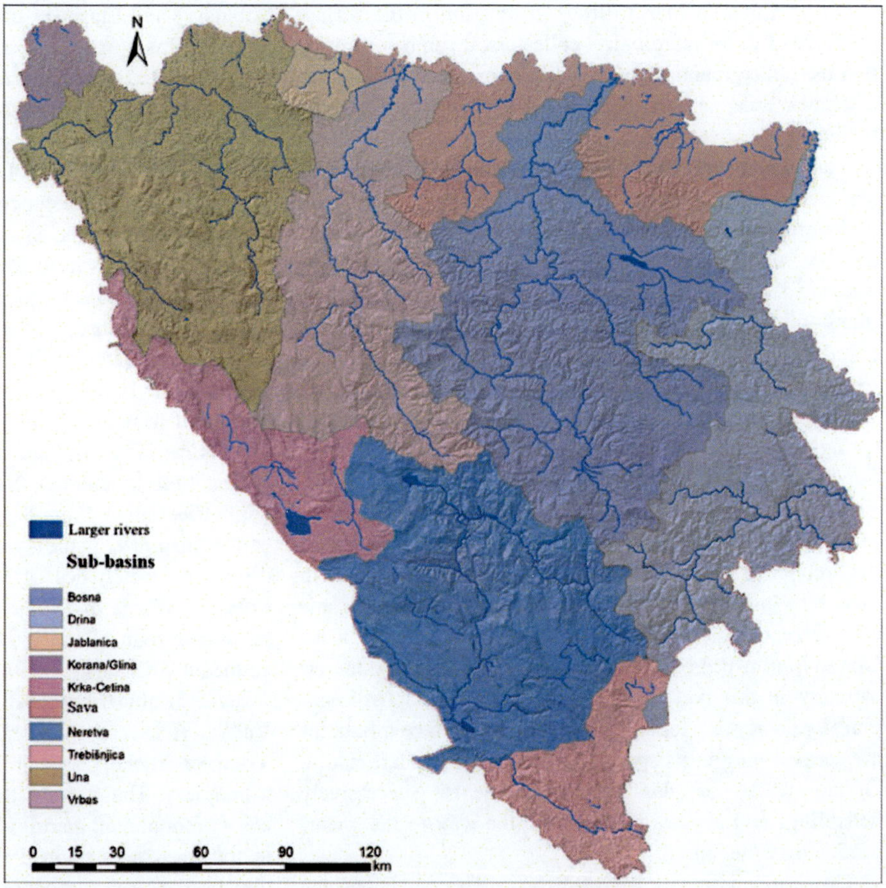

Fig. 6.3 Sub-basins in Bosnia and Herzegovina

unreliable, so it is not possible to determine them with certainty. For this reason, for presenting the amount of water on the outlet profiles of Glina and Korana rivers to the Republic of Croatia, they are given in summary, based on the hydrological regionalization of the entire basin (Encyclopedia of Yugoslavia 1983).

6.2.2 Una River

Una river basin covers the northern and northwestern part of the Dinaric karst zone, i.e., Mesozoic limestones, due to which an underground runoff was formed in the karst part of the basin. In the zones where the limestone and limestone-dolomite parent rock substrate dominates, a zone of hydrogeological collectors has been formed., which conditioned the absence of surface river network or leachate of rainwater through cracks, fissures, and channels into the karst underground. In clastite zones, which usually form the hanging walls of the river valleys and basin widening of the Una and Sana rivers and their larger tributaries, they have the characteristics of hydrogeological insulators, which had a positive effect on the survival of the surface water network (Đug et al. 2008). In accordance with the presented facts, it can be said that the Una basin area is characterized by a large percentage of groundwater runoff, which certainly had a negative impact on the more even spatial distribution of the network of surface watercourses. It can be generally concluded that in the zones of Triassic, Jurassic, and Cretaceous limestones, the percentage of filtered rainwater into the karst underground is very high, so in such areas, the surface network of watercourses has not even developed (Drešković and Mirić 2017).

The predominant groundwater runoff takes place through fissure systems, which are mostly connected to fault zones. Groundwater in the form of springs of varying degrees of water yield emerges on the surface, mainly at the contacts of karstified limestone and impermeable dolomite or poorly permeable clastic deposits. Considering the basic geological-geotectonic characteristics of the basin, it was determined that the Una, Velika, and Mala Neteka springs and a series of smaller springs along the Una river (upstream from Martin-Brod) are fed from the abyss in Brezovac in the southwestern part. In the southeastern area, from the Vagan and Resanovci abysses (near Grahovo), the source of the Krka river (east of Begluk) and the source of Bastašica along the Unac (in the western part of the Drvarsko polje) are fed, with the same water occurring at the very strong Crno

Vrelo on Unac (at Martin-Brod). In relation to the surface presence of these hydrological collectors, it can be estimated that the underground runoff developed on about 40% of the catchment area; i.e., the coefficient of surface runoff is about 62.6%.

The largest watercourse and chirographic backbone of the entire area is the river Una, with a total length of 212 km. Due to the predominant limestone structure and dynamic relief, it can be said that the Una River is hydrographically very weakened. More specifically, the distribution of carbonate rocks—limestone-dolomite structures—significantly influenced the morphology of the river valley, and thus the reduction of the surface river network. The spring of the river Una consists of a large number of more abundant karst springs. The main spring of Una river is below the mountain Stražbenica (the village of Donja Suvaja), with two more significant in this zone in terms of water yield: Velika and Mala Neteka. Immediately downstream from these springs, the Srebrenica stream flows into the river Una. After 12 km of flow, Una (partly as a border river) becomes a river in Bosnia and Herzegovina. The most important right tributaries of the Una to Martin-Brod are Krka and Unac. Unac, as the largest tributary in this part of the Una basin, is also hydrographically poorly developed (considering the surface tributaries are formed only in its spring, as well as on the left side to Drvar, while in downstream segments successively bedsprings and abysses alternate, all the way to the water-producing Crno Vrelo).

Downstream from the Martin-Brod extension to Bihać, the valley of the river Una has mostly gorge features so that the tributary system is very scarce, and the replenishment of the flow takes place directly from karst springs (mainly concentrated around Kulen-Vakuf). Only on the stretch from Ripče to Bihać do significant tributaries appear Račićki potok, Ripački potok, Hrgarski potok and Orljanski potok, as right tributaries, and Dobrenička rijeka and Drobnička rijeka (Zavaljska and Skočajska draga), as left tributaries. A very significant right tributary is the river Klokot with the tributary Mresnica, which flows into the Una in the city of Bihać. In the area of the Brekovica village, Brekovački izvor flows as a left tributary into the Una. A more significant left tributary is the river Krušnica, which flows into the Una in the zone of the town of Bosanska Krupa. Although the river Krušnica is short-flowing, it is a water-flowing stream, considering that it originates from a very abundant Krušnica spring.

The next significant left tributaries are the rivers Bastra, Glodina, and Ivanjska, which affect the Una in the zones of the settlements of the same name. The most significant right tributary in this flow segment is the river Vojskova, which flows into the Una in the zone of the village of Rudice. The next significant tributaries are the Bukovska rijeka (left tributary) and Ljusinski potok (right tributary), which flow into the Una upstream (Fig. 6.4) from the village of Otoka.

In the area of the town of Novi Grad, its largest tributary, the Sana river, flows into the Una river. Sana river originates from three water-producing karst springs in the karst plateau zone near the villages of Donja Pecka and Jasenovi Potoci. All three springs after ephemeral streams (about 1.5 km) form a single stream of the Sana. Given the slight anthropogenic pressures, Sana river, in the part of the flow to Ključ, has the characteristics of a clean river. In the part of the flow to Sanski Most, anthropogenic pressures increase, which causes a decrease in water quality. In the village of Vrhpolje, Sana is joined by its largest tributary—Sanica. Besides it, the most important tributaries are Banjica, Čajića rijeka, Milošnica, Sokošnica, Kozica, Tomina, Kijevska rijeka, Glamašnica, Sasina (left tributaries), Ključki potok, Dabar, and Zdena (right tributaries). From Sanski Most, the longitudinal profile of the Sana river is quite harmonized, which is why it takes on the characteristics of a lowland river. Upstream from Prijedor (village of Gomjenica), Sana is joined by a water-producing right tributary—Gomjenica. In addition, the most abundant tributaries include Pilješka Rijeka, Bukovica, Ježevac (right tributaries), Bliha, Dugovača, Majdanuša, Velika (Dževarska) river, Utvinac, Potok Oštra luka, Vranjska rijeka, Kostajnica, and Brdarska rijeka (left tributaries). The flow of the Sana to the town of Prijedor has a predominant meridional direction, from where it flows in the form of a trapped meander in a prevailing parallel direction. A large number of tributaries flow into this segment of the Sana, namely, Puharska rijeka, Svetinja, Vragolovača, Prljugovac (with Jesić river), Skakavac, Petkovačka rijeka, Krivoglavski izvor (left tributaries), Ljubija, Volar, Trgovištki potok, Crna, Velika, Japra, and Suća (right tributaries).

Sana river flows into the Una river in Novi Grad. Downstream from the mouth of the Sana, as part of the left tributary system, the following rivers flow into the Una: Tuinica, Strižna (with Vodićevska rijeka), Dobski potok, Strigova, Tavija (in the village of Kostajnica), Vučijak, Slabinja, Mlječanica, Binjačka rijeka (in the Kozarska Dubica), Moštanica, Virovska, and Rakovica (in the zone of the confluence of the Una and Sava rivers near the village of Jasenovac). The formation of the longitudinal profile of the Una riverbed was directly influenced by neotectonic movements, which, along with the specific lithological composition of the terrain, reflected the formation of numerous geological-geomorphological-hydrological phenomena in the riverbed such as travertine waterfalls, cascades, and rapids. Thus, for example, along the transverse faults, the Martin-Brod and Štrbački buk waterfalls were formed (Figs. 6.5 and 6.6), where the increased fall of the river

Fig. 6.4 Travertine barriers on the Una River upstream from the confluence of the Unac. (Photo by Drešković Nusret)

Fig. 6.5 Martin-brod Waterfall. (Photo by Drešković Nusret)

Fig. 6.6 Štrbački Buk Waterfall. (Photo by Drešković Nusret)

caused its substantial vertical incision and increased deep erosion into the parent rock substrate. It should certainly be noted that, in addition to the vertical incision, it is very intense and lateral, which causes a more pronounced meandering of Una in the zones of floodplains (Đug et al. 2017).

The dominant type of travertine deposits represent biogenic travertine that develops in certain segments along the river Una, particularly in the area upstream from the town of Bihac. Research has also established that the basic hydrological and physical-chemical parameters of water have acceptable ecological values and favor the further development of natural travertine deposits (Đug et al. 2017). In its annual water supply, Una river has a characteristic nival-pluvial regime that causes elevated flows during spring (late spring maxima) and late autumn and early winter (with early winter maxima). Minimum flows occur in the second half of summer and the first half of autumn (Fig. 6.7).

In the annual course of the flow, the first period of maximum flows occurs during the spring with the values of periodic averages of 220 m^3/s (HS "Martin-Brod") and 140 m^3/s (HS "Kralje"). The second, late autumn, and early winter periods of elevated flows have 175 m^3/s and 165 m^3/s, respectively. The period with minimal flows occurs in the second half of summer and the first half of autumn with periodic averages of 48 m^3/s and 55 m^3/s, respectively. Certainly, in relation to the stated averages, there are years during which extreme values and low and high flow occurred.

6.2.3 Vrbas River

The backbone of the river network of this area is the Vrbas river with its developed surface and underground tributary system (Fig. 6.8). In the upper part of the watercourse (up to Donji Vakuf), Vrbas has a developed surface river network, i.e., a larger number of tributaries, some of which show karst features. A large number of springs rich with abundant water is the primary characteristic of the spring catchment area of the Vrbas river. More specifically, the source of the Vrbas river is the spring of Vrbas, which is located on the southwestern foothills of the mountain Zec, at about 1730 m above sea level. Among the more significant flows in the zone of the spring catchment area are also the Sikir and Deral streams (Drešković and Mirić 2017).

The first significant right tributary of the Vrbas in this part is the river Desna, which flows into Vrbas after the first 16 km of flow from the source. After approximately 1 km, the river Tuščica flows into the Vrbas as a left tributary. In the

Fig. 6.7 Annual course of mean monthly flows on the Una River from 1980 to 2010. (HS—hydrographic station)

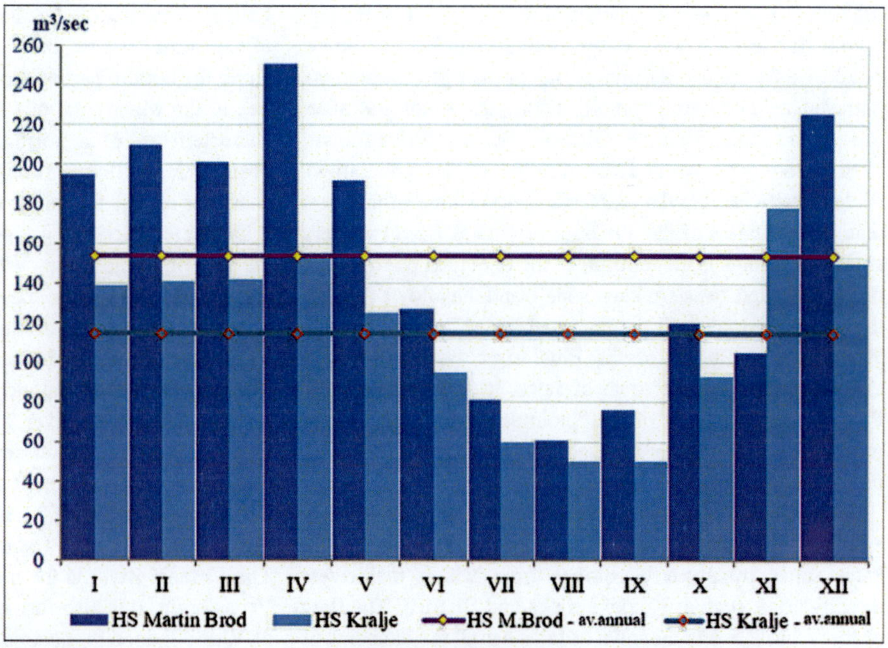

Fig. 6.8 River Vrbas in Ključ city. (Photo by Aida Bidžan-Gekić)

downstream segment of the flow to the town of Donji Vakuf, the following rivers flow in Vrbas: Crnodol, Kruščica, Bistrica (with Suhodolski potok), Ričica, Zaneski potok, Odžački potok, Goruški potok, Vileški potok, Vitina (town of Bugojno), Rovanjski potok, Drvetinska rijeka, Ćehajićka rijeka, Volušnica, Oboračka rijeka (town of Donji Vakuf) (as right tributaries), Trlica, Trnovača (with Trunta), Barski potok, Mrkodolski potok, Cipalska rijeka, Bunta, Duboka rijeka (with Poričnica) (town of Bugojno), Lubovski potok, Prusačka rijeka (town of Donji Vakuf) (as left tributaries).

The density of the surface river network in the downstream segment of the Vrbas flow (up to the town of Jajce) is slightly decreasing due to the increase in karst relief and underground water sinking. The main streams that make up the tributary system in this segment of the Vrbas are: Barićki potok, Sandžački potok, Babin potok, Goduša, Sokolinski potok, Bila voda, Rika (town of Jajce) (as right tributaries), Semešnica, Podvodski potok, Đihanjski potok, Kračec, Rasadnički potok, Hornjak, Glasinac, Sedinac, Pliva (downstream from the town of Jajce—Jezero) (as left tributaries). Pliva river (with Jošavka and Crna rijeka) is one of the most aquatic tributaries of the Vrbas. It originates from two very abundant springs in the foothills of the Pljevljski podovi (part of the so-called "western karst zone") and thus flows through the town of Šipovo, all the way to Jajce. After entering the Pliva lake, the water of the Pliva river further flows in two ways: through the town of Jajce, it flows through the Pliva Waterfall as part of the "biological minimum," while most of the water flows through the tunnel to Hydropower plant (hereinafter: HPP) Jajce I.

Part of the mentioned watercourses has (as already emphasized) a distinct karst character; i.e., it is formed in karst zones by powerful springs from which real rivers directly arise, such as Kozićka rijeka and Bistrica. The flow of the Vrbas River to the town of Jajce has all the elements of a natural flow regime, with downstream from Jajce; there are strong modifying influences caused by the operation of HPP Jajce I, and (immediately downstream) the impact of HPP Jajce II. Immediately downstream from HPP Jajce II, the Vrbas river flows through a canyon river valley—the Vrbas canyon, which ends upstream from the city of Banja Luka. Significant tributaries in this segment of the flow are Lučina, Komotinski potok, Ugar, Bregovski potok, Svrakava (with Rijeka) (as right tributaries), Crna rijeka (Mrkonjić-Grad), Krupa, Rekavica (as left tributaries). Ugar river is also one of the most water abdundant flows in the basin. It occurs along the bottom part of the northwestern foothills of Vlašić, i.e., in the wider zone of Ljuta greda (1739 m), and has the characteristics of a normal hydrographic catchment area. From here, it flows through a valley of primarily gorge character except for minor morphological extensions where it has the character of a typical river valley. Upstream from the city of Banja Luka, Vrbas takes on plain features with very small slopes on the longitudinal profile. The most important tributaries in this part of the Vrbas river are: Vrbanja (cities: Banja Luka, Čelinac, Kotor-Varoš), Brusina, Turjanica (with Slatina and Vukešnica), Crkvena, Gumjera, Velika (Šeškovačka) river, Savica (with Povelićka Rijeka and Krivaja) (as right tributaries), Rujišnjak, Surtolija, Crkvina (city of Banja Luka), Rudnička rijeka, Ivaštanka (with Dragočajska rijeka), Bukovica, Klašnička rijeka, Mahovljanska rijeka, Osorna Kurjača (as left tributaries). The Vrbanja River is the most watery tributary in the entire Vrbas basin. It is also formed in the wider northern foothills of Vlašić mountain, upstream from the village of Pilipovina (at an altitude of 1520 m), in the region of Prelivode—a watershed with the Ugar river. It has a length of about 85.3 km and flows through a river valley which only in places (in the upstream part of the stream) has gorge features.

In the hydrological regime, in the upstream segments of watercourses, the nival-pluvial water supply regime dominates, noting that the main flow occurs in the second half of spring, i.e., in the hydrological April–May period (Fig. 6.9). The periodic average of the main hydrological

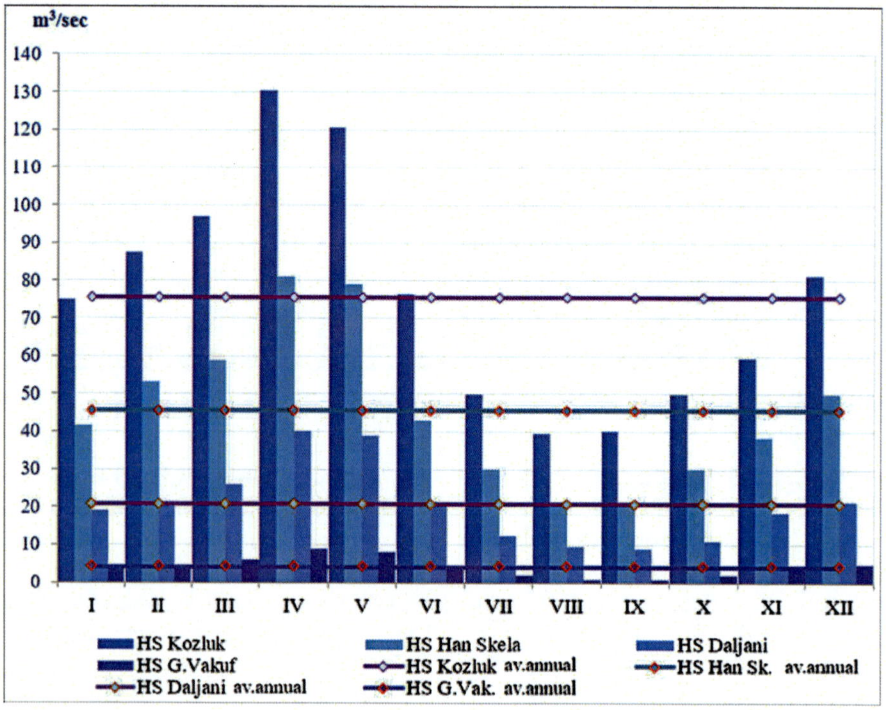

Fig. 6.9 Annual course of medium monthly flows on the river Vrbas from 1980 to 2010. (HS—hydrographic station)

maximum is 125 m^3/s (HS "Kozluk") and 80 m^3/s (HS "Han-Skela"). The second hydrological maximum occurs in the period of late autumn and the first half of winter with a periodic average of 70.5 m^3/s (HS "Kozluk") and 45 m^3/s (HS "Han-Skela"). The period with minimal flows occurs in the period of the second half of summer and the first half of autumn, i.e., in the period August–September, with a periodic average of 40 m^3/s (HS "Kozluk") or 21.5 m^3/s (HS "Han-Skela"). The measurements indicate similar hydrological characteristics at the remaining two hydrological stations: Daljani and Gornji Vakuf. More specifically, the main periodic maximum on HS "Daljani" occurs in the period April–May with an average of about 40.0 m^3/s. At HS "Gornji Vakuf" the periodic maximum is 8.5 m^3/s. The situation is similar for periodic minima. The main periodic minimum at HS "Daljani" occurs in the period August–September with an average of 9.5 m^3/s, while the same for HS "Gornji Vakuf" is only 2 m^3/s (Consortium for drafting the water management strategy of the Federation of BiH 2010).

6.2.4 Bosna River

The Bosna River Basin covers the areas of the two macro-regions of Bosnia and Herzegovina; i.e., in the northern part it belongs to the Peripannonian, while the central and southern parts belong to the mountain-valley region of Bosnia and Herzegovina.

Most of the river network, about 90%, of the catchment area, consists of surface watercourses that spread in the northern, eastern, western, and central parts of the area. The main surface watercourses are mostly developed on the main alluvial plateaus and collect the waters of tributary streams whose valleys are formed along the steep slopes. According to the method of origin, most of the surface river network has the characteristics of a normal hydrographic headland, which is why all ranks of watercourses can be distinguished within it: streams, smaller streams, streams, smaller rivers, and rivers. In the karst terrains of the Crnorječka plateau and the wider area around Igman, Bjelašnica and Treskavica in the south of the catchment area, there are all hydrogeological preconditions for the development of an underground hydrological network. Underground flows appear on the topographic surface at the contact of water-permeable and water-impermeable sediments, most often in the form of water-source springs and springs. Some of them are strong enough to directly form the right currents. The surface runoff coefficient (lo) is quite large—about 0.5, which indicates that almost half of the rainwater in the Bosna River Basin evaporates before it reaches the watercourse (Fig. 6.10).

The hydrographic backbone of the river network in the basin is the river Bosna, whose total length is about 270.8 km (excluding the river Željeznica). The hydrographic source of the river Bosna is a drainage karst spring in the foothills of Bjelašnica—the spring of Željeznica, at an altitude of about 1020 m. and impermeable Neogene deposits of the Sarajevo

Fig. 6.10 The Spring of the Bosna river. (Photo by Ranko Mirić)

field. All springs, in relation to the thermal characteristics of water, are differentiated into a total of five groups of springs, of which over 30 of them have a constant water yield. The total water yield of the spring is about 65 m^3/s, with the central spring emitting an average of about 5 m^3/s of water, or 27 m^3/s during the flood season and 1.5 m^3/s during the drier part of the year. In hydrological terms, its main source branch is the river Željeznica, which is the first large right tributary to flow into the river Bosna in Sarajevsko polje, east of the settlement of Otes.

At about 582 m downstream from the confluence with the Željeznica, the river Zujevina flows into the river Bosna from the left. It is formed from the Kradenik spring at about 1140 m above sea level on the northern foothills of the Maiden's Rocks and has a very developed hydrological headland. At about 2.9 km downstream, the river Dobrinja flows into the river Bosna from the right. Its source head is formed by a large number, mostly periodic, flows of the river Sušica, which are formed in the zone between 800 m and 1000 m above sea level. The Miljacka River is the next right tributary, which flows into the Bosna River in Sarajevsko polje, 1.2 km downstream from the previous mouth. It is formed by merging the streams of Mokranjska and Paljanska Miljacka, which originate from several water-generating springs in the foothills of Romania. The source and central parts of the course have all the characteristics of a gorge to the settlement of Bentbaša, from where they take on the characteristics of a normal river valley. The next significant right tributary is the Vogošća River, which flows into Bosna about 7.1 km downstream from the mouth of the Miljacka River, in the area of the Vogošća-Semizovac extension. Further downstream, at a distance of about 2.4 km from the previous mouth, on the right side, the river Ljubina flows into the river Bosna (Đug and Drešković 2008).

The spring headland of Ljubina, which consists of several water-source springs, is located at about 1070 m above sea level. The Misoča River flows into the Bosna River on the right side, 8.9 km downstream from the mouth of the Ljubina, in the zone of the Ilijaš morphological extension. A small town and the municipal center of Ilijaš developed on this river. The next more important right tributary of the Bosna River is the Stavnja River. Its source head was formed in the foothills of the Zvijezda mountain, within the hypsometric range from 1100 m to 1300 m. Downstream, at 7.3 km from the mouth of the Stavnja, the Fojnička River flows into the Bosna River. This watercourse originates in the foothills of Vranica, from the rivers Jezernica and Borovnica. In the zone of the town of Fojnica, it is called Dragača, and it gets its name from the Fojnička river after the river Željeznica flows downstream from the town. It flows into the river Bosna in the area of the town of Visoko. The river Goduša flows into Bosna in the zone of the settlement Mulići, about 3.6 km downstream from Visoko. On the stretch of the Visoko extension, tributaries of Bosna are: Grđevački potok (right tributary), Zimčanica, Radovljanska rijeka (with Orašački potok), and Suvaja (left tributaries).

In the downstream segment of the flow, the following flows flow to the Kakanj extension of Bosna: Trstionica (with Ričica), Zgošća, Ribnica (right tributaries), Papratnica, Jošanički potok, and Modrinjski potok (left tributaries). In the wider zone of the settlements of Katići and Grablje, the river Lašva flows into the river Bosna (Fig. 6.11). The source head of this watercourse is formed in the wider zone of Radalj mountain and Galičica, within the hypsometric zone from 1000 m to 1200 m. Within its larger composite extensions, the following cities developed: Travnik and Vitez. Its main tributaries are the rivers Grlovnica, Bila, Prala, Kruščica, and Kozica. In the downstream segment of the flow, the Zenica morphological extension in the river Bosna is reached by: Mrstava, Stijenčice, Babina rijeka (town of Zenica), Dobra voda, Gračanička rijeka (right tributaries), Suha, Kočevska rijeka (town of Zenica), and Bukovica (left tributaries). In the segment of the Vrandučka gorge (Fig. 6.12) to the Žepa morphological extension, the following reach Bosna: Ostrožnica, Bistrička rijeka, Orahovička rijeka and Bistričak, Starinski potok, Željeznica, Rakovica, (left tributaries), Crni potok, Nemilska rijeka, Velika rijeka, Pepelarska rijeka and Mrčajska rije (right tributaries). In the segment of the Žepa valley, a number of tributaries touch the river Bosna: Ograina, Papratnica, Ljubna, Bljuva (left tributaries), Želećka river, Bukovički potok, Bistrička river, Brisina and Pejića river (right tributaries). Immediately at the beginning of the Zavidovići expansion, the Gostović River flows into Bosna as a right tributary.

In the zone of the town of Zavidovići, the river Krivaja flows into Bosna. It is formed by the confluence of the rivers Bioštica and Stupčanica in the settlement of Olovo, from where it flows with a distinctly inconsistent longitudinal profile which causes the appearance of smaller rapids and waterfalls in the riverbed. The rivers Stupčanica and Bioštica have a source head in the wider zone of Šljemenska planina, which is formed by a large number of contact springs and springs, such as the spring of Bioštica, in the foothills of the mountain Bukovik. Significant tributaries of the Krivaja are the rivers: Tribija, Duboštica, Župeljeva (with Velika Maoča), Mala Maoča, Ribnica, Tatašnica, Vozućica, Maošnica, etc. Rujnica, Bočinja, Jablanica (with Rakovec) (town of Maglaj) (Fig. 6.13) Paklenica, Janjin stream, Velika rijeka, Spreča (town of Doboj) (right tributaries), Dolinski stream, Fojnica, Lješnica (with Stupinska river), Tujnička river, Ošvenski stream, Bistrica, Kosovski potok, Usora (left tributaries).

The source ridge of the river Spreča is formed from the river Papraća (in the village of Papraća), at about 600 m above sea level. The remaining part of the headgear is formed along the northwestern foothills of Velja glava, at about 450 m above sea level.

Fig. 6.11 Plava voda a tributary of the river Lašva. (Photo by Aida Bidžan-Gekić)

Fig. 6.12 Bosna River near Vranduk. (Photo by Ranko Mirić)

The main tributary system of Spreča consists of: Bukovica, Tavna, Sajtovića river, Kalesijska river, Dubnica, Mala Spreča, Rainska river, Gribaja, Gračanica river, Krivača, Lukavička river, Krivača, oskova (with Gostelj), Suha river, Ugar, Mednica, Turija, Jala (town of Tuzla), Šikuljačka river, Dobošnička river, Babin stream, Ponikva, Gnojnički stream, Jadrina, Miričinski stream, Kamenička river, Rašljevska river, Lohinjski stream, Sočkovački stream, Prenja, Sokoluša (town of Gračanica), Brijesnički stream, Crna river, Velika river, and Jošavačka river. By dividing

Fig. 6.13 Bosna River near Maglaj. (Photo by Ranko Mirić)

the river Spreča, an artificial lake Modrac was formed, 12 km long and 700 m wide on average, with a maximum depth of 14 m. It was made for the needs of water supply of the population and industry of the Tuzla region. The Usora River is formed by the confluence of the Velika and Mala Usora in the zone of the Teslić settlement. The length of the stream is about 82 km. The source head of both streams is markedly developed and formed in the foothills of Borje mountain, within the hypsometric range from 800 m to 1000 m above sea level. A part of the spring headland of Velika Usora is also formed in the foothills of Očauš, at about 1200 m above sea level. The tributary system of the Bosna River downstream of the town of Doboj has typical plain features, in which the longitudinal profile is balanced and with very small falls, which causes a reduced speed of water flow. More specifically, the tributary system to the confluence with the Sava River consists of: Lukavica (with Grapska river), Kožuhski potok, Brgula, Velika rijeka, Babošnica, Dusa (town of Modriča) (right tributaries), Rudanka, Foča, Veličanka, Lovnica, Glogovica, Plavuša, Ljubioča, Botajička river, Gnionica, and Jošava (left tributaries) (Hadžić and Drešković 2014).

In accordance with the current situation, it can be stated that Bosna has a natural hydrological regime, but it has recently been less disturbed by the construction of the flow HPP "Vranduk." There are also other less pronounced impacts that primarily relate to water intakes within the tributary system for the needs of water supply of the economy and the population, the waters of the Željeznica. Also, there is the impact of the operation of HPP Bogatići on the river Željeznica, which disrupts the natural flow regime of this river. The technogenic impact is particularly pronounced on the Spreča River, where, due to the impact of the Modrac hydro-accumulation, the flow regime of the Spreča River downstream of the dam is practically artificially regulated. In relation to the existing physical-geographical conditions and hydrographic elements of the river regime, the largest part of the river network in the basin of Bosna has a pluvial-nival type of water regime—a moderate-continental variant. The main characteristic of this type is the appearance of maximum water levels during March and April or April and March, while the minimum water levels are time-positioned in September and August. This is especially true of the hydrographic systems of the main right tributaries whose source ridges to the immediate confluence with the Bosna River have complete mountain features. This type of water regime is maintained downstream from the mouth of the Stavnja, where the elements of the water regime of the Posavina variant are already felt (Drešković and Mirić 2017).

Based on the presented facts, it can be concluded that in the natural water supply regime of the entire river network in the basin, the dominant role is played by rainwater that is secreted during the period of precipitation maxima. Another important but secondary factor is the water formed by the melting of the snow cover which has accumulated during the winter in higher hypsometric levels and which is felt more strongly in the water regime in late spring. All prominent elements of the water regime can be seen on the main watercourse in the Canton—the river Bosna. The basic quantitative indicators of the flow at the hydrological stations of the sub-basin of the river Bosna as well as the intra-annual change of the average monthly flows are presented in Fig. 6.14.

In the annual course of this type of flow, two characteristic periods can be distinguished. The period of the first hydrological maximum is time-positioned on March—May, during which the average on HS "Maglaj" is about 356 m^3/s with a maximum in April of 400 m^3/s. The value of the periodic average on HS "Raspotočje" is about 165 m^3/s, with an April average of 188 m^3/s. The second periodic maximum occurs during late autumn and early winter, i.e., in the period November–December. The periodic average on HS "Maglaj" is about 232.5 m^3/s with a maximum in December of 277 m^3/s. The value of the secondary periodic average on HS "Raspotočje" is about 111.5 m^3/s, with a December average maximum of about 188 m^3/s. The rest of the year is characterized by values below the annual average. The main minimum flow occurs during the second half of summer and early autumn, i.e., in the period July–September. The periodic mean for HS "Maglaj" is about 118.5 m^3/s with an average monthly minimum in August—99.5 m^3/s (Consortium for drafting the water management strategy of the Federation of BiH 2010). Similar qualitative indicators can be determined for the annual flow of flows on HS "Reljevo." Specifically, in the annual course of the flow, two characteristic periods can be distinguished. The period of elevated water levels is timed to December—May, during which the average is 37.5 m^3/s with a maximum in April of 51.3 m^3/s and in May of 39.3 m^3/s. The remaining (second) part of the year is characterized by values below the annual average. The periodic mean is 20.1 m^3/s, with minimal flows during August and September of 18 m^3/s.

6.2.5 Drina

The hydrographic backbone of the basin is Drina river. According to the data from the Regional Geography of Bosnia and Herzegovina I, "the Drina River is formed by the confluence of the Piva and Tara rivers in the Šćepan polje zone"). In the downstream flow segment, about 5 km from the confluences, the first significant left tributary flows into the Drina river – Sutjeska rijeka. The spring catchment area of Sutjeska Rijeka (Fig. 6.15) is formed in the mountainous areas of Zelengora, Volujak, and Lebršnik, within the hypsometric range from 1300 m to 1500 m above sea level" (Drešković and Mirić 2017).

The river valley has a predominant type of canyon valley, often with very steep slopes, relative heights of up to 500 m. At 11 km downstream from the mouth of the Sutjeska rijeka, the next significant left tributary flows into the Drina river — Velika Bjelava river (with Vrbnička rijeka). At the same stretch, but as a right tributary, Birotićki potok flows into the Drina. The next left rich by water tributary, Bistrica river, flows into the Drina river in 3.7 km from the previous confluence of Velika Bjelava river. Bistrica river has a significantly developed source head formed on the slopes of Treskavica mountain on one side and Jahorina mountain on

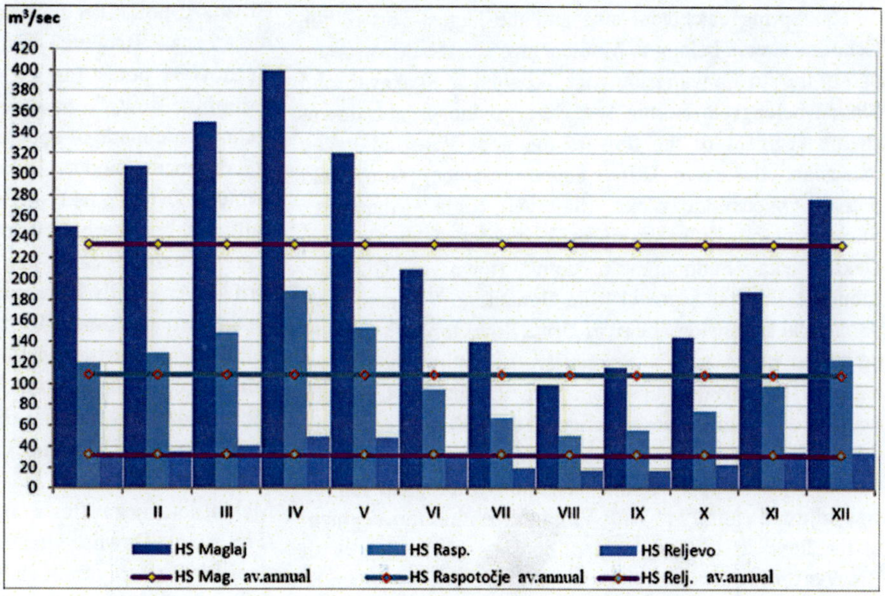

Fig. 6.14 Annual course of medium monthly flows on the river Bosna from 1980 to 2010. (HS—hydrographic station)

Fig. 6.15 Upper stream of Sutjeska Rijeka. (Photo by Ranko Mirić)

the other. It flows through a very deep rocky valley with very steep slopes from the source to the confluence. At about 4.2 km downstream, in those of the town of Foča, rich by water right tributary—the river Ćehotina—flows into the river Drina.

The Spring catchment area and the largest part of the Ćehotina stream belong to Montenegro; i.e., the downstream part of the stream with tributaries Skopotnica, Skakavac, and Slatina belongs to Bosnia and Herzegovina. In the downstream segment of the flow to the next Višegrad valley extension, the river Drina flows: Jošanica, Gunjevićki potok, Oglečevska rijeka, Sastavci (right tributaries), Susješni potok, Kolunska rijeka, Kosovska rijeka, Osanica, Ocka rijeka, Podhranjevski potok (town of Goražde) tributaries). In the segment from Goražde to Višegrad morphological expansion, the river Drina flows through the predominant gorge valley. Practically the entire part of the natural Drina riverbed upstream from Višegrad to Ustiprača was flooded with the construction of the Višegrad reservoir, due to which the flow of Drina river lost all elements of the natural water supply regime; i.e., the hydrological regime was adjusted entirely to HPP Višegrad. Within this segment of the flow, its tributary system consists of: Janjina (with Batovka), Lim, Kruševica, Uzavnički potok, Rzav (town of Višegrad) (as right tributaries), Sopotnička rijeka, Prača, Sušica, Lacanj, Skakavac, Jablanica (town of Višegrad) (as left tributaries). Among the most significant tributaries of the Drina in this segment are Lim and Prača. Lim is a watercourse that also belongs to Bosnia and Herzegovina only in the downstream part. More specifically, Lim springs from lake Plav (at about 900 m above sea level), from where it flows mainly through a composite valley in which valley extensions and gorges narrow alternately. At the town of Prijepolje, it flows from Montenegro to the territory of Serbia; that is, at the place of Uvac, it flows from Serbia to Bosnia and Herzegovina. In Brodar, it flows into the river Drina. Significant tributaries on the territory of Bosnia and Herzegovina are Radojna (with Sućeska) and the Sokolovačka rijeka. The spring catchment area of the Prača river originates in the foothills of Jahorina and Romanija mountains, within the hypsometric zone from 1300 m to 1500 m above sea level. From the village of Renovica, it flows through a distinct canyon valley with very steep sloping sides of relative height up to 500 m. In the zone of the village of Borač, it is joined by the Rakitnica river, where the town of Rogatica was formed. It flows into the river Drina in the zone of the village of Ustiprača. Downstream from the city of Višegrad, the hydrological regime of the Drina river is under another distinct deterministic influence—the operation of HPP "Perućac." More specifically, the deceleration elevation practically coincides with the zero elevation of HPP "Višegrad," which means that these two hydropower plants practically harmonize the regime of discharge and accumulation of water. For this reason, it is not possible to talk about the natural flow of the Drina river within the segment between the two mentioned hydropower plants. The tributary system of the Drina in this part consists of: Banja, Mušićka rijeka, Kuka potok, Klašnički potok, Kamenički potok, Brusnički potok (as right tributaries), Barimski potok, Suvi do, Žepa, Banjevi, Crni potok, Radoševićki potok, Beširevićki potok (as left tributaries). In the downstream part of the flow, Drina river is once again entirely under the technogenic mode of operation, since HPP "Divič" was built at a distance of about 110 km from Peruća, which formed an artificial Zvornik lake and whose deceleration level practically reaches the bottom of the Peruća reservoir. Accordingly, it can be stated that the hydrological regime of the Drina river is fully harmonized with the regime of operation of the two mentioned hydropower plants. In this part, the following waterstreams flow into Drina: Toplička rijeka, Baraković rijeka, Jagodnja, Mlečvanska rijeka, Grabovička rijeka, Loznička rijeka, Saska rijeka, Kriţevica, Glogovska rijeka (town of Bratunac), Slapašnička Rijeka, Polomski potok, Drinjača (with Jadar), Kamenica, Jošanička rijeka, Mekinjska rijeka (town of Zvornik) (as left tributaries) (Fig. 6.16). Downstream from the town of Zvornik, Drina river flows mainly through a gorge valley until it reaches the

Fig. 6.16 Drina River in Zvornik. (Photo by Ranko Mirić)

Fig. 6.17 River Drina in town of Goražde. (Photo by Aida Bidžan-Gekić)

Peripannonian area (in the zone of the village of Karake). Significant left tributaries in the segment of the flow to the confluence with the Sava river are: Hoča, Sapna, Kokića potok, Kozlučka rijeka, Šipulja, Jasenička rijeka, Vrela, Lokanjska rijeka, Pilićka rijeka, Tavna, Brezovica, and Janja river (Nezirović et al. 2015).

Following the stated facts, it can be said that the Drina river, only in the upstream segment of the flow (up to Ustiprača), has a natural hydrological regime of water supply. For this reason, in the analysis of the hydrological regime, only HS "Goražde" was considered, which also has a representative multi-year series of measurements (Figs. 6.17 and 6.18). The hydrological flow regime is dominated by the nival-pluvial regime, with the main hydrological maximum in the second half of spring and early summer. The period average for these 3 months is about 575 m^3/s, with a maximum in May—620 m^3/s. The second, the secondary hydrological maximum, occurs during late autumn and early

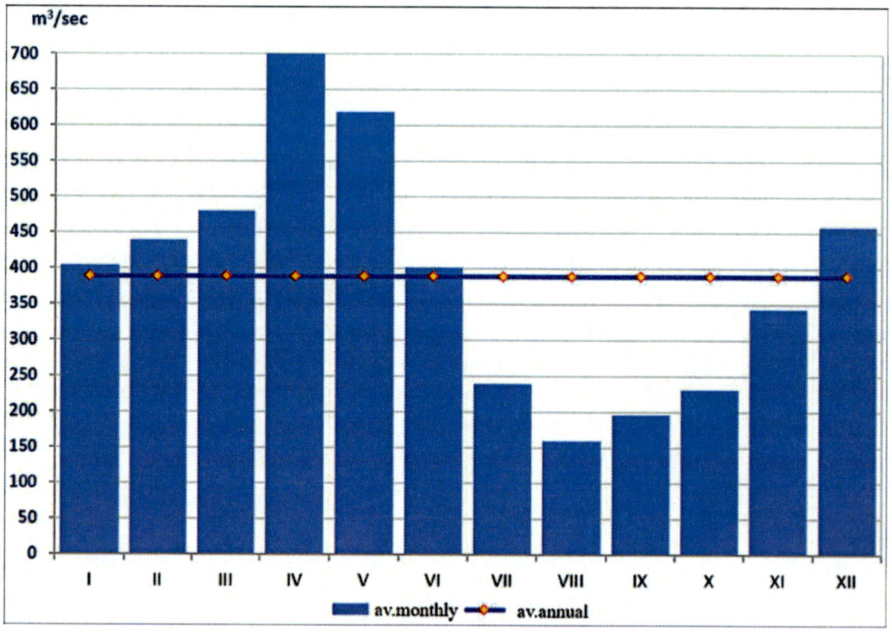

Fig. 6.18 Annual flow on the river Drina (Hydrographic station "Goražde") from 1980 to 2010

winter, i.e., in the period November–December. The value of the periodic average is 405 m^3/s, with a maximum in November—about 460 m^3/s (Consortium for drafting the water management strategy of the Federation of BiH 2010).

Precipitation minimum occurs in the period August–September, with a periodic average value of 180 m^3/s. The minimum flow occurs in August—160 m^3/s—indicating more pronounced continental characteristics of the entire Drina river basin. Changes in the annual flow at the level of distinct segments of the longitudinal profile of the Drina river and its main tributaries can be seen from the following data: Bastasi—164 m^3/s, Foča—212 m^3/s, Višegrad—341 m^3/s, Zvornik—387 m^3/s, a confluence (into the Sava river)—401 m^3/s, Sutjeska (Igoče)—14 m^3/s, Ćehotina (Foča)—16 m^3/s, Prača (Rakitnica)—2.4 m^3/s, Lim (Rudo)—115 m^3/s, Drinjača—6.5 m^3/s.

6.2.6 Immediate Basin of the Sava River

General hydrographic characteristics of the immediate Sava river basin are defined by geological-geotectonic, morphological, climatic, and vegetation specifics. Accordingly, as well as the fact that watercourses directly gravitate to the Sava river, it can be said that the hydrographic systems of these sub-basin areas are primarily undeveloped. Indeed, the backbone of the hydrographic network of the immediate basin and the entire Black Sea basin is Sava river. It is formed in the Republic of Slovenia by two equally developed streams: Sava Bohinjka and Sava Dolinka in the foothills of Triglav mountain. Apart from Slovenia, the Sava river flows through the Republic of Croatia, Bosnia and Herzegovina (as a border river), and the Republic of Serbia. As the border river, Sava in Bosnia and Herzegovina has variable hydrological characteristics that directly depend on the climate season. Summer flow values are significantly lower than in winter when precipitation recharge is maximum.

The mean annual value of mean flows is about 1260.5 m^3/s. In the annual course of medium water levels, time intervals with increased and decreased values can also be distinguished. Both water flow maxima are clearly expressed and with very close values. The first maximum falls on the period April–May with an average of 1614.6 m^3/s, while the second maximum is positioned in November–December with a periodic average of about 1632.9 m^3/s. The main minimum water levels also occur during the second half of summer and early autumn, i.e., August–September. The periodic average of median water level is about 826.4 m^3/s, which is about 25% below the annual value. The second, less pronounced minimum water level was time-positioned in January–February with a periodic average of about 1180.7 m^3/s.

Certainly, in relation to the summary indicators for the Sava river, it is necessary to present the basic hydrological parameters at the level of the sub-basin areas of the immediate Sava river basin. The backbone of the river network of the sub-basin of the river Jablanica is the river Jablanica with its tributary system, which consists of the Vrbaška river (with Golubača and Crna rijeka) and Lubina. In addition to it, other tributary systems of rivers have been formed in the sub-basin area: Virovska, Ribarice (with Pejinovac), Podgračka, Gašnica, Bistrica, Jurkovica (with Vrbaska) (town of Gradiška), Gostinja, Matura, and other smaller rivers.

The backbone of the river network of sub-basin of the river Ukrina is the river Ukrina with its tributary system. The river Ukrina is formed by the confluence of the Velika and Mala Ukrina in the wider zone of the village of Kulaši. In the hydrological sense, the course of Ukrina originates from Velika Ukrina, whose source head is formed in the northeastern foothills of the mountain Uzlomac. The tributary system of this flow consists of Davidovića rijeka and Bistrica (with Lukavac) (in the area of the spring catchment area), Manastirica, Šibovača, Borovica, Vrletnica, Kamenica, Velika Borovica (right tributaries), Čađavica, Jelovac, Kamenica, Sobanjska rijeka, Jelova (left tributaries). The tributary system of Mala Ukrina consists of Bijela rijeka and Planski potok (the spring catchment area), Velika Javorova, Osivica, Radnja (with Vranjak, Rastužica, Ostružnja), Lepenica (right tributaries), Kruševica, and Veliki potok (left tributaries). From the confluence from Velika Ukrina with Mala Ukrina, before the confluence to Sava river, the following rivers flow into Ukrina: Crna rijeka, Lađašnica, Ilova (with Dražana), Poježnica, Malica, Lupljanica, Zelenska rijeka, Lješnica, Urinac (right tributaries), Kremna (with Rakovac), Kremnica, Vijak, Jadova (with Kunova, Živanički potok and Putnjača), Orljak, Agićka rijeka (with Ukovija), Beglučki potok, Crna rijeka (left tributaries). According to the results of instrumental monitoring, the average annual flow of Ukrina (HS "Derventa") is about 17.5 m^3/s.

In the area of the rivers Tolisa, Tinja, Brka, Lukavac, and Dašnica, more rich watercourses define this sub-basin area with their tributary systems that flow directly into the Sava river. Accordingly, in the area of the sub-basin, the sub-basins of the following watercourses have been singled out: Tolisa, Tinja, Brka, Lukavac, and Dašnica. Tolisa river has a developed spring catchment area formed along the northern slopes of the mountain Trebavac, in the wider zone of Zelinjsko brdo (Jezero—499 m). The left part of the tributary system of Tolisa is quite poorly developed (Desna and Lijeskova rivers) in relation to the right, which consists of: Srnava, Korničica, Gradašnica (town of Gradačac), Kruševica, and Briježnica (with Prvakovac).

Tinja river includes a hydrographic system that was formed in the eastern part of this sub-basin. The spring catchment area of the Tinja river was developed and formed in the southern foothills of the Majevica mountain. Of the larger streams in the spring catchment area, the following streams can be singled out: Dragunjska rijeka and Jasenički potok. In the downstream part of the flow, the tributary system consists of: Kugićka rijeka, Urvenica, Dubokovac, Bistrica, Bijeli potok, Klinar, Podarevac, Tinjica, Bukovac, Lukavac (right tributaries), Caška rijeka, Slanska rijeka, Lužnički potok, Moštanica, Mala Tinja (with Lukavac, Rajska rijeka, Zelinjska rijeka, Međička rijeka and Vućkovac), Lomnica, and Smrdulja (left tributaries). Given the fact that the hydrological regime of all watercourses in the eastern part of the sub-basin is very similar, the analysis of flow indicators for the river Tinja was performed. The average annual flow determined on HS "Srebrenik" is about 2.3 m^3/s. There are periodic maximums and minimums in the annual course of the flow. The main periodic maximum occurs in the period of the second half of winter and the first half of spring (February–March), with a periodic average of 3.73 m^3/s. The periodic minimum occurs in the second half of summer and the first half of autumn (August–September), with an average of only 0.95 m^3/s.

Brka river is more abundant with water in the area of this sub-basin. The source headland is formed in the northern and northeastern slopes of the Majevica mountain, within the hypsometric zone from 500 m to 600 m above sea level. The tributary system in the zone of the spring catchment area is formed by the Rašljanska, Rahićka, and Štrepačka rijeka. Brka originates from the confluence of the rivers of Rašljanska rijeka and Rahićka rijeka, in the zone of the village of Maoča. The downstream tributary system of the river Brka is mainly connected with the river Zovičica (which is formed by the confluence of the Štrepačka rijeka and Lužnica), which flows into Brka in the wider zone of the town of Brčko.

Lukavac river is formed by the confluence of the rivers Lukavac and Lukavac (Gnjica) in the zone of village of Glavičorak. The spring catchment area of the watercourse was formed in the area of the northern and northeastern slopes of the eastern part of the Majevica mountain, at about 600 m above sea level. The source headland consists of: Drijenovačka rijeka, Erićka rijeka, Šibošnica, Jablanička rijeka, and Gnjica. The most important tributary, the river Gnjica (with Bijela rijeka and Bukovica), flows into the river Lukavac in the zone of the village of Vršani (Drešković and Mirić 2017). In addition to these watercourses in this sub-basin area of the Sava River, several other rich by water watercourses have been formed, such as Dašnica, Bistrik, Široki Resnik, and Prugnjača (in the extreme western part of the basin).

6.2.7 Neretva River

Hydrographic specifics in the Neretva basin are a direct consequence of the geological structure and relief of the terrain as well as climatic specifics. The total length of the permanent surface river network in the Neretva basin, with flows that have a permanent hydrological function, is about 1948 km. The average density of the surface river network is 0.27 km/km^2, which is below the average for the continental part of Bosnia and Herzegovina. However, from the aspect of the amount of water obtained at the outlet profiles of watercourses, it can be said that the amount of water is above the value of most other catchment areas of Bosnia and Herzegovina. Neretva river basin is characterized by numerous mountain lakes, almost a regular hydrological phenomenon on all major mountains. Among the most famous are: lakes on Zelengora mountain (Kotlaničko, Orlovačko, Crno, Bijelo, Donje bare, Gornje bare, Štirinsko and Kladopoljsko), lakes on Treskavica mountain (Veliko, Platno, Crno and Bijelo), lake on Crvanj mountain (Uloško), lake on Prenj mountain (Boračko), lake on Čvrsnica and Vran Mountains (Blidinje). Especially important from the aspect of protected nature is the area of Deransko lake (blato), which in 2001 was declared a Ramsar site. In the area of the Neretva basin, a large number of springs have been formed with a very large water yield, which is sufficient to form real rivers directly. The Buna river spring, Krupačko vrelo, Bregava river spring, and the like are among the most famous (Drešković and Mirić 2016).

The backbone of the river network of the entire basin is the river Neretva (Narona—Roman name for the Neretva). The Neretva river is the longest and most abundant with water in the Adriatic Sea basin on the Balkan Peninsula. The total length of the Neretva river is about 225 km, with the largest part of the Neretva river belonging to Bosnia and Herzegovina—about 204 km. The remaining part, downstream from Metković, belongs to the Republic of Croatia. The spring catchment area occupies the area of the highest Dinaric mountains of the southeastern highlands of BiH, i.e., the mountains of Zelengora, Lelija, and Treskavica. The main part of the source of the Neretva river is in the area of the southern and southwestern slopes of Zelengora (Ravno brdo) (1299 m) and Grdelj (1227 m) mountains, or in the area of Borača, at about 1095 m above sea level, while in the southeast it drains water from the northern and northeastern slopes of Živanj (1696 m), Kuk (1584 m), and Mjedena glava (1602 m). To the village of Ulog, several shorter but rich by water streams, on both the left and right valley sides, flow into Neretva. The characteristics of the river valley in this flow segment are steep sides which give the valley the characteristics of a gorge with an average height of 600 m to 700 m and southeast-northwest direction. From the village of Ulog, Neretva changes the direction of flow to the north, and from here, it has mostly developed only the right tributary system formed on the eastern foothills of the mountain Crvanj. The most important tributary in this part is the Krupački potok, which flows downstream from the Ulog extension, and which is sufficiently abundant with water for the Neretva to take on the characteristics of a real river from the place of its inflow (Fig. 6.19).

Downstream from the village of Obalj, in the zone of lake expansion, the most important tributary in this part of the stream flows into the Neretva river–Jezernica river, which is considered an integral part of the Neretva spring catchment area. Jezernica mostly drains water from the southern and southwestern foothills of Treskavica mountain and, to a

Fig. 6.19 The canyon of the river Neretva upstream from Ulog. (Photo by Drešković Nusret)

lesser extent, from the northern and northwestern slopes of Lelija mountain. From the confluence of Jezernica, Neretva changes its direction of flow by about 90° and passes into the predominantly western direction, which it retains all the way to the village of Glavatičevo. The river valley in this stream segment takes on all the characteristics of a gorge with very steep valley sides, whose relative height exceeds 700 m in some places.

The most important right tributaries are Ljuta and Rijeka on the right and Živanjski potok and Krupac on the left valley side. Ljuta river is primarily known for its hydromorphological specifics, which formed its practically canyon valley between the mountain morphostructures of Treskavica and Visočica. From Glavatičevo, Neretva river bends for about 45° to the northwest and keeps such a flow direction all the way downstream from the town of Konjic. The basic morphological features are the canyon type of the valley with predominant steep valley sides, relative height up to about 500 m. The tributary system in this part has similar hydromorphological features as in the previous segment; i.e., the main tributaries are abundant with water, short in length, and flow through gorge-canyon valleys. The most important right tributaries are Razićki potok, Rakitnica, (Konjic's), Ljuta and Trešanica, or Šištica and Bijela on the left side of the valley.

This watercourse segment is known for its rafting water trail, about 23 km long, characterized by several waterfalls and rapids and a total drop of about 86 m. and over 1000 m. Since, due to the slower vertical fluvial incision, it lags behind Neretva, Rakitnica flows into the Neretva in the hydroform of a waterfall. It is similar to the island of Boračko jezero, the river Šištica, which has the so-called hanging valley so that it flows into the Neretva in the form of a waterfall of the relative height of about 30 m. This part of Neretva until the town of Konjic is hydromorphologically very often referred to as the upper or upstream part. Downstream from Konjic, the first hydro-accumulation was formed—"Jablanica"—by blocking the flow of the Neretva upstream from Jablanica (village of Slatina). With the construction of a 75 m high dam, the Neretva riverbed was submerged from downstream from Konjic as well as the Rama riverbed to the Gračac village, and the Jablanica hydro accumulation was formed, with an area of about 14.5 km^2 and a volume of about 320 million m^3 (average length about 25 km, the average width is approximately 0.6 km; that is, the average depth is about 22 m) (Drešković and Đug 2010).

The water from the hydro-accumulation is directly transferred through a tunnel about 3 km long to the turbine plants, so that the old riverbed, from the dam to the village of Jablanica (between the Tovarnica massif (823 m) and Ćeharski vrh (827 m), has almost lost its hydrological function). In the zone of the Jablanica hydro-accumulation (Fig. 6.20), the right tributary system forms several smaller

Fig. 6.20 Jablanica hydro-accumulation near Čelebići. (Photo by Drešković Nusret)

rivers, among which the largest are Kraljuščica, Neretvica with the stream Blučica, and Rama with the stream Volujak. The left tributary system consists of Baščica, Javorski potok, and several other streams of lower water yield. A water reservoir was also built on the Rama tributary in 1986. The length of this water reservoir is about 9 km, the average width is about 2 km, and the area is 1.8 km^2, with a maximum depth of 95 m. At a medium water level, about 450 million m^3 of water is stored in this reservoir. Immediately downstream from the dam, the flow of the Neretva turns retrograde for about 90° in the south direction, and Neretva keeps that direction, with minor angular deviations, all the way to the mouth of the river Buna (Drešković and Mirić 2017).

The basic hydromorphological features of the Neretva flow downstream from the dam, all the way to Bijelo polje (northern part of the Mostar valley), are gorge features of the valley, steep valley sides (with relative heights of over 1000 m), two reservoirs, and a very abundant with water tributary system. The most important right tributaries are Doljanka, Grabovica, and Drežnica, while from the left valley side, the Jablanica stream, Ravančica with tributaries, and Bijela flow into the Neretva. In this part of the course, Neretva is deeply cut into the limestone rock substrate, so that it built an imposing canyon valley, which is wedged between the mountain morphostructures of Prenj on the left and Čvrsnica on the right, stretching from Donja Jablanica to Bijelo polje. In the area of the village of Grabovica in 1981, the Grabovica dam was built, which flooded the Neretva riverbed and stream all the way to Donja Jablanica, as well as part of the Grabovica river (Fig. 6.21). The total area of the Grabovica hydro-accumulation is about 0.4 km^2, and about 198 million m^3 of water is accumulated in it. In the same period, the hydro-accumulation Salakovac was formed, with an area of about 2.55 km^2 and with a volume of about 68 million m^3. In this way, the river Neretva has entirely changed the natural hydrological specifics; Neretva has completely lost natural flow characteristics in the flow segment.

The first sizeable composite extension in the Neretva valley is the Mostar valley, which in morphological terms can be divided into three smaller morphological units: Bijelo polje (field) (in the north), Mostar valley (in a narrower sense), and Mostarsko or Bišće polje, which are interconnected by shorter morphological narrowings of the sutjeski type. In Bijelo polje, immediately upstream from Mostar, the fourth reservoir on the Neretva was formed—the Mostar hydro accumulation, with an area of about 1.54 km^2—and was built for the water supply needs of the Mostar industry.

The surface tributary system in this morphological segment is very poorly developed due to the predominant limestone structure of the terrain and high average annual temperatures. The exception is the river Buna with Bunica, which originate from very water-producing karst springs and

Fig. 6.21 Grabovica hydro-accumulation. (Photo by Drešković Nusret)

are very short-flowing, about 8 km, which with more widespread alluvial (watertight) sediments is enough to preserve water supply throughout the year.

Significant, but underground inflow is also achieved by the waters of the river Lištica which sink in Mostarski blato and near the village of Bačevići, partly over the surface course of the river Jasenica and partly underground infiltrate into the watercourse of the Neretva river. In relation to the mentioned morpho-hydrological differentiation of the flow and the Neretva valley, the central part ends up to Mostar, i.e., the third downstream segment starts from the Mostarsko polje to the confluence into the Adriatic Sea. Downstream from the confluence of the Buna river, Neretva cuts into the karst-corrosion Brotnjansko-dubravska plateau, so that in this segment of the flow, to the Čapljina extension, the Neretva has a gorge-type valley with relative heights of valley sides of about 250 to 350 m. The flow is also a change of direction from the meridional to the predominant southwest direction. There are no surface tributaries in this morphological segment of Neretva. Thanks to the development of rock masses with the properties of hydrological insulators (Pliocene-Eocene flysch and Quaternary alluvium) in the Čapljina-Hutovo extension-delta, a bilateral surface tributary system was formed. The most important left tributaries are Bregava and Krupa rivers on the left and the river Trebižat on the right side of the valley. These are watercourses with a primarily stable hydrological regime, except for the river Bregava, which, during the period July–September, sinks into the karst underground downstream from the town of Stolac, although most of these waters drain the Neretva from the underground. The width of the Neretva river increases in this part and reaches 100 m, in some places up to 150 m, while the depth is on average about 1.5 m, and in parts of the vortex, it is up to 20 m.

The basic parameters that define the water yield of the Neretva and its river regime, during the year or in certain climatic seasons, are determined by geological-geotectonic terrain, spatial, and temporal dynamics of the principal climatic elements, but also anthropogenic pressures, especially hydrotechnical interventions both in the basin and in the flow itself. In connection with the above, it can be said that the monitoring and analysis of the basic indicators of the natural river regime of the Neretva are reasonable and only possible in the segments of the flow to Konjic and downstream from Mostar. For this reason, hydrological stations in Glavatičevo, Konjic, and Žitomislići were taken as the basis for presenting the hydrological features of the Neretva river.

For annual and monthly water flows, based on the appearance of their graphs, it can be concluded that they largely correlate with water level equivalents (Fig. 6.22). The mean annual flow over a multi-year period at HS. Glavatičevo is about 39.5 m^3/s, HS. Konjic—49.2 m^3/s, while on HS Žitomislići flow rate is 293.4 m^3/s. The difference in the amount of water flowing through the Neretva riverbed in the upper and lower part of the stream is very noticeable; i.e., in the delta zone, the amount of water is about nine times higher than in the Glavatičevo zone (Consortium for drafting the water management strategy of the Federation of BiH 2010). This difference can be explained by the fact that the profile of Žitomislići shows water content over 3/4 of the

Fig. 6.22 Annual flow on the Neretva River (hydrographic stations: Glavatičevo, Konjic and Žitomislići) from 1980 to 2010

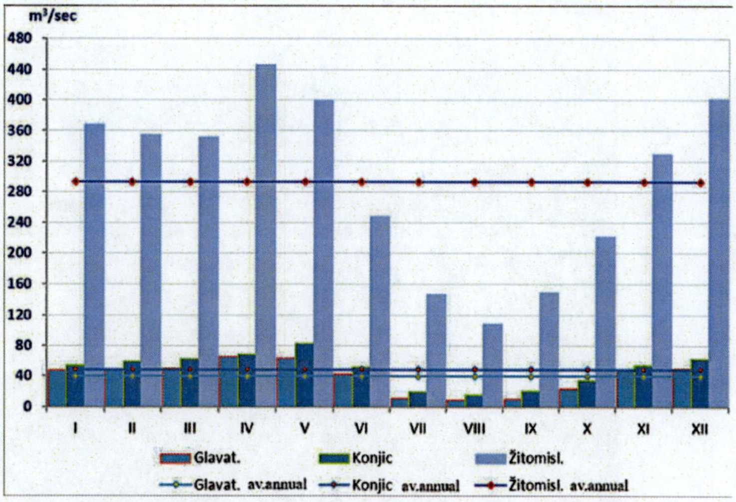

entire basin, while the profile at Glavatičevo water yield is quite low due to the relatively small number of tributaries. Similar to water levels, there are marked irregularities in the annual course of flows, so distinct extreme periods with increased and decreased flows can also be distinguished. The flow's main maximum is April–May with the following values: HS Glavatičevo—64.5 m^3/s, HS Konjic—76.2 m^3/s, and HS Žitomislići—422.4 m^3/s.

The flow's main minimum occurs during the period July–September, when the flow values average about 25% of the maximum or, more specifically, HS Glavatičevo–10.5 m^3/s, HS Konjic—17.5 m^3/s, and HS Žitomislići—127.5 m^3/s. Secondary extreme periods fall at the same time intervals and with aliquot quantitative relationships with water level equivalents. The secondary maximum of the flown is December–January, with values: HS Glavatičevo—49.5 m^3/s, HS Konjic—58.2 m^3/s, and HS Žitomislići—384.4 m^3/s. The secondary minimum flow was positioned in February–March with the following average values: HS Glavatičevo—49 m^3/s, HS Konjic—60.5 m^3/s, and HS Žitomislići—352 m^3/s (Fig. 6.23).

The Neretva Valley is dominated by typical Mediterranean influences. The water yield and characteristics of the river regime of other watercourses in the Neretva basin also correlate well with climatic characteristics and geological structure. The most important indicators for the water yield of larger flows in the basin are given in the following table. The water yield of watercourses in any return period of maximum flows directly correlates with the rank of watercourses; i.e., the highest water yields have the highest flows and vice versa.

6.2.8 Trebišnjica River

Before completing the first phase of hydromelioration interventions, the Trebišnjica river basin area was known for the longest karst underground stream in the world—the Trebišnjica river, with a length of about 90 km. After completing the mentioned works, the course of Trebišnjica was redirected to the surface; that is, an artificial accumulation was built in its upstream part—Bilećko jezero, which holds about 1.28 billion m^3 of water. Before the construction of the hydro system, Trebišnjica flowed from the spring near Bileća (330 m above sea level) along the natural riverbed to the south, all the way to the village of Lastva, from where it flowed to the west, i.e., to Trebinje (at 272 m above sea level) and further to Popovo polje, where it plunged into the karst underground through numerous cracks at the bottom of the depression. Measurements also proved that Trebišnjica had direct underground hydrological connections with Hutovo blato, Neretva, and the Adriatic Sea (Drešković and Mirić 2017).

Numerous researches conducted to use the hydro potential during the 70s and 80s of the last century have established that the waters of the Trebišnjica catchment area drain underground watercourses and the Neretva basin. More specifically, it was determined that the surface waters of the river Zalomka, after sinking, flow into the Buna, Bunica, and Neretva, respectively, from the area of Nevesinjsko polje. Groundwater flowing down the Fatnica river from the surface of the Fatnica polje partly flows into Bregava and the other into Trebišnjica. The surface waters of Gatačko polje (Opačka rijeka) and Cerničko polje flow underground to

Fig. 6.23 Neretva river in the zone of the Počitelj fortress. (Photo by Nusret Drešković)

Trebišnjica, with a smaller part flowing into Fatničko plje. The surface waters of the Ljubinjsko polje and Ljubomirsko polje flow into the course of Trebišnjica by underground runoff. Instrumental monitoring determined that the average annual flow of the river Trebišnjica near Trebinje is 96.1 m^3/s and 81.9 m^3/s near Grnčarevo. The variation of flowing waters largely depends on the climatological seasons, so that the share of winter flow was over ¾, and the share of summer flow was below ¼. More specifically, during the period of hydrological minimum (August), the surface flow drops to only 3–5 m^3/s, while the amount of water during the late autumn maximum can reach values up to 900 m^3/s.

Energy utilization of the waters of the Trebišnjica basin is defined by the water management plan from 1967, and the Base Project from 1976, which planned to build a total of seven hydropower plants and five reservoirs. The realization of works defined by the project of the first water management master plan from 1958 started with the construction of HPP "Dubrovnik" (126 MW) in 1965, HPP "Trebinje I" (168 MW)in 1968, PHE "Čapljina" (420 MW) in 1979, and HPP "Trebinje II" (8 MW) in 1988. During 1979 and 1980, the Upper Horizons project planned three remaining hydropower plants: HPP Nevesinje (61 MW), HPP Dabar (160 MW), and HPP Bileća (160 MW). The total installed capacity of all planned hydropower plants in the Trebišnjica system would be 1075 MW, with a total annual production of 3900 GWh. The planned water use system in the Trebišnjica basin and the entire eastern Herzegovina has become questionable primarily due to changed planning conditions and extremely negative impacts that the planned water intakes have on the downstream part of the Neretva River and delta, especially Hutovo Blato (Drešković and Đug 2010).

6.2.9 Cetina and Krka

The most important watercourses in Cetina and Krka basin are Kriva jaruga and Ribnik in Glamočko polje, the river Milač in Kupreško polje, the river Šuica in Duvanjsko polje and the rivers Bistrica, Sturba, Žabljak and Ričina in Livanjsko polje. The waters of the natural accumulation formed in Glamočko polje (Pučina retention) flow into Livanjsko polje through the Dragnić abyss. Accumulated waters from Duvanjsko polje through the Kovači abyss flow to the spring of Ričina, while in Livanjsko polje, the waters gravitate to Buško jezero. Natural retentions in the northwestern part of Livanjsko polje, through the Kazanci and Čaprazlije abysses, flow towards Cetina. In Livanjsko polje, there is a strong influence from the complex hydro-technogenic system that collects and drains water to HPP "Orlovac" in the Republic of Croatia. Part of that system is Buško jezero, Mandak Reservoir, Drinovački Kanal (which collects the waters of Bistrica, Sturba, and Žabljak), and finally, lake Lipa, where all these waters are gathered and drained through the tunnel to HPP "Orlovac."

In accordance with the above, it can be said that Buško jezero is the main accumulation basin in which water is transferred from the upstream segments of the basin. Buško blato with Bijelo polje is a separate spatial unit of Livanjsko polje. The bottom of the field and the peripheral area of the lake are covered with a series of marly-clayey-sandy, in some places gravelly Neogene fractions, whose thickness is not unique and varies from a few meters in the peripheral segments of the field to ten or more meters in zones of widespread alluvial plateaus in the central part.

Buško blato is formed at altitudes from 700 m to 710 m. It is supplied with water from several Voklija springs, which dry up in summer. The most important tributary is Ričina, which flows from the eastern side of the lake for about 5 km. The most significant number of abysses is located on the western side of the lake valley. In the 70s of the last century, Buško blato was transferred to a techno-aquatic complex by hydro-technical interventions for electricity production and intensification of agricultural production. The accumulation of water was carried out in the area of Buško blato, which until the formation of the artificial hydro-accumulation represented the wetland-marsh system—Buško blato. The maximum amount of water at the highest water level is 800 million m^3 of water, which makes Buško jezero one of the largest freshwater reservoirs in Europe. The surface streams of Ričina, Sturba, Žabljak, Bistrica, Jaruga, and Plovuća are of the most significant importance for the water supply of Buško jezero. In addition to these permanent watercourses, there are several smaller streams, especially in the foothills of Tušnica, which deliver additional larger amounts of water during the precipitation season and the snowmelt period. With the construction of Buško jezero, the bed of the river Ričina was flooded; i.e., there is only a small part which is practically a continuation of the river Šujica from Duvanjsko polje. Lake water levels vary significantly, which is directly dependent on the pluviometric regime. The height of the oscillations is affected considerably by the reversible channel that connects the Lipa compensation basin with Buško jezero. Maximum water levels appear during early and late spring. In addition to water consumption for energy purposes, the negative components of the hydrological balance include the infiltration of about 5 m^3 of water/day into the rock underground (Consortium for drafting the water management strategy of the Federation of BiH 2010).

In order to define the basic features of the hydrological regime in the basin, an analysis of the annual flow on the river Cetina (HS "Sturba") was performed. The average annual flow is 4.5 m^3/s, with significant variations compared to climatological seasons. More specifically, periods of increased and decreased flows can be distinguished in the annual course of flows.

The flow's main maximum is time-positioned at the end of winter and the first half of spring, i.e., in the period March–April. The periodic average is about 6.3 m^3/s, with a maximum of 6.6 m^3/s in April. It is important to emphasize that this period of high flows lasts up to 5 months, which is certainly of primary importance for general hydrological relations in the basin. The secondary maximum of the flow is time-positioned in the period of late autumn–early winter, i.e., November–December. The value of the periodic average of the flow's secondary maximum is identical to the previous one—6.3 m^3/s. This indicates the fact that the hydrological maxima at the level of the basic periods are quite uniform. Flow's periodic minimum occurs in the second half of summer and the first half of autumn (August—September), with a periodic average of 1.9 m^3/s.

References

Consortium for drafting the water management strategy of the Federation of BiH (2010) Strategija upravljanja vodama Federacije Bosne i Hercegovine 2010–2022 [Water Management Strategy of the Federation of Bosnia and Herzegovina 2010–2022]. Agencija za vodno područje rijeke Save, Agencija za vodno područje Jadranskog mora, Sarajevo, Mostar

Drešković N, Đug S (2010) Rijeka Neretva – potamološke karakteristike [Neretva River – potamological characteristics]. Zbornik radova I međunarodnog simpozijuma ribarstva i akvakulture, Sarajevo

Drešković N, Mirić R (2016) Bathymetric chart of Lake Derane as the basis for defining the hydro-ecological optimum of the Deransko wetlands. Annales Ser Hist Nat 26(2):235–244

Drešković N, Mirić R (2017) Regionalna geografija Bosne i Hercegovine I [Regional Geography of Bosni and Hercegovina I]. Faculty of Science, University of Sarajevo, Sarajevo

Đug S, Drešković N (2008) Prirodno naslijeđe Kantona Sarajevo [Natural heritage of Canton of Sarajevo]. Kantonalni zavod za zaštitu kulturno-historijskog i prirodnog naslijeđa. Arch Design, Sarajevo

Đug S, Drešković N, Jovanović Ž, Jauz R, Mujkić H et al (2008) Prostorni plan područja posebnih obilježja od značaja za Federaciju BiH, Sliv rijeke Une 2008–2028 [Spatial plan of the area of special features of importance for the Federation of BiH, Una River Basin 2008–2028]. Urbanistički zavod Bosne i Hercegovine, Sarajevo

Đug S, Drešković N, Okerić Š (2017) Sedronosne naslage rijeke Une – postojeće stanje i katastar sedronosnih naslaga [Tufa deposits of the Una River – current status and cadastre of tufa deposits]. In: Drešković N (ed) Zbornik radova četvrtog kongresa geografa Bosne i Hercegovine, Geografsko društvo u Federaciji Bosne i Hercegovine. Prirodno-matematički fakultet Univerziteta u Sarajevu, Sarajevo

Encyclopedia of Yugoslavia (1983) Socijalistička republika Bosna i Hercegovina. Jugoslovenski leksikografski zavod, Zagreb

Hadžić E, Drešković N (2014) Analiza uticaja temperaturnih i padavinskih oscilacija na riječne protoke u Sarajevskoj kotlini [Analysis of the influence of temperature and precipitation oscillations on river flows in the Sarajevo valley]. Vodoprivreda 46:65–75

Hydrological data (1980–2010) Hydrological yearbooks of Bosnia and Herzegovina from 1980 to 2010. SHMZ Beograd, FHMZ, Sarajevo

Nezirović S, Drešković N, Pobrić A, Mirić R (2015) Geographical aspects of river flows in the region of northeast Bosnia. In: Serban G (ed) Air and water components of the enviroment conference proceedings, pp 80–88

7 Pedogeography of Bosnia and Herzegovina

Abstract

Following the classification system of soils in Bosnia and Herzegovina, the genetic-evolutionary principle of classification of soil types is applied, mainly based on pedogenetic processes and soil properties they cause, considering soil development by developmental stages. The differentiation of soil types has a hierarchical character with a starting point in the basic classification unit—soil type, which is further ranked in higher or lower classification units according to the following classification categories: division (department), class, type, subtype, variety, and form. Based on the results of previous scientific research in Bosnia and Herzegovina, a total of three soil divisions have been identified and differentiated. These are terrestrial (or automorphic) soils, hydromorphic soils, and subhydric soils. Terrestrial soils of Bosnia and Herzegovina are presented with a total of seven classes and 34 types from this division. Hydromorphic soils consist of 5 classes (9 types) and subhydric of 2 classes with 4 soil types. Within these divisions and related classes, many subtypes stand out, which show a very complex pedological structure of Bosnia and Herzegovina. The chapter presents the main pedological types within the associated classes, with basic chemical and physical characteristics and distribution in Bosnia and Herzegovina. The most common soil types are Dystrict cambisol on acidsilicate rocks (27.95%) and Calcocambisol and mosaic calcomelanosol—calcocambisol on limestone (20.68%).

Keywords

Soil types · Calcocambisol · Dystrict Cambisol · Mosaic · Physical Geography

7.1 National Classification and Soil Types

The starting point in the national classification system is represented by the type of soil, which consists of a hexagonal three-dimensional land area (which is arranged in the form of honeycombs) whose area can be from 1 m^2 to 10 m^2. The mentioned basic pedological form is called pedon in the classification pedosystematics, which is characterized by: predominant one-type structure of profiles, characteristic order of horizons and their one-type physical-chemical composition, one-type processes of movement and transformation of organic and mineral matter, etc.

National classification systems are, in principle, adaptable to the WRB International Soil Classification System (World Reference Base for Soil Resources), which has the function of supporting a better understanding of basic pedosystemic and their subsystem categories at the level of different countries. The WRB reference system was established in 1992 to make the UNESCO/FAO revised soil map the basis for establishing a future international reference base for soil classification.

In accordance with the mentioned Western European classification concept, the genetic-evolutionary principle of classification of soil types is applied in Bosnia and Herzegovina, mainly based on pedogenetic processes and soil properties they cause, taking into account soil development by developmental stages. The differentiation of soil types has a hierarchical character with a starting point in the basic classification unit—soil type, which is further ranked in higher or lower classification units according to the following classification categories: division (unit), class, type, subtype, variety, and form. Divisions (or departments) are classification units of the highest rank that unite soil groups according to the related character of wetting and the chemical characteristics of the

H. Gekić et al., *The Geography of Bosnia and Herzegovina*, World Regional Geography Book Series,
https://doi.org/10.1007/978-3-030-98523-3_7

water used for wetting. Classes in the classification system represent divisions that include all soil types with the same order of characteristic horizons. Soil type is a basic pedosystematic unit with the same order of horizons and similar processes of transformation and migration of pedosubstrate. A subtype is a lower subcategory unit whose differentiation within soil types is based on the specifics of pedogenetic processes, the similarity of the parent substrate, soil depth, etc. Variety is defined based on the type of parent substrate within a particular subtype. The shape is distinguished based on variations in the textural composition of the variety (Resulović et al. 2008).

In Bosnia and Herzegovina, a total of three soil divisions have been identified and differentiated: terrestrial (or automorphic) soils, hydromorphic soils, and subhydric soils.

7.1.1 Terrestrial (or Automorphic) Soils of Bosnia and Herzegovina

The essential feature of the distribution of terrestrial (or automorphic) soils is the regular hydrological regime of wetting through rainwater that is not retained in the soil, where the percolation of infiltrated water is free. In the area of Bosnia and Herzegovina, a total of seven soil classes from this division have been singled out. As a consequence of the absence of stagnant water and continuous wetting in this division of the soil, the processes of reduction or gleying are entirely excluded.

7.1.1.1 Underdeveloped Soils and Terrestrial Regosoles

Soil types from this class have a humus horizon (A horizon) and a mineral complex in the initial phase of formation, which is why the organo-mineral complex has not developed either. The process of formation of the mineral complex is in the initial phase of very weak chemical decomposition of the parent substrate with very low production of organic matter. For this reason, the initial horizon mainly develops on a loose substrate in which the root system of higher plants has just begun to develop. On a compact parent rock substrate, the decomposition process is started by species from lichen and moss communities, which is why the soil in such cases is very shallow and undeveloped (Resulović & Ćustović 2002).

Within the class of undeveloped and underdeveloped soils in Bosnia and Herzegovina, there are four basic types of soils: limestone-dolomite lithosols (either rocky or lithosols), regosoles on loose rocks, colluvial soils (or colluvium), and aeolian soils (or arenosols).

Lithosols are extremely young and undeveloped soils formed on a solid, rocky base, on which a poorly developed humus layer of only up to 2 cm of thickness is formed. The process of formation of the organo-mineral complex in the humus layer is possible, although only in the initial phase of development. The sparse vegetation that develops on the rock substrate accumulates small amounts of organic residues, which are very easily washed away through the systems of cracks and pores within and between detritus, which is why the accumulation of humus is very weak and sporadic. Depending on the type of geological substrate, different subtypes can develop: on acid rocks, on neutral and basic rocks, on peridotites, and on limestones and dolomites. This type of soil is most common in the zones of Herzegovinian rocky landscapes, i.e., in areas where limestone and dolomite rock detritus is in the form of fragmented and "unformed" rock material of decimeter to meter dimensions. The complexes of this rock material mentioned in the geomorphological literature are known as "rocky landscapes." In addition to the above, lithosols are formed in the mountainous area of Bosnia, mainly in the higher hypsometric zones where the rock masses of Mesozoic (Jurassic and Triassic) limestones dominate, which are decided by more pronounced primary cracking (Drešković and Mirić 2017).

Sierozems (Russian: "raw soil") on loose rocks are undeveloped or poorly developed soils on loose rock detritus (except for alluvial, deluvial and aeolian recent deposits), which, as a rule, are not skeletal. They are formed by the erosion of previously formed soils or initial processes of pedogenesis, which has not yet resulted in the creation of the A horizon, as a result of which they represent the initial stage of soil development. The parent substrate, which consists of igneous rocks, marls, marly limestones, and dolomites (loose clayey substrate), is subject to more pronounced mechanical destruction and chemical decomposition. More plants that inhabit this type of habitat with their root system promote mechanical destruction and decomposition of primary silicate minerals into secondary ones, which reduces the fragmentation of detritus and the formation of mechanical soil particles below 2 mm in diameter. In relation to the character of the rock mass, the following can be formed: carbonate, marly, silicate, eutric, etc., subtypes of sierozems. The geographical distribution of this type of soil in Bosnia and Herzegovina is of a local character and is marked in small areas where deposits of shale regoliths, sandy dolomites, granular eruptions, and zones of moraine deposits have been developed (Encyclopedia of Yugoslavia 1983).

The main characteristic of *colluvial soils* is under development or poor development, which is why they have only (A) or Ap horizon. The pedogenetic processes that cause the formation of this type of soil are related to fluviodenudation processes on the rocky slopes of mountain morphostructures by which the destroyed rocky detritus from the higher hypsometric floors is transferred by gravity creeps to the flatter parts of the slopes. These deposits usually have a proluvial character, considering that they are associated with climatological periods with precipitation maxima. Accumulated rock detritus both along the slopes and in the bottom parts of them is not completely rounded, which is why it has the characteristics of proluvial deposits that are deposited in accordance with the size of the material; along the slopes and edges, a coarser material, while in the central parts of accumulation zone, a material of greater degree and fineness of grain fineness is deposited. In accordance with the predominant mineral composition, this type of soil can be differentiated into several subtypes: eutric and silicate (on neutral and basic minerals), district silicate (on acidic minerals), carbonate (on carbonate rock minerals), and colluvial-prolivial subtypes. In relation to the mentioned characteristics of the pedogenetic process, it can be said that the geographical distribution of this soil type is primarily related to the mountain morphostructures of the mountain-valley macroregion of Bosnia and Herzegovina and its transition zones into neighboring geographical macroregions.

Arenosols or "living" sands are formed under the direct influence of the aeolian (wind) exogenous geomorphological process by which, in the immediate geological past, powdery and sandy particles were transferred from the source areas to the Pannonian Plain. In relation to the predominant mineral structure, they can be quartz (> 95% SiO2), silicate (<95% SiO2), carbonate (> 95% carbonate minerals), and mixed (quartz-silicate-carbonate). Due to the prevailing powdery-sandy structure, these are soils characterized by more pronounced mobility, except in cases when different types of forest vegetation stabilize them. The geographical distribution of these soil types is primarily related to the Peripannonian area of the macroregion of northern Bosnia, which belongs to the extreme southern edge of the Pannonian Basin.

7.1.1.2 Humus-Accumulative Soils

Soils from the humus-accumulative class are characterized by the presence of a developed humus horizon, which, directly or via the AC horizon, passes into the parent rock base. In accordance with the general geological, geomorphological, hydrological, and climatic characteristics in Bosnia and Herzegovina, no typical soils from this class have been formed—only some less developed types are represented: black soils on limestones and dolomites (or calcomelanosols), humus silicate soils (or rankers), humus carbonate soils (or rendzina), chernozem, and vertisols.

Black soils on limestones and dolomites (or calcomelanosols) are formed on a compact limestone and/or dolomite substrate containing less than 1% of dry residue formed in karst karstification processes. The structure of the undissolved substrate is dominated by the minerals sesquioxide (R2O3), which represent the mineral basis for the formation of black soils. In accordance with the time dynamics of development for black soils of this type, the following chronological phase varieties can be distinguished: organogenic black soil, organogenic black soil, organogenic-mineral black soil, browned (brownized) black soil, brown soil on limestone (or dolomite), and loess soil with limestone (or dolomite). The intensity of this development process largely depends, in addition to the mineralogical-petrographic composition of the parent rock substrate, on the general climatic conditions of a particular area. The geographical distribution of this type of soil is in line with the basic characteristics of the pedogenetic process. More specifically, limestone-dolomite black soils are formed in a very wide range of climatic and morphological variations within the mountain-valley macroregion of Bosnia and Herzegovina, especially in the zones of mid-mountain to high-mountain terrain (example of mountains Kalin and Rudina near Bugojno) (Gekić 2011).

Rankers represent humus soil formed on hard (Ah—mC) substrate and, less frequently, on crumbly, silicate rock substrate. This type of soil can have Amo (molecular) and Aum (umbric) or Aoh (ochric) horizon that is formed on the parent rock, on whose mineral-petrographic composition their characteristics also depend. More specifically, humus silicate soils can be formed on several different types of rock substrate, such as basic and ultrabasic rocks (from which eutric rankers are formed) and silicate rocks (from which acid rankers are formed) (Resulović et al. 2008). The main factor of intensive accumulation of humus is the unfavorable hydrothermal conditions of mineralization of organic matter, which is why this type of soil usually has a skeletal structure. The geographical distribution of this type of soil in Bosnia and Herzegovina is related to the distribution of basic and ultrabasic mountain massifs on the one hand and silicate mountain massifs on the other, mainly in the transition zone from mountain-valley to macro-region of northern Bosnia. Large areas under the ranker are east of the river Vrbas: starting from Komar through Kaštel, Kamenjaš, Skakavac, including the area of Ravno Rostovo east of Kalin and Rudina. To the south, the ranker spreads in the area of Vranica. These lands are spread in the extreme northwest all the way from Pelakovac to Babin potok and in the southwest in the area of Pandurica, Hum, Planinica. They are overgrown with forest and meadow-pasture vegetation (Gekić 2016).

In general, *rendzina* is formed on loose carbonate sediments (especially on scattered dolomites, marly limestones, marls, and carbonate sandstones), cemented products of proluvial-colluvial processes—conglomerates and breccias, as well as on moraine deposits. The average depth of the humus horizon is 30–40 cm, although it can be larger. Rendzinw generally has a molecular (Amo) horizon over which they gradually pass into the parent rock substrate. Rendzina is formed by the accumulation of mature (mull) humus rich in $CaCO_3$ (10–50%), which turns into a humus-gley complex under the influence of moisture. Under the influence of the humid climate from the mentioned complex, carbonate wear and leaching occurs, which, under the additional influence of H_2CO_3, leads to the decomposition of silicates and clay formation. In this process, $Fe_2O_3 \times nH_2O$ is released, forming a brownish B horizon consisting of a brownish rendzina. Due to the fact that they have a favorable regime of aeration and wetting, rendzinas are mostly agricultural lands that, with favorable slopes and terrain exposures, are mainly used as fruit or vineyard lands. In accordance with the above, rendzina occurs in most of the hilly and mountainous regions of Bosnia in conditions of humid (Cf) climate.

The main characteristic of *chernozem* is the powerful A horizon, which in the areas of Bosnia and Herzegovina can be from 40 to 80 cm, while the power of the transitional AmoC is up to 30 cm. The basic parent substrate on which, under the influence of arid or semiarid climate, chernozem forms is carbonate loess (with 20–30% $CaCO_3$). In addition to the above, chernozem can also develop on other types of substrates, such as sandy loess, aeolian sands, or precipitated alluvial deposits. In chernozem on typical loess, the structure is clayey, on a sandy base sandy-loamy, while on loess terraces, the structure is clayey-loamy.

The dominant vegetation covering chernozem is grass vegetation (in natural conditions) or cereals or fodder crops in agricultural production conditions. A large amount of organic matter reaches the organic horizon. In a drier climate, part of the organic matter mineralizes, while the remaining part mixes with the pedosubstrate A of the horizon under the influence of pedofauna (moles and earthworms). In this way, there is an increase in the mineral component in the A horizon, i.e., an increase in agricultural productivity of this type of land. Considering the characteristics of the parent substrate, chernozem is geographically connected with the macroregion of northern Bosnia, i.e., the Peripannonian area, where it can appear on the following relief macroforms: loess plateaus (90–120 m above sea level)–typical loess; loess terraces (75–90 m above sea level)—loess in conditions of increased substrate humidity; old river terraces (up to 70 m above sea level) —fluctuated sedimented loess (Resulović et al. 2008).

Vertisols are formed on substrates containing at least 30% of tertiary lake clay of predominantly montmorillonite type (such as marly clays or carbonate clay shales) but can also be formed on the elution of certain basic rocks whose decomposition results in a larger amount of the same mineral (such as basalt, gabbro or diabase) or on Quaternary sediments in which a larger amount of montmorillonite is present. Due to the dominance of montmorillonite clay, pedoturbation (lat. Verto-turn) occurs, which conditions the homogenization and deepening of the A horizon. Pedosubstrate in its composition contains a higher amount of $CaCO_3$ and $MgCO_3$ as a consequence of formation on basic rocks. Amo horizon is formed under conditions of terrestrial pedogenesis and has a predominant thickness of over 30 cm. The AC horizon has a predominant capacity of 20–30 cm. However, due to poor internal drainage, it acquires the predominant properties of hydro-morphic humus. Although they have high porosity (40—60%), the presence of large pores is very small (below 5%), so these soils have a very unfavorable water-air regime (Resulović & Ćustović 2002). Due to that, they are very hard and cracked during the dry period, while during the precipitation periods, they turn into a very clastic, slippery, and difficult-to-pass substrate. Vertisols are formed in arid and semiarid climates within a hypsometric range of 200–600 meters above sea level. The predominant area of occurrence in Bosnia and Herzegovina is the macroregion of northern Bosnia, especially in the zone of transition from hilly to mountainous terrain.

7.1.1.3 Cambic Soils (or Brown Soils)

The basic characteristic of cambic soils is the presence of a cambic (B) horizon that is formed between A (or O) and C (or R) horizons. Their formation is conditioned by more intensive pedogenetic processes that take place in favorable hydrothermal conditions, as a result of which the biological and chemical decomposition and wear of the parent substrate are enhanced. More specifically, soils of this class can be formed in the processes of chemical dissolution of the limestone parent substrate during which residue is formed (designation Brz) or in the processes of decomposition of silicate minerals contained in silicate parent rock substrate, which form clay minerals in the subsurface horizon (designation Bv). The accompanying effect of these processes is increased oxidation or release of iron oxide, which results in the soil's yellowish or reddish color. The final process in soil formation is argylosynthesis, which conditions the brownization of soils from this class (Resulović et al. 2008).

Within the class of cambic soils in Bosnia and Herzegovina, the following basic types are represented: brown soils on limestones and dolomites (or chalcocambisols), eutric brown soils (or eutric cambisols), acid brown soils (dystrict brown soils or dystrict cambisols), red soil or terra rossa, terra fusca, and pelosols (or vertical cambisols).

The main characteristic of *brown soils on limestones and dolomites (or calcocambisols)* is the presence of a residual (Brz) horizon primarily formed on the limestone (and somewhat less often on the limestone-dolomite) parent rock substrate. The thickness of the soil is, therefore, quite small and varies in the range from 30 cm (on limestone) to about 60 cm (on dolomite). More specifically, this type of soil is formed from limestone mulches as of the moment when their thickness reaches 30 cm. The dominant area within which residual soils are formed (Brz) is the area of Herzegovinian karst, where, in the conditions of semiarid climate, a very slow process of chemical dissolution of limestone mass takes place. The geographical distribution of this type of soil is conditioned by the mentioned petrographic substrate and is related to the macroregion of central (mountain-valley) Bosnia and Herzegovina, i.e., to mountain morphostructures that predominantly form limestone or limestone-dolomite rock masses (Đug and Drešković 2008).

Eutric brown soils (or eutric cambisols) are, as a rule, formed on the predominant basic parent substrate, where the saturation of the absorption horizon with bases is above 50% or pHHOH> 5.5. The parent substrate can also be carbonate, with these soils never forming on limestone or dolomite. Soils from this class are characterized by the presence of a molecular or ochric-humus-accumulative horizon (20–30 cm thick), which usually lies above the identification, cambric (Bv) horizon. The Bv horizon mainly occurs under intensive conditions of the parent carbonate substrate metamorphosis under in situ conditions. In the early stages of the formation of brown soils, the power of this horizon is lower—about 10–20 cm, while in the later stages of the development of these soils, its potency increases from 80 to 100 cm. The transformation of organic matter occurs with abundant CO2 production, which accelerates the decomposition of minerals, especially at depths of 30–50 cm, where there are more favorable hydrothermal conditions and lower biological intensity of base accumulation. More specifically, part of the decomposition products enters the process of argylosynthesis (synthesis of secondary aluminosilicates) in *in situ* conditions, while the other part that absorbs the root system of plants by physiological processes is returned to the surface. The color of the Bv horizon is predominantly brown or yellowish-brown, depending on the degree of hydration of the released iron (Fe). Depending on the characteristics of the parent substrate and pedogenetic processes, this type of soil has several subtypes and varieties.

Acid brown soils (dystrict brown soils or dystrict cambisols) are primarily formed on acidic quartz-silicate substrates where the degree of base saturation is below 50% or pHHOH <5.5. They can also form on compact carbonate rocks as well as on a loose series of sediments. The surface (A) horizon is ochric, except in cases of the formation of these soils at higher altitudes when it has umbric characteristics. Its capacity is generally small and amounts to about 10 cm. Since it is formed in a humid forest climate, the humus layer is formed from decomposed or semi-decomposed organic matter. It may have the characteristics of mature or semi-crude (when decomposition processes are not complete). The basic Bv horizon was formed in similar pedogenetic processes of weaker argylosynthesis, where, due to the lack of iron minerals, the accumulation of iron oxide is weaker, which is why the color of the Bv horizon varies—from light brown to ocher. The thickness of the basic horizon is up to 50 cm, and in its content, clay particles are present in a higher concentration. In the process of organic matter transformation, fulvic acids dominate in the basic horizon; i.e., decomposition takes place in the direction of the formation of the ochric A horizon. Acid-brown soils originate from Rankers and, depending on the characteristics of the climate in Bosnia and Herzegovina, the following subtypes can be developed: typical, podzolic, and leached acid-brown soils. Most of these soil types are covered by various forest communities, especially phytocenoses of beech, beech-spruce, and beech-spruce-fir forests. The geographical distribution of acid-brown soils is related to the mountainous areas of the macroregion of mountain-valley Bosnia, in the predominant hypsometric range from 700 to 1,500 m above sea level. Subtypes of these soils can also be formed in the zone of mountainous relief of the macroregion of northern Bosnia, in the wider zone of transition to the macroregion of central Bosnia (from 400 to 1000 m above sea level) (Fig. 7.1).

The main characteristic of *terra rossa* is the building mineral of iron—hematite (Fe2O3), which gives this type of soil a characteristic red color, according to which it got its local name—red soil (crvenica). It is common for red soil to bind to scarce residual soils occurring in a chemical dissolution of limestone or limestone-dolomite compact rock substrate in sub-Mediterranean climates. This process is called rubification and involves the dehydration and crystallization of iron oxides absorbed from the surface of clay minerals. Dehydration occurs during the dry season and implies good drainage of the substrate and crystallization of hematite. During the wet period, the mineralization of organic matter accelerates, resulting in the formation of cryptocrystalline ferric hydride, which gradually turns into hematite. The surface Ah horizon is mainly clayey clay, whose thickness is about 15 cm. The ratio of dry residue to the basic unit of dissolved mass is only 0.2%, which means that in natural conditions, the process of forming red soil takes a geochronologically very long time. SiO2 and R2O3 (1.3: 1.5) are almost equally present in the residue structure. Different geomorphological methods have demonstrated that it takes up to 10,000 years to form 1 cm3 of dry residue. The potency of the basic Brz horizon in the Mediterranean and sub-Mediterranean climate conditions is a maximum of 70 cm. The origin of rubella has not been fully elucidated; today, in essence, there are two theories about its origin. According to the first theory, red soil is a product of the dissolution of the limestone substrate and the formation of a dry residue, which was subsequently exposed to the mentioned process of rubification. According to another theory, redness was formed as a product of the decay of limestone rocks, which were subsequently transformed into residue or red soil. The geographical distribution of red soil in Bosnia and Herzegovina, in addition to the region of Herzegovina, is also related to the limestone "islands" of Triassic and Jurassic

Fig. 7.1 Mountainous soils in rural areas near Bugojno. (Photo by Haris Gekić)

limestone in the macroregion of mountain-valley Bosnia (Drešković and Mirić 2017).

Terra fusca is formed on carbonate and rocks that contain gypsum. The humus-accumulative Ah horizon is less powerful and varies in the range between 10 and 30 cm. The basic Bv horizon is characterized by a more pronounced clay content, i.e., lower water permeability due to low capillary porosity. Its acidity is moderate, and acidophilic vegetation mainly develops on this type of substrate, but vegetation that is also characteristic of carbonate soils can also develop. More specifically, the dominant forest vegetation inhabiting this type of pedosubstrate is beech and beech-fir forests and pasture vegetation that develops in more humid climates. The most important geographical area in which terra fusca is continuously developed is the area of northwestern Bosnia and the region of Bosanska Krajina, with the proviso that in more limited areas, it can occur in other areas of the macroregion of mountain-valley Bosnia and Herzegovina.

Pelosols are soils formed on a parent substrate containing higher concentrations of clayey substrate and marly clays from which clayey loams and clays are formed. The surface Ah horizon contains decomposed organic matter whose power is generally small—up to 20 cm. In the period of wetting, the clay matter (illite clay) swells in the basic P horizon and then becomes completely waterproof. On the contrary, the clay matter dries out during dry periods, resulting in noticeable vertical cracks forming in the basic horizon, which can be extended to the parent C horizon. By this property, pelosols are very similar to vertisols. Very often, a transitional subsurface horizon is formed between the Ah and P horizons, which contains a large amount of clayey matter up to 10 cm thick, which has a rounded surface and which, in fact, reflects a vertical property. For this reason, these soils are very unstable, and during the rainy seasons in the hilly and mountainous relief conditions, landslides are very often formed on them. Pelosols are widespread in the wider area of the mountain-valley and mountainous relief of Bosnia and Herzegovina, where marly-clay lots of sediments form the parent substrate (Resulović et al. 2008).

7.1.1.4 Eluvial-Illuvial Soils

The basic feature of this soil class is a pronounced process of eluvial-illuvial migration, which results in the leaching of humus and decomposition of clay minerals into silicates and sesquioxides and transfer of sesquioxides to the basic horizon. More specifically, in the conditions of good rainwater drainage, the process of moving undamaged clay particles from the surface layer (eluviation) and their deposition in deeper soil horizons (illuviation) is intensified. These processes have led to the emergence of a particular E horizon, which refers to these leaching products (eluare—to wash off). Within the class of eluvial-illuvial soils in Bosnia and

Herzegovina, differentiation has been made into the following types: luvisols, acrisols, podzols, and brown podzolic soil or Brunipodzol.

Luvisols are formed in the conditions of a semihumid climate on a flat or hilly relief on a parent substrate that can be constructed of igneous, metamorphic, and sedimentary rocks. The surface Ah humus-accumulative layer contains mature humus (up to 3%) and has a thickness of 10–60 cm. In it, the processes of enrichment of the surface substrate with clay take place. The basic Bt horizon is carbonate-free in its composition and is built of loam or clay loam whose thickness averages 30–40 cm, or a maximum of 80 cm. In relation to chemical reactions, the pedo-substrate of the basic horizon has moderately acidic properties with a pH value of 6.0 to 6.5. The Ah the orizon has a slightly lower acidity of 5.5–6.0. The degree of base saturation averages 50–70%. Under these conditions, the process of peptization of clay minerals takes place, which elute from the surface into the basic Bt horizon. The intensity of elution of clay minerals depends on a number of factors, among which are more important: mechanical sorption, coagulation of clay minerals, pH value of the substrate, and others (Resulović et al. 2008).

The geographical distribution of this soil type is related to the Peripannonian area of Bosnia and Herzegovina, i.e., in the zone of flat or hilly relief within the hypsometric zones from 100 to 600 m.

Acrisols are significantly similar to luvisols in terms of their mechanical structure and vertical structure of the pedosubstrate. The parent pedosubstrate on which these soils were formed are red soils, on which a woody granular-powdery material is most often applied by aeolian, which is very similar to the clay substrate that is formed in the E horizon. More specifically, the E horizon is formed in the process of moving clay minerals from the surface horizon, with the washed fractions partially retained in this eluvial and mostly in the basic Bt horizon. The E horizon has a lower base saturation—up to 30%. The surface Ah horizon usually has umbric properties with a mediocre content of mature humus. The basic Bt horizon can be of greater power (up to 100 cm depth) if the acrisols are gleyic; i.e., there is less power in situations when the basic horizon is more humus (humus acrisols). In Bosnia and Herzegovina, this type of soil dominates in the subregion of northwestern Bosnia, but they can also be found in Herzegovina, i.e., in zones where compact limestones and dolomites form the parent rock substrate.

Podzols are soils formed by the decomposition of clay minerals into sesquioxides (R2O3) and silicate residue (SiO2) and the transfer of sesquioxides to deeper (E and B) horizons. These soils have a developed pedological profile that includes humus-accumulative (Ah), eluvial (E), basic B (with two subtypes: Bh and Bs), and parent loose (IC) horizon. The E horizon has a predominant sandy structure and contains a small percentage of clay particles—up to 10%. Its power varies from 10 (in poorly developed podzols) to 60 cm (fully developed podzols). According to the chemical reaction, together with the Ah horizon, it belongs to acidic soils with a pH value of up to 4.

The basic B horizon, according to the content, can be highly humus (Bh), iron (Bs), or humus-iron (Bsh). The specific ratio of iron-humic matter depends on the conditions of podzolization, primarily on the content of fulvic acids, the intensity of leaching of bases, acidification, the intensity of movement of sesquioxide, climate, etc. Bh subhorizon is rich in humus originating from decomposed organic matter within forest cover. The composition of the Bs subhorizon is dominated by iron oxides (Fe) and is characterized by intense sedimentation of sesquioxides, while the Bsh subhorizon has the characteristics of both the previous ones. Podzols are soils that are typically formed in the humid or perihumid climate of Central and Northern Europe. In Bosnia and Herzegovina, they are limited to the high mountain areas of the macroregion of mountain-valley Bosnia and Herzegovina, at altitudes above 900 m, where the thermal characteristics of the Df climate type dominate (Drešković & Mirić 2017).

Brunipodzols are soils that, unlike podzols, have a mixed accumulative-humus (A) and eluvial (E) horizon, which can be organic or umbric and which, as a rule, is formed under the influence of decomposed organic substrate subsequently partially mineralized. The humus content is relatively low and varies in values up to 10%, which results in poorer elution or separation of the entire eluvial horizon. Below the mixed horizon is developed B horizon with its subhorizons (Bh and Bs) with a significantly higher share of humus in them (30% <v <40%). This A/E horizon is actually an early stage that precedes the formation of a separate eluvial horizon.

It is built mainly of sandy-loam particles, the thickness of which is usually not large and is from 10 to 15 cm. Its capillary porosity is relatively large, which causes good water permeability and good aeration of the pedosubstrate. Its acidity is somewhat more pronounced and averages 4.0–4.5. The B horizon is characterized by a higher capacity—from 50 to 70 cm, with the fact that its share in the structure increases due to the movement of surface clay materials. The degree of acidity of the B horizon is somewhat lower and amounts to 4.5–5.0, with a content of mature humus up to about 2% (Resulović et al. 2008).

7.1.1.5 Soils with Stagnant Upper (Surface) Water—Pseudogley Soils

The main characteristic of this soil class is the continuous or occasional stagnation of precipitation water in the upper (surface) segments of the pedological profile. This property arises from the fact that the pedosubstrate of these soils is built of water-impermeable or very poorly water-permeable, mostly clay minerals, as a result of which water from the surface can penetrate only into the shallow parts of the upper horizons. For this reason, these soils show a lot of similarities with gley soils. More specifically, there are mainly three horizons in the pseudogley class of soils: humus (Ah), pseudogley (S or B), and waterproof. The humus horizon is built mainly of decomposed organic matter whose composition and capacity depend on the type of vegetation inhabiting these pedoenvironments. If it is an association of beech forests with thermophilic to mesophilic species of woody vegetation, then it is formed by the decay of the leaf cover, and then it has ochric properties primarily. It can also be inhabited only by herbaceous (meadow communities), which cause a faster transformation of decomposed organic matter, which is why it has a slightly higher capacity—up to 30 cm. Beneath this, a pseudogleyic horizon is formed whose genesis depends on the specific conditions within which the glazing process developed. If it is a pseudogley that has developed on impermeable or very poorly permeable clayey or predominantly clayey soil layers, then it has the characteristics of a developed pseudogley horizon whose thickness is up to 50 cm. This pedogenetic type of pseudogley is very often singled out as a separate type called primary pseudogley (Sw or g1 horizon) (Saneković 1958).

This horizon is developed on a parent substrate built of metamorphic rocks (marble) whose thickness can be up to several meters (Sd or g2 horizon). Suppose the pseudogleyic horizon was formed, the so-called by a secondary path, i.e., by moving colloidal clay particles from the surface to the subsurface (Eg) horizon. In that case, it is a process of ilimerization, i.e., a continuation of the process from the formed luvisols. The waterproof layer (Bgt) is clayey and compact with a thickness of less than 1 m. This type of pseudogley stands out as another type—secondary pseudogley. Regardless of the power and genesis, the common feature of both types of pseudogley soils is the longer retention of precipitation waters in the pseudogley horizon, based on which a horizon was formed, which partly has the characteristics of hydromorphic soils. Pseudogleyic soils are more widespread in Bosnia and Herzegovina. Their geographical distribution is related to the mountainous morphostructures of the macroregion of northern Bosnia, on which it builds a pedosubstrate up to about 600 m above sea level. Pseudogley is especially present in two large belts in the central part of the Uskopaljska valley along the Vrbas River and is connected to the fluvisol belt. The first belt west of the Vrbas begins north of Kutanja and extends south all the way to Udurlije and Vedro Polje. The second pseudogley belt is almost twice as large and covers the area east of Vrbas, from Malovina, Jemanlići, Čehajići in the north through Drvetine and Rosulje (Fig. 7.2) in the south all the way to Odžak in the length of about 18 km and an average width of about 3.5 km (Gekić 2016).

7.1.1.6 Anthropogenic Soils—Anthrosols

Anthrosols are a class of soils formed under pronounced anthropogenic influences so that the degree of modification conditioned the formation of practically new soil types in relation to the natural pedosubstrate. In most cases, all soils in this class have arable (Ap) and parent (C) horizons. The type and degree of modification are directly dependent on the purpose or forms of farming, and in this connection, the following four types from this soil class can be distinguished: acrisols (or aric anthrosoli or arable soils), hortisols (or garden soils), regosols, and soils of covered spaces (or greenhouse soils).

Acrisols (or arable soils) are cultivated soils whose primary purpose is the cultivation of agricultural arable crops. Such a purpose of these soils implies that the entire arable layer is mechanically processed and adapted to the cultivation of certain crops. To improve productive abilities, many different types of organic and inorganic stimulants have been introduced into these soils, which is why their mineral structure and physical-chemical composition, in general, have been completely adapted to production. They are significantly enriched with humus and nitrogen-phosphate compounds that improve the mineral nutrition content of arable crops. Acrisols can be further differentiated into subtypes primarily according to the depth of the arable layer to shallow (up to 20 cm), medium-deep (20 to 40 cm), and deep (above 40 cm). These soils are widely distributed in Bosnia and Herzegovina on flat or slightly sloping terrain on which field and vegetable crops are grown, as well as in plantation cultivation of fruits and vines.

Hortisols are anthropogenic soils in which the arable (Ap) horizon is slightly deeper and is about 80 cm. Their purpose is defined by the cultivation of garden crops, especially vegetables and flowers, as a result of which these soils are highly humized and enriched with phosphate-nitrogen compounds. These soils also have significantly improved water-air properties. They are mainly represented in populated places.

Regosols are a unique form of anthropogenic adaptation of the natural pedosubstrate. More specifically, mechanical processing (trenching or deep ridging) and physicochemical stimulation, and numerous indigenous soils (such as rendzina, eutric soils, colluvial soils, etc.) have been almost wholly modified and adapted for production purposes. The arable layer (Arig) is very deep and is 50 to 100 cm. These soils are less supplied with humic substances (from 1 to 3%), which is why they have the character of moderately acidic soils. The geographical distribution of this type of soil is also associated with lowland and hilly relief, and they have the purpose of growing field-vegetable and plantation fruit crops.

Greenhouse soils represent a unique form of cultivated soils because they are mechanically closed, which is why they have significantly changed microclimatic conditions in

Fig. 7.2 Fertile soils in Rosulje near Bugojno. (Photo by Haris Gekić)

relation to the ambient climatic characteristics. They are intensively cultivated and have a wholly altered mechanical structure, formed under the influence of intensive digging and mechanical comminution of the pedosubstrate. They contain a large amount of introduced organic matter (especially fertilizers) as well as mineral additives which improve the total yields of mainly vegetable and berry crops. Areas under these forms of anthropogenic soils in Bosnia and Herzegovina are increasing today, which is in line with the national strategy for improving agricultural production.

7.1.1.7 Technogenic Soils—Technosols

Soils of this class were formed under the influence of various technological processes that include numerous aspects of exploitation of mineral raw materials and disposal of discarded products of production processes in the surface layer of the soil. Accordingly, the technogenic soil class can be differentiated into the following subclasses and soil types: I subclass—soils on altered natural substrate (deposols; recultisols), II subclass—soils on various waste (cinerosols—soils on fly ash and slag; rhodic technosols—soils on red mud; garbisoli—soils on communal waste), III subclass—altered natural soils under the impact of fire (pyrosols or combustosols), IV subclass—natural soils into which various materials have been imported (necrosols or cementery soils), V subclass—soils of urban areas (urbisols), and VI subclass (glasshouse soils).

Deposols are technogenic soils that were removed as roofing material in the process of exploitation of mineral resources and on which no planned reclamation measures have been taken. Such soils are usually allochthonous because they have been transferred from the place of exploitation to specific disposal locations. It is a tailing whose structure is different depending on the type of pedological substrate; i.e., they can be of silicate or carbonate character. It is common for a part of the deposited tailings material to be covered with natural vegetation, which can be woody and grassy.

Recultisols are, in fact, deposols subjected to planned reclamation measures, and in connection with that, additional planning measures are implemented on them, including vegetation arrangement. Under the influence of reclamation, there are changes in physical and chemical properties over time, which primarily depends on the type of tailings material that is deposited.

Cinerosols are soils formed on landfills of combustion products of different types of coal in thermal power plants, i.e., soils formed on deposited ash and slag. Physico-chemical characteristics depend on the type of coal used, and above all, on the sulfur content in it. They are mainly formed in wider zones around thermal power plants. They are challenging to recultivate due to the very unfavorable mechanical structure and a large amount of pollutants present.

Rhodic technosol is formed in the technological process of alumina production from bauxite ore, during which large quantities of waste material are generated—red sludge (after which they got their name). The composition of this material is dominated by aluminum and soda, which have a toxic effect on the surrounding vegetation, which require special and functional measures for the remediation of this type of soil. The most famous sites and exploitation basins in Bosnia and Herzegovina are in the vicinity of Mostar and Zvornik.

Garbisols are formed in the zones where sanitary landfills have been formed to dispose of municipal and partly industrial waste. They represent the applied layers of soil that are directly interacting with the disposed waste, as a result of which they change their primary physical properties very quickly and permanently. These are mainly protective layers of soil that cover the surface of the disposed waste. Most often, under the influence of various chemical reactions that occur in the deposited material, their physicochemical properties change entirely in accordance with the intensity of these chemical processes.

Pyrosols (or combustosoils, burnt soils) were formed under the influence of burning the surface (Ah) horizon or internal fires, which resulted in the modification of the remaining horizons under the influence of high temperatures. Their physicochemical characteristics vary greatly, which directly depends on the same as on the textural composition and mineralogical-petrographic characteristics of the pedosubstraate before combustion. For this reason, the pH value and reactions to bases vary widely. The content of humic matter is also directly dependent on the character of the same before burning. These soils have developed in zones of relict or recent fires, especially in forest areas with coniferous vegetation.

Necrosols are a type of soil formed at selected locations that have the function of areas intended for cemeteries. They develop through the interaction of natural soils with human corpses and other imported grave objects that are subject to subsequent rot and decay. Under the influence of decay and decomposition of the introduced organic and inorganic matter, the surrounding natural soil enters into chemical interaction with it, and in this way, they are entirely or partially modified. Their physicochemical composition is also affected by certain herbaceous crops planted on these surfaces.

Urbisols are soils formed within urbanized areas and can be preserved in a completely natural state (which is less common) or are entirely anthropogenic. The construction of various infrastructural facilities mainly conditions their anthropogenic modification and usurpation for urban development needs. Given the very intensive recent urbanization processes, it can be said that most of the pedosubstrate in urban areas is completely modified in several ways: complete or partial removal of one or more horizons and their coverage by entirely different types of pedosubstrate depending on construction interventions. In accordance with the above, the following subtypes can be singled out in this group of soils: natural soils, anthropogenic soils, and damaged soils. A particular form of urbisol is horticultural areas or green areas in cities adapted to certain forms of vegetation cultivation. In the cities, most of the land is covered by structures. Buildings, roads, and artificially surfaced areas are associated with vegetated areas and bare soils, which occupy discontinuous but significant surfaces (Drašković et al. 2016).

Glasshouse soils are characterized by wholly altered basic mechanical and physicochemical properties (similar to greenhouse soils). More specifically, to maximize production capabilities, this type of soil is continuously and systematically supplied with a large amount of organic and inorganic stimulants, which has led to a complete change in its natural properties. They are mechanically very unstable due to the fact that they are constantly dug up and drained mechanically. They have essentially the same production function as greenhouse soils; i.e., they are primarily adapted for the vegetable production of berries.

7.1.2 Hydromorphic Soils

The basic feature of the division of hydromorphic soils is the constant or occasional surplus wetting of the whole or a particular part of the pedological profile. Excessive wetting is the excessive filling of the pores with water that moves slowly or stagnates in them, which in the initial stages results in reduction processes of iron, manganese, and sulfur compounds, or the final gleying of the pedological profile. Reduction processes develop under anaerobic conditions, which results in the final gleying or formation of the gley (g1) horizon. The surplus water in the profile either originates from rainwater or has the character of the so-called complementary waters (i.e., catchment, flood, or groundwater). In Bosnia and Herzegovina, five soil classes from this division, with a total of nine separate soil types, have been singled out.

7.1.2.1 Undeveloped Hydromorphic Soils

Fluvisols are soils formed on river sediments; i.e., the primary agent that affects their formation and structuring is river water. Given that this is a dynamic process that lasts continuously for fluviosols, it can be stated that they are primarily young soils and that, due to the recent formation, they do not have consolidated or clearly differentiated horizons. More specifically, a surface (Ai) layer can be singled out on the vertical profile below which the layers formed in accordance with the phases of fluvial development of watercourses continue. Their structure, i.e., granulometric composition and characteristics, correspond to the general paleo-geographical conditions within which the material was deposited.

In essence, the layered structure corresponds to the degree of fragmentation of gravelly-sandy material and fine rock detritus; i.e., the accumulation of material is adequate to the transport energy of watercourses. In the upstream segments of watercourses, where the energy of water is higher, the largest rock detritus is deposited and accumulated. In contrast, towards the confluence, more fragmented and finely processed gravelly-sandy material accumulates. In other words, the degree of comminution and processing of rock detritus is also consistent with water-flow energy. In accordance with the structure of the pedosubstrate, the basic water and aeration properties are defined; i.e., the loamy and sandy-loamy structure of the pedosubstrate is the most favorable (Resulović and Ćustović 2002). Considering that the mentioned fluvial process is taking place in the recent period, it can be said that fluvisols are developing continuously, with the recent deposits being the youngest and mechanically the least consolidated. Since there are no clearly separated horizons, the stated stratification of fluviosols is marked in chronological order by Roman numerals, with each number corresponding to one layer. Similar differentiation and structuring are present in fluvisols that develop in the pre-plane planes. Further classification of this type of soil into subtypes can be performed based on the mineralogical-petrographic composition of pedosubstrate, primarily from the aspect of carbonate content in their composition.

The geographical distribution of fluvisols is defined by the network of watercourses and their rank, i.e., the most developed fluvisols are formed in the alluvial zone of downstream segments of larger watercourses, i.e., their development decreases in upstream segments of especially smaller watercourses, where they are practically present only in fragments, in zones of morphological gorges (Drešković and Mirić 2017).

7.1.2.2 Stagnogley Soils

Stagnogley is a type of soil formed in conditions of pronounced or excessive moisture, especially under the influence of rainwater or groundwater, which is why it has specific pseudogleyic properties. Due to excessive humidity, a highly glazed horizon (Btg or g2) appears, which develops under anaerobic conditions, very marble, and with noticeable participation of clay matter. Under the influence of precipitation or catchment waters, clayey colloidal particles are transferred to the transitional (E or g1) horizon, which is less developed in relation to typical eluvial soils. The surface (Ah) horizon is rich in humus, which is why its thickness can be up to 30 cm. The geographical distribution is characterized by the distribution in the form of smaller isolated areas within the mountain-valley macroregion of Bosnia and Herzegovina (Đug and Drešković 2008).

7.1.2.3 Semigley Soils—Meadow Soils

Humofluvisols are soils primarily formed in wide alluvial plateaus of larger rivers or the lowland-hilly terrain of alluvial terraces—relict forms of former river plateaus. As a rule, meadow soils are overgrown with various meadow communities that are occasionally replaced by hygrophilous forest communities (which develop along watercourses), which, in addition to pedogenetic processes, are primarily influenced by the climatic factor. The surface (Aoh) horizon has a pronounced humus character, given that it is formed under the influence of decaying meadow or forest vegetation, which is why, with intensive wetting by rainwater, its capacity can be several tens of centimeters. There is an (AC or Gs) horizon in the surface substrate formed by different types of alluvial deposits whose mineral-petrographic composition varies between carbonate and non-carbonate. Its mechanical composition is, as a rule, granular with varying types of fragmentation—from sandy-clayey to fine-grained. The lower part of the basic horizon is under the pronounced influence of groundwater, which is why to a certain extent, it has gley properties (hydrogenation process—Gr horizon). Considering that the groundwater level during the year varies considerably, the carbon horizon can have a capacity of over 1 m (Gs horizon). Due to the more significant presence of clay material in the pedo-substrate, the surface and upper segments of the basic horizon of these soils in the dry period of the year are arid, which causes pronounced cracks that can affect the entire profile to groundwater levels (Resulović and Ćustović 2002). During intensive wetting with rainwater, capillary water carries a lot of oxygen, so this part of the profile has predominant terrestrial properties. This type of soil is widely represented in the area around the Pannonian basin in Bosnia and Herzegovina and partly in the transition zone with hilly relief.

7.1.2.4 Gley Soils

Humogley is a type of soil characterized by a pronounced depth of the profile, which is usually above 1.0 m deep. The surface (Ao) horizon has distinct humus properties and high thickness (from 50 cm to 1.0 m), which are a consequence of the decay and decomposition of a large amount of mostly marsh vegetation. Under the influence of rainwater, a large amount of primary mineral matter decomposes (argylogenesis) and moves to the next horizon. The movement of decomposed minerals occurs under the influence of rainwater, which often comes into contact with groundwater. Depending on the groundwater level in the upper parts of the basic or transitional horizon, aerobic or anaerobic conditions are formed, leading to differently directed matter flows. In the conditions of continuous or predominant groundwater retention (above 1.0 m from the soil surface), the processes of mineral decomposition are very difficult; i.e., reduction conditions occur, which results in increased participation of clay substances in the transition (AC) and basic (CG) horizon (Resulović and Ćustović 2002). Humogley, as a rule, appears in the dead zones and meanders of watercourses in the Peripannonian part of Bosnia and Herzegovina and other areas of flat or concave relief on the surface of which standing water can remain for a long time.

Eugley is a muddy swamp soil in which most of the profile was formed in conditions of long-term wetting of groundwater—the process of hydromorphism. More specifically, in these soil types, groundwater covers the baseline horizon (Gr) and part of the transition horizon (Go) throughout the year, resulting in complete anaerobic conditions for the decomposition of organic matter. This results in a more intensive transformation process of the primary minerals iron and manganese and the formation of the organo-mineral complex and sulfur compounds. In the transition horizon, there are conditions of long-term wetting; i.e., a saturation of pores through groundwater lasts most of the year, while in the remaining period, there are anaerobic conditions in which oxidation processes take place under the influence of oxygen and whose products are transferred to the basic horizon. The surface (Ah) horizon is very rich in humus matter that results from the decomposition of wetland vegetation (herbaceous—reed, sedge, bulrush, or woody—willow or alder) taking place in the predominant aquatic environment. Depending on the content of carbonate minerals (above all CaCO3), these soils can be carbonate or non-carbonate; i.e., their pH value varies in the values of acid soils, which are primarily saturated with bases. The geographical distribution of these soils is related to the distinct plain relief, i.e., to the wide zones of the confluence of the main tributaries of the Sava river, as well as in morphodepressions with lake water. Eugley is represented, for example, on only 0.3% of the Uskopaljska valley. It spreads into two smaller areas. The first is located in the central part of the Uskopaljska valley in the area of Čaušlije and Lug, and the second about 10 km south in the area of Donje Crniče near Bugojno (Gekić 2016) (Fig. 7.3).

Peat-gley soil is formed in an environment that is hydrologically shaped and with more clearly differentiated segments of the profile within which the processes of decomposition and decomposition of matter take place. In the surface (nH) horizon with a capacity of up to 30 cm, intensive decomposition processes of a large amount of hygrophilous woody vegetation (oak, ash, cottonwood, alder, and willow) take place in conditions that are very often anaerobic. The basic (nHG) horizon is dominated by a mineral component whose decomposition takes place in typical anaerobic conditions (with complete absence of oxygen), determined by excess groundwater wetting. Depending on the presence of carbonate minerals (especially CaCO3, which can vary from 1% to 20%), these soils can occur in the subtype of carbonate or non-carbonate soils or their transitional varieties in relation to the degree of acidity and pH (Resulović & Ćustović 2002). The geographical distribution of these soil types is related to the Peripannonian morphological region in parts of the remaining meanders (also known as "old water").

7.1.2.5 Peat Soils—Histosols

Low peat was formed under the dominant influences of the fluvial process, which causes excess wetting within the flattened mainly lowland relief, especially in the zone of wide alluvial plateaus of large rivers. The process of peat formation takes place under the influence of groundwater, the level of which fluctuates significantly during the year. This is reflected in the process of only partial formation of anaerobic conditions; i.e., the decomposition of the organic component is only partial and, as such, is deposited in special humus

Fig. 7.3 Agricultural soils near Bugojno. (Photo by Haris Gekić)

(nH) horizon in the base of which is the basic (usually gley) mineral horizon. The continuous change of drier and wetter periods leads to the backfilling of semi-decomposed wetland layers with terrigenous material, where a new process of wetting with marsh vegetation takes place again. In this way, a layered structure is formed, i.e., a larger number of subhorizons within a unique pedological profile. The basic physical and chemical characteristics of the humus horizon largely depend on the type of decomposed wetland-marsh vegetation, on which its capacity and production characteristics largely depend. In Bosnia and Herzegovina, they are mainly related to certain zones along the bottom of karst fields in Herzegovina, among which the most famous are Livanjsko (northwestern part), Hutovo blato, Glamočko, Kupreško, and other karst fields. Also known are sites in the area of the Bardača wetland and other similar areas along the Sava River (Đug and Drešković 2010).

High peat occurs in mountainous areas on highly acidic and silicate or parent substrates with low base saturation and areas of mountainous terrain dominated by low-temperature conditions and humid climate (high rainfall and high humidity). The formation of peat deposits is associated with *Sphagnum* mosses, whose decomposition is limited by a larger amount of rainwater in the prevailing terrestrial conditions. This unfavorable thermal-hygienic regime affects the reduced development of microorganisms, which further causes incomplete decomposition of organic matter. In this way, the phases of development and partial decomposition of moss and other accompanying herbaceous substrate change continuously, which causes a layered structure and growth of the humus horizon. If the humus horizon develops on extremely acidic podzolic soils, it further reduces the process of organic decomposition. In this connection, larger amounts of raw moss material accumulate, increasing the humus layer's capacity, which further reduces organic processes. In this way, a distinctly acidic oligotrophic humus is formed, on which mainly moss stands of the stated order develop.

A large number of sites with high peat have been registered in Bosnia and Herzegovina, especially in the zone of the mountains Zvijezda, Konjuh, Vranica, and other pictorial massifs of the mountain-valley macroregion of Bosnia and Herzegovina.

Wetlands are soils that develop fully under the influence of water that can appear in the soil as running, groundwater, or catchment water. Under such conditions in morphodepressions of mainly lowland areas, water is continuously retained throughout the profile, originating from several other primary hydromorphic soil types. The most famous type of wetlands in Bosnia and Herzegovina is Deransko blato (Gornje blato), which today is only partially preserved in significantly changed conditions of the former natural hydrological regime of water supply.

7.1.3 Subhydric Soils

In addition to the above-mentioned main soil types, which have continuous and separated surface areas, the sporadic presence of certain other soil types that do not belong to the above-mentioned A and B divisions has been established in Bosnia and Herzegovina. More specifically, these are soils whose genesis and development are related to stagnant waters, i.e., which are located at the bottom of lake basins, ponds, or marine coastal areas. The process of pedogenesis of these soils is combined with the process of sedimentation that takes place in the reducing conditions of freshwater or saltwater environment. Soils that develop under the above pedogenetic processes belong to a special division of soils—subhydric soils. In relation to the basic physical and chemical characteristics of the aquatic environment as well as the intensity of the pedogenetic process, this soil class is differentiated into two basic classes: undeveloped subhydric soils and developed subhydric soils.

7.1.3.1 Undeveloped Subhydric Soils

Protopedon represents the initial stage of pedosubstrate development on clastic sediments along the bottom of shallow lake basins or wetlands. Underwater lower algae initiate the pedogenetic process through the production of crude humus in the initial stage of decomposition. The process of sedimentation of mainly terrigenous material occurs in combination with the deposition of humus material, which results in the formation of a thinner surface layer of pedosubstrate.

7.1.3.2 Developed Subhydric Soils

Sapropel or rotten sludge is formed on the entire Bosnia and Herzegovina along the bottom of the basin with standing waters—wetland ecosystems. Wetland habitats are characterized by a more developed process of eutrophication, which results in a reduced concentration of oxygen in the water. Considering that the decomposition of organic matter takes place by bacterial decomposition in anaerobic conditions, the process of putrefaction or formation of pedosubstrate in the form of rotten sludge of different thicknesses containing metal sulfides and gaseous substances (H2S, H2, etc.) is pronounced (Đug and Drešković 2010).

Dy or muddy brown soil also develops in aquatic conditions with very poor aeration. The humus horizon is formed by the coagulation of humus colloids contained in standing water. It is very poorly biologically active due to the pronounced acidity of the aquatic environment.

Gyttja or muddy gray soil is characterized by a more developed humus horizon compared to the previous two types. This is a consequence of greater air permeability of the water mass, i.e., higher presence of oxygen and reducing bacterial organisms. As a consequence of this relationship, the decomposition process of organic and mineral components is intensified; i.e., the surface horizon is characterized by a greater power. In addition to mineral particles, coprogenic compounds of predominant olive green color are also present in the sludge content.

At the end of this chapter, there is detailed Table 7.1 and Fig. 7.4. about the type of soils in Bosnia and Herzegovina and their share in the total area of the country.

Table 7.1 Type of soils in Bosnia and Herzegovina

No.	Type of soil	Area (%)
1	Lithosol, calcomelanosol, and mosaic lithosol—calcomelanosol on limestone and dolomite	2.88
2	Calcocambisol and mosaic calcomelanosol—calcocambisol on limestone	20.68
3	Mosaics calcomelanosol—luvisol, calcocambisol—luvisol, calcomelanosol—calcocambisol—luvisol and rendzina-calcocambisol—luvisol complex on limestone or dolomite	16.60
4	Mosaic calcomelanosol—terra rossa on limestone	0.28
5	Mosaic calcocambisol—terra rossa on limestone and dolomite	0.81
6	Complex calcocambisol—eutric cambisol on a series of limestone and marl	0.10
7	Terra rossa on limestone and dolomite	1.04
8	Rendzina and lithosol—rendzina complex on dolomite	3.10
9	Complex rendzina—chalcocambisol on dolomite and mosaic calcomelanosol—calcocambisol on limestone	0.37
10	Complex rendzina—chalcocambisol on dolomite	0.19
11	Acrylic luvisol on limestone	1.01
12	Rendzina on moraine material	0.63
13	Regosol—rendzina complex on flysch-like sediments	0.15
14	Rendzina on marl	0.40
15	Rendzina—vertisol complex on marl	0.22
16	A series of rendzina—calcocambisol on marl and limestone	0.24
17	A series of rendzina—eutric cambisol on marl	0.18
18	Complex rendzina—district cambisol—pseudogley on flysch-like sediments	0.96
19	Complex rendzina—pseudogley on flysch-like sediments	0.09
20	Rendzina—eugley complex on tertiary sediments	0.16
21	Lithosol and ranker on serpentinite and peridotite	0.07
22	Eutric cambisol and series ranker—cambisol and cambisol—luvisol on serpentinite and peridotite	3.21
23	Series of ranker—eutric cambisol on bases rich in silicate rocks	0.05
24	Eutric cambisol on bases rich in silicate rocks	1.14
25	Complex ranker—dystrict cambisol on acid silicate rocks	0.04
26	Dystrict cambisol on acid silicate rocks	27.95
27	Complex dystrict cambisol—eutric cambisol on a series of carbonate and silicate rocks	0.16
28	Complex dystrict cambisol—luvisol complex on acid silicate rocks	0.23
29	District cambisol—luvisol—pseudogley complex on acid silicate rocks	0.17
30	Luvisol on silicate rocks	1.26
31	Luvisol—pseudogley complex on tertiary sediments	0.24
32	Pseudogley	6.03
33	Vertisol, pelosol, and vertisol—pelosol complex on clay and marl	1.28
34	Fluvisol	4.29
35	Complex fluvisol—pseudogley	0.04
36	Eugley	1.05
37	Eugley—semigley complex	1.19
38	Semigley	1.47
0	Water area	0.03

Source: Based on Drešković and Mirić 2017 and field research of autorsh

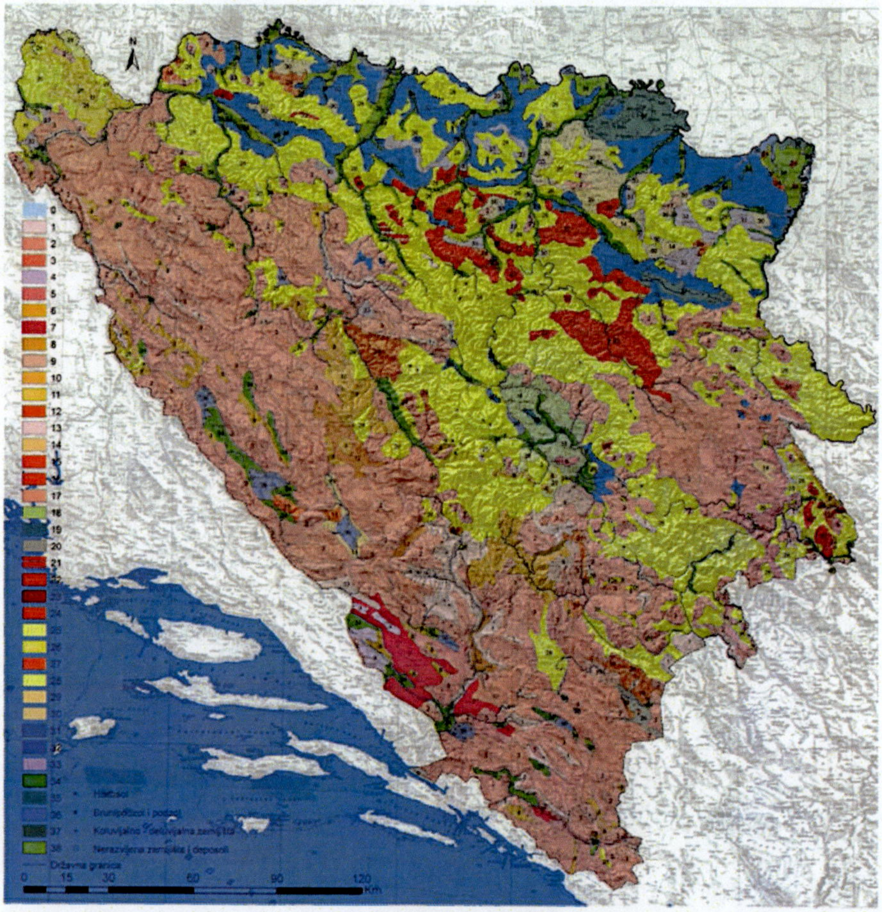

Fig. 7.4 Main types of soils in Bosnia and Herzegovina. (Based on Drešković & Mirić 2017 and field research of authors; Reference numbers for the legend are explained in the Table 7.1)

References

Drašković B, Drešković N, Mirić R (2016) East Sarajevo (Bosnia and Herzegovina) twenty years later –changes in land use. Geogr Pannonica 20(3):161–167

Drešković N, Mirić R (2017) Regionalna geografija Bosne i Hercegovine I [Regional Geography of Bosni and Hercegovina I]. Faculty of Science, University of Sarajevo, Sarajevo

Đug S, Drešković N (2008) Prirodno naslijeđe Kantona Sarajevo [Natural heritage of Canton of Sarajevo]. Kantonalni zavod za zaštitu kulturno-historijskog i prirodnog naslijeđa, ArchDesign, Sarajevo

Đug S, Drešković N (2010) Zaštita prirode u Bosni i Hercegovini: stanje i perspektive [Nature protection in Bosnia and Herzegovina: state and perspectives]. Zbornik radova, Federalno ministarstvo okoliša i turizma, Sarajevo

Encyclopedia of Yugoslavia (1983) Socijalistička republika Bosna i Hercegovina. Jugoslovenski leksikografski zavod, Zagreb

Gekić H (2011) Društveno-geografske determinante kao osnova prostornog plana općine Bugojno [Socio-geographical determinants as the basis of the spatial plan of the municipality of Bugojno]. Master thesis, Department of Geography, Faculty of Science, University of Sarajevo, Sarajevo

Gekić H (2016) Geografske osnove poljoprivredne proizvodnje u Uskopaljskoj kotlini [Geographical bases of agricultural production in the Uskopalj valley]. Doctoral thesis, Department of Geography, Faculty of Science, University of Sarajevo, Sarajevo

Resulović H, Ćustović H (2002) Pedologija, opći dio [Pedology, general part]. Poljorivredno-prehrambeni fakultet Univerziteta, Sarajevo

Resulović H, Ćustović H, Čengić I (2008) Sistematika tla/zemljišta, nastanak, svojstva i plodnost [Soil/soil systematics, origin, properties and fertility]. Poljoprivredno-prehrambeni fakultet Univerziteta u Sarajevu, Sarajevo

Saneković Đ (1958) Nekoliko riječi o pedološkom sastavu Bosne i Hercegovine [A few words about the pedological composition of Bosnia and Herzegovina]. Poljoprivredni Pregled 3-4:12–26

8 Biogeography of Bosnia and Herzegovina

Abstract

A significant factor for the general biogeographical features of the Dinarides of Bosnia and Herzegovina, as the dominant spatial unit, is their belonging to parts of the subtropical and temperate climate zone, i.e., contact with the Adriatic Sea in the south, southwest, and southeast, and connection with the Eurasian mainland in the west, north, and east. Another important factor is the orographic structure of the terrain (presence of all six energy classes of relief and pronounced morphological disintegration on the vertical profile from 0 to 2600 m above sea level). In Bosnia and Herzegovina, three groups of sectors can be singled out on the horizontal profile, including hardwood evergreen, temperate-deciduous, mixed deciduous-coniferous coniferous, and high-mountain forest vegetation, which spread over three biogeographical regions. Within each region, the spatial differentiation of the 19 most important climatogenic ecosystems is explained with their specificities regarding dominant vegetation species and communities with an overview of general physical-geographical features and faunal composition. Based on all the specifics of physical and geographical factors, with previously established protected areas and the application of criteria according to the UICN in the modern period, Bosnia and Herzegovina has only 2.6% of the territory under a certain degree of protection (far below regional and European level). The preservation of natural heritage requires establishing a more efficient system of long-term protection of the most valuable and important natural areas characterized by a high degree of biological diversity and following the European Union's modern ecological criteria and standards.

Keywords

Protected areas · Flora · Fauna · Ecosystem · Physical Geography

8.1 Structure and Dynamics of the Ecosystem

Primary differentiation and spatial variation of natural-climatogenic biocenoses, i.e., ecosystems in Bosnia and Herzegovina, arose as a product of long-term interaction of different physical-geographical factors, especially: mineralogical-petrographic composition of rock masses, climatic specifics, specifics of the hydrographic system, and pedological biology, while the subsequent modification of their structure and dynamics was significantly influenced by various anthropogenic factors, among which the uncontrolled and excessive exploitation of bioresources can be particularly noted. More specifically, under the influence of anthropogenic factors on primary-climatogenic ecosystems, their basic production property changes, i.e., the ability to produce green mass and biomass as a whole. According to the degree of exploitation and negative impacts, it can be said that the decrease in biomass production of primary biocenoses is practically proportional to the increase in the degree of degradation, i.e., the anthropogenic transformation of climatogenic ecosystems. Suppose we mark the biomass production of primary ecosystems from 100%. In that case, the production of secondary ecosystems falls to 66.6%, tertiary ecosystems to about 33.3%, while production in technogenic systems (anthropogenic deserts) falls to 0%.

The most important consequence that accompanies the change in the structure of primary-climatogenic ecosystems refers to the change in the number and structure of communities and species that inhabit them. The results of field phytocenological research of primary, secondary, and tertiary vegetation of particular ecosystems on the vertical profiles of the mountain-valley region of Bosnia and Herzegovina clearly indicate this statement. For example, out of a total of 233 species found in three different primary-climatogenic macrophytocenoses in the Sarajevo-Zenica Mesoregion, 197 have the character of industrial, medicinal, vitamin-producing, and edible plants or about

H. Gekić et al., *The Geography of Bosnia and Herzegovina*, World Regional Geography Book Series,
https://doi.org/10.1007/978-3-030-98523-3_8

85%. In four macrophytocenoses of secondary vegetation, a total of 176 species were found, of which 112 (or about 64%) belong to the category of industrial, medicinal, vitamin-producing, and edible plants, which indicates the significant fact that secondary vegetation has not only lower biomass production but also its production quality for economic exploitation significantly reduced. A total of 250 species appeared in six macrophytocenoses of tertiary vegetation, of which 178 (or about 71%) from the mentioned categories, indicating a significant decrease in the percentage of biomass production compared to the primary macrophytocenosis. The increased presence of industrial, medicinal, vitamin-producing, and edible plant species in phytocenoses of tertiary vegetation in relation to secondary vegetation, apart from the specific structure of tertiary vegetation, can be explained by better adaptation of this vegetation to anthropogenic influences and increasing needs for their physicochemical and biochemical characteristics, which affected their additional cultivation (Drešković et al. 2011).

The mentioned researches of various indicators at the level of populations and species in primary, secondary, and tertiary biocenoses on numerous vertical profiles of the mountain-valley region of Bosnia and Herzegovina resulted, among other things, in defining the degree of preservation or degradation of ecosystems based on which the following vegetation categories can be singled out: primary ecosystems, primary-secondary ecosystems, secondary-primary ecosystems, secondary ecosystems, secondary-tertiary ecosystems, tertiary-secondary ecosystems, tertiary ecosystems, tertiary-primary ecosystems, and primary-tertiary ecosystems.

Indicators of primary-climatogenic vegetation include all species that are part of the primary-climatogenic biocenoses, i.e., macrophytocenoses, and which in the spatial sense generally do not occur outside them. Indicators of primary-secondary vegetation indicate species that find their developmental biological optimum in primary ecosystems, with the proviso that they can be found as accompanying species in secondary ecosystems that have emerged due to the degradation of the respective primary ecosystems. Secondary-primary vegetation indicators indicate species or populations that find the optimum in secondary vegetation and characterize the biocenoses of secondary vegetation, and may also occur in primary vegetation, which is formed by the programming of the corresponding secondary vegetation.

Secondary vegetation indicators indicate species that find their developmental biological optimum in secondary vegetation and generally do not occur outside this vegetation. Indicators of secondary-tertiary vegetation indicate species optimally found in the secondary vegetation of meadows and pastures, with the proviso that they can sporadically appear as accompanying species in macrophytocenoses of tertiary vegetation of rangelands, fields, landfills, etc.

Tertiary-secondary vegetation indicators indicate species that are optimally found in anthropogenic ecosystems of urban and rural areas, and as ancillary species, they also appear in secondary ecosystems that are created by the programming of appropriate tertiary ecosystems. Tertiary vegetation indicators indicate species in tertiary ecosystems and are contained in their macrophytocenoses, i.e., biocenoses. Tertiary-primary vegetation indicators indicate rare species that find the optimum in tertiary vegetation ecosystems and may sporadically occur in the nitrified forest or some primary ecosystems.

Primary-tertiary vegetation indicators indicate species that find the optimum in primary-climatogenic vegetation ecosystems, while they can only occur sporadically in tertiary vegetation ecosystems. Indeed, in addition to the above basic categories, such species and populations can connect all three types of vegetation—primary, secondary, and tertiary, noting that their importance is an indicator of the degree of degradation is generally small, which is why this type of differentiation applies only on a large spatial scale. From the aspect of degree of integration and degree of ecological variation of primary, secondary, and tertiary vegetation, primary, then secondary, and lowest corresponding tertiary macrophytocenoses have the highest degree of integration of macrophytocenoses, while the degree of their ecological variation is inversely proportional to their degree of integration.

8.2 Basic Rules of Horizontal and Vertical Ecosystem Distribution in the Dinarides

The spatial differentiation of ecosystems and their communities in the Dinarides is primarily defined by the general astronomical and physical-geographical position of the Dinaric geostructure. The wider area of the Dinarides of the Balkan Peninsula occupies the area between the so-called Narbe (scar) zone in the northeast (the area north of the Ljubljana valley), Peć geofracture in the southeast, the tectonic trench of the Sava in the northeast, and Adriatic masses in the southwest. The Dinaric geostructure extends within 42° and 46.5° N and 14° and 20.5° E within the stated limits.

A particularly important factor for the general biogeographical features of the Dinarides is the spatial affiliation to parts of the subtropical and temperate climate zone, i.e., contact with the Adriatic Sea in the south, southwest, and southeast of the area and land connection with the Eurasian mainland in the west, north, and east. Due to the presence of these belts and their spatial distribution in the Dinarides during certain climatological seasons, air masses with significantly different physical characteristics are intertwined and mixed. More specifically, this area is dominated by moisture-rich air masses of the western and southwestern circulation alternating with dry air masses from the continental east and northeast, and polar maritime and continental air from the European north and subtropical air masses from the African north. The mutual relations of these air masses define the existing radiation-circulation processes, which in the wider region results in the formation of a larger number of local climates that significantly contribute to the formation and development of a large number of very different life forms.

An important factor of biogeographical diversity is the orographic structure of the terrain, which is characterized by the presence of all six energy classes of relief, due to which there are prerequisites for local ecosystem specifics to appear at the local spatial level. The pronounced morphological disintegration of the relief influenced the establishment of multiple differentiation of genetic systems and their communities on the vertical profile of the Dinarides (from 0 to 2600 m). More specifically, in the area of the Dinarides, which is under the influence of the Adriatic Sea, a particular arrangement of genetic and biocenological systems has been formed, which differs significantly in relation to the spatial part, which is under the dominant continental influences.

According to the prominent orographic-climatic specifics, in Bosnia and Herzegovina, orographic units can be distinguished at the zonal level, characterized by a separate structure and spatial distribution phylogenetic and biocenological systems.

The first (and at the same time typically coastal) orographic unit in Bosnia and Herzegovina is represented by the mountain Orjen (1895 m), which has a prominent position and other mountain massifs of this area and represents the first morphological barrier to humid and warm air masses. The Mediterranean is moving towards the north, i.e., towards the mainland of the Balkan Peninsula. The next orographic unit in the Mediterranean area includes the area along the northern edge of the lowest karst fields and consists of mountain massifs (Herzegovinian) Bjelašnica, Crvanj, Velež, Čabulja, Čvrsnica, Prenj, Vran-mountain, and others. According to the continental area, the transitional orographic unit in the zone of the central Dinarides of Bosnia and Herzegovina consists of Vranica, Bitovnja, Bjelašnica, Treskavica, Jahorina, Zelengora, Lelija, Maglić, Volujak, and others. Typical continental orographic units in the area of the inner Dinarides of Bosnia and Herzegovina include Grmeč, Srnetica, Klekovača, Vitorog, Manjača, Ljubišnja, Romanija, Ozren, Zvijezda, Ravan-planina, Čemerska planina, Vlašić, Sljemenska planina, Devetak and others. These four groups of mountains have differently structured and height-positioned ecosystems on the vertical profile, which, in addition to the number of species that make them up, also differ in the overall degree of biodiversity.

8.3 Biogeographical Differentiation of Dinarides

Considering the primary physical-geographical specifics, it can be stated that, in the area of Bosnia and Herzegovina, three groups of sectors can be distinguished on the horizontal profile, including hardwood evergreen temperate-deciduous, mixed deciduous-coniferous, coniferous, and alpine forest vegetation. More specifically, under the influence of the Adriatic climate, along the Adriatic coast, formed hardwood evergreen forests of the class Quercetea illicis. In the immediate coastal zone, in the sub-Mediterranean area, temperate deciduous xerothermic forests of the order Quercetalia pubescentis have been developed, while in the extreme areas of this zone, mesophilic temperate deciduous xerothermic forests of sessile oak and hornbeam (Querco-Carpinetum celaduna ilyricum) and Hungarian oak and Turkey oak (Quercetum conferte-cerris) have been formed. The main vegetation belt in the Dinarides in BiH is formed by temperate deciduous forests of the Querco-Fagetea class, whose altitude continuity consists of the ecosystem of dark coniferous forests (Piceion abietis) and the ecosystems of pine juniper (Pinion mugi).

In accordance with the highlighted facts in the area of the Dinarides of the Balkan Peninsula (and Bosnia and Herzegovina), three biogeographical regions can be distinguished on the macroplane (Fig. 8.1): Mediterranean biogeographical region, Euro-Siberian-Boreo-American biogeographical region, Alpine-High Nordic biogeographical region.

8.4 Mediterranean Biogeographical Region

The Mediterranean biogeographical region includes the southern, southwestern, and southeastern parts of Bosnia and Herzegovina under the dominant influence of the Adriatic (Mediterranean) climate, i.e., the area that, in the historical-geographical context, is spatially differentiated as Herzegovina. From the morphological point of view, this macro-regional unit is differentiated into the following three sub-regional areas: Lower Herzegovina, Western Herzegovina, and Eastern Herzegovina. The Mediterranean biogeographical region in Bosnia and Herzegovina is represented by the Adriatic province, which includes a very narrow belt between the seacoast up to about 400 m above sea level in the south and up to about 300 m above sea level in the north-exposed parts of the terrain. In addition to the aforementioned immediate coastal area, the Adriatic province

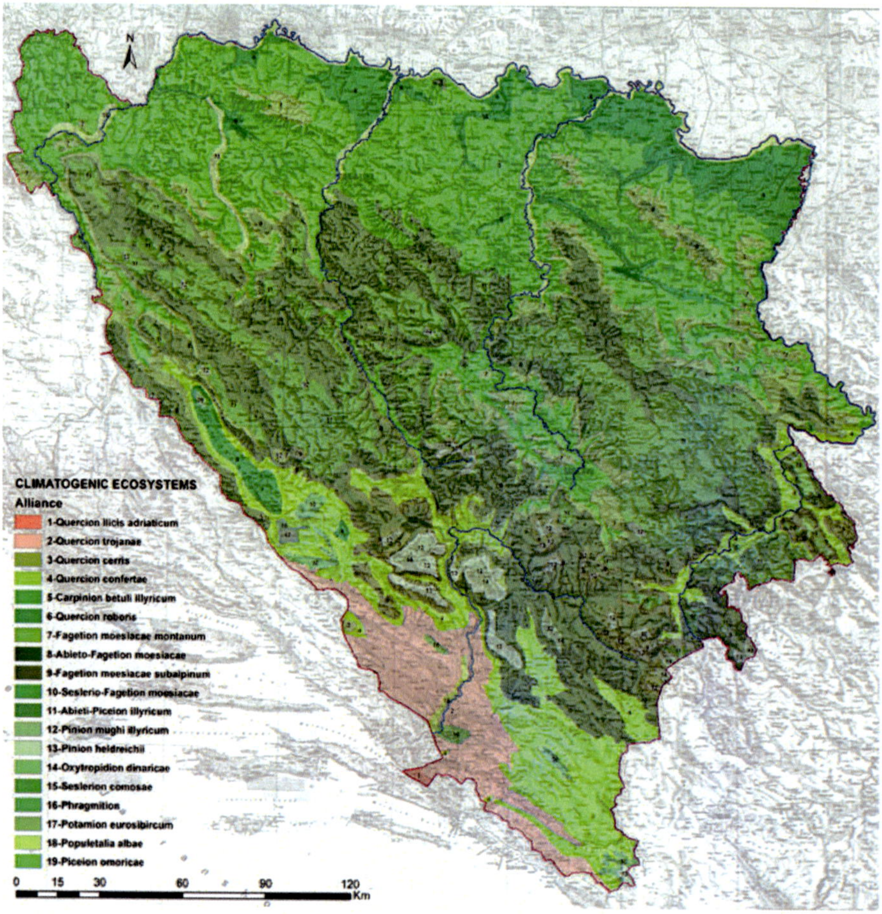

Fig. 8.1 Climatogenic ecosystems of Bosnia and Herzegovina

has the area of low Herzegovina, i.e., the unique area of the Neretva river valley to Mostar's Bijelo polje with Dubravsko and Brotnjansko areas and Popovo polje. From the geological-geomorphological aspect, it is an area with typical karst features, considering that the mineral-petrographic structure of the rock substrate structure is dominated by Cretaceous limestones with a high percentage of $CaCO_3$ (which in some places is over 90%).

The Mediterranean biogeographical region primarily encompasses the Adriatic province in which the hardwood evergreen forests of Orno-Quercetum illicis are climatogenically represented. Circum-Mediterranean species such as Quercus illex, Quercus coccifera, Phyllirea latifolia, Viburnum tinus, Myrtus communis, Pistacia terebinthus, Lauris nobilis, and others play a dominant role in this vegetation. The ecosystems of the Adriatic province are characterized by an almost continuous year-round vegetation period. The maximum of biodynamics is in the spring period, i.e., from April to July, while in the period from July to September, there is a decrease in biodynamics caused by summer drought and high temperatures. The maximum autumn dynamics lasts from September to November, after which low temperatures cause a period of slow biodynamics (from November to March) (Drešković et al. 2011).

8.4.1 Ecosystems of Adriatic Province

Given that the climatogenic forest along the entire Adriatic coast has been significantly degraded under strong anthropogenic pressures (mainly related to deforestation and logging), it has recently been replaced by low thickets or macchia or, even lower, thickets developed in rocky or garrigue. Rocky pastures correspond to the highest degree of forest degradation, while in extreme degradation conditions, the vegetation of Eumediterranean limestone ridges or rocks is formed.

Macchia are thickets containing almost all the elements of climatogenic Eumediterranean forests of evergreen oak and even the evergreen oak itself (Quercus illex), and we include them in the same community with real Eumediterranean forests—Orno-Quercetum Illicis Ht. et Horvatić. Mediterranean garrigues are low thickets dominated by low thickets of the genera Erica and Cistus, after which the order of garrigues on the Adriatic coast was named Cisto-Ericetalia Horvat. The primary producers of organic matter in these developmental stages of Mediterranean vegetation are: Cistus villosus, Erica arborea, Cistus salviefolius, Erica verticillata, Erica multiflora, Juniperus oxycedrus, Juniperus phoenicea, Rosmarinus officinalis, Asparagus acutifolius, and others.

Mediterranean rocky areas most often have semi-desert characteristics, given that the height of vegetation averages from 15 to a maximum of 60 cm, and the vegetation cover per 1 m^2 falls below 50% on average. The primary producers of Eumediterranean rocks are species from the genera: Chrysopogon, Cymbopogon, Andropogon, Brachypodium, Bromus, Koeleria, Festuca, Stipa, etc.; and therefore this vegetation approaches the vegetation of semi-deserts and steppes of the subtropical and tropical zone of the Northern Hemisphere. In addition to species from the family Gramineae (Poaceae), a large number of species from other families, i.e., genera, also occur in rock communities: Teucrium, Salvia, Trifolium, Medicago, Doricnium, Chelychrysum, Ornithogalum, Crocus, Colchicum, Romulea, and others (Encyclopedia of Yugoslavia 1983).

The vegetation of cracks in the limestone rocks of the Eumediterranean belt belongs to the Centaureo-Campanulion Horvatić connection. Considering that the biomass production is very low and that the degree of vegetation cover is below 10%/m^2, the cracks of limestone rocks have practically desert vegetation characteristics. Dominant soils are cracked limestone Mediterranean regosols or cracked Mediterranean black soils, which are most often organomineral. The main producers of this community are species from the genera: Centaurea, Campanula, Crithmum, Limonium, Sesleria, Scorzonera, Seseli, Alyssum, Allium, Inula, Asplenium, and others. Tertiary vegetation consists of species from the following genera: Chenopodietalia, Secalinetalia, Plantaginetalia majoris, and others.

The main producer species of the biocenosis of hardwood evergreen forests of the Eumediterranean belt in Bosnia and Herzegovina are: Quercus illex, Arbutus unedo, Pistacia terebinthus, Viburnum tinus, Rosa sempervirens, Rubia peregrina, Carex distaehya, Asplenium onopteris, Festuca heterophylla, Phyllierea latifolia, Myrthus communis, Rhamnus alaternus, Lonicera implexa, Smilax aspera, Cyclamen repandum, Oryzopsis virescens, Viola reichenbachiana, and others (Stefanović et al. 1983).

Azonal vegetation of the Eumediterranean belt in low Herzegovina includes several communities of class or order, of which we will mention only the class of marsh vegetation Fragmitetea, which covers the wider area of the Neretva delta, and the class of temperate deciduous forests which, through floodplains of Popultelia albea, a belt of hardwood evergreen vegetation (Fig. 8.2).

The most important consumers and reducers of this ecosystem are:

- Mammal: golden jackal (Canis aureus), European rabbit (Oryctolagus cuniculus).
- Birds: Spanish sparrow (Passer hispaniolensis), western black-eared wheatear (Oenanthe hispanica), black-headed bunting (Emberiza melanocephala), olive-tree warbler (Hippolais olivetorum), eastern olivaceous warbler (Hippolais pallida), melodious warbler (Hippolais polyglotta), western yellow wagtail (Motacilla flava), warblers (Sylvia melanocephala, S. cantillans, S. hortensis), and others.
- Reptiles: sheltopusik (Ophisaurus apodus), lizards (Lacerta fiumana, L. melisellensis, L. sicula), Mediterranean house gecko (Hemdactylus turcicus), racer snake (Colluber quator linneatus), common wall gecko (Tarentola mauritanica), horned viper (Vipera ammodytes), and others.
- Amphibians: Rana dalmatina, Bufo viridis, Hyla arborea, Clemmys caspica, and others.
- Insects and arthropods: Egyptian grasshopper (Anacridium aegyptium), oleander hawk-moth (Deilephila nerii), cicadas (Cicada orni, Lyristes plebejus), Italian scorpion (Euscorpius italica), red-headed centipede (Scolopendra morsitans), spiders (Tarantula sp., Lycosa sp., etc.), and soil organisms from the group Collembola: Orchesella biforciata, Odontella lamelifera, Folsomides parvulus, Neanura gneiweri, Odontella armatea and Symphylella vulgaris from the group Symphylla, and others.

8.5 Euro-Siberian-Boreo-American Biogeographical Region

The areas in contact with the extreme northern parts of the Adriatic province belong to the wider transition zone of the Mediterranean to the Euro-Siberian-Boreo-American biogeographical region.

Continuous surface coverage in the spatial distribution of thermal type—warm summer is disturbed in the zones of mountain morphostructures whose altitudes are above 1000 m. More specifically, in the region of Mediterranean Bosnia and Herzegovina, spatial continuity of hot summer disturbs the high Herzegovinian mountains: Prenj, Čvrsnica (Pločno 2225 m), Čabulja (Velika Vlajna 1776 m), Vran (Veliki Vran 2074 m), Velež (Botin 1968 m), Crvanj (Zimomor 1920 m), Šiljevac (1297 m), Sniježnica (Komolj 1263 m), Sitnica (Kobilja glava 1419 m), Radimlja (1136 m), Viduša (Veliki Tisac 1328 m), Ruda planina (1273 m), Orjen (Orjen 1893 m), Hrgud (Veliki Perić 1109), Bjelašnica (Motka 1395 m), Lib-planina (Svinjar 1481 m), Štitar-planina (Štitar 1426 m), Oštrc (1304 m), Midena (Sinjal 1222 m), Zavelim Kolokovac (1347 m), Velika Golija

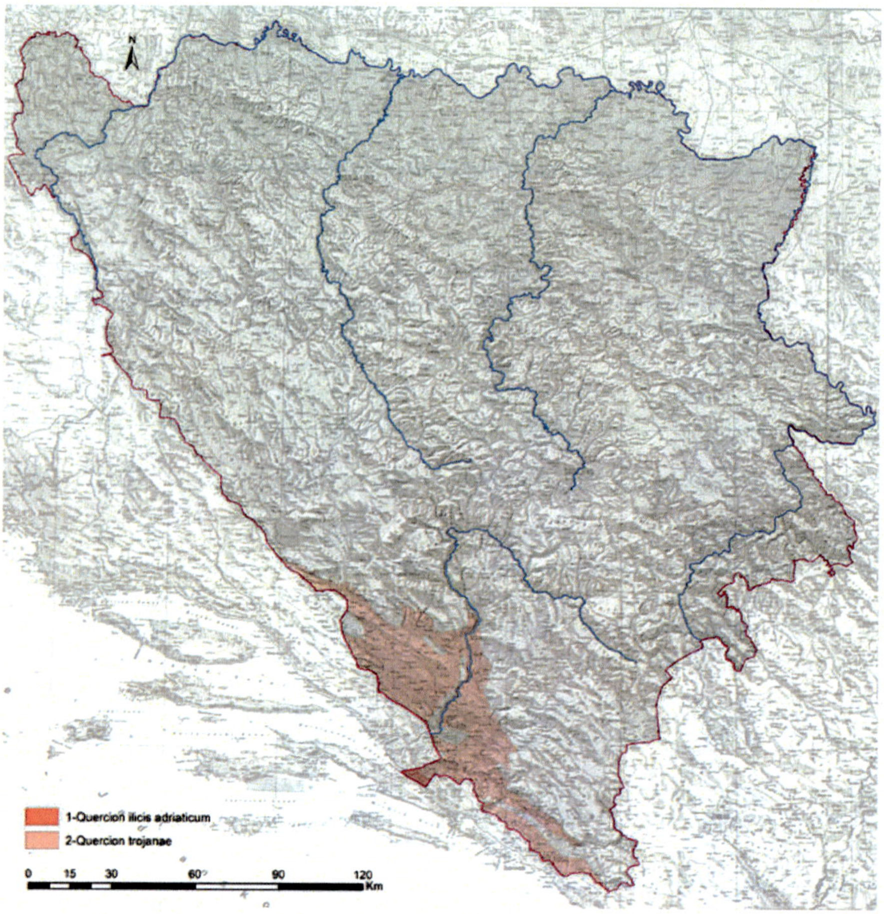

Fig. 8.2 Geographical distribution of climatogenic ecosystems of the Adriatic province

(Veliki vrh 1890 m), Krug-planina (Gagić 1333 m), Košutija glava (1148 m), Broćanac (Kušanovac 1117 m), Jastrebinka (1138 m), and other less pronounced morphostructures. On the mentioned mountain, morphostructures, on the one hand, and deeper towards the continental interior, on the other hand, the climatogenic biocenoses of the Euro-Siberian-Boreo-American region have a dominant vegetation role.

The Euro-Siberian-Boreo-American biogeographical region in Bosnia and Herzegovina is represented by two biogeographical provinces, namely, the Illyrian and Moesian provinces.

The Illyrian province covers the largest part of the Bosnian Dinarides; i.e., it covers a continuous area from the upper border of the hardy evergreen vegetation of the Adriatic province to the lower edge of the High Dinaric province of the Alpine-High Nordic region.

It is differentiated on the vertical and horizontal profile of the Dinarides into a large number of vegetation belts and sub-belts, such as:

- Belt of xerothermic deciduous forests and thickets of sub-Mediterranean parts (Ostryo-Carpinion orientalis adriaticum), with two sub-belts: sub-belt of forests and thickets of oriental hornbeam (Carpinetum orientalis adriaticum) and sub-belt of black hornbeam forests and thickets (Seslerio-Ostryetum carpinifoliae).
- Belt of mesophilic oak-hornbeam forests (Querco-Carpinetum betuli illyricum).
- Belt of beech forests (Fagion illyricum) with sub-belts: thermophilic beech forests (Seslerio-Fagetum illyricum), mountain mesophilic beech forests (Fagetum illyricum montanum), mountain beech-fir forests (Abieto-Fagetum ilyricum), and subalpine beech forests and thickets (Aceri-Fagetum illyricum).
- Belt of xerothermic pine forests (Pinion heldreichii and Pinion peucis).
- Belt of dark coniferous spruce and fir forests (Vaccinio-Piceion) with two sub-belts: sub-belt of mounted dark coniferous forests (Piceetum abietis montanum, Abieto-Piceetum, Blechno-Abietum, Calamagrosti-Abietum) and sub-belt of subalpine dark coniferous forests (Piceetum abietis subalpinum).
- Belt of low pine juniper thickets (Pinetum mugi illyricum).

The Moesian province covers the southeastern parts of Bosnia and Herzegovina. The following vegetation belts and sub-belts are differentiated on the vertical profile of this province:

- Belt of thermophilic forests of Hungarian oak and Turkey oak (Quercion fanetto).
- Belt of mesophilic sessile oak forests (Quercion petraeae).
- Belt of beech forests (Fagion moesiacae) with sub-belts: sub-belt of mounted Moesian beech forests (Fagetum moesiacum montanum), sub-belt of Moesian beech-fir forests (Abieto-Fagetum moesiacum), and sub-belt of subalpine Moesian beech forests (Aceri-Fagetum moesiacum subalpinum).
- Belt of light coniferous forests on peridotites and serpentines (Pinion austriacae).
- Belt of thermophilic spruce forests (Piceion omorikae).
- Belt of mesophilic spruce and fir forests (Vaccinio-Piceion) with two sub-belts; sub-belt of mounted dark coniferous forests (Piceetum abietis montanum, Abieto-Piceetum, Blechno-Abietum, Calamagrosti-Abietum) and sub-belt of subalpine dark coniferous forests (Piceetum abietis subalpinum).
- Belt of Moesian juniper pine (Pinetum mugi moesiacum).

8.5.1 Ecosystems of the Illyrian and Moesian Provinces

The Ecosystem of sub-Mediterranean xerothermic temperate-deciduous forests (Quercetalia pubescentis Br.-Bl.) is characterized by specific production dynamics that reach the optimum in spring and early summer. Increased biodynamics of the communities of this system (especially in the stratum of herbaceous plants) is also present during the autumn period and is caused by a large amount of precipitation and moderate air temperatures. Therophytes and geophytes play the dominant production role from the following genera: Crocus, Colchicum, Euphrasia, Odontites, Cyclamen, and others. The lowest biodynamics characterizes late summer to early autumn due to extremely high air temperatures and intense drought. The vegetation of this ecosystem belongs to the class Querco-Fagetea No.-Bl. et Vlieger, or its order Quercetalia pubescentis Br.-Bl. Xerothermic oaks of temperate deciduous forests cover two quite demarcated areas and two quite clearly demarcated habitat types and are phytocenologically differentiated into two connections: Ostryo-Carpinion orientalis Horvat et Horvatić and Quercion farnetto Horvat.

The vegetation communities of both connections include xerothermic forests and thickets of the sub-Mediterranean and Mediterranean-mountain belt from about 200 to about 1200 m above sea level. The most important edificatory species in vegetation communities on limestone soils of this ecosystem are: Carpinus orientalis, Petteria ramentacea, Quercus pubescens, Fraxinus ornus, Coronilla emoroides, Ostrya carpinifolia, Acer monspessulanum, Quercus macedonica, Quercus cerris, Punica granatum, Palirus aculeatus (australis), and others (Bakaršić 1957).

Soils on a dominant dolomite substrate are characterized by less pronounced differentiation of developmental phases, absence of ilimerized soil, the rare occurrence of brown dolomite soil, and less variation of hydrothermal, mineralogical, and pH regimes. Most of the soil is water permeable, which, depending on the pH value, results in a basic or neutral reaction, and which causes the appearance of more xerothermic and basophilic phytogenetic and biocenological systems on them, such as Erico-Pinetalia pine forests, which precede thermophilic temperate-deciduous forests of the Carpinion orientalis connection.

Within this ecosystem, the most important edificatory species of dolomite soils are Pinus nigra ssp. austriaca, Pinus nigra ssp. dalmatica, Pinus silvestris, Erica carnea, Fraxinus ornus, Cotinus coggygria, Peucedanum austricum, Peucedanum oreosolinum, and endemic species Euphorbia herzegovina, Thymus aureopunctatus, Alyssum moelendorfianum, Dianthinos onjuius, and Dianthinos onjuius.

The hydrothermal regime on silicate soils is quite favorable, which is why more mesophilic forests have been developed on these lands, such as chestnut (Caastanea sativa), Hungarian oak (Quercus farnetto), sessile oak (Quercus petraea), and others. The most important indicator species on these soils are Castanea sativa, Quercus cerris, Quercus farnetto, Quercus petraea, Genista tinctoria, Crytisantus nigricans, Crytisantus hirstutus, Crytisantus hirstutus, Luzula sp., and others. Under the influence of pedogenetic factor in acidic sub-Mediterranean silicate soils, there are forests of chestnut, sessile oak, and Hungarian oak with several acidophilic elements in the stratum of herbaceous plants, such as species from the genera Luzula, Deschampasia, Festuca, Veronica, Pteridium, and others.

Diverse thermal-pluviometric regime conditions further vegetation differentiation on the vertical and horizontal profile of the sub-Mediterranean belt on practically the same geological base and the same soil types differentiation into two subbands (vegetation connections): sub-belt of eastern hornbeam forests and thickets (Capinion orientalis Bleč. et Lakušić) and sub-belt of hop hornbeam forests and thickets (Orneto-Ostryon Tomažić).

The sub-belt of eastern hornbeam forests and thickets differs in height into a series of associations, which, in contact with the evergreen forests of the Adriatic Province, begin with the association Rusco-Carpinetum orientalis Bleč. et Lakušić 67. The height continuity is represented by the association Aceri-Carpinetum orientalis Bleč. et Lakušić

67, which most often occurs in the form of smaller "vegetation islands" in the belt of mesophilic oak forests.

The sub-belt with hop hornbeam is differentiated on the vertical profile into several associations, the most important of which are: the Mediterranean-Montane community (Seslerio angustifolii-Ostryetum carpinifoliae Lakušić 70), which predominantly develops in river canyons in continental Bosnia and Herzegovina. These areas are inhabited mainly by steep and north-exposed limestone slopes, with a slope of over 45°. According to the genesis of the soil, they are most often organogenic or organomineral and are represented by Mediterranean-montane limestone black soil. The average annual temperatures in its habitat are around 8 °C, and the average annual relative humidity is between 65% and 70%. The main edificatory species of this community are Ostrya carpinifolia, Fraxinus ornus, Quercus daleschampi, Cytisus bosniacus, Sesleria angustifolia, Cotoneaster tomentosa, Amelanchier ovalis, Dianthus kitaibelii, Scabiosa leucophylla, and others. The autumn hornbeam sedge community (Seslerio autumnalis-Ostryetum carpinifoliae Horvat et Horvatić) develops in average annual temperatures of 9 and 12 °C, and average annual relative humidity of 55% and 60%. The communities of sedge and hop hornbeam (Seslerio angustifolii-Otryetum carpinifoliae Lakušić 70) develop in this sub-belt in thermally pluviometrically slightly modified conditions, which requires a certain vegetation modification.

Geological and pedological factors cause the differentiation of the vegetation of thermophilic deciduous forests to: vegetation connections on limestone soils represented by: (Carpinion orientalis Bleč. et Lakušić) and (Orneto-Ostryon Tomažić), vegetation bundles on dolomite soils represented by: (Orno-Ericion dolomiticum Horvat), vegetation ties on limestone soils represented by: (Quercion frametto Horvat, Castanion sativae, Quercion robori-petraeae).

Each of these connections is characterized by a series of vegetation associations formed on soils formed on an adequate geological basis. Facies of temperate-deciduous thermophilic forests with black and white pine facies develop on sierozem and shallow rendzina, while thermophilic forests and scrub bushes on warmer and black hornbeam and autumn thickets develop on deeper rendzina and brown dolomite soils.

Vegetation of thermophilic temperate-deciduous forests of the sub-Mediterranean belt on silicate rocks is differentiated into four habitat types based on the hydro-thermal regime: warmest and wettest with pedunculate oak forests (Quercion scutariensis), the warmest and driest with carp forests (Carpinion orientalis), the coldest and driest with forests of black hornbeam and black ash (Orneto-Ostryon), the coldest and wettest with pedunculate and sessile oak forests (Quercion robori-petraeae). Each of these communities forms a separate climatogenic complex with a vegetation structure primarily defined by a specific hydro-thermal regime.

The secondary vegetation of the ecosystem of sub-Mediterranean xerothermic temperate-deciduous forests consists of rocky areas belonging to the order Scorzonero-Chrysopogonetalia Horvatić and which, depending on the hydro-thermal regime of the habitat, are further differentiated into several connections and associations. The most widespread are associations belonging to Satureion subspicatae Horvat et Horvatić and Scorzonerion villosae. The main edificatory species in sub-Mediterranean and Mediterranean-montane rocks associations are Satureia subspicata, Satureia cuneifolia, Teucrium montanum, Carex humilis, Globularia cordifolia, Bromus erectus sl., Salvia officinalis, Stipa pennata, and Scovzonera Feather, Scorzonera.

Tertiary vegetation of ecosystems of sub-Mediterranean xerothermic deciduous forests depending on the concentration of nitrates in the habitat and the intensity of anthropogenic pressures belongs to the orders: Chenopodietalia, Planteginetalia majoris, and Secalinetalia. The main edificatory species of nitrophilous vegetation of the order Chenopodietalia are Chenopodium spp., Carthamus lanatus, Urtica dioica, U. urens, U. pillulifera, Capsella rubella, and others. The vegetation of trampled habitats of the order Plantaginetalia majoris in this zone is dominated by species Plantago major, Polygonum aviculare, Lolium arvense, Agropyrum mediteraneum, and others. The vegetation of the habitat of the order Secalinetalia is dominated by the species Trifolium arvense, Convolvulus arvense, Galeopsis sp., Rhynanthus sp., Veronica sp., and others.

The consumer-reducing component of this ecosystem is characterized by significant wealth from various plant and animal groups. Of the plant components, the species from the genera stand out: Orobanche, Limodorum, and Hymantoglossum. Of the faunal consumers, the following groups of animals play a significant role. Mammals: European badger (Meles meles), hedgehog (Erinaceus europaeus), yellow-neced mouse (Apodemus flavicolis), wolf (Canis lupus), fox (Vulpes vulpes), the European mole (Talpe europaea), the lesser white-toothed shrew (Crocidura suaveolens), forest dormouse (Dryomys mitedula) etc., and of domestic animals' donkey (Equs asinus), and goat (Capra domestica). Birds: the sombre tit (Parus lugubris), Syrian woodpecker (Dendrocopus syriacus), the Levant sparrowhawk (Accipiter brevipes), the European turtle dove (Streptopelea tutur), the Eurasian collared dove (Streptopelia decaocto), the European golden oriole (Oriolus oriolus), the European green woodpecker (Picus viridis), common nightingale (Luscinia megarhynchos), the Eurasian scops owl (Otus scops), the house sparrow (Passer domesticus), European magpie (Pica pica), jackdaw bird (Coleus medula), rock dove (Columba livia), and many others. Reptiles: This group is very numerous in the belt of sub-Mediterranean and Mediterranean-mountain forests and thickets. The most important species are: lizards (Lacerta trilineata, L. praticola, L. viridis i L. agilis), sheltopusik (Ophisaurus apodus), horned viper (Vipera amodites), Aesculapian snake (Elaphe longissima). Amphibians: turtles (Testudo graeca, T. hermani), frogs (Hila arborea and Bufo viridis), salamander (Salamandra maculosa), and others. Insects: Insects are present in the area of climatogenic thermophilic deciduous forests with a large number of families and genera. Of the species that occur exclusively in this ecosystem or find the optimum in it, we will list the following: the ant lion (Myrmeleon formicarius, Ascalaphus sp.), the common ant (Myrmica ruginoides), scarce swallowtail (Papilio podalirius), the sloe emperor moth (Saturnia pyri i S. spini), the oak processionary moth (Thaumatopoea processionea), cockchafer (Melontha pectoralis), the European stag beetle (Lucanus cervus), the great capricorn beetle (Cerambyx cedro, C. scopoli), and others (Fig. 8.3).

The ecosystem of mesophilic temperate-deciduous oak and hornbeam forests (Carpinion betuli illyricum Horvat) is predominantly represented in the foothills of the central and continental Dinarides of BiH. More specifically, the vegetation communities of this belt were developed in the hypsometric belt of 300 and 800 m above sea level, where the Cfb climate type dominates. The lower vegetation boundary of the biocenoses of this belt consists of floodplain forest communities, while the upper boundary is mixed with the belt of montane beech forests. Fragments of sessile and hornbeam vegetation also occur azonally, in the belt of sub-Mediterranean vegetation, on moist habitats and deep soils (Querco-Carpinetum submediteraneum). The biocenosis of Sessile oak and hornbeam is the most organized biocenosis in Bosnia and Herzegovina, in which a large number of plant and animal species live in very close and complex interactions that often accompany the occurrence of parasitism, semi-parasitism, symbiosis, and other forms of association.

The dominant type of geological substrate in the belt of mesophilic deciduous forests consists of rock masses built of dolomite and limestone, marl and clay, to peridotite, diabase, melaphyre, quartz porphyrite, and quartz sandstone. The pronounced mineralogical-petrographic variation of the parent rock substrate also conditioned the more pronounced

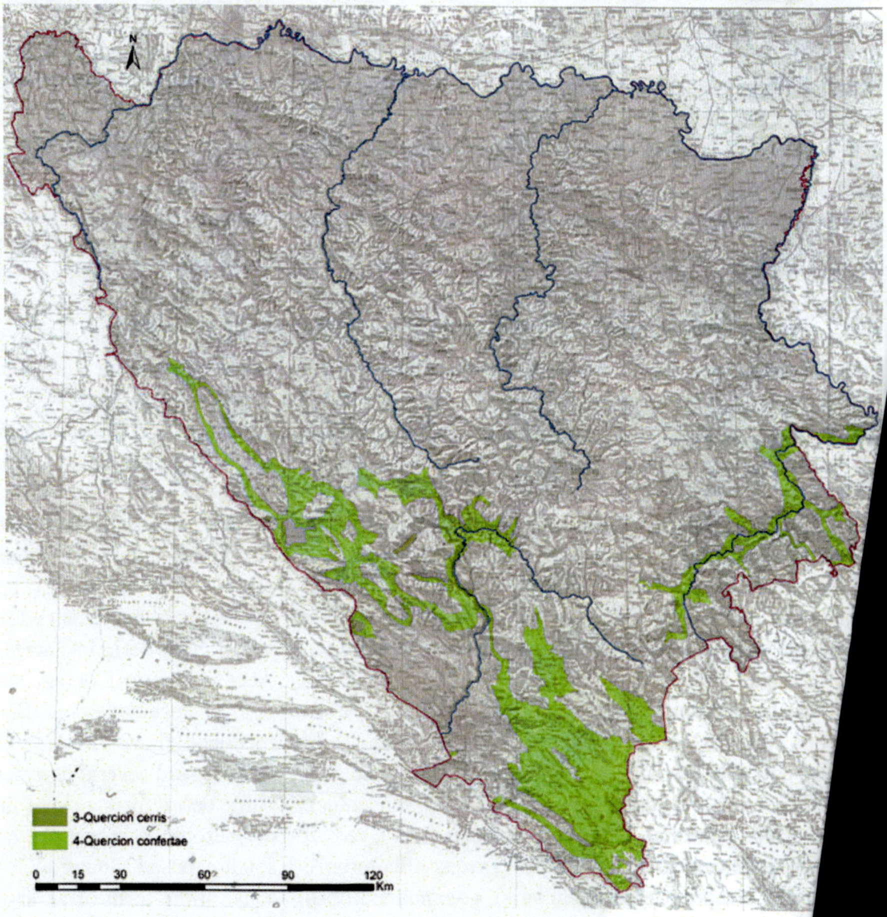

Fig. 8.3 Geographical distribution of climatogenic ecosystems of sub-Mediterranean xerothermic deciduous forests

variation of the parent pedological substrate, which is then accompanied by an adequate variation of biocenosis differentiations. On dolomite and shallow limestone soils in this belt, continental esclaves of sub-Mediterranean vegetation of the Orno-Ostryon Tomažić connection were formed on acidic and moist soils of pedunculate and Sessile oak forests. On moderately moist soils of limestone and marl series, i.e., on hilly brown limestone soil types, on hilly brown soils on marls and clays, and mountain parapodzol (or pseudogley), and tertiary sediments (marls, clays, shales) and common hornbeam (Querco-Carpinetum illyricum Horvat).

The main edificatory species of phytocenosis of mountain mesophilous forests are Quercus petraea, Carpinus betulus, Acer campestre, Prunus avium, Acer tataricum, Corylus avellana, Ligustrum vulgare, Viburnum lantana, Anemone nemorosa, Primula vulgaris, Pulmonaria officinalis, Symphytum tuberosum, Potentilla micrantha, and others. The secondary vegetation of the ecosystem of mesophilic deciduous forests is represented by the valley mesophilic meadow connections Arrhenatherion elatioris Br.-Bl. sativa, Knautia arvensis, Holchus lanatus, and others. It is an ecosystem of mowed meadows created and maintained under the 'nfluence of continuous mowing or intensive grazing. Thanks ı the presence and a high number of species of nutritious ısses, this ecosystem is one of the most useful for the 'elopment of animal husbandry (Lakušić 1981).

'he tertiary vegetation of this ecosystem was formed ' very intense anthropogenic influences on the primary ːcondary vegetation, respectively. Characteristic ter- 'egetation units in this ecosystem are Centaurion 'onvolvulion arvensis, Agropyro-Rumeio, Polygonin ', Bidention, Menthion pulegii, Senecion fluviatilis, ın, Onopordion acanthii, and others.

ısumer-reducing component of the biocenosis of deciduous forests is characterized by an excep- h of plant and animal species. Among the are numerous edible and poisonous fungi and 'uch as Lathrea squamaria, Neotia nidus avis, and others. Of the faunal consumers, the ɔs of animals play a significant role. /ulpes vulpes), wolf (Canis lupus), hedge- 'opaeus), badger (Meles meles), European (Glis glis), yellow-necked mouse 'is), wild boar (Sus scrofa), roe deer), red deer (Cervus elaphus), European aea), bank vole (Cleothrionomys use (Muskardinus avellanarius), wild- d of domestic animals: horse (Equs), sheep (Ovis domesticus), and goat : redhead (Erithrocus rubeculus), blue tit (Parus coreuleus), redtail blackbird (Turdus ericotorum), a familiaris), common Eurofish Columba oenas, Coccothrantes coccothrantes, Buteo buteo, Picus canus, Tardus merula, and others. Reptiles: Lacerta agilis, Anguis fragilis. Amphibians: Hyla arborea, Rana fragilis. Insects: the Old World swallowtail (Papilio machaon), the small emperor moth (Saturnia pavonia), the gypsy moth (Lymantria dispar), the common cockchafer (Melolontha melolontha), longhorn beetle (Leptura virens, Morimus funereus, Prionus cariarius, Rosalia alpina), locusts in secondary vegetation (Gonophocerus rufus, Chorthippus longicornis, Ch.dosatus, Odontopodisma sp.), and others (Fig. 8.4).

Structure and biodynamics, horology, and chronology of *beech forest belt ecosystems* (Fagion moesiacae Bleč. et Lakušić) represent continuity in the evolution and dynamics of oak-hornbeam forests. Still, due to somewhat reduced ecological conditions, biomass production in this zone is reduced. More specifically, the biocenosis of climatogenic beech-fir forests has the highest biomass production of this belt, with the same declining in the sub-belt of mounted beech forests and, further, towards higher hypsometric zones (Lakušić 1981).

Under the influence of different hydro-thermal regimes, the ecosystem of the beech forest belt in Bosnia and Herzegovina is differentiated into the following four sub-belts: sub-belt of thermophilic Mediterranean-mounted beech forests (Seslerio-Fagetum moesiaceae), sub-bet of mesophilic montage forests of beech (Fagetum moesiacae montanum), sub-belt of beech and fir mountain forests (Abieto-Fagetum moesiacae), and sub-belt of subalpine beech forests (Aceri-Fagetum subalpinum) (Stefanović et al. 1983).

Thee ecosystem of the sub-belt of thermophilic Mediterranean-montane beech forests (Seslerio-Fagetum moesiacae Horvat, Aceri obtusati-Fagetum moesiacae Fabijanić, Fukarek et Stefanović) covers a wide area of the outer and central Dinarides of BiH. Its lower vegetation boundary is formed by the sub-belt of black hornbeam and autumn sedge (Seslerio autumnalis-Ostryetum carpinifoliae Horvat et Horvatić). In contrast, its upper vegetation is formed by the sub-belt of subalpine Bosnian pine forests (Pinion heldreichii Horvat). The community of Aceri Obtusati-Fagetum moesiacae represents the continental exclave of this sub-belt (Fig. 8.5).

The most significant edificatory types of this sub-belt, i.e., associations Seslerio-Fagetum moesiacae su: Fagusmoesiaca, Acer hyrcanum, Corylus colurna, Ostrya carpinifolia, Viburnum maculatum, Sesleria autumnalis, Melitis melissiphyllum, Calamintha officinalis, Clinopodium vulgare, and several others. Continental exclaves of thermophilic beech forests occur in the belt of mesophilic montane or mesophilic mountain beech-fir forests on steeper and south-exposed slopes with shallow limestone or dolomite soils, in the form of an association Aceri obtusati-Fagetum

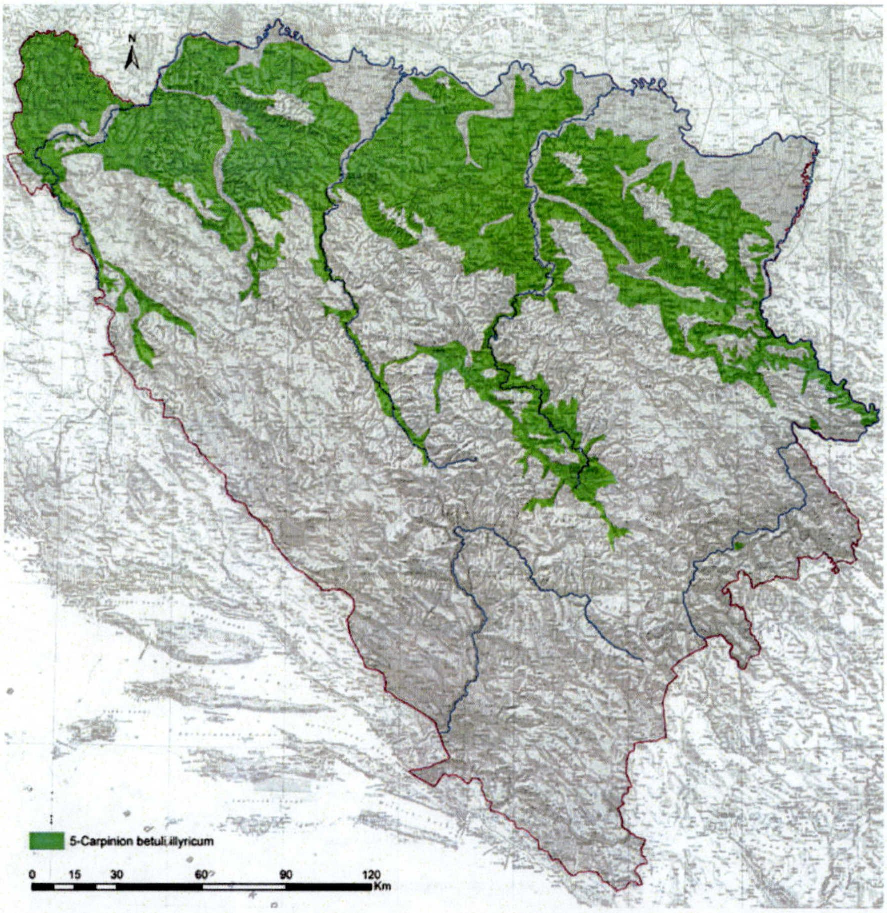

Fig. 8.4 Geographical distribution of climatogenic ecosystems of mesophilic deciduous oak and hornbeam forests

moesiacae. The edificatory types of this association are: Acer obtusatum, Fagus moesiaca, Ostrya carpinifolia, Acer campestre, A. pseudoplatanus, A. platanoides, Fraxinus ornus, Viburnum lantana, Ligustrum vulgare, Crataegus monogyna, and others. Depending on the type of geological base, both associations of thermophilic beech forests are differentiated against the dominant geological base in two variants: limestone (Seslerio-Fagetum moesiacae calcicolum and Aceri obtusati-Fagetum moecum moesiacae calcicolum) and dolomite (Seslerio-Fagetum moesiacae dolomiticum and Aceri obtusati Fagetum moesiacae dolomitikum).

The sub-belt of mesophilic beech forests (Fagetum moesiacae montanum Horvat) covers a large area of the central and continental Dinarides of BiH, in a wider morphological zone of transition from the upper parts of the mountain to the lower parts of the mountain belt. The main edificatory species of this ecosystem are: Fagus moesiaca, Acer pseudoplatanus, Fraxinus excelsior, Corylus avellana, Crataegus sp., Asarum europaeum, Sanicula europaea, Cardamine bulbifera, Cardamine enneaphyllos, Anemone nemorosa, and others. Depending on the parent rock substrate and soil, the montage beech forest ecosystem is differentiated into two geological variants: Fagetum moesiacae montanum calcicolum and Fagetum moesiacae montanum silicicolum. The ecosystem of Fagetum moesiacae montanum calcicolum is characterized by a carbonate structure on which brown mountain limestone soils are formed, a favorable hydrothermal and pH regime, and a much more diverse biocenosis that has a large number of species (Encyclopedia of Yugoslavia 1983).

The ecosystem of Fagetum moesiacae montanum silicicolum differs markedly from the previous ecosystem in terms of the structure and dynamics of the biocenosis and physicochemical properties of the parent rock substrate and soil types, which is why it is often classified as a particular vegetation class Quercetea robori-petraeae, i.e., order Quercetal Luzulo-Fagion connection. However, taking into account the fact that the differentiation is based on the geological structure of the terrain, these two types of ecosystems are characterized by their biocenological and other specifics, due to which they are recognizable and adequately typified (Fig. 8.6).

The ecosystem of the sub-belt of mountain beech-fir forests (Abieto-Fagetum moesiacae Horvat) was developed

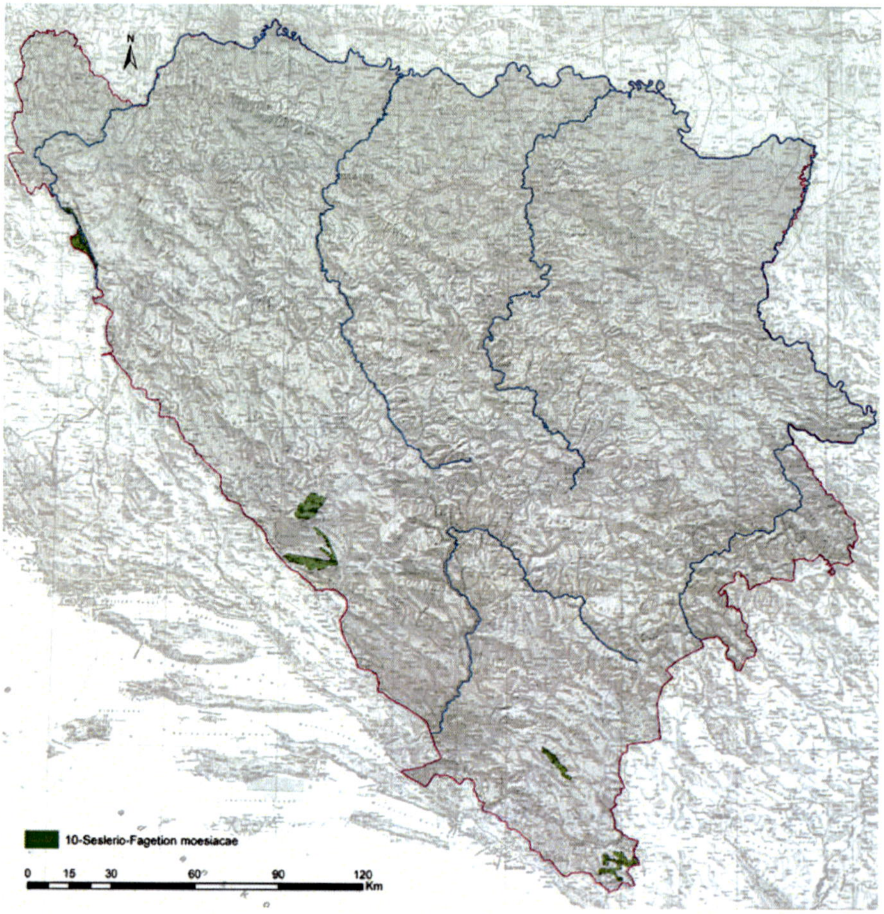

Fig. 8.5 Geographical distribution of ecosystem of the sub-belt of thermophilic Mediterranean-montane beech forests

in the central and continental Dinarides in BiH (most often between the ecosystems of mountain beech and subalpine beech forests). The climatic specifics of this area indicate the characteristics of the Cfc climate type—moderately warm and humid climate with fresh summer. The average annual temperatures in it vary between 5 and 7 °C, while the absolute temperature extremes reach values of −35 °C (absolute minimum air temperatures) and 30 °C (absolute maximum air temperatures). The average annual relative humidity varies between 75% and 85%, with the hygienic regime (similar to the previous vegetation sub-zone) ranging from 50–80% to 100% (from October to June).

The heterogeneity of the geological and pedological substrate has led to a marked differentiation of the beech-fir forest ecosystem into two variants: Abieto-Fagetum moesiacae calcicolum and Abieto-Fagetum moesiacae silicicolum. Both variants of the optimum are found in similar ecological conditions, which reduces their phytocenological differentiation.

The main edificatory species of this vegetation sub-zone are: Abies alba, Fagus moesiaca, Acer platanoides, Ulmus montana, Fraxinus excelsior, Lonicera coreulea, Daphne mezereum, Sambucus racemosa, Festuca drymeia, Prenanthes purpurea, Aremonia agrimonoides, Elymus europaeus, Brachypodium silvaticum, Festuca silvatica, and others. In the limestone variant, there are a large number of species from the same variant of the sub-zone of mounted beech forests; i.e., in the silicate variant, many acidophilic species from the same variant of the previous vegetation zone also appear: Luzula silvatica, L. pilosa, L. nemorosa, L. luzulina, Veronica officinalis, Vaccinium myrtillus; then various species of mosses from the genera Polytrichum, Rhytidiadelphus, Hylocomium, Mnium, Bryum, and others (Drešković et al. 2011) (Fig. 8.7).

The subalpine beech forest sub-belt ecosystem (Aceri-Fagetum moesiacae subalpinium) represents the altitude continuity of the beech-fir forest ecosystem, mainly in the central Bosnian Dinarides. Only in places within smaller isolated areas, this ecosystem appears in the mountainous regions of the outer Dinarides. The pine juniper ecosystem forms the upper limit of this sub-belt (Fig. 8.8).

Mean annual air temperatures in the subalpine beech forest ecosystem area vary in the range of 3–6 °C and are in line with the general thermal characteristics of the Dfc climate.

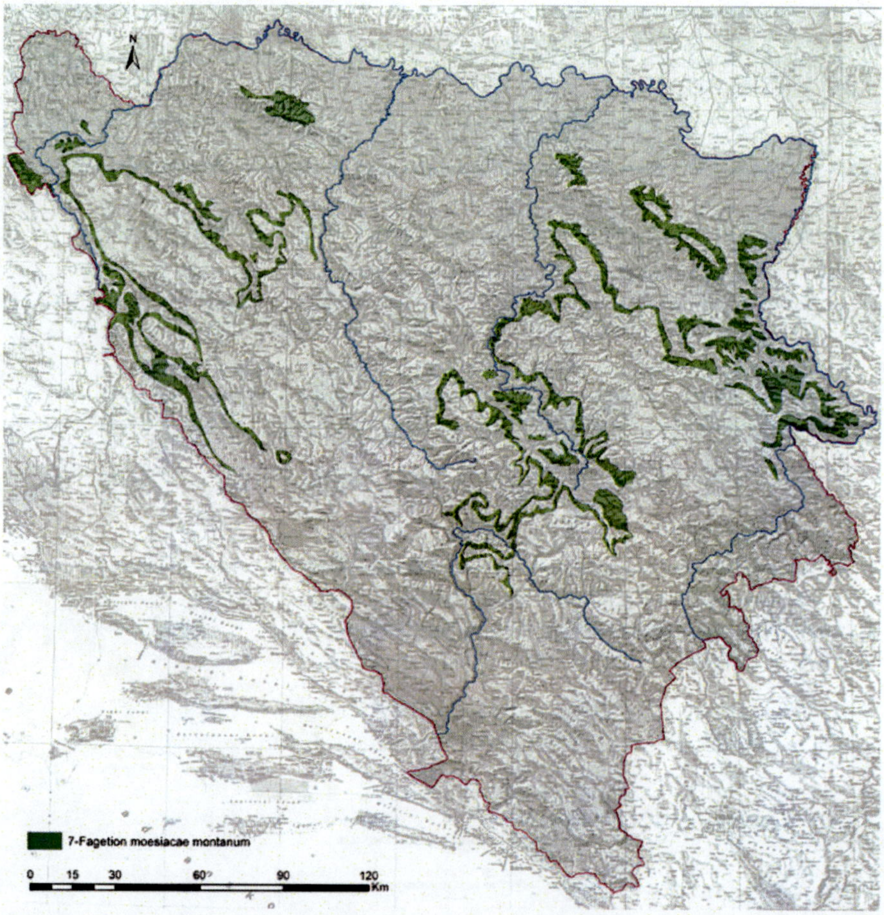

Fig. 8.6 Geographical distribution of ecosystem of the sub-belt of mesophilic beech forests

Absolute maximum temperatures are up to 25 °C, while the minimums reach up to −35 °C. The average annual relative humidity is from 70% to 80%, with the vegetation period lasting shorter and characterized by physiological drought, which has a negative impact on the production of total biomass. The heterogeneous geological structure of the terrain in this sub-belt conditioned the differentiation of this sub-belt into two separate mineralogical-petrographic spatial units, i.e., areas with dominant limestone and volcanic acid silicate rock masses, between which there are transitional varieties such as basic volcanic or acidic sedimentary rocks (Dalmatin et al. 2008).

Compared to the geological base, soil types are characterized by less diversity, which is also conditioned by unfavorable pedogenetic factors. More specifically, the most common soils of this sub-belt are subalpine limestone soils (organogenic and organomineral, and brownized), subalpine humus silicate, subalpine acid humus silicate, and subalpine acid brown soils. Depending on the specific geological and pedological relationships, subalpine forests and thickets are differentiated into two ecosystems, which actually represent different biocenoses: Aceri-Fagetum moesiacae subalpinum calcicolum and Aceri-Fagetum moesiacae subalpinum silicicolum.

The ecosystem of moesiacae subalpinum calcicolum develops on a limestone rock substrate and adequate subalpine limestone black soils and subalpine rendzina on moraines. The main edificatory species of this phytocenosis are Fagus moesiaca f. subalpina, Acer pseudoplatanus f. subalpina, A. heldreichii var. pančićii, Rhamnus fallax, Lonicera alpigena, Daphne mezereum, Cardamine savensis, Valeriana tripteris, Ranunculus platanifolius, R. lanuginosus, Sorbus aucuparia f. glabrata, and others.

The ecosystem of Aceri-Fagetum moesiacae subalpinum silicicolum develops on silicate soils formed on basic and acidic parent rock substrate. In accordance with the above, geological-pedological subvariants can be produced in this ecosystem on subalpine humus silicate and subalpine humic acid silicate soils. This system's highly organized communities are less likely to develop on subalpine acidic brown or subalpine podzolic soils. The leading edificatory species of this phytocenosis are Fagus moesiaca f. subalpina, Sorbus aucuparia f. glabrata, Acer heldreichii, Lonicera

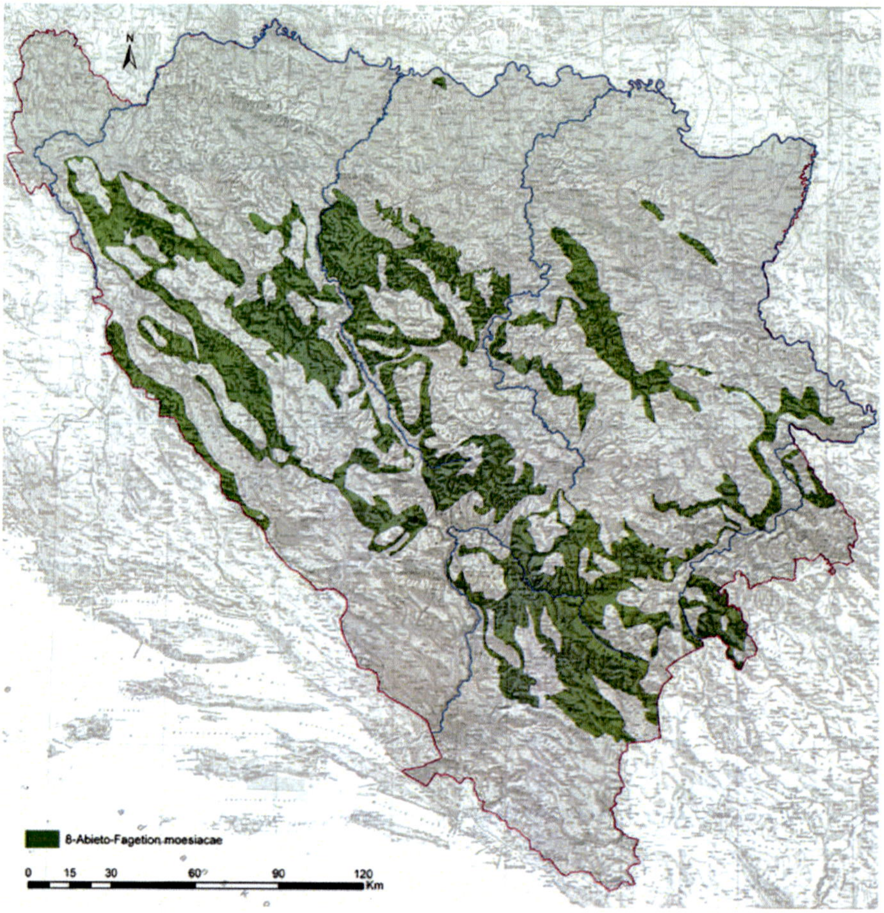

Fig. 8.7 Geographical distribution of climatogenic ecosystems of the sub-belt of mountain beech-fir forests

borbaschiana, Daphne mezereum, Rubus idaeus, Vaccinium myrtillus, Luzula pilosa, L. nemorosa, and others.

Following the prevailing hydro-thermal regime, subalpine beech forests are differentiated into two sub-associations: Aceri-Fagetum moesiacae subalpinum silvicolum and Aceri-Fagetum moesiacae subalpinum patulectorum.

Aceri-Fagetum moesiacae subalpinum silvicolum sub-association develops in the lower part of the subalpine belt in warmer and wetter habitats, which is why it shows significant biocenological similarity to the Aceri-Fagetum moesiacae calcicolum association. In drier and colder habitats, the Aceri-Fagetum moesiacae subalpinum patulectorum sub-association develops in the middle and upper part of the subalpine belt and, consequently, shows significant floristic similarity with the Pinetum mugi illyricum calcicolum community (Lakušić 1981).

The secondary vegetation of the beech forest belt is mainly represented by mesophilic meadows belonging to the Pančićion Lakušić association in the southeastern part of the Dinarides, and the Cynosurion Tx, and Bromion erecti association in the northwestern Dinarides. The main edificatory types are: Pančićia serbica, Poa alpina, Phleum alpinum, Polygonum bistorta, Trisetum flavescens, Alchemilla vulgaris s. l., Knautia sarajevensis, Anthoxanthum odoratum, Festuca fallax, Lilium bosniacum, Carum carvi, Silene sendtneri, Viola elegantula, Orchis sambucina, Primula intricata, Trifolium pratense, T. repens, T. ochroleucum, Crepis conyzifolia, Leontodon autumnalis, Centaurea jacea, C. crnogorica, Campanula patula, C. glomerata, Rhynanthus rumelicus, Pedicularis hermanniana, and others.

Tertiary vegetation in the hypsometrically lower areas of this belt consists of associations of the order Chenopodietalia, Plantaginetalia majoris, and Secalinetalia. In the central hypsometric parts of the belt, associations are represented: from the order Chenopodietalia (alliance Chenopodion subalpinum Br.-Bl.), from the order Plantaginetalia majoris (alliance Plantaginion reniformis), and from the order Secalintalia (alliance Centaurion cyani Lakušić). In the hypsometrically highest parts of this belt, the most widespread associations from tertiary vegetation are from alliance Chenopodion subalpinum and alliance Plantaginion reniformis.

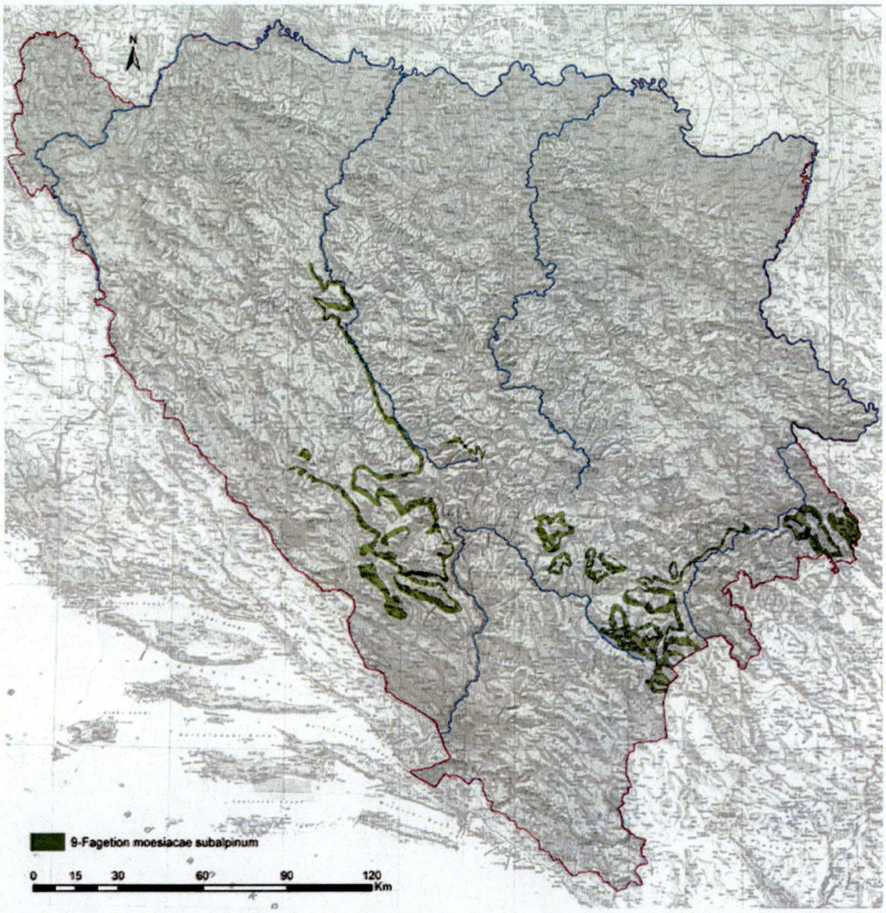

Fig. 8.8 Geographical distribution of climatogenic ecosystems of subalpine beech forests sub-belt

The consumer-reducing component of the biocenosis of beech forests is very similar to those from the biocenosis of mesophilic oak-hornbeam forests.

Of the faunal consumers, the following groups of animals play a significant role: Ursus arctos (opt.), Sus scrofa (opt.), Glis glis (opt.), Felis silvestris (opt.), Columba oenas, Erithacus rubecula, Picus canus, Buteo buteo, Anguis fragilis, Lacerta fragilis, Hyla arborea, Rana agilis, Papilio machaon, Saturnia pavonia, Leptura virens, Rosalia alpina, and others. The dynamics of the biocenosis are differentiated into three main aspects: spring, summer, and autumn. It is best expressed during the summer in the ecosystem of the sub-zone of mountain beech forests and the weakest in the ecosystem of subalpine beech forests and thickets.

The ecosystem of dark coniferous forests of the Balkan Peninsula (Piceion abietis) is represented by the southernmost exclaves of the Euro-Siberian taiga zone. More specifically, in the central Dinarides of Bosnia and Herzegovina area, they include the northern exposures of the mountain and subalpine belt. They are most widespread within the hypsometric zone from 1000 to 1800 m above sea level, with the proviso that, in some places, they can also occur within the lower (up to 800 m) or higher (up to 2000 m) hypsometric floors (Stefanović et al. 1983).

Considering the general morphology and physiological system, it can be said that the vegetation period of dark coniferous forests is functionally short, from May to September, with the photosynthetic process essentially lasting throughout the year. Consequently, the vegetation period in the layer of herbaceous plants lasts differently: late spring, summer, and early autumn in mountain forests to late spring and summer in subalpine spruce forests.

Climatic specifics, especially the thermal-pluviometric regime of the area, influenced an evident differentiation of ecosystems of dark coniferous forests of the Dinarides of BiH at the level of two climatic subtypes that dominate in the biocenoses of the mountain and subalpine spruce forests. Climatic characteristics of mountain spruce forests are defined by the average annual air temperature of up to 6 °C, with the average annual temperature amplitude being quite pronounced—about 19 °C; i.e., the average January temperatures range from −4 to −5 °C, while the average July temperatures are around 15 °C. Absolute temperature extremes range from −30 °C (absolute minimums) to about

30 °C (absolute maximums). The average annual humidity is about 75%, with a marked increase in the late autumn pluviometric maximum to about 90%.

The dominant production function of this ecosystem is mainly carried out by species Picea abies. Besides it, the following species can also be found at tree stratum: Abies alba and Pinus silvestris, more below Lonicera alpigena, L. coerulea, L. nigra, Daphne mezereum, Rhamnus fallax, Vaccinium myrtillus, and others. The stratum of herbaceous plants is built of acidophilic species: Luzula nemorosa, L. pilosa, L. maxima, Veronica officinalis, V. urticifolia, Lycopodium annotinum, Oxalis acetosella, and many species of moss from the genera: Polytrichum, Dicranum, Mnium, Bryum, Hylocomium, Rhytidiadelphus, Ctenidium, and in wetter habitats even Sphagnum (NEAP 2003).

Of the floristic consumers, the following plant species play a significant role: Neottia nidus avis, Corallorhyza trifida, Monotropa hypopitis, and others. Of the faunal consumers, the following groups of animals play a significant role. Mammals: Ursus arctos, Sus scrofa, Vulpes vulpes, Canis lupus, vjeverica, kuna zlatica, Lynx lynx, puh, srna, Apodemus flavicola, A. silvaticus, and others. Birds: Regulus regulus, R. ignicapillus, Turdus viscivarus, Parus cristatus, P. ater, Fringilla fringilla, Nucifraga caryocatactes, and many other bird species from the beech forest ecosystem. Herpetofauna: representatives of the herpetofauna are very rare in the biocenosis of dark coniferous forests. Of the reptiles, the following are rare: Elaphe longissima, Vipera ursini, V. berus, Lacerta vivipara, Salamandra atra, Triturus vulgaris, and less frequently other species. Reducents are made up of numerous species of bacteria, fungi, and various animals in the soil. The ecosystem of dark coniferous forests is both in the past and today under strong anthropogenic influences, especially as a basis for the development of the wood industry. As a function of adapting this ecosystem for livestock development in the mountain belt of this ecosystem, meadows of Pančićion Lakušić on the southeastern Dinarides were formed, and in the subalpine pasture of mountain ores on limestones of Festucion pungentis Horvat, i.e., on silicates of Jasionion orbiculatae (Drešković et al. 2011) (Fig. 8.9).

Around the pens in the belt of this ecosystem, tertiary vegetation develops, namely, Chenopodion subalpinii Br.-Bl., Plantaginion reniformis, and Rumicion alpinii Br.-Bl., whose structure and dynamics and biomass production

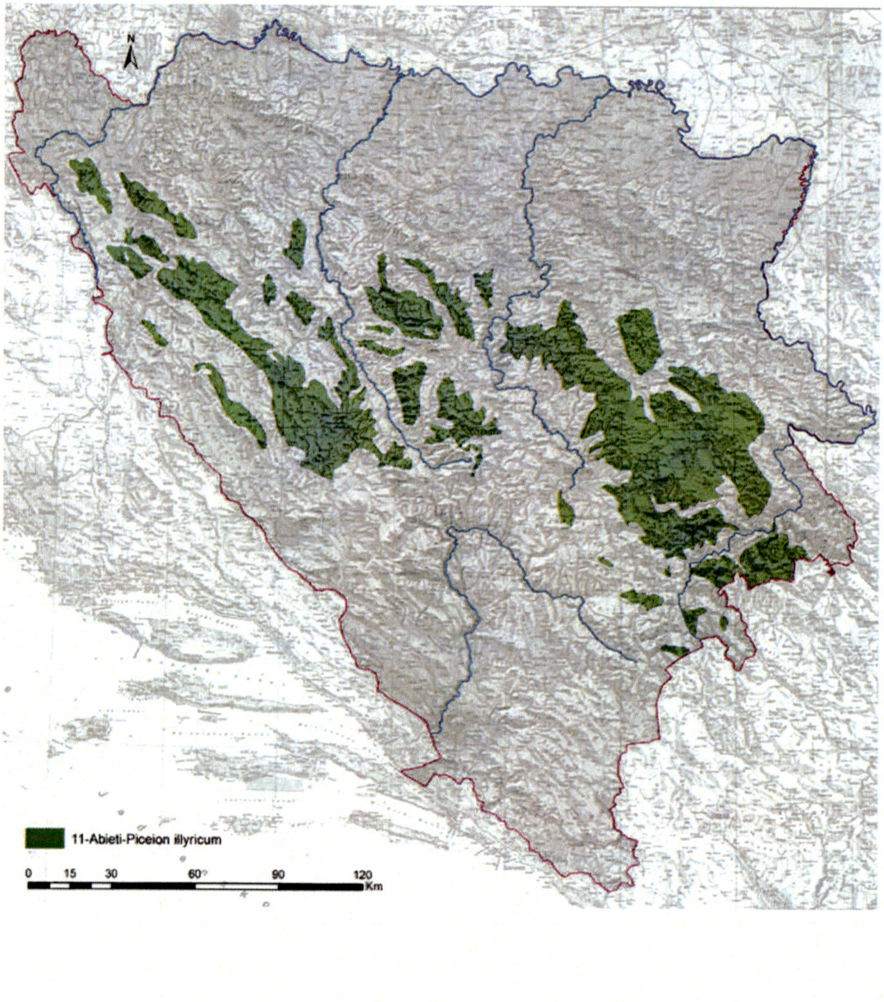

Fig. 8.9 Geographical distribution of climatogenic ecosystems of dark coniferous forests

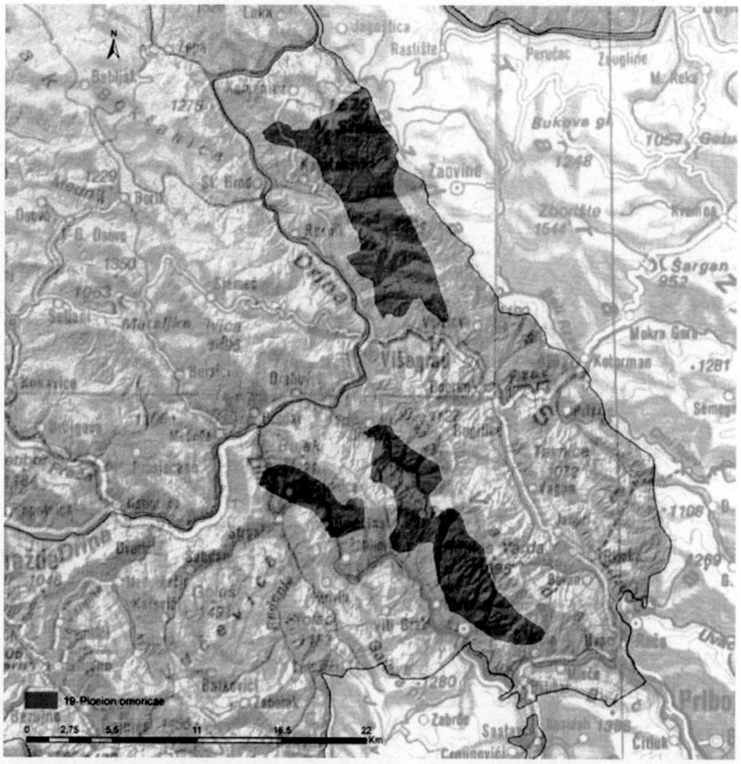

Fig. 8.10 Geographical distribution of climatogenic ecosystems of coniferous forests of Pančić spruce

indicate a high degree of anthropogenic degradation of the climatogenic ecosystem.

The ecosystem of relict and endemic forests of Pančić spruce (Piceion omorikae) developed under the influence of the orographic structure of the terrain and other ecological conditions within the hypsometric zone from 800 to 1600 m above sea level, in the area of the upstream part of the catchment area of Drina river and the downstream portion of the catchment area of the Lim river. The predominant locations where Pančić spruce develops are the south-exposed slopes of the said belt (Fig. 8.10).

The average annual air temperatures in the habitats of the Pančić spruce ecosystem usually vary between 5 and 9 °C. Absolute temperature extremes depend on altitude within the belt. Consequently, minimum absolute air temperatures of −35 °C occur in the habitats of subalpine populations. In contrast, absolute maximum temperatures up to about 32 °C appear in the habitats of the downstream part of the mountain belt. The average annual relative humidity ranges from about 60% to about 70% (in mixed forest habitats) to about 90% (in black alder habitats).

The geological background in the habitats of this spruce forest ecosystem consists mainly of Triassic limestones and rarely serpentinites. Soils in Pančić spruce habitats are most often undeveloped phases of the calcareous series of soils, such as sierozem, black soils, brown limestone, and somewhat less frequent soils on serpentine deposits such as wetland gley soil or hydrogen black soils in the hygrophilous ecosystem with black alder (Piceo omorikae-Alnetum glutinosae Čolić et Gigov). The dominant producer floristic species of the Pančić spruce ecosystem among the trees are Picea omorika, Pinus nigra ssp. austriaca, Picea abies, Acer pseudoplatanus, Fagus moesiaca, Ostrya carpinifolia, and other. Among thickets, the following are dominant: Daphne blagayana, Amelanchier ovalis, Cotoneaster tomentosa, Spirea chamaedryfolia, Lonicera alpigena, Daphne mezereum, Corylus avellana, Erica carnea, Vaccinium myrtillus, Rosa pendulina, Lonicera xylosteum, and others. In the stratum of herbaceous plants, the producer role is carried out by Orthilia secunda, Laserpitium krapfi, Goodyera repens, Gentiana asclepiadea, Hieracium transsilvanicum, Euphorbia amygdaloides, Lathyrus vernus, Luzula silvatica, Calamagrostis varia, Valeriana montana, Veronica chamaedrys, Doronicum columnae, Libanotis sibirica, Galium schultesii, Succisa pratensis, Lilium martagon, Platanthera bifolia, and others.

Of the faunal consumers, the following groups of animals play a significant role. Mammals: Vulpes vulpes, Lepus europaeus, Canis lupus, and Meles meles. Birds: Nucifraga caryocatactes, Columba livia, Parus ater, Sitta europaea, Dendrocopus major, Turdus merula, and others. Reptiles: Vipera ammodytes, V. ursini macrops, Lacerta vivipara, L. viridis, Anguis anguis, Elaphe longissima, and others. Amphibians: Testudo graeca ibera, Rana graeca, Hyla

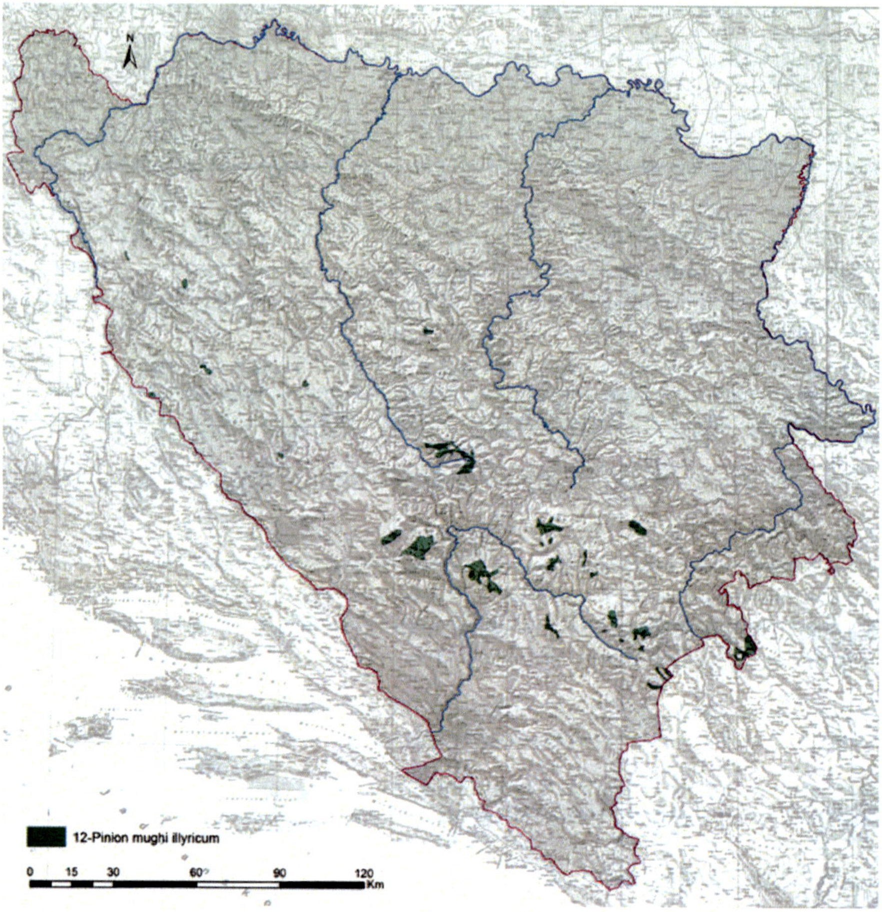

Fig. 8.11 Geographical distribution of climatogenic ecosystems of pine juniper conifers

arborea, and others. Insects: Pyrgo-morphella serbica, Parnasius apolo, and many other species including numerous endemics and relics of the Balkan Peninsula.

The communities of juniper pine ecosystems (Pinion mugi) are in the form of exclaves of different surfaces formed in the circum-Mediterranean mountainous area of the upper part of the subalpine Dinarides belt (Fig. 8.11). The most important biocenoses that depend on the physical and geographical specifics of the area that appear in this ecosystem are Pinetum mugi Illyricum, Pinetum mugi croaticum, Pinetum mugi montenegrinum, etc. The vegetation period of communities of this ecosystem lasts from late May or early June and ends in September. The period of intensive biomass production lasts during the short summer period. According to the flowering period of some plants from the herbaceous floor, it is differentiated into early and late summer (when most species are in bloom) (Stefanović et al. 1983).

Geological formations, forming the basis of the pine juniper ecosystem, form rock masses of different mineral-petrographic compositions such as limestone, dolomite, quartz porphyrites, rhyolites, dacites, etc. Depending on the geological background, this ecosystem is differentiated into geological variants such as Pinet calcicolum, Pinetum mugi dolomiticum, and Pinetum mugi silicicolum.

The soils of the pine juniper ecosystem represent the primary development phases of the limestone, dolomite, and silicate series of soils, i.e., soils such as subalpine calcomelanosol and subalpine district melanosol, i.e., rankers, dominate. Only exceptionally, the structure of this ecosystem includes acidic brown soils, i.e., subalpine cambisol. A common feature of all soils of pine juniper ecosystems is the richness of humus and nitrogen with good air and a weaker water regime.

The main producer plant species is Pinus mughus. Of the low thickets, among the most important producer species are Vaccinium myrtillus, V. vitis idaea, Rhododendron hirsutum, Erica carnea, Calluna vulgaris, and others. The floor of herbaceous plants is dominated by different plant species, which is directly related to the type of geological substrate. Acid volcanic rocks are dominated by: Luzula nemorosa, L. maxima, L. pilosa, Deschampsia flexuosa, Oxalis acetosella, Homogyne alpina, Gentiana asclepiadea, and others.

Fig. 8.12 Viola elegantula. (Photo by Aida Bidžan-Gekić)

Limestone soils are dominated by herbaceous plant species: Valeriana tripteris, Dentaria savensis, Polygonatum verticillatum, Symphytum tuberosum, Anemone nemorosa, Viola elegantula (Fig. 8.12), Asarum europaeum, and other species from the secondary vegetation of mountain foothill limestone grasslands on limestones. On the floor of mosses and lichens, the most common species are Dicranum, Polytrichum, Tortella, Hylocomiurri, Rhythidiadelphus, Ctenidium, Cetraria, Cladnia, Peltigera, and others.

Of the faunal consumers, the following groups of animals play a significant role. Mammals: Ursus arctos, Sus scrofa, Canis lupus, Lepus europaeus, Vulpes vulpes, Apodemus flavicola, Dolomis sp., Talpa sp., and others. Birds: Tordus torquatus, Prunella modularis, P. collaris, P. merula, Phylloscopus collybita, Purrhula pyrulla, Regulus ignicapillus, Erithacus rubecula, Poenicurus poenicurus, Parus cristatus, P. coeruleus, Fringilla coelebs, Carduelis carduelis, C. chloris, Loxia curvirostra, Silvia atricapilla, and others. Reptiles: occasionally coming from neighboring ecosystems: Vipera ursini macrops, V. berus bosniaca, Coronella austriaca and Lacerta vivipara, and others. Amphibians: Salamandra atra and Triturus alpestris reiseri (on Vranica mountain).

Of the faunal consumers, the following groups of animals play a significant role. Mammals: Ursus arctos, Sus scrofa, Canis lupus, Lepus europaeus, Vulpes vulpes, Apodemus flavicola, Dolomis sp., Talpa sp., and others. Birds: Tordus torquatus, Prunella modularis, P. collaris, P. merula, Phylloscopus collybita, Purrhula pyrulla, Regulus ignicapillus, Erithacus rubecula, Poenicurus poenicurus, Parus cristatus, P. coeruleus, Fringilla coelebs, Carduelis carduelis, C. chloris, Loxia curvirostra, Silvia atricapilla, and others. Reptiles: occasionally coming from neighboring ecosystems: Vipera ursini macrops, V. berus bosniaca, Coronella austriaca and Lacerta vivipara, and others. Amphibians: Salamandra atra and Triturus alpestris reiseri (on Vranica mountain).

The ecosystem of xerothermic pine forests (Erico-Pinetea Horvat) belongs to the biocenosis of light coniferous forests and develops on the profile of the outer Dinarides of BiH, from the belt of hardwood evergreen forests (on the shores of the Adriatic Sea) to the juniper pine belt, mountain heaths Bruckenthalion spiculifoliae Horv. em. Lakušić and mountain ores on limestones and silicates (up to about 2000 m above sea level). In the earlier stages of development, pine forests, except in the sub-Mediterranean and subalpine zone, had a zonal character in other climatic zones but under the influence of global climate change in these areas were preserved only in habitats with pronounced extreme variations of primary ecological factors.

Today's pine forests of the Mediterranean and sub-Mediterranean belts and pine forests of the hilly and mountainous belts in the Dinarides have a secondary character. This genetic category also includes Aleppo pine forests in the eumediterranean and sub-Mediterranean belt, Dalmatian black pine forests, and Illyrian black pine forests on limestone. Illyrian pine forests on dolomites and serpentines and Bosnian pine (munika) forests in the subalpine zone are much better developed, considering that they have continuously developed from xerotherms to the present day. Aleppo pine forests (Pinion halepensis) represent recent developmental stages of garrigue and maquis in the eumediterranean and lower sub-Mediterranean belt and for a relatively short historical time (from several hundred years ago) have experienced succession with climatogenic vegetation of pine, kermes oak, or downy oak. The situation is similar with the forests of Dalmatian black pine (Pinion dalmaticae) and Illyrian black pine (Pinion austriacae).

Bosnian pine forests (Pinion heldreichii) are developing as zonal vegetation of the upper part of the Mediterranean-Montan and subalpine belt of the coastal and central Dinarides of BiH. In the mountains with the less favorable hydro-thermal regime, as is the case with the mountains of the coastal Dinarides (Orjen, Velež, Hercegovačka Bjelašnica, etc.), they are developed in the subalpine zone and have a structural form in the form of savanna forests or forest-steppes with scattered trees, while the majority of vegetation consists of communities of cliffs, rocks, and screes (Encyclopedia of Yugoslavia 1983).

In the littoral mountain belt of the Dinarides, these forest communities develop in the hypsometric zone up to 1800 m above sea level, with a more pronounced biocenological development in the zone from 1200 to 1500 m above sea level. Above them, the vegetation of limestone foothill limestone grasslands, screes, and rocks usually develops, while the juniper pine belt is missing. In the group of mountains at the transition to the central Dinarides (Prenj, Čabulja, Čvrsnica, Vran-planina, etc.) with a more favorable hydrothermal regime, these forests develop in the upper part of the mountain and subalpine belt and form associations: Pinetum heldreichii mediteraneo- montanum, Pinetum heldreichii hercegovinum, and others. These associations are mainly formed on steep limestone or dolomite terrains and represent the surface continuity of communities from the Seslerio autumnalis-Fagetum moesiacae belt or even on the Seslerio autumnalis-Ostryetum carpinifoliae in the littoral Dinarides.

The very unfavorable water regime had a negative effect on the very reduced production of both the main edificatory species Pinus heldreichii and other species within this biocenosis. The main production types, apart from Bosnian pine, are: Rhamnus fallax, Viburnum maculatum, Lonicera glutinosa, Senecio visianianus, Gentiana amblyphylla, Daphne blagayana, D. laureola, and others.

Xerothermic pine forests develop on volcanic mountain massifs (Pinion austriacae serpentinicum) of the continental Bosnian Dinarides such as Konjuh, Stolac, and others. Forest vegetation is the primary vegetation, which has been maintained since the xerotherms in the form of permanent developmental stages of the forest in this area, thanks, above all, to the steeper slopes on the peridotites and serpentines. Forest vegetation habitats on volcanic rocks are characterized by continuous fluviodenudation soil erosion and a weak hydrothermal regime, characterized by lack of water and large temperature fluctuations both on a daily and monthly basis.

The geological background of the terrain is primarily formed by deposits of Paleozoic age, which are represented by mineralogical-petrographic varieties of rhyolite, metariolite, quartzite, porphyrite, quartz porphyry, and other rock masses characterized by a high degree of metamorphosis. Brown and ilimerized soils (from the more developed phase of the serpentine series) have much more favorable both water regime and mineral nutrition regime, as a result of which much more organized phytocenoses of leaf-deciduous forests of Hungarian oak, Hungarian oak and Turkish oak, Turkish oak and Sessile oak, Sessile oak, beech or beech and fir in the mountain belt, which leads to the conclusion that the vegetation of pine forests is a permanent stage.

Secondary vegetation in the habitats of the coastal pine forest ecosystems of Pinion halepensis and Pinion dalmaticae consists of communities of rocky pastures of the order Cymbopogo-Brachypodietalia H-ić. Secondary vegetation in Pinion austriacae habitats is formed by communities of rocky pastures on limestones of the order Scorzonero-Chrysopogonetalia H-ić, while communities of the order Halacsyetalia Ritter are developed on serpentines. Secondary vegetation in the habitats of Pinion heldreichii Horvat communities is formed by associations of mountain pastures on limestones of the Dinarides of the order Crepidetalia dinaricae, i.e., the subalpine association Festucion albanicae.

Tertiary nitrophilic vegetation in the eumediterranean belt, i.e., in the forest belt of Pinion halepensis and Pinion dalmaticae, consists of associations of Chenopodion muralis H-ić, while the vegetation of well-trodden habitats consists of the association Polygonion avicularis. Similar associations occur in the habitats of pine forests of Pinion austriacae, while in the belt of subalpine pine forests of Bosnian pine, there are associations of Plantaginion reniformis Lakušić and Chenopodion subalpinum Br.-Bl.

The consumer-reducing component has a high percentage of endemic Dinaric or Balkan species. The producer component has similar endemic characteristics, which is why this ecosystem stands out in a particular province of Balkan xerothermic pine forests, which, from the ecological aspect, connects the Mediterranean with the Euro-Siberian-Boreo-American and Alpine-High Nordic biogeographical region.

Of the faunal consumers, the following groups of animals play a significant role. Mammals:

- In coastal pine forests: Canis aureus, Vulpes vulpes, Oryctolagus cuniculus, and others.
- In subalpine pine woods: Rupicapra rupicapra, Vulpes vulpes, Lynx lynx ssp. pardina, Microtus nivalis ssp., Dolomys bogdanovi, D. grebenscikovi, Talpa caeca, T. hercegoviensis, Pitymys sp., and others.

Representatives from the groups of birds, herpetofauna, and insects in coastal pine forests are the same as in the ecosystem of Mediterranean evergreen forests of evergreen oak or in the ecosystem of thermophilic deciduous forests of the order Quercetalia pubescentis. The following representatives of the fauna exist in the ecosystem of subalpine pine forests of the Pinion heldreichii order: Birds (Eremophila alpestris balcanica, Alectoris graeca, Carduelis citrinella, Pyrrhocorax pyrrhocorax, Serinus canaria, Nucifraga caryocatactes, Parus ater, Dendrocopus major, Menticola saxatilis, and others), reptiles (Vipera ammodytes, V. ursini macrops, Lacerta mosorensis, L. vivipera, and others), amphibians (Rana graeca, Salamandra atra), and insects (Psorodonotus, Pyrgomorphella, Oropodisma, and others). The great faunal diversity and endemicity indicate the specificity of this ecosystem. They are also the basis for its separation into a particular biogeographical province of the Balkan Peninsula's endemic and relict pine forests (Drešković et al. 2011).

8.6 Alpine-High Nordic Biogeographical Region

The Alpine-High Nordic biogeographical region in Bosnia and Herzegovina includes the alpine climate zone that extends to the highest mountains of the central Bosnian Dinarides. More specifically, in the southeastern Bosnian highlands, it contains a continuous vegetation belt at altitudes above 2000, formed on the highest mountains of this area: Maglić and Volujak. On other high mountains of the central Dinarides and mountains of the transition zone such as Zelengora, Lelija, Treskavica, Bjelašnica, Vranica, Prenj, Čvrsnica, Vran, and others, the vegetation of this biogeographical province occurs in the form of discontinuous areas, most often of juniper pine. However, occasionally (on slopes under shades), it can appear even in the belt of subalpine forests of munika, beech, or spruce. Given the small and limited surface areas caused by relatively low altitudes (Maglić 2386 m), ecosystems in Bosnia and Herzegovina in the Alpine-High Nordic biogeographical region appear from only one biogeographical province—the High Nordic Province.

8.6.1 Ecosystem of High Nordic Province

Per the specific geographical position of the high mountains of the central Bosnian-Herzegovinian Dinarides, certain ecological factor specifics cause some differences in the structure and dynamics of biocenoses of the High Nordic Province. Such positional characteristics conditioned the need to analyze the general environmental conditions and characteristics of plant communities at the level of both vertical and horizontal biogeographical profiles.

On the vertical profile of the mountains of the High Nordic province, three vegetation belts can be distinguished: discontinuous belt of subnival vegetation, belt of alpine foothill limestone grasslands, and belt of subalpine foothill limestone grasslands of secondary origin. On the horizontal profile of the mountains, this province is differentiated into six sectors: High Orjen sector, High Prenj sector, High Maglić sector, and High Vranica sector.

High Orjen sector covers the area of high mountains of the Bosnian-Herzegovinian coastal Dinarides (Orjen, Hercegovačka Bjelašnica and other), which have a developed belt of foothill limestone grasslands above or within the belt of the munika, belonging to a special order of Seslerion robustae. The vegetation of the rocks of this sector belongs to the order Amphoricarpion neumayeri, while the vegetation belongs to the community of Drypeion jacquinianae. Endemic species that characterize this sector are Amphoricarpus neumayeri, Edraianthus lovćenicus, Edraianthus wettsteinii, Lamium lovćenicum, Satureia horvatii, Edraianthus caricinus and others, while the vegetation units are Seslerietum robustae, Seslerietum juncifoliae mediteraneo-montanum, Amphoricarpion neumayeri, Drypetum jacquinianae, and others.

High Prenj sector includes the high mountains of Hercegovina: Velež, Prenj, Čvrsnicu, Čabulju, and Vran. Characteristic vegetation units of this sector are Oxytropidion prenjae, Carici-Dianthetum freynii, Gentianetum dinaricae hercegovinicum, Seslerietum juncifoliae hercegovinicum, Amphoricarpi-Campanuletum hercegovinae, Elyno-Edraianthetum serpyllifolii hercegovinicum, Festucetum pungentis hercegovinicum, Saxifragetum prenjae hercegovinicum, and others. Within this sector, there are also numerous endemic plant species such as Dianthus prenjus, Euphorbis hercegovina, Dianthus freynii, Saxifraga prenja, Amphoricarpus autoriatus, Campanula hercegovina, Gentiana dinarica, Edraianthus hercegovinus, Oxytropis prenja, Leontopodium nivale-hercegovinus, and others.

High Maglić sector includes the high mountains of the southeastern Bosnian highlands: Volujak, Maglić and Lebršnik. In Bosnia and Herzegovina, it is differentiated into two: the Maglić sector and the Volujak-Zelengora sector. The Maglić subsector includes the belt from the upper border of juniper pine and subalpine spruce forest, i.e., the belt of juniper pine and subalpine beech forest on slopes with southern exposures. The Volujak-Zelengora subsector continues to the juniper belt of pine and subalpine beech forests. At the same time, on the south-exposed slopes, it also descends to lower hypsometric levels, practically to the belt of thermophilic subalpine beech forests. Characteristic plant species of the High Durmitor sector are Valeriana braunii-blaquetii, Plantago durmitorea, Daphne malyana, Edraianthus sutjeskae, Edraianthus glišićii, and others. Among phytocenoses more significant are: Elyno-Edraianthetum serpyllifolii, Stachydi-Festucetum pseudoxanhinae, Euphorbio-Valerianetum bertisceae, Edraintho-Potentilletum clusianae, Edraiantho-Daphneetum malyanae, Senecio-Festucetum spadiceae, and others.

The high Vranica sector of the high Dinaric province includes the area of high mountains of the mountain group Vranica, Vlašić (Fig. 8.13), Ljubišnja, etc. This sector contains a discontinuous belt to which, at the lower border, juniper pine and subalpine spruce forest continues. The characteristic and differential types of this sector are: Edraianthus niveus, Rhodiola rosea, Gnaphalium supinum-balcinicum,

Fig. 8.13 Vlašić mountain. (Photo by Aida Bidžan-Gekić)

Rhododendron hirsutum, Rananculus crenatus, Lycopodium alpinum, Jasione orbiculata-bosniaca, Saxifraga caesia, Leucanthemum rotundifolium, Veronica fruticans, and others. Characteristic and differential phytocenoses are Meo-Festucetum spadiceae, Carici-Edraianthetum nivei, Festucetum variae, Festucetum halleri, Lycopodietum alpini, Caricetum curvulae-bosniacum, Junceto-Primuletum glutinosae, and others.

The ecosystem of foothill limestone grasslands on limestones (Crepidetalia dinaricae) is limited to the high Dinaric province of the Alpine-high Nordic region, i.e., the southeastern Dinarides. It occurs primarily in the alpine climate zone of the southeastern Dinarides; i.e., it developed between belts of juniper pine or subalpine forests on the one hand and subnival vegetation of the order Selicetalia retusae serpyllifoliae on the other. Because of the anthropogenic influences on subalpine forests and thickets, it also develops in the subalpine zone at 1500 and 2500 meters of altitude. The habitats of this ecosystem are in the form of exclaves, which primarily include the peaks of the highest Dinaric mountains, such as Maglić and Volujak in Bosnia and Herzegovina (Lakušić 1981).

Climatic characteristics within this and all other ecosystems of this province have the attributes of ET climate or tundra climate. In contrast, in the zones where these ecosystems descend below 2000 m, Dfd climate type characteristics are also encountered. Average annual temperatures in the ecosystem of mountain minerals on limestones range from 0 to 2 °C. Absolute minimum air temperatures drop to −45 °C, while summer absolute maximums reach values up to 30 °C. The average annual relative humidity is 50–70%, with a maximum of 80% reached in the second half of spring. A minimum relative humidity of 50% occurs during July.

The geological basis of the ecosystem of mountain ores of the order Crepidetalia dinaricae is primarily formed by more compact limestones and dolomites of the Triassic age, while only in some places Jurassic and Cretaceous limestones appear. The pedological substrate of the ecosystem of mountain ores of the order Crepidetalia dinaricae is calcomelanosol, while carbonate lithosol or calcocambisol are less common (Fig. 8.14).

The soils are characterized by a shallow humus profile with 5–25 cm of depth. The percentage of humus is usually high and amounts to 10% and 30%, which gives these soils partial peat characteristics. The pH value in shallow calcomelanosols is neutral, in more developed calcomelanosols slightly is acidic, and in the deepest calcomelanosols, it is acidic. Due to the predominant limestone base, the soils have a weak water regime, which harms the biodynamics of this ecosystem.

The primary biocenosis of this ecosystem is made up of herbaceous plants from the families: Poaceae, Asteraceae, Fabaceae, Lamiaceae, Scrophulariaceae, etc. Under anthropogenic influences, this biocenosis is differentiated into two subbiocenoses belonging to different community: primary-alpine and alpine. The alpine subbiocenosis belongs to the Oxytropidion dinaricae community, while the subalpine subbiocenosis belongs to the Festucion pungentis community.

The most important producers of organic matter within the Oxytropidion dinaricae community are Oxytropis dinarica, Festuca paniiciana, Sesleria tenuifolia, Carex laevis, Allysum montanum, Crepis dinarica, Gentiana verna, Dryas

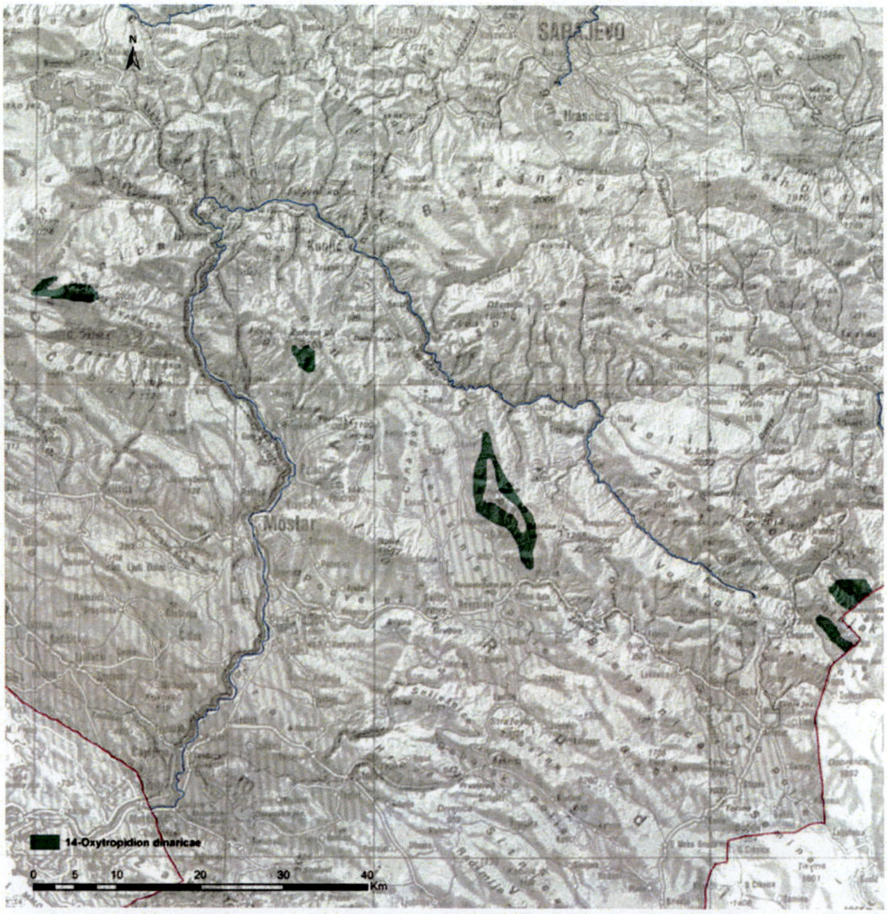

Fig. 8.14 Geographical distribution of climatogenic ecosystems of foothill limestone grasslands on limestones

octopetala, Helianthemum seridianus alidrostus H. montenegrinus, E. hercgovinus, E. murbekii, E. croaticus, etc. The most important producer species within the Festucion pungentis community are: Festuca pungens, Onobrychis montana, Sesleria robusta, Helianthemum grandiflorum, S. gigantea, Potent ilia heptaphyl crantzii, Cerastium lanigerum, Minuartia juniperina, Festuca glauca, and F. amethystina.

Of the faunal consumers, the following groups of animals play a significant role. Mammals: Dolomys sp., Talpa sp., Apodemus sp., Lepus sp., Sus scrofa, Vulpes vulpes, Rupicapra rupicapra, and others. Birds: Montifringilla nivalis, Anthus spinoletta, Poenicurus ochrurus, Oenanthe oenanthe, Anthus trivialis, Eremophia alpestris, Falco tinunculus, Aquila chrysoetes, Gyps fulvus, Pyrrhocorax graculus, Pyrrhocorax graculus, Carduelis colabina, Carduelis comlabina. Reptiles: They are very rare in this ecosystem and are almost the only species of Vipera ursini macrops, Coronella austriaca, and Lacerta vivipara. Amphibians: Rare specimens of Salamandra atra and Triturus alpestris appear only during long summer rains.

The ecosystem of foothill limestone grasslands on silicates (Seslerietalia comosae) only secondarily covers the highest areas of high silicate mountain massifs, most often above the juniper pine and mountain heather ecosystem; i.e., in Bosnia and Herzegovina it occurs only on high habitats of Vranica mountain (where Pinetum mugi silicicolum ecosystem was primarily developed).

The dominant geological basis of this ecosystem is formed by rhyolites and metariolites of the Lower Paleozoic age, which are supplemented in places by quartzites, dacites, and porphyrites. Highly acidic soils with a shallow profile and a large percentage of humus (15–30%) were formed on this geological base.

The climatic specifics of this ecosystem are quite similar to the previous one. More specifically, average annual temperatures in this ecosystem also range from 0 to 2 °C. Absolute minimum air temperatures drop to −35 °C, while summer absolute maximums reach values up to 35 °C. The average annual relative humidity is about 70%, with a maximum of 80% reached in the second half of spring. A minimum relative humidity of 60% occurs during July. In the alpine and subalpine zone, shallow soils develop on acidic volcanic rocks—district melanosols or rankers—and district cambisols much less often. Basic soil types have better water and air regime, which affects slightly better community biodynamics. The average soil pH is 3.5–5.5.

In structural terms, the biocenosis of this ecosystem is most similar to the biocenosis of the ecosystem of mountain minerals on limestone. The leading producers of organic matter in this ecosystem are Carex curvula, Jasione orbiculata, Festuca halleri, Agrostis rupestris, Sieversia montana, Phyteuma confusum, Poa violacea, Nardus stricta, Achilllea lingulata, Festuca varia, Juncus trifidus and Primula glutinosa, Lycopodium alpinum i dr. The consumer-reducing component of this biocenosis consists of almost the same species that participate in the biocenoses of mountain mineral ecosystems on limestones.

The ecosystem of subnival vegetation around the highest peaks (Salicetea herbaceae) in Bosnia and Herzegovina develops only extra zonal (in the zone of mountain ores on limestone or silicate) in depressions on the northern exposures of the highest peaks of Maglić, Volujak, Zelengora, Lelija, Vranica, Prenj, Čvrsnica, and Vran mountains. Areas with this vegetation are very small and usually amount to several tens to several hundred square meters, which is why they typically have to be mapped disproportionately (Đug and Drešković 2008).

The biodynamics of this ecosystem is characterized by long phases of prolonged changes, caused, among other environmental factors, by a short vegetation period. More specifically, the vegetation period lasts from July to September, which causes a low level of biocenosis organization and a low degree of species association. The climatic specifics of this ecosystem are essentially not significantly different from the previous two, except that the thermal regime has somewhat more extreme features. The primary producer component of the biocenosis of this ecosystem is composed of hamephytes and hemicryptophytes from the genera Salix, Ranun culus, Carex, Poa, Cerastium, Alopecurus, Trifolium, Plantago, Soldanella, Crepis, Taraxacum, Leontodon, etc. Depending on the composition of the parent substrate of this ecosystem; i.e., from the physicochemical properties of the soil, the producer component is differentiated into two vegetation orders: the order Salicetalia retusae-serpyllifoliae and the order Salicetalia herbaceae. The order Salicetalia retusae-serpyllifoliae includes communities on acid volcanic rocks, while the order Salicetalia retusae-serpyllifoliae includes associations on limestones and dolomites. The consumer-reducing component consists of rare species of arctic-alpine distribution such as Montifringilla nivalis, Eremophila alpestris, and others.

8.7 Protected Natural Areas of Bosnia and Herzegovina

Bosnia and Herzegovina is one of the countries in Europe with the greatest diversity of species of plants and animals. Genetic diversity makes Bosnia and Herzegovina an extraordinary asset. Of particular importance is that which is contained in several hundred endemic species of plants, animals, and fungi. This makes the gene pool invaluable. That Bosnia and Herzegovina ranks third in Europe. Genetic and species diversity is contained in various forms of ecological diversity (populations, biocenoses, ecosystems, biomes, biocycles, biosphere) (Redžić 2012). Vascular flora accounts for about 5000 confirmed taxa (a taxon—plural taxa—is a group of one or more populations of an organism or organisms seen by taxonomists to form a unit) of species, subspecies, and variety and form levels. As much as 30% of the total endemic flora in the Balkans (1800 species) is contained within the flora of Bosnia and Herzegovina. BiH has many unique aspects of biodiversity such as natural and preserved forest ecosystems (virgin forests, of special importance for genetic diversity), large areas of karst with many unexplored speleological structures, relict-refugial types of ecosystems as endemic centers of development and occurrence of new species, and numerous endemic and relict species.

Relict-refugial types of ecosystems (ecosystems in the crevices of carbonate, dolomite, volcanic and silicate rocks) and ecosystems of screes (predominantly represented in the canyons of the Una, Vrbas, Drina and Neretva rivers) are of special value to the biodiversity of BiH. These ecosystems are especially important as endemic centers of development and occurrence of new species. Relict and refugial species are surviving unfavorable climate conditions in shelters suitable to their ecological niche and available in the nature of BiH. Additionally, glacial relicts are found in ecosystems of snow patches, subalpine and alpine meadows as well as in ecosystems of rock crevices in the high-mountain area (USAID BiH 2016).

The high level of endemic and relic qualities gives it significance on a global level. Exactly 5134 taxa (Thallophytes and Cormophytes) have been identified so far putting Bosnia and Herzegovina among the richest in flora in Europe. The most specific characteristic of Bosnia and Herzegovina's flora is a large number of paleo, neo, and steno endemic species, tertiary and glacial relicts maintained in refugial habitats, such as cliffs, canyons and mountain cirques. The most part of endemic taxa are recognized within the systematic group of Cormophytes, and it is currently estimated to 450 endemic taxa. These endemic species have been found in Herzegovina, within the mountain complex

Prenj-Čvrsnica-Velež and canyons of the Neretva River, and its tributaries. These areas are recognized as endemic centers.

Fungi inhabit both terrestrial and aquatic environments. It is estimated that between 15000 and 20000 mushroom species can be found in Bosnia and Herzegovina. However, only 552 species of higher fungi have been identified. There are many species of Macromycetes which have high economic potential and value. They are an important source of income for local communities. The most important among them are: Morel, King Bolete, Chanterelle, Macrolepiota sp., and Saffron Milk Cap. Lichens, on the other hand, are an insufficiently studied group of organisms in Bosnia and Herzegovina. More than 300 species of lichens have been recorded so far, and their biodiversity is estimated to around 1000. However, for the territory of BiH, no assessment of the degree of threat according to the IUCN criteria has been made so far.

The country's fauna is characterized by the occurrence of refugia and development centers (in unfavorable geological periods, canyons and gorges had favorable combination of ecological factors for survival of certain species, and relicts, and those species became endems after long periods of time), and by the most unique fauna of karst areas, mountains and canyons. Limnofauna of invertebrates is very diverse due to the diversity of aquatic habitats. Fifty species of annelids have been identified belonging to 19 genera, and 8 species of leaches belonging to 7 genera. Freshwater crustaceans include 31 species, of which 16 are endemic. Freshwater ecosystems are characterized by rich and diverse aquatic insects. The order Ephemeroptera (mayflies) comprises 58 species belonging to 20 genera, out of which five are Dinaric, Balkan, or Dinaric-alpine endemic species. In the order of Plecoptera (stoneflies), 74 species from 15 genera have been described. The order of Trichoptera (Water moths) is plentiful with 215 detected species from 78 genera. The most interesting is genus Drusus. Very interesting life forms are endemic species of caves in southern Bosnia and Herzegovina area.

The total number of species of vertebrates in Bosnia and Herzegovina is: fish 119, amphibians 20, reptiles 38, birds 340, and mammals with 87 species. Within ichthyofauna (fish) the highest diversity is recognized within the family Cyprinidae (51 species) and Salmonidae (8 species).

The diversity of amphibians consists of 7 genera, 21 species, and 22 subspecies. Among tail-less (Anura) most abundant is genus Rana with 7 species and among tail (Caudata) amphibians this is genus Triturus with 5 species. Reptiles inhabit freshwater, ponds, marshes, and almost all terrestrial ecosystems (especially rocky grassland). They belong to 40 species (45 sub-species) from 12 families. The highest reptile diversity in Bosnia and Herzegovina is evident in the Mediterranean region and supra-Mediterranean belt. Birds: Bird diversity comprises 340 species within 189 genera, 68 families and 22 orders.

The fauna of mammals comprises 87 identified species within 51 genera and 19 families. Most species inhabit terrestrial habitats, while a small number of species inhabit aquatic ecosystems secondarily or only occasionally. High abundance and diversity characterize the family Vespertiolinidae (bats) with 20 identified species from 8 genera. A new species of bat, pond bat (Myotis dasycneme), has been recently discovered in the Cave of Bijambare. Bosnia and Herzegovina is the most southern location where this species is found in Europe. In the same caves another new species of bat were registered for BiH, the mountain long-eared bat (Plecotus macrobullaris).

A very important step in the protection of endangered species is the introduction of red lists and red data books. According to the Law on Nature Protection of Federation of Bosnia and Herzegovina the "Red List—list of endangered species/subspecies" is the official document containing the list of endangered wild species/subspecies, arranged by category of threat of extinction. These categories are determined in line with the methodology of the International Union for Conservation of Nature (IUCN). The Red Lists of Wild Species and Subspecies of Plants, Animals and Fungi of Federation of Bosnia and Herzegovina were developed in January 2014. They contain 659 endangered flora taxa, 284 fauna taxa, and 86 fungi taxa. In Republika Srpska, 7 Red lists for 7 taxonomic groups are in place. Republika Srpska Government adopted the Decree on the Red List of Endangered Species of Flora and Fauna in 2012. The Red List contains endangered species of flora and fauna in this Bosnian entity and consists of several different lists according to different taxonomic groups. The methodology used for development of red lists in the entity of Republika Srpska is different than the methodology used in Federation of Bosnia and Herzegovina. Therefore, data are not comparable or suitable for compilation and presentation at national level (USAID BiH 2016).

Nature protection in Bosnia and Herzegovina, as in other countries in the region, began in the second half of the nineteenth century with the protection of certain game species. The first officially protected area in Bosnia and Herzegovina was established in June 1954, in the SW part of the country, in the basin of the Perućica stream. Only 2 months later, in August 1954, the National Institute for the Protection of Cultural Monuments and Natural Rarities of the People's Republic of Bosnia and Herzegovina protected Prokoško jezero on Vranica Mountain as a natural rarity under the 1947 Law on the Protection of Cultural Monuments. The first national park, Sutjeska, was established in 1969, and Perućica became the nucleus of this park (Fukarek 1970).

Spatial Plan of Bosnia and Herzegovina for the period 1981–2000 envisioned the protection of 8062 km^2 or 15.03% of the country's territory through an integrated approach. Until 1990, in accordance with the Law on Nature Protection and the Law on Protection of Cultural, Historical, and Natural Heritage, only 0.55% of the territory of Bosnia and Herzegovina was protected (253 areas—28127 ha) (NEAP Directorate 2003).

Structurally, the stated nature protection plan at the level of individual categories has the following indicators: national parks—1730 km^2, regional parks—4500 km^2, especially valuable natural areas—1200 km^2, memorial parks—100 km^2, forest and other reserves, natural monuments, etc.—10 km^2, water surfaces, coastal area, natural and artificial lakes—440 km^2, other forms of protected nature—about 350 km^2. The first newly protected areas, based on IUCN criteria, have been established in Sarajevo Canton. These are the natural monuments of Vrelo Bosne and Skakavac and the protected landscape of Bijambare (Fig. 8.15) (Drešković and Đug 2008).

The Blidinje Nature Park (Fig. 8.16) was established in the Herzegovina-Neretva Canton in 1995, while the Prokoško jezero Nature Monument was established in the Central Bosnia Canton in 2005 (Đug and Drešković 2012). The natural monument Tajan (2008) was established in the Zenica-Doboj Canton in 2009, and the Konjuh Protected Landscape in the Tuzla Canton in 2009. In May 2008, the Una National Park was established as the first national park in the Federation of Bosnia and Herzegovina (Dalmatin et al. 2008).

In the entity of Republika Srpska, there are 21 protected areas with a total area of 24632 ha (Ljubojević and Marčeta 2007). There are two national parks here: Kozara (3494.5 ha) (Fig. 8.17) and Sutjeska (17250 ha) (Fig. 8.18), as well as two rainforest reserves: Janj (195 ha) and Lom (297.75 ha). The current state of other protected areas is very problematic, as they are under a very pronounced anthropogenic influence.

The Spatial Plan of the Republika Srpska in its first phase (1996–2015) set a very ambitious goal to protect 15–20% of the territory of this entity by establishing 11 new national parks and 11 nature parks (Kadić and Marković 2006). Despite these intentions, even the existing protected areas are endangered. For example, 12 reserves of Pančić spruce (Picea omorica (Panč.) Willk.), with an area of 320 ha, are threatened due to poor management and lack of adequate protection (Ljubojević and Marčeta 2007).

Despite all efforts, Bosnia and Herzegovina has managed to protect only 2.6% of its territory, which is far below the regional and European levels. The results of the analyses show that 134440.8 ha (1344.4 km^2) are protected in Bosnia and Herzegovina. Ramsar sites in Bosnia and Herzegovina are Hutovo blato—1972, Ramsar Site Number 1105; Močvara Bardača—February 2007, Ramsar Site Number 1658; and Livanjsko polje—2008, Ramsar Site Number 1786.

Habitat fragmentation and destruction, together with the overexploitation of natural resources, led to a dramatic increase in environmental degradation in the post-war period. Therefore, the need to establish new and re-valorize existing protected areas in the country was emphasized. The protection of natural heritage requires establishing a more efficient system of long-term protection of the most valuable and vital natural areas characterized by a high degree of biological diversity and following the European Union's modern ecological criteria and standards. In achieving this goal, developing and implementing a methodology that will enable permanent monitoring of the situation and changes in protected areas and their environment is necessary. The highest priority should be given to developing a comprehensive information system as a crucial tool for effective monitoring and prompt and objective insight into the state of the natural heritage. In this way, further degradation of natural resources could be prevented, and harmonization of development activities with nature protection could be enabled. It is also necessary to plan infrastructure very carefully in environmentally sensitive areas by taking into account the visual identity and values of the landscape. Nature protection should have the same status as other sectors in spatial planning (NEAP 2003).

In addition to the three existing national parks (Sutjeska, Kozara, and Una), it is necessary to effectively protect the following areas as critical zones in the network of protected areas: Olympic mountains—Igman and Bjelašnica, mountains of the endemic center in Herzegovina—Prenj, Čvrsnica, and Čabulja, Vranica mountain, Konjuh—Zvijezda—Tajan mountain complex, Livanjsko polje as the most extensive karst field in the world, Šator mountain and Grmeč mountain. Other areas proposed in the spatial entity plans will serve as islands and corridors in the network of protected areas of Bosnia and Herzegovina. For example, National Park Una does not provide great opportunity for the development of high number of activities, like mass tourism, except in certain limited locations (Hrelja et al. 2015). Similar situation is with other protected areas.

Although putting in place a monitoring system on ecosystems is an obligation in accordance to the laws on

Fig. 8.19 The Wilson promenade in Sarajevo. (Photo by Aida Bidžan-Gekić)

carriage connected visitors with the Great Alley to Vrelo Bosne in 1985 (The Great Alley in Ilidža 2018). It is suitable for walking, cycling, and jogging. It leads to the spring of Bosnia, where swans live on small lakes, the most recognizable symbol of this area. The Great Alley is an inseparable whole with the Spring of Bosnia. This area is equipped with park furniture for staying in nature, and a playground for children. Alley and Vrelo Bosne are unique in appearance, location, and content, so it is not surprising that it is a favorite resort of citizens of Sarajevo, but also of many tourists who visit this city. This alley is suitable for the development of sports-recreational, excursion, educational and cultural-manifestation tourism.

Wilson's promenade (Fig. 8.19) was built by the Austro-Hungarian monarchy, and it was arranged and put into operation at the beginning of the last century under the name "Kalaj Promenade" after the name of the politician, Minister of Finance and head of Bosnia and Herzegovina, Benjamin Kalaj. The Austro-Hungarians planted linden trees in it (480 in total), most of which have survived to this day, because during the war (1992–1995) no destruction was spared in Bosnia and Herzegovina.

After the fall of the monarchy, the street was named after the 28th President of the United States, Thomas Woodrow Wilson. In the period from 1941 to 1945, the name Wilson's Promenade was changed to "Mussolini's Promenade" or "Mussolini Promenade." When the war ended, the promenade was given the name it deserved, and that is "Wilson's Promenade." Already in 1960, the name of the promenade was changed again, and it was named "Youth Promenade." After that, again, for the third time, the name Wilson's Promenade is returned to him. It is important to emphasize that in addition to numerous street name changes, the name Wilson's Promenade has not changed to this day (The Wilson promenade 2020).

The Botanical Garden of the National Museum of Bosnia and Herzegovina is located in the central part of Sarajevo and is part of the National Museum complex (Fig. 8.20). It is the only monument of its kind in Sarajevo and Bosnia and Herzegovina. It was founded in 1913, and its founder was the prominent botanist, Carlo Maly. The area of the botanical garden is 14270 m^2 and more than 2000 plant species are grown in it. The garden contains very valuable collections: endemic plants, alpine species, the most important medicinal plants, ferns, succulents, wetlands and aquatic plants, flora of xerothermic habitats "Mediterraneum," flora of the Miljacka river valley, and flora on dolomites, limestones, serpentines, etc. Numerous exotic species of trees can be seen in one place, such as: Japanese cherry (Prunus serrulata), mammoth (Sequoia), ginkgo (Ginkgo biloba), etc. A special attraction is the fountain in the central part of the garden, surrounded by lush vegetation, which contributes to the visitor's impression of a romantic flower park (Đug and Drešković 2008). This garden is suitable for the development of cultural-event and educational tourism.

Fig. 8.20 The Botanical Garden of the National Museum of Bosnia and Herzegovina. (Photo by Aida Bidžan-Gekić)

References

Bakaršić S (1957) Prevoj Ivan [Ivan Pass]. Geografski pregled 1:66–84

Dalmatin M, Drešković N, Đug S (2008) Zaštićena područja Bosne i Hercegovine [Protected areas of Bosnia and Herzegovina]. Ekološka udruga Lijepa naša, Čapljina

Drešković N, Đug S (2008) Geografski informacioni sistem za Nacionalni park Una [Geographic information system for Una National Park]. Zaštićena područja u funkciji održivog razvoja, Federalno ministarstvo okoliša i turizma, Sarajevo

Drešković N, Đug S, Stupar V, Hamzić A, Lelo S et al (2011) NATURA 2000 u Bosni i Hercegovini [Natura 2000 in Bosnia and Herzegovina]. Centar za okolišno održivi razvoj, Sarajevo

Đug S, Drešković N (2008) Prirodno naslijeđe Kantona Sarajevo [Natural heritage of Canton of Sarajevo]. Kantonalni zavod za zaštitu kulturno-historijskog i prirodnog naslijeđa, Arch Design, Sarajevo

Đug S, Drešković N (2012) Nature protection in Bosnia and Herzegovina: state and perspectives. Rev Geogr 7(1):69–80

Encyclopedia of Yugoslavia (1983) Socijalistička republika Bosna i Hercegovina. Jugoslovenski leksikografski zavod, Zagreb

Fukarek P (1970) Šumske zajednice prašumskog rezervata Perućice u Bosni [Forest communities of the Perućica rainforest reserve in Bosnia]. Posebna izdanja Odjeljenja prirodnih i matematičkih nauka ANUBiH 15(4):157–262

Hrelja E, Drešković N, Mirić R, Avdić B (2015) Geoecological evaluation of terrain in National park Una. Zbornik radova I Međunarodne konferencije iz turizma i menadžmenta u ugostiteljstvu, Sarajevo, pp 75–83

Kadić J, Marković B (2006) Zaštićena područja prirode u prostornom planu RS 2001–2015 [Protected areas of nature in the spatial plan of RS]. Zbornik Jahorina, Istočno Sarajevo

Lakušić R (1981) Klimatogeni ekosistemi Bosne i Hercegovine. Geografski pregled 25:41–69

Ljubojević S, Marčeta D (2007) The level of forest exploitation in protected areas of Republic of Srpska. Glasnik Šumarskog Fakulteta 7:23–50

NEAP (2003) Akcioni plan za zaštitu okoliša BiH [Action plan of environment protection of B&H]. Direktorat NEAP, Sarajevo

Redžić S (2012) Biodiverzitet Bosne i Hercegovine—stanje, mogućnosti upotrebe i neophodnost održivog upravljanja [Biodiversity of Bosnia and Herzegovina—state, ability to use and the need for sustainable management]. In: Redžić S (ed) Drugi međunarodni kolokvijum Biodiverzitet—teorijski i praktični aspekt, Zbornik radova/Proceedings 22. Posebna izdanja/Special Editions CXLVIII of Department of Natural Sciences and Mathematics of ANU BiH, Sarajevo, pp 47–70

Stefanović V, Beus V, Burlica Č, Dizdarević H, Vukorep I (1983) Ekološko-vegetacijska rejonizacija Bosne i Hercegovine [Ecological and vegetation regionalization of Bosnia and Herzegovina]. Šumarski fakultet Sarajevo, Sarajevo

The Great Alley in Ilidža (2018). Available at https://www.navigator.ba/#/p/aleja-na-ilidzi. Accessed 11 Sept 2018

The Wilson promenade (2020). Available at http://sarajevski.ba/vilsonovo-setaliste-mjesto-za-svako-doba-dana/19025. Accessed 22 Nov 2020

USAID BiH (2016) Country biodiversity analysis: Bosnia and Herzegovina, actions necessary for biodiversity protection. USAID/Bosnia and Herzegovina, Sarajevo

Part III

Human Geography of Bosnia and Herzegovina

9 Historical Geography of Bosnia and Herzegovina

Abstract

An overview of the historical and geographical development of Bosnia and Herzegovina starts from the Old Stone Age—Paleolithic—continues through ancient times with the Illyrians and the rule of Rome, the arrival of the Slavs, the great Bosnian kingdom in the Middle Ages, more than 400 years of Ottoman rule, to the modern and independent state of Bosnia and Herzegovina in the twenty-first century. It is important to point out that the territory of Bosnia and Herzegovina, unlike neighboring countries, is poorly explored archaeologically, and quite significant discoveries are expected in the future. The oldest settlements belong to the Mousterian culture in northern Bosnia from the Middle Paleolithic period.

In prehistoric and ancient times, the Illyrian peoples left their mark on historical and geographical development until the arrival of the Roman Empire. A significant period of prosperity began during the Bosnian kingdom from the second half of the fourteenth century. Over the last 500 years, Bosnia and Herzegovina has preserved its territorial borders significantly and is undoubtedly one of the most specific countries in the world. Some of the most important battles in the Second World War, crucial for the formation of socialist Yugoslavia, were fought on the territory of Bosnia and Herzegovina. Modern Bosnia and Herzegovina, after leaving the socialist and communist system of the former state of Yugoslavia by declaring independence and joining the UN, suffered great material and human destruction and atrocities in the period 1992–1995 that had not been seen in Europe since the Second World War. In the third decade of the twenty-first century, Bosnia and Herzegovina aims to become full member of the European Union.

Keywords

Mousterian culture · Illyrians · Kingdom of Bosnia · Bosnian eyalet · Communism · Republic of Bosnia and Herzegovina

9.1 Historical Geographies Before the Twentieth Century

9.1.1 Paleolithic and Mesolithic

The area of today's Bosnia and Herzegovina (hereinafter: BiH) has been inhabited since the Old Stone Age—Paleolithic. Traces of the habitat in Makljenovac near Doboj discovered in 1949, then in Kulaši near Prnjavor and Sokolac, are a witness to this. Roughly worked stone tools were found here—pebbles attributed to *Homo erectus*. Many discoveries in the river valleys of northern Bosnia followed, especially Ukrina and Usora. Discovered sites of the Old Stone Age in northern Bosnia can be included in open habitats mainly located on low hills.

Unlike the Old, the Middle Paleolithic, with a duration of 100,000–35,000 years before the present, left more traces. Habitats of *Homo sapiens neaderthalensis*, both in the open and in caves, have been discovered in several places, especially in northern Bosnia. A wide variety of tools and weapons were found in them, most often stone blades and scrapers.

The Middle Paleolithic industry from the Rastuša site is characterized by artifacts of mainly small dimensions, a high share of scrapers, and a small presence of perforated tools, and the appearance of Charente elements was registered in the form of a bilaterally retouched limace scraper tool (Jovanović et al. 2014).

H. Gekić et al., *The Geography of Bosnia and Herzegovina*, World Regional Geography Book Series,
https://doi.org/10.1007/978-3-030-98523-3_9

The results of the typological and technological analysis of materials from the Zobište site near Derventa confirmed the assumption of the existence of a regional tradition but also indicated a noticeable continuity of settlement and occupation of these sites over time (Baumler 1987). The connection of the Middle Paleolithic of northern Bosnia with other parts of Europe is still not sufficiently clarified. The situation is similar to the establishment of connections and similarities between the localities in northern Bosnia.

There are numerous archeological discoveries in the river valleys of northern Bosnia, especially Ukrina and Usora. The cultural remains are deposited in layers of light ocher clay, originating from the Wurm Ice Age and interspersed with gray veins of dusty matter, applied by the wind to the cracked soil of the Dryas. The Late Paleolithic is thus already significantly represented. The bearer of that time, *Homo sapiens sapiens*, used more advanced tools technically. In addition, he left the first traces of his spiritual life. In 1976, for the first time in BiH, an engraving on a rock was discovered in the Badanj cave near Stolac (Fig. 9.1), representing an extraordinary work of Paleolithic man in this area. It is a remarkable phenomenon and the first case of this type of Paleolithic art in southeastern Europe (Benac et al. 1984). It shows an animal similar to a horse seen from the right side. The culture from the uppermost layer of this site belongs to the late Epigravettian and is dated, together with the drawing, to the geological period of the Younger Dryas. The engravings in Pećine pod lipom near Sokolac also stand out, as well as the stone head of a bird from Mala Gradina from Kulaš near Prnjavor and the bone whistle from the Crvena Stijena cave at the spring of Trebišnjica. Thousands of stone tools, engraved bones, and jewelry in the form of necklaces were found in the Badanj cave, and they were made of pierced snails, small shells, and deer teeth (Basler 1976).

The oldest settlements found in BiH belong to the Mousterian culture. They were discovered in the lower layers at the sites of Kamen and Lonđa in Makljenovac near Doboj, Markovac and Zobište in Donji Detlak near Derventa, Kadar in Donji Svilaj near Odžak, Mala Gradina in Kulaši, and elsewhere. Luščić in Kulaši and the younger strata on Kamen and Lonđa belong to the Aurignacian culture. The finds of the upper layers in Kadar date from the Gravettian era (around 26,000 to 21,000 years before the present).

A series of thermoluminescent dates from individual sites support the chronological framework of the artifacts found: the Mousterian from the Zobišta site near Derventa dates to a period of about 87,000 years before the present. This could be classified as an early stage of the last glaciation (Montet-White 1992).

Fig. 9.1 Engraving of the back of the animal's body in Badanj cave near Stolac. (Photo by Aida Bidžan-Gekić)

Archaeological site Danilovića brdo is located in the village of Podnovlje, near Doboj. This site was recognized in 1959. The large number of upper Paleolithic artifacts found on this site during the early 1960s suggests that during this period, occupation of the site was more intensive than during Middle Paleolithic times, but that can also be a sign of a techno-economic behavior in the Upper Paleolithic was more intensive than in the Middle Paleolithic. Considering that the stratigraphic situation on northern Bosnia's sites is not promising site Danilovića brdo, and other sites in this region have an immense potential for future research of the Middle and Upper Paleolithic in this region, especially in the research of open-air sites and reviewing patterns in settlement caves and open-air sites in the narrow and wider region (Dragosavac 2016).

Due to the nature of the soil of the Quaternary layers of northern Bosnia, i.e., its specific acidity, osteological findings

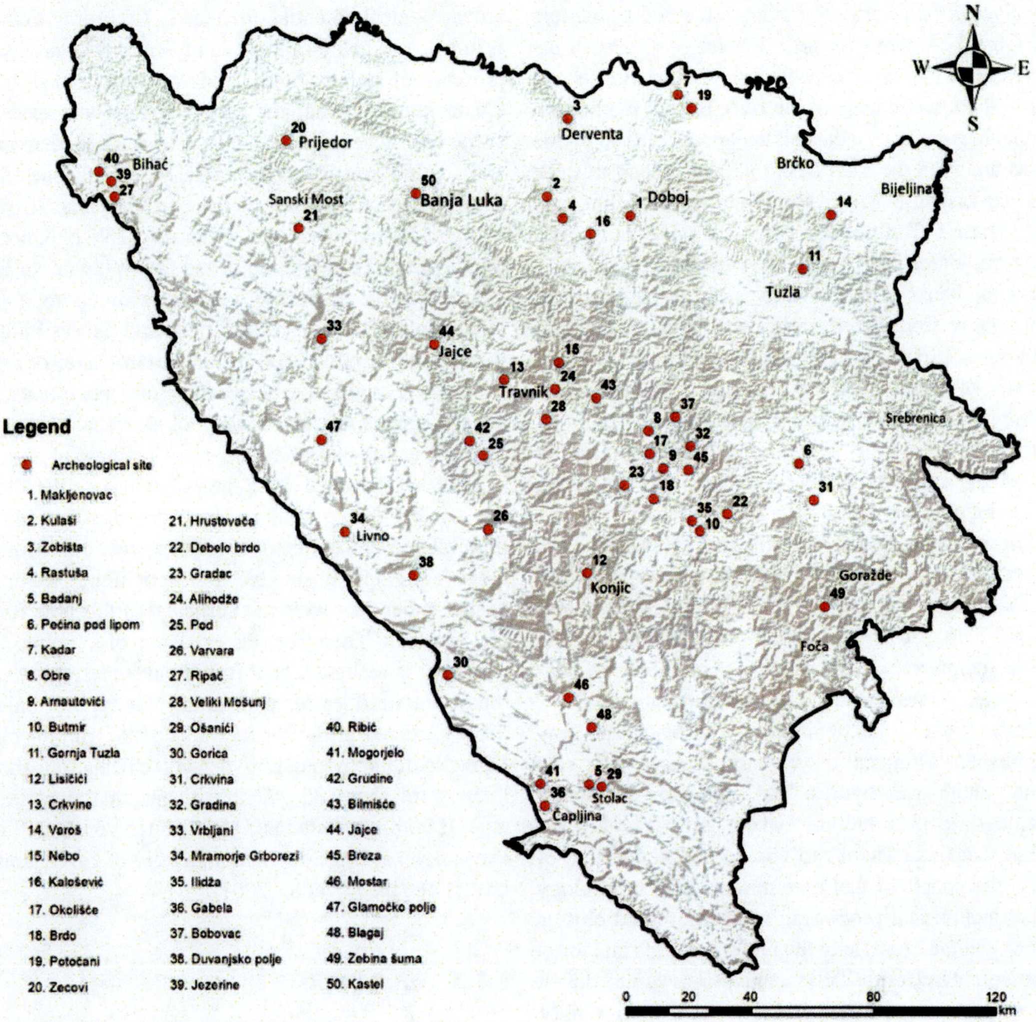

Fig. 9.2 Archaeological sites mentioned in the text

have not been best preserved, so the remains of an animal or human bones are rarely found at archaeological sites. In the area of northern Bosnia, according to the example of Rastuša cave located on the slopes of Hrnjino brdo near Teslić (Fig. 9.2), the following species of animals were most represented: cave bear, red deer, aurochs, steppe rhino, giant deer, wolf, and cave lion (Pandžić 2014).

It is reasonable to expect that Neanderthals were also present in northern Bosnia about 130,000 years before the present when the Betal cave in Slovenia and the Krapina cave in Croatia were inhabited. Finds of Middle Paleolithic material older than 290,000 years in Italy provide an opportunity to find Middle Paleolithic material of similar age in northern Bosnia in the future (Mussi 2002).

The Mesolithic, or the last phase of the Ice Age, did not leave significant traces in BiH. Traces of habitat have been found only in eastern Herzegovina near Gacko and Crvena Stijena. Characteristic finds are tiny precision-made stone tools and weapons that served as deadly tips for sticks and harpoons (Basler 1979).

9.1.2 Neolithic

This period in BiH is primarily characterized by a number of settlements in the valley of the river Bosna and its northern tributaries, i.e., in the valley of the Neretva river. There are two primary cultural zones: the Adriatic in Herzegovina and the Pannonian-Balkan in Bosnia. The upper basin of Bosnia is a transitional area between these zones. All three Neolithic phases are represented in BiH. The settlements in Gornja Tuzla (Starčevo culture), Zelena pećina near Mostar (Impresso culture), and Obre I near Kakanj (mixed culture) belong to the Older Neolithic. The most important representative of the Middle Neolithic is the Kakanj culture (Kakanj, Obre I, Arnautovići near Visoko—artifacts in the form of vessels with bell-shaped legs and cult rhytons stand out here) in the upper course of the river Bosna; in northern Bosnia, the bearers of Vinča culture appear. Most of the settlements are from the Late Neolithic and belong to three groups: Hvar-Lisičić (Lisičići near Konjic, Peć Mlini near Tihaljina), Butmir (Butmir, Okolište, Obre II), and Vinča (Gornja Tuzla, Varoš near Koraj, Kaloševič on Usora, etc.). Only along the banks of the Sava river can the influence of the Sopot culture from Pannonia be felt. Research on Neolithic cultures in BiH has notably contributed to the study of the genesis of Neolithic cultures and Mediterranean-Pannonian relations in this period (Benac 1979).

The most famous Late Neolithic culture was Butmir, named after the site Butmir near Sarajevo. The following sites also belong to this culture: Gornje Polje near Kakanj, Okolište near Visoko, Brdo in Kiseljak, Nebo in Bila, Crkvina near Turbet, and others. This culture gave significant findings that indicate the rich spiritual life of the inhabitants of that time. It is read through small anthropomorphic and zoomorphic sculptures, among which it is possible to distinguish European, Armenoid, and Negroid elements.

The Vinča culture is recognizable by its black and gray quality polished ceramics of various shapes with rich decorations, altars, and figures. The settlement in Gornja Tuzla belongs to this culture, where salt was already exploited at that time. Thanks to mining and knowledge of metallurgy, the people of that time mastered the technology of gold production and processing (numerous gold objects were found). On the example of the findings of rich and long-lasting settlement in Gornja Tuzla, conclusions can be drawn about the importance of salt deposits and salt springs since the Middle Neolithic period (Tasić 2007).

The discovery of the site and the Neolithic settlement in Butmir happened by accident during the construction of the Agricultural unit in Butmir. Archaeological excavations that followed immediately after the discovery carried out between 1893 and 1896 investigated the total area of the settlement of 5305 m^2. During this research, a large amount of pottery, stone tools, and anthropomorphic figurines were discovered, which intrigued the scientific community of Europe from the end of the nineteenth century to such an extent that, in 1894, the Austro-Hungarian administration organized the World Congress of Archaeologists and Anthropologists (Čović 1988a; Kaljanac 2014).

Shortly afterward, the Butmir site was marked as an eponymous site of the eponymous culture of the Late Neolithic, which remains to this day the most famous cultural and historical monument of BiH. In September 2016, in the national monument area—the archaeological area of the prehistoric settlement in Butmir—rescue archeology was conducted to determine the actual area of distribution of the archaeological potential on the site. According to the overall results, specifically of the area of Sections 2 and 4 where the presence of archaeological finds was determined, it is possible to conclude that the archaeological site extends on a surface of 4 hectares and that it is positioned towards east and south in relation to the initially defined space where the Butmir culture was once discovered (Kaljanac 2016).

In the Visoko basin (the settlement of Visoko, not far from Sarajevo) at the beginning of the fifth millennium BC, there were fortified, organized villages with up to a thousand inhabitants. Traces of fortifications and houses built in rows were found. Around 4800 BC, in the entire area of the Visoko basin, there could have been about 3500 inhabitants, which is an average population density of 31 inhabitants per km^2. Reconstruction of arable land gives a picture of an organized system that stretched along the valley of the river Bosna. The use of this fertile land implied, among other things, the possibility of the outbreak of conflict situations. Mountainous areas and land unsuitable for agriculture were used as pastures, but they were not sufficient in this area for stationary livestock. Therefore, the existence of a mobile livestock economy is assumed, as it is still encountered in these parts today. Possibilities for performing this type of animal husbandry existed in the broader area of the Butmir culture. A Late Neolithic population density calculation shows that there were about 32,000 inhabitants in the entire Butmir area. If both populated and uninhabited zones are taken into account, an average population density of 1.8 inhabitants per km^2 is obtained (Müller 2007).

9.1.3 Eneolithic

New developments in the area of Bosnia arose in the middle of the fourth millennium BC when there were several waves of migration. They were connected with the steppe nomadic peoples—cattle breeders made mobile by horses and carriages with carriages. They are also recognizable by the manner of their burial. Specially built mounds of earth and later also of stone were dug (piles or mounds).

They are the bearers of the Lasinja, Badanj, Kostolac, and Vučedol cultures spread from the Pannonian-Danube area. Their settlements were located along the Vrbas, Bosna, and Drina rivers in areas rich in ores. The site of the Kostolac culture, Potočani near Odžak, yielded extremely beautiful, high-quality, and diverse ceramics decorated with rich geometric motifs filled with white inlay (a mixture of resin and ground snail houses).

The most significant Eneolithic culture is Vučedol, which spreads from the neighboring areas of today's Slavonia and Srijem in eastern Croatia. The most famous localities are Zecovi near Prijedor, Hrustovača near Sanski Most, Debelo Brdo, Fortica near Sarajevo, Gradac in Kiseljak, and

Alihodže in Travnik. This culture was known for its advanced copper metallurgy (molds for casting tools and weapons have been preserved). In Herzegovina, this period was marked by the Adriatic culture that emerged from the late phase of the Hvar-Lisičić culture. New changes occurred at the end of the third millennium BC, with a wave of new settlers from the lower Danube region. They also brought the spread of bronze (copper and tin alloys).

Quite exact data on social stratification during the Eneolithic in the territory of BiH is not extensive, but judging by the data and analogies from other Eneolithic sites outside BiH, they are sufficiently indicative. In that sense, the main house on Zecovi near Prijedor stands out, which probably belonged to the head of the settlement, and the oval building in Pivnica, which, had it not belong to a person with a distinct status in the settlement, certainly had a special purpose. The Eneolithic is an age that brings a complete and definitive break from the Neolithic age. This process is conditioned by the gradual degeneration of Neolithic cultures, which is felt in all spheres of life. However, a much more significant cause of these changes is the gradual but pronounced population shift, which has affected all parts of Southeast Europe, leading to the Indo-Europeanization of these areas. Neolithic cultures, both physically and spiritually, could not withstand the intense pressure from the East. One part of them is entirely layered over and remained almost without any share in later development (Butmir culture), while the other part only managed to biologically endure a short period and partially maintained the continuity of its development during the transition from Neolithic to Eneolithic (Hvar-Lisičić culture). The already developed Eneolithic phases do not belong to the Neolithic tradition (Tasić et al. 1979).

9.1.4 Bronze Age

Around 1800 BC, the youngest Eneolithic cultures disappeared, which was caused by the penetration of the last waves of Indo-European migration and the drowning of Eneolithic populations in the new Early Bronze Age cultures. These waves include the bearers of the Perjámos culture that occupy northeastern Bosnia (Gornja Tuzla I) and the bearers of rough pottery and stone battle-axes found in Glasinac, southern Bosnia and northern Herzegovina (Debelo Brdo, Kotorac, Pod, Varvara). In southern Herzegovina, there is a symbiosis and fusion of Older Eneolithic elements with early bronze newcomers (ceramics of the Cetina-Kotorac type).

The Early Bronze Age (1800 to 1550 BC), as a whole, is mainly characterized by the settlement of mountainous areas and the development of hillforts. The funeral rites are dominated by mounds (tumulus), often with a contorted position of the dead. During the Middle Bronze Age (1550 to 1250 BC), the development that began in the Early Bronze Age (fortified settlements and burial under tumulus) continued, and in the lower Podrinje, the influences of the Central European burial mound culture were observed. The Late Bronze Age (1250 to 850 BC) brought significant changes: the bearers of the urnfield culture penetrated northern Bosnia and founded many settlements and necropolises. Central Bosnia culture is developing in central Bosnia, characterized by a specific ("Western Balkan") geometric style on ceramics and metal products. At Glasinac, in eastern BiH, continuous development on the old foundations (the so-called Proto-Illyrian period) could be observed. A large number of garden settlements were founded, and domestic bronze metallurgy began (Čović 1976).

The Bronze Age reflected a significant extent on the economy, spiritual culture, and social relations. Although cattle breeding and agriculture remained priority occupations, metallurgy gradually became increasingly important, as evidenced by the workshop centers in Ripče, Donja Dolina, Pivnica, Varvara, Debelo Brdo, etc. In spiritual culture, the cult of the dead was observed through burials beneath tumulus and flat graves with grave goods. Changes in social relations are essential because of the processes that began to influence the formation of tribal communities, which will nevertheless complete their full affirmation in the Iron Age. The social and political organization and the way of life in the Bronze Age of the population in BiH can be presented only in the form of quite general assumptions. If nomadic or seminomadic cattle breeding as a basic form of economy and way of life is taken as a realistic precondition for the Early and Middle Bronze Age, relatively small, quite mobile communities and without firmly demarcated territories would have to be taken into account. However, some phenomena (e.g., strongly fortified settlements with double or triple dry stone walls, forts with some kind of acropolis—Veliki Gradac na Privali, Trostruka—and some richer or with special attention equipped tumulus) indicate the existence of social discipline and a particular hierarchy within these communities, and to hold, at least for several generations, certain territories in possession.

The Late Bronze Age, during which relatively clearly demarcated larger territorial units (known as certain cultures or cultural groups) are formed, is probably the time when the formation of larger tribal communities begins, some of which will, during the Iron Age, grow into complex political and ethnocultural organisms (e.g., Japodian, Maezaei, Delmatae, Daesitiates, and others). Leading these communities is already a strong layer of the early clan-tribal aristocracy, whose existence is reflected, for example, in the hoard from Veliki Mošunj or some of the wealthy tombs of the Late Bronze Age from Glasinac (Čović 1988b).

9.1.5 Iron Age

The rich iron ore deposits in northwestern and central Bosnia enabled the development of iron metallurgy as early as the tenth and nineth century BC, but the true Iron Age can only be observed from the middle of the eighth century BC. In material culture, northern Bosnia was strongly connected to the western Pannonian area throughout the Iron Age and was visibly separated from the southern, Illyrian parts. The Pannonian peoples—Maezaei, Osseriates, and Breuci—lived there, while the surroundings of Bihać belonged to the easternmost settlements of Japodians. Daesitiates lived in central Bosnia. The material culture of the Delmatae characterizes southwestern Bosnia. The Glasinac culture in eastern Bosnia can be attributed to the Illyrian Autariates. Eastern Herzegovina, and perhaps the areas along the right bank of the Neretva river, belonged mainly to the Illyrian Daorsi.

The settlements were located predominantly on the tops of hills. The most famous settlements of this type are Pod in Bugojno and Gradina above Ošanić near Stolac. Stilt house settlements were also discovered in Bosnia, on the rivers or along the rivers (Gradina in Donja Dolina near Gradiška, along the right bank of the Sava river, and Otok in Ripče near Bihać). The inhabitants of the Iron Age were primarily buried in earthen or stone mounds (the most famous necropolis is Glasinac, East of Sarajevo). In the northern parts of Bosnia, only flat cemeteries were discovered, without any markings on the surface (the most famous necropolises are Jezerine near Bihać and Donja Dolina near Gradiška).

One of the most important sites of the Iron Age is the settlement of Pod near Bugojno. The Iron Age period includes five phases of settlements in Pod. This period lasted continuously from about 750 (725) to the transition of the fourth to the third century BC. It consists of five phases. Life in the settlement continued because in the Pod C phase's oldest layer, some of the Pod B phase's ceramic forms, i.e., the Late Bronze Age, appear. The compact layer from the Iron Age in Pod has two distinct residential horizons with significant innovations in the organization of the settlement, the defense system, and the characteristics of the movable inventory of ceramics and metals.

The first iron objects were imported, but from the eighth century, the exploitation of domestic iron deposits and the export of iron objects to the Glasinac area and Herzegovina began. Iron weapons did not represent any particular advantage over bronze ones, but the availability of iron ore and mass production led to mass armaments. The Iron Age prioritized areas that had accessible iron ore deposits, such as central Bosnia, and mining and metallurgy became a vital force for overall progress.

Economic progress led to both the social stratification on the one hand and inter-tribal connections on the other, thus creating a large tribal community known as the Daesitiates. Special strata in this community and the Glasinac area were the warriors, who were strong supporters of the tribal aristocracy, which was led by the duke of all the Daesitiates. The result was establishing a solid hierarchy, which would eventually enable Baton to oppose Rome for many years in the conquests of the Illyrian lands.

The oldest written monument in the territory of BiH was found on the Pod (near Bugojno) hillfort, and it is a vase from the sixth century BC with an inscription in the Etruscan-Umbrian language. It is kept in the National Museum of BiH. It is a transcript dedicated to the gods Jan and Juturna. These letters were engraved on other vessels as well. Their appearance testifies to the strong and lasting ties of the settlement of Pod with the Greek colonies in the south and the Etruscan settlements in the north of Italy. The basis of these connections is a well-organized store. With developed artisanal production, able-bodied residents created a small but well-organized settlement, well-fortified, with a proper layout of houses, narrow streets maintained under strict communal laws, and finally, a square and suburbs. It was not a village, but a city, at least in an embryonic sense. Other settlements of Daesitiates were not at this level of development, but centers such as Debelo brdo and Pod would pull forward the entire economic development of the country, encouraging production, trade, movement of people, and ideas, and affect the overall cultural development (Čović 1976).

The exploitation of iron ores and iron production in central Bosnia began in the nineth century BC at the latest. The basic cultural-geographical zones in the Iron Age largely coincide with those from the previous period. The exceptions are Herzegovina, sharply divided into the western part in the Early Iron Age, culturally closely related to southwestern Bosnia (and central Dalmatia), and the eastern part, which first entered the circle of the Glasinac culture and was then exposed to strong Hellenization.

Iron production was developing in the whole area around the rivers Vrbas, Sana, and Una; trade is intensifying, and a particular cultural and ethnic differentiation can be observed: Japodes live in Pounje, Osseriates around Donja Dolina, Maezaei from Sana to Vrbas, and Breuci in northeastern Bosnia. As of the fourth century BC, there were strong Celtic influences, reflected in imported or taken over pieces of jewelry and costumes, the appearance of Celtic weapons, and especially numerous gray Celtic pottery. The Celtic Scordisci may have occupied the extreme northeast (Semberija).

Forts as the primary form of settlement and burial under tumulus, with various rites and often rich contributions (especially in the so-called princely graves), are characteristics of Glasinac culture, one part of which can undoubtedly be attributed to the Autariates and others to other related peoples.

After its culmination, the Glasinac culture in the fourth century BC weakened, only to disappear entirely at the beginning of the third century, leaving the area empty, which can be associated with the famous migrations of the Autariates. Fortified towns and fortifications around larger karst fields are typical for southwestern Bosnia, and more than 400 of them have been recorded. Some of them were very strongly fortified, and a certain number also had suburbs. Burial is also characteristic: flat graves, exclusively with inhumations.

The economy of the Iron Age was more diverse than in the previous period, and the differences were more significant among the areas. Data for northern and central Bosnia speak of several types of wheat and barley, oats, millet, and several legumes. Livestock also played an important role, especially in mountainous areas. Based on iron production, the production of weaponry was developing to a much greater extent, which has undoubtedly had multiple social and political consequences. It is quite possible that iron was exported from some parts. Other crafts also developed: pottery, bone and horn processing, various types of fabrics, and footwear. Hunting has undoubtedly become a secondary way of providing food. Jewelry, metal utensils, parts of military equipment, and fine ceramics are also imported.

Social and political development during the Iron Age led from the strengthening of local (clan and tribal) communities to the creation of larger communities, peoples, many of whom are already known to us from ancient written sources (Japodian, Maezaei, Osseriates, Breuci, Daesitiates, Delmatae, Autariates). The cult of the dead still plays an important role in the field of religion. In some areas, incineration dominated, including northern Bosnia, while in some, it gradually strengthens towards the south. Certain religious understandings and practices break through and spread and transcend tribal and national characteristics. In addition to the small clay plastic created at the end of the Late Bronze Age, which is still developing, it is joined by the so-called columnar idols (like those found in Pod near Bugojno).

Finds from the area of Glasinac culture testify to the missions of the Illyrian peoples who visited famous Greek shrines (in Sparta, Olympia, Corinth, etc.), which indicates the existence of a system of deities similar to that of Hellenes. The discovery of two sanctuaries gave further very important information about religious life. One of them was discovered in a layer from the time around 600 BC in Pod near Bugojno and is a kind of autochthonous type with a shrine—"an altar"—and a special kind of cult plastic ("columnar idols"). The second sanctuary is in Gorica in southwestern Herzegovina. It is a small temple of the Mediterranean type, and the time of its founding (seventh century BC) coincides with the time of the rise in the construction of early archaic temples in Greece (as well as in Etruria).

High art, in the form of stone sculpture and relief, developed only in the areas inhabited by Japodes, with the strongest center around Bihać. In its essence, it is part of the circum-Adriatic tomb sculpture (stelae, sarcophagi, urns), and in terms of style, it is mainly dependent on the mature Greek archaic. Although some of the works of this art date back to Roman times, they belong to the pre-Roman era, and the stele from Golubić and probably some other of these monuments were undoubtedly created during the period of Japodian independence (Čović 1988c).

The material legacy of the Autariates not only shows the peculiarities of general Illyrian culture but is also an important representative of that culture. The rumored culture of the Early Iron Age is the only culture that can be considered unequivocally Illyrian without any doubt, and it can be rightly said that it represents the classical period of Illyrian material production (Mesihović 2014).

9.1.6 Early Antiquity

The Illyrian regions, south and southeast of the Neretva river, were during the last centuries BC under the strong influences of the Hellenistic world. Therefore, this period of early Illyrian history can be called "Illyrian Hellenism." Extremely strong permeation of autochthonous material and spiritual culture with Hellenistic influences in the area of contemporary BiH was established only in eastern Herzegovina, where the Illyrian people Daorsi lived (Fig. 9.3). From around 300 BC to the middle of the first century BC, Hellenism of the southern Italian type was manifested in eastern Herzegovina, best evidenced by several finds of pottery of the Gnathia type. Vast quantities of fragments of amphorae from the Daorson area speak of a developed maritime trade with Greece itself and Greek settlements in southern Italy and trade with more distant provinces. In the second century BC, the Daorsi also minted money.

An important factor during the early ancient period was the Illyrian state. The founding of the Illyrian state took place in the second half of the fifth century BC under the influence of the Greek world. The common interests united several ethnically related peoples into an alliance that grew into a complete state formation with its nobility, army, and rulers at the end of the fourth century BC. Along with the Illyrian

Fig. 9.3 Illyrian peoples at the beginning of our era on the territory of modern Bosnia and Herzegovina and the surroundings. (Megistias, public domain, modified by Haris Gekić)

people of Ardiaei, which are supposed to have been its founder, there were some other Illyrian peoples from the south of BiH, such as the Daorsi and the Pleraei.

The high level of cultural development led to the early emergence of literacy in eastern Herzegovina (the use of the Greek alphabet). Judging by the numerous miniature metal figures and reliefs on tiles found in the ruins of the town of Daorson, the center of the people of the same name (Ošanići near Stolac), it can be concluded that the beliefs of the local population in the Hellenistic period largely corresponded to Greek beliefs. The end of the Hellenistic period in eastern Herzegovina falls in the first century BC, when Romanization, and with it the Latin language, replaced elements of Greek culture in this area (Marić 1977).

The wars that began on the eastern coast of the Adriatic Sea in the third century BC were the prelude to the long-running battles waged by the Illyrian peoples (Ardiaei, Dalmatae, Daorsi, Japodes, etc.) against the Roman invaders. The Romans finally conquered the Illyrians during the Dalmatian-Pannonian uprising at the beginning of our era. Since then, the parts of contemporary BiH have been part of the province of Dalmatia (before that Illyricum). The new government built many roads, which were of strategic and economic importance because they connected the Adriatic Sea with the Danube, and in addition, built military camps and fortifications, founded new cities, and inhabited them by Italics and veterans (Pašalić 1960). Roman emperors gradually carried out Romanization, dividing the civic population. New economic relations are being established: agriculture and crafts are being developed, and special attention was paid to mining. The Roman religion also spread, and with the frequent immigration of the oriental element, eastern cults also spread. Illyrian cults have long been held in parallel with the official religion of the conquerors. Despite intensive Romanization, the Illyrians managed to preserve many features of their indigenous culture. In the third century, they also had a greater political influence because emperors (Diocletian and others) also had the Illyrian background. After the collapse of the Western Roman Empire (476), the territory of BiH experienced new changes and changed rulers (Eastern Goths, Byzantium) (Pašalić 1984).

Unfortunately, to date, no urban settlements from this period have been found and explored in BiH, although they can be anticipated in several places in Srebrenica, Ilidža, Stolac, Bileća, Gacko, Kiseljak, Tomislavgrad, Travnik, Banja Luka, Gradiška, and others. In addition to urban settlements, rustic villas were also important—rural farmsteads that were registered throughout BiH and Roman agricultural estates that were awarded to veterans as a reward for their military service. Gradually, the personal names of the people changed with the settlers, so foreign Roman names are also found on the monuments (Fig. 9.4).

Religion-wise, the Roman Empire specifically introduced the cult of the three supreme deities of Jupiter, Juno, and Minerva, and they were also tolerant of domestic cults. Over time, there was an equalization of domestic and Roman deities (e.g., Bind—the deity of water is equated with Neptune, etc.). Cults of individual deities spread with merchants,

Fig. 9.4 Tombstone found in Borčani (Tomislavgrad) from the second century dedicated to Manes, mother Dastona and sister Paiona. (Kept in National Museum of BiH in Sarajevo; photo by Aida Bidžan-Gekić)

colonists, the army, and from distant lands. This is especially true in mining areas where there were many foreigners. Among these deities, the cult of Mithras of the invincible Sun, which arrived even from distant Persia, stands out. The two most important monuments of this cult were found in Jajce and Konjic.

At least in the villages, the Illyrian peoples spoke their language until the fourth century, remained faithful to their customs (costumes visible on monuments), and preserved their names. Changes would occur but only much later in late antiquity.

The great Illyrian uprising against Rome (from 6 to 9 AD) was named the Baton War (Bellum Batonianum) after one of the champions of the Desitiates, the people which, along with Breuci, Pirustae, and Delmatae, were one of the most powerful in Illyria. The uprising broke out in the land of the Daesitiates and was joined by Pirustae, Delmatae and Maezaei, Breuci, and others who were dissatisfied with how they collected tribute and recruited their youth into the Roman army abroad. The Romans pacified the conquered land by enslavement, population displacement, mass recruitment, and confiscation of land and natural resources. The repressive measures were so thorough that the Illyrian and Pannonian peoples never rebelled again. Most of contemporary BiH belonged to the province of Dalmatia, and the wide belt south of the Sava River belonged to the province of Pannonia. Over a thousand different inscriptions from this period (epigraphic monuments) are known to be from BiH. One of the most important archeological sites is the Japodian necropolises near Bihać (pre-Roman and partly Roman): Jezerine and Ribić (Bojanovski 1988a).

Moreover, based on this very pronounced sedentary culture of human existence in the ancient period, it is quite normal to expect Latin literacy to penetrate to a considerable extent and rise among the population that gave a stamp to the ancient development of the Sarajevo area. Thus, they left a whole series of testimonies of Latin, ancient literacy in the Sarajevo area. These are mostly epigraphic monuments. To this day, a significant number of such inscriptions from the ancient period (beginning of the first century—beginning of the seventh century) have been found in today's Sarajevo area, which shed considerable light on the early history of Sarajevo. Within this group, most of the epitaphs are left behind by members of the upper class. To a lesser extent, these are solemn dedications to high state officials (emperors, individual senators) and deities (Mesihović 2008).

The Romans showed great interest in the territory of Bosnia both because it was geopolitically and strategically very important and because it was an area known for its natural resources. There lay large deposits of iron, silver, and other metals and other treasures, and a greater number of slaves could easily be found. Even in the early days of Roman rule, several urban-type centers and settlements arose. Some received the rank of municipalities already in the first century, i.e., self-governing city municipalities with an area stretched to the surrounding area and its villages. The majority of the local, autochthonous population remained outside the municipal administration and continued to live in the organization it had at the time of its independence. The indigenous village existed alongside the currents of urban civilization and only partially and superficially adopted some of the new urban ways of life. Until later time—namely, 212, when by Caracalla's constitution all the subjects of the empire received the right of Roman citizenship—most of the indigenous people remained in the status of peregrines and provincials. It should be emphasized that Rome exercised constant supervision in rural municipalities and parishes.

Forums, mansions, baths, temples, and beautiful residential buildings were built in this period in the cities, following the example of Italic and other developed centers. Remains of such buildings and structures can be found in municipalities, primarily those that existed on the site of today's Ilidža, Stolac, Gradina near Sas, Tomislavgrad, etc. Characters from tombstones are almost the only source from which we can learn how local people dressed, what jewelry and ornaments they wore, etc. (Pašalić 1984).

9.1.7 Late Antiquity

The earliest written record that mentions the presence of a new religion, Christianity, on the territory of today's BiH comes from a well-known early Christian worker from the fourth century, St. Jeronimo, a native of BiH. In his biography, he literally states that he was born into a Christian family and that in his native Stridon (probably located somewhere in northwestern Bosnia), there was a Christian community. In 374, he sent an epistle to the deacon there, while, in another one, he mentions the priest of the city of Lupicin. The first reliable information about the organized church comes from the sixth century.

Archaeological finds, unlike written sources, provide far more information about early Christianity. At the end of the nineteenth century, an early Christian oratory (prayer place) with an altar plate was discovered in Vidoštak near Stolac. Similar remains were discovered in Borasi near Vitina, Panika near Bileća, etc. All these finds date back to the invasion of the Western Goths in this area, i.e., before the fourth century (Imamović 1998).

From the beginning of the fourth century to the end of the sixth century, ancient culture changed significantly. These processes were caused by external events such as the penetration of the Germanic and Mongol peoples and changes in the attitude of the bearers of ancient society towards the state, society, religion, and art. Until 480, BiH mostly belonged to the province of Dalmatia, and later (until just after 535), it was a part of Odoacer's and then the East Gothic state for a short time; later, it nominally belonged to the Eastern Roman Empire, but the frequent invasions of the Avars and Slavs hindered its development.

The fortified villa on Mogorjelo near Čapljina (Fig. 9.5) dates back to the beginning of the fourth century, similar to Diocletian's Palace in Split, but less and more modestly equipped. A large number of fortifications were built on the hills, from the fourth to the first third of the sixth century, with an irregular ground plan.

The most famous are Ključ on Sana river, Blagaj on Buna river, parts of the town of Stolac, Koštur on Dabrica, Vrbljani near Ključ, Crkvina at the mouth of the Usora, parts of Bobovac, Debelo Brdo in Sarajevo, etc. Quite numerous early Christian basilicas, from the fifth and sixth centuries, are modest by size. Their interior was usually richly and distinctively decorated. The art of late antiquity is attributed to the local Romanized population. Influences from Syria are also undisputed, and it is unknown to what extent have the Eastern Goths also influenced its formation (Basler 1972).

The ancient and late antiquity period in BiH is still nowhere near sufficiently explored. Mainly residential buildings were investigated, only a few fortifications, a small number of sarcophagi, stelae, wall tombs, and early Christian basilicas. At the very beginning of late antiquity, Christianity and Mithraism coexisted in terms of religion. The most important and most reliable sources for studying BiH in late antiquity are archaeological finds that can be used to explore the cultural, historical, and other manifestations that occur in the society of that time.

Due to the war situation in the fourth century, new fortifications (castella) began to be built, restoring the old ones from the time of pacification (first–third centuries) and building shelters (refuges, strongholds) for the people. This category of sites includes Gradac in Homolj near Kiseljak, Gradina in Dabravine near Vareš, and Gradina in Karaula near Kakanj. Significant archeological sites are represented by iron mines and metallurgical workshops of that time. This is the area of northwestern Bosnia (Crkvina in Blagaj near Novi Grad, Majdanište in Maslovare, etc.). The mines welcomed the invasions of the Avars and Slavs, and they were mainly exploited by the Roman army. Late antiquity stelae are also important, especially for the study of spiritual culture. They point to art and the ritual of burial. Except for one from Donji Ribnik near Ključ, all belong to central Bosnia: Zenica (Bilmišće), Višnjica (Pod Gracem), and Putićevo. The reliefs on the stelae are without refined shaping, and this art is accompanied by a pronounced symmetry, the appearance of proportionality of individual limbs, etc. Incineration as a burial method existed in the fourth century, although it is evident from material evidence that inhumation prevailed in the whole area. Thus, from the fourth century onwards in BiH, we do not find a stela as the main bearer of the rite of incineration.

The mausoleum from Šipovo is a representative of the ritual of inhumation, as are the grave constructions covered with slabs (Jajce, Grudine in Bugojno, Sarajevo). The Mithraic religion was also the state religion at the beginning of the fourth century and as such, was in constant conflict with Christianity until it became a free religion in 313, and a little later, the religion of the Roman Empire. From the fourth to the sixth century, large goods were formed in agriculture—latifundium (Mogorjelo, Suvaja). Thus, the slave-owning system gradually disappeared and moved into feudalism (Paškvalin 1988). The area of the upper course of the Neretva river, the unexplored part of the Roman province of Dalmatia, is closely related to the cultural, political, and economic development of one, from a scientific point of view poorly explored Illyrian people, Narensi (Šačić 2012).

Ancient settlements in BiH have not been sufficiently researched, and it is not possible to say with certainty where the location of any of them is. An archaeologically extremely rich area in central Bosnia, with numerous ancient

Fig. 9.5 Mogorjelo villa rustica near Čapljina from the fourth century. (Photo by Aida Bidžan-Gekić)

settlements, is today's territory of the municipality of Bugojno. Unfortunately, it has not been sufficiently researched either. The largest ancient settlement is presumably in the west of today's town of Bugojno (in Čipuljić). Extensive archaeological research at the Grudine site in Čipuljić has established four cultural strata: prehistoric, Roman, late antiquity, and medieval, a large medieval cemetery with over 750 excavated graves. Finds such as the early Christian Basilica and money, among others, dating from the second to the sixth century (Bojanovski 1988b), also support the location of the ancient settlement of Bistue Nova.

9.1.8 Middle Ages

The short-lived penetration of the Western Goths affected only a small, southern part of Herzegovina, so the traces of their demolition are quite insignificant in relation to other areas through which they passed. The Eastern Goths briefly passed through BiH in the middle of the fifth century.

In this period, two phases stand out in BiH: the older, already marked by the penetration of the Eastern Goths in 459, and especially the inclusion in their state (from 489 to about 536/540), and the younger, characterized by invasions of Avars and Slavs from the mid-sixth century, the fall of Salona (around 614), and the final stabilization of the Slavs during the following centuries. Archaeological monuments of the older period are the findings of runes (Breza), jewelry (Potoci, Mihaljevići), Sarajevo, Gornje Pećine, Rakovčani), or money (Vinica, Bihać). Apart from sporadic specimens of jewelry from the Avar circle (Velika Kladuša, Čebe, Duvanjsko polje), the oldest Slavic settlements with semi-dugouts (Mušići, Batković, Ostojićevo), remains of settlements on forts or ancient complexes, forts, traces of incineration, and a rich inventory of many necropolises and graves, with the characteristics of the Frankish cultural circle (Mogorjelo, Mostar, Stolac, Bugojno, etc.), belong to the younger phase. Monuments of sacral architecture and sculpture, along with frescoes, show the characteristics of Byzantine and early Romanesque art (Glamočko polje, Sarajevsko polje, Zavala, Trebinje, Humac) (Lovrenović 1998).

The peace established during the reign of the Eastern Roman Empire was short-lived. Avars and Slavs began to gather on the borders of the empire. In 602, they crossed the Sava river (the northern border of today's country) and invaded the territory of BiH. They headed inland through quality Roman roads and river valleys. In the resulting whirlwind, all the achievements of ancient civilization in this area finally disappeared. It is difficult to find out what happened on the soil of today's BiH during that time because we lack written reports. Relying on the archaeological material, it can be assumed that the autochthonous population, to a large extent, suppressed the penetration of the Slavs. They achieved this due to the good weapons supplied from their rich mines and the hilly and mountainous configuration of the country, all of which were unfavorable to the invaders. Those who did manage to penetrate (primarily the Avars) quickly drowned in the indigenous mass, so it can be said that the areas of today's BiH entered the Middle Ages with a compact indigenous (Illyrian) population. Such a situation did not last long because the country was in a complete Slavic environment. Slavification followed, to such an extent that Bosnia seemed to be a Slavic country in the Middle Ages, although, in essence, it was not (Imamović 1998).

The Slavs, mixed with the Avars, did not occupy the Balkan Peninsula all at once but for a longer period of time (sixth–seventh centuries). The Slavs did not settle everywhere in compact tribal groups but mixed with each other.

Thus, in the seventh century, there were 25 peoples in the European part of the Byzantine area under the common name "Slavae"—mixed with the Avars, who gave the Slavic masses political leadership by directing its movements and attacks on the centers of Byzantine rule.

Until the Slavic settlement, the Illyrian ethnos had a millennium and a half of continuous life on the territory of today's BiH. During the four and a half centuries of Roman rule, the upper stratum of Illyrian society was largely Romanized, while most retained their language and established a way of life. With the arrival of the Slavs, the process of merging a large number of autochthonous non-Romanized and semi-Romanized Illyrian ethnos and Slavic newcomers, especially in some more isolated areas, was intense, and the Illyrian autochthonous element played a vital role in forming the cultural, somatic, and psychological characteristics of the population. The symbiosis of two diverse ethnic elements is evidenced by the relatively well-preserved pre-Roman and Roman toponymy (names of rivers and mountains) and numerous linguistic borrowings that have become permanently ingrained in the vocabulary of the settled Slavs; the name of the river Bosna seems to be a translation of the Illyrian name "Bathinus" (Basanius).

At least a century before the mention of the first name, the famous Bosnian Ban Borić (1154–1163), the idea of Bosnia as a unique early feudal state was created in the surrounding world. Even its frequent mention until the end of the eleventh century shows that Bosnia in that period was a territorial-political organism whose borders in the east-west direction included the area around the upper Drina to the upper Vrbas. The then interstate relations in the western part of the Balkan Peninsula were in the same line as Raška and Croatia.

The strong Avar influence on the social development of the Bosnian Slavs is evidenced by the ruling title urun (urum), undoubtedly of Avar origin meaning gentleman, which is found carved in the Humac plate in the tenth/eleventh century, next to the name of the founder of the church of St. Michael. Also, the Bosnian medieval ruling title ban originates from the Avar basis bajan (rich, possessor) (Lovrenović 1998).

The first mention of the name Bosnia appears in work *De Administrando Imperio* by Constantine Porphyrogenitus in the middle of the tenth century. He mentions Bosnia (Βοσωνα/Bosona) as a "small country" inhabited by Bosnians (*Bošnjani*). The parishes of Pset (Bosanski Petrovac) and Pleba (Pliva) are also mentioned, as well as many inhabited towns that were probably fortifications and centers of certain parishes on the territory of today's BiH. In the original Bosnia, he mentioned only two cities, Desnik and Katera, for which, despite research, the exact location is not known. Historical and archaeological information about early medieval Bosnia is very scarce. There are no reliable data on the sociopolitical system in BiH in most of the early Middle Ages, nor on the cultural and geographical conditions among the settled Slavs.

After the collapse of the first Avar Khaganate, caused by the defeat in 626 in Constantinople, the Avars were forced to leave the Balkan Peninsula and the territories under the nominal rule of Byzantium. From that moment, the merging of cultural, ethnic, and political units into a new core, known in history as the Sclavinia, began. The newly arrived south Slavs and the Illyrian indigenous-Romanized population took part in this process. The Slavs prevailed, imposed their language, and became the ruling social class. Regardless of the number, which was by no means negligible, the natives had a secondary role, but their characteristics still entered the genetic characteristics of the new national-political phenomena. In the sixth century, the Slavs inhabited the entire Western Balkans, including Bosnia (Fine 1991).

It was established that the medieval Bosnian polity stretched from the Sarajevo valley in the south to the Zenica valley in the north, while the eastern border was the Prača valley towards the Drina and the western went along the Lepenica and Lašva valleys. It is believed that the nucleus in which the earliest Bosnian state developed is located in the Visoko valley (Filipović 2017). It is known that the first known ruler of Bosnia, Prince Stjepan Bosanski, ruled the present-day area of the Sarajevo and Visoko valleys.

In the Middle Ages, mainly in today's BiH but also partly in today's Croatia and Montenegro, the state of Bosnia existed, which at the peak of its power was one of the strongest Balkan states (Malcolm 2011). The peak of her power was in the time of King Tvrtko I Kotromanić. It is believed that the medieval Bosnian state was constituted at the end of the eighth century and is "apparently the oldest among its neighbors" (Klaić 1994). Medieval Bosnia lost its independence to the Ottomans in 1463.

Through its internal development and the influence of more developed and advanced external social and political factors (mostly Byzantium) and the Franks, the old clan-tribal society began to disintegrate in Bosnia, and elements of early feudal society developed. By the end of the nineth century, Bosnia had become a state with developed feudal characteristics. The Bosnian feudal system is based on the noble heritage that is common in Western European countries. It is interesting that until the fourteenth century, slavery was very developed in Bosnia.

In the developed Middle Ages, the ruler in Bosnia had the title of ban. From the coronation of Ban Tvrtko I Kotromanić as king (1377), Bosnian rulers would carry the royal title until the country lost its independence, i.e., until the fall of Bosnia to the Ottomans in 1463. Neighboring countries played a significant role in the political life of Bosnia, mostly Byzantium and the Kingdom of Hungary. These two powers fought for supremacy on the Balkan Peninsula and had a significant influence on the history of medieval Bosnia.

Although the institution of a ban in Bosnia is mentioned earlier, the first Bosnian ban known by name was Ban Borić (1154–1164) who sided with the Hungarian king.

After Ban Borić, the famous Kulin Ban (1180–1204) came to power in Bosnia. Ban Kulin came to power before 1180 as a Byzantine vassal. With the death of the Byzantine emperor Manuel Comnenus, the Byzantine presence disappeared, and his independent rule over Bosnia began. In its further development, Bosnia was more inclined to Hungary, with whose presence it built its own independent expression. Ban Kulin militarily strengthened his state and expanded it to the parishes of Soli and Usora so that Bosnian rule was established over the entire basin of the river Bosnia. The territorial expansion of Bosnia is an expression of the previous internal development of the state. This was manifested by economic indicators, as well as international respect. Wanting business cooperation with neighbors, Ban Kulin concluded a trade agreement with Dubrovnik. The resulting document, the Charter of Kulin Ban of August 29, 1189, shows several segments of the growing Bosnian state and is one of the most famous documents in the history of BiH. The Charter of Kulin Ban was written in Bosnian Cyrillic, widely known as Bosančica (an extinct variant of the Cyrillic alphabet that originated in medieval Bosnia) in a language in which the Old Stokavian vernacular is mixed with Church Slavonic forms, with parts in Latin language written in Latin alphabet (Fig. 9.6).

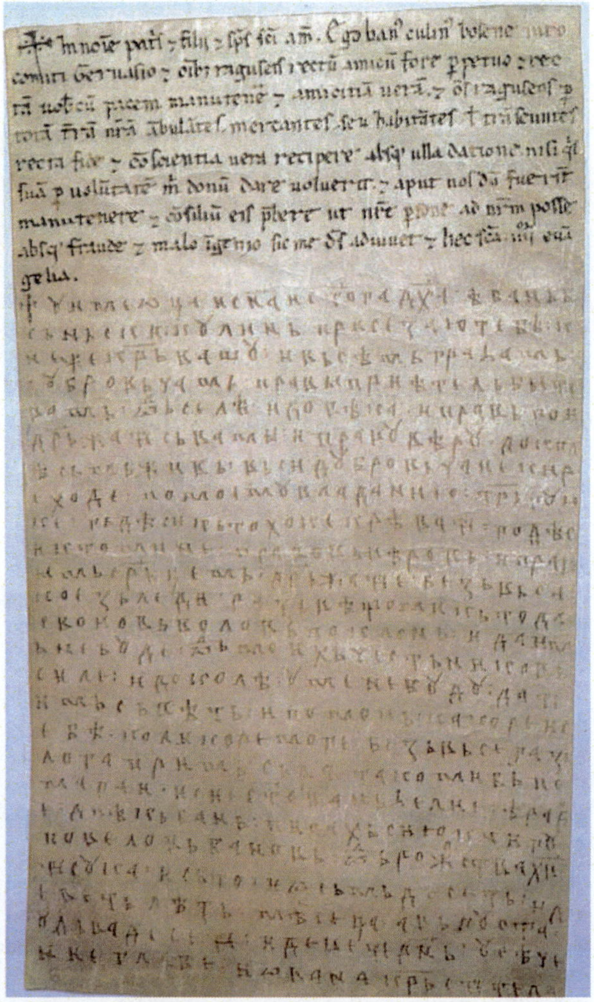

Fig. 9.6 Charter of Kulin Ban. (Photo by Aida Bidžan-Gekić from Državni arhiv in Dubrovnik)

At the end of the twelfth century, accusations began that Kulin Ban was protecting heretics. This did not suit Pope Innocent III, who wanted to destroy the heretics. He called on the Hungarian-Croatian king to crusade against what he said were the infidels in Bosnia. Realizing the danger that threatened Bosnia, Kulin Ban, in the presence of a large number of subjects and believers in front of the papal envoy Kazamaris, on Bilino polje near Zenica in 1203, publicly accepted the teachings of the church as required by Rome. He managed to avoid a great conflict, but the religious situation in Bosnia, with the presence of a different religious understanding, gained a new dimension that would mark its medieval existence and a recognizable reflection in the then Western European world. Kulin Ban left a significant mark on the rise of Bosnia among the most important entities on the eastern Adriatic coast. His time is remembered as a time of Bosnian independence and unmistakable identity. That is why, not without reason, the time of Kulin Ban in the history of medieval Bosnia is recorded by a folk proverb that reads "ever since the Kulin Ban and good old days" (Nakaš et al. 2018).

In the thirteenth century, religion will get its organizational form and character as the Bosnian Church with a hierarchy and teachings that will take strong roots among the population. Under the pretext of destroying heresy and intending to conquer Bosnia, Hungary would undertake a series of crusades against Bosnia. These conflicts strengthened the unity between the new faith and the nobility, and the result was the expulsion of the organization of the Bosnian Catholic Church (Bosnian Diocese) to Hungary (Đakovo). The Bosnian Church became the only "heretical" church ever to become an earthly church, the Bosnian state religion.

At the beginning of the fourteenth century, Bosnia became stronger again and, over time, grew into the most powerful state in the Balkans. It was not long until Ban Stjepan II Kotromanić added the area of Hum to Bosnia and the territory from the Neretva to the Cetina, which became part of Bosnia. Ban Stjepan II Kotromanić also annexed Bosnia and a large part of Dalmatia, from Dubrovnik to Split, and Bosnia also annexed the lower regions, Usora and Soli. He established very good relations with foreign powers, primarily with Hungary, and assisted the Bosnian king with the Bosnian military troops in quelling the revolts of the

Hungarian nobility. Ban Stjepan II Kotromanić then concluded a trade agreement with Dubrovnik in 1334, and in 1335 with Venice. In his Charter to the People of Dubrovnik in 1332, he wrote: "If a citizen of Dubrovnik has any claims towards a Bosnian (*Bošnjanin*), let him be summoned before ban. . . ."

This is a very important document because it says that Bošnjani (or Bosnians) live in Bosnia, Bošnjani as inhabitants of the medieval Bosnian state. This is evident in all official documents of the Bosnian state and in many unofficial ones, such as glosses on *krstjanskim* (Christian) religious books and inscriptions on stećak tombstones, where the inhabitants of medieval Bosnia call themselves Bosnians or Bošnjani (Nakaš et al. 2018).

Ban Stjepan II Kotromanić supported the Bosnian Church, whose hierarchy stood firmly with the Bosnian rulers and nobles, which brought him into conflict with the Pope. In order to appease the Pope, he agreed in 1340 to give a great deal of freedom to the Franciscans in Bosnia, while converting to Catholicism in April 1347, when he sent a letter to the Pope, in which he asked him to increase the number of trained Catholic priests in Bosnia, who according to the requirements should know the Bosnian language. In 1350, the Serbian Empire invaded Bosnia. The army, led by Emperor Dušan, targeted the area of Hum. However, the Bosnian army, led by Stjepan II Kotromanić, wisely managed to expel the Serbian army from the Bosnian country completely.

The Franciscans soon formed the Bosnian Franciscan Vicariate with its seat in Mile (near Visoko), which expanded to include a much larger territory than the Bosnian state, extending all the way to Romania, so that, as early as 1385, it had 35 monasteries, of which only 4 were in Bosnia. These are the Franciscan monasteries in Visoko, Lašva, Kraljeva Sutjeska, and Olovo. By 1463, 12 more Franciscan monasteries were being built throughout Bosnia, mostly in the strong economic development centers of the medieval Bosnian state. A tombstone of Radovan Pribilović (fourth century) was photographed and recorded during a tour of the museum of the Franciscan monastery of Kraljeva Sutjeska by Aida Bidžan-Gekić. The tombstone is one of the valuable monuments of the Bosnian alphabet Bosančica (Fig. 9.7).

Fig. 9.7 A tombstone of Radovan Pribilović (fourteenth century) is a valuable monument of the Bosnian alphabet Bosančica kept in the museum of the Franciscan monastery of Kraljeva Sutjeska. (Photo by Aida Bidžan-Gekić)

The reign of Ban Stjepan II Kotromanić was marked by strong economic development based on mining and trade, especially in exporting the highly sought-after domestic raw materials of silver, copper, and lead. The Bosnian dinar was being minted—Denarius, groschen. New urban settlements, caravan stations, and road communications were being developed. There was an increasing number of foreigners who find their economic interest in Bosnia. It also strengthened the domestic economic base. Through the caravan trade, carried out by armed escort Ponosnici (i.e., the proud ones) under the leadership of Kramari, Bosnian products were exchanged with coastal towns. Numerous trade unions, operating in the wider area, were beginning to emerge. This contributed to the development of customs in medieval Bosnia. By marrying his daughter Elizabeth in 1353 to the Hungarian King Louis I of Anjou, at the time one of the most influential European rulers, Stephen II raised the reputation of Bosnia and the Kotromanić dynasty, which came close to the highest levels of European politics at the time.

Bosnia will experience its greatest rise during the time of its last ban and first king, Tvrtko I Kotromanić (1353–1391). In 1353, Ban Stjepan II Kotromanić died and was buried in the Franciscan monastery in Visoko. He left behind a stable Bosnian state; independent, economically advanced, and militarily very strong. He left behind an advancing power of his vassals, who began to participate in the politics of the central government as a significant political force. After the death of Stjepan II, Tvrtko I Kotromanić, the son of Stjepan's brother Vladislav, only 15 at the time, took the title of ban, while his mother, Jelena, aided him with state affairs. He initially had big problems within the country, and Hungary exploited that.

In 1363, a conflict broke out between Tvrtko and Louis I of Anjou, in which Bosnia emerged victoriously. The young ban wreaked into conflict with Hungary, gaining the trust and prestige of Bosnian magnates. This enabled him to make new territorial and political changes in Bosnia.

After stabilizing power in the country, Tvrtko expanded Bosnia on its southeastern borders. In alliance with the Serb regional lord, Prince Lazar Hrebeljanović, he defeated the Serb regional lord Nikola Altomanović and annexed his land to Bosnia in 1373. Bosnia expanded to parts of the upper Podrinje and Polimlje. At the beginning of 1377, Tvrtko expanded Bosnia to Trebinje and the parishes of Konavle and Dračevica. Thus, Bosnia reached the Adriatic coast and south of Dubrovnik in the area of the Bay of Kotor (Šehić and Tepić 2002).

The advanced economic development, strengthening the country's military force, territorial expansions, and the general situation in a broader context enabled the Bosnian ruler a new ideological and political position of power. Tvrtko decided to declare Bosnia a kingdom. Tvrtko was crowned King of Bosnia in 1377 in Mile near Visoko (today's place of Arnautovići). It was also the coronation place of Bosnian kings to come. According to the charter of April 10, 1378, with the coronation of the king of Serb land, Bosnia, Pomorje, and the western lands, the Bosnian rulers acquired the tribute given on St. Demetrius Day in the amount of 2000 perpers. Along with the earlier tribute of Ston, these are the most important tributes that the people of Dubrovnik had to pay to the Bosnian rulers. Tvrtko's later conquests in Dalmatia and Croatia enabled him to be called the king of Croatia and Dalmatia. To pursue an independent economic policy on the eastern Adriatic coast, King Tvrtko I founded a trading port on the northern side of the Bay of Kotor called Novi (today's Herceg Novi) (Fig. 9.8).

During the reign of King Tvrtko, the Bosnian was the most modern army in the entire Balkans. Bosnia was the first country in the Balkans to introduce firearms, which was then a revolutionary step in military technology at the time. The first use of firearms in the Balkans was made by the Bosnian army on August 13, 1378, during the attack of the Venetian fleet on the city of Kotor. On that occasion, the Bosnian army used three bombers (cannons) and successfully defended the city. After the death of the Hungarian king in 1382, riots broke out in the areas of the Dalmatian coast, and King Tvrtko used this opportunity to send the Bosnian army to that part of Dalmatia. He thus took complete control over the whole of Dalmatia, together with the islands, and annexed Split, Trogir, and Šibenik and the islands of Brač, Korčula, and Hvar to Bosnia. The exception remains Zadar, which was under the rule of Venice, making Bosnia by far the most powerful state in the Balkans.

In the last decade of his reign, King Tvrtko faced Ottoman incursions into Bosnia, first in the autumn of 1386 and then in the summer of 1388 near Bileća, when the Bosnians, led by Duke Vlatko Vuković, defeated the Ottomans. Shortly afterward, the Bosnian army, again under Duke Vlatko Vuković, participated in a powerful military coalition composed, in addition to Bosnians, of the leaders of the coalition of Serbs, Hungarians, Croats, Albanians, and Greeks, which opposed the Ottoman Empire in 1389. While Tvrtko boasted that the Christian coalition won, the Ottomans increased their penetrations even more, and they became the first Bosnian neighbors on the southeastern borders of Bosnia through the vassal estates of Vuk Branković (Kurtović 2019). Although the Ottoman presence in medieval Bosnia can be traced since 1386 and in numerous other incursions of varying intensity and significance, only the events of the mid-second decade of the fifteenth century marked their more active political engagement. Medieval Bosnia developed its cultural and artistic expression in a wide range of diverse currents that covered the eastern Adriatic coast throughout the millennium. The strong outlines of Illyrian, Roman, and Germanic culture merged with Slavic culture through direct contacts in the early Middle Ages. This diversity was greatly complicated by the presence of a strong Frankish, Byzantine, Hungarian, and Mediterranean culture, which, through the development of the medieval Bosnian state and society, had a significant interference and influence on Bosnian territory.

In this diverse encounter and application of various cultural and artistic influences, Bosnia was also developing a domestic Bosnian cultural heritage which, in addition to the application of highly developed foreign influences, also created the features and peculiarities of its own cultural expression and style. Recognizably, Bosnian-style products, made in the Bosnian way, find their application beyond the borders of the medieval Bosnian state in Venice and other European countries.

The development of literacy in Bosnia, where Bosnian, Greek, Latin, and Italian and Glagolitic and Bosnian Cyrillic (*Bosančica*) script were used, in addition to the usual transcriptional activity at the time, apocryphal texts expressing criticism of Christianity and a different view of the world, also offered the pieces of domestic Bosnian literature. The plaque of Humac reflects early medieval literacy, and extraordinary examples such as Hval's manuscript show the interest of the Bosnian feudal community in the exceptional artistic and aesthetic values of the time. The Knights' Tournament in Buda in 1412, the affiliation of the Bosnian rulers to the famous Order of the Dragon, and the development of musical culture show the dedication to the most important currents of knightly and courtly culture in Europe.

The development of the Bosnian heraldry created extraordinary examples of coats of arms and seals of the Bosnian rulers and nobles. This is most faithfully shown through the seals of Bosnian rulers and paintings of Hrvoje Vukčić Hrvatinić. The architecture of medieval Bosnia finds expression in defensive fortresses (Fig. 9.9), courts of rulers (like the magnificent Bobovac, Blagaj, and Jajce), and summer and winter residences of nobles (like Kuknje and Samobor) but also in the usual residential culture of everyday life imbued with the increasing use of stone as a building material, which shows an advanced status of the area. As in other European environments, Romanesque and Gothic styles predominate.

Fig. 9.8 Medieval Bosnian Kingdom expansion. (Optimus Pryme, Creative Commons Attribution-Share Alike 3.0 Unported)

Fig. 9.9 The Počitelj fortress one of the famous medieval fortresses in Bosnia and Herzegovina. (Photo by Aida Bidžan-Gekić)

The most important status symbol of medieval Bosnia's cultural and artistic expression is recognized in the stećak tombstones, the bearers of the still alive, exciting, and enigmatic expression in the culture and art of BiH. The stećak tombstones are medieval tombstones. Prominent necropolises of stećak tombstones are located in Blidinje, Radimlja, Boljuni, Grborezi, and Popovo. A vast number of necropolises with over 50,000 preserved stećak tombstones

Fig. 9.10 Stećci in urban area of Konjic. (Photo by Aida Bidžan-Gekić)

of various shapes, symbols, decorations, epigraphic messages, and their artists still testify to this timeless time.

The decorations on the stećak tombstones are social symbols (coats of arms, animal figures, lilies, and rosettes), religious symbols (cross, solar signs, crescent moon, star, swastika), figural representations (portraits of the deceased, scenes of circle dances, battle games, hunting), and pure ornaments (architectural, plant, and geometric motifs) (Fig. 9.10). The stećak tombstones belong to Romanesque and Gothic art. They are characterized by regional diversity.

The stećak tombstone, a unique example of medieval tombstones, is declared a UNESCO (United Nations Educational, Scientific and Cultural Organization) World Heritage List. This was decided on July 15, 2016, by the UNESCO Committee at its session in Istanbul. The list includes a total of 30 necropolises, 22 of which are located on the territory of BiH.

Inscriptions on stećak tombstones sometimes reveal the confessional composition of individual necropolises, i.e., the affiliation of the deceased to one of the three church organizations present in medieval Bosnia and Hum—the Orthodox, Catholic, and Bosnian Church. This interconfessional character of stećak tombstones is one of the most striking features of the religious life of that time (Lovrenović 2009). Stjepan Tvrtko II Kotromanić (born before 1382, died in November 1443) was a Bosnian king who lived and ruled in a very turbulent period of Bosnian history, where the influences of two great regional powers (Hungary on the one hand and the Ottoman Empire on the other) alternated. He ruled in three periods: from 1404 to 1409, 1421 to 1433, and 1435 to 1443 (Fig. 9.11).

In 1443, the Bosnian nobility appointed King Stjepan Tomaš to power at a session of the medieval Bosnian parliament called rusag. The most powerful Bosnian nobleman in southeastern Bosnia, Stjepan Vukčić Kosača, initially refused to recognize Stjepan Tomaš as the new king of Bosnia, resulting in an internal conflict in Bosnia, which would end in 1446 when his daughter, Katarina Kosača, married the Bosnian king. However, Stjepan Vukčić Kosača continued to have a very strong influence on his estate, which is confirmed by the fact that in 1448 he took the title of *Herceg* (Herzog, duke), after which this part of Bosnia would later be called Hercegovina (*Herzegovina—land of Herzog*) (Filipović 2017).

The Ottoman expansion forced king Tomaš to seek foreign aid in 1450 and address the Pope on the matter, using the fact that the pope showed a great interest in Bosnia, primarily because of the Franciscans, who at the time, led by the Vicar of Bosnia named Jakob Markijski, sought to establish Catholic domination in Bosnia since King Tomaš converted to Catholicism in 1445, which the Pope wanted to take full advantage of. Also, the Pope was obsessed with the issue of the Bosnian Church, accusing it, for who knows how many times, of heresy, so that in 1459, King Tomaš agreed to the Pope's request to start persecuting "heretics" in Bosnia.

King Tomaš convened the elders of the Bosnian Church and offered them a very limited choice: to accept Catholicism or be expelled from Bosnia. According to the Pope's later records, about 2000 of them agreed to convert to Catholicism, and of the others who did not accept, about 40,000 took refuge in southern Bosnia, in Hum, and found protection under the Stjepan Vukčić Kosača, who also belonged to Bosnian Church. This was described in detail in records of the then Pope, Pius II, who later, in his Comments (Commentarii), also writes:

> The King of Bosnia, to wash away the stain left by surrendering Smederevo (a city in today's Serbia) to the Ottomans and prove his Christian faith, or, as many believe, motivated by greed for

Fig. 9.11 King Tvrtko II coins kept in the National Museum of BiH. (Photo by Aida Bidžan-Gekić)

money, forced the Manicheans (their real name is *krstjani*—followers of the Bosnian Church), which were numerous in the kingdom, to emigrate, leaving their possessions, should they not receive the baptism of Christ. About two thousand were baptized, and about forty thousand or a little more, who stubbornly clung to delusion, fled to Stephen (Vukčić Kosača), his friend in disbelief.

That Bosnia was full of "Manicheans" even before the fall is proved by the report of Bishop Nikola Barbuči to the Pope in May 1456, in which he states that King Tomaš justified himself by saying that he could not fight the Ottomans because of the "Manichaeans," i.e., *krstjani*—followers of the Bosnian Church, who prefer the Ottomans over the Christians and who are in large numbers in his country (Ćirković 1964). It should be noted that the thesis that the Bosnian Church was dualistic or heretical has not yet been scientifically proven, and this was the position of the predominantly Catholic Church led by the Pope.

Catholics viewed the Bosnian faith as paganism, which meant a dualistic religion. The members of the Bosnian Church did not remain indebted to other churches either, calling them "Latin heresy" and "Greek heresy." Members of the Bosnian Church called themselves *krstjani* (Christians). Representatives of the Bosnian Church did not build religious buildings similar to those in Catholicism and Orthodoxy. They had no possessions either.

In 1461, Stjepan Tomašević, the son of King Stjepan Tomaš, was elected the new king of Bosnia. He faced a serious threat from the Ottoman invaders and their pretensions to Bosnia. He appealed in vain for help in defending himself from the Ottoman Empire. After the Hungarian king promised to help Bosnia in 1462, King Stjepan Tomašević refused to pay tribute to the Ottoman Empire. This endangered the peace in Bosnia. Thanks to a strong foothold in the parish of Vrhbosna, the Bosnian Frontier, the Ottomans were able to penetrate Bosnia very quickly. A year later, in 1463, a large Ottoman army, under the command of Mehmed II, conquered Bosnia. In a strong Ottoman raid, the Bosnian royal residence of Bobovac (near Vareš) fell on June 20, 1463. King Stjepan Tomašević fled west to the Ključ fortress. Surrounded by Ottoman forces, he decided to surrender, after which he was taken to Jajce and executed before Sultan Mehmed II the Conqueror. Thus, the medieval Bosnian state ceased to exist (Kurtović 2019).

In 1448, Stjepan Vukčić Kosača sided with the Ottomans and the Serbian despot Đurađ Branković in the attack on the Bosnian king and adopted the title of duke, which was addressed to him by Frederick III, emperor of the Holy Roman Empire. During his third marriage with the Bavarian princess Cecilia, he used in correspondence the German transliteration of his title of the duke—Herzog (which, in the vernacular, he evolved into "herceg"). The territory under his rule stretched from Lim to Cetina and from Rama to the Bay of Kotor. This was the area where he ruled independently.

Herceg Stjepan had several residences in two cities: Blagaj during the summer and (Herceg) Novi during the winter. The aspirations for economic independence were also expressed in his efforts to develop the city of Novi into a maritime and trade center (in 1449, he founded a broadcloth mill). At the same time, the war with Kotor and Dubrovnik broke out, which lasted from 1451 to 1454. The Second Konavle War, in which the Herzog's opponents also got involved (King Stjepan Tomaš, the Herceg's son Vladislav Hercegović Kosača, the noble Vlatković family, and some of the dukes' vassals). The Ottoman military aid sent to the Herceg shook his opponents, so in 1453 he reconciled with his son Vladislav and then, in 1454, with Dubrovnik. As the Ottomans invaded his country, he reconciled with the new Bosnian king, Stjepan Tomašević (1461). After the fall of Bosnia in 1463, his lands were gradually conquered by the Ottomans (Hum and Podrinje) and the Venetians (Neretva region) from 1465. Towards the end of his life, his power was reduced to a narrow coastal area with the town of Novi, which the Ottomans occupied in 1482. Herceg Stjepan Vukčić Kosača died in 1466. After the Ottoman conquest, the land once owned by Herzog was named Herzegovina (Filipović 2017).

Contrary to popular belief, the Middle Ages were a time of clear rules and norms of behavior, conditioned primarily by the biblical understanding of the world and its correct order. Only a certain number of families had the "divine authority" to rule the kingdoms of this earthly world, and it was impossible to enter that circle without authorization. All theories that the Bosnian nobility rejected this order because they belonged to the dualistic heresy have been proven wrong today, and, in the future, it is necessary to look at the actions of that nobility through the prism of norms and behavior of the nobility in the near and far surroundings (Dautović 2014).

The Bosnian Church as an institution developed much later than the custom of Bosnian-style marriage and is only associated with the perseverance of marriage practice and by no means with its origin and evolution. Bosnian-style marriage and interconfessional marriages were among the most interesting features of the period. They differentiate Bosnia from contemporary societies in adjacent and more distant countries. The Bosnian style of marriage demonstrates, unlike any other type of contemporary source, the difference in practice between the Bosnian and the Roman churches. Even though the Bosnian Church did not institute this custom, it nurtured it at a time when the rite of marriage was slowly but steadily establishing its place in the theology of the Roman Church (Dautović 2019).

The Bosnian Church (in Latin Ecclesia bosniensis) was an autochthonous Christian church in Bosnia that existed during the Middle Ages. The Bosnian Church was a community of heterodox Christians, which appeared in the late twelfth century in the medieval Bosnian state. It represents the most important identity and expression of the Bosnian state of that time. The members of this Church were called Krstjani (Christians, "adherents to the cross"). Church Slavonic was used in the rites, and the bishop was chosen from among the monks (Filipović 2017).

9.1.9 Period of the Ottoman Rule

For almost a century, the Ottomans maintained constant military, economic, and political pressure on Bosnia, undermining the sovereignty and the established system, weakening the power of rulers and peoples, and, at the end of 1463, organized in a military manner the last episode of Bosnian-Ottoman relations. Despite that, certain territories of the former Bosnian state were maintained even after the famous 1463, when neither the state nor the institution of rulers existed. The end of the medieval Bosnian state must be observed in the context of the events of that time in Southeast Europe, where many states suffered the same fate at the same time, i.e., the penetration of the most powerful empire of the world at that time. These would be well-founded historical facts, and the question remains to what extent would the man of today, in forming their attitude towards the past, pay attention to fundamental scientific achievements based on many years of research into demanding historical sources.

It was a showdown between the most potent power in the world (the Ottoman Empire), which invaded Europe and which, before the destruction of the Bosnian state, had already wiped out the mighty Byzantine Empire and conquered Constantinople (later Istanbul), and threatened the powerful Central European states coming close to Rome and Vienna. Due to the fact that they did not help Bosnia at the time, a myth was built in Europe soon after the fall of the Bosnian state about how the Ottomans conquered Bosnia thanks to the weak resistance of the Bosnians. It was an attempt to "wash the conscience." In essence, this is also happening today, with the attempts to construct the story that the Ottomans in the mid-fifteenth century were not conquerors but liberators, which is absolutely meaningless. It is a myth that could be further talked about in detail.

The Ottoman period of Bosnia lasted from 1463, when the Bosnian Kingdom collapsed, to 1878, when Austro-Hungary occupied the Bosnian pashalik. The arrival of the Ottomans was an important event in the history of BiH and had enormous religious, linguistic, cultural, political, and military consequences for the area of present-day BiH. During the Ottoman rule, Islam and Judaism appeared significantly in Bosnia for the first time, while Orthodoxy spread to the entire area of today's state. Politicians and thinkers of Bosnian descent have influenced the course of history far beyond the borders of their homeland. Cities like Sarajevo and Mostar have been founded in Bosnia, and with them, impressive architectural structures such as the Old Bridge and the Ferhadija Mosque. The modern borders of BiH also date from this era.

Before the final fall of the Bosnian state in 1463, the Ottomans had already subjugated large parts of the territory to their rule. Those areas that had not yet been subjugated were in a vassal relationship with the Ottoman sultan. Bosnia was a real battlefield for more than 50 years to come. In the conquered area, the Ottomans organized a new administrative unit, the Sanjak of Bosnia, with its headquarters in Vrhbosna (close to the present-day Sarajevo). It was included in the Rumelia eyalet. As Herceg Stjepan showed intentions to regain the lost territories of BiH, the Ottomans occupied the whole of Herzegovina, except for small parts of the coast. In that area, in 1470, Sanjak of Herzegovina was formed. It included all areas from Prijepolje (today's Serbia) to Makarska (today's Croatia) that were once under the rule of Herceg Stjepan. That Sanjak was also attached to Rumelia eyalet. After an inevitable stalemate in the early sixteenth century, the Ottomans again undertook large-scale military actions. They reoccupied the Srebrenica Banovina in 1512 and formed a special Sanjak of Zvornik, while Sanjak of Klis was formed after the occupation of Klis in 1537. In July 1580, all this territory, along with the Sanjak of Krk-Lika, Sanjak of Prizren, and Vushtrri, was united into one administrative-political and military unit, the Bosnian Pashalik or Eyalet.

The Ottoman Empire was essentially a military organization, interested only in new conquests and tax collection, and its administrative system was designed to provide primarily two things: manpower for the army and money to pay for it. The first tax register of the Ottoman Empire was established in Bosnia as early as 1468. From there, it can be seen that, in the first 5 years, not a large number of the population converted to Islam. According to the tax register from 1520, the number of Muslims exceeded 45% of the total number of inhabitants, while the tax register from 1624 tells us that in Bosnia there live about 150,000 Catholics, about 75,000 Orthodox, and 450,000 Muslims. Thus, the main process of conversion to Islam took about 100 years and did not happen overnight. Throughout the Ottoman Empire, Bosnia had a privileged status in relation to all other countries which were a part of it. Bosniak feudal lords not only retained their estates but also had the right to inherit them within their families, and, as we will see later, Bosnians in Bosnia, over time, completely took the local power into their own hands, which was not the case in other countries which were the part of the Ottoman Empire. From the already mentioned tax registers, one can clearly follow another process, and that is the spread of Orthodoxy in Bosnia. Orthodoxy spread among Bosnians during medieval Bosnia, with almost no results, except in Herzegovina, where some Bosniaks accepted Orthodoxy peacefully, while, around the Drina river, it spread violently. However, after Bosnia became part of the Ottoman Empire, Vlachs (Orthodox cattle breeders) immigrated to Bosnia, and through them, another part of Bosniaks accepted Orthodoxy. Like the Bosnians who had previously accepted Orthodoxy, they became Bosniak Orthodox, and a part of the Vlachs became Bosniaks (Šabanović 1982).

The Ottomans were initially dissatisfied with the conquered areas in northwestern Bosnia. Therefore, they constantly attacked the lower course of the Una and Bihać rivers, which they finally occupied in 1592. The inclusion of Bihać and the Bihać region in the Bosnian eyalet completed the Ottoman conquests in that area. The territory of eyalet was therefore completed, and its formation brought the territory of the medieval Bosnian state as a whole into a new administrative unit. Thus, the conquest of Bosnia lasted more than a century and a half. During the period of temporary and, at times, permanent Ottoman rule, certain areas were given to the management of prominent soldiers, usually beys. These larger or smaller areas were called vilayets. The permanent system of government is characterized by the establishment of the judiciary (qadi) and the related administrative division. The largest administrative unit within the eyalet, where the entire administrative and judicial power takes place, was the Sanjak.

In the conquered BiH, the Ottomans, following their ideology and practice, destroyed the previous government and its bearers and the higher nobility and introduced a Sipahi-Timar system based on giving privileges to the members of the Islamic religion. Society was divided into two basic classes: the ruling feudal and subordinate rayah, which created material resources through labor. Soon a large population of all social and found religious groups converted to Islam, so the people of the Islamic religion became the most numerous. The weakness and non-unity of the church organization in most of Bosnia facilitated the conversion to Islam. It is very important to point out that there was no forced Islamization as a practice in BiH.

Progress in the development of the Ottoman Empire by the end of the sixteenth century was also reflected in Bosnia. The stability of the system and legality were felt. The land was cultivated, and the benefits were regular and strictly controlled. Cities and their economy began to develop rapidly. Thus, Sarajevo became one of the largest cities in the European part of the Ottoman Empire, and Mostar, Banja Luka, Foča, Livno, etc. also developed. Cities were built mainly by autochthonous inhabitants. In the first instance, sacral buildings were built, followed by trade and craft shops, barracks and other economic structures, and residential areas in their immediate or distant surroundings (Zlatar 1998).

Very early, Bosnia became a border area towards the Austrian and Venetian possessions in Croatia and Dalmatia and therefore exposed to all the troublesome characteristics of such regions. The incursions on both sides were frequent, with the aim of looting, killing, and enslaving individuals. Due to that, a system of captaincies as military institutions was established on the borders of Bosnia, which gradually expanded towards the interior. The crisis in the Empire began to be felt more strongly in the late sixteenth century and continued throughout the seventeenth century. It was created after the end of the Ottoman conquests when the sources of income began to dry up. At the end of the seventeenth century, the so-called great wars finally ended with the Ottoman Empire's defeat, leaving Bosnia directly exposed to attacks by Austria and Venice (Fig. 9.12).

In the Bosnian eyalet, over 75% of the population was of the Islamic religion, and this share was later steadily declining as the Ottomans brought over Vlachs (livestock population of predominantly Orthodox and Catholic faith from neighboring countries) as labor forces. During the seventeenth century in BiH, urban development flourished, especially in the urban economy and culture. It is the age of appearance of prominent creators in every scientific and artistic discipline. A special place is occupied by Hasan Kafi Pruščak, whose works have been translated into numerous languages around the world (Pelidija 1998).

The general crisis in the Empire is a consequence of the disruption of agrarian relations, which is reflected in the emergence of the process of chiftlicization. This process means the suppression of the military strata of Sipahi and the creation of the strata of noncombatant possessors, and the attack on the productive population and its security. The defeat of the Ottomans under Vienna (1683) and the following years in Hungary, then the deep penetrations of the Austrian army into Bosnian territory greatly damaged Bosnia

Fig. 9.12 Eyalet of Bosnia around the year 1600. (Modified by Haris Gekić from Šupuk 2016; Creative Commons License Attribution-ShareAlike 3.0 Unported)

(the penetration of Eugene of Savoy to Sarajevo in 1697; first he occupied Doboj and Maglaj without resistance on 16 and 17 October 1697, and on the same day he plundered and set fire to Visoko, and then quickly penetrated to Sarajevo, captured the city, and plundered and burned it; due to the coming winter, Eugene of Savoy withdrew on October 25, 1697, taking a large number of inhabitants into slavery).

Partly under the influence of their churches, and partly due to the weakening and deformation of the system, the Christian population, which was in a position of rayah until the end of the seventeenth century, showed growing sympathy for Christian states and readiness to participate in wars against the Ottoman rule. The first Serbian uprising (1804) caused a lively movement among rayah in BiH. After the victory of the Serbian insurgents in the Battle of Mišar (1806), where many Bosnian beys and aga were killed, the self-awareness and rebellion of the rayah increased throughout Bosnia. Muslims considered the Ottoman Empire their own, because it was the bearer and protector of the Islamic religion. However, there was no complete unity between them. There were class differences among Muslims, and they were further intensified by chiftlicization. When the classical Ottoman system disintegrated and open conflicts between Bosnian Muslims and Sublime Porte arose, their unity came to the fore.

Captain Husein Gradaščević (1802–1834), better known as the Dragon of Bosnia, was a military leader in the service of the Ottoman Empire and later a Bosnian captain who fought for Bosnian autonomy in the Ottoman Empire. In 1831, he was elected leader of the movement for Bosnian autonomy and took over the title of vizier. He quickly expelled the Ottoman vizier and other opponents of Bosnian autonomy and thus became the de facto ruler of the Bosnian pashalik. The uprising was aimed at achieving autonomy, lasted for 3 years, and included a demand for the abolition of Ottoman loyalties, mainly in Herzegovina. Among the notable achievements is that on July 18 of the same year, when Gradaščević led the Bosnian army, which defeated the Ottoman army under the command of the Grand Vizier near Štimlje in Kosovo. He decided not to continue the campaign and returned to Bosnia, where his soldiers proclaimed him the new vizier on September 12. However, after 3 years of fighting, the uprising failed, while all captaincies were abolished by 1835.

Gradaščević had to flee to the Habsburg Monarchy on May 31. From there on, he negotiated his return to the Ottoman Empire with the Sultan, which granted his wish, but he could not return to Bosnia. He moved to Belgrade and then to Istanbul, where he died of cholera on August 17, 1834. Gradaščević received the honorary title of "Dragon of Bosnia," and even today, he is considered a Bosniak national hero and one of the most prominent figures in the history of BiH.

At the beginning of the last quarter of the nineteenth century, an armed uprising broke out in BiH on a larger scale. With the entry of Serbia and Montenegro into the war against the Ottoman Empire, social problems were pushed into the background, and political elements imbued with national ideas and with strong religious accents emerged. The development of events was significantly influenced by the fact that Austria-Hungary showed a particular interest in BiH. The uprising began in June 1875 in Herzegovina with the revolt of Orthodox and Catholic peasants and then spread to Krajina (north-western Bosnia), parts of eastern Bosnia, and Sanjak (today parts of south-western Serbia and north-western Montenegro). The uprising was forcibly suppressed. BiH changed governor with a later political game and international settlements at the Berlin Congress in 1878. The new governor brought a well-established policy of exploitation, a strong political system, and a determination to oppose the national liberation of the Bosnian population. Two neighboring Slavic states (Serbia and Montenegro) did not assist the people of BiH because their interests did not coincide (Tepić 1998).

9.2 Contemporary Historical Geographies

9.2.1 Period of the Austro-Hungarian Rule

The Austro-Hungarian occupation of BiH began on July 29, 1878, and was finally completed on October 20. The occupying army was met with great resistance from the Bosnian population, which was broken only after 2 and a half months.

Unlike other provinces in the Austro-Hungarian Empire, BiH was not part of either the Austrian or the Hungarian part but was the place where Austria and Hungary ruled together. For the entire period of Austro-Hungarian rule, military and civilian power were not completely separated. Given the Sultan's formal sovereignty over BiH, its international and domestic position remained unresolved, making it difficult to consolidate new power. The annexation was a permanent goal of the Austro-Hungarian Empire. It was proclaimed after extensive preparations on October 5, 1908, definitely making BiH a part of the Austro-Hungarian Empire. In 1910, Emperor Franz Joseph declared the national Constitution in BiH, after which numerous other laws were passed, and numerous institutions were introduced.

In the period 1878–1908, from occupation to annexation, BiH was formally an Ottoman province, but the real power belonged to Austro-Hungary. According to the Berlin Agreement and the Constantinople Convention, sovereignty over BiH belonged to the Sultan, but in reality, that right was nudum ius because the actual exercise of power was in the hands of the Austro-Hungarian Empire. After the annexation, BiH formally became a province, but it also retained a unique position as a separate body within the Austro-Hungarian Empire. The inhabitants of BiH were not legally Austrian or Hungarian citizens but were treated as Bosnian nationals. As a corpus separatum within the Monarchy, Bosnia was not equal to its other two states. BiH did not have any self-governing rights, nor could it participate in the so-called common affairs in the Monarchy, so in political terms, it was in an entirely dependent position in relation to Austria and Hungary.

The Austro-Hungarian administration advocated a pluralistic and multi-confessional Bosnian nation. The Joint Imperial Minister of Finance and the Vienna Governor Béni Kállay thus supported Bosnian nationalism, aiming to instill a sense of belonging to a great and mighty nation in BiH. They saw Bosnians as a nation that 'speaks Bosnian and is divided into three religions with equal rights." Before them, between 1861 and 1869, Topal Osman Pasha, the Ottoman Grand Vizier, sought to do the same. On the one hand, this policy aimed to try to isolate BiH from its irredentist neighbors (Orthodox Serbia, Catholic Croatia, and the Muslim Ottoman Empire) and to marginalize the already established ideas of Serb and Croat nationality among Bosnian Orthodox and Catholic communities, supported by the neighborhood (Serbia and Croatia) for decades. Most Orthodox and Catholic politicians eventually ignored or opposed such a policy. At the same time, Austro-Hungarian officials actively promoted BiH as a new and prosperous country in their ruling crown. Habsburg officials published several exhibits on Bosnian history, folklore, and archeology, with artists such as Alphonse Mucha presenting the Bosnian pavilion at an exhibition in Paris in 1900.

Fra Antun Knežević, Bosnian Franciscan, historian, and writer, was one of the many main proponents of Bosniak nationhood, and he fiercely advocated against imminent Croatization of Bosnian Catholics on one side and imminent Serbianization of Bosnian Orthodox people on the other (led by ideologists and missionaries from neighboring countries), as he called them Catholic Bosniaks and Orthodox Bosniaks in his work. His position and doctrine were that all Bosnians or Bosniaks are one people of three faiths and that up to the late nineteenth century, no Croats and Serbs lived in Bosnia and Herzegovina (Truhelka 1930). He identified himself as a Bosnian and being a supporter of multi-faith Bosnian nationalism.

The idea of a united South Slavic state (usually expected to be led by the independent Kingdom of Serbia) became a popular political ideology in the region at the time, including BiH. The newspaper *Bošnjak* was published in BiH at that time. These newspapers provoked heated debates in neighboring Croatia and Serbia. The paper supported the politics of Béni Kallay, whose goal was to strengthen the Austro-Hungarian rule in BiH. Until 1905, nationalism was an integral factor in Bosnian politics, with national political parties corresponding to the three groups that dominated the elections.

In administrative-territorial terms, BiH was divided into six districts, and these into districts and county offices. District and county heads, i.e., offices, conducted all political and administrative affairs in their area under the direct control of the Provincial Government. In addition to this state administrative structure, there was substantial self-government in Bosnia in rural and urban municipalities. There was also a particular organization of the judiciary in BiH. The first instance was composed of 48 district courts. Six district courts exercised the appellate authority. There existed a Supreme Provincial Court in Sarajevo. All district courts also had sharia courts for the waqf, family, and inheritance affairs of the Muslim population. The Supreme Sharia Court (Imamović 1997) acted as an appellate court to the Supreme Court in Sarajevo.

On February 20, 1910, the Constitution for BiH was proclaimed. The Constitution did not bring about almost any changes to the administration of BiH. The supreme administrative authority remained under the jurisdiction of the Joint Ministry of Finance in Vienna and its agency—the clerical Provincial Government in Sarajevo. The Constitution and its accompanying laws introduced Diet into the country's political life with very limited legislative rights. The Diet of BiH consisted of 72 elected deputies and 20 virilists. The Diet is organized based on the social and national curial system. However, unlike other obsolete representative bodies in the Monarchy, the Bosnian Diet was not found on any particular domestic tradition. Citizens were divided into three curiae by national groups, providing each group with a certain number of delegates in proportion to its size. There were special curiae within the national electorate on a social basis: urban, rural, landowner, and intellectuals curia.

After the occupation in 1878, Austro-Hungarian Empire worked on a new system of governance to improve the political, economic, and cultural life of BiH. There was a modernization of the outdated Ottoman order in all these areas, which made BiH a backward region of this part of Europe. Based on the provision from January 1, 1880, according to which BiH was introduced into the Austro-Hungarian customs area, the rapid economic development began (Imamović and Hadžibegović 1998).

The main task was to cover the costs from own revenues. Therefore, Austro-Hungarian Empire sought to support the country's economic potential, focusing its policy on exploiting BiH's economic wealth. In the first decade of the twentieth century, raw materials accounted for 83% of all Bosnian exports. Wood industry products were the most important item. By 1885, by direct settlement and auction, BiH sold about 144,000 oak trees. Based on a contract, the largest wood company Morpurgo Parente, from Trieste, cut down 1,300,000 trees with 2,600,000 m^3 of wood. This provided the basis for the development of the forestry industry, and several new companies were established. It was systematically invested in the forests of BiH, so at the beginning of the twentieth century, the export of Bosnian wood became a serious competitor to the Austrian one. Timber worth 30–35 million crowns is exported from BiH annually. Along with the forestry industry development, there is also the construction of railways, which were the basis for connecting raw material basins.

After forestry, mining developed—the exploitation of coal, iron, and salt. The new venture required significant capital, so the state secured loans from large banks. In BiH, the exploitation of iron was especially advanced, which brought over the first companies in the chemical industry. However, despite the significant investment in mining, it developed gradually, limiting itself to a few important items (coal, iron, magnetite, and salt). In 1913, 847,289 tons of coal were produced in the country, and 2669 people were employed. Of the iron mines, the one in Vareš stood out in particular, which was used during the Ottoman rule but was significantly modernized during the Austro-Hungarian period, so ore production increased from 2305 tons in 1890 to 219,113 tons in 1913. In 1893, a private forge for the production of steel, rolled and barred iron, rolled wire, and mining rails was opened in Zenica with the private capital of the Provincial Government. Towards the end of the Austro-Hungarian administration, in 1916, the exploitation of the wealthiest deposit began in Ljubija near Prijedor.

Among other industries, a special place was taken by the construction material industry. With several companies, the chemical and food industries also stood out. With the opening of the refinery in Bosanski Brod in 1893, the era of refining crude oil from imported oil began. The company Solvay was opened in Lukavac, which produced ammonia, caustic, and crystal soda. Dry distillation of wood was performed in Teslić, calcium carbide, ferrosilicon, chlorinated lime and chloride in Jajce, cellulose in Drvar, and turpentine and rosin in Višegrad.

Despite significant results, BiH has not experienced full industrialization, remaining a predominantly agrarian country. According to incomplete data from 1907, 46,593 state and private companies and shops were registered in BiH, which employed 101,664 people or 5.7% of the total population of BiH. During the Austro-Hungarian Empire, the first two scientific institutions were formed: the National Museum in Sarajevo in 1888 and the Institute for Balkan Research in 1904.

In addition to the annexation of BiH, the Young Turk revolution caused other changes in the Balkans. The uprising of Albanians in Albania, Kosovo, western Macedonia, and the Sanjak of Novi Pazar, and the Italian-Ottoman war in 1911 prompted the Balkan states (Serbia, Montenegro, Bulgaria, and Greece) to take advantage of the new situation and divide the remaining Ottoman territory in the Balkans. In October 1912, the Balkan allies declared war on the Ottomans and inflicted a heavy defeat. The Sanjak of Novi Pazar, inhabited by Bosniaks, was occupied by the Serbian and Montenegrin armies. After the assassination of Franz Ferdinand on June 28, 1914, carried out by members of the "Young Bosnia" organization, Austro-Hungarian Empire declared war on Serbia on July 28. The Sarajevo assassination served as the occasion for the outbreak of the First World War. Namely, Gavrilo Princip, a member of the secret organization "Young Bosnia" which aimed to overthrow the Austro-Hungarian government in BiH through riots, uprisings, and armed assassinations, assassinated the Austro-Hungarian heir to the throne and Archduke—Franz Ferdinand and his pregnant wife Sofia on June 28, 1914.

With the coup in Serbia and the coming to power of the former French officer Petar I Karađorđević in 1903, the political goals began to change, and one of the most important goals became the "liberation of the enslaved Serb brothers" in Austro-Hungary. The foreign support for this policy was found in new, friendly, allied relations with Russia and France, while the cooling with Austro-Hungary led to a customs war (1906–1911), while the annexation of BiH in 1908 almost led to an actual war. Serbia's victory in the Balkan Wars (1912 and 1913) and its annexation of Kosovo and Macedonia did not hinder the ambitions to free the enslaved brothers in Western countries, growing strong constantly.

On the other hand, in the Balkan areas of the Austro-Hungarian Empire, the idea of creating a South Slavic state union in which the rights of Serbs, Croats, and Slovenes will not be oppressed by Germans and Hungarians is strengthening. This dissatisfaction of the Slavs became apparent at the beginning of the twentieth century to the younger generation of civil servants in Vienna, whose leading representative became Crown Prince Franz Ferdinand, who began to advocate the reform of the dual Austrian-Hungarian union into a triple Austrian-Hungarian-Slavic union.

The central role in the assassination will be played by the secret Serbian terrorist organization "Black Hand," officially founded on June 10, 1910, although its members had been involved in terrorist operations and political assassinations, such as the assassination of Serbian King Aleksandar Obrenović, even before. The Commander-in-Chief of the "Black Hand" was also the Commander-in-Chief of the Serbian Military Intelligence Service, Colonel Dragutin Dimitrijević. In Bosnia, the "Black Hand" controlled the "Young Bosnia" organization, whose members will carry out the assassination. The assassination involved seven people deployed along the route from the field where the exercises were held to the city hall (Fig. 9.13).

Fig. 9.13 Museum of Sarajevo "1878–1919" in front of which the Sarajevo assassination took place. (Photo by Aida Bidžan-Gekić)

Gavrilo Princip and other assassins down the street heard a bomb explode and left their seats. By an unusual set of coincidences, Gavrilo Princip took the opportunity when the Archduke's car briefly passed by him and pulled out a revolver. The first shot hit the Archduke, and the second, intended for Oskar Potiorek, hit Duchess Sofia. He, like the other assassins, was captured and tried from October 12 to 28, 1914. Immediately upon receiving the news of the assassination, the Austro-Hungarian monarchy accused Serbia and its government of organizing coups outside its territory and conducted only a formal investigation. The word about the assassination came just at the right moment for Austro-Hungarian Empire, which was looking for an excuse for war with Serbia, convinced that it would subdue and destroy it with one blow. The intentions of Austro-Hungary are best seen in the ultimatum sent to Serbia.

The situation for Austro-Hungarian Empire after the assassination was quite simple. The heir to the throne was killed by members of the "Young Bosnia" association, trained and armed in Belgrade. Also, the Austro-Hungarian police found a document on the organization of the assassination signed by the President of the Council of Ministers of Serbia, Nikola Pašić, on June 5, 1914 (23 days before the assassination), containing the nickname of one of the assassins. On July 28, 1914, Austro-Hungarian Empire declared war on Serbia. Shortly afterward, Russia declared war on Austro-Hungarian Empire, and Germany declared war on Russia and France, thus beginning the First World War.

The Bosnian regiments became part of the sixth Austro-Hungarian army, crossing the Drina river on August 12. Earlier, during the preparations for the war, a general mobilization was carried out, cattle were procured for towing, the army was supplied with essential food items, and transportation was mobilized for military needs and goals. With the capitulation of Serbia in December 1915 and Montenegro in January 1916, BiH ceased to be a war hinterland. The cessation of direct warfare did not mean the end of suffering for the population of BiH. The war mobilization left the villages and the cities without labor, agrarian incomes were twice as low, and thorough economic restrictions completely impoverished the population.

Already in the first days of the outbreak of war, difficult times began throughout the country. With the collapse of Serbia and Montenegro in October 1915, the fronts of the First World War moved far from its borders, but many Bosniaks, mobilized in the Austro-Hungarian regiments, died and perished on those distant fronts. Bosnian regiments fought mainly on the Russian and Italian fronts. On the Russian front, Bosniaks took part in numerous battles. Although far from large battlefields, the consequences of the war, primarily in the form of mass famine, were felt everywhere in BiH. After the capitulation of Serbia and Montenegro, the Bosnian regiments continued to fight on other battlefields: in the Carpathians, Galicia, and Brest Litovsk against the Russian army and the Sochi river and South Tyrol against the Italian army. According to some estimates, the number of Bosnians killed and missing in the First World War is over 100,000. The country welcomed the end of the war in great misery and poverty.

Culturally, the area of BiH was still very backward. About 80% of the population was illiterate. There were few primary schools, while secondary schools were rare. There were no higher education institutions until 1937 when the Higher Islamic Sharia-Theological School in Sarajevo was founded and, a little later, the Faculty of Agriculture. Due to the lack of higher education institutions, Bosnian high school students studied in Zagreb or Belgrade. During this period, many newspapers and magazines began to appear, the founders of which were mainly political parties and religious institutions. Theaters were established in Tuzla and Sarajevo.

At its session on October 2, 1915, the Hungarian government officially adopted the plan to join BiH to Hungary, requesting that all forms of autonomous political life in BiH be abolished before the position of the Monarchy, i.e., Hungary, was strengthened. The idea that it was possible to join BiH to Hungary was abandoned after Yugoslav-oriented, Orthodox, and Catholic Bosnian politicians delivered a memorandum to Hungary on November 20, 1918, in Sarajevo, seeking a solution to the Bosnian question based on the principle of self-determination, in the form of equal and politically independent peoples. Simultaneously with these turmoil in Bosnia and around Bosnia, on the other hand, there was an ongoing process of resolving the Yugoslav question in the form of creating a common Yugoslav state, uniting the Kingdom of Serbia (and Montenegro) with the South Slavic countries of the Habsburg Monarchy.

Among Bosnian Muslim (Bosniak) politicians, Parliament Speaker Safvet-beg Bašagić and Šerif Arnautović, then one of the most influential leaders of the United Muslim Organization, were the first to speak. On July 17, 1917, they handed over a memorandum to Tsar Charles, with a proposal that the state-legal position of BiH be resolved by its accession to Hungary, with guaranteed autonomy. In contrast, younger Bosniak politicians and intellectuals in Bosnia gradually embraced the Yugoslav orientation (Imamović 1997).

By the act of unification on December 1, 1918, BiH became part of a united Yugoslav state, but it retained some elements and forms of its statehood and autonomy for some time. The Main Board of the People's Council for BiH ceased to function on December 31, 1918, but the People's Government for BiH continued to perform its regular duties. As early as the beginning of January 1919, the People's Government ceased to function, and on January 31, the Provincial Government for BiH was formed. Its competencies were very limited, as most were transferred to the central government in Belgrade. Thus, only four commissions were retained: internal affairs, judiciary, agriculture, and education (Imamović and Hadžibegović 1998).

9.2.2 Period of 1919–1945

The Kingdom of Serbs, Croats, and Slovenes would later become the Kingdom of Yugoslavia. At that time, political life in Bosnia was marked by social and economic conflicts over the issue of property ownership and the creation of more political parties. The dominant ideological conflict of the Yugoslav state was between Croatian regionalism and Serbian centralism, while the Bosnians approached it depending on the overall political atmosphere. Although the state was divided into 33 areas that erased traditional geographical boundaries, Bosnian politicians such as Mehmed Spaho ensured that BiH preserved its territorial integrity.

The establishment of the Kingdom of Yugoslavia in 1929 brought about the redrawing of borders into administrative units (banat) that supposedly bypassed all ethnic and historical lines yet erased every trace of the Bosnian entity. The famous Cvetković-Maček agreement created the Banovina of Croatia in 1939, practically dividing Bosnia between Serbia and Croatia, which they had long expected. However, that period was marked by the beginning of the Second World War in 1939 in Europe, while Yugoslavia would become a target only after March 25, 1941, when it joined the Triple Alliance, but soon left when a coup d'etat was carried out. This infuriated Hitler, who promised the complete destruction of the Yugoslav state, which was attacked on April 6, 1941.

Shortly after Nazi forces conquered Yugoslavia, the country was divided into occupation zones. BiH was in the German and Italian zones, and the border ran along the line Bosanski Novi-Mount Šator-Mount Bjelašnica border with Montenegro. At the same time, the Axis Powers are creating a pro-fascist Independent State of Croatia (NDH), which included today's BiH. During the occupation, the NDH authorities systematically expelled and killed the Jewish and Serbian populations. This prompted many Serbs to join the Chetnik movement, which, in addition to the initial fight against the occupiers, which was later stopped, began committing crimes against Bosniaks, especially in eastern Bosnia.

After the creation of the Independent State of Croatia on April 10, 1941, Bosnian Muslims (Bosniaks) were declared Croats of the Muslim faith. The territory of BiH is divided into 22 administrative units (large parishes) whose names coincided with the names of the parishes of the medieval Bosnian state. The Muslims of BiH were considered the "flowers" of the Croatian people and BiH the "soul and heart" of the NDH. Some Muslim politicians accepted cooperation with the Ustasha authorities. A good part of the population was utterly indifferent to the new government, and most chose resistance and joined the partisans, making Bosnia the main scene of war operations during the Second World War on the territory of the former Yugoslavia.

As the NDH authorities were unable to protect Muslim villages from Chetnik and partisan attacks adequately, a movement has emerged among some Bosniak representatives in power since 1942 that sought to secede or at least maximize the autonomy of BiH within the NDH. The signing of an agreement on cooperation between the NDH authorities and some Chetnik dukes in mid-1942 further contributed to the distrust of Muslims towards the NDH. Some Muslim politicians sought the idea of autonomy by trying to win over Hitler by offering him the service of Muslims in SS units. In that case, Bosnia would gain autonomy under the German governor; eastern Herzegovina would be joined to Montenegro, while western Herzegovina would remain in the NDH. Hitler disagreed with the idea of Bosnian autonomy, but he accepted the idea of founding a Muslim SS unit, and in 1944, the 13th SS Khanjar Division was formed.

In early 1941, the Yugoslav Communists, led by Josip Broz, organized a multiethnic resistance movement simply called the Partisans. In addition to fighting Nazi forces, the Partisans were opponents of the Chetniks. On November 25, 1943, the Anti-Fascist Council of Yugoslavia established BiH in Jajce as a Republic within the Yugoslav Federation, with historical borders dating back to the Ottoman era. This decision was preceded by a decision made on November 25 in Mrkonjić Grad at the first session of ZAVNOBiH (State Anti-Fascist Council for the National Liberation of Bosnia and Herzegovina) and confirmed at the second session of ZAVNOBiH in Sanski Most.

From 1943, the allies also began to support the Partisans, and the Chetniks have since lost all external support. Since then, the Partisans have achieved several military successes and, in the end, became the most successful anti-guerrilla group that will liberate the whole of Yugoslavia by the end of the war. At the end of the war, a new Constitution was adopted in 1946, officially recognizing Bosnia and Herzegovina as a republic in the new Yugoslav state.

The Communist Party of Yugoslavia took advantage of the internal conflicts in BiH, calling on its members to revolt on June 22, 1941, when Germany invaded the Soviet Union. After heavy defeats in Serbia in the fall of 1941, the Headquarters of the Partisan forces, led by Tito, moved to the territory of BiH and remained there until 1944. The partisan movement was supported by most Serbs, who were endangered and were not in the Chetnik ranks, then Bosniaks, anti-fascist Croats, Jews, and others. In BiH, partisan

Fig. 9.14 The monument "Battle of Sutjeska," built in 1971, is dedicated to the famous Battle of the Sutjeska from 1943. (Photo by Aida Bidžan-Gekić)

liberated areas were created, including the so-called Republic of Bihać in 1942, where the first session of the Anti-Fascist Council of the People's Liberation of Yugoslavia (AVNOJ), i.e., the first Partisan parliament, was held.

At the end of 1943, the second session of AVNOJ was held in Jajce, which announced a new Yugoslavia and created its first government. Thus, BiH became not only the center of the Partisan movement but also the area of the fiercest civil war and the focus of future Yugoslavia. Of the seven anti-partisan offensives led by Germans, Italians, units of the Independent State of Croatia, and others, six were in BiH. The most famous were the Battle of the Neretva in early 1943, the Battle of Sutjeska in the late spring of 1943, and the Drvar landing in May 1944. The battle of Kozara from June and July 1942 is well-known, which became a symbol of resistance, courage, and suffering (Fig. 9.14).

Parallelly with the new authorities for the whole of Yugoslavia, they emerged during the struggle that the partisans called the People's Liberation War and the beginnings of BiH, as one of the federal units within the other Yugoslavia. Already in Bihać, at the end of 1942, the authorities of the future republics were being established, but this was postponed due to the decisive battles on the Neretva and Sutjeska rivers. On the eve of the second session of AVNOJ in Jajce at the end of 1943, the first session of the National Anti-Fascist Council of the People's Liberation of Bosnia and Herzegovina (ZAVNOBiH) was held in the neighboring town of Mrkonjić Grad. It was the inception of the Bosnian parliament, which at the subsequent two sessions made decisions on BiH as a separate republic in that, another Yugoslavia.

Deprived of political integrity and included in the NDH by the will of the Germans and Italians, where it was allowed to exist only in the past as a geopolitical term, during the war, through the fight of its people, BiH restored its historical political integrity and eventually secured the status of a federal unit within Yugoslavia. Democratic centralism limited and determined its political sovereignty. Being a part of NDH at the beginning of the war, BiH could be presented on the geopolitical map as a historical political entity with a renewed existence by the end of the war. An important element of the victory of the national liberation movement in Bosnia and Herzegovina was the reality of the oridinary life of its peoples, which could not be prevented by plans of ethnic homogenization by methods of biological eradication – ethnic cleansing and genocide. The historical merit of the national liberation struggle is that it provided Serbs, Croats, and Muslims (Bosniaks) with the right to life (Redžić 1998).

9.2.3 The Period of Communism

On April 28, 1945, the People's Government of Bosnia and Herzegovina was appointed at the session of ZAVNOBiH. According to the AVNOJ agreements and Yugoslav constitutions, a joint territory of Serbs, Croats, and Bosniaks (then Muslims), indivisible according to the principles of the historical facts of the existence of the Bosnian state, the

Republic of Bosnia and Herzegovina was defined as one of the six constituent federal units of socialist Yugoslavia, under the name Bosnia and Herzegovina, and later the Socialist Republic of Bosnia and Herzegovina. It was recognized as the most diverse republic within the former Yugoslavia. It had its uniqueness because it consisted of three main South Slavic peoples and other citizens. Josip Broz Tito often emphasized that Bosnia was indivisible and that, at the same time, it belonged to Serbs, Croats, and Muslims.

BiH experienced its greatest industrial expansion during this period. Therefore, factory chimneys were on its coat of arms as a symbol of modern BiH based on industrial development.

Upon entering the new state, the territory of BiH was the subject of strong repression and reprisals. The Law on Agrarian Reform and Colonization from August 1945 limited the area of arable land to 10 hectares, and the remaining land was confiscated and handed over to the state. The following year, a law on nationalization was passed, based on which banks, roads, and all economic goods became state property. The colonization process was underway in Herzegovina so that several tens of thousands of inhabitants from the war-torn area moved to the regions of Vojvodina and Slavonia. Through youth labor actions, several railways were built: Brčko-Banovići, Šamac-Sarajevo, and Doboj-Banja Luka. The work of the Islamic religious community was approved in 1946, but in the spirit of anti-religious ideology, the new government hampered its work. In the late 1940s, there was a great persecution of members of the Young Muslims. With the law from September 1950, a widespread campaign of abolishing the veil and niqab began, disturbing the position of women in society. Two years later, Islamic kuttab schools were closed, and Sufi orders were banned. Until 1964, it was forbidden to publish Islamic literature. The image of Islam began to change only in the mid-1950s, due to Tito's policy of the Non-Aligned Movement, whose members mainly were Islamic. The Catholic Church in BiH has been treated harshly due to its close cooperation with the Ustasha criminal fascist movement. Numerous churches were destroyed and monasteries closed. The Orthodox Church had a better treatment by far.

Serbs and Croats in BiH have systematically imposed their language and suppressed the Bosnian language for a full 100 years. BiH had to be governed from Belgrade, directly from the Central Committee of the CPY, led by Tito. Muslims were not given the right to be equal people or a people in general, so many Muslims were liquidated, imprisoned, and persecuted for various reasons. Leading and high positions in all spheres of life in socialist BiH were held mainly by ethnic Serbs and Croats (economy, schools, army, etc.)

After the Second World War, the first real rebellion against the new socialist order broke out in BiH. In 1950, Muslims rebelled, as did Serbs and Croats on the border between western Bosnia and Croatia. The Cazin uprising (or Kordunaš uprising, as it was called in Croatia) was suppressed in the blood by the police and the army, and it was kept as a top-secret by the state for years to come.

In 1960, economic reforms in Yugoslavia, aiming at decentralization and de-statization, gave certain results but were stopped due to the very nature of the communist system. Should they had been carried out to a logical end, it would mean the end of one-party totalitarianism and Yugoslavia based on it. Since the pro-Serbian leadership of the League of Communists was removed in 1966, more free political thinking began in BiH as well. Similar processes in Croatia and other republics have contributed to this.

The first step in easing in was the decision of Tito's leadership to recognize the status of the people in 1971 and the population of the majority of the Islamic religion, which until then had to choose to declare themselves as Serb, Croat, Yugoslav, or neutral. Since then, Muslims in BiH have been written with a capital letter as a people. As a result, the 1971 census saw a sharp decline in the number of Serbs, Croats, and Yugoslavs in favor of Muslims. Thus, for example, the number of Serbs decreased from 42.9% in 1961 to 37.2%, and the number of Muslims in the same period increased from 25.7% to 39.6%. Then, for the first time, Muslims in official data became the relatively most numerous people in BiH and, at the same time, were recognized as one of the equal peoples in the former Yugoslavia. Bosnian Muslims later became one of the fiercest supporters of the Yugoslav state.

Since the mid-1970s, BiH has increasingly made economic, cultural, political, and sporting progress. Sarajevo is becoming a metropolis, and there, as well as in Tuzla, Zenica, Mostar, Banja Luka, Velika Kladuša, and several factories, combines are being built, finding their place on the world market (Energoinvest, Energopetrol, Agrokomerc, Željezara Zenica, SOKO, etc.) This progress was made by removing the old partisan leadership and bringing in new cadres at the head of political and economic institutions.

On May 4, 1980, in Ljubljana, Josip Broz Tito, the undisputed leader of the SFRY, died. Within the League of Communists, after Tito died in 1980, there was a stratification into centralist and federalist forces. This is also happening within the League of Communists of BiH, where the commitment among Serbs towards a Greater Serbia is strengthening, while Muslims are striving for federalism, autonomy, and reformism.

Due to its central geographical position within the Yugoslav Federation, post-war BiH was marked as the basis for developing the military defense (Fig. 9.15). This meant a great concentration of the Yugoslav People's Army and the development of a dedicated industry. A few years after the end of the Second World War, the reconstruction

Fig. 9.15 Central position of Bosnia and Herzegovina in former Yugoslavia. (UN Cartographic Section 2007, public domain)

and construction of the devastated country began, with new railways, roads, factories, and hydroelectric power plants being built. There was an accelerated development of industry, especially metallurgical, special-purpose machine; food; textile; production of medicines; construction of electric power system, hydroelectric power plants, new cities and factories, schools and higher education institutions, hospitals, and health centers; the establishment of health and social protection systems for the entire population. All this accelerated economic and transport development means the expansion of existing cities and the emergence of new ones, as well as an increase in the number of inhabitants in urban areas and large migrations from rural areas. However, far less was invested in the development of BiH compared to the neighboring republics of Croatia and Serbia. The importance of the Federal Republic of Bosnia and Herzegovina in the SFRY and the world was confirmed by the election of Sarajevo as the host of the 1984 Winter Olympics.

9.2.4 Period After 1990

At the end of 1988, the constitutional order in the SFRY began to collapse. In 1990, parliamentary elections were held in all Yugoslav republics. In the first multiparty elections in BiH, on November 18, 1990, three national parties won the most votes: the Party of Democratic Action (SDA), the Serb Democratic Party (SDS), and the Croat Democratic Union (HDZ). Alija Izetbegović was elected the first President of the Presidency of BiH. As all attempts to reorganize Yugoslavia failed, on June 25, 1991, Slovenia and Croatia declared independence and seceded from the Yugoslav state union. The decision to call a referendum on the independence of BiH was preceded by the adoption of the Act on the Reaffirmation of the Sovereignty of the Republic of Bosnia and Herzegovina in the same institution on October 15, 1991. The Republic of Bosnia and Herzegovina was first recognized by Bulgaria on January 31, 1992.

At the same time, the International Community convened a conference on Yugoslavia, with the task of helping resolve political relations between the republics in the process of dissolving Yugoslavia. At the Council of Ministers of the European Community in Brussels on December 16, 1991, the European Community and its member states adopted the Declaration on Yugoslavia and the Declaration on Guidelines on the Recognition of States in Eastern Europe and the Soviet Union. In response to the above guidelines, on December 20, 1991, the Government of BiH adopted the Decision on the Recognition of the Statehood or Independence of Bosnia and Herzegovina.

Following Opinion No. 4 of the Arbitration Commission, the Assembly of the Socialist Republic of BiH passed a decision to call a referendum, according to which it was held on February 29 and March 1, 1992. Out of a total of 3,253,568 persons with the right to vote, 2,073,568 persons, or 63.73%, went to the referendum. Two million sixty-one thousand nine hundred thirty-two voters, or 99.44%, voted for a sovereign and independent BiH, a state of equal citizens, the people of BiH—Muslims, Serbs, Croats, and members of other peoples living in it.

The international recognition of BiH followed at the session of the Council of Ministers of the European Community in Brussels on April 6, 1992, with the entry into force the following day. With this recognition, BiH formally ceased to be part of the SFRY. The political-administrative borders of the former republic, as one of the six federal units of the SFRY, became international state borders. On May 22, 1992, Bosnia and Herzegovina was admitted to the United Nations as the 177th member, together with Slovenia and Croatia. Beginning of 1992, the historical name Bosniak started to return as the name of the ethnic group, which replaced the previous religious designation "Muslim."

Immediately after the country declared independence and international recognition in April 1992, the war against the Republic of Bosnia and Herzegovina officially broke out. Although the Republic of Bosnia and Herzegovina was admitted to the United Nations on May 22, 1992, the ruthless war continued (Kurtćehajić 2017). Today, there is numerous documentation on the planned division of BiH, including the most important transcripts from Franjo Tuđman, but also the testimonials of Stjepan Mesić, former President of the Presidency of SFRY, Ante Marković, former Prime Minister of SFRY, and many other witnesses of the time.

Armed conflicts began as early as November 1991, when the Yugoslav People's Army (JNA), the official army of Yugoslavia, attacked and destroyed the village of Ravno in the municipality of Trebinje (as part of actions of JNA against Croatia). At the same time, numerous military organizations were created, offering resistance to the JNA and soon to the newly established paramilitary Army of the Republika Srpska (VRS). The Army of the Republic of Bosnia and Herzegovina was the official armed force of the Republic of Bosnia and Herzegovina during the war in BiH. The army was founded on April 15, 1992. In addition to the already created so-called Republika Srpska in 1992 (originally called the Republic of the Serb People of Bosnia and Herzegovina), the so-called Croatian Republic of Herceg-Bosna was created in 1993, practically dividing BiH into three ethnic parts, two of which are entirely separatist in nature.

The Yugoslav People's Army, along with Bosnian Serb paramilitary authorities, surrounded BiH's capital, Sarajevo, and kept it under siege for 1425 days, the longest siege in the history of the war (Fig. 9.16). Sarajevo Roses (Fig. 9.17) is a term used to describe holes in the asphalt made of grenades and shrapnel, later filled with red paint. About 470,000 grenades fell on Sarajevo during the siege, or 330 a day, and it used to happen that over 3000 grenades fell in 1 day. Ten thousand six hundred fifty people were killed, including 1601 children. Those places where grenades fell are remembered in Sarajevo for the abundance of blood of citizens killed or wounded by grenades. The blood would disappear after a while, and painful memories of the places of

Fig. 9.16 Monument to 7-year-old boy killed by aggressor sniper in 1994 in front of the Faculty of Philosophy in Sarajevo. (Photo by Aida Bidžan-Gekić)

suffering were awakened by the characteristic holes of grenades and flower-shaped shrapnel, which would remain on the asphalt. The holes were filled with red after the aggression, so this color contributed to making it look like a rose in memory of the dead. This was done in every place where a grenade killed three or more citizens of Sarajevo. As you walk through Sarajevo, you will often notice the Sarajevo Rose, which symbolically and with dignity reminds of the citizens of Sarajevo, killed while going for water or to work or reminiscent of children killed while playing carefree (Halilović 2008).

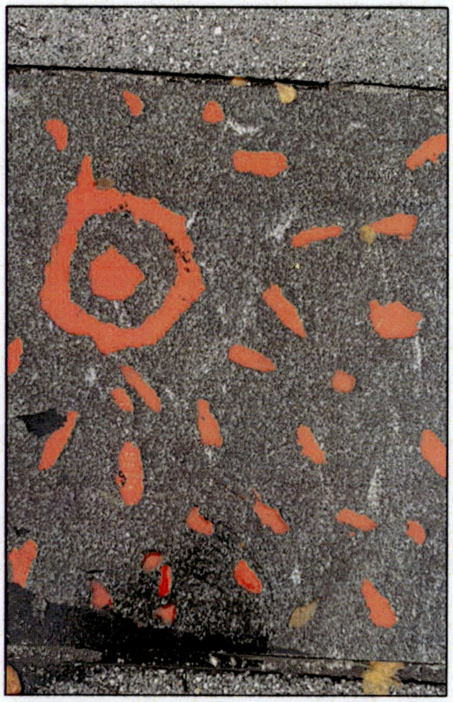

Fig. 9.17 A Sarajevo Rose. (Photo by Aida Bidžan-Gekić)

The Army of Republika Srpska occupied many towns in BiH at the very beginning of the aggression: Foča, Višegrad, Bijeljina, Banja Luka, Prijedor, and Doboj. This was where Serb military formations massacred and abused Bosniaks and Croats. At the same time, in the territory under the control of the Croatian Defense Council (HVO), the HVO clashed with units of the Army of the Republic of Bosnia and Herzegovina, especially in Mostar, Travnik, Livno, Ljubuški, Čapljina, and Gornji Vakuf. In the autumn of 1993, Fikret Abdić seceded the municipality of Velika Kladuša, causing the conflict between Bosniak military units; established autonomy; and proclaimed the Autonomous Province of Zapadna Bosna (Western Bosnia) (Čekić 2012).

In July 1995, with logistical assistance from Serbia and special units of the Serbian Ministry of internal affairs, the "Scorpions," and other formations from Serbia, the Army of Republika Srpska committed genocide against the Bosniak civilian population in Srebrenica, in eastern Bosnia (Fig. 9.18). The list of those killed in Srebrenica is not final; there are currently 8372 names. The war lasted almost until the end of 1995, with Bosniaks suffering the most, genocide and ethnic cleansing being committed, but with both Serbs and Croats also suffering losses. At least 6925 children aged 0–18 were killed in BiH (Tokača 2012). Almost 105,000 people died or were killed in the war. Numerous mass graves (with more than 100 victims) have been found across the country and are still being discovered. The largest so far was found in the village of Tomašica near Prijedor, with victims of Bosniak and Croat ethnicity. All three peoples in the country experience the war differently, seeing it as a threat to their national interests. So for many Serbs, it is a "patriotic" war, and for many Croats, a "homeland" war, although it is true that the war was started for the sake of the goals of great-state projects of neighboring countries. With the successful resistance of the citizens of the Republic of Bosnia and Herzegovina to the aggressive war and the commitment of the Government of the Republic of Bosnia and Herzegovina to a multiethnic and democratic

Fig. 9.18 Part of Srebrenica-Potočari Memorial and Cemetery for the victims of the 1995 genocide. (Photo by Aida Bidžan-Gekić)

state of equal peoples, BiH has preserved its statehood and historical continuity. The war ended on November 21, 1995, with the signing of the Dayton Peace Agreement in Dayton (Ohio, USA), where great credit goes to the USA.

In the American city of Dayton on November 21, 1995, all warring parties in the territory BiH signed a peace agreement (the agreement was signed as parties to the war by the presidents of independent and internationally recognized states: Alija Izetbegović (Bosnia and Herzegovina), Franjo Tuđman (Croatia), and Slobodan Milošević (FR Yugoslavia, president of Serbia) and witnesses from the USA, France, the UK, Germany, Russia, and the European Union, which unofficially ended the war (Summary of the Dayton Peace Agreement 1995). The final agreement was signed in Paris on December 14, 1995. It was signed without the knowledge of the citizens of BiH. The Dayton Agreement confirmed Bosnia and Herzegovina as an independent and sovereign state in the European family of states. According to the agreement, BiH consists of two administrative units: the Federation of Bosnia and Herzegovina and Republika Srpska. A few years later, the Brčko District was established on March 8, 2000, 1 year after the international arbitration for Brčko, and thus the territory of the Brčko District has a special status. It does not belong to any entity, but the state of BiH has sovereignty over it.

The Constitution was adopted based on the Dayton Peace Agreement. Under the Constitution, Bosnia and Herzegovina is recognized as an independent, sovereign state with full legal personality based on the values of peace and reconciliation, prepared for Euro-Atlantic integration. Under the Dayton Peace Agreement, the High Representative is responsible on behalf of the international community (OHR) for overseeing the implementation of the civilian aspects of the Peace Agreement in BiH. The High Representative also has the task of coordinating the activities of international civil society organizations and agencies operating in the country (Dayton Accords 1995).

It is very important to emphasize that Radovan Karadžić, former Bosnian Serb political leader who served as the president of Republika Srpska during the war, was convicted of genocide, crimes against humanity, and war crimes by the International Criminal Tribunal for the former Yugoslavia in Hague in 2016, and in appeal process by International Residual Mechanism for Criminal Tribunals in 2019 was sentenced to life in prison (IRMCT 2019). Also, Ratko Mladić, a war criminal, former commander and general of the Army of Republika Srpska between 1992 and 1995, was sentenced by the International Residual Mechanism for Criminal Tribunals in the Hague to life in prison for genocide, war crimes, and crimes against humanity in 2021 (IRMCT 2021). Despite that denial of genocide and mass crimes and the celebration of war criminals are a daily occurrence in Bosnian society.

The findings of Srebrenica genocide denial report (2021) indicate that genocide denial, including the glorification of war crimes and criminals, remains widespread in both Bosnia and Herzegovina (BiH) and neighboring countries. Over the course of the past year, quantitative analysis identified 234 instances of genocide denial in the regional public and media discourse. The vast majority of these instances occurred in Serbia (142), followed by BiH (60), and Montenegro (19). The three most common rhetorical tactics used in genocide denial remain, disputing the number and identity of victims, conspiracy theories which challenge the rulings and integrity of international courts and nationalist historical revisionism. Within these general categories, common practices include conspiracy theories and "alternative histories" which contradict the judicially established facts of the Srebrenica genocide; attempts to silence or disparage public discourse about the genocide; portraying the recognition and memorialization of the genocide as an attack on the Serbian people; opposition to the official recognition or condemnation of the genocide by states, local governments, and institutions; and obstruction of the adoption of legislation banning genocide denial (Srebrenica genocide denial report 2021).

Dismantling the existing culture of denial is by no means an easy undertaking. Although civil society in particular has so far been spearheading initiatives about truthful approaches to dealing with the past, it has been weakened by the uniformity of views between the incumbents and opposition regarding denial and left to its own devices by international actors. As the top political echelons of Republika Srpska are gripped by denial, tackling it may necessitate not only political courage and but also capitalizing of windows of opportunity that crises may offer by exposing the levels of governance mishaps and by increased international shaming. At the same time, top-down messaging as well as opening truthful and critical debates about the past will need to be cultured and introduced in the media, universities, schools, and public institutions. Without these debates and more democratic deliberation, Bosnia and Herzegovina will remain in a limbo, caught between the past and the present, without any viable future prospects for its citizens. These are important lessons with policy implications for those interested in helping divided societies build a positive future (Barton Hronešová 2021).

The records and transcripts from the Assembly of the Serb People in Bosnia and Herzegovina sessions offer tremendous insight into various aspects of the war that the Bosnian Serb and Serbian political class waged on Bosnia and Herzegovina. For example, Dobrica Ćosić's recommendation that the Vance-Owen Peace Plan (the proposal involved the division of Bosnia and Herzegovina into ten semiautonomous regions) be accepted on the grounds that it promised an end to the very "concept of that state of Bosnia and Herzegovina" reveals, among other things, that dismantling the state of Bosnia and Herzegovina was one of the primary aims of the Serbian nationalist project. It also indicates that the real center of gravity for the Serb elites in Bosnia and Herzegovina was located in Belgrade. The tensions that began to develop between Karadžić and Mladić in the autumn of 1994, additionally indicated a previously unknown degree of control that Belgrade maintained over the Bosnian Serbs throughout the 1990s (Suljagić 2021).

The involvement of neighboring countries of Serbia and Montenegro (in that time FR Yugoslavia) in the war against BiH was confirmed in 2021 by the verdict of the UN International Residual Mechanism for Criminal Tribunals in the Hague. According to judgment "The Trial Chamber finds that the common criminal objective, as defined, was shared by the senior political, military and police leadership in Serbia, SAO Krajina, SAO Slavonia, Baranja and Western Srem and Republika Srpska". Among those people was Slobodan Milošević, the former president of Serbia and FR Yugoslavia was part of a joint criminal enterprise (JCE) in BiH and Croatia. This is the essence of the verdict of the International Residual Mechanism for Criminal Courts in The Hague against former leaders of the State Security Service of Serbia, Jovica Stanišić and Franko Simatović (UN 2021).

Also, the court in the Hague handed down long-term verdicts for war crimes to the leaders of the creation of the so-called Herceg Bosna: Jadranko Prlić, Bruno Stojić, Slobodan Praljak, Milivoj Petković, Valentin Ćorić, and Berislav Pušić. The Appeals Chamber of International Tribunal for the Prosecution of Persons Responsible for Serious Violations of International Humanitarian Law Committed in the Territory of the Former Yugoslavia since 1991 states that in the first instance, the decision remains that Croatia was involved in the war in BiH and that the first President of the Independent Republic of Croatia, Franjo Tuđman, was part of a joint criminal enterprise (JCE) together with Herceg-Bosna leaders (UN 2017).

The Bosnian war was driven by fantasies of geopolitical and demographic engineering which led to ethnic cleansing and the separation of Bosnia's ethnic qua national communities from each other. The post-war peace opened up the possibility of returns, most especially the politically sensitive cross-boundary return of Bosniaks to their old homes in Republika Srpska and Serbs to the residences in Federation of Bosnia and Herzegovina (Ó Tuathail and Dahlman 2006). The reception of refugees and returnees across Bosnia-Herzegovina has been varied; the presence of the international organizations and officials—UNHCR, OHR, OSCE, and EUFOR officials—has nudged the process towards completion and managed to diffuse potential flashpoints; however, although some communities are mixed again, this is no guarantee of interethnic harmony. In plenty of communities, returnees are reminded of their second-class status as minority returnees and not as ordinary pre-war residents of their own hometowns (Ó Tuathail and O'Loughlin 2009).

In the period 1990–2000, the position of BiH was determined by two important factors: aggression and the war caused by aggression (1992–1995) and the 5 years of the Dayton Peace Agreement (1995–2000). This decade has no parallel in the entire history of BiH. It was only in this period, in the last decade of the twentieth century, that BiH came to the brink of ruin and a point of no return. However, we testified undoubtedly that history in this decade was not inclined to BiH, that it was "about the head of BiH," and that BiH was left to fate. To survive, BiH itself must understand its position. Regardless of the future attitude of the international community, the survival of BiH is primarily in its hands, mind, and practice. In any case, BiH owes history its gratitude for its knowledge and plausible lessons. From the Congress of Berlin (1878) to the Dayton Accords (1995), the international community showed interest in determining the position and abolishing the independence of small states by dividing them into spheres of influence.

According to the historical tradition, especially after the Second World War, in interventions in crisis areas, it adheres to the practice of dividing crisis territories into East-West or North-South. There are no indications that the International Community intends to abandon and revise this practice. In this decade, BiH was a victim of aggressive hegemonic nationalisms and its own inability to formulate and follow the imperative of the common BiH interest to ensure and strengthen state independence and autonomy. As an independent state, BiH is possible if the ideology and policy on the great national power in Serbia and Croatia and "national homelands" among Serbs and Croats in BiH is overcome. Bosniaks are obliged to give up their aspirations for primacy that BiH is first theirs, and then Serb and Croat. These are external and internal assumptions whose realization has the significance of a categorical imperative for the survival and perspectives of BiH.

All historical processes have their beginning and end. The thin Dayton vertical of BiH's reintegration is a fundamental point in both the theory and practice of BiH's existence and future. In the last decade of the twentieth century, BiH has been the focus of world attention more than ever in its history, which has not been to its advantage. BiH's integration into the European Union, outside of which the independent state of BiH is impossible in the modern European constellation, can be of crucial importance in the process of state reintegration of BiH. BiH is a long-lived phenomenon, one of the oldest states in the region, which is the East to West and West to East, but its "age" is not a sign and imperative that the time of its death has come, but longevity is proof of its historical roots, reasons, and ability to testify to the understanding of the legality of history through its nature and existence (Redžić 2003).

The state of Bosnia and Herzegovina has two major national holidays: Independence Day and Statehood Day. BiH's Independence Day is a national holiday celebrated on March 1 every year, celebrating BiH's independence from the Socialist Federal Republic of Yugoslavia. On February 28, 1995, the state parliament decided to start observing March 1 as the Independence Day of BiH.

Bosnia and Herzegovina Statehood Day is a national holiday in BiH celebrated every November 25 (adopted by the Parliament in late 1995). On that day in 1943, at the first session of ZAVNOBiH in Mrkonjić Grad, the Resolution of ZAVNOBiH was adopted, expressing the determination of the people of BiH to make their country a community of brotherhood in which full equality of all citizens and equality of the Republic of BiH within the Yugoslav federation, with historical boundaries dating back to medieval Bosnia, will be ensured. Contemporary BiH aims to become a member of the European Union.

References

Barton Hronešová J (2021) Ethnopopulist denial and crime relativisation in Bosnian Republika Srpska. East European Politics

Basler Đ (1972) Arhitektura kasnoantičkog doba u Bosni i Hercegovini [The architecture of late antiquity in Bosnia and Herzegovina]. Veselin Masleša, Sarajevo

Basler Đ (1976) Ein Viertel-Jahrhundert der Palaolithforschung in Nord bosnien, Yearbook ANU BiH, vol 11. Centar za balkanološka ispitivanja, 11, Sarajevo

Basler Đ (1979) Paleolitsko i mezolitsko doba u Bosni i Hercegovini [Paleolithic and mesolithic in Bosnia and Herzegovina]. In: Benac A (ed) Praistorija jugoslavenskih zemalja, I Paleolitsko i mezolitsko doba. Veselin Masleša, Sarajevo, pp 309–350

Baumler M (1987) Core reduction sequences: an analysis of blank production in the Middle Paleolithic of Northern Bosnia. University of Arizona, Tucson

Benac A (1979) Prelazna zona [Transition zone]. In: Benac A (ed) Praistorija jugoslavenskih zemalja, II, Neolitsko doba. ANUBiH, Centar za balkanološka ispitivanja, Sarajevo, pp 363–473

Benac A, Basler Đ, Čović B, Pašalić E, Miletić N, Anđelić P (1984) Kulturna istorija Bosne i Hercegovine [Cultural history of Bosnia and Herzegovina]. Veselin Masleša, Sarajevo

Bojanovski I (1988a) Doba Rimskog Carstva (Principat) [Age of Roman Empire (Principate)]. In: Čović B (ed) Arheološki leksikon Bosne i Hercegovine, I. Zemaljski muzej B&H, Sarajevo, pp 29–35

Bojanovski I (1988b) Bosna i Hercegovina u antičko doba [Bosnia and Herzegovina in ancient times]. ANUBiH, Centar za balkanološka istraživanja, Knjiga 6, Sarajevo, p 160

Čekić S (2012) Genocid i istina o genocidu u Bosni i Hercegovini [Genocide and the truth about genocide in Bosnia and Herzegovina]. Univerzitet u Sarajevu, Institut za istraživanje zločina protiv čovječnosti i međunarodnog prava, Sarajevo

Čović B (1976) Od Butmira do Ilira [From Butmir Culture to Illyrians]. Veselin Masleša, Sarajevo

Čović B (1988a) Naučna djelatnost iz oblasti prahistorijske arheologije [Scientific activity in the field of prehistoric archeology]. In: Palavestra P (ed) Spomenica stogodišnjice rada. Zemaljskog muzeja Bosne i Hercegovine, Sarajevo, pp 74–95

Čović B (1988b) Bronazno doba [Bronze age]. In: Čović B (ed) Arheološki leksikon Bosne i Hercegovine, I. Zemaljski muzej B&H, Sarajevo, pp 22–23

Čović B (1988c) Željezno doba [Iron age]. In: Čović B (ed) Arheološki leksikon Bosne i Hercegovine, I. Zemaljski muzej B&H, Sarajevo, pp 24–26

Ćirković S (1964) Istorija srednjovekovne bosanske države [History of the medieval Bosnian state]. Srpska književna zadruga, Belgrade, p 312

Dautović D (2014) Slavenska varijanta feudalizma (analiza na primjeru bellatores bosanskog srednjovjekovlja) [Slavic variant of feudalism (analysis on the example of bellatores of the Bosnian Middle Ages)]. Bosna Franciscana 41:133–165. Sarajevo

Dautović D (2019) The papacy and marriage practices in medieval Bosnia. In: Medieval Bosnia and south-east European relations. Arc Humanities Press, Leeds, pp 113–135

Dayton Accords (1995). Available at https://2009-2017.state.gov/p/eur/rls/or/dayton/index.htm. Accessed 19 Nov 2020

Dragosavac S (2016) Tehnološka analiza artefakata od okresanog kamena sa paleolitskog lokaliteta Danilovića brdo u Podnovlju kod Doboja [Technological analysis of painted stone artifacts from the Paleolithic site Danilovića brdo in Podnovlje near Doboj] Godišnjak/Jahrbuch 2017 46:19–42 ANUBiH, Centar za balkanološka ispitivanja, Sarajevo

Filipović E (2017) Bosansko kraljevstvo, Historija srednjovjekovne Bosanske države [Kingdom of Bosnia, history of the medieval Bosnian state]. Mladinska knjiga, Sarajevo

Fine J (1991) The early medieval Balkans: a critical survey from the sixth to the late twelfth century. University of Michigan Press, Ann Arbor

Halilović J (2008) Reakcije: Nestaju Sarajevske ruže [Reactions: Sarajevo roses are disappearing]. Available at https://www.islamskazajednica.ba/vijesti/vijesti-arhiva/5281-reakcije-nestaju-sarajevske-rue

https://commons.wikimedia.org/wiki/File:IllyricumAD6RomanConditionofTribes.png (Map of Illyrian nations at the beginning of our era on the territory of modern Bosnia and Herzegovina and the surroundings), Megistias, public domain, (modified by Haris Gekić). Accesed 5 June 2020

https://commons.wikimedia.org/wiki/File:Medieval_Bosnian_State_Expansion-en.svg, Creative Commons Attribution-Share Alike 3.0 Unported, (Medieval Bosnian Kingdom expansion). Accessed 6 July 2020

https://hr.wikipedia.org/wiki/Datoteka:Bosanski_pa%C5%A1aluk_1600._godine.png (Šupuk A. 2016. Map of Eyalet of Bosnia around year 1600) Creative Commons License Attribution-ShareAlike 3.0 Unported. Accessed 5 July 2020

Imamović M (1997) Historija Bošnjaka [History of Bosniaks]. Preporod, Sarajevo

Imamović E (1998) Prostor BiH u prahistoriji i antici [The area of B&H in prehistory and antiquity]. In: Imamović E (ed) BiH od najstarijih vremena do kraja Drugog svjetskog rata. Bosanski kulturni centar, Sarajevo, pp 36–44

Imamović M, Hadžibegović I (1998) Bosna i Hercegovina u vrijeme Austrougarske vladavine [Bosnia and Herzegovina during the Austro-Hungarian rule]. In: Imamović E (ed) BiH od najstarijih vremena do kraja Drugog svjetskog rata. Bosanski kulturni centar, Sarajevo, pp 223–297

IRMCT (2019) International Residual Mechanism for Criminal Tribunals, Case of Radovan Karadžić. Available at https://www.irmct.org/en/cases/mict-13-55. Accessed 22 Oct 2020

IRMCT (2021) International Residual Mechanism for Criminal Tribunals, Case of Ratko Mladić. Available at https://www.irmct.org/en/cases/mict-13-56. Accessed 22 Oct 2020

Jovanović I, Pandžić I, Preston M (2014) The Chipped Stone Industry from Rastuša Cave. Banja Luka, Faculty of Philosophy, University of Banja Luka

Kaljanac A (2014) Historija arheologije – U potrazi za prošlošću [History of archeology – in search of the past]. University of Sarajevo, Sarajevo

Kaljanac A (2016) Rezultati arheološke prospekcije 2016. godine na lokalitetu Butmir [Results of the archaeological prospecting in 2016 at the Butmir site]. Godišnjak/Jahrbuch 2017 46:19–42. ANUBiH, Centar za balkanološka ispitivanja, Sarajevo

Klaić N (1994) Srednjovjekovna Bosna [Medieval Bosnian state]. Eminex, Zagreb

Kurtćehajić S (2017) Prilozi za politički sistem Bosne i Hercegovine [Contributions to the political system of Bosnia and Herzegovina]. Futur art, Sarajevo

Kurtović E (2019) Kratka historija Srednjovjekovne Bosne [A short history of Medieval Bosnia]. Dobra knjiga, Sarajevo

Lovrenović D (1998) Od slavenskog naseljavanja do bana Kulina [From the Slavic arrival to Ban Kulin]. In: Imamović E (ed) BiH od najstarijih vremena do kraja Drugog svjetskog rata. Bosanski kulturni centar, Sarajevo, pp 43–55

Lovrenović D (2009) Stećci – Bosansko i humsko mramorje srednjeg vijeka [Stećci – Bosnian and Hum marble of the middle ages]. Rabic, Sarajevo

Malcolm N (2011) Bosna – kratka povijest [Bosnia – short history]. Buybook, Sarajevo

Marić Z (1977) Arheološka istraživanja akropole ilirskog grada Daors..a na Gradini u Ošanićima kod Stoca [Archaeological excavations of the acropolis of the Illyrian town of Daors ... and on Gradina in Ošanići near Stolac]. GZM 30–31:5–99. Sarajevo

Mesihović S (2008) Inscriptiones latinarvm saraevonensis.... Historijska traganja 1:9–68

Mesihović S (2014) Historija autarijata [History of Autariates]. Filozofski fakultet u Sarajevu, Sarajevo

Montet-White A (1992) Paleolithic settlement patterns in Northern Bosnia. Preistoria Alpina 28:91–102

Müller J (2007) Demografski varijabiliteti u kasnom neolitu na području Bosne i pitanje gustine naseljenosti u neolitu jugoistočne Evrope [Demographic variability in the late Neolithic in Bosnia and the issue of population density in the Neolithic of Southeast Europe] Godišnjak XXXVI 34:17–33. ANUBiH, Centar za balkanološka ispitivanja, Sarajevo

Mussi M (2002) Earliest Italy, an overview of the Italian paleolithic and mesolithic. Kluwer Academic Publishers, New York

Nakaš L, Župarić D, Lalić A, Dž D, Kurtović E (2018) Codex Diplomaticus Regni Bosnae, Povelje i pisma stare Bosanske države [Charters and letters of the old Bosnian state]. Mladinska knjiga, Sarajevo

Ó Tuathail G, Dahlman C (2006) The 'West Bank of the Drina': land allocation and ethnic engineering in Republika Srpska. Trans Inst Brit Geogr 31(3):304–322

Ó Tuathail G, O'Loughlin J (2009) After ethnic cleansing: return outcomes in Bosnia-Herzegovina a decade beyond war. Ann Assoc Am Geogr 99(5):1045–1053

Pandžić I (2014) Arheološki dnevnik paleolitskih istraživanja [Archaeological diary of Paleolithic researches]. Muzej Republike Srpske, Banja Luka

Pašalić E (1960) Antička naselja i komunikacije u Bosni i Hercegovini [Ancient settlements and communications in Bosnia and Herzegovina]. Zemaljski muzej B&H, Sarajevo

Pašalić E (1984) Period rimske vladavine do kraja III vijeka naše ere [The period of Roman rule until the end of the 3rd century AD]. In: Benac A (ed) Kulturna istorija Bosne i Hercegovine. Veselin Masleša, Sarajevo, pp 191–258

Paškvalin V (1988) Doba kasne antike (Dominat) [Late Antiquity (Dominate)]. Zemaljski muzej B&H, Sarajevo, pp 36–39

Pelidija E (1998) Bosanski ejalet od 1593. do Svištovskog mira 1791. godine [Bosnian eyalet from 1593 until the Peace of Svishtov in 1791]. In: Imamović E (ed) BiH od najstarijih vremena do kraja Drugog svjetskog rata. Bosanski kulturni centar, Sarajevo, pp 135–173

Redžić E (1998) Bosna i Hercegovina u Drugom svjetskom ratu [Bosnia and Herzegovina i World War II]. Oko, Sarajevo

Redžić E (2003) Bosna i Hercegovina u posljednjoj deceniji 20. vijeka [Bosnia and Herzegovina in the last decade of the 20th century]. In: Redžić E (ed) Proceedings of Conference „Istorijska nauka o Bosni i Hercegovini u razdoblju 1990–2000", Sarajevo, pp 7–13

Šabanović H (1982) Bosanski pašaluk – postanak i upravna podjela [Bosnian pashaluk – origin and administrative division]. Svjetlost, Sarajevo

Šačić A (2012) Kulturno-historijski razvoj ilirskog naroda Naresa [Cultural and historical development of the Illyrian people of Nares], vol 41. Centar za balkanološka ispitivanja, Godišnjak/Jahrbuch, Sarajevo, p 109

Šehić Z, Tepić I (2002) Povijesni atlas Bosne i Hercegovine [Historical atlas of Bosnia and Herzegovina]. Sejtarija, Sarajevo

Srebrenica genocide denial report 2021. Memorial Center Srebrenica. Available at https://srebrenicamemorial.org/en/investigations/srebrenica-genocide-denial-report-2021/8. Accessed 11 Oct 2021

Suljagić E (2021) Genocide by Plebiscite: the Bosnian Serb assembly and social construction of "Turks" in Bosnia and Herzegovina. J Genocide Res 23(4):568–587

Summary of the Dayton Peace Agreement on Bosnia-Herzegovina (1995). Available at http://hrlibrary.umn.edu/icty/dayton/daytonsum.html. Accessed 18 Oct 2020

Tasić N (2007) Neolit u senci – još jedan osvrt na starčevačka naselja u Bosni [Neolithic in the shadows – another review of Starčevo settlements in Bosnia]. Godišnjak XXXVI 34:15–16. ANUBiH, Centar za balkanološka ispitivanja, Sarajevo

Tasić N, Dimitrijević S, Jovanović B (1979) Eneolit [Eneolithic]. In: Benac A (ed) Praistorija jugoslavenskih zemalja III, eneolit. Centar za balkanološka ispitivanja, Sarajevo, pp 137–251

Tepić I (1998) Bosna i Hercegovina od kraja 18. vijeka do austrougarske okupacije 1878. godine [Bosnia and Herzegovina from the end of the 18th century until the Austro-Hungarian occupation in 1878]. In: Imamović E (ed) BiH od najstarijih vremena do kraja Drugog svjetskog rata. Bosanski kulturni centar, Sarajevo, pp 175–222

Tokača M (2012) Bosanska knjiga mrtvih [Bosnian book of dead]. Istraživačko dokumentacioni centar, Sarajevo

Truhelka Ć (1930) Jedno zanimljivo pismo bosanskog historičara fra Ante Kneževića [An interesting letter from the Bosnian historian Fra Ante Knežević]. Narodne starine 9(2):227–233

UN (2017) International tribunal for the prosecution of persons responsible for serious violations of International Humanitarian Law committed in the territory of the Former Yugoslavia since 1991, public summary of judgement Prosecutor v. Prlić, Stojić, Praljak, Petković, Ćorić, Pušić. Available at https://www.icty.org/x/cases/prlic/acjug/en/171129_Prlic_Summary_of_Judgement.pdf. Accessed 12 June 2020

UN (2021) International Residual Mechanism for Criminal Tribunals, Judgement, Persecutor v. Stanišić and Simatović. Available at https://www.irmct.org/sites/default/files/case_documents/20210806-Judgement-Stanisic_Simatovic.pdf. Accessed 15 Aug 2021

UN Cartographic Section (2007) Map of former Yugoslavia, public domain. Available at https://www.un.org/Depts/Cartographic/map/profile/frmryugo.pdf (Central position of Bosnia and Herzegovina in former Yugoslavia). Accessed 11 May 2020

Zlatar B (1998) Bosna i Hercegovina u okvirima Osmanskog Carstva (1463–1593) [Bosnia and Herzegovina within the Ottoman Empire (1463–1593)]. In: Imamović E (ed) BiH od najstarijih vremena do kraja Drugog svjetskog rata. Bosanski kulturni centar, Sarajevo, pp 99–134

10 Political Geography of Bosnia and Herzegovina

Abstract

According to the area it occupies, Bosnia and Herzegovina belongs to the category of small countries. Accordingly, it is the 125th country globally and the 27th in Europe, right after Croatia and ahead of Slovakia and Estonia. Bosnia and Herzegovina has a highly favorable, compact shape of the territory and exits to the Adriatic Sea in the area of Neum in 29.6 km of the coast length. Its post-conflict nature, multicultural and multiethnic population, and the overcomplicated political and administrative structures among others all create a unique political geographical environment. The current geopolitical position of Bosnia and Herzegovina, however, cannot be assessed as favorable. For the last 30 years, the influences of the neighboring countries, namely, the Republic of Croatia and the Republic of Serbia, as well as the interest influences of the great world powers of the USA and the Russian Federation, have intertwined throughout the territory of Bosnia and Herzegovina. The political system of Bosnia and Herzegovina is one of the most specific globally, with a Constitution and an electoral system that discriminates against its citizens and a vaguely defined structure. Bosnia and Herzegovina is specific for the presence of the Office of the High Representative and the European Union Special Representative for Bosnia and Herzegovina as the leading organization for the civilian aspect of peace implementation. The modern political-geographical structure, established by the Dayton Peace Agreement, is dysfunctional and has led to the disintegration and division of numerous natural-geographical and socio-geographical objects and systems. The significance and activity of conservative ethnonational political parties and elites are highly significant. Despite all the current problems, Bosnia and Herzegovina aims to become a member of the European Union.

Keywords

Modern political-geographical development · Political system · Political-territorial division · Political parties · Elections

10.1 Political System of Bosnia and Herzegovina

The modern political-geographical development of Bosnia and Herzegovina (hereinafter: BiH) begins with political-territorial independence, which represents the will of the citizens of BiH expressed in the Independence Referendum on February 29 and March 1, 1992, and is verified by the European Union (April 6, 1992) and the UN (May 22, 1992).

Immediately after the international recognition of the Republic of Bosnia and Herzegovina, a war of terrible proportions followed, with far-reaching political, economic, and social consequences. The war lasted until the end of 1995, when the Dayton Peace Agreement was concluded and signed, thus establishing a complex country of BiH with a specific political-territorial organization, which, according to legal experts, is atypical and has no foundation in any European state-legal-historical experience. The implications of the political and geographical structure of BiH on the geographical structure and development of the country are very complex and numerous and manifest themselves in a complex and multidimensional way in all segments of Bosnian society and determine all aspects of slow social, political, economic, and cultural development, as well as the stagnation of integration of BiH into modern global and Euro-Atlantic trends. The current political and geographical structure of BiH makes this impossible.

H. Gekić et al., *The Geography of Bosnia and Herzegovina*, World Regional Geography Book Series,
https://doi.org/10.1007/978-3-030-98523-3_10

BiH is a democratic state that functions under the law and is based on free and democratic elections (Constitution of BiH, Article 1). BiH consists of two entities: the Federation of Bosnia and Herzegovina and the Republika Srpska and the Brčko District of Bosnia and Herzegovina. In addition to having the characteristics of a high-degree state-legal decentralization and representing an atypical state organization, BiH has an asymmetric political-territorial structure at the entity level because the entity of Republika Srpska (hereinafter: RS) is politically and territorially organized as unitary and centralized, while the entity of Federation of Bosnia and Herzegovina (hereinafter: FBiH) is decentralized, in which many state functions are reduced to the cantonal level (Fig. 10.1).

Dayton acknowledged the continuity of the Republic of Bosnia and Herzegovina as an independent and sovereign state. There is no form of state structure anywhere in the Constitution of BiH which creates space for different interpretations that move from the country being a Union to that of being a unitary, highly decentralized state. According to the majority of the features from the Constitution, it seems that there is a federal state structure. However, by definition, BiH is not a federation. From a detailed analysis given by Kurtćehajić, one can conclude that BiH in terms of state regulation lies between the federal and highly decentralized state (Kurtćehajić 2017a). BiH is internationally recognized without entities and as such was admitted to the UN on May 22, 1992. The entities have emerged as a result of an international mistake as well as the current electoral system that is discriminatory (Kurtćehajić 2017b).

As a result of the modern political-territorial organization, there is a slow establishment of a coherent state-political system that would enable the full and efficient functioning of the country. In such conditions, the Inter-Entity Boundary Line (hereinafter: IEBL) is a disintegrating factor in establishing a unified state-political-territorial system of BiH.

Maintaining and even disrupting the existing state of the undeveloped constitutional and legal functioning country of BiH, which as such corresponds to the secessionist forces in the country and the radicalization of geopolitical and geostrategic relations in the western part of Southeast Europe, objectively feed the citizens' fears for its sovereignty and territorial integrity, which have proven to be realistic in the recent past.

The tardiness in gaining the political affirmation of BiH in Europe and the world is a direct consequence of its political-territorial structure. Namely, almost 30 years since gaining independence, it is still not a member of the most important European and global organizations (such as the EU, NATO, etc.). BiH has not yet claimed the fundamental geopolitical basis for comprehensive inclusion in the European and global system, primarily due to many problems in the cooperation of political subjects at the national level. In the atmosphere of the existing political-territorial structure of BiH, the incoherent attitude of political actors regarding the solution of border problems with neighboring countries, etc. seems logical.

History will remember the definition of BiH which was determined in the resolution of the first session of ZAVNOBiH (State Anti-Fascist Council for the National Liberation of Bosnia and Herzegovina), and that contains the denial and affirmation, which says that BiH is neither Serbian nor Croatian nor Muslim, but Serbian, Muslim, and Croatian. This definition has touched on the very core of the survival of BiH, to which academician Redžić says: "Negation does not allow Bosnia to be divided between Serbs, Muslims, and Croats; affirmation states that Bosnia belongs equally to Serbs, Muslims, and Croats. Negation and affirmation are not confronted but are the expression of unity in diversity, they are mutually conditional, confirm one another, one contains the other." This is proof that the fundamental form of political representation during the Second World War and after it should have been ethnic. As a result of historical events, after the Second World War, it comes to the establishment of a worker in relation to national political representation, and a noncompetitive political and electoral system is established. After the war events of the 1992–1995 period, in the first plan comes the ethnic representation by the Dayton Peace Agreement (Arnautović 2013).

The General Framework Agreement for Peace in Bosnia and Herzegovina, initialed in Dayton and signed in Paris on December 14, 1995, contains the Constitution of Bosnia and Herzegovina as Annex 4. The Peace Agreement was signed by the heads of state of the Republic of Bosnia and Herzegovina (Alija Izetbegović), the Republic of Croatia (Franjo Tuđman), and the Federal Republic of Yugoslavia (Slobodan Milošević, President of the Republic of Serbia, by authorization). The USA, Britain, France, Russia, Germany, and the European Union are witnesses to the agreement. This International Agreement reached consent on the political will of the signatory parties for everything contained in it and therefore for what is contained in Annex 4. Annex 4 was signed by authorized representatives of the Republic of Bosnia and Herzegovina, the FBiH, and the RS. This has exhausted all constitutional issues, as well as political issues, which led to the new Constitution of Bosnia and Herzegovina, as an expression of the political will of its three peoples (Bosniaks, Serbs, and Croats), expressed in the Preamble of the Constitution, which states that these three peoples, together with others, and the citizens of BiH determine the Constitution of Bosnia and Herzegovina.

> Bosniaks, Croats, and Serbs as constituent peoples (together with others) and citizens of Bosnia and Herzegovina determine here that the Constitution of Bosnia and Herzegovina is as follows . . .
>
> The Republic of Bosnia and Herzegovina, whose official name from now on will be 'Bosnia and Herzegovina,' will continue its legal existence under international law as a state

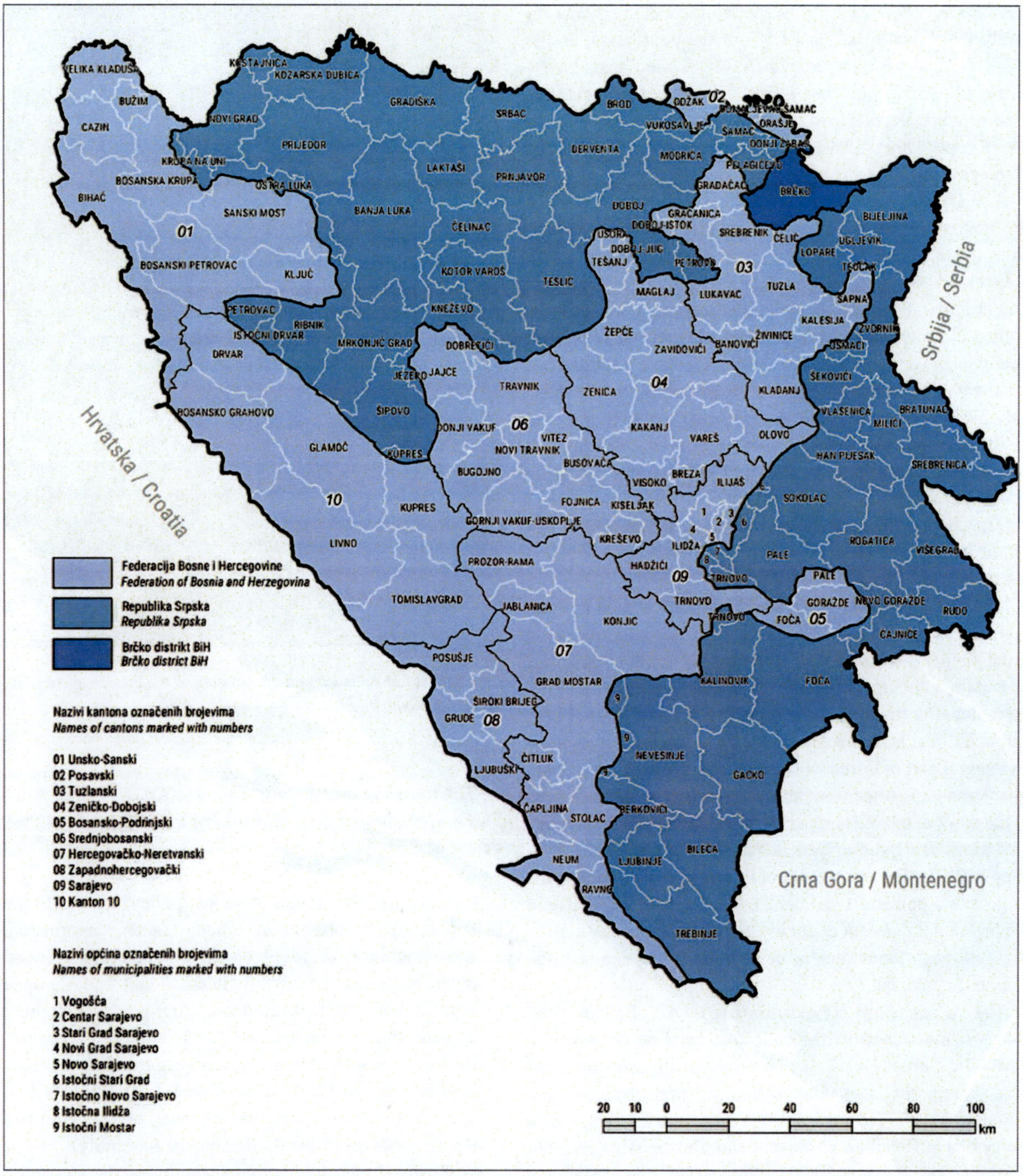

Fig. 10.1 Administrative and territorial structure of Bosnia and Herzegovina. (Agency for Statistics of Bosnia and Herzegovina 2016)

with an internal structure modified as defined here and with the existing internationally recognized borders. It will remain a member state of the United Nations and may, like Bosnia and Herzegovina, retain or apply for membership in organizations within the UN system and other international organizations. (Constitution of BiH, Article 1)

According to the population census from 2013, BiH has a population of 3,531,159, of which 62.9% live in the FBiH, 34.8% in the RS, and 83,516 or 2.3% in Brčko District of Bosnia and Herzegovina. The majority in the ethnic composition of the population have Bosniaks with 50.1%. Serbs participate with 30.8%, Croats with 15.4%, and "Others" with 2.9%, and 0.8% did not specify their ethnicity. In the entity RS, Serbs make up 81.5% of the population, and in FBiH Bosniaks participate with 70.4%, while Croats have a share of 22.4%. In the Brčko District of Bosnia and Herzegovina, Bosniaks make up 42.4%, Serbs 34.9%, Croats 20.7%, and "Others" 2% (Agency for Statistics of Bosnia and Herzegovina 2016).

The Dayton Agreement as an imposed agreement which the government of BiH had to sign that violated five laws from international relations: the right to state integrity, state sovereignty, the use of the army, not permitted for change of border under the United Nations Charter, and genocide—the United Nations Convention on Genocide requests to prevent and punish genocide crimes, while RS was awarded for genocide, right to return of refugees. According to the international law, refugees have the indisputable right to return to their former homes. By the Dayton Agreement, this right was confiscated and replaced by the right to return of property in which an entity has the right to pay to the former owner (irrespective of his/her wishes) so that he/she would not return to his/her home from which he/she was expelled (Daadler 2014).

The Constitution of Bosnia and Herzegovina, as an integral part of the General Framework Agreement, contains the usual elements inherent in the highest legal and political act of a state, although it is partly in content an act of international law (by the method of adoption; manner of adoption; and additional agreements—conventions on human rights—an integral part of the constitutional text; members appointed by international organizations and agencies are also involved in the work of a large number of institutions), and some of its provisions are more agreed in nature than normative ones (Begić 1997). The complexity of this document, which is the subject of a political agreement, can be seen more clearly in the fact that it does not explicitly mention the state system of BiH, although terms such as democratic, sovereign, and the like are mentioned.

The Presidency (*Predsjedništvo*) of Bosnia and Herzegovina represents the collective head of the complex state. The current Presidency continues a Yugoslavian tradition, given the fact that collective presidencies were introduced at the federal level in Yugoslavia in 1971 and in the Socialist Republic of Bosnia and Herzegovina in 1974 (Bieber 2008; Pejanović 2005). Since 1995, the Presidency has been an executive body consisting of one member from ranks of Bosniaks, one member from ranks of Croats and one member from ranks of Serbs (three members in total).

The powers of the Presidency (Fig. 10.2) speak to a combination of the elements of a parliamentary and presidential political system (Sahadžić 2011). In addition to the usual representative responsibilities, the office also comes with important political duties.

Fig. 10.2 Main entrance into the building of the Presidency of BiH in Sarajevo. (Photo by Aida Bidžan-Gekić)

The most important are as follows (Art. V Par. 3, Art. V Par. 4, Constitution of BiH; Rules of Procedure of the Presidency of BiH 2004):

- Conducting the entire foreign policy of BiH (i.e., appointing ambassadors and other international representatives; deciding on the foreign policy course; representing the country in international and European organizations and institutions; arranging international agreements, and denouncing; and, with the consent of the Parliamentary Assembly, ratifying treaties).
- Nominating the Chair of the Council of Ministers (who shall take office upon the approval of the House of Representatives of the Parliamentary Assembly).
- Proposing (upon the recommendation of the Council of Ministers) an annual budget to the Parliamentary Assembly but also other legislation.
- Civilian commanding of the armed forces (until the unification of the armed forces, the Presidency had authority over the two separate entity armed forces).
- Appointing five members to the Governing Board of the Central Bank of Bosnia and Herzegovina and of the Commission for the Protection of National Monuments.

The Parliamentary Assembly of Bosnia and Herzegovina is the highest legislative body of BiH, consisting of two houses—the House of Representatives and the House of Peoples—and all legislative decisions are made by adoption in both houses of the Parliamentary Assembly of BiH. The term of office of elected members of the BiH Parliamentary Assembly until 2002 was 2 years, and after the 2002 elections, they are elected and delegated for 4 years.

The Parliamentary Assembly of BiH enacts laws, approves the budget and decides on the sources and amount of revenues for financing BiH institutions and BiH international obligations, confirms the appointment of the chairman and members of the Council of Ministers, approves ratification of international treaties, and decides on other issues necessary of the execution of its powers or granted by agreement of the entities. The Houses of the Parliamentary Assembly of BiH perform legislative, informational, and control activities and regulate the internal organization and their work. In accordance with Article 4 of the Constitution of BiH, the Parliamentary Assembly is competent to enact laws necessary for the implementation of decisions of the Presidency or the implementation of its functions under this Constitution; decide on the sources and amount of funds for the work of the institutions of BiH and the international obligations of BiH; and approve the budget for the institutions of BiH; decide on the consent for ratification of the treaty and other matters necessary for the exercise of its functions, or those assigned to it by mutual agreement of the entities.

As follows from Article 4 of the Constitution of BiH, the role that is automatically linked to the Parliamentary Assembly (Fig. 10.3) is the legislative role. In performing this role, the Parliamentary Assembly acts in areas defined by the Constitution of BiH as the competencies of the state level of government. According to Article 3 of the Constitution of BiH, these are foreign policy; foreign trade policy; customs policy; monetary policy; financing of institutions and international obligations of BiH; immigration, refugee, and asylum policy and regulation; implementation of international and inter-entity criminal law, including relations with Interpol; establishment and functioning of typical and international means of communication; regulation of inter-entity transport; and air traffic control (Constitution of BiH).

All functions and powers that are not explicitly entrusted to the institutions of BiH by the Constitution of BiH belong to the entities. However, it is also envisaged that BiH, i.e., the state level of government, will take over all issues in which the entities agree, as well as the matters provided for in Annexes 5–8 of the Dayton Agreement or in matters necessary to preserve the sovereignty, territorial integrity, political independence, and international subjectivity of BiH.

In addition to the legislature, the Parliamentary Assembly also has a financial role, given that it is responsible for deciding on the sources and amount of funds for the work of BiH's institutions and BiH international obligations and approving the budget for BiH institutions. Without going into the essence, it is enough to emphasize that the same provisions are applied to regulate the basic legislative procedure when adopting the budget.

Directly related to the financial role of the Parliamentary Assembly is its supervisory role, which is reflected both in the obligation of the executive to submit a report on budget execution to the Parliamentary Assembly and in the right of the Parliamentary Assembly to accept or reject that report.

Fig. 10.3 Building of the BiH Parliamentary Assembly in Sarajevo. (Photo by Aida Bidžan-Gekić)

The Parliamentary Assembly is also authorized to either deny the executive power the consent to ratify an international treaty or agreement, i.e., cancel it. This authority indicates that the Parliamentary Assembly, in addition to the above, has a controlling role. The control role of the Parliamentary Assembly is also expressed through the work of its House of Peoples, whose task is to protect the interests of the three constituent peoples of BiH, i.e., to take care not to violate the so-called vital national interests of one of the three constituent peoples. The control role of the Parliamentary Assembly is reflected in its authority to confirm the appointment of the Council of Ministers of BiH (performed by the House of Representatives), monitor and control its work through various mechanisms, and vote no confidence if deemed necessary. In applying these powers, as it can be concluded from the abovementioned, the control role is supplemented by the electoral role, which was especially evident with the establishment of the Commission for the Preparation of Elections of the Council of Ministers of BiH in mid-2006. Its existence is determined by the Law on Amendments to the Law on the Council of Ministers of BiH, while its work is prescribed by the Rules of Procedure of the House of Representatives of the Parliamentary Assembly—which imposes the conclusion that the Commission operates within the House of Representatives (Mirić 2011).

The number of members of the Election Commission is not constant and is determined for each convocation of the House, with the stipulation that the Commission consists of one representative of each political party, coalition, and list of independent candidates, as well as all independent candidates who hold parliamentary seats. The Commission is competent to examine and analyze all available information submitted to it by the Central Election Commission and the State Investigation and Protection Agency to assess and determine the suitability of candidates for the position of chairman and members of the Council of Ministers of BiH, i.e., the same responsibilities previously held by the Office of the High Representative for BiH (OHR). In its opinion, the Commission makes a recommendation to the House for confirmation of candidates appointed as chairman and members of the Council of Ministers of BiH, although, it should be noted, its opinion is not binding. Finally, when it comes to the roles of the Parliamentary Assembly, it should be mentioned that it is the only one authorized to give authentic interpretations of the acts it adopts, which means that it also has the role of an interpreter of the current legislation.

The 42 representatives of the House of Representatives are directly elected, two-thirds in the FBiH and one-third in the RS (Art. IV Par. 1, Constitution of BiH). The number of mandates is not defined by ethnic criteria but is divided between the two entities on a territorial basis. Each entity is then subdivided into smaller electoral districts. Mandates are allocated according to the proportional representation formula per Sainte-Laguë method, with the application of the so-called compensation seats as a corrective.

The House of Peoples consists of 15 indirectly elected members: 5 Bosniak, 5 Croat, and 5 Serb delegates. While the five Serbian delegates are elected by the National Assembly of the Republika Srpska, the Bosniak and Croat members are elected by the Bosniak and Croat members' delegates of the House of the Peoples of the Federation (Art. IV Par. 1, Constitution of BiH). Thus, the formation and election of the two chambers are based on the principle of parity, namely, the parity of the entities (1/3 from RS, 2/3 from FBiH, in both chambers) and peoples (in the House of Peoples). Similar to the Presidency elections, the election for the House of Peoples excludes Serbs in the FBiH and Croats and Bosniaks in the RS, as well as all others in both entities. Moreover, Croat and Bosniak delegates are exclusively elected to the House of Peoples by Croats or Bosniaks, which prohibits Serb and any other delegates in the House of Peoples of the Federation from actively participating in their election. However, in the RS, all members of the National Assembly appoint the five Serb members of the state-level House of Peoples.

The competences of the Parliamentary Assembly of Bosnia and Herzegovina (PA BiH) are vaguely defined in the Constitution. This constitutional gap has been filled by the two chambers through the adoption of their Rules of Procedures. The main functions comprise the following: adoption of constitutional amendments and legislation, ratification of international treaties, approval and execution of the state budget, the appointment and oversight of the Council of Ministers, and the appointment of different independent institutions (Marković and Sahadžić 2013).

Unlike the House of Representatives, the House of Peoples is represented by three fixed caucuses: Bosniak, Croat, and Serb delegates (all per five). The House of Peoples holds its constitutional session usually later then the House of Representatives, as members are not directly elected, but appointed by the entity parliaments.

The Council of Ministers (Vijeće/Savjet ministara) of Bosnia and Herzegovina represents the state-level government. The Presidency appoints the Chairman of the Council of Ministers, who will take office with the approval of the House of Representatives. The Speaker shall appoint the Minister of Foreign Affairs, the Minister of Foreign Trade, and other ministers as necessary, all of whom shall assume office upon the approval of the House of Representatives. The chairman and ministers together form the Council of Ministers and are responsible for carrying out policies and decisions of BiH in various areas as set out in the Constitution of BiH and report to the Parliamentary Assembly (including, at least annually, expenditures by Bosnia and Herzegovina). A maximum of two-thirds of all ministers may be appointed from the territory of the FBiH. The Speaker will also appoint Deputy Ministers (who will not be from the same constituent people as their ministers), who will take office after approval by the House of Representatives. The Council of Ministers will resign if the Parliamentary Assembly votes no confidence in it.

The Council of Ministers is a body of executive power of BiH that performs its rights and duties as a government function, under the Constitution of BiH and laws and other regulations of BiH. The seat of the Council of Ministers is in Sarajevo. The mandate of the Council of Ministers is the same as the mandate of the Parliamentary Assembly of Bosnia and Herzegovina. The Council of Ministers of BiH consists of the Chairman and ministers from the following areas: foreign affairs, foreign trade, economic relations, finance and treasury, communications and transport, civil affairs, human rights and refugees, justice, security, and defense (Constitution of BiH).

The Constitution puts a strong emphasis on human rights and the prohibition of discrimination. However, it remains in breach of the European Convention on Human Rights, as stated in the Sejdić-Finci ruling of the European Court of Human Rights. The constitution prevents citizens who do not declare themselves as one of the country's three constituent peoples from standing as candidates for the Presidency and the House of Peoples of the Parliamentary Assembly. The constitution has not created a democratic state. B&H is not based on civic principle as it only recognizes the constituent peoples (Serbs, Croats, and Bosniaks). The citizens who do not belong to the three constituent peoples do not have the same rights. The constitution entirely sacrificed efficient and rational decision-making to the principle of involving representatives of each constituent people in decision-making processes. Therefore, the decision-making is not based on rational, professional, real, and democratic principles but on ethnicity. The so-called ethnic principle in public administration favors further ethnic divisions (Branković 2016).

Even though there was an ad-hoc parliamentary committee on the implementation of the Sejdić-Finci case established in 2011, none of the models entered parliamentary procedure or was voted on in the Parliamentary Assembly, keeping the discussion on the reform of the Presidency outside the legislature, among party elites and without broader social debate (Banović et al. 2021).

Discrimination through ethnic segregation is especially visible in BiH's education system, where the so-called two schools under one roof continue to segregate children by ethnicity in 56 schools. In these schools, children of different ethnicities are physically separated from each other and learn from different curricula. Final and binding decision of the FBiH Constitutional Court from 2014 found there was discrimination and ordered the Herzegovina-Neretva Canton to establish unified, integrated, and multicultural schools which has not been implemented yet (OSCE 2018). Discriminatory practices in the education sector are also evidenced through the situation surrounding Bosniak students in RS, who are not offered the right to call their language "Bosnian" even though this right has been confirmed by the BiH Constitutional Court. Social exclusion and discrimination against Roma, BiH's largest national minority remains widespread, particularly in housing, employment, education, and health services (OSCE 2019).

LGBT persons face discrimination in numerous fields, including employment, education, housing, and health care. According to CSOs, discrimination on the grounds of sexual orientation, gender identity, and sexual characteristics is most prevalent in the employment sector, as well as in accessing goods and services. Persons with disabilities face discrimination in accessing education, employment, social protection, and health care. The failure to implement relevant regulations contributes to the exclusion of children and young people with disabilities from the educational system. Most public buildings and institutions, as well as public transport infrastructure, remain inaccessible to persons with disabilities (OSCE 2019).

The Bosnian constitution delegates the organization and responsibilities for the judicial system to the entities and Brčko District of Bosnia and Herzegovina (Art. II 3(a), Constitution of BiH). The BiH Constitutional Court is the only judicial institution whose responsibilities, competences, and organization are mentioned in the state Constitution. Others like the High Judicial and Prosecutorial Council, the Court of BiH, the state Prosecutor, and the Public Attorney will be later on regulated by the state law. Apart from courts that work on dispute resolution, the justice system of BiH includes other institutions that enable the application of the law, such as (1) ministries of justice at all levels in BiH, (2) prosecutors' offices, (3) general attorneys' offices, (4) law firms, (5) notary public offices, and the European Court of Human Rights, considering that BiH has been a member of the Council of Europe since July 2002 (Smailagić and Keranović 2009). One might include three constitutional courts in BiH into the justice system as well, as those courts are special courts whose primary objective is the protection of principles of constitutionality and legality—the BiH Constitutional Court, the FBiH Constitutional Court, and the RS Constitutional Court. The establishment of the Court of BiH (Sud BiH) in 2007, which has state-level jurisdiction, was of particular importance to the BiH judicial system. Its tasks are comprised of the protection of effective implementation of the central state's competences and the protection of human rights and rule of law (Art. 1 Law on the Court BiH). The Court consists of departments for administrative law, civil law, criminal law, and appeal.

10.2 Contemporary Political-Geographical Structure of Bosnia and Herzegovina

Ever since gaining international recognition (session of the Council of Ministers of the European Community in Brussels on April 6, 1992), BiH (the Republic of Bosnia and Herzegovina from April 8, 1992, to the adoption of the Dayton Constitution of Bosnia and Herzegovina in late 1995) has experienced significant quantitative-qualitative changes in its internal political-territorial organization due to war period 1992–1995.

The historical genesis of the middle level of government in Bosnia shows that throughout the history of BiH, there hasn't been any governmental structure that is constituted along the national or ethnic lines and connected to the national/ethnic territories. However, during the war, a compromise was reached between the relevant entities of this period (the Republic of Bosnia and Herzegovina, the Republic of Croatia, and FR Yugoslavia) through the peace process to form ethnicity-based territorial units.

BiH consists of two entities, the Federation of Bosnia and Herzegovina (FBiH) and the Republika Srpska (RS), and one district, the Brčko District of Bosnia and Herzegovina.

The Constitution of BiH has laid the foundations for a new constitutional and legal structure of BiH as a decentralized state with two multiethnic entities and central institutions vital for the functioning of an internationally recognized state, which had far-reaching and radical reflections on the political-territorial organization of our country. Brčko District of Bosnia and Herzegovina (hereinafter Brčko District) is a separate administrative unit in BiH that does not belong to either the entity of FBiH or the entity of RS, but the state of BiH. Brčko District was created by a decision of the International Arbitration Court and was established on March 8, 2000.

10.2.1 Political-Territorial Division of Bosnia and Herzegovina at the First Level

As stated above, BiH is, legally and constitutionally, a peculiar country in the world. According to the Constitution of BiH, Article 1, item 3—Composition, the territorial-political organization of the territory of BiH during the initial division was performed at the first (primary) level on two entities: FBiH and RS.

Through the further transformation of the political-territorial organization of BiH, by the arbitral award of the High Representative of the United Nations for BiH (in accordance with paragraph 9 of the Final Arbitral Award for Brčko of March 5, 1999), the provision on the establishment of the Brčko District, covering the area within the municipality of Brčko from 1992, entered into force. The Brčko District is the unique administrative unit of local self-government under the sovereignty of BiH. With this decision, the territorial-political organization of the territory of BiH was expanded in the initial division, because the Brčko District, according to its Statute (Statute of Brčko District of Bosnia and Herzegovina 2010), is under the direct jurisdiction of the entity and state institutions of BiH.

Dayton acknowledged the continuity of the Republic of Bosnia and Herzegovina as an independent and sovereign state that it acquired on April 6, 1992. There is no form of state structure anywhere in the Constitution of BiH which creates space for different interpretations that move from BiH being a Union to that of being a unitary, highly decentralized state. According to the majority of the features from the Constitution of BiH, it seems that there is a federal state structure. However, by definition, BiH is not a federation. From a detailed analysis of the author, it can be concluded that BiH in terms of state regulation lies between the federal and highly decentralized state (Kurtćehajić 2017a).

There is no constitutional impediment to BiH being considered a decentralized state with two autonomous multi-ethnic entities, which, in terms of the extent of its competencies and the constitutionally defined presumed competencies of the entities, brings BiH closer to a federation. However, according to the constitutional definition, BiH is not a federation, much less a confederation or a union, because it does not have elementary theoretical assumptions to be a union: there are no two or more sovereign states that it unites. As it is known, entities are not sovereign. Only BiH has the sovereignty recognized by the international community.

There are at least two legal bases in the current Constitution for BiH to be considered a simple decentralized government. One legal basis lies in the constitutional definition of BiH, and the other in the representation of peoples, not entities, in the institutions of BiH. Nowhere in the Constitution can a legal basis be found for BiH to be a complex state. Article I-1 of the Constitution of BiH reads: "The Republic of Bosnia and Herzegovina, the official name of which shall henceforth be 'Bosnia and Herzegovina,' shall continue its legal existence under international law as a state, with its internal structure modified as provided herein…". The internal structure modified by this Constitution is already contained in the same article, namely, point 3: "Bosnia and Herzegovina consists of two entities: the FBiH and the RS (hereinafter: 'the entities')." It is no coincidence that both the FBiH and the RS are called entities. Precisely to preserve BiH internal structure and the non-division of a sovereign state. Therefore, BiH is not composed of two states but two entities. If the constitution-maker wanted BiH to be composed of two states, he would say so. The constitution-maker did not see the FBiH or the RS as separate states but as entities of a single state, and that state is BiH. Not to be understood as states by any chance, they are considered entities of a single state, BiH. That is why it is not a coincidence that point 3, in parentheses, states that both the FBiH and the RS will be "hereinafter referred to as entities."

Another legal basis for BiH to be considered a simple decentralized state is that the constitutional term "institutions of BiH" must not be mistaken with the already domesticated and unconstitutional phrase "joint" institutions of BiH. The Constitution of BiH does not recognize "joint institutions of BiH," only the institutions of BiH. What does that mean from the point of view of defining the form of state organization in BiH? It means that the institutions of BiH are not in fact "joint institutions" of the entities, which is often tried to impose, especially from the political representatives of the RS, but that they are, against the will of the entities, the originary institutions of BiH as a sovereign state, whose legal existence, competencies, and functions are ensured by the Constitution itself, not by the political will of the entities. This approach to the interpretation of institutions of BiH, as not being derived from the will of the entities, but as originary, original constitutional institutions of BiH, is supported by the content of Article 9.3 of the Constitution of BiH, which reads: "Officials appointed to positions in the institutions of BiH shall generally be representatives of the peoples of BiH," i.e., not the peoples in the entities, but the peoples of the entire BiH.

What does this approach mean? It means that BiH is not defined in this Constitution either as a federation of peoples or as a federation of entities. Therefore, a distinction should be made between the sovereignty of the state and the sovereignty of the people as an ethnocultural communities such as BiH. It is not necessary to homogenize the people on a physically unbroken territory in relation to the entire state of which they are citizens, because they exercise political sovereignty both as a people and as citizens in the entire area of the state of which they are citizens. Hence, in this case, entity citizenship must not be an obstacle to the exercise of rights from the citizenship of the state of BiH on its entire state territory (Ibrahimagić 2012).

Officially, the territory of BiH covers the area of 51,209.2 km (out of which 12.2 km^2 is the sea area; Bosnia and Herzegovina in numbers 2020). At the second instance of administrative division, and according to the statistical institutes of the respective entities, the area occupied by the entity of FBiH is 26,110.5 km^2 (51% of the total area of BiH; Statistical Yearbook 2020), the entity of RS 24,641 km^2 (48% of the total area of BiH; Statistical Yearbook of Republika Srpska 2020), and Brčko District 493 km^2 (about 1% of the total area of BiH; Statistical data of Brčko District 2020).

According to the data stated above, there are certain discrepancies, i.e., deviations from the total area of BiH in these different sources. According to the Agency for Statistics of BiH, the area of BiH is 51,209.2 km^2, while the statistical institutes of entities, when reporting on the area of the entities and the area of Brčko District in sum, provide a surplus of 35.3 km^2. The cause of this discrepancy in the data should also be sought in the inconsistency of certain parts of the IEBL so that some data have more significant ranges.

According to the provisions of Article 3 of the Constitution of BiH and the Constitution itself, the foundations have been laid for a new constitutional-legal and state-legal structure of BiH as a decentralized state. The entities in BiH, FBiH, and the RS, according to their constitutions, have a broad range of competencies (hence, BiH is specific on a global scale), which include establishing parallel (special) relations with neighboring countries, followed by the administration, judiciary, police, education, health, housing policy, spatial planning, transportation, communication, etc.

It is important to mention that in addition to being the capital of BiH as a state, Sarajevo is the capital of both the FBiH and RS entities (Article 1 of the Constitution of BiH; Article 4 of the Constitution of FBiH; Article 9 of the Constitution of RS).

10.2.2 Political-Territorial Division of Bosnia and Herzegovina at the Second Level

The political-territorial organization of BiH in the second level is a consequence of a series of political processes that confirmed their legal framework since the Washington Agreement, signed on February 29, 1994, when the Constituent Assembly of FBiH adopted the Constitution of FBiH at a session on March 30, 1994, which entered into force on the same day. This marked the beginning of the process of constitutional and legal transformation of the former Republic of Bosnia and Herzegovina.

In the normative-legal and political part, this process ends with signing the General Framework Peace Agreement in BiH in Dayton (USA), i.e., Paris (France—December 14, 1995), which resulted in the Constitution of BiH. With Washington and then the Dayton Peace Agreement, FBiH was established as a complex entity with ten cantons on which its political-territorial and administrative structure rests. According to the provisions of the Law on Federal Units (cantons-counties) published in the *Official Gazette* No. 3/9 of 1996, cantons are constituted considering the ethnic, economic-functional, natural-geographical, and communication principle.

In the FBiH, ten federal units, cantons have been formed. Their constitutions determine the names and seats of the cantons, and the name of the canton is determined exclusively by the cities that are the seat of the cantonal government or by regional geographical characteristics. In most cases, this principle was respected, which is evident from

Table 10.1 Cantons of the FBiH and their share in the total area of FBiH and BiH

Canton	Seat	Area (km^2)	Share in FBiH (%)	Share in BiH (%)	Number of local self-government units (municipalities/cities)
Una-Sana	Bihać	4125.0	15.8	8.1	8
Posavina	Orašje	324.6	1.2	0.6	3
Tuzla	Tuzla	2649.0	10.2	5.2	13
Zenica-Doboj	Zenica	3343.1	12.8	6.5	12
Bosnian-Podrinje Canton Goražde	Goražde	504.6	1.9	1.0	3
Central Bosnian	Travnik	3189.0	12.2	6.2	12
Herzegovina-Neretva	Mostar	4401.0	16.9	8.6	9
West Herzegovina	Široki Brijeg	1362.2	5.2	2.7	4
Sarajevo	Sarajevo	1276.9	4.9	2.5	9
Canton 10	Kupres	4934.1	18.9	9.6	6

Adapted from Statistical Yearbook (2020)

the name of the canton, with some federal units changing their names (Table 10.1).

From the constitutional and legal aspects, FBiH is defined as a community of cantons on the principle of decentralization of power and precise division of competencies between cantons and FBiH. According to the Section 3 of the Constitution of FBiH, Division of responsibilities between the Federation government and the cantons (Article 1), exclusive responsibilities of FBiH are conducting foreign affairs; organizing and conducting the defense of the FBiH; and protecting its borders, including the establishment of a joint command of the military forces, the supervision of military production, and the conclusion of military agreements; citizenship; determining economic policy, including planning and reconstruction and land use policy at the federal level; trade regulation, including customs, international trade and finance, trade within FBiH, industrial property rights, product standardization, securities, and communications; regulation of finances and financial institutions, including the establishment and control of the currency, the establishment of monetary and fiscal policy, and the establishment of a central bank; combating international and inter-cantonal crime, in particular terrorism, drug trafficking and organized crime, as well as cooperation with Interpol; allocation of electronic frequencies for radio, television, and other purposes; determining energy policy, including the distribution between cantons and providing and maintaining the necessary infrastructure; and financing the activities of federal authorities and institutions established by the federal government through taxation, borrowing, or other means.

Cantons, as territorial-political units in FBiH, have a significant scope of competencies in different areas. According to the Constitution of the FBiH (Section III, Article 4), cantons have all competencies that are not explicitly entrusted to the entity government. The particular competencies of the cantons are in establishing and supervising police forces that will have uniform federal uniforms with cantonal insignia; establishing education policy, including the enactment of education regulations and the provision of education; establishing and implementing cultural policy; determining housing policy, including the adoption of regulations concerning the arrangement and construction of housing; establishing policies regarding the regulation and provision of public services; enacting regulations on local land use, including zoning; enacting regulations on the improvement of local business and charitable activities; enacting regulations on local energy production facilities and ensuring their availability; establishing policy regarding the insurance of radio and television, including the adoption of regulations on the insurance of their work and construction; implementation of social policy and establishment of social protection services; and creation and implementation of cantonal tourism policy and development of tourist resources financing the activities of the cantonal government or its agencies by taxation, borrowing, or other means (Constitution of FBiH 1994).

The political-geographical structure of the entire state, as well as the entities, is further complicated by the cantons. Each canton has its own Constitution, which ensures the functioning of the legislative, executive, and judicial authorities, as well as other institutions and bodies, and the protection of rights and freedoms, established by the Constitution of FBiH.

According to Articles 2 and 3 of the Constitution of FBiH (Section III—Division of Competences between the Federal and Cantonal Authorities), the joint competencies of the FBiH and the cantons are in the following areas: guaranteeing and enforcing human rights, health care, environmental protection policy, communication and transport infrastructure, and social policy; enforcement of laws and other regulations on citizenship; immigration and asylum; tourism; and the use of natural resources (Constitution of FBiH 1994).

Table 10.2 Mean values of isochrons and distances for the cantons in the FBiH

Canton	Mean isochronous value relative to cantonal center (in minutes)	Mean value of distance from the cantonal center (in km)
Una-Sana	64.7	56.7
Posavina	28.5	37.0
Tuzla	48.7	36.7
Zenica-Doboj	71.7	62.6
Bosnian-Podrinje Goražde	22.5	46.0
Central Bosnian	70.4	46.7
Herzegovina-Neretva	62.7	63.7
West Herzegovina	44.7	33.0
Sarajevo	24.2	19.8
Canton 10	68.8	59.7

Adapted from Mirić (2009)

These competencies may be exercised jointly or separately or by the cantons in coordination with the federal government, where the cantons and the federal government agree on these competencies on an ongoing basis. In cases of laws and other binding regulations throughout the FBiH, the federal government takes into account cantonal competencies, different situations in individual cantons, and the need for flexibility in implementation.

This political and legal framework of BiH, at the first and second level of political-territorial division, is very important due to the competencies of the entities, districts, and cantons, which, in geographical terms, is reflected primarily in nodal-functional, economic-geographical, and administrative-political transformation of the area of BiH. The legal and constitutional framework on which the entities regulate their mutual relations, and relations with the state of BiH, and inter-cantonal relations and relations between the canton and FBiH represent the basis for a more detailed study of complex geographical and regional-geographical processes and conditions.

In an analytical review of the political-territorial structure of the FBiH, there is a joint assessment that the cantonal division is unsustainable from the aspect of the criteria on which it is constituted. One of the principles of constituting cantons in the FBiH is the natural-geographical approach. Analyzing the attitude towards the physiognomic regions of BiH (Nurković and Mirić 2005) and the cantonal structure, the conclusion is that this principle is very little respected because most cantonal units from this aspect represent a heterogeneous space. This is most evident in the examples of Una-Sana, Herzegovina-Neretva, and Zenica-Doboj cantons.

Regarding the economic-functional principle of constituting cantons, it should be emphasized that in spatial reality, the division into entities caused a decrease in spatial interaction by 40% in cities in FBiH. The cantonal division of the FBiH has further accelerated the disintegration of the economic space, creating explicit, very difficult-to-maintain spaces. It should be emphasized that, with the dissolution of the single economic space of BiH and the economic "collapse" of large industrial and commercial enterprises in the period 1992–1995, this problem further complicated the possibilities of sustainable economic development, which was the basis of economic and functional connection of specific areas.

In terms of the primacy index and gravity models, the cantonal division shows major shortcomings in two respects: one is the low primacy index of the cantonal seats in relation to the corresponding local self-government units (municipalities/cities), and the other is reflected in the overlapping gravitational influences of neighboring regional centers.

Examples of a low degree of primacy are particularly pronounced in Posavina (1.1), West Herzegovina (1.1), and Canton 10 (1.1). The gravitational interaction and gravitational area models are closely related to the communication principle of canton separation in the FBiH. When analyzing the values of these parameters at the cantonal level, Mirić (2009) points to the exact inclination in many cases (Table 10.2).

It is noticeable that the communication principle is extremely fulfilled for the areas of Sarajevo Canton and Tuzla Canton, comparing the elements of the spatial size of these administrative-political units. A different case is expressed in Zenica-Doboj and Central Bosnian cantons, which is even more complicated by the existence of a large number of municipalities in them (12 each) and realistically much weaker communication infrastructure (quality of the road network in the first place). These parameters were also used to analyze the spatial relationship of contact municipalities to neighboring cantons based on the communication principle to see the functional components and possibilities of a more rational structure of the FBiH.

Fig. 10.4 The building under construction is in the entity of RS (municipality of Istočno Novo Sarajevo), and parking is in the entity of FBIH (municipality of Novi Grad Sarajevo). (Photo by Aida Bidžan-Gekić)

There are many examples in which the affiliation of individual municipalities to certain cantons is questioned, taking into account the above indicators and the communication principle in constituting administrative units in the FBiH. The most striking examples are the municipalities in Zenica-Doboj and Central Bosnian cantons neighboring the Sarajevo Canton. This primarily refers to the municipalities of Kiseljak, Breza, and Olovo, although the parameters indicate the presence of a larger number of cases of distances and isochrones with lower values compared to neighboring cantonal centers compared to the home (e.g., Busovača, Vareš, or Drvar) (Mirić 2009).

Considering the stated principles, according to which the cantons are differentiated within FBiH, the resolute conclusions point to the inevitability of restructuring and reorganization of the central level of administrative-political structure and thus the functioning of government. In doing so, the possible models in the current political and socioeconomic moment should be developed in several directions. One of them is the predominant spatial transformation of the current structure, which should be based on the exact and rational principles of differentiation of the scope of the central government. In this regard, the starting point should be based on spatial-functional rationalization, which necessarily requires a reduction in the number of cantons or their reorganization according to some of the already mentioned principles (primarily economic-social and communication).

The most rational model would be reducing the number of cantons by the process of their integration, i.e., merging, and, in the next step, joining individual municipalities to newly formed cantons exclusively on the principles of communication, economic-social, and natural-geographical parameters, or complete abolition of cantons and formation of territorial units in FBiH on the principles of NUTS (Nomenclature of Territorial Units for Statistics) regions as in the European Union, as a prerequisite for the establishment of a European model of regionalization in the entire territory of BiH.

10.2.3 Political-Territorial Division of Bosnia and Herzegovina at the Third Level

The third-level political-territorial division is based on the existence of the lowest territorial-administrative units—municipalities/cities (hereinafter: municipalities). In 2021, their total number was 79 in the FBiH and 63 in the RS. Four municipalities are part of the city of Sarajevo, and six are part of the city of Istočno Sarajevo. The Brčko District is a special territorial unit that we partially analyze at this level due to its size and importance.

Based on the administrative-territorial division from 1992, the number of municipalities increased by 34 new local self-government units, representing an increase of approximately 31%. Most of the newly formed local self-government units were constituted by the disintegration of municipal territories in 1992, as a direct consequence of the formation of the IEBL (example of the disintegration of Sarajevo, Fig. 10.4). However, several newly formed municipalities are not constituted by the division of pre-war municipalities that disintegrate the entity border, namely, the municipalities of Bužim, Kostajnica, Milići, and Stanari.

Within the entity of FBiH, 13 new municipalities have been constituted. In the entity of RS, the newly formed municipalities are spatially related mainly to the 1992 disintegration of municipal territories, except for the municipalities of Milići, Kostajnica, and Stanari. Out of a total of 63 municipalities, 21 are newly formed municipalities on the territory of the RS after the war in 1995.

IEBL, in addition to causing the emergence of new municipal territories due to its spatial character, has also led to minor but significant changes in the spatial structure of existing municipalities in BiH. This is primarily reflected in relatively minor changes in the reduction or increase in the area of a significant number of municipalities.

In BiH, only 47 of the 109 municipalities have retained their territorial area compared to 1992, as the direct result of the existence of the IEBL. Having said this, it is necessary to mention minor but numerous significant changes in the territory of municipal borders that are directly related to the IEBL. These changes have a lesser territorial coverage, and entity laws regulate their affiliation.

Furthermore, constituting local self-government units with the status of a city in the last decade and a half has introduced another level into the political-territorial organization of BiH. There are 25 such cities, of which only 2 cities mentioned earlier make up more than 1 municipality (Sarajevo and Istočno Sarajevo). All the other cities have grown from a single municipality into a city and do not have any significant benefits because of that, and from that aspect, this change is mostly cosmetic.

The sheer quantity of administrative units making up the tertiary division of BiH in terms of territory and politics suggests that such a division is not the product of efforts that would result in a better spatial reorganization of the inherited municipal network. The recent situation with the number of municipalities/cities, which is 142 (if we count the Brčko District, it is 143, but in this case, we omit it from this number only because of its political and legal subjectivity), is the result of compromise at higher political and territorial levels. Such solutions have led to, in some cases, "spatial miniaturization" of individual municipalities.

This spatial miniaturization is primarily a characteristic of a large number of newly formed municipalities, which, unlike the pre-Dayton municipal network in which there was also "miniaturization" (justified by the high degree of concentration of population—e.g., Sarajevo municipalities of Novi Grad, Centar, or Novo Sarajevo, which are small in area but large in population), neither have considerable population potential nor do they cover larger areas in space.

There are many such examples, but geographically unique municipal areas such as Istočni Mostar, Kupres and Istočni Drvar (RS), Dobretići, Sapna, Teočak, and Ravno (FBiH) stand out. Such municipalities are characterized by a small area and an underdeveloped municipal center, which does not have all the functional features of the city, or such a center does not exist, but the functions are dislocated to larger urban centers to which these municipalities gravitate (e.g., Istočni Mostar and Nevesinje).

The enormous fragmentation of recent political-territorial differentiation in the third level is evidenced by the fact that, in 1992, the average territory of a municipality in BiH was 469 km^2, while, as of 2021, the average territory of a municipality was 358 km^2. Among political-territorial units of local self-government (municipalities or cities), the territory area ranges from 9.9 km^2 (municipality of Novo Sarajevo) to 1239 km^2 (Banja Luka). At the same time, as many as 12 municipalities have an area of less than 50 km^2, and their total area is less than the average area of the municipality in BiH in 1992. The second smallest municipality in BiH, Doboj-Jug, with an area of 10.2 km^2, is a direct consequence of the establishment of the IEBL and an example of illogical modern political-territorial division that has a destructive effect on the system of the overall socioeconomic development. In addition to the above, special emphasis should be placed on the size of some newly formed municipalities, with an underdeveloped municipal center that did not have the status of an urban settlement before 1995, which are characterized by a very small area. These are the municipalities of Istočna Ilidža (29 km^2), Teočak (29 km^2), Istočno Novo Sarajevo (36 km^2), Doboj-Istok (41 km^2), Domaljevac-Šamac (44 km^2), Kupres (RS, 45 km^2), Donji Žabar (49 km^2), and Usora (49.8 km^2). The largest local self-government units are Banja Luka (1239 km^2), Mostar (1175 km^2), Konjic (1169 km^2), Foča (RS, 1135 km^2), and Glamoč (1034 km^2) (Table 10.3; Fig. 10.5).

There are numerous problems within the process of transformation of the municipal network, unresolved issues on defining the spatial coverage of individual municipal territories, unorganized cadastre and its harmonization between the entities, and the absence of professional service or Ministry of geodetic affairs at the level of BiH. Consequently, there are numerous difficulties in the exact determination of the spatial scope. Therefore, we can only talk about the cross-section of the situation on this issue because the time dynamics of change are very pronounced.

Even though municipalities are at "the bottom" of the hierarchical structure, they perform very important functions in the areas of housing, education, health, security, etc. Financially speaking, in 2015, budgetary incomes for all municipalities in BiH were ca. 1.6 billion convertible marks (comparable, state level had for that year a budgetary income of somewhat less than a billion) (Huškić 2017).

Spatial units of local self-government (municipalities) are organized into local communities and then into populated places (settlements). The number of local communities in BiH is 2640, while the number of settlements is 6141. Within a municipality, local communities are established as a

Table 10.3 Classification of municipalities/cities by area in Bosnia and Herzegovina in 2021

Category	Area in km^2	Number of municipality/city	Area in % from total area of country
Micro	Under 50	12	0.8
Very small	50–149.99	23	4.4
Small	150–249.99	26	10.2
Medium	250–499.99	46	31.5
Large	500–699.99	18	20.8
Very large	700–899.99	10	15.5
Huge	More than 900	8	16.8

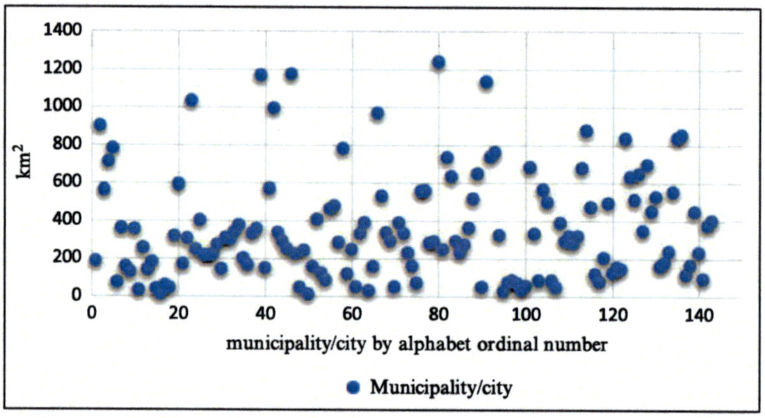

Fig. 10.5 Municipalities/cities in Bosnia and Herzegovina in relation to the area in 2021

mandatory form of local self-government enabling the citizens to participate in decision-making on matters within the scope of self-government and local affairs that directly and daily affect their lives and work. Unlike populated places, a local community has the status of a legal entity within the rights and duties determined by the statute and the decision on the establishment. The local community is entered in the register of local communities kept by the competent municipal service. The Municipal Mayor decides on entry in the register of local communities, and a decision of the Municipal Council determines the registration procedure and the manner of keeping the register. The local community has a statute, its own Council, the President of the Council, and the Citizens' Assembly. Municipal councils organize local elections for the Council and the President of local communities at the municipal level.

10.2.4 Degree of Correlation of Political-Territorial Organization with Major Physical-Geographical, Socio-geographical, and Economic Factors

From the idiographic analysis of the political-territorial organization of BiH, and especially its division by the IEBL and the differentiation of the larger entity into cantons, to the particularization of the municipal network, arises the complexity of its reflection on the complex geographical reality. In the following text, we will look synthetically at some segments that arise from this state of spatial territorial division of BiH. Their quantitative-qualitative dimension points to the most important problems that arise when studying the relationship of regional structure to political-territorial division.

As the relief structure is one of the leading elements of regional-geographical differentiation of space, there is a significant discrepancy between the largest geomorphological-geotectonic units and the IEBL in BiH. First of all, it refers to the differentiation of the territory of BiH into three geotectonic complexes (belts), namely, the outer, central, and external Dinarides. They run in the NW-SE direction, while the IEBL, for the most part, has the N-S direction of provision, except for the southwestern part of BiH. Due to such spatial features, spatial disintegration is highly present at the level of lower geotectonic units (zones), which are part of these complexes.

In the outer Dinarides complex, an example of the dissolution of the southeastern part, i.e., the lower zone of the High Karst, is illustrative, while in the central Dinarides, disintegration is fully expressed in the dissolution of the zones that form this complex (e.g., Mesozoic limestone and dolomite zones with central Bosnian shale mountains in the core, Bosnian flysch zone—Sarajevo-Banja Luka flysch zone).

As IEBL in the northern parts of BiH which runs extremely oriented to the N-S direction, in this part of the central Dinarides, there is also a pronounced disintegration of the entire complex and associated zones. The most obvious example is the inner ophiolite zone (Savsko-Vardarska), which is disintegrated into several smaller spatial units.

In terms of hypsometric features, which give a significant dimension to the physiognomic differentiation of space, the relief of BiH is classified in the category of middle-mountain ranges. In the North, BiH occupies the peripheral parts of the Pannonian Plain. Lower altitudes are dominated by many basins and depressions and the appearance of lower mountains in the form of ridges and massifs. This physiognomically very homogeneous area is decomposed to a great extent, where the particular emphasis should be placed on the disintegration of depressions such as Sprečka and Bosansko Brodska and the ridge and massif the areas of Majevica, Ozren, and Trebovac.

In the central mountain-valley area, the IEBL intersected a large number of dominant mountain morphostructures of this area, such as Vitorog, Vlašić, Bjelašnica, and Treskavica, and from the synform area of the Sarajevo valley. Somewhat further South in the area of Visoko karst, we should point out the disintegration of the mountain massifs of Velež, Crna Gora, and Hrgud and the area of Popovo polje.

The analysis of the spatial manifestation of IEBL shows a high degree of disintegration of river basins and river networks in BiH in quantitative and qualitative aspects. This relationship is manifested in the dissolution of 8 of the 11 isolated river basins.

As much as 92% of the catchment area in BiH, separated by river basins, is disintegrated by the IEBL. River flows of the highest rank (Bosna, Neretva, Vrbas, Una, Drina, and Trebišnjica), which also have the largest catchment areas, intersect with the IEBL (flows of Bosna and Drina twice). In the further ranking, at the level of tributaries of the first rank, the IEBL disintegrated 26 river flows (mainly in the Bosna basin (seven) and in the Drina basin (six)). At the level of second-tier tributaries, the number of river flows divided by the IEBL is 15, and disintegration is again most pronounced in the Bosna basin (six river flows). At the level of third-tier tributaries, the disintegration of the river network and basins has quantitatively come to the fore, as the IEBL divides as many as 96 flows. Bearing in mind that the hydrographic network and river basins are one of the important factors in the methodology of separating regional units of different ranks in BiH, these indicators are necessary for an exact consideration of the relationship between modern political-territorial division and regional-geographical structure of BiH.

The spatial manifestation of the IEBL has led to a high degree of disintegration and a unified electric power structure of BiH. As a consequence of the political-territorial structure and competencies of the entities in terms of resource management, there is a division of the power system. Despite that, BiH satisfies its electricity needs entirely by producing electricity from its power plants, using the available hydro potentials and domestic coal. The state of domestic consumption also enables the export of part of the produced electricity.

Progress has been made with the establishment of the State Electricity Regulatory Commission (SERC) to regulate electricity transmission and create conditions for the development of international electricity trade, with the establishment of the state Independent System Operator (NSO, with the task of managing the electricity transmission system to ensure the continuous supply for consumers according to defined standards of quality and reliable operation), and the company Elektroprijenos, which is responsible for the development, construction, operation, and maintenance of the transmission network in BiH. However, the competence of Elektroprijenos as a state-owned company over the transmission network is organizational, while the actual owners of the network are entities. Of the three power companies in BiH, two are in FBiH—JP Elektroprivreda BiH d.d. Sarajevo and JP Elektroprivreda Hrvatske zajednice Herceg-Bosna d.d. Mostar—and one in RS (Elektroprivreda RS in Trebinje). Unique river basins and streams and potentials for energy profitability through hydropower plants have been disintegrated so that the functioning of energy systems is very problematic and complicated. The Spatial Plan of the RS in the electricity segment was developed as an independent system without influence and interconnection with the Spatial Plan of the FBiH. Projects that cover the territory of both entities and which were defined by long-term spatial plans, spatially protected, have now been significantly changed, and mutual harmonization will be necessary. Development projects have an exclusive entity-level character.

The total number of settlements and municipal centers in BiH is systematically organized only based on the administrative-territorial hierarchy. For the spatial interaction of the centers, there are no indications from which it could be concluded that these centers in BiH have the characteristics of systemic connectivity. This is a real consequence of the administrative division of BiH into two entities and one district. Before 1992, BiH operated with systemic links between central settlements. On the first day of the war in 1992, these systemic links were severed, and the current administrative structure did not re-establish them. The settlement system has evolved over a long time. From the area in which the system functioned, the division into parts that did not previously function as a subsystem does not automatically form systemically articulated spaces. Such is the case of the two entities with the establishment of the IEBL, which is a direct consequence of the 1992–1995 war lines. With the IEBL, in administratively and functionally fragmented municipalities, 304 settlements were divided. The degree of disintegration is spatially different, which is a consequence of the unequal population density of certain parts of BiH and the spatial character of the IEBL. Spatially, the process of disintegration is most pronounced in the northeastern parts (valleys of Usora and Spreča, and the lower course of Bosnia), then in the metropolitan area of Sarajevo, then in Goražde and Jajce. A unique effect of these divisions is the formation of new municipalities in which the choice for the function of a central settlement is reduced to a rural settlement. Such settlements do not objectively constitute a collectivity that can generate spatial interaction processes.

In socio-geographical terms, the IEBL has dismantled the settlement system formed over many years under specific historical and socioeconomic conditions. The result of such long-term development is forming a specific spatial structure of settlements of different hierarchical rank, conditioned by the spatial distribution of the population, economic contents, and accompanying infrastructure. These processes also formed a network of urban settlements of different hierarchical ranks, leading to the spatial system of "urban reinforcement" as the dominant factor of socioeconomic and other aspects of BiH development. This urban armature is reflected in the spatial existence of the primary direction (Šamac-Doboj-Zenica-Sarajevo-Konjic-Mostar-Čapljina) and lateral directions from east to the north (Goražde-Sarajevo-Travnik-Jajce-Banja Luka-Gradiška) and in the northern part from west to east (Bihać-Prijedor-Banja Luka-Doboj-Tuzla-Zvornik). Comparing such urban networks with the network of road and railway routes, their apparent spatial connection can be noticed. Urban systems of the highest rank (the example of Sarajevo, as a metropolitan area) have lost significantly in their demographic concentration. A similar example is the urban areas of Doboj, Goražde, and Jajce.

On the other hand, the Dayton Peace Agreement, which "primarily took into account the ethnic and military-political criteria," divided BiH into two parts, breaking the hierarchical-functional, infrastructural-economic, and all other relations between large urban centers, centers of lower hierarchical rank, and their hinterland from their resource potentials, production, and other contents (Bublin 2000).

The assumptions about establishing the systemic functioning of central settlements in each entity separately are very questionable. If the strategic commitment is to form entity settlement systems, then this process would take a very long time, and the partial, nonintegrated infrastructure network irrational and costly and the entire state of BiH would not be compatible with the EU system.

Polarization in urban development, strengthening the role of the cities of Sarajevo, Banja Luka, Mostar, and Tuzla, and weakening the role of Trebinje, Bijeljina, Bugojno, Doboj, Zvornik, etc. The problem of a sort of ruralization of cities is an objective consequence of political and ethnic homogenization at the entity and cantonal levels.

The IEBL also conditioned the disintegration of transport infrastructure to a significant extent, separate development and reconstruction due to the consequences of the war. The IEBL cuts the highway in 24 places, the regional roads in 51, and the local roads in as many as 92 places. Prioritization and significant investments in the reconstruction of road communications are slowing down the integral development of this vital sector in BiH. A similar example is with the railway traffic, which significantly depended on the functioning of the economy. There is a particularly pronounced disintegration in the area of the intersection of the railway routes East-West (Zvornik-Tuzla-Doboj-Banja Luka-Novi Grad) and North-South (Šamac-Doboj-Zenica-Sarajevo-Mostar-Čapljina).

Reményi et al. (2016) in their paper aimed to present how much regional development (especially transportation development) in BiH is implemented according to an ethno-territorial logic. In a small country lacking resources, which is also hindered by internal separatist tensions, transportation development implemented along the lines of ethnic thinking results in the establishment of parallel structures and a waste of resources. The underlying issue is due to the conflict of the ethno-nationalist elites, but the institutions which enable this were created by Dayton. According to them it would be desirable to change this, and a state-level integrated regional policy would be an important pillar of a stable BiH, but based on the failures of the previous attempts, they have no illusions that this will occur in the near future.

The centuries-old multiconfessionalism, multiculturalism, and multiethnicity of BiH are being called into question due to the confessional, cultural, ethnic, and political homogenization that is taking place in the atmosphere of its political-territorial structure. This homogenization, which began during the war (1992–1995), is a severe threat to the centuries-old character of the territory of BiH. This is reflected in establishing ethnically homogeneous areas such as those in the East and far West of BiH, which were not the case before. There is a pronounced polarization in the modern demographic development of BiH in the light of the modern political-territorial structure and implied by ethnic and political homogenization.

The issue of the return of refugees and displaced persons to their settlements after the war is prolonged and extremely difficult, especially in the entity of RS. This is especially evident in creating a regionally different environment for sustainable return and is in accordance with the modern political-territorial organization of the state, which in turn affects the creation of specifics in population and other elements of socioeconomic development (Nurković and Mirić 2005).

The problem of brain drain from BiH is directly related to the extremely slow and difficult implementation of economic, social, and political reforms in the atmosphere of the current political-territorial structure. This problem is evident, and the assessment of the competent services shows that around

530,000 people left BiH in 2013–2019, of which 80% were young (Labor Force Survey 2019).

A particular problem is the unequal treatment (discrimination) of BiH citizens in neighboring countries with regard to ethnic and entity affiliation. This is a direct consequence of its political-territorial structure, i.e., the right of the entities to establish parallel relationships with neighboring countries (FBiH with Croatia, RS with Serbia) and the double-natured attitude of these countries towards BiH. Related to that is the problem of controlling the entry and treatment of foreign persons in BiH, which affects regional and European security. This problem is seen as a severe obstacle to removing barriers to the movement of BiH citizens in European countries.

The political-geographical structure of BiH also has its economic-geographical implications. First, it is important to point out the nonunified and disintegrated economic system of BiH at the entity level. There are very few macroeconomic policy instruments available to BiH. The disintegration of the spatial organization and valorization of natural resources of BiH at the entity and cantonal level (forests, ores, fertile land, water, etc.) is a direct consequence of its political-territorial structure (the natural resources are in one entity and processing capacities in the other one, for example, metallurgical resource-production complex). BiH's transport and infrastructure systems at the entity level have been disintegrated. It is especially important to mention the disintegration of road infrastructure and railway infrastructure. The disintegration of transport infrastructure, in turn, implies the objective existence of problems that occur in the maintenance and safe operation of transport on these communications. The disintegration of the electric power system at the entity level, which in the FBiH has the characteristics of disintegration at the ethnic level, is reflected in the dissolution of the pre-war generation and transmission power network into three separate power systems in BiH, as we previously mentioned (Elektroprivreda BiH, Elektroprivreda HZ Herceg-Bosne, Elektroprivreda RS).

The political-territorial structure of BiH established by Dayton in economic-geographical terms is, in modern times, the critical barrier to its inclusion in global, regional, and Euro-Atlantic integrations (Kukić 2001), because integration into international economic flows implies a comprehensive sustainable economic system; complete legislative and institutional harmonization with developed countries, especially those in the EU; etc. The current state of the economy in BiH is not a sufficient guarantee for sustainable economic development because the unstable political situation and undeveloped legal and institutional infrastructure of the entire state structure prevent even the most basic, which is defining goals of economic policy of BiH for its sustainable development.

10.3 Electoral Geography

Joint coexistence in BiH mixed up the members of the most numerous three of its peoples through marriages, friendly ties, etc. so that the conflicts in BiH had additional dramatic consequences. BiH has been a Yugoslav-oriented republic. However, the flow of events imposed by other republics, which shook the foundations of Yugoslavia, demanded that its political representatives be more actively involved in the events that followed. So political changes began in BiH. It was the SDA (Party of Democratic Action) that first formed in mid-1990 which was supposed to articulate and protect the interests of Bosniaks and BiH. Two months later, the SDS (Serb Democratic Party) was founded by Bosnian Serbs, and immediately after that, Croats created their own HDZ (Croat Democratic Party) branch in BiH. Although deeply confronted by many issues, three national parties agreed on a closed meeting to act uniquely against then the main opponents—communists who had in the meantime transformed into the League of Communists, the Party of Democratic Change (SK-SDP), to be transformed later into the Social Democratic Party (SDP).

In the parliamentary elections held on November 18, 1990, the national parties SDA, SDS, and HDZ won, winning 84.6% of seats in parliament. National parties were surprised by a great victory, but most of the population voted for them fearing that the members of the other would win and endanger their survival. This problem would be less pronounced then but also now if the danger over BiH has not been present because of the pretensions of its neighbors towards it. Thus, everyone most trusted to their national party. In response to the request for recognition of the independence of BiH, Badinter's commission proposed a referendum to which all the citizens of Bosnia and Herzegovina would be invited to participate in under international control (Trnka 2000).

Bosnian Serbs who were dissatisfied with the proposal and wished to remain in the mainly Serb Yugoslavia, aware that the referendum would show that citizens would want independence, on January 9, 1992, declared a proto-state formation "Republic of the Serb People of Bosnia and Herzegovina" later renamed to "Republic of Srpska." Unlike Bosnian Serbs, Bosnian Croats received instructions to vote on the referendum for independence, because they would leave Yugoslavia that way and it would be easier to join Croatia later. In order to realize these goals, the elected representatives of Bosnian Croats, on November 18, 1991, at a session in Grude, made a decision on the establishment of the Croatian Community "Herceg-Bosna," a proto-state formation that should have at the right time be annexed to Croatia.

During all these combinations, the referendum for independence was held on February 29 and March 1, 1992, under the control of international observers who later confirmed that it was implemented according to all international democratic standards. According to the official results, 64.31% of the 3.25 million registered voters came out to the referendum. 99.44% of those who came out voted "for." This was a necessary condition for the international recognition of BiH. Consequently, the European Community and the Member States recognized the Socialist Republic of Bosnia and Herzegovina (SRBiH) as a sovereign and independent state within the existing borders at Luxembourg session on April 6, 1992. The USA recognized BiH as an independent state on April 7, 1992 (they recognized Slovenia and Croatia on the same day). As early as April 8, the Presidency of SRBiH issued a decree on changing the name of SRBiH to the Republic of Bosnia and Herzegovina (RBiH) and the decision on declaring a direct threat to war. Later, the JNA army (National Army of neighboring country Yugoslavia), assisted by paramilitary forces of the Bosnian Serbs, started aggression against the independent state of BiH, and at its session held on June 20, 1992, the Presidency of BiH issued the decision on proclaiming the state of war. The war for the survival of BiH and Bosniaks lasted until the signing of the Dayton Peace Agreement on December 15, 1995.

10.3.1 Electoral Law and Electoral System

In contemporary BiH, elections are held every 2 years alternately, general and local, respectively; thus, the term of office between the elections lasts 4 years. This system has been established by the Election Law and the General Framework Agreement for Peace in Bosnia and Herzegovina and has been in force since 1996 (General Framework Agreement for Peace in Bosnia and Herzegovina, Annex 3).

In 2001, the Parliament Assembly (PA BiH) adopted the BiH Electoral Law. This law made no significant changes to the temporary electoral system setup through the DPA.

The basic characteristics of the permanent electoral system are as follows (Trnka 2006):

- Election of the PA BiH and members of the BiH Presidency have been further regulated.
- Conditions for acquiring the right to vote have been also regulated.
- Voters have the right to vote in person or in absence in the municipality where they had residence according to 1991 population census.
- A citizen of BiH who has dual citizenship has the right to vote only if BiH is his or her permanent residence.
- Persons accused or sentenced by the International Criminal Tribunal for the Former Yugoslavia (ICTY) for serious violations of humanitarian law cannot be elected for any function.
- The principle of incompatibility is regulated.
- Mandates belong to the candidate, not to the party or coalition of which the candidate is a member.
- Permanent terms for the conduction of elections were stipulated.

The authorities responsible for the conduct of elections in the permanent electoral system, according to the Electoral Law, are the Central Election Commission, Municipal Election Commissions, and the Polling Station Committees. The composition of these authorities should be multiethnic, reflecting the population of the constituent peoples and the others in accordance with the most recent population census.

The electoral system of BiH uses different models and principles of mandate distribution. There are elements of both direct and indirect elections applied: majoritarian, proportional, and compensatory method in mandate distribution with the system of one-member and multimember constituencies. Majoritarian system in the general election in BiH is implemented in the course of electing the members of the Presidency of BiH and the President and two vice President of the RS.

On the other hand, proportionate representation system is implemented in the course of electing the representatives for the House of Representatives of the PA BiH, the House of Representatives of the Parliament of the FBiH, the NA RS, and the cantonal assemblies. When it comes to the distribution of mandates, in the proportionate representation system, it is performed by implementing the Sainte-Laguë method for converting the votes into mandates, dividing the total number of votes realized with the odd numbers 1, 3, 5, 7, 9, and 11 as far as it is necessary to distribute mandates.

As a correction, the electoral system uses the compensatory mandate system (Sahadžić 2009). The compensational election system has a goal of ensuring political representation for all three constituent peoples and reflects the core nature of the Bosnian political system. The compensation mandates are awarded to those political subjects that, in reference to the realized election results, won a disproportionate number of regular mandates (or won none) and that crossed legally determined threshold of 3% in the calculation of votes in all multimember electoral units by applying Sainte-Laguë method. The role of the compensation electoral system is to represent the scattered voices generated by the existence of constituencies.

In the distribution of compensatory mandates, only the political parties and coalitions can be involved, but not independent candidates. Moreover, the electoral system adopted the system of semi-open candidate lists. Meaning, the registered voters have the right, other than a political party, to vote for a particular candidate on the list. At the same time, citizens are not able to decide on the list of candidates. It is up to the political parties to decide how to form candidate lists. But once a candidate is being elected, he/she holds a mandate not depending on the membership to a particular political party or a coalition. It has been perceived as a more democratic but connected with difficulties of votes counting (Sahadžić 2009).

Electoral system applies the threshold of 3% for the political parties, meaning that the political subjects may not participate in the distribution of mandates if they do not win more than 3% of the total number of valid votes in the electoral unit in which they are running for a certain governmental body. Comparing to the other countries, this has been considered to be low. Comparatively, Spain, Greece, Romania, and Ukraine have the same threshold as BiH, whereas, for example, Croatia has 5%, Russia 7%, Turkey 10%, etc. (Venice Commission 2010). Also, there is no compulsory voting: the minimum age for voting and age for running is 18.

The Electoral Law of BiH prefers large parties, while nonparliamentary parties and independent candidates have a worse position in terms of funding as well as in providing support for candidacy and access to media space. Inequality in terms of funding is reflected in the fact that large political parties have more budgetary resources since the funds are allocated according to the number of seats won at all levels of government (Law on Financing Political Parties of BiH, Article 7).

Electoral units are clearly defined by the Election Law of BiH (eight units, Fig. 10.6). Mandates are distributed as follows: for the House of Representatives of the Parliamentary Assembly of BiH, 21 direct mandates from 5 multimember constituencies from the territory of the FBiH and 9 direct mandates from 3 multimember constituencies from RS territory. In addition, seven Compensation Mandates for the territory of the entire FBiH and five Compensation Mandates for the territory of RS are also allocated (Election Law of Bosnia and Herzegovina, Articles 9.2 and 9).

One of the problems that arise due to the poor application of electoral legislation is that in some cases the elections result in a different voice value in some constituencies. This is particularly evident in the elections for the House of Representatives of the Parliamentary Assembly of BiH, where we have significant differences in the number of voters enrolled in the electoral roll and the number of direct mandates from the respective constituencies. For example, for the 2018 general elections in BiH, in the electoral constituency 512 (area of West Herzegovina and Herzegovina-Neretva Cantons), the average was 89,547 registered voters per direct mandate from that unit, while in the electoral constituency 513 (area of Canton of Sarajevo), that average was 110,338 voters enrolled in a voter list for one direct mandate. Although the Election Law of BiH foresees the re-examination of the electoral constituencies and the number of mandates, this has not happened so far (Electoral Law of Bosnia and Herzegovina, Article 10.9).

There is discrimination on the basis of ethnicity in the Constitution of BiH because people who do not declare themselves as members of one of the three constituent ethnic groups cannot apply for the Presidency of BiH. The same discrimination affects all those who claim to be Croats or Bosniaks and have their permanent residence in RS or declare themselves as Serbs and have residency in the FBiH, as well as people who do not declare on ethnicity. Access to candidacy is also prohibited in the case of the completion of the House of Peoples of the Parliamentary Assembly of BiH. The European Court of Human Rights rendered verdicts in the cases of "Sejdić and Finci," "Zornić," "Pilav," and "Šlaku" versus BiH regarding the possibilities of elections for the Presidency of BiH and the House of Peoples of the Parliamentary Assembly of BiH. All these verdicts have established that the fundamental human right of every citizen to elect and be elected from the European Convention on Human Rights was compromised. Also, in the House of Peoples of BiH, the constituent peoples are not represented in proportion to the total number of inhabitants of their proper ethnic/national group. Thus, Bosniaks are discriminated against, because they make up more than half of the total population of BiH and have only a third of the representatives in the House of Peoples. While, for example, Croats are overrepresented, as they make up only 15.4% of the total population according to the last population census from 2013.

People are not yet accustomed to making free democratic decisions in BiH. The electoral law has been drafted to maintain the status quo, to make possible for everyone to vote for their national heroes, and to politically categorize BiH into the Bosniak, Croatian, and Serbian electoral bodies where the elected ones will be responsible for and only take care of their ethnic group. For a citizen voter, his/her free will and free choice simply have nothing to do there. Political representatives of ethnic groups behave quite comfortably because they know they have the guaranteed number of votes of their national populous. If they were elected by all the citizens of BiH regardless of their nationality and that they respond for their work to the ones who elected them, it would be another situation and story. A special reason for the victory of national parties is disunity and inability to gather the opposition itself. If a democratic block was created on a clearly defined basis, the results of the elections would be far more favorable for the opposition civil parties. The media largely contribute to the election results, of which the most powerful medium of television is under the monopoly of the ruling national parties. It is prevalent in the public opinion that the Social Democrats or Liberals, etc., Bosniaks, Serbs,

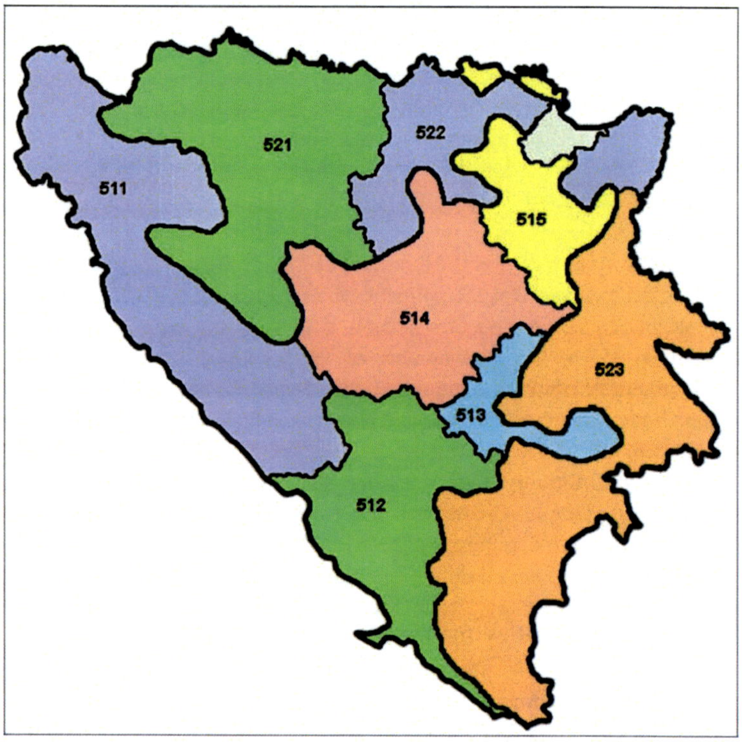

Fig. 10.6 Electoral units of Bosnia and Herzegovina

and Croats are less valuable and that they are anti-national and not true members of the mentioned ethnic groups (Duraković 1999). Not much has changed in the last 22 years.

Expecting that political will already exist as clear and concrete and manifested in a Bosnian citizen would mean a complete misunderstanding of the state in which his political life lies. People who do not go to the polls are not politically disinterested persons but those who are dissatisfied with the election offer. In BiH there is a cognitive pattern of belief in the inability to change the state in which the society is found, a scheme of the so-called learned helplessness according to which even when we can act, due to the perceived and learned experience that our commitment cannot change anything (because we are the product of collectivist culture and our individuality is not cared for or developed in any of the segments of political socialization, from family, school to society in general), we remain in an unchangeable state. Citizens who are disoriented in this turmoil of unclear expectations, vague values, and unarticulated own desires seem to have entered the vicious cycle of learned helpless behavior, even when they have the opportunity to act with the power of their will remain in that cliché but already confirmed, proven, tried, and the least (currently) painful behavior.

In today's BiH the trust relies not on the state or state institutions but on the ethno-political parties and the religious groups. According to a survey published by the Sarajevo-based think tank Analitika in 2013, governmental and parliamentary bodies enjoy significantly less trust (between 22.1% and 25.7%) than do religious institutions (+50%) (Jusić and Ahmetašević 2013).

In general terms, the BiH youth remain unengaged and marginalized from political debate and decision-making. In terms of electoral participation, the 2018 elections saw a 51.6% voter turnout for 18- to 30-year-olds, which was slightly below the overall turnout average. Three out of four persons aged 16–30 claim not to be interested in politics. Regarding other forms of participation, such as membership of political parties or youth organizations, direct contact with politicians, or participation in public debates or online political discussions, youth involvement is generally low (Commission for the Coordination of Youth Issues BiH 2008).

Finally, the repercussions of such a political culture and related developments on the stability and legitimacy of the political system are not to be underestimated. Although support for democratic values in the country is quite strong, it is continuously undermined by the contrasting political and territorial aspirations of the individual ethnic groups. To make matters worse, disillusionment and lack of confidence in political actors, as a result of endemic corruption, are increasing political disaffection and lack of trust in the democratic system. Therefore, the obvious impact of this rudimentary democratic political culture is especially high dissatisfaction, disenchantment, and apathy, which were clearly manifested in the low turnout for the 2018 elections with 54% participation, which has been consistently and slowly decreasing since the elections in 2010 (56.6% and 54.5%) (Banović et al. 2021).

The majority of experts label BiH political life as ethnicized and consider it its determinative aspect. As a consequence, and also due to the overcomplicated political and electoral system, only a few papers have been published on the electoral geography of the country. It is almost an axiomatic truth that the majority of voters cast their ballot in BiH according to their ethnicity on parties organized along ethnic lines. On the one hand, it can be seen as the consequence of the war and the post-conflict characteristics of the society or as an unwanted side effect of the Dayton Peace Accords (e.g., McMahon and Western 2009; Kartsonaki 2016).

To better understand electoral geography of BiH on the basis of data of population census from 2013, we calculated and analyzed the typology of ethnic distribution of municipalities where we used modified Bougarel's typology mentioned by Žíla (Bougarel 1992; Žíla2013) which defines six basic patterns of ethnic configuration in Bosnia (with our adaptation, see Fig. 10.7).

1. A homogeneous community: With a prevailing ethnic group representing 80% of the total population of a particular municipality. In this category are 68% of the municipalities (97) with 65% of total Bosnian population. With Bosniak majority, there were 41, with Serb majority 42, and with Croat majority 14 municipalities. The largest municipalities in this group are Banja Luka, Novi Grad Sarajevo, Zenica, Bijeljina, Cazin, Ilidža, and Bihać (all between 60,000 and 120,000 inhabitants).
2. A majority biethnic community: The dominant ethnic group was represented by 60–80% of the total population, or the local majority ethnic group was twice as high in numbers as the numerically second largest ethnic group. The third ethnic group did not exceed 10% of the total number of people of a particular municipality. The number of municipalities in 2013 were 21 with 17% of the total Bosnian population. This type had more than 60% representation of Bosniaks in 6, Serbs in 12, and Croats in 3 municipalities. The largest were Tuzla, Doboj, Novo Sarajevo, and Travnik (all between 50,000 and 111,000 inhabitants).
3. A biethnically balanced community: The prevailing ethnic group reached a of maximum 65% of the total population of an administrative unit, or its population did not exceed twice the number of people of the second numerically strongest ethnic group. The third ethnic group did not exceed 10% of the total number of people of a particular municipality. The number of municipalities of this type were 21 with 15% of the total Bosnian population. The largest municipalities in this group are Mostar, Prijedor, and Zvornik (all between 50,000 and 110,000 inhabitants).
4. A heterogeneous biethnic community: None of the ethnic groups exceeded 60% of the total local population. The numerically most powerful ethnic group did not exceed the second largest ethnic group twofold. The local numerically strongest ethnic group exceeded the third ethnic group twofold, which was, however, represented by at least 10% of the total number of people of a specific municipality. Only Vukosavlje municipality was in this type (4667 inhabitants).
5. A heterogeneous ethnically balanced community: Individual ethnic groups are relatively balanced in numbers. The largest ethnic group is not represented by more than 50%, and the second numerically strongest ethnic group does not exceed the third ethnic group twofold. The third ethnic group is at the same time represented by at least 15% of the total number of people of a specific municipality. Only Brčko (District) and Glamoč municipality were of this type (2.5% of the total Bosnian population).
6. A heterogeneous community with a majority ethnic group: The largest ethnic group was represented in a municipality by at least 50% and not more than 60%. At the same time, it has twice the number of people as the second-ranked nation. It also has to be true that the second and third ethnic groups were represented by at least 10%. There were no municipalities of this type in 2013.

Constitution and election law were based on the international convention of UN and EU about human rights, equality of all citizens, and on wide regional decentralization in favor of local administrative units inside the country. The Constitution of BiH does not contain a principled position on the acceptance of the principle of division or the principle of unity of government. The analysis of the constitutional provisions leaves no doubt that it is a specific one, so far in a comparative contemporary constitution an unknown model of division of power.

One of the specifics of BiH from the point of view of this analysis, besides the mere existence of ethnoterritorial units (entities and cantons) and their influence on electoral geography, is the existence of entity citizenship which has no significance in international relations as opposed to the citizenship of BiH. Bosnia's Constitution allows its citizens to have dual citizenship if there is a bilateral (international) contract with a particular state, which further influences election results.

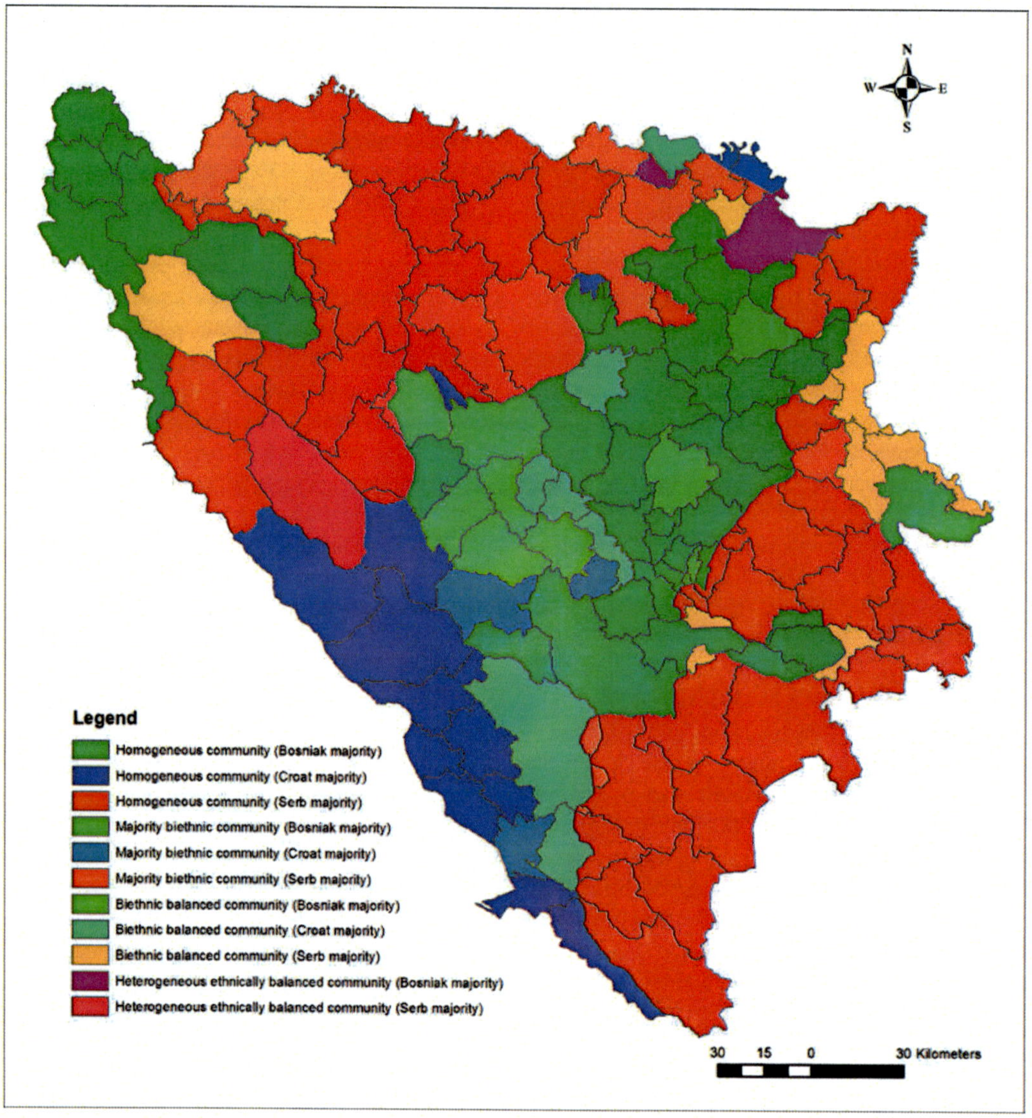

Fig. 10.7 Typology of ethnic distribution of municipalities of Bosnia and Herzegovina in 2013

10.3.2 Political Parties

The Constitution of BiH does not address the question of political parties, nor does a specific state-level law exist regarding this matter. The only state-level legislations are the Law on Party Financing from 2012 and the Electoral Law of 2001. Political parties must have 50 (in FBiH), 500 (in RS), and 300 (in BD) founding members before registering as an association with the relevant municipal court (Stjepanović 2016).

The exact number of active political parties varies from election to election. For example, after verification of the registration documents for the 2018 general elections, the state presidency was contested by 15 candidates, while 37 candidates stood for the president of the RS. For parliamentary elections, lists of 22 parties and 6 coalitions appeared on the ballot for the BiH House of Representatives; 23 parties and 5 coalitions competed for the FBiH House of Representatives; and 21 parties, 6 coalitions, and 4 independent candidates—for the RS National Assembly (ODIHR 2019).

The Party of Democratic Action (Stranka demokratske akcije or SDA), founded by Alija Izetbegović, later the first president of the independent Republic of Bosnia and

Herzegovina, is the largest Bosniak party of Bosnia and Herzegovina. The party ideologically has its roots in the old Yugoslav Muslim Organization, a conservative Bosniak party in the Kingdom of Yugoslavia. Since the beginning of 1992, SDA has always been a member of the central state government, with the exception of the period of the Alliance for Change (2000–2002) and from 2011 until 2015. In December 2004, the party was granted observer status in the European Peoples Party. During his time as Chairman, Alija Izetbegović achieved equal legal status for the three constituent peoples in the entirety of BiH through a constitutional complaint. The party advocates constitutional changes and hence strengthening central state institutions. Former Presidency member and son of Alija Izetbegović, Bakir Izetbegović, leads the party since 2014. He succeeded Sulejman Tihić, party president from 2001 until 2014, which has undertaken several reforms and acted as a moderate politician, especially in the context of constitutional reform discussions (Topić 2007).

The Union for Better Future (Savez za bolju budućnost or SBB) is a center-right, conservative political party with Bosniak voters base, founded in 2009 by the Bosniak media magnate Fahrudin Radončić. Immediately in the 2010 elections, the party has positioned itself as the second biggest Bosniak political party, and it has maintained this position until today. As DF, the party is very much based on the cult of the party leader. Even though Radončić repeatedly before the 2018 general elections stated that he does not intend to form a coalition with the corrupt SDA, he did not keep his promise. As DF, SBB was a coalition partner of the biggest party, securing the SBB leader a seat in the Council of Ministers (Balkan Insight 2019a, b).

The 25 years after Dayton have also confirmed that the current political setting does not leave too much space for multiethnic parties. Multiethnic alternative political movements are the exception and without country-wide relevance (Banović et al. 2021). Multiethnic and normal citizen ideas (e.g., as in EU countries) are not yet widely supported in Bosnia, and national issues still dominate everyday politics in the whole country. Social Democratic Party of BiH (SDP), officially a multiethnic party, was established in 1992. It is a successor of the former state communist party. The SDP was the largest body of opposition, until they managed to win the parliamentary elections in 2010. Nevertheless, they were not able to fulfil the hopes of the citizens as they made significant compromises with nationalistic parties, resulting with Željko Komšić, the member of the Presidency from Croatian people, leaving the party and founding the Democratic Front. Since 2014, the party is in opposition on state and entity level, run by Nermin Nikšić. The party is associated with the Party of European Socialists.

The Democratic Front (Demokratska Fronta or DF) is a social-democratic political party in BiH founded by Željko Komšić in 2013, at that time member of the Presidency from Croatian people, in protest to the coalition of SDP with nationalistic parties. The fundamentals of the party are mostly former SDP member and political activists from the political left. It has diverse ethnicity members and the orientation of party is center-left. Željko Komšić was followed by other SDP members, leaving the party and joining him. Today, the party lives predominantly on the old fame of the party leader, who managed to get re-elected to the Presidency for a third term in 2018. For a second time in row, the party is also in coalition with SDA, besides strong promises of the party president that this will not be the case after the 2018 general elections (Balkan Insight 2019a, b).

The Naša Stranka (Our Party) is one of the rare political parties which was not created as a split from existing political parties, but as a totally new political movement. What used to be a small, regional party in the Canton Sarajevo and few municipalities around the country is today a relevant player on the political scene in the FBiH. Since the 2018 general elections, the party is also represented in the state-level parliament. Ideologically the party self-identifies as socio-liberal and multiethnic, being part of the Alliance of Liberals and Democrats in Europe. It is the only bigger party in the FBiH which is run by a Serb in 2021, Predrag Kojović.

The Croat Democratic Union (Hrvatska Demokratska Zajednica or HDZ BiH), conservative nationalist party, claims the exclusive right to protect Croat interests in BiH. Local elections in 2008 and general elections 2010 confirmed HDZ as the strongest Croat party. This has been also reassured by the election results of the 2014 and 2018 General Election results. In the last two mandates, HDZ is the only political party which holds three ministries in the Council of Ministers. Similar to the SDA, the HDZ has observer's status in the European People Party. Dragan Čović is party president since 2005. Currently, party cooperates closely with SNSD on the state level.

The Serbian Democratic Party (Srpska demokratska stranka or SDS), founded by Radovan Karadžić, war criminal, convicted of genocide, crimes against humanity, and war crimes by the International Criminal Tribunal for the former Yugoslavia (ICTY) in 2016, is the war-time party of Bosnian Serbs, but it lost its primate position in 2006 to SNSD. After the failure of the parliamentarian majority of SDS-HDZ-SDA in 1991, the SDS left the central state parliament and began constructing a separatist Serb para-state. At that point, the party was an all-encompassing movement with the support of the majority of the Bosnian Serb population, a trend that continued in post-war years. Yet internal conflicts eventually led the party to fragment. Today the SDS is the key opposition party in entity of RS, a status confirmed in the 2018 general election. Regularly, on the state level, SDS collaborates with SNSD, putting the interests of entity of

RS first whenever needed. Mirko Šarović is running the party since 2019.

The strongest party in the entity of RS is the SNSD (Alliance of Independent Social Democrats or Savez nezavisnih socijaldemokrata). SNSD is a Serb political party in BiH (mainly Serb members; Serb center-right party) which is founded in 1996. Their politics succeeded politics of SDS in the last 10 years as main Serb nationalist party. The party was founded in 1996 by a group of opposition politicians in the entity of RS. Its Chairman, Milorad Dodik, assumed the responsibilities of the government chairperson from 2005, following his initial term as Prime Minister from 1998 to 2001. From 2010 until 2018, Dodik was the President of the entity of RS, and since 2018, he is a member of the state-level Presidency. The SNSD strongly advocates the preservation of the entity, its central institutions, and competences. On the platform of planning to hold a referendum on independence of entity, the party mobilized Serb voters, leading to regular victories in the 2006, 2010, 2014, and 2018 elections. The party governs in cooperation with several smaller parties. As a result of his incontestable position, Dodik has been confrontative towards the EU and OHR. In January 2017, the US Department of the Treasury's Office of Foreign Assets Control imposed sanctions on Milorad Dodik. The SNSD was expelled from the Socialist International in 2012 for continuing to espouse a nationalist and extremist line. Ideologically, the party developed into the opposite direction, and on several occasions, representatives of the SNSD signed cooperation agreements with representatives of Putin's United Russia. All of this will not prevent the party to continue being the leading party in the entity of RS for four election rounds in a row (Banović et al. 2021).

In addition to the large parties mentioned above, there are several smaller parties including the Movement of Democratic Action (PDA), People and Justice (NiP), Independent Bloc (NB), Independent BiH List (NBL), Party of Democratic Activity (A-SDA), Socialdemocrats of BiH (SD BiH), Party for BiH (SBiH), Democratic People's Union (DNS), the Party of Democratic Progress (PDP), Peoples Democratic Movement (NDP), the Socialist Party of Republika Srpska (SP), and United Srpska (US) and several local parties Laburists of BiH (Laburists), Croatian National Movement (HRP), Zenica Iniciative (ZIN), and People Movement (NRP). These parties receive some votes in each election but have not been able to make a name for themselves except for SBiH prior to 2010.

Most political parties, the larger ones, in particular, can be described as "charismatic leader organizations" (Jovanović 2002). Orientation towards a leading figure is a legacy from Communist times and is especially evident when a new party is in the process of being formed. Today, the parties are more characteristic of patronage associations. A strong cult has shaped around the Chairman of the SNSD, Milorad Dodik. The large parties are able to maintain their membership due to the fact that economic resources, industry, energy, and party media are under their control. In BiH there is no institutionalized process for intraparty programmatic discussion; decisions are made on behalf of the party exclusively by the party leader. The absence of a culture of discussion has caused most of the rifts. In the case of diverging opinions, there is no attempt to achieve a compromise. Instead, new small parties form with new leading figures. However, these splinter parties rarely evolve into a successful opposition and usually fail to have significant political influence.

10.3.3 General and Local Elections

As before in the general elections of 2010, 2014, and 2018, voting occurred in eight constituencies. In 2010, the number of registered voters in the general elections for the Parliament of Bosnia and Herzegovina was 3,126,599, of which 1,769,247 (56.6%) went to the polls (number of voters living in Croatia was 7970 and in Serbia 1098). There was 1,641,569 (92.8%) valid ballots, while 127,678 (7.2%) was invalid. There were a total of 39 political parties, 11 coalitions, and 13 independent candidates in elections with 8242 candidates (36.8% of candidates were females). The winner of the elections was the SDP on the one hand and SNSD on the other, besides the newly formed SBB. The following political parties received the most votes in percentage: SNSD (16.4%), SDP (16.2%), SDA (12.1%), SDS (8.4%), SBB (7.6%), and HDZ (6.8%) (Table 10.4).

In the 2014 general elections for Parliament Assembly, the total number of voters increased by 4.9%, so the total number rose to 3,278,908 (number of voters living in Croatia was larger by 17% and those from Serbia by 27%) of which 1,787,449 (54.5%) participated in the elections. There was 1,630,920 (91.2%) valid ballots and 156,529 (8.8%) was invalid. There were a total of 65 political parties, 24 coalitions, and 24 independent candidates (7748 total candidates, of which 42.3% were females). SNSD remained the strongest party in the RS; however, SDS was able to increase its votes. In the FBiH, SDP weakened while SDA was gaining ground, as well as the newly formed social democratic party DF. The following political parties received the most votes: SDA (18.7%), SNSD (15.6%), SDS (9.8%), DF (9.2%), SBB (8.7%), and HDZ (7.5%). The biggest loser was the SDP (6.6%), which received about 59% fewer votes than in 2010 (the main reason is the disappointment of voters and the emergence of a new political party, DF, from the SDP). Conversely, these voters voted overwhelmingly for the SDA, which received about 54% more votes than in 2010.

The general elections of 2018 saw a 2.3% rise in the number of voters compared to 2014 and 7.3% compared to

Table 10.4 Largest political parties with votes won in general elections in 2010, 2014, and 2018

	2010		2014		2018	
Party	Number	% of total valid votes	Number	% of total valid votes	Number	% of total valid votes
DF			150,767	9.2	96,174[a]	5.8
HDZ (1990)	49,549[a]	3	40,113	2.5		
HDZ BiH	112,115	6.8	123,022[a]	7.5	149,872[a]	9.0
Naša Stranka					48,401	2.9
PDP	40,070	2.4	50,516[a]	3.1	83,832[a]	5.1
SBB BiH	124,114	7.6	142,003	8.7	68,993	4.2
SDA	197,922	12.1	305,394	18.7	281,731	17.0
SDP	266,023	16.2	108,501	6.6	150,454	9.1
SDS	137,844	8.4	211,562	13.0	162,414[a]	9.8
SNSD	269,009	16.4	255,024	15.6	265,593	16.0

Source: calculated from data on www.izbori.ba
[a]In coalition with one or several smaller parties

2010. The number of voters living in Croatia was 2.5 times larger and those living in Serbia 10 times larger than in 2014 (results of actions of conservative nationalist parties HDZ and SNSD). The total number of voters was 3,355,429, but only 1,812,960 or 54% participated in the elections. The number of valid ballots was 1,656,516 (91.4%), while 156,444 (8.6%) was invalid. A total of 69 political parties, 36 coalitions, and 34 independent candidates participated in the elections with 7497 candidates (41.6% candidates were females). Party support did not change much since the previous elections. SDA remained the leading party in FBiH, as SNSD in the RS. Previously established new parties (SBB, DF) were losing some support, while SDP gained more votes. Croatian coalitions led by HDZ experienced one of their best results. The most votes by percentage were received by SDA (17%), SNSD (16%), SDP (9.1%), and HDZ (9%). The biggest loser was the DF (5.8%), receiving about 36% fewer votes than in 2014 (the main reason being the loss of voter confidence in the new postelection coalition agreements with the SDA). Vice versa, these voters voted overwhelmingly again for the SDP, which received 38.7% more votes than in 2014. Big losers were also SBB (51.4% fewer votes than in 2014, mainly due to earlier unfavorable postelection coalition agreements with other national conservative parties) and SDS, which received about 23% fewer votes (their voters mainly voted for SNSD) (Fig. 10.8).

The general elections in 2014 for the House of Representatives of the BiH Parliamentary Assembly saw the national conservative political parties winning again by far the most votes. Most of the municipalities voted for SNSD (43 of them), SDA (41), HDZ (25), and SDS (23). The only socialist and liberal political party that received the most votes in the municipalities is the DF in the municipalities of the capital of Sarajevo: Centar Sarajevo, Novo Sarajevo, Stari Grad, and Novi Grad Sarajevo. This arose due to voters' disappointment in the second-largest party in terms of the number of votes from 2010, SDP, and the formation of a new party, DF, from the ranks of this party (Results of general elections 2010, 2014) (Fig. 10.9).

The general elections of 2018 for the House of Representatives of the Parliamentary Assembly of BiH saw the national conservative political parties winning again by far the most votes from municipalities/cities. Most of the municipalities voted for SNSD (46 of them), SDA (36), HDZ (26), and SDS (15). Socialist and liberal political parties received the most votes in the following municipalities/cities: Centar Sarajevo, Novo Sarajevo (Naša Stranka, NS), Lukavac, Doboj Jug, Gradačac, Gračanica, Maglaj, and Tuzla (SDP) (Results of general elections 2018) (Fig. 10.10).

In the elections for members of the Presidency in the last three election cycles (2010, 2014, and 2018), seven candidates received the most votes to enter the Presidency. If we look at all three election cycles, only two candidates had more than 20% of the total number of valid votes: Milorad Dodik (21.8% in 2018 elected a member of the Presidency from the ranks of the Serbian people) and Željko Komšić (20.7% in 2010 elected a member of the Presidency from the ranks of the Croatian people). Of all the elected members of the Presidency, the lowest percentage of valid votes was won by Dragan Čović (7.8% in 2014 elected a member of the Presidency from the ranks of the Croatian people) and Bakir Izetbegović (10% in 2010 elected a member of the Presidency from the ranks of the Bosniak people) (Table 10.5).

Like all other institutions at the state level, the functionality of the Presidency depends greatly on the general political climate and the relations between the main political parties of the three ethnic groups (Banović et al. 2021).

In addition to the usual parity of the three ethnic groups, the election of the Presidency also has a territorial component outlined in the election rules. The eligible voters (citizens over the age of 18) in the FBiH directly elect one Bosniak and one Croat, while voters in the RS vote for the Serbian member (Article V, Constitution of BiH). As in elections for

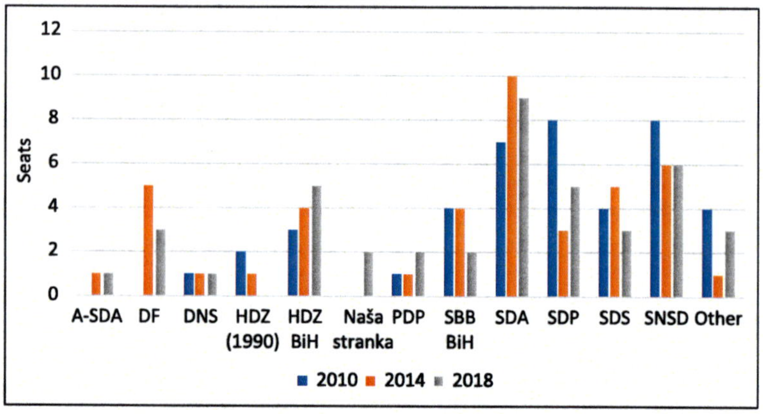

Fig. 10.8 Distribution of seats in the state-level House of Representatives of Bosnia and Herzegovina in 2010, 2014, and 2018

Parliament of BiH voters in Brčko District can choose to vote for preferential candidate or party from RS or FBiH, but not from both. Serbs from the FBiH and Croats and Bosniaks from the RS, as well as all other citizens that do not belong to one of the three largest ethnic groups (the "Others"), do not have a passive electoral right and are excluded from the highest executive body (Bieber 2008). December 2020 marked the 11th anniversary of the Sejdić and Finci judgement of the European Court for Human Rights, in which this arrangement has been assessed as discriminatory and not in line with the European Convention for Human Rights. This judgement was followed by judgements in the Zornić, Šlaku, Pilav, and Pudarić cases, in which the court stated that in order to implement these judgements, BiH would have to introduce constitutional and legislation amendments, ensuring every citizen passive and active voting rights. Regarding local elections, we will focus on the analysis of the previous elections from 2020. At its session held on May 23, 2020, the Central Election Commission of Bosnia and Herzegovina announced the local elections in 2020, initially for October 4, 2020, and then postponed to November 15, 2020, due to the Covid-19 pandemic and lack of funds, since the politicians from SNSD and HDZ elected at the state level did not want to approve funds for holding local elections in time. Mayors and municipal councils were elected in 120 municipalities, mayors in 22, and city councils in 21 cities, and the Brčko District Assembly (Central Election Commission 2020). A total of 3,283,380 BiH citizens were registered to vote. The turnout was 50%. Of the total, 95% of ballots were valid. Of the 425 candidates for mayor, only 29 were women (6.8%). In the 2020 local elections, 17 political parties won the most votes in municipal/city councils. The best result was achieved by SNSD (most votes in 53 municipalities/cities), followed by SDA (37), HDZ (22), and SDS (12) (Fig. 10.11).

These elections confirmed that there is still a distinct dominance of national conservative political parties at all levels of government. At the local level in 2020, the exceptions were municipalities where the SDP and SDBiH won (Doboj Jug, Gradačac, Lukavac, Maglaj, Zavidovići, and Tuzla) (Results of local elections 2020). The elections were held amid a pandemic a year after the delayed formation of the state-level government. National coalition partners, supported by the majority of Bosniaks (SDA), Serbs (SNSD), and Croats (HDZ), defended most of the seats won in the previous local elections in 2016. In Mostar, these elections were held for the first time since 2008 on December 20, after the electoral amendments were adopted in July 2020, and the decision of the European Court of Human Rights in Baralija vs. BiH was implemented.

According to Abazović (2015), in BiH, people are faced with the fact that leading political actors cannot be held accountable, and very often, the practice of "shifting responsibility" to someone else (the international community, etc.) occurs mainly when it comes to the so-called sensitive and "complex" issues. Judging by domestic policies and governments and the people themselves, social capital in BiH will remain only at the level of a certain set of informal values or norms among group members that enable cooperation between them for a long time. Nevertheless, there is a hope that the network of social relations that determines the behavior of individuals and thus influences economic development, if social capital is taken seriously, can result in a culture of tolerance, in which extensive networks of voluntary associations grow and emerge.

The characteristics of the electoral geography of BiH are similar to the Central and Eastern European ones. The most significant difference is made by the ethnonational divisions of BiH and its constitutionalization in the Dayton Peace Accords. Therefore the ethnicization of the electoral geography of the country is not instead of but on the top of the other factors, which, in the majority of places, thus become of secondary importance, but in some places eventually can be stronger than the ethnic ones. If and when it can be overcome, party support would be defined by other social variables.

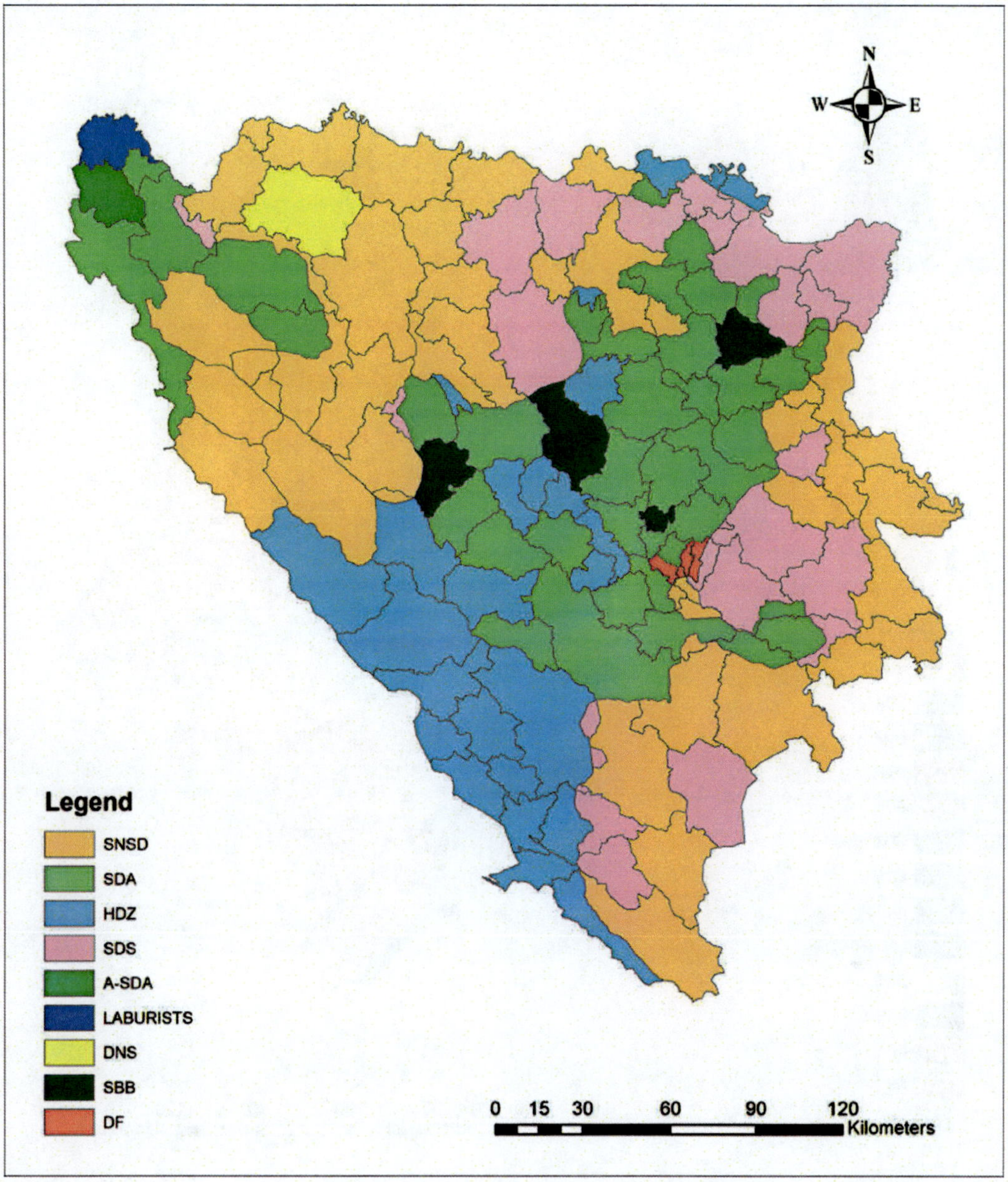

Fig. 10.9 Political parties with most votes for the House of Representatives in general elections in 2014

10.4 European Union and BiH

When creating the Dayton Constitution of BiH in 1995, the prospect of full membership in the EU did not exist. As the perspective of full membership for BiH became more realistic, so did the legislative and institutional frameworks and practices for achieving that perspective. As of June 2000, when the Ministry of European Integration was established, two periods can be observed until today. The first period lasted from 2000 until the end of 2010. This period was marked by actors who developed a narrative about the need for coordination and mutual cooperation to achieve the perspective of EU membership and were ready to take concrete steps, partly due to the stronger influence of the international community and partly due to pressure from the domestic public and their expectations.

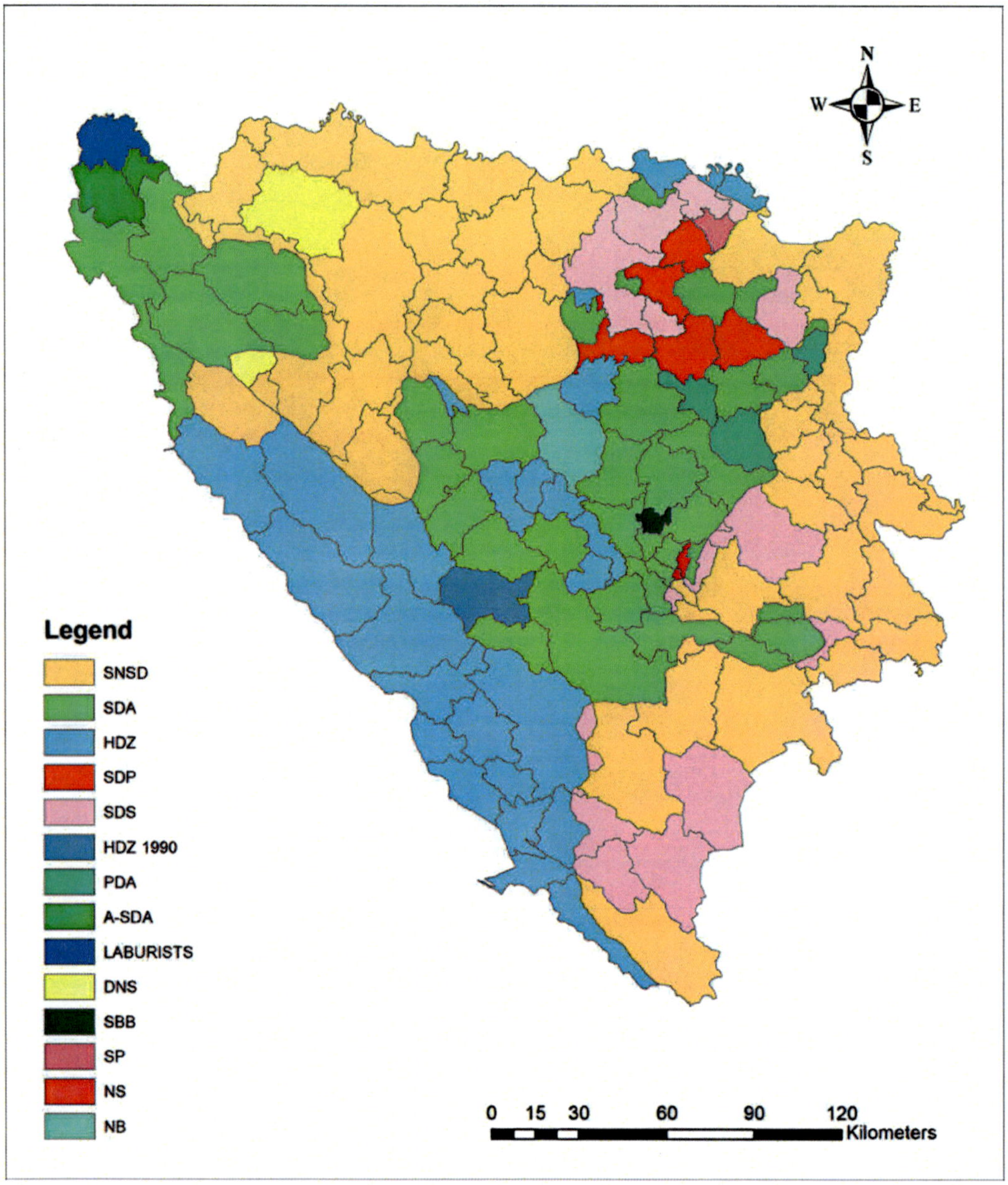

Fig. 10.10 Political parties with most votes for the House of Representatives in general elections in 2018

Table 10.5 Elected members of Presidency with the number and percentage of valid votes in 2010, 2014, and 2018

	2010		2014		2018	
Surname and name	Number	% of total valid votes	Number	% of total valid votes	Number	% of total valid votes
Čović Dragan			128,053	7.8		
Dodik Milorad					368,210	21.8
Džaferović Šefik					212,581	12.6
Izetbegović Bakir	162,831	10.0	247,235	15.0		
Ivanić Mladen			317,799	19.3		
Komšić Željko	337,065	20.7			225,500	13.3
Radmanović Nebojša	295,629	18.2				

Calculated from data on www.izbori.ba

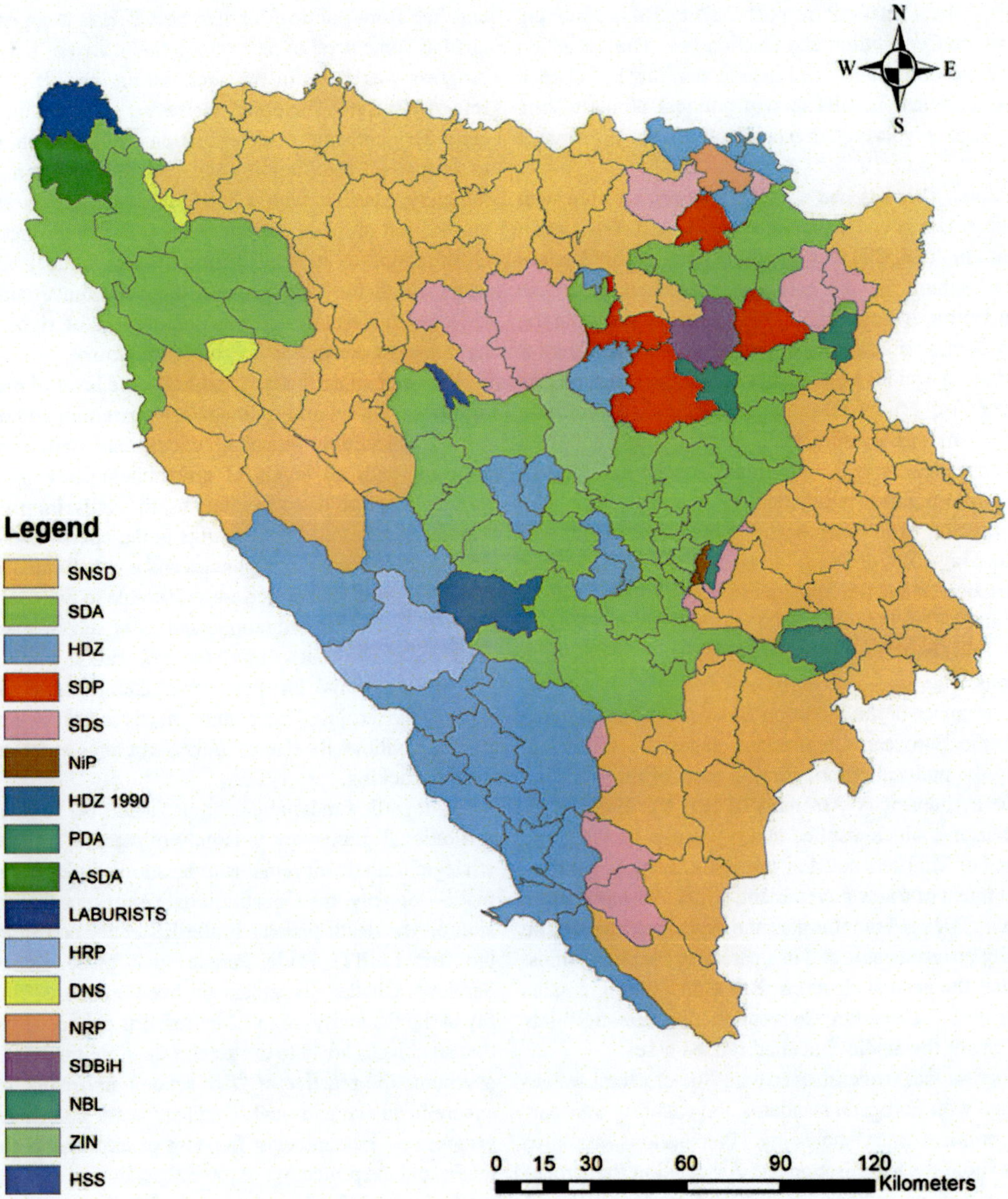

Fig. 10.11 Political parties with most votes in local elections in 2020

From the beginning of 2011, BiH is in the second period, marked by a narrative of reduced political will for mutual cooperation and coordination of policies and activities. The European Commission's reports have since become increasingly negative. The 2011 Report states that BiH "lacks a common vision of the country's future and institutional framework" for EU integration (Ćerimagić 2015), which was even more negatively reported in the Report for 2021 (Bosnia and Herzegovina 2020a, b Report).

There are several reasons for such developments. From the aspect of EU integration, we can talk about the relationship between what the EU offered and demanded from politicians in BiH in that period. For submission of credible application for EU membership, the EU sought an agreement on implementing the European Court of Human Rights ruling in the Sejdić and Finci case. A credible application for EU membership does not mean much from the aspect of the long road to full EU membership, and the process of agreeing on Sejdić and Finci developed into a discussion on ethnoterritorial political relations in post-Dayton BiH, assuming changes to the Constitution and election law. The debate on changes to the Constitution and the election law has opened space for politicians in BiH to win political points among voters through advocating irreconcilable views (Ćerimagić 2015).

Since 2008, BiH has had a contractual relationship with the European Union under the Stabilization and Association Agreement. In 2016, BiH was the last country in the Western Balkans to submit an application for membership in the European Union. After receiving the questionnaire from the European Commission, it took 14 months for BiH to answer the first 3242 questions and another 8 months to answer an additional 655 questions. Despite the established coordination mechanism for EU issues, the authorities could not agree on providing answers to 22 questions: 1 from the political criteria, 4 from regional policy, and 17 from education policy. The opinion of the European Commission (Avis) was received in May 2019, defining 14 priorities divided into four areas—democracy and the functioning of institutions, rule of law, fundamental rights, and public administration reform—which BiH needs to meet in order to start full membership negotiations.

In the summary of the report on Bosnia and Herzegovina for 2020, the European Commission states the following. Regarding the political criteria, during most of the reporting period, the Parliamentary Assembly of BiH was blocked for political reasons, which resulted in a legislative backlog, and the Council of Ministers acted in caretaker function, pending the appointment of a new government, which only took place in December 2019, 14 months after the general elections. The FBiH entity government is still in a caretaker function almost 3 years after the general elections. BiH Constitution remains in breach of the European Convention on Human Rights (ECHR), as per the Sejdić-Finci and related cases.

No progress was made in improving the electoral framework in line with European standards and ensuring transparency of political party financing. Preparations are also ongoing to hold the Stabilisation and Association Parliamentary Committee, as well as develop and adopt a national program for the adoption of the EU acquis. No progress was made in ensuring an enabling environment for civil society. Meaningful and systematic consultations with civil society remain to be ensured.

BiH is at an early stage with public administration reform, and there was no progress on ensuring a professional and depoliticized civil service and a coordinated countrywide approach to policy making. All levels of government adopted the strategic framework on public administration reform and now need to adopt the related action plan. A political body steering the coordination of public administration reform is not yet established. To guarantee a professional civil service, civil service procedures must be based on merit principles and free from political interference. BiH is at an early stage and has some level of preparation in the area of judiciary. No progress was made in this area during the 2019 and 2020. Integrity reforms have met resistance from within the judiciary. Evident signs of deterioration require urgent measures to strengthen the integrity and regain citizens' trust in the judiciary, starting with a credible and rigorous system of verification of financial statements of judicial office holders. Obstructions to judicial reforms from political actors and from within the judiciary and the poor functioning of the judiciary undermine citizens' enjoyment of rights and the fight against corruption and organized crime.

BiH is at an early stage and has some level of preparation in the prevention and fight against corruption and organized crime. Corruption remained widespread and is a serious concern, with all levels of government showing signs of political capture directly affecting the daily life of citizens. There are systemic shortcomings in the operational cooperation of law enforcement agencies and a very limited exchange of intelligence. The police are vulnerable to political interference. Financial investigations and asset seizures are largely ineffective. The contact point for cooperation with Europol is not yet operational. No steps were taken to establish cooperation with Eurojust. The country needs to improve its capacity and continue its efforts in the fight against terrorism and drug trafficking.

As regards fundamental rights, while the legislative and institutional framework is largely in place, a comprehensive strategic framework remains to be adopted. Some steps were made—notably, the Constitutional Court repealed the provision on the death penalty in the RS entity constitution, and the first LGBTI Pride Parade took place peacefully in Sarajevo. Challenges remain on freedom of assembly, notably in the RS entity. Significant reforms are needed to ensure that all citizens are able to exercise their political rights and to overcome the practice of "two schools under one roof" and ensure inclusive and quality education for all. There was no progress on guaranteeing freedom of expression and of the media and the protection of journalists by ensuring the appropriate judicial follow-up to cases of threats and violence against journalists and media workers and ensuring the financial sustainability of the public broadcasting system.

BiH is overall at an early stage and has some level of preparation regarding its level of preparedness and ability to take on the obligations of EU membership and needs to significantly step up the process to align with the EU acquis and implement and enforce related legislation. Limited to no progress was made on the different EU acquis chapters during the reporting period. Particular attention should be paid to the areas of free movement of goods, right of establishment and freedom to provide services, information society and media, agriculture and rural development, fisheries, transport policy, energy, economic and monetary policy, statistics, social policy and employment, enterprise and industrial policy, regional policy and coordination of structural instruments, education and culture, consumer and health protection, and financial control (Bosnia and Herzegovina 2020a, b Report).

According to Pejanović (2016), it is imperative to emphasize that accelerating the integration of BiH into the European Union also means strengthening the stability of the state in all aspects of its development. Accelerating BiH's integration into the European Union means the possibility of preventing the danger for BiH that comes from changing geopolitical trends in Europe and the world and especially from the dangers posed by terrorism and the strengthening of the Russian Federation's influence in the Balkans. One of the new possibilities that can prevent the dangers of negative geopolitical forces towards BiH in the internal plan of BiH's political development is the introduction of a model of a broad coalition of parliamentary parties to dedicate all political energies to accelerating BiH's integration into the European Union.

10.5 The Office of High Representative (OHR)

The Office of High Representative (OHR) is an ad-hoc international institution responsible for monitoring the implementation of the civilian aspects of the Dayton Peace Agreement (DPA). The mandate of this institution arose from Annex 10 of the DPA, which states its competences as follows: monitor the implementation of the peace settlement; maintain close contact with the parties to the agreement, to promote their full compliance with all civilian aspects of the agreement; and coordinate the activities of civilian organizations and agencies in BiH to ensure the efficient implementation of the civilian aspects of the peace settlement. The High Representative shall respect their autonomy within their spheres of cooperation while, as necessary, providing general guidance about the impact of their activities on the implementation of the peace settlement; facilitate, as the High Representative judges deem necessary, the resolution of any difficulties arising in connection with civilian implementation; participate in meetings of donor organizations; report periodically on progress to the United Nations, European Union, the USA, Russian Federation, and other interested governments, parties, and organizations; and provide guidance to the United Nations International Police Task Force.

The DPA declares the High Representative as the final authority responsible for interpreting the agreement on the civilian implementation of the peace settlement. Nevertheless, since the adoption of this document, the tasks of the OHR have changed according to the developments of political life in BiH, as well as according to the decisions of the Peace Implementation Council (PIC) (OHR 2021).

One of the most remarkable and innovative resolutions affecting the work of the OHR was the adoption of the so-called Bonn powers, which stemmed from the PIC Conference in Bonn, Germany, in December 1997. These powers were built upon the previously mentioned Annex 10 of the DPA and were thought to empower the OHR to be able to remove from office those public officials who violate legal commitments and/or the DPA, as well as to impose laws as he/she understands are needed in order to further the development of BiH when BiH institutions fail to do so. The High Representative imposed 112 laws within the legislative competences of the PA BiH during the period from 1996 to 2007. These interventions pertained mostly to the field of judiciary reform, followed by the fields of citizenship, personal and travel documents, public property, privatization, the electoral system, and telecommunications, with the HR acting as a legislator. Additionally, it is worth mentioning that from 1997 to August 2012, the High Representative dismissed over 200 officials using the Bonn powers in approximately 900 cases (Banović et al. 2021).

The use of the so-called Bonn powers by different High Representatives has solved some political blockades through the imposition of a series of laws relating to varying fundamental areas, such as a single currency, the anthem, the flag, and the election law. Decisions have been primarily on state symbols, the judiciary, removal/suspension from office of state officials, property laws, and war crime processing. The PIC was established at a conference held in December 1995, after signing of the DPA, and its main purpose was to mobilize international support for the agreement. The PIC was founded by 55 member states and agencies and has a fluctuating number of observers supporting the stabilization and development of BiH through different means: financially and politically and providing troops for the international military forces and as consultants on various subjects.

During this conference, a Steering Board for the PIC was established, which works under the leadership of the High Representative. The members of the Steering Board are Canada, France, Germany, Italy, Japan, Russia, United Kingdom, the USA, the Presidency of the European Union, the European Commission, and the Organization of the Islamic Conference, represented by Turkey. The aim of the Steering Board is to provide the High Representative with political guidance. This Steering Board meets weekly in Sarajevo with Ambassadors, and three times a year with political directors.

At their meeting in Brussels on the 26th and 27th of February 2008, the Political Directors of the Peace Implementation Council Steering Board set out the need to be met by BiH authorities prior to the closure of the OHR, known as the "5 + 2" plan. The objectives that need to be met by the BiH authorities prior to OHR closure are acceptable and sustainable resolution of the issue of apportionment of property between the state and other levels of government;

acceptable and sustainable resolution of defense property; completion of the Brčko Final Award; fiscal sustainability (promoted through an Agreement on a Permanent ITA Coefficient methodology and establishment of a National Fiscal Council); and entrenchment of the rule of law (demonstrated through the adoption of a National War Crimes Strategy, passage of the law on aliens and asylum, and adoption of a National Justice Sector Reform Strategy).

In addition to these objectives, the PIC Steering Board agreed that additional two conditions need to be fulfilled prior to OHR closure: signing of the Stabilisation and Association Agreement (SAA) and a positive assessment of the situation in BiH by the PIC Steering Board based on full compliance with the Dayton Peace Agreement.

As of today, there is still a long way ahead in order for BiH to achieve the completion of the 5 + 2 agenda. According to the OHR itself, at the state level, there is complete stagnation, evidenced by the poor legislative output of the BiH Council of Ministers and the BiH Parliamentary Assembly. During the current mandate, apart from budget acts, only one new law has been fully adopted. This is part of a general decline in legislative output over the past several mandates but is nonetheless a new low. In such an environment, it should come as no surprise that the BiH authorities have made no progress in implementing the 5 + 2 agenda, with the exception of the continued positive trend in Brčko District. Not only is there no progress on the other objectives, there are frequent attempts to roll back existing reforms in crucial areas (OHR 2021).

10.6 National Symbols

The flag of BiH is the state symbol of BiH. It was proclaimed on February 4, 1998, and adopted by the Parliamentary Assembly of BiH in 2001. It replaced the flag of the RBiH, which has been used since the declaration of independence in 1992.

The original light blue color on the flag was the blue color of the United Nations. However, dark blue ended up being chosen, corresponding to the blue color of the European Union flag. The yellow triangle represents the geographical appearance of BiH. The stars represent the European environment of BiH (Fig. 10.12).

In early 1998, the High Representative for BiH, Carlos Westendorp, made several neutral proposals. After the delegates in the Parliamentary Assembly failed to constitute the necessary majority to adopt any proposal, with the support of all delegates of SDS and HDZ BiH, Westendorp declared the current flag on the third of February of the same year (Official Gazette of Bosnia and Herzegovina 2001).

Fig. 10.12 National flag of Bosnia and Herzegovina

Fig. 10.13 Coat of arms of Bosnia and Herzegovina

The coat of arms is very similar in graphic elements to the flag: It is divided diagonally into upper and right yellow and lower and left blue parts. In the blue part along the border with the yellow field, there are seven white five-pointed stars, of which the uppermost and lower ones are shown only partially. There are several explanations for the symbolism of the coat of arms; the yellow triangle should symbolize the geographical shape of BiH, its three sides, the three constituent peoples, and the yellow color—the sun itself. The blue color symbolizes the sky or international organizations (UN or European Union), and the stars are also a symbolic link to the EU flag (Fig. 10.13).

According to Article IV(4)(a) of the Constitution of BiH, the Parliamentary Assembly of BiH (House of Representatives on June 6, 2001, and House of Peoples on May 23, 2001) adopted the Law on the Coat of Arms of BiH in 2001. The Law was published in the Official Gazette of Bosnia and Herzegovina on August 3, 2001 (Official Gazette of Bosnia and Herzegovina, Year V, No. 19, Sarajevo, 2001).

The Law on the National Anthem of BiH was passed by the High Representative of the International Community in BiH on June 25, 1999. The anthem of BiH was officially performed for the first time at the seventh session of the House of Representatives of the Parliamentary Assembly of BiH on July 15, 1999, and at the sixth session of the House of Peoples of the Parliamentary Assembly of BiH on July 19, 1999.

The Law on the National Anthem of BiH, with the musical part "Intermezzo" (authored by Dušan Šestić from Banja Luka), was adopted at the fourth session of the House of Representatives on February 10, 1999, and at the third session of the House of Peoples on May 23, 2001, and published in the Official Gazette of BiH No. 19 from 2001 (National flag, coat of arms, and anthem of Bosnia and Herzegovina). The anthem is one of the few in the world that does not have lyrics because the leading political parties still cannot agree on it.

References

Abazović D (2015) Sociološka čudovišnost političke volje u BiH [Sociological monstrosity of political will in BiH]. In: Abazović D (ed) Kratki problem u uvod političke volje: Studija slučaja Bosne i Hercegovine. FES, Sarajevo, pp 58–71

Agency for Statistics of Bosnia and Herzegovina (2016) Census of population, households and dwellings in Bosnia and Herzegovina 2013. Final results, Sarajevo. Available at https://www.popis.gov.ba/popis2013/doc/RezultatiPopisa_BS.pdf. Accessed on 21 Sept 2019

Anketa o radnoj snazi [Labor Force Survey] (2019) Agency for Statistics of Bosnia and Herzegovina, Sarajevo

Arnautović S (2013) Izborni sistem u Bosni i Hercegovini prema Deklaraciji o pravima građana Bosne i Hercegovine [The electoral system in Bosnia and Herzegovina according to the declaration on the rights of citizens of Bosnia and Herzegovina]. Historijska traganja 12:49–82

Balkan Insight (2019a). https://balkaninsight.com/2019/05/16/bosniak-parties-hatch-coalition-agreement-in-bosnia/. Accessed on 11 June 2021

Balkan Insight (2019b). https://balkaninsight.com/2019/12/21/reform-program-fails-to-clarify-bosnias-nato-ties/. Accessed on 11 June 2021

Banović D, Gavrić S, Barreiro M (2021) The political system of Bosnia and Herzegovina institutions—actors—processes. Springer

Begić K (1997) Bosna i Hercegovina od Vanceove misije do Daytonskog sporazuma [Bosnia and Herzegovina from the Vance Mission to the Dayton Agreement]. Bosanska knjiga, Sarajevo

Bieber F (2008) Bosna i Hercegovina poslije rata: Politički sistem u podjeljenom društvu [Bosnia and Herzegovina after the war: a political system in a divided society]. Buybook, Sarajevo

Bosnia and Herzegovina (2020a) Report, Commission Staff Working Document, 2020 Communication on EU Enlargement Policy. European Commission, Brussels, Belgium, pp 5–77

Bosnia and Herzegovina (2020b) Report, European Commission, 2020 Communication on EU Enlargement Policy, Brussels. Available at https://ec.europa.eu/neighbourhood-enlargement/sites/near/files/bosnia_and_herzegovina_report_2020.pdf. Accessed on 21 Jan 2021

Bougarel X (1992) Bosnie-Herzégovine: anatomie d'une poudrière. Hérodote 67:84–147

Branković A (2016) Administrative structure of Bosnia and Herzegovina. In: Oğurlu Y (ed) Bosnia and Herzegovina: law, society and politics. International University of Sarajevo, Sarajevo, pp 10–28

Bublin M (2000) Prostorno planiranje [Spatial planning]. Univerzitetska knjiga, Sarajevo

Ćerimagić A (2015) The imaginary world of competencies. In: Kapetanović A (ed) The legacy of peace. FES, Sarajevo, pp 223–244

Commission for the Coordination of Youth Issues BiH (2008) Analysis of the position of young people and the youth sector in BiH. Sarajevo. Available at http://www.mladi.gov.ba/pdf/Mladi_politika-EN_final.pdf. Accessed 10 Aug 2020

Constitution of Bosnia and Herzegovina. Available at https://www.ustavnisud.ba/public/down/USTAV_BOSNE_I_HERCEGOVINE_engl.pdf. Accessed on 16 July 2020

Constitution of the Federation of Bosnia and Herzegovina (FBiH). Available at https://www.predstavnickidom-pfbih.gov.ba/bs/page.php?id=103. Accessed on 14 Apr 2021

Daadler I (2014) Getting to Dayton: the making of America's Bosnia policy. Brookings Institution Press, Washington, DC

Duraković N (1999) Bosanski izazovi [Bosnian challenges]. DES, Sarajevo

Election Law of Bosnia and Herzegovina, Articles 9.2 and 9. Available at https://www.izbori.ba/Documents/documents/ZAKONI/BiH_Election_Law_last_consolidated_version_2018.pdf. Accessed on 23 June 2020

General Framework Agreement for Peace in Bosnia and Herzegovina, Annex 3, Article 4, Regularity of Democratic Elections in BiH. Available at https://www.osce.org/files/f/documents/e/0/126173.pdf. Accessed on 22 June 2020

Huškić J (2017) Lokalna samouprava u Bosni i Hercegovini [Local self-government in Bosnia and Herzegovina]. Pravni vjesnik: časopis za pravne i društvene znanosti 33(1):105–123

Ibrahimagić O (2012) Bosanski identitet i suverenitet [Bosnian identity and sovereignty]. University of Sarajevo, Sarajevo

Jovanović M (2002) Die politischen Parteien in Bosnien-Herzegowina. Master thesis, Bonn

Jusić T, Ahmetašević N (2013) Media reforms through intervention: international media assistance in Bosnia and Herzegovina. Analitika—Center for Social Research, Sarajevo

Kartsonaki A (2016) Twenty years after Dayton: Bosnia-Herzegovina (still) stable and explosive. Civil Wars 18(4):488–516. https://doi.org/10.1080/13698249.2017.1297052

Kukić S (2001) Bosanskohercegovački apsurdi kao brana uključenja u zajednicu demokratskih država [Bosnian absurdities as a barrier to inclusion in the community of democratic states]. In: Kukić S (ed) Bosna i Hercegovina i evroatlantske integracije. Sarajevo, pp 43–55

Kurtćehajić S (2017a) Prilozi za politički sistem Bosne i Hercegovine [Contributions to the political system of Bosnia and Herzegovina]. Futur Art, Sarajevo

Kurtćehajić S (2017b) Pravno-politička promišljanja o Bosni [Legal-political considerations on Bosnia]. Futur Art, Sarajevo

Law on Financing Political Parties of BiH, Article 7. Available at https://www.izbori.ba/documents/documents/zakoni/ppf_law_new.pdf. Accessed on 22 June 2020

Marković G, Sahadžić M (2013) Legal and political framework of the work of the parliamentary assembly of BiH. In: Orlović S (ed) Comparative analysis of democratic performances of the parliaments of Serbia. BiH and Montenegro, Faculty of Political Sciences, Belgrade, p 47

McMahon P, Western J (2009) The death of Dayton. How to stop Bosnia from falling apart. Foreign Aff 88(5):69–83

Mirić R (2009) Prostorno- funkcionalna analiza savremenog političko-teritorijalnog ustroja Federacije Bosne i Hercegovine: kantonizacija ili regionalizacija [Spatial-functional analysis of the modern political-territorial structure of the Federation of Bosnia and Herzegovina: cantonization or regionalization]. In: Spahić M (ed) Zbornik radova II kongresa geografa Bosne i Hercegovine. Geographical Society of FBiH, Sarajevo, pp 651–664

Mirić R (2011) Kompleksnost geografskih implikacija političko-teritorijalne organizacije Bosne i Hercegovine u svjetlu savremenih evropskih integracijskih procesa [Complexity of geographical implications of the political-territorial organization of Bosnia and Herzegovina in the light of modern European integration processes]. PhD thesis, Faculty of Science, University of Sarajevo, Sarajevo

National flag, coat of arms and anthem of Bosnia and Herzegovina. Available at http://www.mvp.gov.ba/dobro_dosli_u_bih/zastava_grb_himna/?id=7262. Accessed 06 July 2020

Nurković S, Mirić R (2005) Implikacije političko-teritorijalnog ustroja Bosne i Hercegovine na njenu regionalno-geografsku strukturu i razvoj [Implications of the political-territorial structure of Bosnia and Herzegovina on its regional-geographical structure and development]. In: Spahić M (ed) Zbornik radova Prvog kongresa geografa Bosne i Hercegovine. Geographical Society of FBiH, Sarajevo, pp 26–45

ODIHR (2019) Bosnia and Herzegovina. Election Observation Mission Final Report. Available at https://www.osce.org/odihr/elections/bih/409905?download=true. Accessed on 14 June 2021

OHR (2021) 59th Report of the High Representative for Implementation of the Peace Agreement on Bosnia and Herzegovina. Available at http://www.ohr.int/59th-report-of-the-high-representative-for-implementation-of-the-peace-agreement-on-bosnia-and-herzegovina-to-the-secretary-general-of-the-united-nations/. Accessed on 15 July 2021

OSCE (2018) Two schools under one roof: the most visible example of discrimination in education in Bosnia and Herzegovina. OSCE Mission to Bosnia and Herzegovina, Sarajevo

OSCE (2019) Assessment of the work of Bosnia and Herzegovina institutions in combating discrimination. OSCE Mission to Bosnia and Herzegovina, Sarajevo

Pejanović M (2005) Politički razvitak Bosne i Hercegovine u postdejtonskom periodu [Political development of Bosnia and Herzegovina in the post-dayton period]. Šahinpašić, Sarajevo

Pejanović M (2016) Bosna i Hercegovina i geopolitičke promjene u Evropi i svijetu na početku XXI stoljeća [Bosnia and Herzegovina and geopolitical changes in Europe and the world at the beginning of the 21st century]. In: Cvitković I (ed) Geopolitičke promjene u svijetu i Evropi i položaj Bosne i Hercegovine. ANUBiH Department of Social Sciences (8), Sarajevo, pp 42–58

Reményi P, Végh A, Pap N (2016) The influence of ethnic policies on regional development and transport issues in Bosnia and Herzegovina. Revue belge de géographie 2016(1):1–21. https://doi.org/10.4000/belgeo.18991

Results of general elections (2010). Available at http://www.izbori.ba/Finalni2010/Finalni/PredsjednistvoBiH/Default.aspx. Accessed on 19 Nov 2020

Results of general elections (2014). Available at http://www.izbori.ba/Potvrdjeni2014/Finalni/PredsjednistvoBiH/Default.aspx. Accessed on 20 Nov 2020

Results of general elections (2018). Available at http://izbori.ba/rezultati_izbora?resId=25&langId=1#/1/0/0/0/0. Accessed on 19 Nov 2020

Results of local elections (2020). Available at https://www.izbori.ba/Rezultati_izbora/?resId=27&langId=1#/8/0/0. Accessed on 18 Nov 2020

Rules of Procedure of the Presidency of BiH (2004). Available at http://predsjednistvobih.ba/nadl/default.aspx?id=18265&langTag=en-US. Accessed on 17 July 2020

Sahadžić M (2009) The electoral system of Bosnia and Herzegovina: a short review of political matter and/or technical perplexion. Contemp Issues 2(1):61–78

Sahadžić M (2011) Priroda političkog sistema u Bosni i Hercegovini [The nature of the political system in Bosnia and Herzegovina]. In: Banović D (ed) Država, politika i društvo u Bosni i Hercegovini. University Press Magistrat, Sarajevo, pp 15–37

Službeni glasnik Bosne i Hercegovine (2001) [Official Gazette of Bosnia and Herzegovina] Year 5, No. 19, Sarajevo

Smailagić N, Keranović N (2009) Pravnosudni sistem [Judicial system]. In: Gavrić S (ed) Uvod u politički sistem Bosne i Hercegovine—izabrani aspekti. Sarajevski otvoreni centar/Konrad Adenauer Stiftung, Sarajevo, pp 234–263

Statistical data of Brčko District BiH (2020) Agency for Statistics of Bosnia and Herzegovina. Branch of Brčko, p 54

Statistical Yearbook (2020) Institute for Statistics of FBiH. Sarajevo, p 57

Statistical Yearbook of Republika Srpska (2020) Institute for Statistics of Republika Srpska. Banja Luka, p 15

Statute of the Brčko District of Bosnia and Herzegovina (2010). Available at https://skupstinabd.ba/images/dokumenti/ba/statut-brcko-distrikta.pdf. Accessed on 22 Feb 2021

Stjepanović D (2016) Analiza procesa i uslova registracije političkih subjekata u državama članicama OSCE-a [Analysis of the process and conditions of registration of political entities in the OSCE participating States]. Pod lupom, Sarajevo

Topić T (2007) Hibridni režim i političke partije u Bosni i Hercegovini [Hybrid regime and political parties in Bosnia and Herzegovina]. In: Puhalo S (ed) Socio-psihološki profil glasača i apstinenata u Bosni i Hercegovini između Opštih izbora 2006. godine i Lokalnih izbora 2008. godine. Art Print, Banja Luka, pp 13–52

Trnka K (2000) Konstitutivnost naroda [Constitutivity of the people]. Congress Council of Bosniak Intellectuals, Sarajevo

Trnka K (2006) Ustavno pravo [Constitutional law]. University of Sarajevo, Faculty of Administration, Sarajevo

Venice Commission (2010) Report on thresholds and other features of electoral systems which bar parties from access to parliament (II). Strasbourg. Available at https://www.venice.coe.int/webforms/documents/default.aspx?pdffile=CDL-AD(2010)007-e. Accessed on 21 Oct 2020

www.izbori.ba. Accessed on 01 Feb 2021

www.parlament.ba. Accessed on 11 Aug 2020

Zakon o federalnim jedinicama [federal units law]. Available at http://www.fbihvlada.gov.ba/bosanski/zakoni/1996/zakoni/id%20o%20kantonima.htm. Accessed on 11 Apr 2021

Žďla O (2013) Ethno-demographic development in Bosnia and Herzegovina in 1971–1991 and its propensity for ethnic conflict. Acta Universitatis Palackianae Olomucensis – Geographica 44(1): 5–25

11 Population Geography of Bosnia and Herzegovina

Abstract

According to the total number of inhabitants, Bosnia and Herzegovina belongs to the category of small countries. Accordingly, it is the 132nd country globally and the 28th in Europe, right after Croatia and ahead of Albania and Lithuania. The population of Bosnia and Herzegovina is in a post-transition stage of population development in which the rates of natural change are negative. The total fertility rate is among the lowest in the world in the last several years. It is a kind of paradox that Bosnia and Herzegovina is not a geographical area of such a level of development that the post-transition (stationary) population would be a normal expression of socioeconomic conditions. The modern demographic picture has been strongly influenced by external factors from past periods, such as strong emigration and losses in the last war (1992–1995). They shook and significantly accelerated the "normal" demographic dynamics. This is a significant obstacle to the optimal socioeconomic development of Bosnia and Herzegovina. According to estimates, 3.6% of the total population has emigrated permanently since 2013. The population of Bosnia and Herzegovina is characterized by the advanced process of demographic aging, i.e., a decline in the number and share of young people and an increase in the share of older people in the total population. The main factors of demographic aging are unfavorable processes in demo-reproduction (decline in birth rates), emigration, war losses, and the lack of quality stimulating population policy. Estimates of future demographic changes in Bosnia and Herzegovina are extremely negative.

Keywords

Spatial distribution · Natural change · Migrations · Population aging · Educational structure

11.1 Contemporary Characteristics of Population and Spatial Distribution

According to the latest census (2013), Bosnia and Herzegovina had a population of 3,531,159. According to the statistical institutes' estimates, there was a total of 3,403,754 inhabitants in Bosnia and Herzegovina at the end of 2020, showing a decrease of 3.6%. As much as 18% of the active population is engaged in agricultural activities, and more than half of the total population lives in rural areas (BHAS 2016; FZS 2021; ISR 2021).

Bosnia and Herzegovina (hereinafter: BiH) is facing a depopulation process, which is mostly caused by emigration, decreasing birth rates, increasing death rates, and population aging. Unfortunately, there is no institutionalized population policy in this country which could lead to demographic revitalization (Kadušić and Suljić 2018). According to Halilovich (2012), Bosnians nowadays represent one of the most widespread emigrant communities from the Balkans.

Only the capital of BiH, Sarajevo, has more than 250,000 inhabitants. The population is aging very fast, and life expectancy is increasing, which raises concerns about the demographic trends. Negative migration rates have been characteristic since the late nineteenth century; not even the rapid industrialization and urbanization in the period 1965–1980 stopped the emigration of the Bosnian population. Emigration increased to an unprecedented scale during the 1992–1995 war. BiH is an ethnically and religiously heterogeneous country. According to the estimate of the national statistical institutes, the number of inhabitants after the war (1992–1995), in 1996, was lower by 16.7% (732,033). According to the data from the most recent (2013) census, the population in BiH was 19.3% lower than at the penultimate census in 1991—bearing in mind that the last census was made according to the de facto population concept compared to the previous censuses, which used the de jure population concept (BHAS 2019).

H. Gekić et al., *The Geography of Bosnia and Herzegovina*, World Regional Geography Book Series,
https://doi.org/10.1007/978-3-030-98523-3_11

The geographical area of BiH consists of parts of physiognomically different and functionally diverse units. Its space is defined by complex social (ethnic, socio-cultural, religious and economic, and settlement) processes and military-political events and constellations in the long past of Southeast Europe. Therefore, today's inhabitants and population of BiH reflect the historical-demographic process and its distributive, dynamic, and structural differences.

The population of BiH, with its distribution, density, population growth, migration, composition, and other characteristics, shows extreme complexity. The primary source of population analysis of data in this chapter is the population census of BiH from 2013. Due to the lack of an adequate population register, later estimates of population and certain other characteristics (other than vital statistics) should be taken with caution.

Differences in population density are a consequence of complex processes but with an unequal natural (geographical) basis and, later, a social superstructure as fundamental components of the general valorization of the space of modern society. The population density of each area is a function of social processes and value systems that also direct the spatial mobility of the population.

So far, these processes have made BiH a country with a medium population density of 66.5 inhabitants per km^2 of an area in 2020. European countries such as the Republic of Ireland, Ukraine, and Bulgaria have similar population densities. At the level of smaller spatial units, the differences are significant. Today's population density results from population growth and migration, i.e., general population growth over the past few generations. Due to BiH's very dynamic terrain and dissected relief in the central area and flatlands in the northern and southern parts, its population is unevenly distributed.

The specificity of the BiH area is the distribution of the total population by hypsometric levels. In 2013, the biggest number of inhabitants lived in the hypsometric level of 201–500 m AMSL (height above mean sea level), 40.1% of the total population of BiH or 1,415,995 inhabitants. 30.4% of the total population of BiH or 1,073,472 inhabitants lived at AMSL of up to 200 m. The third level in terms of population was 501–1000 m AMSL, with 29.2% of the total population of BiH or 1,031,098 inhabitants living there. The least number of inhabitants lived at the level of 1001–1500 m AMSL, only 0.3% of the total population of BiH or 10,594 inhabitants (most of them in the municipalities of Kupres-FBiH, Kalinovik, and some mountain settlements in Prozor-Rama and Gornji Vakuf-Uskoplje).

From 1961 to 2013, the number and share of the population living in higher hypsometric levels decreased, depending mainly on the possibilities for settlement in lower areas and employment in industrial activities in urban and suburban areas. The share of the population at the level of

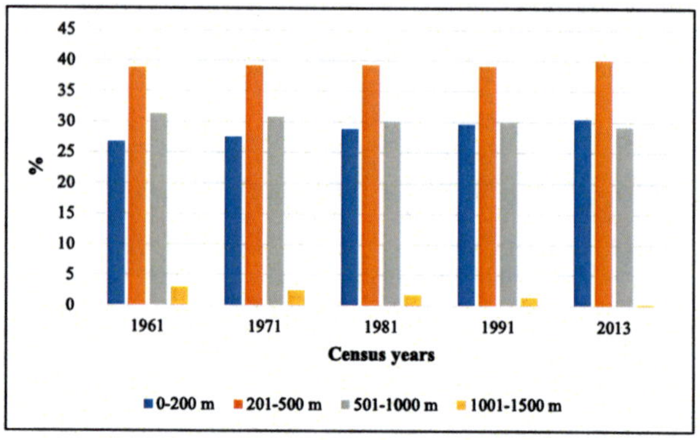

Fig. 11.1 Population distribution in relation to altitude from 1961 to 2013

1001–1500 m AMSL decreased from 3.0% in 1961 to only 0.3% in 2013 (the number of inhabitants at this level decreased from 99,347 to 10,594 or by 89.3%). The increase in the number of inhabitants was the highest at the level up to 200 m AMSL, from 875,053 in 1961 to 1,294,103 in 1991 (by 47.9%). Due to the extensive depopulation during and after the 1992–1995 war, the population of this level decreased to 1,073,472 in 2013 or 17.0%. A characteristic of the height distribution of the population in BiH over the last 60 years is the constant decrease in the number of inhabitants at the highest levels in favor of lower levels. The population of the area above 1000 m AMSL in BiH in the last 30 years is meager and very rare (Fig. 11.1).

The population of BiH in the last almost 150 years is the result of the influence of many factors and extremely complex historical and geographical development. This means that the reasons for different populations of certain parts of the country should not be sought only in natural preconditions for population but also in specific general circumstances of its socioeconomic development in the past, which have proven to be an important factor in the population of BiH in the past and today.

It should be noted that the population of BiH and, subsequently, the distribution of the population as its indicator is very uneven. Differences in the spatial distribution of the population are very pronounced.

During the last 60 years, the influence of natural and geographical factors on the movement and development of the population of BiH has significantly weakened. At the same time, the impact of socio-geographical determinants of population development has intensified, especially economic changes in a society driven by the processes of industrialization, deagrarization, deruralization, and urbanization.

Under the direct influence of morphological-hypsometric characteristics of space, the impact of climate on population in BiH is twofold. In areas up to 1000 m AMSL, the climate is generally favorably affected by moderate air temperatures

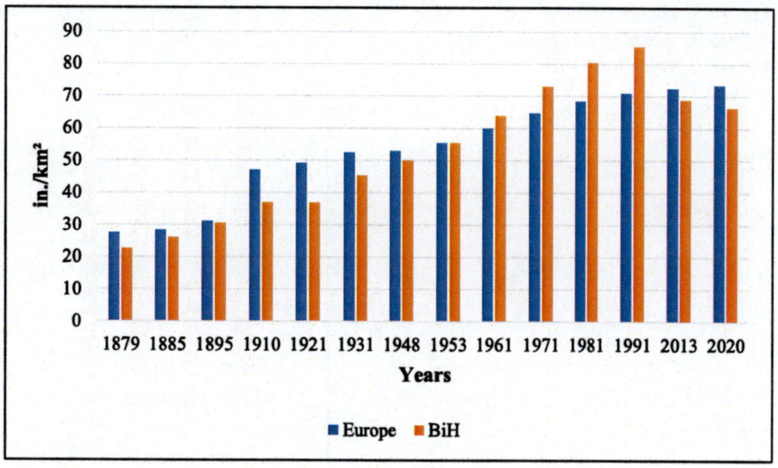

Fig. 11.2 Population density in Europe and Bosnia and Herzegovina from 1879 to 2020

and humidity and an even distribution of precipitation and their effective rhythm throughout the year. At this height level in 2013, 99.7% of the total population of the country was concentrated. The favorable temperature-precipitation complex is an essential and undeniably important element of the population of BiH.

The climate has a highly unfavorable effect on the population in mountainous areas above 1000 m AMSL due to its lower temperatures and cold. Relief appears in these areas as a climate modifier. Only 0.3% of the population, according to the 2013 census, lived in these climate-unfavorable areas. Also, it is evident that the sunny areas (areas facing the sun) and the foothills are significantly more populated than the shady areas (areas sheltered from the sun). This is especially noticeable on the slopes of numerous mountains in BiH, such as Kalin and Rudina in the municipality of Bugojno, Vranica in the municipality of Gornji Vakuf-Uskoplje, Vlašić in the municipality of Travnik, and others.

It is clear that the areas of BiH characterized by the presence of at least one major surface watercourse are significantly more populated than those with no such flows. Thus, primarily mountainous regions and areas are almost uninhabited without a significant surface river network. All this proves that the presence of water that is necessary for the water supply of the population, settlements, industry, agricultural land, etc. is one of the essential factors in the population of BiH. The impact of water on population development is far more favorable in the northern, central, eastern, and southern part of the country compared to the western, southeastern, and partly southwestern, which due to geological structure and relief, does not have such an abundance of watercourses as the previously mentioned parts of BiH.

The impact of soils on population is reflected in its distribution and fertility. The population is very pronounced where the soils are fertile, such as along the great rivers Bosna, Vrbas, Drina, Neretva, and Una and the lower reaches of their tributaries. With the change of distribution from fertile to less fertile soils, the population density also changes. That is why the areas in the mountainous regions where rankers and district cambisols dominate the soil are almost uninhabited or are together with lithosols as in the west, southeast, and southwest.

The strengthening of the country's economy towards industry and service activities, mainly located in larger urban centers and their suburban areas, has led to the relocation of the population from rural to urban settlements, resulting in spatial relocation in the population structure. Due to the high population growth and occasionally very intensive immigration, the population density of BiH increased almost continuously from 1879 to the beginning of 1992.

Compared to the European continent, the population density in BiH was lower until the early 1950s. Since then, primarily due to improved health care of the population (especially infants), the population is growing rapidly, increasing population density. Already in 1961, the population density in BiH was higher than the European average (64 in./km^2 compared to 60 in./km^2), and in 1991, this ratio was 85.5 in./km^2:71.0 in./km^2 (Fig. 11.2). Before the war, BiH was 20.4% more densely populated than the European average. Unfortunately, the great war suffering and emigration of the population associated with natural depopulation have led to the fact that, according to estimates from 2020, BiH is 9.5% less populated compared to Europe.

We supplemented the analysis of the relative population density with the analysis of the index of the average distance of one inhabitant from another, which starts from the assumption that the distribution of the population is uniform. Beginning in 1879 until 1991, following the general trends of demographic and socioeconomic development, the index of the average distance of the population in BiH decreased from 225.9 to 116.2 (Fig. 11.3). In this analysis, extreme events such as the 1992–1995 war disrupted this trend. According to 2013 data, the average population distance index increased

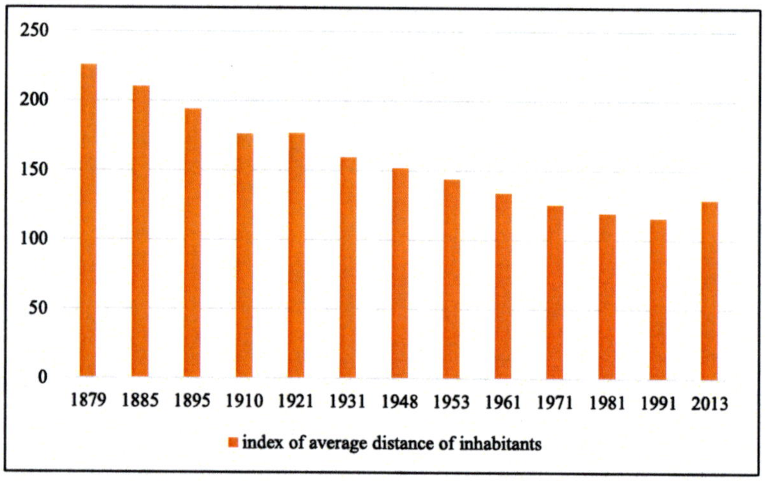

Fig. 11.3 Index of average distance of inhabitants in Bosnia and Herzegovina from 1879 to 2013

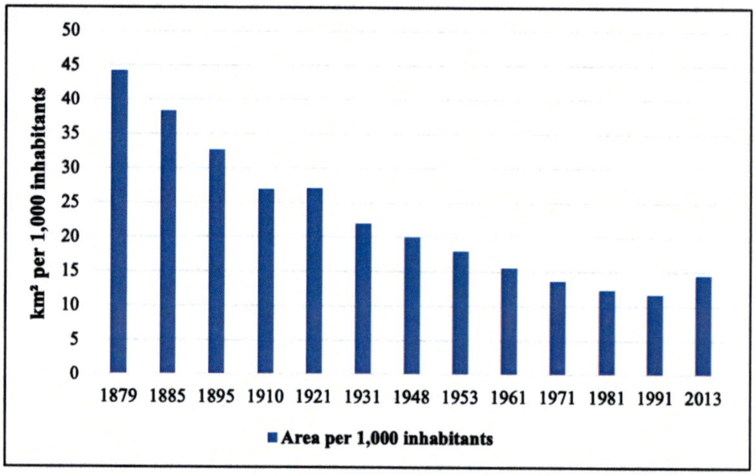

Fig. 11.4 Area per 1000 inhabitants in Bosnia and Herzegovina from 1879 to 2013

by 11.4% to 129.4%, confirming that this is a period of demographic regression and depopulation.

Also, the territory area per 1000 inhabitants decreased in BiH until 1991, from 44.2 km^2 in 1879 to only 11.7 km^2 in 1991. In 2013, due to a significant decline in the total population, this area increased to 14.5 km^2 or by 23.9% and was almost at the level of the early 1960s (Fig. 11.4).

The main areas of settlement were the central parts of the valleys (Sarajevo-Zenica, Mostar, Banja Luka, Tuzla, Bihać, Uskopaljska, etc.), especially in urban settlements that became the center of economic development such as Sarajevo, Zenica, Mostar, Bihać, Prijedor, Banja Luka, Bijeljina, Brčko, Tuzla, Bugojno, etc. The increase in the total population is necessarily reflected in the rise in the population density. From 1961 to 1992, BiH was part of a group of above-average inhabited European areas.

There were significant population distribution and population density changes during the last inter-census period (1991–2013). Due to the great war destruction in 1992–1995 and depopulation, the general population density in BiH decreased from 85.5 in./km^2 to 69.0 in./km^2 or by 19.3%. Due to the reduction mentioned above, BiH became a below-average populated country in Europe, as the general population density was 3.5 in./km^2 lower than the European average.

Compared to the national average in 2013 (69.0 in./km^2), 65 municipalities/cities (hereinafter: municipalities) had an above-average population density in which the population density ranged from as much as 6547 in./km^2 in Novo Sarajevo (the smallest in area of municipalities in BiH with only 9.9 km^2) up to 69.4 in./km^2 in Bratunac. In the remaining 78 municipalities, the average general population density ranged from only 1.1 in./km^2 in Istočni Drvar to 67.9 in./km^2 in Gradiška.

If we consider the average general population density of BiH in 2013, then its municipalities can be differentiated into five groups. The first group includes municipalities with a very low density (less than 20 in./km^2). There are 30 of them

(21% of all municipalities in BiH), covering 25.5% of the state's total area. Slightly more than a quarter of the territory of BiH is with very low population density, and only 4.2% of the total population of BiH lives in this area.

It is necessary to emphasize here that this is the general population density within the administrative units. It is quite clear that there are substantial uninhabited areas in which unfavorable geographical conditions prevail within the area of almost all municipalities. Characteristic poles from this group are Istočni Drvar with 1.1 in./km^2 and Krupa na Uni with 19.0 in./km^2. These are areas of unfavorable mountain and karst on the west, east, and southeast. The population is concentrated mainly in valley areas up to 500 m AMSL.

The second group includes municipalities with a low population density (20–39.99 in./km^2), which in 2013 were 22 (15.4% of all municipalities in BiH), covering 19.7% of the total area of BiH. 8.3% of the total population lived in low-populated areas. If we add the previous group, we can conclude that almost half of the total area of BiH territory is characterized by a low population density (low and very low). The poles in this group are Jezero with 20.4 in./km^2 covering an area of 56 km^2 and Srbac with 38.8 in./km^2 covering an area of 453 km^2. The municipalities of this group cover large areas that extend towards mountain peaks or are located in geographically unfavorable karst areas.

The third group consists of municipalities with a medium population density (40–99.99 in./km^2), which are the most numerous, namely, 39 (27.3% of all municipalities in BiH) covering 27.8% of the total area, and in which 23.1% of the total population of BiH lives. These are municipalities located near major urban and economic centers in BiH, such as Mostar in the south (Posušje, Široki Brijeg, Grude, and Stolac), in central Bosnia around Sarajevo (Ilijaš, Hadžići, and Pale-RS) and somewhat further west (Bugojno, Donji Vakuf, Gornji Vakuf, Jajce), in eastern and northeastern Bosnia in the near and far vicinity of Zvornik (Vlasenica, Milići, Bratunac) and Tuzla and Doboj (Petrovo, Stanari, Pelagićevo, Zavidovići, Čelić, Derventa, Prnjavor, Teslić), in the vicinity of Banja Luka (Gradiška and Srbac), and near Prijedor and Bihać (Sanski Most, Kozarska Dubica, Bosanska Krupa, Ključ, Novi Grad, and Kostajnica).

The fourth group consists of municipalities with a high population density (90–149.99 inhabitants/km^2). There are 23 of them (16.0% of all municipalities in BiH), covering 14.3% of the total area, and 21.4% of the total population of BiH lives in them. These are some of the most developed municipalities in BiH but also those located in their immediate vicinity of places such as Mostar in the south and, next to it, Ljubuški, Čitluk, and Čapljina; then Velika Kladuša in the far northwest; a somewhat east Prijedor and Laktaši; many municipalities near Tuzla (Ugljevik, Banovići, Lukavac); then Doboj; near Zenica (Žepče, Maglaj) and Sarajevo (Kakanj, Kiseljak); and somewhat further west Busovača, Novi Travnik, etc. The poles in this group were Mostar with 90.0 in./km^2 and Usora with 132.6 in./km^2 (rural municipality with only 49.8 km^2 area).

The last group consists of municipalities with a very high population density (150 and more in./km^2), of which there are 29 (20.3% of all municipalities in BiH), covering 12.7% of the total area, and in which lived as many as 43.0% of the total population of BiH. The poles in this group were Bijeljina with 146.8 in./km^2 and Novo Sarajevo with 6547 in./km^2. In addition to Novo Sarajevo, the most densely populated are the other three municipalities, which are administratively part of the city of Sarajevo (Novi Grad Sarajevo with 2522 in./km^2, Centar Sarajevo 1672 in./km^2, Stari Grad Sarajevo with 725 in./km^2). In the municipalities adjacent to the city of Sarajevo, the density is also very high, namely, Istočna Ilidža 527 in./km^2 and Ilidža 467 in./km^2. Very high population density also characterizes the most economically developed areas in BiH, Banja Luka, Zenica, Brčko, Tešanj, and Tuzla, and its immediate surroundings in the northeast, where the population density of municipalities ranges from 160 in./km^2 in Srebrenik to 377.5 in./km^2 in Tuzla (Fig. 11.5).

Regional differences in the spatial distribution of the population are very prominent. The existence of regional differences in population is one of the fundamental demographic characteristics of BiH. The largest population is located in the central part of BiH (27.8% of the total population), characterized by an uneven population due to the characteristic mountain-valley conditions. The capital Sarajevo is located in this part of BiH. The most densely populated are the valleys, hilly and mountainous areas less frequently, while mountainous and high-mountainous regions and areas are very sparsely populated.

24.2% of the total population lives in the northern part of BiH. It is predominantly a lowland area with fertile agricultural land, and the largest portion of the population is located in Banja Luka. 19.5% of the total population lives in the northeastern part of BiH, and the largest portion of the population is in Tuzla and its surroundings. It is a mining, industrial, and agricultural area. 10.9% of the total population lives in the south of BiH, mostly in Mostar and its immediate surroundings. The Neretva river valley has a much higher population than other areas. 7.9% of the total population lives in the northwestern part of BiH, and the largest portion of the population is concentrated in the valley of the Una river. Seven percent of the total population of BiH is concentrated in the east, and the population is significantly declining in the direction from Pale to Goražde to the far east of the country due to the presence of high mountain massifs. In the west of BiH, there is only 2.7% of the total population. It is mainly a karst high-mountainous area with far unfavorable geographical living conditions than other BiH parts.

Fig. 11.5 Population density in Bosnia and Herzegovina in 2013

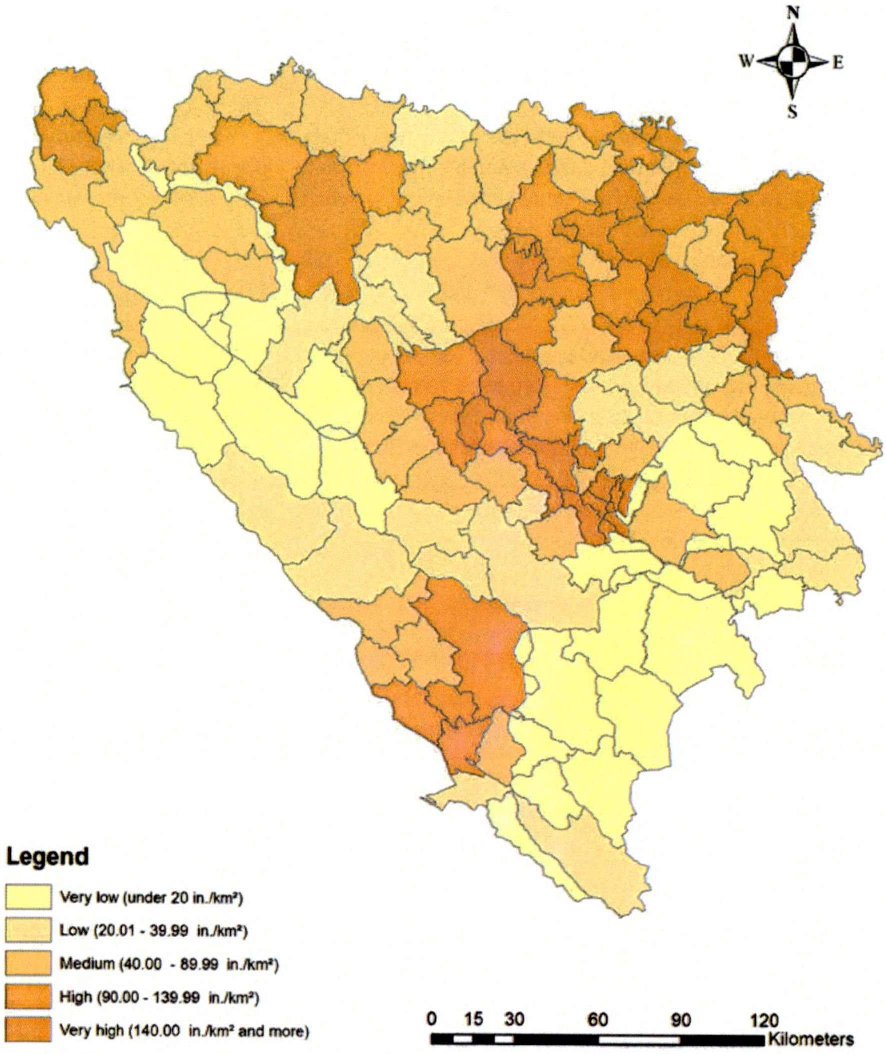

Physiological population density is an important indicator of the agrarian-geographical potential of the area. The value of physiological density of 500 inhabitants per 100 ha of arable land is considered to meet the population's nutritional needs. Any value above the one stated previously indicates that residents in an area cannot produce enough food for their needs and must therefore buy it (Husain 1996). The physiological population density of BiH increased from 170 inhabitants in 1953 to 275 inhabitants per 100 ha of total arable land until 1991. After that, due to the decrease in the number of inhabitants, it decreased to 220 inhabitants per 100 ha of total arable land in 2013.

It should be noted that arable land in BiH has been constantly reduced in favor of other categories of land use (forests, construction land, etc.). From 1953 to 2013, arable land was reduced by almost 108,000 hectares or about 6.3% (Ministry of Foreign Trade and Economic Relations of BiH 2018). At the same time, the population is higher by 24%. Therefore, BiH is a country that, with regard to arable land, can meet the needs of twice as many inhabitants as it has, i.e., it is one of the countries with a moderate physiological population density.

A significant feature of the distribution of the population of BiH is the unequal population by municipalities. Given the size of the contingent in the total population, municipalities are classified into seven categories. After 1995, due to establishing the inter-entity borderline (hereinafter: IEBL), numerous municipalities were formed with a very small population and now form mainly the first group of municipalities with a size of up to 2000 inhabitants. There are 11 of them (7.7% of the total number of municipalities in BiH), and only 0.3% of the total population of BiH lives in them. They are also called miniature municipalities. The smallest municipality in BiH in terms of population is Istočni Drvar, with 79 inhabitants in 2013. Slightly larger than this municipality are Istočni Mostar (257), Kupres-RS (300), and

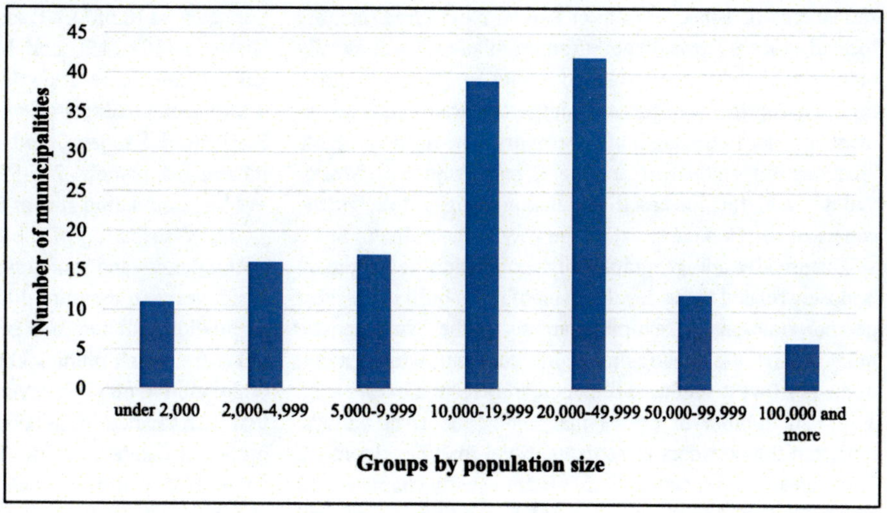

Fig. 11.6 Classification of municipalities by population in 2013

Petrovac (361). The largest municipality in this group in terms of population is Foča (FBiH), with 1933 inhabitants.

The characteristic of the second-largest group of municipalities (2000–4999 inhabitants) is its location mainly in the karst area and the populating of naturally geographically scarce regions of the west, south, and southeast of BiH. There are a total of 16 such municipalities (11.2% of the total number of municipalities in BiH), and only 1.5% of the total population of BiH lives in them. These are municipalities in the extreme south of BiH, such as Neum on the seaside and has only 4653 inhabitants, then Ravno 3219, slightly north of them Ljubinje (3511) and Berkovići (2114), Kalinovik (2029), and Bosansko Grahovo (2449) and Glamoč (3860) in the far west, among others (Fig. 11.6).

The third group includes municipalities with 5000–9999 inhabitants, which in 2013 were 17 (11.9% of the total number of municipalities). 3.3% of the total population of BiH lived in them. Similar to the previous group, these are mainly municipalities located in karst, mountainous, and naturally geographically scarce areas for populating, such as Kupres (FBiH) (5057), Drvar (7036), and Bosanski Petrovac (7328) in the west and northwest of BiH, followed by Rudo (7963) and Gacko (8990) in the east and southeast, and numerous rural municipalities formed by the establishment of IEBL such as Pelagićevo (5220), Osmaci (6016), Ribnik (6048), Petrovo (6474), Usora (6603), Šekovići (6761), and Teočak (7424).

If we merged the first three groups into one, we would conclude that in the area of almost one-third of all municipalities in BiH live only 5.1% of the total population. These are the areas with the most pronounced depopulation in which the number of inhabitants is estimated to be rapidly decreasing.

The central part of BiH is a mountainous area intersected by inter-mountain and intra-mountain depressions (valleys) bounded by heights and mountains. These are mainly highly populated river valleys of the upper parts of the watercourse of Bosna and Vrbas rivers and their tributaries. Numerous valley extensions are located there, and the largest, most populated, and economically most developed is the Sarajevo-Zenica valley with the centers Sarajevo and Zenica. This represents the most important direction of populating, i.e., the axis of population development in BiH, parallel to the direction of the Dinaric mountain system. Downstream, the river Bosna receives Lašva as her left tributary, in whose valley are located the towns of Busovača, Vitez, Novi Travnik, and Travnik, which is also a significant direction of populating. A little further north is the valley of the river Krivaja, which connects the towns of Zavidovići and Olovo. A very important direction of populating and the second most important in BiH, parallel to the Dinaric mountain system, is related to the Spreča tectonic depression in which the urban settlements Gračanica, Lukavac, Živinice, Kalesija, and Tuzla in the Tuzla valley are located.

Gornji Vakuf, Bugojno, and Donji Vakuf, and northwest of them Jajce, are located in the inter-mountain and intra-mountain depressions that form Vrbas valley in its upper part of the course, such as the Uskopaljska valley. The most important transversal routes of population settlement in BiH are connected in the north to the Una river valley downstream from Bihać, Sana from Ključ to Prijedor, Vrbas, downstream from Jajce, and in the east along the entire length of the Drina river, followed by the Neretva valley downstream from Konjic in the south.

In the area of the Bihać valley, the valley of Una river occupies the southwest-northeast direction, and the urban settlements of Bihać, Bosanska Krupa, Novi Grad, etc. are located here. Downstream from Jablanica, tectonically predisposed, the Neretva valley occupies a north-south direction, making this area the most important transversal settlement corridor in the south of the country. The most favorable locations for settlement are the Bijelo Polje and Mostar valleys, and the increased population density follows the course of the Neretva river, all the way to the Adriatic Sea.

Historically and geographically, population development is closely related to the development of roads built in the most geomorphologically favorable areas. In the characteristic relief of BiH, the Dinaric mountains stand out, which extend in the northwest-southeast direction and create a barrier with a limited number of passes that, along the river valleys, represent the corridors of settlement and road development. The plateaus and mountains of BiH are usually characterized by an isolated and dispersed population with a very small number of inhabitants in specific areas (Ahmetbegović 2012).

There were 39 municipalities in the fourth group (10,000–19,999 inhabitants), which represents 27.2% of the total number of municipalities, and 16% of the total population of BiH lived in them. It is a diverse group of municipalities located in all parts of BiH but mostly near major city centers or the most important state roads. In the south of BiH, they are Stolac (14,502), Grude (17,308), and Čitluk (18,140); in the east Sokolac (12,021) and Foča-RS (18,288); in the north Brod (16,619) and Šamac (17,273); in the west and northwest Šipovo (10,293) and Ključ (16,744); and in central BiH they are Fojnica (12,356) and Donji Vakuf (13,985), as well as many others.

In the fifth group (20,000–49,999 inhabitants), there were 42 municipalities, which makes for 29.4% of the total number of municipalities, and 36.1% of the total population of BiH lived in them. It is the most populous group of municipalities in terms of population. It is also a diverse group of municipalities located in all parts of BiH, and their characteristic is more extensive economic development than in previous groups. In the south of BiH, they are Čapljina (26,157) and Ljubuški (28,184), in the east Bratunac (20,340) and Goražde (20,897), in the north Modriča (25,720) and Prnjavor (35,956), in the west and northwest Livno (34,133) and Sanski Most (41,475), in central BiH they are Bugojno (31,470) and Visoko (39,938), and many others (Fig. 11.7).

The sixth group includes 12 municipalities (8.4% of the total number of municipalities) with a population of 50,000–99,999. 21.8% of the total population of BiH lived in them. The largest municipalities in this group are Prijedor (89,397), Brčko (83,516), Ilidža (66,730), Cazin (66,149), and Novo Sarajevo (64,814). The seventh group consists of the most economically potent municipalities that continue to be attractive to migrants Banja Luka (185,042), Novi Grad Sarajevo (118,553), Tuzla (110,979), Zenica (110,663), Bijeljina (107,715), and Mostar (105,797). The capital of BiH, Sarajevo, as a single administrative spatial unit with 4 city municipalities (Novi Grad Sarajevo, Novo Sarajevo, Stari Grad Sarajevo, and Centar Sarajevo), has the most inhabitants, namely, 275,524. In BiH, 1 municipality had an average population size of about 24,693 inhabitants in 2013.

The distribution of the population by municipalities has become more and more uneven over the decades. One of the easier ways to measure differences in the distribution of the population and the area of the territory is to establish the value for which municipalities should relocate the country's population to obtain an even population among them. We did that with the help of the population concentration index. The higher this index, the more concentrated the population. A value of 0 would mean an even distribution throughout the country, and a value of 100 would mean a concentration of population in a single municipality. The population concentration index quantifies the distribution of the population at a particular time (according to Nejašmić 2005). In 2013, the population concentration index in BiH was 38.1%, which means that it is necessary to relocate more than 1/3 of the total population to obtain an even distribution of the population by municipalities.

The territorial distribution of the population of BiH is conditioned by the vertical fragmentation of the terrain, the geomorphological structure of the terrain, the richness and quality of arable land, climatic conditions, and the quality of roads. A very complex and close interdependence of relief morphological and morphometric characteristics of space and population distribution can be observed in BiH. This population-geographical characteristic of the distribution of BiH population is a consequence of the complex natural-geographical basis and socio-geographical factors.

11.2 Population Dynamics

11.2.1 Total Number of Population

Due to the lack of reliable data on the population of BiH, until the first, officially recognized and methodologically correct census in 1879, conducted by the Austro-Hungarian authorities, it was not possible to give more reliable estimates of the population or some other characteristics of the total population. Demographic and socioeconomic development of BiH took place in strong interdependence with relatively favorable natural-geographical preconditions for settlement (prominent traffic-geographical position, richness of quality forest, numerous rivers, fertile land) and complex political events, which often directed and modified population-settlement processes. In the environment of favorable natural-geographical conditions for settlement and complex

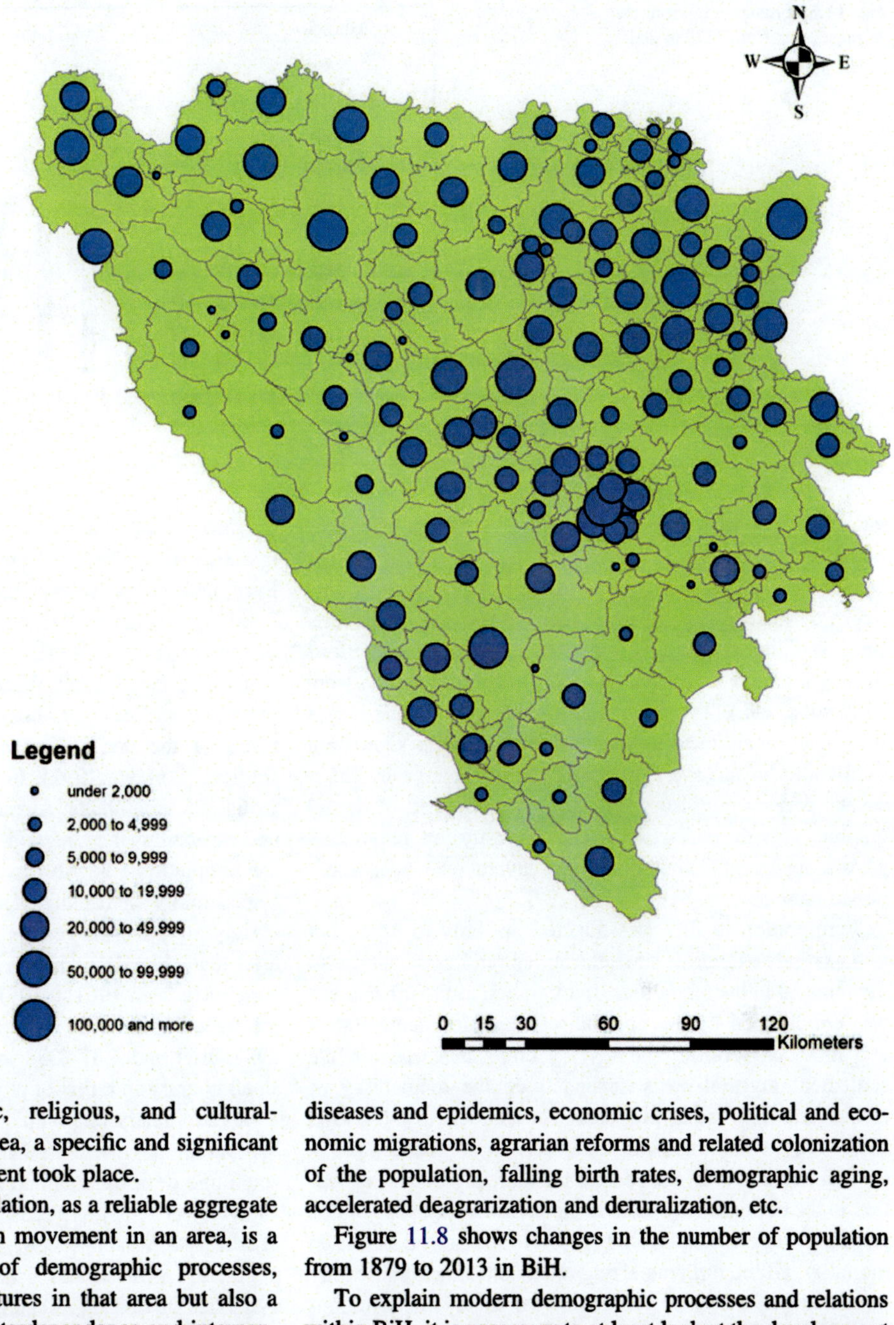

Fig. 11.7 Size structure of municipalities in Bosnia and Herzegovina in 2013

historical, political, economic, religious, and cultural-civilizational relations in this area, a specific and significant numerous population development took place.

The change in the total population, as a reliable aggregate indicator of the total population movement in an area, is a clear indicator of the level of demographic processes, conditions, relations, and structures in that area but also a precise argument proving the interdependence and interpenetration of different historical, political, social, economic, and territorial developments and changes in structural processes in population development. The movement of the total population of BiH in the entire observed period (1879–2013) took place in broader development conditions in which the development and movement of the population were dominated by destabilizing "internal" and "external" factors of demographic growth, such as the First and Second World War, the aggressive war against BiH between 1992 and 1995, diseases and epidemics, economic crises, political and economic migrations, agrarian reforms and related colonization of the population, falling birth rates, demographic aging, accelerated deagrarization and deruralization, etc.

Figure 11.8 shows changes in the number of population from 1879 to 2013 in BiH.

To explain modern demographic processes and relations within BiH, it is necessary to at least look at the development trends in the past. In this chapter, we have considered the numerical movement of the population from the first modern census in BiH in 1879 to the last in 2013, partially taking into account the last estimate of the population from 2020. This period is long enough to assess the fundamental long-term trends in the movement of the population of the observed area (a total of 141 years).

The development of the population of BiH in the last 141 years has characteristics of the so-called natural-

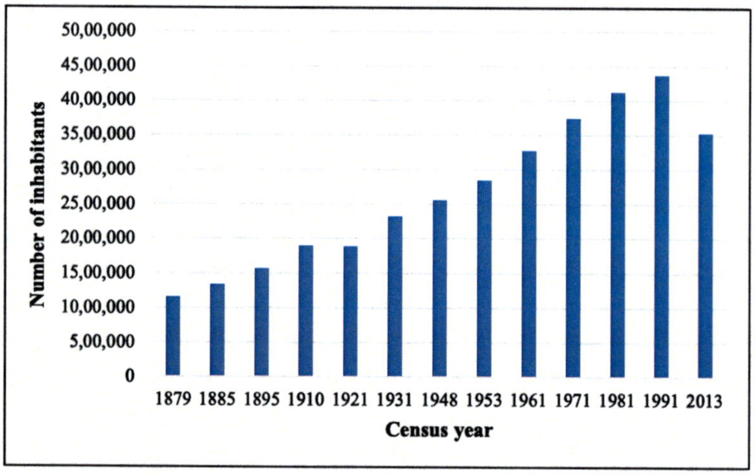

Fig. 11.8 Changes in the number of population from 1879 to 2013

migratory type of population formation. Although migratory currents from the nineteenth century until the 1990s did not take on the features of mass migrations from earlier historical stages of societal development, their numbers and character surpassed the constantly present individual migrations (immigration of Catholics from Dalmatia, Orthodox from Vojvodina, etc.). The movement of the total population of BiH in the mentioned period was uneven, with significant oscillations in the average annual rate of change by inter-census periods but also with a pronounced trend of demographic growth until 1992. The intensity of population growth depended on the influence of several previously mentioned factors.

Immigration to BiH was directed not only to urban but also to rural settlements since the agricultural population also participated in the migrations. Immigration to the urban areas was conditioned by the possibility of employment in state and financial services and less by crafts and trade, while settlement in rural areas depended on the availability of arable land and vacant properties. In addition to the period 1910–1921, when there was a decrease in population due to the First World War, the second far more significant exception in the continuous growth is the period after 1991. In this period, the aggression on BiH (1992–1995) caused increased mortality, falling birth rates, migration, and later significantly disrupted population development, so there was a demographic regress from 1991 to 2013 (−0.9% on average per year or even −19.3% in total).

In the period 1879–1991, the population growth of the country was 204.8%, from 1,158,440 to 3,531,159. During the Austro-Hungarian rule, population growth rates were moderately high. Thus, the population growth rate in the period 1879–1885 was 2.6% on average per year, so, in 1885, BiH had 1,336,091 inhabitants. Already in 1910, the number of inhabitants was 1,898,044, which was 42% more than in 1885. The First World War brought a decrease in the number of inhabitants according to the 1921 census by 0.04% or 7604 inhabitants. During this war, about 360,000 inhabitants, or 19% of the total population of BiH, lost their lives, and the most devastating fact is that, out of the stated number, as many as 100,986 children, younger than 10, died of starvation.

During the Second World War, BiH lost 10.1% of its pre-war population, or about 316,000 inhabitants, as a direct result of the war, of which 174,000 were civilians, not soldiers (Žerjavić 1992). In 1948, BiH had a population of 2,564,308 inhabitants. Already after the Second World War, and especially since the mid-1950s, with the improvement of the population's health care, the number of inhabitants has been rapidly increasing. The growth rates were different, mainly in the inter-census period 1948–1953, 2.2%, and after that, they decreased to 1.9% in the period 1953–1961, then 1.4% from 1961 to 1971, and 1% from 1971 to 1981. In the period 1981–1991, the population growth was 0.6%. In 1991, BiH had 4,377,033 inhabitants and was, therefore, a small country in terms of population.

It can already be pointed out that the periods of stronger demographic growth in BiH are also periods of stronger economic development and in connection with immigration and vice versa. A more pronounced increase after the Second World War was felt until 1992, primarily due to natural population growth and immigration. However, population dynamics slowed significantly in the 1981–1991 inter-census period. Population growth in the period 1981–1991 was 0.6% on average per year. This is a consequence of the strengthening of rural exodus and emigration, the reduction of natural dynamics, and the announcement of reproductive and generational depopulation, especially the deterioration of economic flows and the weakening of immigration. Therefore, the numerical development of the population of BiH between 1981 and 1991 could be considered to be slowing down.

The immigration of the population was especially encouraged by the Austro-Hungarian authorities from 1879 to 1914. This is a period in which the occupying powers are increasingly interested in exploiting indigenous natural resources and agricultural products. The development of manufacturing and industry in the Austrian part of the empire has stimulated the import of food products and raw materials from BiH. The exploitation of forest resources began with the opening of numerous sawmills, and the main activity remained agriculture. It is estimated that about 300,000 inhabitants, mostly Muslims, emigrated from BiH during the Austro-Hungarian rule. At the same time, about 115,000 inhabitants, mostly Catholics and Orthodox from Croatia, Vojvodina, Serbia, and Montenegro and then from Germany, Poland, the Czech Republic, and other countries, moved to BiH. The cause of the migration of the Muslim population was insecurity and status within the Christian state and deep religious, cultural, and emotional ties with Constantinople. For almost the entire period of Austro-Hungarian rule, the emigration of Muslims continued, who sold their property for nothing and went into uncertainty (Ramčilović 2019).

There are no reliable data on the natural population growth in this period because the authorities did not record vital events at that time. Still, suppose we assume the high birth rate and high mortality resulting in moderate population growth, it can be assessed that the previously mentioned types of migration processes were decisive for population growth.

The impact of natural population growth and partly migration on the change in the population of BiH was even more significant and more pronounced after the Second World War. However, since the census was conducted 3 years after the war, the losses of the Second World War remained partly "hidden," as they were largely replaced by post-war immigration. Demographic growth was, of course, significantly conditioned by the post-war compensation period (1946–1954), during which the negative demographic consequences of the Second World War and the post-war period in this area were somewhat mitigated due to the increase in birth rates. Residents from Croatia, Serbia, and Montenegro immigrated to BiH the most.

After the agrarian relocations after the end of the Second World War, immigration of the population followed, stimulated exclusively by the economic (primarily industrial) development of the area, which resulted in an increased tendency of population growth in the period of 1948–1991. After agrarian colonization, economic immigration of labor dominated because, although relatively late, intense industrialization still attracted the population, which, in search of jobs (more) and land (less), simply "overflowed" into this area. Residents from Croatia and Serbia immigrated the most. The ample supply of relatively cheap arable and construction land in the urban areas and suburbs attracted part of the immigrant population.

Substantial populating and natural population growth have been favored by industrial expansion reflected in large economic enterprises and the highly advanced population's health care since the 1960s compared to the time before the Second World War. From the 1950s, strong development of the industry began (manufacturing in particular), namely, metal and machine, military, food, construction, textile, chemical, leather and footwear and wood, and transport and trade. The existence of relatively cheap land and a large workforce, along with planned planning, has enhanced the industrial development of BiH.

The military, metal, and machine industry had a particularly concentrating effect on the population, which, thanks to the large and constantly growing needs for labor, strongly stimulated the immigration of the population to main urban centers and its suburbs. Since the 1960s, most immigrants have been "external" migrants, i.e., immigrants from more economically passive areas of Serbia and Croatia (Vojvodina and Dalmatia). They mainly were qualified and highly qualified people which BiH was missing but also the poor, able-bodied, and reproductively vital people, who managed to strengthen and expand the population base of the state, influencing the population trends, birth dynamics, population structures (primarily age and sex), and the spatial redistribution of population. Immigrants settled in urban areas of strong industrialization and along important roads.

The inter-census period between 1971 and 1981 can be considered the final period of real dynamic demographic growth in BiH. Since 1981, population growth has slowed. From 1948 to 1981, the number of inhabitants of BiH increased by as much as 60.8% or an average of 1.8% per year and from 1981 to 1991, 6.1% or 0.6% on average per year, which is three times less. The absolute increase in population between 1981 and 1991 was only one-sixth of the demographic growth of the period 1948–1981. It is evident that the negative changes in development flows and socioeconomic dynamics have adversely affected the change in demographic development.

For almost 10 years before 1992, BiH was no longer as attractive in terms of migration or natality as it used to be. There was a slowdown in the dynamics of economic development, rise in the emancipation of the female population, reduction in the number of children born per woman in the fertile period, etc. Large companies and factories experienced a significant crisis in the production and marketing of products, which was maintained to reduce labor demand from the earlier period. However, this did not slow down or stop the relocation of the population on the rural-urban line within the country.

There has been a slow deterioration of natural population growth, decline of vital demographic structures, and the actual emergence of the process of spatial redeployment (rural exodus) caused by modern socioeconomic processes of industrialization, deagrarization, and urbanization. The deterioration of the status of agriculture and the lack of funds necessary for technical and technological improvement of agricultural production, especially on individual farms, whose holders became increasingly lonely and older, i.e., more powerless and incapable of successful, market-oriented agricultural activity, contributed to this. In such circumstances, poverty became an increasing feature of most rural households, so emigration to cities or abroad was imposed as almost the only way out, especially for younger generations.

Although BiH had a dynamic movement of the total population for European conditions until the beginning of the 1990s, the reduced biodynamics and the growing importance of emigration, especially since the mid-1980s, undoubtedly indicate a worsening of demographic trends. The last inter-census period brought the most drastic changes in the movement of the population of BiH, which had a significant impact on the enormous overall demographic growth from 1879 to 2013. Due to demographic losses and the consequences of war, the total population of BiH between 1991 and 2013 was reduced by almost a fifth (−19.3%). Spatial differentiation or polarization of BiH demographic growth is evident at the municipal level. In the last pre-war inter-census period, there was a decline in the overall population growth rate of BiH and a demographic depopulation of an increasing number of municipalities.

In this part, our goal was to show the extent of the process of overall depopulation as a consequence of inherited destabilizing factors of demographic development (two world wars, emigration, declining birth rates, demographic aging, etc.) and highly adverse effects of armed aggression on BiH, (depressed birth rate, refugees/emigrants) affected BiH. For this analysis to be as precise as possible, we had to adjust the data from the 1991 census to the new administrative-territorial division created by the Dayton Peace Agreement. Population reduction of more than 50% in the period 1991–2013, with the type of extinction, was recorded in 20 municipalities. Their common feature is accommodation in far less favorable geographical areas than other municipalities like the west, east, and southeast of BiH. The following stand out in particular: Bosansko Grahovo (−70.5%), Pelagićevo (−70.1%), Glamoč (−69.3%), Srebrenica (−67.8%), Vareš (−60.0%), and Derventa (−54.1%).

A total of 37 municipalities were characterized by a type of very strong population regression (30–49.99% decrease) and 28 municipalities by a type of strong regression (20–29.99% decrease), which is a total of 65 municipalities that lost from 20 to 49, 99% of its population. This is 45.5% of the total number of municipalities. The municipalities that had more than 30,000 inhabitants in 1991 and lost the largest share of their population in this group stand out, namely, Bratunac (−44.6%), Konjic (−43.2%), Jajce (−38.2%), Teslić (−37.2%), Goražde (−36.9%), Kozarska Dubica (−34.4%), Bosanska Krupa (−33.8%), Maglaj (−32.6%), Zavidovići (−31.3%), and Centar Sarajevo (−30.4%). In the municipality of Bugojno in 2013, compared to 1991, the total population decreased by 32.9% or 15,419 persons. The municipality of Bugojno is among those municipalities in BiH that, between 1991 and 2013, recorded one of the largest rates of decline in the total population (Gekić 2021). The capital of BiH, Sarajevo, with its 4 city municipalities, has lost slightly more than 80,000 of its inhabitants or 23% according to the 2013 census compared to 1991 observed in today's administrative boundaries.

Compared to 1991, the population increased in 21 municipalities. The most significant increase in population, the type of strong progression (20% and more increase), had six municipalities. By far, the most significant increase in population is in the territory of today's municipality of Istočno Novo Sarajevo, which, although located in the entity of Republika Srpska, primarily thanks to the proximity of the capital Sarajevo and its economic impact through accelerated housing construction increased by 156.1% and 6302 inhabitants. Other municipalities in this group are Ravno (an increase of 81.8%), Istočna Ilidža (an increase of 69.1%), Pale-RS (an increase of 41.2%), Istočni Drvar (an increase of 26.2%), and Čitluk (an increase of 20.3%) (Fig. 11.9) (BHAS 2016; RIS 1948–1991).

According to the last census from 2013, BiH had 3,531,159 inhabitants, of which 2,219,220 (62.8%) lived in the entity of the Federation of BiH (hereinafter: FBiH), followed by 1,228,423 (34.8%) in the entity of Republika Srpska (hereinafter: RS) and 83,516 (2.4%) in the Brčko District. In the mid-2020, BiH had 3,403,670 inhabitants, which is 127,489 fewer than in 2013 or 3.6%. There were 1,136,274 (33.4%) inhabitants in the entity of RS, 2,184,680 (64.2%) in the entity of FBiH, and 82,716 (2.4%) inhabitants in the Brčko District. It should be noted that, according to the estimates of the entity statistical institutes in BiH at the municipal level from 2013 to 2019, the trend of total depopulation in 127 municipalities continued, which is 88.8% of the total number (Fig. 11.10).

The municipalities that had the most significant decrease in the previous period stand out in particular and now have a type of moderate regression of population decrease (decrease of 10–19.99%). These are Oštra Luka (−18.1%), Novo Goražde (−14.6%), Drvar (−13.2%), Trnovo-FBiH (−12.4%), Kalinovik (−11.8%), Krupa na Uni (−11.8%), Vareš (−11.7%), Bosanski Petrovac (−11.7%), Kupres-RS (−11.6%), Bosansko Grahovo (−11.4%), Kneževo (−11.4%), Pelagićevo (−10.9%), and Istočni Drvar (−10.4%). Municipalities achieved a significant increase in population with types of weak, moderate, and strong progression (over 5% increase), namely, Petrovac (49.6%, from

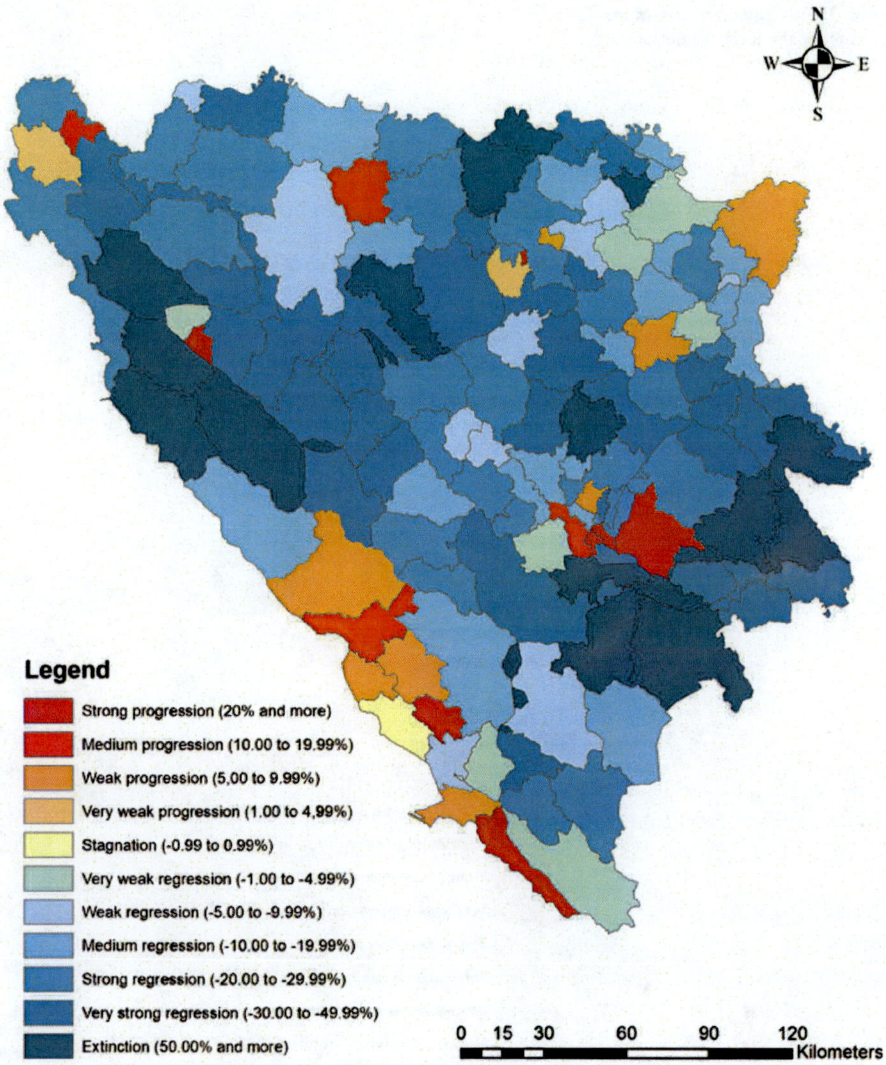

Fig. 11.9 The size of the change in total number of population from 1991 to 2013

357 to 534 through various economic incentives for immigrants to this rural municipality), while in the municipalities of Istočno Novo Sarajevo (13.8%), Vogošća (8.3%), Ilidža (5.7%) and Ilijaš (5.1%) the agglomeration influence of the capital of Sarajevo with new housing construction attracted immigrants (Fig. 11.10) (FZS 2021; ISR 2021).

11.2.2 Natural Movement of Population

During the Austro-Hungarian rule in BiH, vital statistics were not kept (only the church authorities, primarily Catholic, kept precise books of married, born, and deceased members of their religion). During the former Kingdom of Yugoslavia, books on vital events were kept. However, these data mainly refer to urban areas (rarely to rural ones). Even if they exist, they are very unreliable, so it is not possible to monitor in more detail the impact of natural and mechanical components on the overall population movement in BiH (according to Suljić 2011).

Our assessment (based on the structure of the population in later periods, the structure of the fertile female population, the population growth of the population of the wider region—county—etc., taking into account migration flows and prevailing living and development conditions, and total population growth) during the Austro-Hungarian rule, the territory of BiH had high birth rates ranging from 40.1‰ in 1879 to 38.2‰ in 1910, while mortality rates were also high and went from 29.1‰ in 1879 to 25.2‰ in 1910. This resulted in moderate natural population growth rates ranging from 11‰ to 13‰.

In 1921, we estimated that the natural population growth rate was moderate at 14.2‰, resulting from a high birth rate of 36.4‰ and a high mortality rate of 22.2‰. Birth rates and

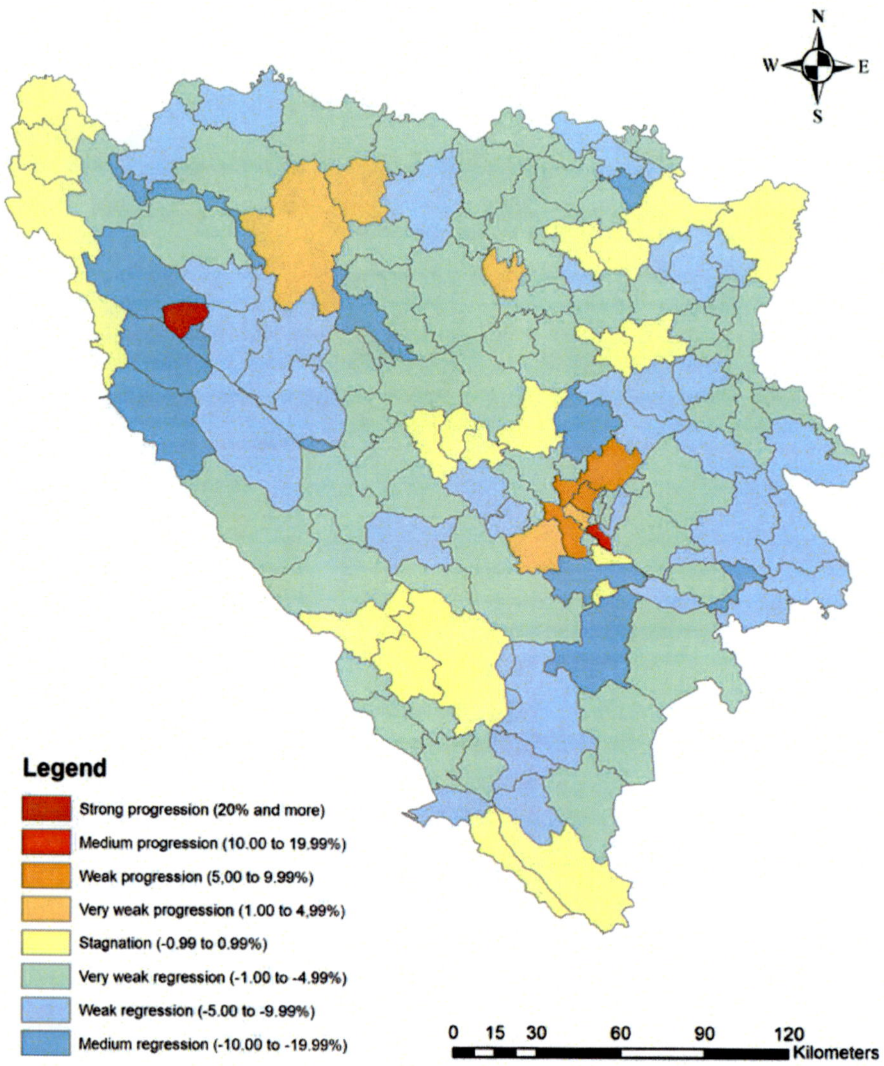

Fig. 11.10 Estimated size of the change in the total number of population from 2013 to 2019

population growth rates increased until the Second World War (from the 1930s, birth rates rose from 44‰ to 52‰ and population growth from 24 to 27, while the mortality rate remained more or less equal to 20–25‰).

The development of the population of BiH in the modern period (especially after 1971) is characterized by increasingly unfavorable natural dynamics as a consequence of specific social and economic development. The dynamics of decreasing natural population growth in BiH is going much faster than its economic development, making BiH already in the post-transition stage of population development in which today are highly developed countries. The former was affected not only by general socioeconomic processes (modernization, industrialization, urbanization, etc.) but also by specific factors, the most important of which are war losses and numerous emigrations abroad. This future transition is a clear reflection of the instability of economic development, as well as the political and military (non)opportunities brought about by the world wars, especially the Second World War and the last war (1992–1995). To this should be added the insufficient involvement of society/state in solving the accumulated demographic difficulties, especially when it comes to the population growth, i.e., the increase in birth rates and the number of children in the family.

After the Second World War, the birth rate declined in line with the transition of the active population to other sectors outside agriculture. Married couples, therefore, wanted fewer children; longer education process and increased female employment were important factors as well. In the year before the last war, in 1991, the number of live births reached 65,430, and it has been in decline ever since. For the period 1992–1995, there is no valid information on this trend. The number of live births in 1996 was 28.8% lower than in the pre-war year of 1991.

At the end of the 2000s, BiH, given the biodynamics of the population, entered a period of pronounced natural depopulation that deepened and lasted continuously for a decade, with no significant indications that it could be stopped or slowed down. Thus, based on the official data of vital statistics of BiH, only between 1981 and 2019, the birth rate in BiH decreased from 71,031 to 28,192 live births or by as much as 60.3%. At the same time, the number of deaths increased from 26,222 to 38,829 or 48%. The "surplus" of deaths over live births amounted to 10,637 in 2019, while the ratio was reversed in 1991 and amounted to 34,019 in favor of live births (which is a negative reversal by 130%), with a potential further tendency to increase "surplus" deaths (BHAS 2019; RIS 1983, 1994).

Between 1996 and 2019, the number of live births decreased by as much as 39.5% or 18,402. The lowest number of live births was 28,192 in 2019 (BHAS 2019). This represents a critical turning point and a worrying phenomenon for the population of BiH. The total fertility rate is 1.23 children born per woman. It is one of the lowest in the world, immediately after Singapore and South Korea (Central Intelligence Agency 2021).

Mortality rates only began to decrease below 10‰ after the 1950s due to improvements in health care and living conditions. It was the lowest in 1974, at just 6‰, and then slowly grew. After 1995, due to increasing population aging, the mortality rate increased to 11.4‰ in 2019. The death toll today is almost equal to that in the 1950s. At the same time, life expectancy at birth has increased: men born in 1961 lived to be 59 years old and women 63, whereas men born in 2019 have a life expectancy of 76 years and women of 80 years (RIS 1948–1991; BHAS 2021). The average age of the female population at the time of death in 2018 was 77 years and of the male population 71. In BiH, just over 22% of deaths occur to people younger than 65 years old. Death before the age of 65 is much more common among male residents (as much as 28% of total male deaths), while it is significantly less in the female population (16%). On average, 90% of the population die from circulatory system diseases, neoplasms, and endocrine and metabolic disorders (BHAS 2019, 2021).

Longer life expectancies at birth, declining birth rates, and an aging population have led to extreme changes in the rate of natural change in the population. This was all exacerbated by the great wave of emigration and casualties during the 1992–1995 war. The total number of casualties in the war in BiH is estimated by the UN-verified ICTY (International Criminal Tribunal for the former Yugoslavia) experts to be 104,732 (2.4% of the total population in 1991), of which 40% are civilians, by gender, 90% are males and by ethnicity, 65% are Bosniaks. Ethnic cleansing processes and more than 2 million people displaced resulted in a major shift in the development of the population of BiH (Zwierzchowski and Tabeau 2010; Tokača 2012).

In BiH, natural population change has been declining since the mid-1950s, when it moved from high rates (above 25‰) to moderate ones. This rate continued to decline and in the early 1970s reached a low level. In the mid-1980s, natural change rate dropped to below 10‰ as a result of a large decline in the birth rate. After 1995, the rate of natural change continued to decrease from 5.9‰ in 1996 to the first negative rate of −0.3‰ in 2007, after which it continued to decline. In 2019, the rate of natural change in the population of BiH was negative (−3.1‰) (Fig. 11.11). The average maternal age at birth of the first child has increased from 22 years in 1991 to almost 28 years in 2019. Marriage rates have been declining gradually since 1948, when it was 13‰; then in 1991, when it was 6.6‰; and to 5.8‰ in 2018 (BHAS 1948–1991,

Fig. 11.11 Natural change rates in BiH from 1948 to 2019

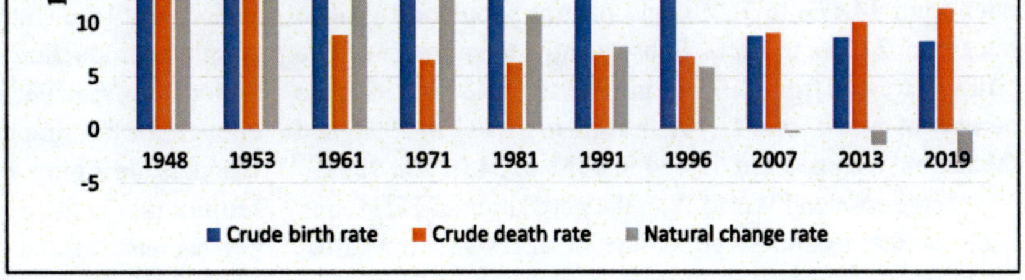

2019). Compared to 1948, the number of marriages decreased by 41%, while the number of children born out of wedlock in 2019 is still very low compared to the countries of the European Union. In 2017, it was around 11% and in 2019 around 6% (BHAS 2019).

On average, the declining reproduction level of the BiH population is a direct consequence of numerous destabilizing determinants of population development, with particular emphasis on the problem of emigration and suffering of the population during the 1992–1995 war, and related disturbed old age structure and polarized population development of towns and villages. The natural decrease of the population affected almost all parts of BiH. In the last 10 years, the problem of population emigration has again become a significant factor in the decline of the population of BiH.

Urban settlements took over the leading role in demo-reproduction from rural ones, which are in a very difficult situation, although there is a strong tendency of declining birth rates and population growth in urban settlements as well. In many parts of BiH today, there is a problem of simple reproduction of the population (for each family to have at least two children). Reducing birth rates and population growth has short-term and long-term effects. In other words, weak natural dynamics do not reflect negatively only on the current total population movement but also in very unfavorable structural-dynamic consequences in 20–25 years when now-born generations (especially women) enter the fertile (reproductively most fertile) period of their life. For almost 15 years, BiH has been characterized by a depopulation process (the "single" or "white plague" system) in the reproduction of the population.

The long-term demo-reproduction trend in the modern period has resulted in an accelerated weakening of natural dynamics in BiH in the 1980s, while, by the beginning of 2007, it was on the verge of a population growth decline. Obviously, the consequences of weakening the population's age structure during the war came to the fore (1992–1995).

Although BiH had a positive population growth movement until 1991, characterized by a very high to moderate natural population growth, the trend of its deterioration is still apparent. The total number of inhabitants increased by 53.7% from 1953 to 1991, the number of live births decreased by 40.7%, and the number of deaths by 23.7%, resulting in the total population growth decrease by 50.8%. Accordingly, birth rates were reduced from 38.5‰ to 14.9‰, mortality rates from 14.4‰ to 7.2‰, and natural population growth rates from 24.2‰ to 7.8‰. Subsequently, the number of live births decreased from 110,373 in 1953 to 65,430 in 1991, the number of deaths from 41,199 in 1953 to 31,411 in 1991, and the natural change from 69,174 in 1953 to 34,019 in 1991.

After the Second World War, the population of BiH found itself in the second stage of the demographic transition, which was characterized by high birth rates and moderate mortality rates, which gave a very high population growth. This stage represents a period in which, under the influence of strong socioeconomic and health progress, there is a transition in the field of mortality and birth rate, i.e., their transition from high to low level. Very high rates of population growth (above 20.0‰), i.e., "demographic expansion" (early transitional substage), lasted until 1966. The reduction in mortality after a certain period has led to a decrease in fertility rates. Thus, both components of population growth found themselves in a transition to a lower level; mortality, however, had a steeper decline, so population growth rates were still high (central transition substage). This substage lasted until 1978.

In the final phase of the transition (late substage), the decline in mortality stops, and the birth rate continues to decrease. This also resulted in relatively low rates of population growth (below 10‰), with a tendency to further decrease. Although we do not have data for the war years (1992–1995), it is known that the difference between the number of live births and deaths is highly negative. In 1996, the post-transition stage in the reproduction of the population began in BiH because, that year, the birth rate was lower than the limit of 14‰ for the first time (it was 13.2‰). The birth rate continued to decrease, so in 2019, it was only 8.7‰.

Slowed demographic dynamics have reflected on the disturbances in the age structure of the population, one of the fundamental determinants of natural dynamics. By reducing the number and share of young people (potentially biodynamically vital), the fertile contingent of the population has narrowed down.

Official data from BiH's vital statistics show that 28,192 children were born in 2019, representing a decrease of 74.5% compared to the number of live births in 1953 and a decrease of 57% compared to the number of live births in 1991. Between 2010 and 2019, the number of live births decreased by 16%. The birth rate in 2019 of 8.3‰ is almost twice lower than in 1986.

After the Second World War, there was a steady decline in the total fertility rate (TFR) in BiH over the last almost seven decades (from 1953 to 2019). In 1953, the population of BiH was still in the first phase of demographic transition, with a total fertility rate of 4.8 children per woman in the fertile period (15–49 years). In 1961, it was 3.8, and in 1971. it was 2.8. In 1981, the total fertility rate decreased to 2.0 and, already in 1991, to 1.9 live births per woman in the fertile period. Until the end of the 1980s, this rate was sufficient for simple reproduction (two children per woman). After this period, the rate decreased so much that it was no longer enough for the simple reproduction of the population (population is declining in numbers by birth alone). The total fertility rate in 2013 was only 1.3, and in 2019, 1.2 children per woman in the fertile period. For the entire period 1953–2019, the total fertility rate decreased by 75%, which is undoubtedly one of the largest reductions in the world (BHAS 2021).

Since the early 1990s, there has been a significant reversal in the circumstances of the reproductive behavior of the BiH population—from more or less stable socioeconomic conditions to increasingly impoverished and politically unstable conditions (especially in the period 1992–1995). The decline in overall fertility rates in the twenty-first century results from several specific factors. Among them are the capacities of the state in unstable national development and opportunities for employment and advancement in education. There is no clear and detailed state-level strategy to improve this rate or general population movement.

The current demographic situation and fertility in BiH after 1995 has some similar characteristics as those mentioned in the theory of the second demographic transition of highly developed European countries (Lesthaeghe and Neels 2002; Lesthaeghe and Surkyn 2002, 2004; van de Kaa 1994, 2001, 2002; van Bavel 2007). This demographic transition, in which the birth rate falls below replacement levels, refers to the constant changes in family and fertility patterns associated with structural and cultural changes marked by modernization, expansion of higher education, increasing secularization, growing individualistic values, and the importance of self-fulfillment.

In recent years, the overall decline in the general fertility rate has been affected in particular by constant changes in the family pattern and various social and economic changes. The gradual decline of number of marriages under the age of 30 continues. The increased age of marriage, mothers' age at first birth, and the reduction of the higher-order birth rate have undoubtedly influenced and contributed to lower fertility rates.

Post-war economic hardships added to the strain of family life; young couples find it very difficult to find affordable housing; unemployment rates are very high; there is a high dependence on both part-time and low-wage jobs; there is no guarantee of continued employment for women who leave the workforce to have children. Also, the stronger decline in fertility in the last decade is much more noticeable than in the years immediately after the end of the war and partly reflects the absence of any systemic population policy.

The population growth of BiH is characterized by significant spatial differences in the characteristics of the changes that have occurred, which is especially visible in the period 2014–2019. There are particularly significant differences between eastern and western Bosnia, on the one hand, and other areas of BiH on the other.

At the entity level in 2019, the FBiH had 18,019 live births (rate 8.2‰), 22,024 deaths (rate 10.1‰), and a natural change of −4005 (rate −1.9‰). Entity of RS had 9274 live births (rate 8.1‰), 15,081 deaths (rate 13.2‰), and population growth −5807 (rate −5.1‰), while Brčko District had 899 live births (rate 10.9‰), 1132 deaths (rate 13.7‰), and population growth 233 (rate −2.8‰) (BHAS 2019).

The overall improvement in health and hygiene and economic conditions in BiH in the late 1960s led to a significant reduction in overall mortality and infant mortality. The general mortality rate decreased to 6.1‰ in 1979 when it started to rise again. In 1991, it was 7.2‰ and in 2001, 8‰. In 2011, the overall mortality rate increased to 9.1‰, and in 2019, it was 11.4‰. Spatial differentiation in terms of the birth rate in BiH is dominated by municipalities with low rates (7.1–10.0‰), namely, 64 or 44.8%. This group of municipalities is very heterogeneous and is represented by municipalities from almost all parts of the country except for the far west and east, where the population is predominantly older (e.g., Istočni Stari Grad and Vlasenica in eastern Bosnia). This group of municipalities fits within the BiH average of 8.3‰ (Fig. 11.12).

The second most widespread group of municipalities are those with a very low birth rate (4.1–7.0‰) and account for 26.6% of the total number of municipalities. These are municipalities with a higher proportion of the elderly (more than 20% aged 60 years or more); since the 1970s, these have mainly been municipalities from which the population emigrated for economic reasons (e.g., Vukosavlje and Kupres FBiH). The third-largest group has an average birth rate (10.1–15.0‰) and is made up of 21 municipalities or 14.7% of the total number of municipalities. This is a group of municipalities with propulsive development (e.g., Ilijaš 13.6‰ and Trebinje 10.1‰). The fourth-largest group are municipalities with extremely low birth rates (below 4‰), which make up 14% of the total number. These are municipalities that are generally small in population and large in area (FZS 2021; ISR 2021); they have a high emigration rate and a high proportion of the elderly (e.g., Lopare 4.0‰ and Ravno 0.2‰). There are no municipalities without live births.

The overall improvement in health and hygiene and economic conditions in BiH in the late 1960s led to a significant reduction in overall mortality and infant mortality. The general mortality rate decreased to 6.1‰ in 1979 when it started to rise again. In 1991, it was 7.2‰ and in 2001, 8‰. In 2011, the overall mortality rate increased to 9.1‰, and in 2019, it was 11.4‰.

An indicator of the general movement of the mortality rate and an indicator of the general living conditions, i.e., society's standard of living, is the mortality of infants (children up to 1 year of age). Extremely high infant mortality rates in BiH (over 100 infant deaths per 1000 live births or 100‰) were present until the early 1960s. In 1953, 15,796 infants died (14.3% of the total number of live births), and already in 1961, this share decreased to 10.0%. The highest mortality of the population was in the age group up to 1 year of age (38.3% of the total number of deaths). This is a

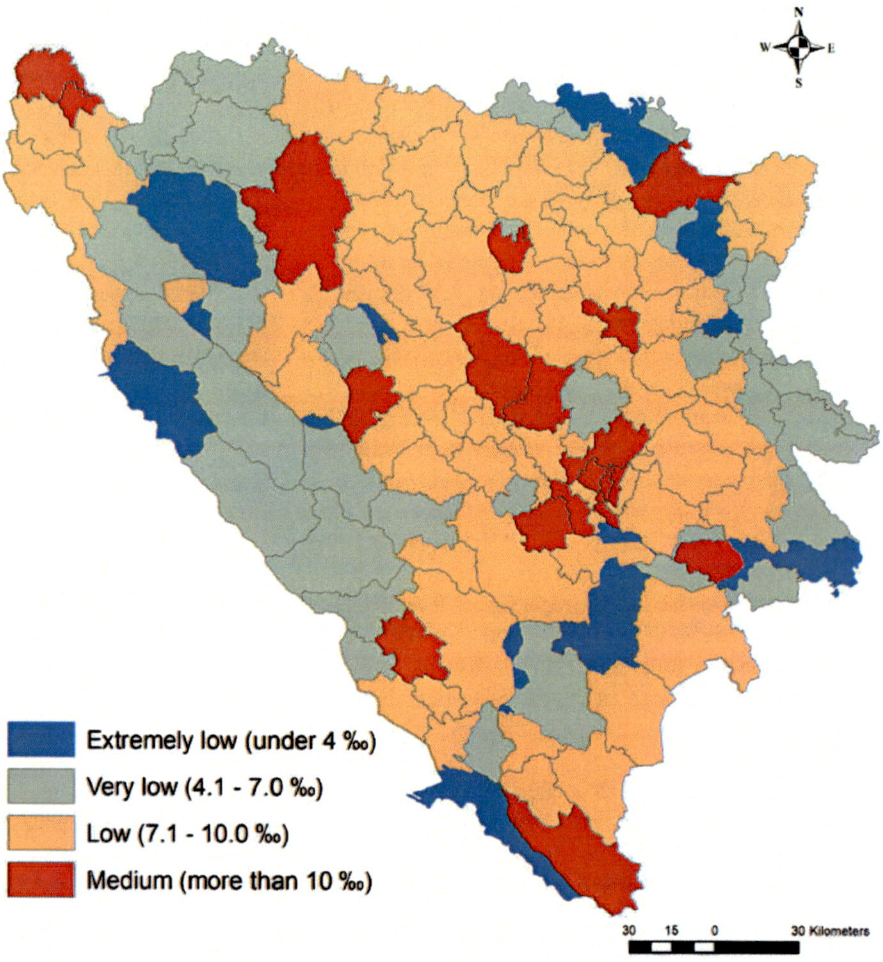

Fig. 11.12 Birth rates in Bosnia and Herzegovina from 2014 to 2019

characteristic of countries with poor health and hygiene protection and conditions that result from poor socioeconomic conditions. Since 1961, infant mortality rates in BiH have been drastically reduced. Thus, already in 1991, this rate was 14.6‰, and in the total number of deaths, infant deaths accounted for 3.4%. In 2019, the infant mortality rate was 6.7‰, and in the total number of deaths, infant deaths accounted for only 0.5%.

Due to the advances in medical science, infant mortality, mainly caused by infectious diseases, has been declining and reaching very low levels, so their effect on life expectancy has been minimal in recent decades. However, other causes have emerged, more or less related to society's economic and social development in the twentieth century: alcohol, smoking, traffic accidents, violence, etc. Apart from these "diseases of modern society," efficiency in combating the other two has not yet been achieved—neoplasms (tumors and cancers) and cardiovascular diseases—which largely determine the duration of human life.

Thanks to the discovery and use of antibiotics, the efficiency of health services in general, maternal and child care before and after birth, and the growing population's education, there has been a sharp decline in infant mortality since the 1960s, and therefore, a decreasing influence on the duration of life. Thus, mortality in the first year of life has become a less critical component of the total mortality of the population in the last 40 years, and it has been replaced by other diseases that cause mortality in other age groups. The population of BiH has a typical death structure according to the causes of death that characterize the phase of advanced mortality transition. The most important causes of mortality in BiH in 2019 are diseases of the circulatory (cardiovascular) system (49%), neoplasms (tumor and cancer) (21%), endocrine and metabolic diseases with eating disorders (5%), diseases of the respiratory system (5%), injuries and poisoning (3%), etc. (BHAS 2019).

The spatial differentiation of mortality rates in BiH is much more uniform than the differentiation according to the birth rate. Two groups, with low and medium mortality rates, account for 84% of all municipalities. Both of these groups of municipalities are very heterogeneous, the difference being that the group of municipalities with an average mortality rate (10.1–15.0‰) is dominated by those in which the older population amounts to over 20% (Fig. 11.13).

The third largest group consists of municipalities with high mortality rates (15.1–20.0‰), which make up 10.5% of the total number of municipalities. These are mostly emigrant municipalities with a high proportion of the elderly (average 25–35%). Six municipalities in BiH form a group with a very high mortality rate, 20.1–30.0‰ (FZS 2021; ISR 2021). Five of them emerged as a result of the Dayton Peace Agreement, by separating some rural areas from the municipalities which were characterized by a higher proportion of the elderly than the young (Oštra Luka, Petrovac, Istočni Stari Grad, Krupa na Uni, and Trnovo FBiH). The extreme group with the lowest mortality rate (up to 5.0‰) includes only two municipalities: Ravno with 3190 inhabitants and Istočni Mostar with 243 inhabitants (Gekić et al. 2020).

The spatial differentiation of the rate of natural change shows the dominance of municipalities with a negative rate of natural change. Of the total number of municipalities, 83.2% have a negative rate of natural change. The largest group, with a rate of −5.0‰ to −0.1‰, is characterized by a very low natural decrease in population (75 municipalities or 52.4%). These numbers are sufficient evidence of the prevalence of natural depopulation in BiH (Fig. 11.14).

While these are very heterogeneous municipalities in terms of size, economic, and urban development, they have a similar percentage of the old population, which is above 20% on average. Municipalities with a low natural decrease make up 19%. These are extremely old municipalities regarding age structure and are located mainly at the outer edge of the country. They are characterized by emigration, a higher

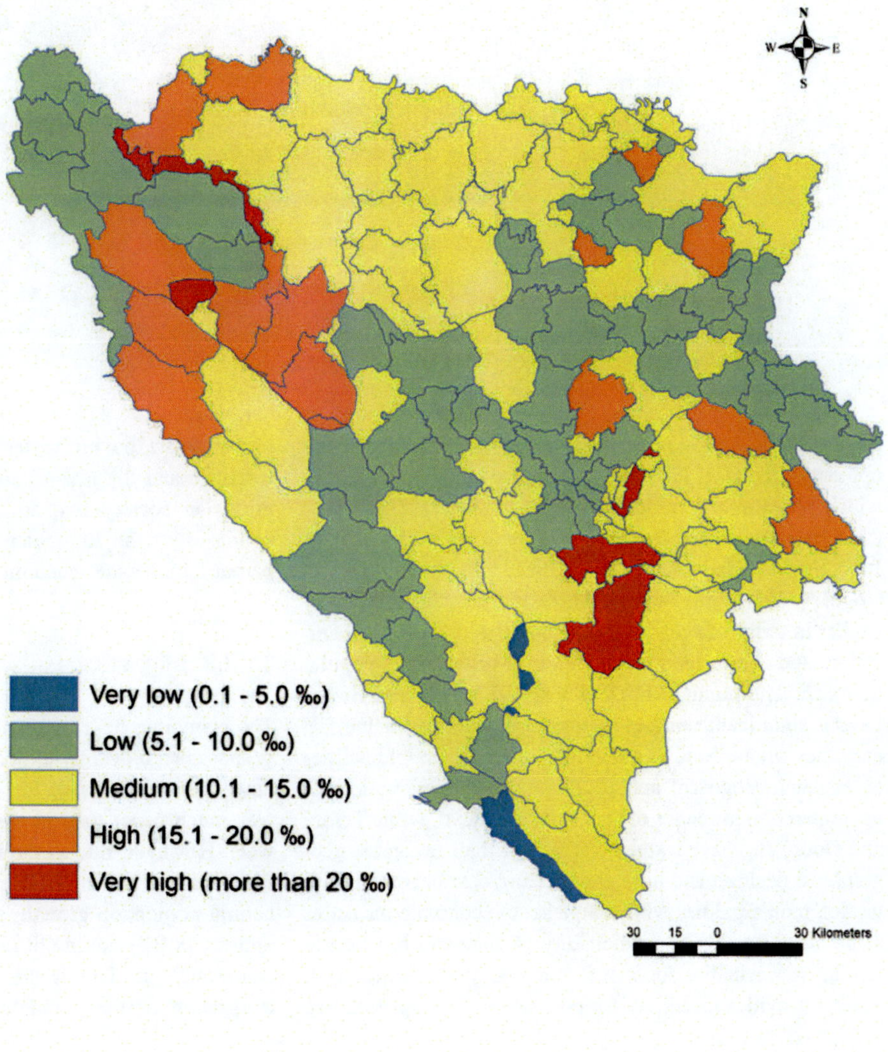

Fig. 11.13 Death rates in Bosnia and Herzegovina from 2014 to 2019

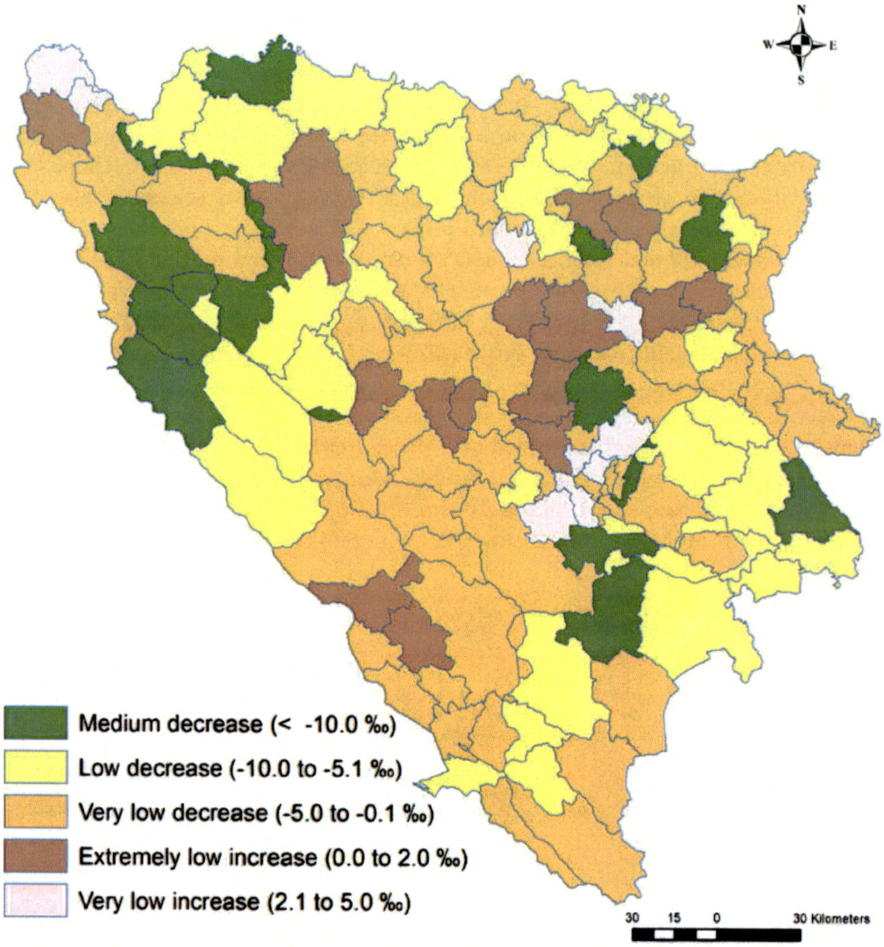

Fig. 11.14 Rates of natural change of population in Bosnia and Herzegovina from 2014 to 2019

proportion of the old population, and poor economic development (FZS 2021; ISR 2021). There are nine municipalities with a very low rate of natural population increase, in which the birth rates exceed the death rate by at least 2‰ (Doboj-Jug, Velika Kladuša, Hadžići, Ilidža, Banovići, Tešanj, Vogošća, Ilijaš, and Bužim). These municipalities are very similar in terms of economic development, degree of urbanization, and age structure, although they differ in the population size. The municipalities of Velika Kladuša and Bužim have a significant number of residents working in the EU countries whose families are left at home. The municipalities of Hadžići, Vogošća, and Ilijaš are developing under the agglomeration influence of the capital Sarajevo, while Tešanj and Doboj-Jug have taken advantage of their favorable geographical position and have even managed to become immigration municipalities with a large set of economic measures at the local level. The municipality of Banovići has been a mining municipality for more than 100 years, where coal mining provides incentives for population development. We can associate low birth rates with the decline and low rates of marriage and the lack of conditions and social and cultural norms in society that are characteristic of the developed countries of the European Union, where forms of single-parent families are much more common.

11.2.3 Migrations

The economic development of BiH after the Second World War was significantly slower than the average of the other, developed republics of the former Yugoslavia. However, it was much faster than in the period between the two world wars. However, due to the inherited low level of economic development and high level of unemployment, with high natural population growth, such a trend of economic development in BiH could not provide employment for all those who sought it. This is the leading cause of very intensive emigration of the population from BiH after the Second

World War until 1992. BiH was a strong emigration area with a negative migration balance of about 693,000 or an average of 16,116 inhabitants per year from 1948 to 1991. Suppose we add to this the number of emigrants by agrarian colonization and emigration until the first census after the Second World War. In that case, we can estimate that in the entire period from 1948 to 1991, about 780,000 more inhabitants emigrated (then immigrated).

From 1945 to 1948, 78,576 people from BiH (Gaćeša 1984) moved to Vojvodina (Republic of Serbia) as part of the planned agrarian colonization, and mainly from the area of western, southwestern, and southeastern BiH. In the period 1953–1961, 74,230 inhabitants immigrated from the territory of Serbia and Croatia to BiH. In the period 1961–1991, the share of the immigrant population in the total increased from 27.7% to 36.7%. According to the 1991 census, immigrants from other settlements in the same municipality (45.3%) and other municipalities in BiH (42.9%) dominated, while the share of immigrants from other countries (10.2%) was small, mainly from Serbia and Croatia (120,397 or a total of 7.5% of all immigrants). About 1.6% of immigrants were from an unknown place of origin. About 75% of the immigrant population was under the age of 30 at the time of the migration.

The main course of migration was in the rural-urban direction. The population left the rural areas and agriculture and moved to urban and suburban areas with employment in industry and other activities. Thus, in 1971, the immigrant population was dominated by immigrants from rural settlements (65.3%).

In 1971, BiH had 137,351 inhabitants or 3.7% of the total population working temporarily abroad, and in 1981, it was 182,940 or 4.4% of the total population. According to the 1991 census, there were 234,213 citizens of BiH temporarily working abroad with their family members, namely, 5.4% of the total population, of which the highest percentage, 48.4, was in Germany, 21.5% in Austria, and 16.4% in Switzerland. In terms of ethnicity, 39.1% were Croats and 25.5% were Serbs. Thus, 12.1% of the total number of Croats before 1992 were abroad.

The overall net migration balance shows that since the Second World War, BiH has been a country with high emigration rates. In the period 1948–1971, approximately 400,000 more people emigrated from BiH than immigrated, mainly to Germany (unqualified workers), Serbia, Croatia, and Slovenia. In the period 1971–1991, the total net migration balance was about −300,000 and in the period 1991–2018, even −1.2 million (RIS 1948–1991; BHAS 2019).

Migrations resulting from the war, as well as the continuing economic ones, have been intensifying changes in the population, in particular in its structure (the most productive young population leaving). Extremely negative changes occurred in all spheres of life in BiH during the period 1992–1995, which in the following 25 years led to a population implosion that rarely occurs in the world. Population emigration has led to a decrease in natural population change, to an aging population, and to a decrease in the natural reproductive potential of the population. This trend continues, without exact and accurate data in the statistical agencies and national and regional governments (Gekić et al. 2020). Unlike in the 1960s and 1970s, when emigrants were mostly unqualified workforce, emigrants today are far more qualified and educated. According to the results of the 2013 census, 9.6% of the population over the age of 15 had completed a higher level of education, which is significantly higher than in 1991, when this share was only 3.7% or in 1981, only 2.2% (RIS 1948–1991; BHAS 2016).

As for sustainable development, the question arises whether future generations will be able to withstand the burdens imposed by the aging of the population in the field of health care, in the provision of funds for pensions, and in the field of social protection (Reynaud and Miccoli 2018). The negative implications of hidden geographies of population implosion are visible in all pores of society, ranging from the pension system, health care, and education systems to the labor market (reduction and aging of the working contingent calling into question the country's faster economic development).

The intensity of migration was constantly increasing so that in a large part of BiH, the negative migration balance prevailed over the population growth. Depopulation areas have thus expanded. In 1953, there were only five depopulation municipalities (Bosanska Dubica, Čitluk, Grude, Ljubinje, and Stolac). Already in 1961, there were 11; in 1971, 21; in 1981, as many as 35; and 34 in 1991. In 2013, the number of depopulation municipalities was 121. Deagrarization as one of the fundamental factors of the process of overall socioeconomic development is the primary factor of mass social and spatial mobility of the population of BiH in the period after the Second World War (Sparavalo 1990) until 1992.

Three migration phases of Bosnian population have stood out in the second half of the twentieth and the beginning of the twenty-first century: a labor force emigration from the former Yugoslavia caused by economic factors, a forced migration in the 1992–1995 period caused by political circumstances, and a post-war migration motivated by political and economic factors.

A weak economic development of BiH in the former Yugoslavia and its economy based on mining and agriculture forced Bosnian population to emigrate to developed countries of the Western and the Central Europe. During the former Yugoslavia, emigration was especially intensified in the 1960s due to a recession of Yugoslavian economy. In the following decades, i.e., before 1991, about 234,213 or 5.4%

of Bosnian citizens emigrated, looking for a temporary employment in abroad (Kadušić and Suljić 2018).

According to the 2013 census, there were 2,158,659 immigrants in BiH (mostly internal migrants), or 61.1% of the total population. The main reason is the war events of 1992–1995 and related migration movements. Immigrants from other municipalities of BiH (34.4%) and other settlements of the same municipality (19.3%) dominated. There is a large share of the population with the unknown municipality of settlement (18.3%) and unknown country of settlement (9.2%). About 9% of immigrants were from the countries of the former Yugoslavia (mainly Croatia and Serbia). In the entity of RS, the share of the immigrant population is as high as 65.4%, while in the entity of FBiH, it is 58.5%, and in the Brčko District, 67.1%.

In the period 1953–1971, larger urban centers experienced rapid growth. The current dynamics of internal migration are significantly lower than in EU countries. In the period 2014–2019, on average per year, about 1% of the total population was mobile in relocations within state borders. The population mostly moved to Sarajevo, Banja Luka, Ilidža, Mostar, Tuzla, Zenica, Vogošća, Bijeljina, Bihać, Brčko, Živinice, Doboj, and Istočno Novo Sarajevo.

Most refugees from BiH during 1992–1995 emigrated to the following countries: Germany 320,000, FR Yugoslavia 277,000 (Federal Republic Yugoslavia comprised the Republic of Serbia and the Republic of Montenegro and existed from 1992 to 2003), Croatia 170,000, Austria 86,500, Sweden 58,000, Slovenia 43,500, Switzerland 24,500, and Turkey 23,500 (Fig. 11.15). Although there were 320,000 refugees in Germany (27.3% of total refugees from BiH), only 14% of that number remained until 2000.

About 1.2 million people were forced to leave the country (Valenta and Ramet 2011), while 1 million citizens were displaced within the country, which, in total, is 2.2 million refugees and displaced people (Ibreljic et al. 2006). Most of the migrants chose neighboring states for their host countries because of geographical proximity (Vrecer 2010) and because of their ethnic background. About 40% of Bosnian citizens immigrated to Serbia, Montenegro, Croatia, and Slovenia between 1992 and 1995. These countries, along with Germany and Austria, hosted 75% of Bosnian refugees. A significant number migrated to other Western European Countries and all over the world—approximately 25% of refugees.

According to 2013 Population Census in BiH, 451,447 citizens returned home from foreign countries. As stated by the United Nations Development Programme (2002), at least 92,000 young people left Bosnia between January 1996 and March 2001. In the first several years of the twenty-first century, there was not any indication that the number of emigrants from BiH was in decline. This process was mostly conditioned by the economic motives and social, political, and other circumstances. Emigrants are mostly highly educated and qualified and are usually between the age of 20 and 40 (Kadušić and Suljić 2018).

BiH has had a negative migration balance since 1996. Only in 2010, it was recorded that 4284 persons left BiH, while only 309 foreigners applied for a permanent residence in this country. Three thousand four hundred ninety-two persons left Bosnia in 2013, 4323 in 2014, 3948 in 2015, and 4019 in 2016. About 36.6% of people who left BiH were between the age of 30 and 39 and 54.1% between the age of 15 and 44. Most of the citizens who left BiH in 2016 went to the following countries: Germany (29.6%), Austria (22.2%), Croatia (21.9%), Serbia (12.1%), and Slovenia (10.5%). Persons who apply for a permanent residence in Bosnia are mostly from China, Croatia, Montenegro, Turkey, North Macedonia, the Russian Federation, Slovenia, Austria, and Syria (Ministry of Security of BiH 2020). The main reasons why Bosnians emigrate are closely related to employment (36%), studying (19%), etc. Emigration outside the state

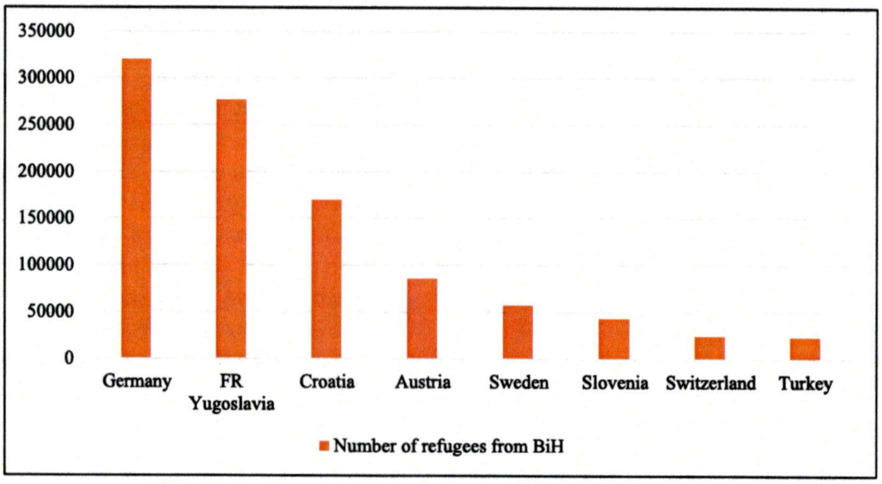

Fig. 11.15 The number of refugees from BiH by leading hosting countries from 1992 to 1995

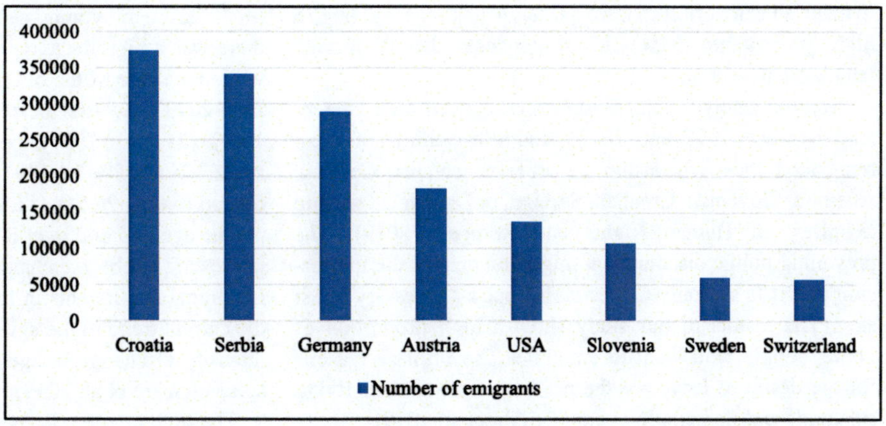

Fig. 11.16 Leading countries by the number of immigrants born in Bosnia and Herzegovina in 2019

borders is a crucial element of the development of the BiH population. Given this, Bosnians are among the European nations with the largest diaspora in the world. The most frequent emigration countries in 2019 were Germany, Austria, Croatia, Slovenia, and Serbia. According to the statistics on the total number of emigrants born in BiH, regardless of their current citizenship, there are 1,807,602 Bosnian emigrants in 52 countries, of which about 60% of them live in 30 European countries (Switzerland, Norway, and the EU-28) (Fig. 11.16).

However, the estimated total number of people of Bosnian origin in the world is between 2 and 2.2 million. This number includes persons born in BiH who have left their country and the estimated number of their descendants born in the receiving countries, regardless of their citizenship. It does not include the so-called old emigration (e.g., emigrants to Turkey before the 1970s).

There has been a continuous trend of increasing remittances from abroad in BiH since 2010, when they amounted to about 1.6 billion EUR. In 2019, remittances from abroad for BiH amounted to about EUR 2 billion, which was 11.4% of total GDP. This amount of remittances in GDP puts BiH among the five countries in Europe. Most remittances came from Germany (21%), Austria (19.7%), the USA (14.5%), Slovenia (10.8%), and Switzerland (8.1%) (Ministry of Security of BiH 2020). Most of the emigrants who send money and goods to BiH reported that they send them to close family members and relatives.

According to estimates commissioned in 2008 by the National Security Council of Turkey, as many as 2,000,000 Turkish citizens are of Bosniak ancestry, but they are not connected with the home country as emigrants from other countries (Germany, Austria, etc.). Migrations of Bosniaks to Turkey are mainly related to the periods before 1970. There are large migrant communities of Bosnians and Herzegovinians living in almost every European country as well as throughout North America and Australia. In many countries they resettled in, they have been regarded as role-model immigrants. Most members of these expatriate communities have retained strong family and other informal social ties both with fellow compatriots in other countries and with those living in BiH. Probably the most cohesive factor in displaced groups of Bosnians and Herzegovinians is their "communal consciousness" and "solidarity" that foster a sense of collective identity and shared responsibility to maintain support in the new countries for their original communities. According to these figures, 43% of all highly skilled individuals from BiH in 2010 resided abroad. The emigration rates from BiH, controlling for the age of entry (including emigrants who moved after a certain age), are not different, suggesting that the majority of emigrants from BiH move after they acquired their education in BiH. Education is also one of the drivers of emigration of youth (Halilovich et al. 2018).

BiH, Croatia, North Macedonia, and Serbia are among the top-ranked countries with the biggest brain drain in the world, according to the Global Competitiveness Report 2019 released by the World Economic Forum. On the scale of 1–7 (1 = all talented people leaving the country; 7 = all talented people staying in the country), BiH scored 1.76, followed by Croatia (1.88), North Macedonia (2.13), and Serbia (2.31).

The fact that emigration implies strong brain drain raises the issue of prioritizing investments within the education sphere (e.g., early childhood development versus higher education), and policies to attract returnees and/or exploit the advantages of the highly educated diaspora come into consideration (Zeneli et al. 2013). A study on youth (Gekić et al. 2020) reported that proportion of young people aged 15–29 who want to leave the country permanently is 46%. The main reasons cited by young people for wanting to leave the country are quality of life (38%), corruption (34%), and the lack of perspective (23%). It is concerning that 46% of young people want to leave the country forever, while 23% plan to stay abroad for more than 20 years. The desire to return has been expressed by 54% of young people. However, they plan to spend the most productive years elsewhere, which has wider social implications. In terms of taking steps towards leaving the country, most have not yet done anything to put their plan into action (57%). Citizens of BiH mostly want to emigrate to Germany (55%), Austria (17%), Switzerland (11%), and Sweden (11%). Most of them are very motivated to make an extra effort to learn a foreign language, although a high percentage (37%) of respondents learnt foreign languages in school.

Approximately 87,500 Bosnians renounced their citizenship between 1998 and 2018. About two-thirds of them renounced their citizenship in order to acquire German, Austrian, Slovenian, Croatian, Serbian, or Danish citizenship (Ministry of Human Rights and Refugees 2019). The prevailing public discourse on migration creates the impression that it is not only a "desirable" but a "necessary" next step. The results of our study show a high proportion of young people who want to emigrate. The primary reasons for the desire to leave are the need to improve their living standards and find employment (Turčilo et al. 2019).

11.3 Population Structure

11.3.1 Age and Gender Structure

Changes in the age structure of the population of BiH also include two comparative global demographic processes: reducing the share of young people while increasing the share of the elderly. In the early 1990s, according to the 1991 census, there were 11.3% of the population aged 60 and over (persons of unknown age were excluded from the total).

According to the data of the census of BiH in 1961, the share of the population aged 60 and over was 6.0%, and the share of young people (0–19 years old) was 46.6%. In the period 1961–1991, apart from the increase in the share of old people in the total population by 5.3%, there was a decrease in the share of young people (0–19 years old) by 14.9%, which is a relevant indicator of demographic aging. Aging has thus become an integral part of the process of demographic transition in its central and late substages, and especially in the post-transition phase. Due to the war in BiH (1992–1995), the demographic aging process has accelerated in addition to other adverse demographic effects (RIS 1948–1991).

The population of BiH is characterized by rapid aging and a high degree of aging. The 2013 population pyramid shows all the characteristics of a regressive population type—a low proportion of young people aged 0–14—so that the base of the pyramid is narrower than the middle part, which indicates a low or negative natural change of population and indicates a depopulation process caused by natural movement.

From 1961 to 2013, there was a clear decrease in the number of young people (up to 34 years old, with a particularly strong decrease in the youngest groups) and, on the other hand, a very high increase in the number of old residents, especially the oldest age groups. In the last intercensus period (1991–2013), there was a significant decrease in the share of young people (especially their number, as there was a high decrease in the total population of 19.3%), while the share of the old population increased. Census data from 2013 show that the share of young people aged 0–14 is 15.4% (24.0% in 1991), the share of young people aged 0–19 is 22.3% (32.5% in 1991), the share of old people aged 60 and over is 20.5% (11.3% in 1991), and the share of old people aged 65 and over is 14.2% (6.6% in 1991). It is very interesting to observe changes in the proportion of individual functional contingents in the total population. They point to the current trend in the aging of the population as well as the intensity of change in age contingents (the so-called partial aging) (Gekić et al. 2019).

The high increase in the proportion of the elderly population, especially the older part of that contingent, indicates that BiH is rapidly approaching the situation in the most developed European countries, where the proportion of the elderly age group of the elderly contingent is growing much faster in the total elderly population. In BiH, this is a consequence of especially the war, and in recent times the economic emigration of the population aged 20–39 associated with the inability to find employment in their country. The average age of the population of BiH is steadily increasing. In 1981, it was 29.6; in 1991, it was 34.0 (calculated from data in the Statistical Yearbook of BiH 1992); to reach a high of 39.5 years in 2013; and is slightly lower than in neighboring countries, e.g., in Croatia (42.2) and Serbia (42.4) (Croatian Bureau of Statistics 2014; Statistical Office of the Republic of Serbia 2014).

The aging index (number of elderly aged 60 and over per hundred young people aged 0–19) in 2013 is 92.2 and is almost half higher than in the previous census (63.3 in 1991), while the aging index calculated from the number of people aged 65 and over per group aged 0–14 is 92.3, and in 1991, it was 27.7. Evidently, in BiH, there are far more older people than in 1991, and this share is almost equal to the share of young people. The comparison of the child base (aged 0–4) and persons aged "75 and over" is very significant for understanding the condition and process of age adjustment. In 1991, these age groups varied significantly, as the number of children was more than three times higher, and in 2013, there were 19.5% more older people than children. A good analytical indicator of age structure and pace of population aging is the age dependency ratio of the elderly, that is, the number of old people per hundred people in working age (shows the workload of the working contingent /15–64/ by the postpartum contingent/65 and more/). According to the census data of 2013, it was 20.2 (in 1991, it was only 9.6). Based on the previous analytical indicators, it is clear that the population of BiH is characterized by a very old age, one of the most unfavorable types of age structure.

According to estimated data for 2018, BiH is in the group of European countries with the smallest share of the elderly

(65 and over) in the total population. The order is as follows (%): Moldova, 11; Albania, 14; Northern Macedonia, 14; Cyprus, 14; Ireland, 14; Belarus, 15; and Ukraine and BiH, 16 each (Population Reference Bureau 2018).

The previously analyzed population structure by age is due to the interplay of several factors that determine the general population movement. In the case of BiH, these factors are decrease in birth rate (decreased birth rate from 14.9‰ in 1991 to 8.3‰ in 2019; reduced total fertility rate from 1.9 in 1991 to 1.2 in 2019; and is one of the lowest in the world), longer life expectancy as a result of lower mortality rates of the population over 30 years old, direct and indirect war losses (about 732,000 fewer inhabitants), and economic emigration of the young population aged between 20 and 39 in the European Union, especially Germany.

The total population of BiH in 2013 was characterized by the fourth type of aging—very old age. At the lower administrative-territorial level, there are significant differences in type. Thus, in the entity of the FBiH, the population belongs to the third type—old age—and in the entity of RS, belongs to the fourth type, very old age, the same as in Brčko District.

Significant changes have taken place in the gender structure of the population. In 1961, the share of the male population in the total population was 48.8%, and the share of the female population was 51.2%. The reason for the higher share of the female population was the suffering of the male population in both world wars and the higher mortality of the male population up to 1 year of age and after the age of 50. At the age of 0–19, the share of men was 50.8% compared to 49.2% of women, while at the age of 60 and over, this ratio was 43.3% of men versus 56.7% of women. In 2013, the share of men in the total population was 49.1% (compared to 1991 lower by 0.8% due to the suffering of men in the 1992–1995 war) and women 50.9%. In 2013, similar relations were maintained in the young and old age groups compared to 1961. The share of men aged 0–19 was 51.3% and women 48.7%. At the age of 60 and over, the share of men in the total population was 43.3%, and the share of women was 56.7% (the main reason is the higher mortality of men in this age group) (RIS 1948–1991; BHAS 2016).

In 1961, the age pyramid showed all the characteristics of a young population (Fig. 11.17).

In 2013, according to the shape of the age-gender pyramid, BiH had a very old population or a very contractual one—in which the share of persons aged 65 and over is 14.2%. The base of the pyramid is exceptionally narrow, and the middle part is convex, so the "pyramid" has the shape of an urn. The share of the young population 0–14 years of age of only 15.4% indicates a very low or negative natural population growth and depopulation processes caused by natural movement.

Compared to 1991, the share of the youngest age group 0–4 years of age decreased from 7.6% to 4.8% (numerically from 332,422 to 174,064). Asymmetry is present in almost all age groups, especially over 60 years of age, where the female population is more numerous by as much as 13.4% (numerically 97,019). There are cuts on the "pyramid" due to the consequences of the last war (1992–1995) in the age group 20–24 years (lower birth rate; 6.5% of the total population) but also the age group 35–44 years for the male population due to suffering and emigration of this population in the war (Fig. 11.18) (BHAS 2016).

To determine the aging degree of a population, especially for the comparison of multi-space units, it is advisable to use a separate model for the evaluation of age structure. Instead of typification based on a simple combination of the "young" and "old," each parameter was scored individually in order to summarize these values to give a point indicator of population aging. Point-based typification has seven types of aging, and each type has a corresponding descriptive characteristic (Table 11.1). It is obvious that the population with more points has a lower degree of aging and vice versa (Nejašmić 2005).

The scoring procedure is as follows: The proportion of young people (aged 0–19) is scored in the range of 0.0–30.0 points, so a higher proportion of young people is scored with more points. The proportion of the elderly (aged 60 and more) is scored in the range of 0.0–70.0 points, so that the higher proportion of the elderly brings a lower number of points (vice versa than in the case of the youth share score); the share value (%) is rounded off to 0.5% (e.g., 25.4–25.5%). Since a higher proportion of young people and a smaller proportion of old people bring a higher final score, it can be argued that a population with a higher score has a lower degree of aging (Table 11.2).

The total population of BiH in 2013 was scored with 72.5 points. This means that it belongs to the fourth type of aging—very old age. All municipalities in BiH are affected by a certain degree of population aging, which is also reflected in the fact that none have type 1 population aging (youth). Only four municipalities are characterized by type 2 (proper aging); the group consists of three municipalities in the far northwest that have more rural than urban population and the highest proportions of the young population in BiH (Bužim, Velika Kladuša, Cazin) and Sapna in northeast BiH with the population aged 0–19 over 25%. The first belt, with type 3 population aging (old age), extends to the northeast of Gradačac and Zvornik, passes through the whole central Bosnia, and ends in western Herzegovina, in Posušje and Ljubuški. The second belt is located in the northwest of the country and covers Bihać, Bosanska Krupa, Sanski Most, and Ključ (a total of 49 municipalities of type 3). Type 4 (very old age) covers 43 municipalities, mainly in Herzegovina, and northern and northeastern Bosnia. The largest cities of BiH

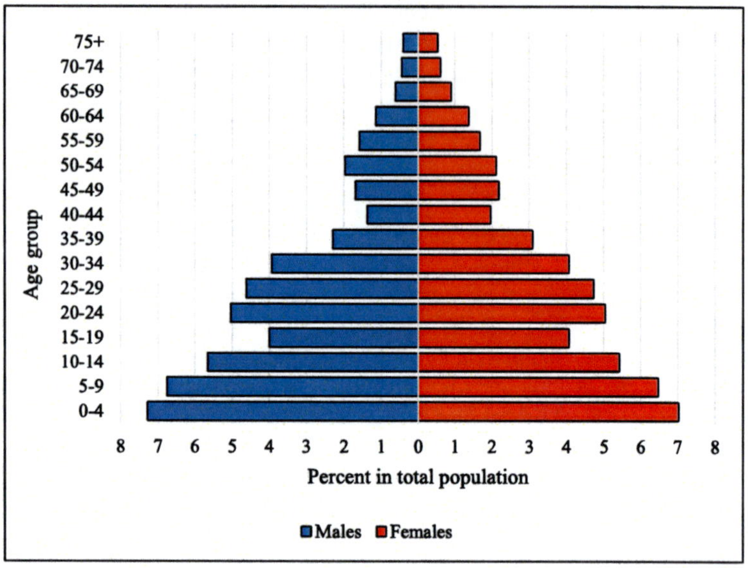

Fig. 11.17 Population pyramid of Bosnia and Herzegovina in 1961

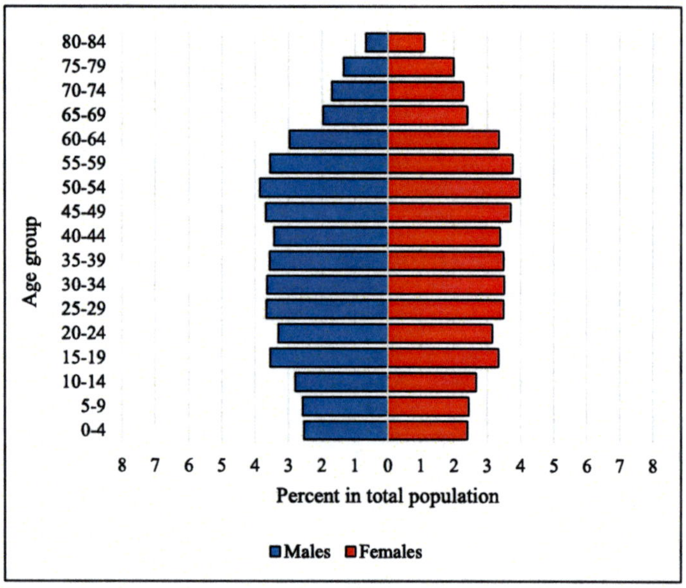

Fig. 11.18 Population pyramid of Bosnia and Herzegovina in 2013

are located in this belt: Sarajevo, Banja Luka, Tuzla, and Mostar. Type 5 (advanced old age) includes mostly municipalities of eastern Herzegovina, eastern Bosnia, and a large belt in the north of the country (41 municipalities). The last type of population aging present in BiH is type 6 (very advanced old age), which covers 6 municipalities characterized by high proportions of the old population and the lowest birth rates in BiH. These are Bosansko Grahovo and Drvar in the west; Kalinovik, Trnovo, and Istočni Stari Grad in the east; and Kupres (Fig. 11.19) (Gekić et al. 2019).

It should be noted that the areas of western and eastern Bosnia and eastern Herzegovina have been areas of intense

Table 11.1 Old age typification based on point values

Point-based indicator of old age	Type	Characteristics
90.5–100.0	1	On the threshold of aging
84.5–90.0	2	Proper aging
73.0–84.0	3	Old age
65.5–72.5	4	Very old age
50.5–65.0	5	Advanced old age
30.5–50.0	6	Very advanced old age
0.0–30.0	7	Extremely advanced old age

Nejašmić (2005)

economic emigration since the 1970s due to the poorer regional economic development, while the vast parts of northern Bosnia and central Bosnia have suffered large population displacements due to the war, with the difference that the area of central Bosnia has been characterized by constantly higher birth rates. Given the metropolitan phase of development of the capital city of Sarajevo, it is to be expected that the situation in the surrounding municipalities (especially in the east) after 1991 would be far more favorable. However, the war events of 1992–1995 disrupted this, and this part, which, by the Dayton Peace Agreement, is mainly in the Republic of Srpska entity, did not continue to develop under the strong agglomeration influence of the Sarajevo metropolis.

From 1950 to 2019, life expectancy at birth was increased from 52.6 to 74.6 years for men and from 54.6 to 79.6 years for women. Higher life expectancy for women and a larger gender gap, which has increased from 2 to 5 years, represent a common trend, although this gap is somewhat smaller among developed countries. For the total population, the increase in life expectancy at birth is from 53.7 to 77.2 years (UNDESA 2017).

Table 11.2 Methodology of age valorization based on point values

Share of young (%)	Points	Share of elderly (%)	Points
0.0–5.0	0.0–5.0	0.0–10.0	70.0–60.5
5.5–10.0	5.5–10.0	10.5–20.0	60.0–50.5
10.5–15.0	10.5–15.0	20.5–30.0	50.0–40.5
15.5–20.0	15.5–20.0	30.5–40.0	40.0–30.5
20.5–25.0	20.5–25.0	40.5–50.0	30.5–20.5
25.5–30 and more	25.5–30	50.5–60.0	20.0–10.5
		60.5–70.0 and more	10.0–0.0

Nejašmić (2005)

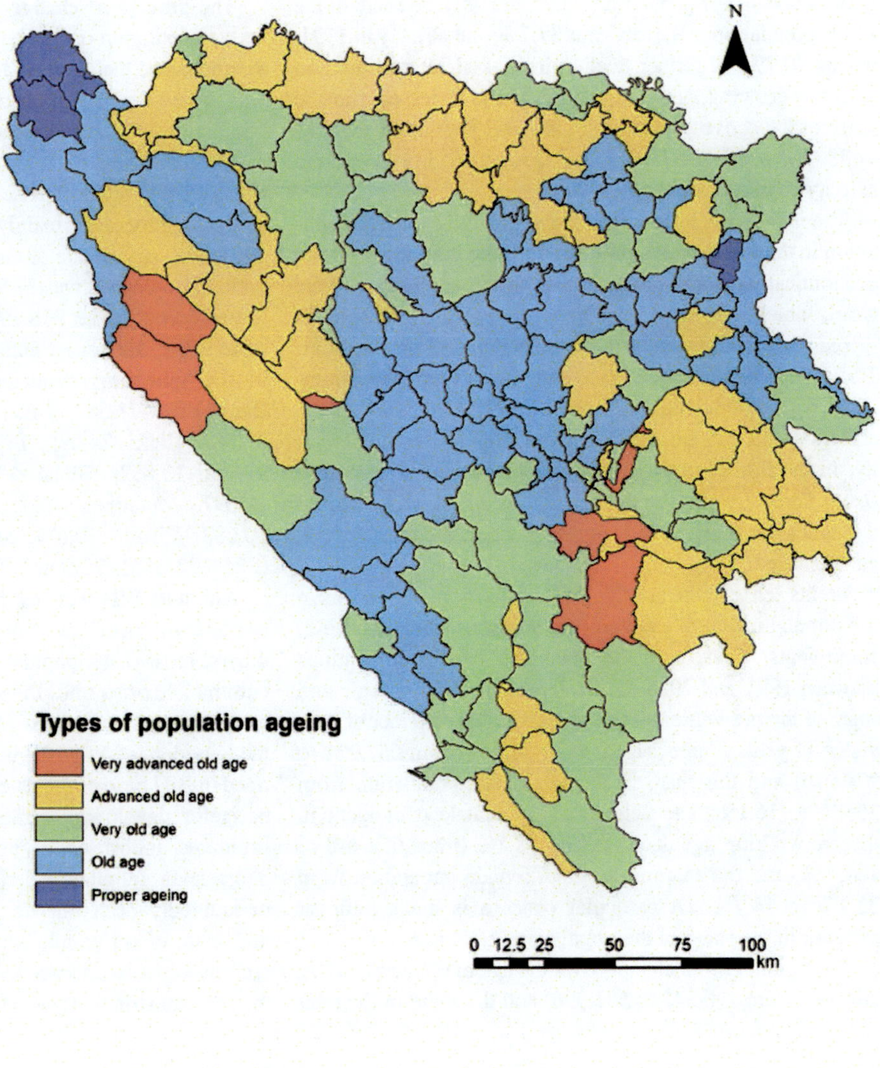

Fig. 11.19 Types of population aging in Bosnia and Herzegovina 2013

11.3.2 Socioeconomic Structure

The economic and social development of BiH after the Second World War, and especially after the 1992–1995 war, had the most direct impact on changes in demographic growth. The most significant changes were in the economic composition, which further transferred to other structures (gender and education) and the total population movement components. Although the average age of the population increased, the share of the active population in the total population of BiH decreased until 1981. The decrease in the share of the active population was influenced by increased education of youth in secondary and higher schools (especially rural youth), then the increased pace of urbanization by rural exodus (especially women who were active before relocation), an increase in the number of retirees, and the share of elderly people in the total population.

From 1953 to 1971, the share of the active population in the total population of BiH decreased from 41.9% to 36.7%. The decline in biological reproduction has led to an increase in the working age population. Thus, in 1981, the share of the active population in the total was 37.7%, and already in 1991, it was 40.7%. A further decline in biological reproduction with the growing aging population led to a decrease in the share of the active population in the total population in 2013 to 38.6%, which can be characterized by the low economic activity of the population.

The increase in the number and share of the elderly population in the total population of BiH resulted in the rise of the economically inactive population, in the broadest sense of the word, which is a very negative trend because it results in increasing social (financial) investments in care mechanisms. The decreasing pressure of the young (i.e., potential economically active population) and the increasing pressure of the elderly on the mature and total population of the municipality, in the future, may cause a lack of own labor resources (labor force), which in the conditions of negative natural dynamics and negative migration balance will not be compensated.

In this sense, there are significant indicators of changes in the population's age structure according to labor (working) contingents. Thus, both the male and female populations between 1971 and 2013 drastically reduced the volume and share of the pre-working age population. The number of men under 14 years of age decreased by 57.6% (from 657,921 to 278,976) and the share in the total male population from 35.9% to 16.1%. The volume of the female contingent in the pre-working age decreased by 58.1% (from 632,448 to 264,743) and the share in the total female population from 33.1% to 14.7%. Of particular concern is the significant increase in the share of the population in old age.

The number of inhabitants over the age of 65 increased by 184.7% (from 176,343 to 501,996) and the share in the total male population of BiH from 4.7% to 14.2%. In addition, in the observed period, aging of the labor contingent took place, which is confirmed by the following indicators: between 1971 and 2013, the share of age groups in the labor contingent from 15 to 30 years of age decreased from 42.2% to 29.1%, and the share of age groups from 50 to 64 years of age increased from 14.5% to 30.5%, when in fact work activity is nearing its end.

The utilization rate of the labor contingent was very variable. In 1971, it amounted to 55.1%, and already in 1991, it reached 50.1% primarily as a result of faster inflow of population into the post-working contingent compared to its activation (the total number of inhabitants in the labor contingent in this period increased by 30.8% and in the post-working contingent 61.3%, while the number of active population increased by 35.2%). The utilization rate of the labor contingent in 2013 increased to 54.8%. However, in 2013, compared to 1991, the total number of inhabitants in the active contingent decreased by 16.1%, while the total number of active inhabitants decreased by 19.3%.

The process of change in the economic structure of the active population of BiH has been accelerated. It is important to emphasize that the average employment of the population was increasing. These changes in the structure of the active population primarily indicate the processes of deagrarization and tertiarization (increase in the number of employees in tertiary activities) as a reflection of the corresponding socioeconomic processes, mainly the process of modernization. In 1991, the activity rate of the population performing an occupation (employed) was 33.9%, while, in 2013, it was 34.6% (based on data from RIS 1948–1991; BHAS 2016).

In 2019, the largest number of employed inhabitants was in the following municipalities/cities: Sarajevo 112,852, Banja Luka 71,584, Tuzla 36,693, Mostar 33,929, Zenica 27,587, Ilidža 24,095, Bijeljina 22,796, Prijedor 14,118, Tešanj 13,852, Bihać 13,806, Travnik 13,796, Doboj 12,697, Živinice 10,911, Gračanica 10,965, Gradiška 10,682, Visoko 10,611, and Laktaši 10,405 (Fig. 11.20) (FZS 2021; ISR 2021).

Although BiH was rich in natural resources, such as various ore of metal and nonmetal, coal, forests, water, and others, most of its population engaged with primarily agricultural activities until the second half of the twentieth century. Previous authorities (Austro-Hungarian Monarchy and the Kingdom of Yugoslavia and Socialist Yugoslavia) developed in BiH base or the so-called dirty industry. It served as a basis for developing ecologically clean and economically profitable industries in neighboring republics of former Yugoslavia (appliances, electric material, etc.). This has undoubtedly been one of the reasons that a large number and share of active inhabitants of BiH as of the 1960s was employed in secondary activities, primarily in industry, mining, construction, etc. (Suljić 2011).

Fig. 11.20 Municipalities/cities with the largest number of employees in BiH in 2019

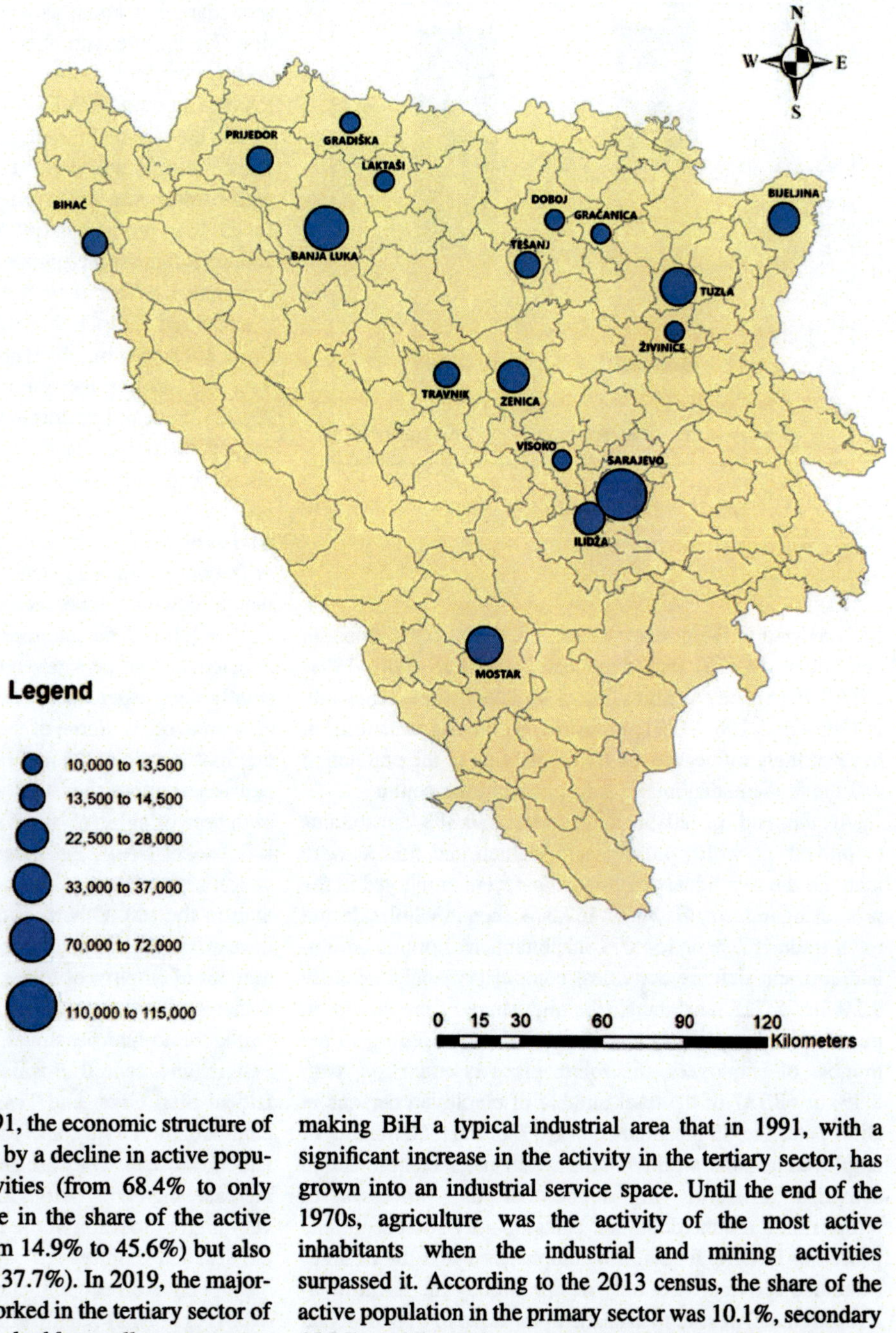

In the period from 1953 to 1991, the economic structure of the population was characterized by a decline in active population engaged in primary activities (from 68.4% to only 16.8%) and a very high increase in the share of the active population in the secondary (from 14.9% to 45.6%) but also tertiary activities (from 10.4% to 37.7%). In 2019, the majority of employed people in BiH worked in the tertiary sector of activities (66.7%), a more than double smaller percentage worked in the secondary sector of activities (31%) and only 2.3% in primary sector of activities.

From 1953 to the mid-1970s, the order of the activity sector on the share of the active population was I-II-III, which is an indisputable indicator of its then agrarian-industrial orientation. In 1981, the share of the active population in secondary activities surpassed a significant share in primary activities by 14%. Relationships between the sectors of the activity have changed, and the order was now II-III-I, making BiH a typical industrial area that in 1991, with a significant increase in the activity in the tertiary sector, has grown into an industrial service space. Until the end of the 1970s, agriculture was the activity of the most active inhabitants when the industrial and mining activities surpassed it. According to the 2013 census, the share of the active population in the primary sector was 10.1%, secondary 46.3%, and the tertiary sector 43.6%. Compared to 1991, the importance of primary activities was reduced by 6.7% of the share, while the importance of the tertiary sector increased by almost 6%. At the same time, the activity of the population in the secondary sector of activity was slightly increased (by 0.7%). The total number of employees in 1971 amounted to 565,957 in BiH, and most were employed in the secondary sector of activity 53.6% (highest industry and mining 214,608) (Fig. 11.21). There were 223,764 or 39.6% in the tertiary sector (mainly in trade and hospitality activities,

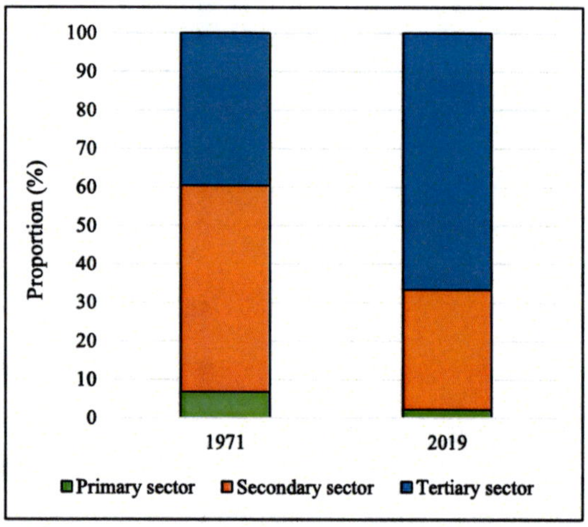

Fig. 11.21 Changes in employment structure between 1971 and 2019 in BiH

58,288), and in the primary sector, 38.178 or 6.8% (most in agriculture 14,479) (based on data from RIS 1948–1991; BHAS 2016). As an expression of social changes, especially in the period 1961–1991, industrialization and urbanization have strongly influenced the transformation of the population and the fundamental image and content of the country.

At the end of 2019, there were 830,383 inhabitants employed, of which 43.3% were women and 56.7% were men. Of the individual activities, most were employed in the processing industry 20.2% or 167,374, then in wholesale and retail trade 18.5% or 154,071 inhabitants and public administration and defense activities, compulsory social security 9.2% or 75,725 inhabitants (the high share of employees in public administration is specific to BiH). According to the number of employees, the fourth place is education, with 8.3% or 68.737 of the total number of employed persons in BiH. Total 54,933 inhabitants, or 6.6% of the total number of employed persons, worked in the health and social protection activities. There is no doubt that the processes of tertiarization are the dominant contemporary socioeconomic processes in BiH. It should be emphasized that the changes described above are not only determined by the usual modernization and post-modernization processes in overall social and economic development but also by war and destruction as a result of the armed aggression in the 1990s. Of the total number of employees in BiH in 2019, 64.8% were in the entity of FBiH, 33% were in the entity of RS, and in the Brčko District 2.2%.

At the end of 2019, there were 401,846 unemployed people in BiH, of which 31.8% were men and 68.2% of women. By the qualification structure of the most unemployed, 141,418 inhabitants (35.2%) were with high qualifications and qualifications, then 123,157 (30.6%) with secondary education, and 119,002 (29.1%) had no qualification. Of the total number of unemployed in BiH in 2019, 76.6% were in the entity of FBiH, 21.7% were in the entity of RS, and in the Brčko District 1.7% (BHAS 2019).

As the population of BiH gets older, more and more elderly people will depend on the pension and social funds, while fewer and fewer employees will contribute to these funds. The problem is more pronounced if we look at the concrete data on the number of employees to the number of pensioners, which in 1991 amounted to 2.4 to 1, and in 2013 it amounted 1.7 to 1. According to the Census of Population from 2013, persons in employment are those aged 15 and over who, during the reference week from September 23 to 29, 2013: worked/performed an activity for at least 1 h to earn money for living (paid work) or if they were temporarily absent from work (because of illness, annual leave, maternity leave, etc.), but had a job to return to and for which they were paid as persons in paid employment or self-employed persons or freelancers or who were involved in a family business for which they did not receive payment.

The ratio of the number of employees to the number of pensioners (workers per retiree ratio) in BiH will be constantly decreasing, which could lead to great difficulties and economic crisis. Some of the miniature municipalities of BiH that have a ratio of the number of employees to the number of pensioners more than two are also characterized by a very high type of aging (5 and 6) such as: Krupa na Uni, Kupres-RS, Istočni Drvar, Pelagićevo, Berkovići, and Donji Žabar, which have a far smaller proportion of the young population than in the rest. Others, such as Tešanj, Cazin, and Bužim, have a favorable type of population aging and the ratio of the number of employees to the number of pensioners. The most difficult situation is in municipalities with less than one employee to one pensioner. These are the municipalities of eastern and central Bosnia (Bileća, Nevesinje, Kalinovik, Istočni Stari Grad, and Trnovo), western Bosnia (Bosansko Grahovo and Drvar), and eastern Herzegovina (Ljubinje and Istočni Mostar). They are characterized by a negative natural increase, high emigration, low proportion of young people, a very high proportion of the elderly population, and otherwise the fifth and sixth type of population aging. For decades, they have been in an economic and social crisis that is not nearly over (Fig. 11.22) (Gekić et al. 2019).

Average monthly salaries rose from 822 convertible marks—BAM (equal to 420 EUR) in 2013 to 932 BAM (476 EUR) in 2019. In the same period, increase in the entity of FBiH was from 841 BAM (430 EUR) to 940 BAM (480 EUR) and in the entity of RS from 803 BAM (410 EUR) to 912 BAM (466 EUR). The minimum monthly salary in BiH rose from 350 BAM (180 EUR) in 2013 to 410 BAM (210 EUR). The total number of pensioners was around 626,000 (17.7% of total population) in 2013 and around 682,000 in 2019 (20% of total population). The

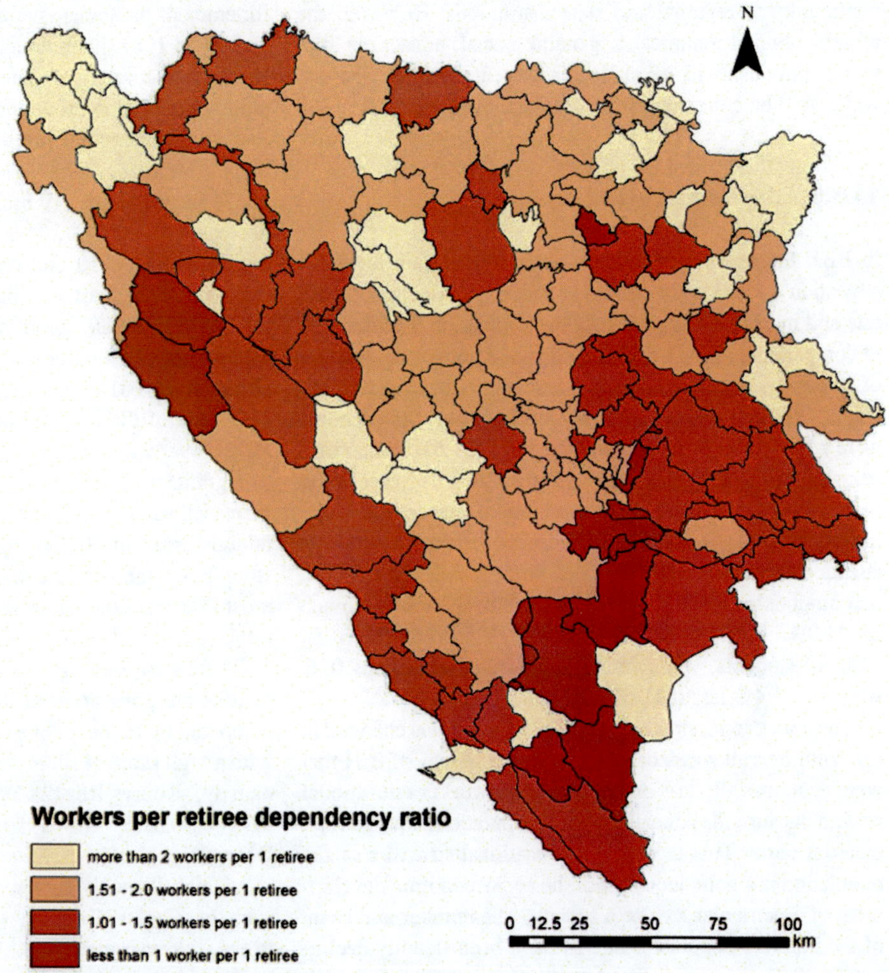

Fig. 11.22 Workers per retiree dependency ratio in BiH in 2013

average monthly pension in BiH rose from 336 BAM (172 EUR) in 2013 (in FBiH 178 EUR; in RS 160 EUR) to 404 BAM (206 EUR) in 2019 (in FBiH 214 EUR; in RS 194 EUR). The minimum monthly pension, for at least 15 years on work, in BiH rose from 250 BAM (128 EUR) in 2013 (in FBiH 158 EUR; in RS 82 EUR) to 302 BAM (154 EUR) in 2019 (in FBiH 190 EUR; in RS 100 EUR) (calculated from data FZMIOPIO 2019; PIO 2019).

All data show that unemployment is the number one problem for young people in BiH. Over 60% of young people are unemployed, of which 13.4% have been unemployed for less than 12 months, half of them for over 2 years and one-fifth for over 5 years. Almost half of the job seekers are young people (15–34 years). Over 50% of children with elementary education from poor families do not continue their education, while 10% of them do not have health insurance (Ministry of Civil Affairs of BiH 2019).

In addition to unemployment, inadequacy of the education system, the depressing political situation in the country, and the lack of prospects for a better future have all caused the trend of mass emigration of young people to other countries. Existential problems of young people, which are mostly the result of long-term unemployment and temporary employment, have a strong impact on having fewer children than desired. Reversing long-term aging and low fertility remains problematic, given that the relevant policies may not pay dividends until the next generation reaches the working age (Rand Europe 2005). BiH will not be able to provide sufficient labor force from its own resources, but will have to focus on importing labor through economic coercion. In this sense, BiH should consider its diaspora as one of the sources of revitalization. The diaspora needs to be supported in making the best out of their lives in the host countries while at the same time being encouraged to keep their identity strong and their relationship with BiH productive (Halilovich et al. 2018). Recent demographic trends in BiH cannot be stopped and reversed in a positive direction without deliberate, systematic long-term state intervention, from the local community to regional and state institutions. However, the aforementioned requires a general social consensus and strong political will of all key actors in Bosnian society, which will be extremely difficult to achieve.

11.3.3 Educational Structure

In BiH, the number of students attending certain levels of education is constantly declining due to the declining birth rate and increased emigration of the population. For the sake of comparison, we will mention the movement of the number of students who attended certain levels of education in 2013 and 2019. In 2013, at the primary level of education, schools were attended by 302,133 students, while in 2019, the number was 280,018 students or 7.3% less (this decrease is far more significant than the estimated total decrease in population in the same period of 3.6%). In 2013, 156,350 students attended secondary education, i.e., high school, and in 2019, that number was reduced to 117,475 students or by as much as 24.9%. Tertiary education in 2013 was attended by 113,290 students, while in 2019, the number was 89,016, which is 21.4% less (BHAS 2016, 2019).

The educational structure of the population has changed in line with overall socioeconomic progress. Areas of BiH that were economically less developed had a weaker educational system because the education of the population required high material costs. That is why the educational structure of the population is a good indicator of the socioeconomic development of a particular area or a society. The number and share of the illiterate population in BiH have been steadily declining. In 1953, there were 853,151 illiterate inhabitants, which is a share of 40.2% of the population aged 10 and over. Almost every other resident aged 10 and over in BiH was illiterate. This data reveals a very underdeveloped economic and social structure in that period.

Changes in the share of the illiterate population indicate intensive changes due to legal regulations making schooling compulsory and very intensive expansion of primary and secondary education and the overall intensive transformation. The majority of the illiterate population in 1953 was over the age of 35, a total of 46.4% of all illiterate inhabitants, while in the age group of 65 and over, illiteracy was around 81%. If we look at the gender structure of the illiterate population, we can notice that there were significantly more illiterate women than men. Of the total illiterate inhabitants, 74% were women, and 26% were men. This is primarily due to the patriarchal way of life in which women mostly ran household activities and men were employed. Women were generally not allowed to become literate (family connection), so 57% of the total number of women were illiterate. As early as 1981, the share of the illiterate population decreased to 14.5% in the total population aged 10 and over and in 1991, to 9.9%. The illiteracy of the female population was rapidly declining, and already in 1981, it amounted to 23.3%. As society's modernization began relatively late in BiH, among the illiterate, the elderly population predominates the female population.

One of the conclusions within this educational structure is that the number of inhabitants aged 10 and over, from 1953 to 2013, increased by 49.8%, while the illiterate population decreased by 89.5%. This means that the increase in the population aged 10 and over was accompanied by almost twice the decline of the illiterate population. From 1953 to 2013, the illiterate population decreased from 853,151 to 89,794 or 89.5%. Illiteracy is still not eradicated in BiH, although in 2013, there were 97.2% of literate residents. The share of the illiterate population in the population aged 10 and over was 2.8%. As in previous periods, urbanization is a significant factor in education development, so the increased share of illiterates is mainly related to rural mountainous areas of BiH compared to urban primarily. In the structure of the illiterate population, 86.4% are women. The most illiterate population is aged 65 and over, 76.2% of the total.

There were also significant differences regarding the share of illiterate population in the population aged 10 and over, compared by municipalities in 2013. Thus, the municipalities with a high share of illiterate inhabitants (over 8%) stand out, namely, Kupres-RS (28.3%), Kneževo (10.9%), Trnovo-FBiH (10.3%), Krupa na Uni (9.8%), Ribnik (8.3%), Dobretići (8.3%), and Istočni Mostar (8.1%), and municipalities with the lowest share of illiterate inhabitants, such as Centar Sarajevo (0.8%), Istočno Novo Sarajevo (0.8%), Novo Sarajevo (0.8%), and Trebinje (0.9%).

The analysis shows that, in 1953, the population of BiH had a very unfavorable educational structure, i.e., level of education. More than a third of the population (67.2%) aged 10 and over were without any or with incomplete primary education. 26.1% had completed primary school. 7.7% of the total population aged 15 and over had completed secondary school or secondary education. Only 0.3% of the population had a tertiary level of education (higher and tertiary education). The share of the female population with any level of education was extremely low (30.9% were women with completed primary education and 19.8% with tertiary education) (BHAS 1948–1991).

The structure of the population by education level is still unsatisfactory, although exceptional results have been achieved in raising the population's educational level. In 2013, 4.9% of the population aged 15 and over had no education, while 9.2% had not completed primary education. 21.4% had completed primary school. Fifty-one percent of the total population aged 15 and over had completed high school or secondary education. The first level of tertiary education (some kind of specialization or completed high school) had 3.9% inhabitants. Only 9.6% of the population had completed tertiary education (higher education). The share of the female population with any level of education was higher than the share of the male population (of those with completed primary education, 57% were women and with completed tertiary education, 51.8%). Most of the population with completed tertiary education is in the age group 25–29 years (20.7%). Considering the share of the population with completed tertiary education in the total population aged 20 and over (average for BiH was 10.4%), there are significant spatial differences in BiH according to the last census in 2013. Six groups of municipalities can be distinguished. In the first group, 22 municipalities has a very low share of the population with completed tertiary education, and those which stand out are Krupa na Uni (1.1%), Kupres-RS (1.5%), Oštra Luka (2.7%), Vukosavlje (2.9%), Pale-FBiH (3.5%), and Dobretići (3.5%) (all municipalities created by the establishment of the IEBL in 1995 and highly rural) (Fig. 11.23).

The most developed and urban municipalities in BiH belonging to the city of Sarajevo had an extremely high share of inhabitants with completed tertiary education: Centar Sarajevo (34.3%), Novo Sarajevo (30.9%), and Stari Grad Sarajevo (20.6%). Several other municipalities were characterized by a significantly high share: Istočno Novo Sarajevo (19.3%), Banja Luka (18.1%), Novi Grad Sarajevo (17.6%), Mostar (17.2%), and East Ilidža (16%) (BHAS 2016).

11.3.4 Ethnical, Religious, and Linguistical Structure

For more than 70 years, Bosniaks have not been given the right to declare themselves under their real name, and it was only in the early 1970s that they were recognized as a separate people by the SFRY authorities under the name Muslims (not Bosniaks). From 1945 to 1991, the communist authorities used various means to force the population to declare themselves under other ethnicities (Serb, Croat, undecided, Yugoslav, etc.). Due to the extremely complex issue of ethnic representation in BiH, especially because, during the time of former socialist Yugoslavia, this entire ethnicity could not give their answer freely, it is difficult to make accurate and especially realistic comparisons. However, we will still list the ethnic structure of the population for the penultimate 1991 census, in addition to the last official one from 2013. Of the total population in 1991, namely, 4,377,033, Muslims accounted for 43.38%, Serbs 31.19%, Croats 17.36%, Yugoslavs 5.53%, Bosnians 0.24%, Montenegrins 0.23%, Roma 0.20%, and others 1.87%. If we recalculate the data from 1991 by municipalities into Dayton administrative-territorial units, we get the following: for the entity of FBiH, Muslims 52.3%, Croats 21.9%, Serbs 17.6%, and others 8.2%, while in the entity of RS, Serbs 55.5%, Muslims 28%, Croats 9.2%, and others 7.3%. In Brčko District ethnic structure of population was: Muslims 44.1%, Croats 25.4%, Serbs 20.7%, and others 9.8% (RIS 1948–1991).

According to the data from the last census in BiH from 2013, slightly more than 99% of the population answered the question of ethnicity/national belonging. Bosniaks make up 50.11% of the total number, Serbs 30.77%, Croats 15.43%, and others 3.69% (primarily Bosnians 1.05%, Roma 0.36%, Muslims 0.34%, and Bosnians and Herzegovinians 0.32%). In the entity of FBiH, the ethnic/national structure was as follows: Bosniaks 70.4%, Croats 22.4%, Serbs 2.5%, and others 4.7%. In the entity of RS, the ethnic/national structure was as follows: Serbs 81.5%, Bosniaks 14%, Croats 2.4%, and others 2.1%. In the Brčko District, Bosniaks made up 42.4%, Serbs 34.6%, Croats 20.7%, and others 2.3%.

Due to the consequences of the aggressive war against BiH and the creation of an inappropriate political-territorial structure after 1995, there was a marked homogenization of political-territorial units, already mentioned in the chapter on the political geography of BiH. It is worth noting here that as many as 92.14% of Serbs live in the entity of RS, while 91.39% of Croats and 88.29% of Bosniaks live in the entity of FBiH.

In 2013, out of the total population, 1,790,454 stated belonging to Islamic religion or 50.7%, 1,085,760 or 30.7% to Orthodox religion, and 536,333 or 15.2% to Catholic religion. Inhabitants also stated that they are Muslim, Roman Catholic, Serbian, Jehovah's Witnesses, and other religions, representing 37,785 or 1.07%. Also, there were 27,853 or 0.8% atheists and 10,816 or 0.3% agnostics. In the entity of FBiH, according to the religious structure, the most inhabitants belong to the Islamic religion or 71.3%, Catholic 22.1%, and Orthodox 2.6%; in the entity of RS, Orthodox makes for 81.4%, Islamic for 14.1%, and Catholic for 2.4%; in the Brčko District, the largest number of inhabitants belong to the Islamic religion, namely, 42.9%; Orthodox 34.5%; and Catholic 20.4%. The spatial distribution of the population by religious structure corresponds to the distribution of the population by ethnic/national structure. It is similar to the population's linguistic structure (BHAS 2016). In the 2013 census, Bosniaks mostly declared themselves to be followers of the Islamic religion and to speak Bosnian; Serbs to be followers of the Orthodox religion and to speak Serbian; and Croats that they are followers of the Catholic religion and that they speak the Croatian language.

In BiH, the population mostly declared that they speak Bosnian, Serbian, and Croatian, which belong to the dialectal continuum of South Slavic languages. Bosnian was the mother tongue of a total of 1,866,585 or 52.9% of the total population, Serbian of 1,086,027 or 30.8%, and Croatian of 515,481 inhabitants or 14.6% (Fig. 11.24). The language structure of the population corresponds almost entirely to

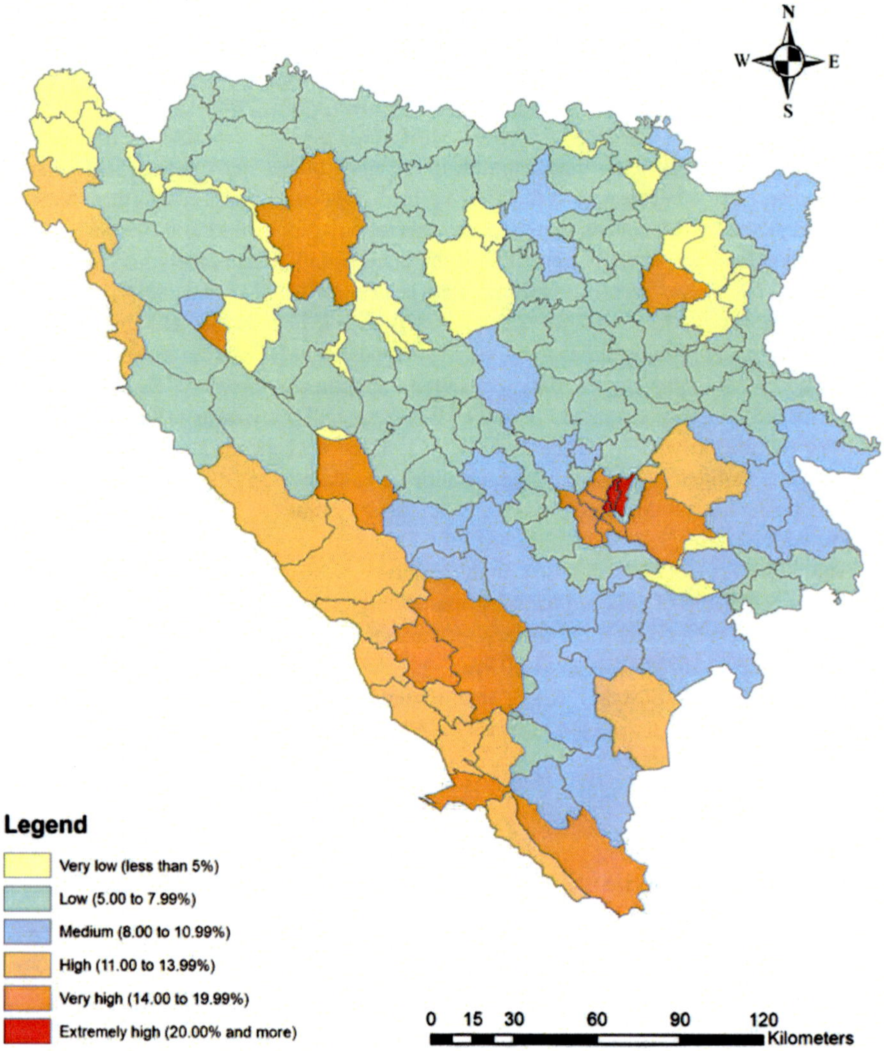

Fig. 11.23 Share of persons with tertiary education in the total population aged 20 and over in 2013

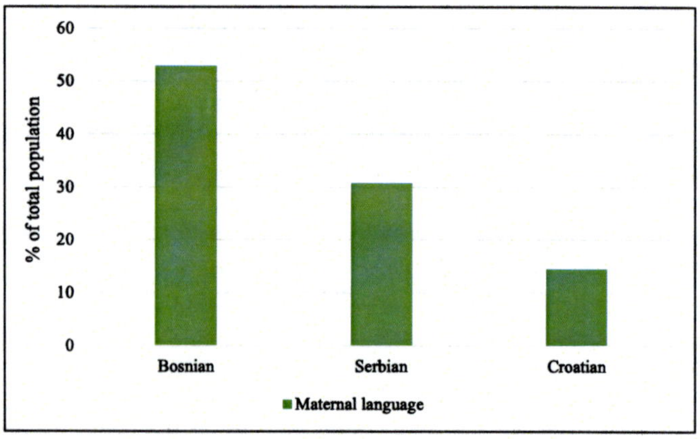

Fig. 11.24 Population by the three largest maternal languages in Bosnia and Herzegovina in 2013

the religious and somewhat less to the ethnic/national structure of the population. It is perhaps the closest indicator of the actual representation of ethnicity in 2013. Of the other languages, Serbo-Croatian is the most represented, with a 0.77% share. In the entity of FBiH, Bosnian is the mother tongue of 74.6% of the population and Croatian of 21.6%, while in the entity or RS, Serbian is the mother tongue of 82.5% and Bosnian of 14.1% of the population. In the Brčko District, 43.5% of the population's mother tongue is Bosnian, 20.1% Croatian, and 34.7% Serbian (BHAS 2016).

11.4 Covid-19 and Future Changes in the Population

In 2020, which was marked by the Covid-19 pandemic, the number of live births in BiH decreased by 1036 or 3.7% compared to 2019 (a larger decrease was, e.g., in 2013 compared to 2012, 5.7%). This is the lowest number of live births in BiH since vital statistics were kept in the entire country. The forecast is that it will continue to decrease due to the aging population and the decrease in the number of women belonging to the fertile contingent. The number of deaths in 2020 in BiH was the highest since 1950 (except for the period 1992–1995, for which we have no exact reliable data) and was 43,808. This represents an increase of 12.8% compared to 2019 and is the largest annual increase since 1955 (again excluding the period 1992–1995). Comparing the annual change in the number of deaths in 2020/2019 is even 9.7% higher than the annual change in the number of deaths in 2019/2018. The internal mobility of the population of BiH in 2020 decreased by 8% compared to 2019 (the sum of internal immigrants and emigrants decreased by 2118).

The health and economic crisis caused by the spread of Covid-19 had significant effects on the labor market as well. Reduced economic activity during the pandemic drastically increased the number of unemployed persons in BiH, especially during 2020. The number of registered unemployed persons at the end of the first month since the coronavirus outbreak in March 2020 was 398,270. By May of the same year, the number of unemployed increased by 23,477 and amounted to 421,747. The number of unemployed persons continued to grow continuously in the following months, reaching 427,593 in August, i.e., 29,323 persons more than in May 2020. Already in the following month, the same number decreased by 9636 people. By the end of 2020, the number of unemployed persons was continuously decreasing, and in December 2020, 413,627 unemployed persons were registered in BiH, which was 2.9% more than in December 2019 (based on data from BHAS 2021).

A total of 111,642 confirmed cases of Covid-19 were registered in 2020, while the total number tested was 514,380. As a result of the Covid-19 disease, 4072 inhabitants died in BiH in 2020, which, given the number of confirmed cases, gives a mortality rate of 3.6%. In the total number of deaths in 2020, those who died from the effects of Covid-19 accounted for 9.3%.

State budget expenditure for pensions and health care for the elderly will almost more than double by 2050. The number of people aged 75 and over will increase by 105% (218,000 people) in comparison with the year 2013; the overall share will rise from 5.9% to 8.1%. According to UNDESA (2019), BiH will have only 1.6 million people in 2100, which is 53% less than in 2019. All of this will put a heavy burden on the welfare and health care funds. With the increase in the number of elderly people, there is a growing need to provide a variety of formal care to this contingent of the population. Due to the decreasing number of children in families, also the number of those subject to social security contributions is decreasing (Gekić et al. 2019).

Depopulation processes are still underway, which undoubtedly affect the negative changes in the age structure of the population in the direction of further aging of the population. This leads to a very unfavorable demographic situation and population development, which can be seen from the estimate for 2050, which says that in the middle variant of population projection, BiH will be characterized by a very old population with an aging index (number of old people aged 60 and over per 100 young people aged 0–19) of as many as 259 or it will have 2.6 times more old than young people.

According to the middle variant of the UN population projection, the working contingent will be reduced by as much as 22,000 a year by 2050 and will be lower by even 930,000 or 37% (UNDESA 2019). Workers are also getting older and older. In 1991, 85.5% of the working contingent was aged 15–49 and 14.5% at the age of 50–64. In 2013, there was a significant deterioration. The share of the labor contingent aged 15–49 decreased to 69.5%, and aged 50–64 increased to 30.5%. In the future decades, the age structure of employees will continue to deteriorate in faster pace (as a reflection of the general demographic aging), and dynamic economic development of the country will be hard to achieve. Furthermore, aging will greatly affect the structure of public and personal consumption. Depopulation and aging of the population as dominant demographic processes in BiH will affect the deterioration of the ratio between the number of employees and the number of pensioners. This raises the question of the financial sustainability of the pension system. As the baby boom generation enters retirement age, this ratio will soon deteriorate further. Unless there is a significant increase in activity rates and employment rates, the dependency ratio in the pension system will be well below the level of sustainability. A slowdown in economic growth can undoubtedly be linked to an aging population. The share of expenditures of the state budget for pensions and health care for the elderly population up to 2050 will almost more than double. The number of people aged 75 and over will be higher for 218,000 (105%) more than in 2013; the share will increase from 5.9% to 8.1% (UNDESA 2019). This will put a heavy burden on welfare and health care funds. With the increase in the number of elderly people, there is a growing need to provide a variety of formal care to this contingent of the population. Due to the decreasing number of children in families, the circle of main care providers for the elderly is decreasing.

References

Ahmetbegović S (2012) Reljef kao faktor okupljanja stanovništva u Bosni i Hercegovini [Relief as a factor of population gathering in Bosnia and Herzegovina]. PhD thesis, Faculty of Science, University of Sarajevo, Sarajevo

BHAS (2019) Bosna i Hercegovina u brojevima 2019 [Bosnia and Herzegovina in numbers 2019]. Agency for Statistics of Bosnia and Herzegovina, Sarajevo

BHAS (2021) Bosna i Hercegovina u brojevima 2020 [Bosnia and Herzegovina in numbers 2020]. Agency for Statistics of Bosnia and Herzegovina, Sarajevo

BHAS 2016, Agency for Statistics of Bosnia and Herzegovina (2016) Census of population, households and dwellings in Bosnia and Herzegovina 2013. Sarajevo

Central Intelligence Agency (2021) The world factbook 2021. Total fertility rates by country. Available at https://www.cia.gov/the-world-factbook/field/total-fertility-rate/country-comparison. Accessed on 12 July 2021

Croatian Bureau of Statistics (2014) Population estimate of Republic of Croatia 2013. Zagreb, pp 2–3

FZMIOPIO (2019) Archives of Federal Institute for Pension and Disability Insurance, Sarajevo

FZS (2021) Data on estimates of population, vital statistics, migrations and employment. Institute for Statistics of Federation of Bosnia and Herzegovina, Sarajevo

Gaćeša N (1984) Agrarna reforma i kolonizacija u Jugoslaviji [Agrarian reform and colonization in Yugoslavia]. Matica srpska, Novi Sad, p 347

Gekić H (2021) Demografska slika općine Bugojno [Demographic profile of the Municipality of Bugojno]. Dobra knjiga, Sarajevo

Gekić H, Bidžan-Gekić A, Mirić R, Drešković N (2019) Spatial differentiation of the age structure of a population in Bosnia and Herzegovina. In: Proceedings, international scientific symposium, new trends in geography, October 3–4, 2019, Ohrid, Republic of North Macedonia, pp 215–226

Gekić H, Bidžan-Gekić A, Mirić R, Remenyi P (2020) Hidden geographies of population implosion in Bosnia and Herzegovina. Eur J Geogr 11(2):47–64

Halilovich H (2012) Trans-local communities in the age of transnationalism: Bosnians in diaspora. Int Migr J 50(1):162–178. https://doi.org/10.1111/j.1468-2435.2011.00721.x

Halilovich H, Hasić J, Karabegović D, Karamehić-Muratović A, Oruč N (2018) Mapping the Bosnian-Herzegovinian diaspora. Ministry of Human Rights and Refugees of Bosnia and Herzegovina, Sarajevo

Husain M (1996) Systematic agricultural geography. Rawat Publications, Jaipur

Ibreljic I, Kulenovic S, Kadusic A, Smajic S (2006) Migration flows in Bosnia and Herzegovina after 1992. 46th Congress of the ERSA, Volos, Greece. Available at http://www-sre.wu-wien.ac.at/ersa/ersaconfs/ersa06/papers/173.pdf. Accessed 11 Sept 2020

ISR (2021) Data on estimates of population, vital statistics, migrations and employment. Institute of Statistics of Republika Srpska, Banja Luka

Kadušić A, Suljić A (2018) Migration and demographic changes: the case of Bosnia and Herzegovina. Eur J Geogr 9(4):75–86

Lesthaeghe R, Neels K (2002) From the first to the second demographic-transition: an interpretation of the spatial continuity of demographic innovation in France, Belgium and Switzerland. European Journal of Population 18:325–360

Lesthaeghe R, Surkyn J (2002) New forms of household formation in Central and Eastern Europe: are they related to newly emerging value orientations? In Economic Commission for Europe, Economic Survey of Europe 2002 (pp 197–216). New York and Geneva: United Nations

Lesthaeghe R, Surkyn J (2004) When history moves on: The foundations and diffusion of a second demographic transition. Paper presented at the seminar on ideational perspectives on international family change. Population Studies Center, Institute for Social Research (ISR), University of Michigan, Ann Arbor

Ministry of Civil Affairs of Bosnia and Herzegovina (2019) Data on youths in Bosnia and Herzegovina. Sarajevo

Ministry of Foreign Trade and Economic Relations of Bosnia and Herzegovina (2018) Strateški plan ruralnog razvoja BiH [Strategic plan of rural development of BiH]. Sarajevo

Ministry of Human Rights and Refugees of Bosnia and Herzegovina (2019) Informations on the status of emigrants from Bosnia and Herzegovina. Sarajevo

Ministry of security of BiH (2020) Migracijski profil BiH 2019 [Migration profile of BiH 2019]. Sarajevo

National Security Council of Turkey (2008) Türkiye'deki Etnik Grupların Dağılım Raporu [Report on the distribution of ethnic groups in Turkey]. Retrieved from https://www.milliyet.com.tr/cadde/turkiyedeki-kurtlerin-sayisi-873452. Accessed on 26 July 2020

Nejašmić I (2005) Demogeografija—stanovništvo u prostornim odnosima i procesima [Demogeography—population in spatial relations and processes]. Školska knjiga, Zagreb

PIO (2019) Archives of pension and disability insurance fund of the Republika Srpska. Banja Luka

Population Reference Bureau (2018) 2018 world population data sheet with focus on changing age structures. Washington. Available at https://www.prb.org/wp-content/uploads/2018/08/2018_World-Population-data-sheet.pdf. Accessed on 14 May 2019

Ramčilović Z (2019) Demografske promjene nakon Berlinskog kongresa (1878) u Bosni i Hercegovini [Demographic changes after Berlin Congress (1878) in Bosnia and Herzegovina]. Historijski pogledi 2:72–84

RAND Europe (2005) Population implosion? Low fertility and policy responses in the European Union. Research Brief: Research and Development Europe. Available at https://www.rand.org/content/dam/rand/pubs/research_briefs/2005/RAND_RB9126.pdf. Accessed on 19 Jan 2020

Reynaud C, Miccoli S (2018) Depopulation and the aging population: the relationship in Italian municipalities. Sustainability 10(4):1004. https://doi.org/10.3390/su10041004

RIS (1948–1991) Population censuses from 1948 to 1991. Republic agency for statistics of Bosnia and Herzegovina, Sarajevo

RIS (1983) Statistical yearbook of Bosnia and Herzegovina 1982. Republic agency for statistics of Bosnia and Herzegovina, Sarajevo

RIS (1994) Statistical yearbook of Bosnia and Herzegovina 1992. Republic agency for statistics of Bosnia and Herzegovina, Sarajevo

Sparavalo J (1990) Deagrarizacija—osnov socijalne i prostorne pokretljivosti stanovništva BiH u periodu poslije Drugog svjetskog rata [Deagrarization—the basis of social and spatial mobility of the population of BiH in the period after the Second World War]. In: Šehić N (ed) Migracije i Bosna i Hercegovina. Institut za istoriju, Sarajevo, pp 365–388

Statistical Office of the Republic of Serbia (2014) Statistical yearbook of Serbia 2013. Belgrade, p 36

Suljić A (2011) Stanovništvo i naselja općine Srebrenica [Population and settlements of the Municipality of Srebrenica]. Geografsko društvo Tuzlanskog kantona, JUNUB „Derviš Sušić" Tuzla, Tuzla

The Global Competitiveness Report (2019) Insight report. World Economic Forum. Available at http://www3.weforum.org/docs/WEF_TheGlobalCompetitivenessReport2019.pdf. Accessed on 26 Aug 2021

Tokača M (2012) Bosanska knjiga mrtvih [Bosnian book of dead]. Istraživačko dokumentacioni centar, Sarajevo

Turčilo L, Osmić A, Šadić S, Dudić A, Kapidžić D, Žiga J (2019) Studija mladi Bosna i Hercegovina 2018/2019 [Youth study Bosnia and Herzegovina 2018/2019]. Friedrich-Ebert-Stiftung, Sarajevo

UNDESA (2017) The world mortality report 2017. United Nations Department of Economic and Social Affairs/Population Division. Available at https://www.un.org/en/development/desa/population/publications/mortality/world-mortality-cdrom-2017.asp. Accessed on 27 June 2021

UNDESA (2019) World population prospects 2019. United Nations Department of Economic and Social Affairs/Population Division, New York. Available at https://population.un.org/wpp/Download/Standard/Population/. Accessed on 21Oct 2020

United Nations Development Programme (2002) Human development report. New York. In: Kadušić A, Suljić A (2018) Migration and demographic changes: the case of Bosnia and Herzegovina. Eur J Geogr 9(4)

Valenta M, Ramet P (2011) The Bosnian diaspora: integration in transnational communities. Ashgate Publishing Limited, Farnham

Van de Kaa DJ (1994) The second demographic transition revisited: theories and expectations. NIDI/CBGS Publication, 30:81–126

Van de Kaa DJ (2001) Postmodern fertility preferences: from changing value orientation to new behavior. In RA. Bulatao & JB Casterline (Eds), Global fertility transition. Supplement to population and development review (Vol 27, pp 290–338). New York: Population Council

Van de Kaa DJ (2002) The idea of a second demographic transition in industrialized countries. Paper presented at the sixth welfare policy seminar of the National Institute of Population and Social Security, Tokyo, Japan

Van Bavel J (2007) Sub-replacement fertility in the west before the baby boom: Past and current perspectives. Popul Stud 64(1): 1–18

Vrecer N (2010) Living in Limbo: integration of forced migrants from Bosnia and Herzegovina in Slovenia. J Refug Stud 23(4):484–502. https://doi.org/10.1093/jrs/feq042

Zeneli B, Cosic E, Dzebo A, Zmejkoski R (2013) Brain gain policies and practices in the Western Balkans. Group 484 Centre for Migration, Belgrade

Žerjavić V (1992) Manipulacije žrtvama drugoga svjetskog rata 1941.-1945.* [Manipulations of the victims of the Second World War 1941-1945*]. Časopis za suvremenu povijest 24(3):149–163

Zwierzchowski J, Tabeau E. (2010) The 1992–95 war in Bosnia and Herzegovina: census-based multiple system estimation of casualties' undercount. International Research Workshop on 'The Global Costs of Conflict' The Households in Conflict Network (HiCN) and The German Institute for Economic Research (DIW Berlin), Berlin, pp 1–25. Available at https://www.icty.org/x/file/About/OTP/War_Demographics/en/bih_casualty_undercount_conf_paper_100201.pdf. Accessed on 07 Nov 2019

12 Urban and Rural Geography of Bosnia and Herzegovina

Abstract

This chapter deals with the basic characteristics of urban and rural settlements in Bosnia and Herzegovina. We also investigate the main factors of settlement development. We predominantly deal with spatial variations of the degree of urbanization, functional typology, and size categories of urban settlements. Bosnia and Herzegovina, considering the size of urban settlements, has a weakly developed urban system. However, the urban settlements of Bosnia and Herzegovina approximately fit into Zipf's law on rank-size at the national level, and therefore Bosnia and Herzegovina differs significantly from the countries in Southeast Europe. Precisely because of that, the centuries-long continuous development of urban settlements within a mostly stable territorial coverage can be observed. By size, the urban system is dominated by urban settlements with less than 5000 inhabitants. More than half of the total population lives in nonurban settlements. Among nonurban settlements, there is domination of settlements by size up to 500 inhabitants (78.9%). During the war period, 1992–1995, a large number of settlements experienced a real urbicide. Thirty-five percent of housing units in Bosnia and Herzegovina were partially or completely destroyed.

Keywords

Urban system · Urban growth · Urban economy · Degree of urbanization · Rural settlements

12.1 Origin and Evolution of Settlements

Urbanization in Bosnia and Herzegovina (hereinafter: BiH) is a process of ancient origin confirmed by the remains of Illyrian hillfort settlements, Roman municipalities and colonies, medieval towns, townships, boroughs, šehers, and Bosnian towns during the Austro-Hungarian rule and between the two world wars. However, in the modern sense, urbanization started developing intensively in BiH only after the Second World War. It differed quantitatively and qualitatively from earlier types of urbanization.

The diversity of settlement types in BiH and their common features result from a long-term mutual, often simultaneous, agency of a number of socioeconomic factors and historical events in a given natural framework. Important Roman settlements sprang up and developed in places and areas known for traces of local, indigenous settlements (in the valley of Sana river in northwestern BiH; in the karst fields in the west, in Uskopaljska valley, the valley of Lašva river, the valley of Neretva river, and southwest BiH). At the time of the arrival of the Romans, the main types of settlements on the territory of BiH were hillforts and stilt houses. Some hillforts were built as fortifications. Stilt houses were built along the rivers, and this type of settlement left its traces on Sava and Una rivers. A hillfort is a settlement on a secluded hill and elevations, even more often on the slopes of hills and on high plateaus. Roman times represented a turning point in terms of settlement accommodation when a large number of them moved to the plains.

In Roman times, most city centers developed on an indigenous basis. We are most familiar with these municipia (settlements with city status) in BiH: Domavia, Bistue Nova, Bistue Vetus, Diluntum, Splonum, Salviae, Pelva, Delminium, etc. Urban settlements are made according to Roman models with forums, baths, basilicas, temples, and water systems. Also, there were villae rusticae throughout BiH as well as numerous agricultural settlements, then mining settlements in eastern (e.g., around Srebrenica), central (Lašva and Vrbas valleys), and western Bosnia (the area around karst fields). There were also settlements next to military camps in naturally suitable places (such as Glamočko polje, Sarajevo, Banja Luka, Mahovljani, Donja Neretva, etc.). Since mineral waters were abundantly exploited around the spas in Roman times, there were several

H. Gekić et al., *The Geography of Bosnia and Herzegovina*, World Regional Geography Book Series,
https://doi.org/10.1007/978-3-030-98523-3_12

significant settlements. The most famous Bosnian thermal springs are Ilidža near Sarajevo (settlement Aqua S, etc.), Laktaši and Gornji Šeher near Banja Luka, and Gata near Bihać.

Roman roads that passed through river valleys and overpasses in BiH had a prominent strategic and economic significance, and significant settlements developed near them. The exploitation of natural resources, especially ores, and the transit of various raw materials; the transport of equipment and food for the army; etc. gave a special role to BiH of that period (Pašalić 1960).

The fortified towns in BiH were built from the twelfth to the fifteenth century, i.e., until the fall of the Bosnian state under Turkish rule. Occupying them, the Turks gave them another purpose, turning them into military watchtowers. A characteristic addition of Turkish fortification architecture to our cities is the addition of tabi', a "bastion" for accommodating cannons. At the end of the fourteenth century, with the introduction of firearms, Bosnian feudal lords and rulers often strengthened the city walls, building new walls next to the old ones, such as on the Jajce citadel.

In the relatively small area of the Cazin region in the northwest of BiH, and the middle and lower course of Una river, the border area of the Hungarian state, numerous fortified towns were built from the twelfth to the fourteenth century, of which Bužim and Bihać stand out. Of the towns along the Sana river, the most famous is Ključ, where the last Bosnian king, Stjepan Tomašević, was captured in 1463.

The hardest and one of the largest cities was Bobovac, which was also the capital of the Bosnian state. Some of the largest cities were located along the Bosna river, namely Doboj, Tešanj, and Maglaj, and at the confluence of Bosnia and Sava rivers, the town of Dobor. Several towns were located along the Drina river, namely, Srebrenica, Kušlat, and Zvonik (Zvornik). In northeastern Bosnia, between Sava and Drina rivers, on the slopes of Majevica, the most fortified town to protect the provinces of Usora and Soli was Srebrenik.

The most important towns along the Neretva river were Valentin, Obalj, Gradac in Glavatičevo, Biograd, Vrbač, Kom, Trešnjevo, Prozor, Mostar, Cim, Gabela, Počitelj, Svitava, Blagaj, Hutovo, etc. The group of towns in eastern Herzegovina consists of Medved, Ključ near Gacko, Torić, Trebinje, Popovo, Biograd near Nevesinje, Stolac, Klobuk, etc. The group of cities in western Herzegovina consists of Gorica near Imotski, Karlovac near Čitluk, Ričani, Gradac near Posušje, Mokro near Lištica, Ljubuški, and Visoko. The cities of all three groups were built on hard-to-reach hills. Most cities had an irregular ground plan adapted to the hilltops, which followed the irregularities of the terrain. In several cities, somewhat more regular ground plans can be seen, which indicates the influence of ancient (Roman and Byzantine) fortifications on the medieval fortifications of this area. We find confirmation for this in the area of Gabela, Stolac, Kostun, and Blagaj (Redžić 2009).

From the formation of the feudal medieval Bosnian state until the middle of the fifteenth century, when there was a rapid development of the commodity-money economy (mainly based on the exploitation of mineral resources: silver, lead, iron, and partly gold and copper), urban settlements are very few and small. There were 25 open urban settlements with the functions of mining, crafts, trade, and administration in the medieval Bosnian state in the fourteenth century and 40 in the fifteenth century. According to the defter of the Ottoman Empire in 1468–1469, it can be concluded that the largest urban the settlement of Bosnia was Fojnica with a population of 1665 residents. In fact, only since the mid-fourteenth century, the accelerated development of the European economy and the conjuncture of metals in its market have created the basic conditions for the emergence of urban settlements in medieval Bosnia. In general, mining played a major role in the economic, political, and cultural development of medieval Bosnia (Kovačević-Kojić 1978).

The rapid development of the commodity-money economy leads to the development of several small market settlements, towns, and mining settlements. The Ottoman conquests stopped this development and steered it in another direction. After a period of stagnation and recovery from the ravages of war and dependent on the overcoming of new directions of strategic and trade communications, economic, administrative, and military needs of the Ottoman Empire, the process of Islamic-oriental urbanization begins with the formation of Muslim settlements as parts of old towns or the establishment of new, primarily open urban settlements with craft-production, trade, cultural-religious, and administrative functions (Kreševljaković 1949).

In the middle of the seventeenth century, Sarajevo was the most important city of the Ottoman Empire in the Balkans with a population of about 80,000 residents. However, in the fundamentally changed circumstances of the political economic and cultural civilization structure of the Turkish period, there is a very rapid and comprehensive establishment of new and development of existing urban centers. Their number has tripled. The intensified urbanization is a consequence of the complex action of numerous factors such as more widespread economic, administrative-political, and cultural activities in the tightly organized Turkish feudal system and the prominent geographical and strategic position of BiH in the territorial part of the Ottoman states. An important factor in this was the strong Islamization when, as recently established, more than two-thirds of the Bosnian population were Muslim in the first half of the seventeenth century (Sućeska 1968).

The fortified towns were classified as palanquins (usually a smaller fortification built of timber, protected on the outside by a dug trench and an earthen embankment), hisari (usually a stone fortification, with or without a tower), and towers. During Ottoman period, the most significant number of today's urban settlements in BiH developed and formed, and the traces of this epoch are still visible today. These processes created, according to the Islamic-oriental pattern (square market, bazaar with artisan quarters and public buildings, separate residential zone, high fences around houses), two basic types of urban settlements: kasaba (town, township) and šeher (city). There was usually a difference, so settlements with a Muslim population were called kasaba. In contrast, the towns with a Christian population were named varoš, and the name was probably used at least two centuries before the arrival of the Ottomans in BiH (Mastilo 1959). During the sixteenth century, due to the development of productive crafts and especially foreign trade, many urban settlements experienced a rise in population and territorial development. Among them, the most important are Sarajevo, Banja Luka, Foča, Mostar, Travnik (Fig. 12.1), Jajce, Zenica, Čajnice, Zvornik, and Livno. At the same time, mining and mining production showed a steady downward trend, which is why the old mining urban centers were stagnating and declining.

A characteristic of many kasabas formed in the second half of the sixteenth and seventeenth centuries is that they did not originate in the old varoš tradition but developed from earlier rural settlements and were formed as exclusively Muslim urban settlements (e.g., the town Četvrtkovište is today's Bijeljina, while kasaba Česta is today's Gornji Vakuf) (Handžić 1977).

It should be emphasized at the outset that the urbanization of BiH in the Ottoman period is a consequence of the complex effects of many factors. It is particularly important to note that urbanization is not only a consequence of better developed productive forces in the better organized feudal system of the Turkish Empire but also a consequence of phenomena emanating from the political, social, and cultural-civilizational structure of the new state (Bakaršić 1961). The motives for urbanization of BiH during the Turkish era are favoring cities according to Islam regulations, the geographical position of Bosnia as the most prominent Turkish province in Europe, the necessity of organizing traffic in a spatially large state, the developed craft for the great needs of numerous Turkish armies, etc. (Bakaršić 1973).

The crisis of the Ottoman Empire, beginning in the seventeenth century, led to the slow development and stagnation of cities. However, they continued to grow in number due to a significant influx of inhabitants from the countryside. Many urban settlements experienced population growth during the eighteenth century with the immigration of Muslim re-emigrants from lost territories. New urban settlements are being built for them (Bosanski Petrovac, Prijedor, Trebinje, etc.).

The period until the end of Ottoman rule (1878) was marked by the stagnation and decline of the city's economy

Fig. 12.1 Ottoman type of houses in the old part of the town of Travnik. (Photo by Aida Bidžan-Gekić)

and thus of urban settlements. Urbanization during the Ottoman era left a network of a large number of small urban settlements (towns of 1000–2000 inhabitants), local economic and administrative centers, and several medium-sized regional centers (about 10,000 inhabitants), with Sarajevo as economic, administrative, and religious-cultural center of BiH. The inhabitants of urban settlements made up just over 10% of the total population.

Traces of the Ottoman period of urbanization are still visible today, both in morphology and in the spatial structure of Bosnian cities: bazaars with craft and trade districts and public institutions and residential zones—neighborhoods (mahala) with the oriental type of houses, mosques, etc. As a result of a very dynamic urbanization, during the Turkish rule in BiH, an urban network was created with smaller urban settlements of local economic and administrative centers and several medium-sized regional Sarajevo as administrative, economic, cultural, educational, and religious centers.

The cities were divided into larger spatial units of congregations and smaller neighborhoods of mahalas. Mahalas were formed mainly around mosques, after which they got their names. Islamic-Ottoman cities consisted of two parts: čaršija (bazaar), the craft and trade part where the economic part of the city was concentrated, and the mahala (neighborhood), residential parts.

Each bazaar is characterized by a square around which numerous streets were developed, having craft shops of the same or related craft. Most of the monumental buildings were built in the area of the bazaar: mosques, hammams, inns and caravanserais, madrasas, and the most important commercial buildings—bedesten (bezistan). A characteristic of residential parts is the abundance of greenery and water. Numerous fountains were erected, and wherever possible, construction was done along the running water. Where there was no such water, backwaters were run through gardens and courtyards.

It was at this time that many cities in BiH emerged and experienced a rise. This is especially true of Sarajevo. Sarajevo is an example of the concept of a planned city-residence, which Gazi Isa-Beg began to build with his endowments on the banks of the Miljacka river. Even before the fall of the Bosnian kingdom, he built a palace-saray, after which the city was named Sarajevo. The following cities were also experiencing a rise: Mostar, Banja Luka, Tuzla, Jajce, Zenica, Travnik, Foča, Zvornik, Trebinje, Višegrad, and other cities in BiH, and traces of this era are still visible today.

At the turn of the nineteenth and twentieth centuries, the cities in BiH developed unevenly, with many local and regional features, whether they were externally visible or internally less visible elements. Some of them became important industrial centers and gradually took the form of a European industrial city (both in social structure and appearance—Tuzla, Zenica), while most remained on crafts, small entrepreneurship (with modern crafts), and increased trade and traffic (Hadžibegović 2004).

According to the 1879 census, 47 places in BiH had the status of the city, and another 27 were registered as trading places or trading municipalities, which were, in fact, small towns in BiH, of which as many as 10 were headquarters of srez (county). On the other hand, some settlements had the status of a city but lost the attributes of a city and an administrative function (Izačić-grad, Gornja Tuzla, Počitelj). In 1879, the largest urban settlement in BiH was Sarajevo, with 21,377 inhabitants. In second place was Mostar with 10,848, in third place Banja Luka with 9560, in fourth place Bijeljina with 6090, in fifth place Travnik 5887, then Tešanj with 5372, and Donja Tuzla with 5119. Only 7 urban settlements had more than 5000 inhabitants, and 43.3% of the urban population of BiH lived in them (Fig. 12.2).

During the rule of Austro-Hungary (1878–1918), the construction of railways was especially emphasized, which encouraged faster development of urban centers and the emergence of new ones in the area of mining resources. Until the second half of the nineteenth century, BiH lived in the lethargy of the Ottoman social system, and it was only with the establishment of Austro-Hungarian rule that significant industrial development began in this region. The industrial enterprises erected from 1878 to 1919 would represent, even in the period between the two wars, the base of BiH's industry, and would significantly determine its development after the Second World War, since until 1948 in BiH were restored only those industries and those industrial enterprises that were built in this region by the Austro-Hungarian authorities (Kamberović 1998).

During the creation of the first industry, railways, and modern roads under the Austro-Hungarian rule, conditions were created for the intensified development of the old and the formation of new urban settlements. New, mostly small, urban centers were growing along with coal mines (Breza, Kakanj, Ugljevik, Kreka) and industrial plants (Zavidovići, Drvar, Teslić, Lukavac, etc.) and along railways and highways (Jablanica, Čapljina, Kalinovik, Han Pijesak, Bosansko Grahovo, etc.). After a long stagnation, numerous older urban settlements in diversified and intensified activities develop and multiply their urban functions, gradually taking on, especially in new parts, the characteristics of Central European cities. Although the urban population has increased in this period, it did not exceed 20% of the total population, and the urban basis still consisted of a larger number of small urban settlements. Between the two world wars, the same tendencies continued but at a much slower pace.

Most rural settlements in BiH were created in the distant past, so the main types and basic layout of settlements are inherited. Some rural settlements date back to the Middle Ages. Most of today's rural settlements originated in the

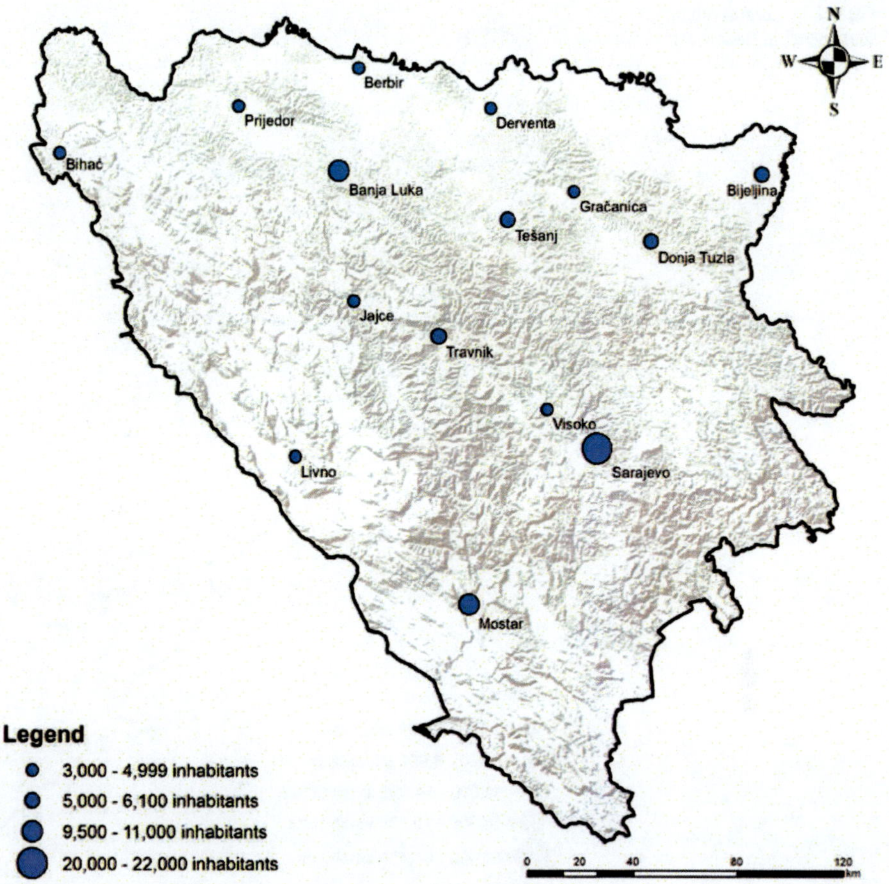

Fig. 12.2 Largest urban settlements in Bosnia and Herzegovina in 1879

Ottoman period, especially in recent centuries. The network of rural settlements later became denser, but until the period of socialist transformation, the rural settlements mostly preserved their old type and physiognomy because the traditional rural economy kept its basic features throughout that time, which determined the type and layout of settlements.

The earlier rural economy was based on seminomadic and extensive cattle breeding (sheep and goats) in mountainous areas, on farming, cattle and pig breeding, and rural crafts in hilly and plain areas (development of dispersed and compact settlements). Especially later, other factors such as the new religion, Islam, in BiH led to the creation of new Muslim rural settlements of a compact type. The settlement of cattle breeders (Vlachs) from the south in the central and northern parts and their transition to agriculture led to the creation of new rural settlements on the clearings, with dispersed houses. Preventing seminomadic livestock movements towards the coast by creating borders led to the permanent settlement of mountain livestock dwellings (katun) of earlier seasonal dwellings. Muslim returnees are settling in towns and rural areas, building new, larger rural settlements with compact houses modeled on towns, especially in border areas. First World War losses and large epidemics create space for multiple new colonizations of the livestock population. Thus, there was a change of population, relocation of settlements, and changes in their appearance and structure (Renier 1931).

With the beginning of industrialization in the Austro-Hungarian era, and although the old feudal relations remained in the countryside, the commodity-money economy put pressure on the countryside. The construction of new communications and opening forest worksites and mines has engaged surplus labor that occasionally or permanently leaves agriculture. The purchase of serfs increases the number of free peasants, the process of dividing large households begins, all leading to the construction of a larger number of new houses so that dispersed settlements become more compact, partially changing the physiognomy. The agrarian reform (1919) accelerated the process of dividing land and large cooperative households; there is a property differentiation. These processes on the rural settlement are maintained in the fragmentation of estates and the construction of new houses, most often alongside the old, common one, so that kinship groups of houses are created. Almost 12,000 km^2 of arable land and forests were confiscated from landowners and handed over to farmers, former tenants. This property was

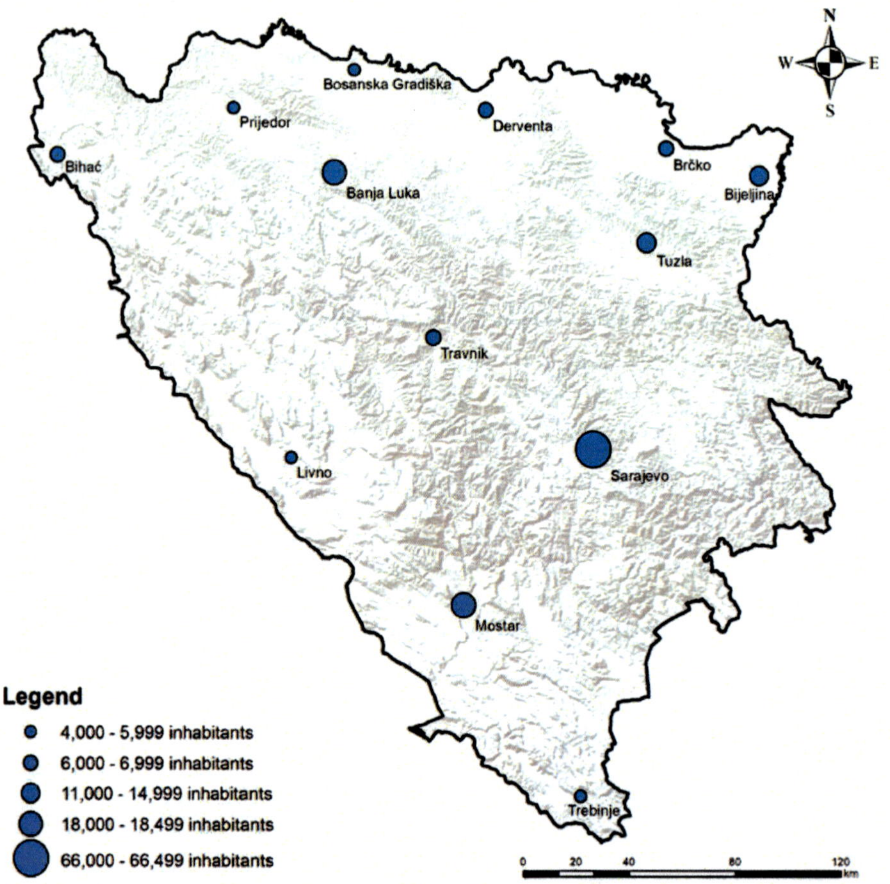

Fig. 12.3 Largest urban settlements in Bosnia and Herzegovina in 1921

confiscated mainly from Bosniaks, beys, and agas, who were never adequately compensated. Between the two world wars, the urbanization process in BiH stagnated (only 2% of the working population worked in industry). In 1921, the largest urban settlements were Sarajevo with 66,317, Mostar with 18,176, Banja Luka with 18,001, Tuzla with 14,036, Bijeljina with 11,013, Brčko with 6832, Bihać with 6372, Travnik with 6334 and Derventa with 6176 inhabitants (Fig. 12.3). There was a total of 11 urban settlements with more than 5000 inhabitants, and in them lived as much as 69.1% of the total urban population.

In the Second World War, many urban settlements suffered extensive damage. After the Second World War, rapid industrialization and urbanization took place. With the rapid and versatile development in the first place of industrial and then tertiary activities in the socialist period, the process of urbanization in all its manifestations was intensified, where the construction of road and railway roads was of particular importance. Although the increase in urban population results from the immigration of rural population into cities, these changes are to a certain extent due to the increased number of urban settlements from 1953 to 1961, due to the city new definition being applied.

After the Second World War, socialist social transformation and widespread intensive industrialization and construction of modern rail and road traffic routes led to the strengthening of the urbanization process, which resulted in establishing new urban settlements—Banovići, Novi Travnik, etc., and increasingly pronounced multifunctionality of other. Many urban settlements have doubled and tripled their population, spreading to the surrounding rural areas. In addition to the old neighborhoods, mostly in new locations, new planned neighborhoods were built on the principles of zoning. For this intensive urbanization process, transforming many rural areas into urban settlements is especially important, primarily along the roads in industrial areas (Bakaršić 1962).

Intensive processes of urbanization, industrialization, and deagrarization of BiH after the Second World War led to the spatial and physiognomic transformation of old and the creation of new villages (rural settlements). This process took place in several stages. In the first phase, after the reconstruction of the destroyed villages, there was a "thickening" of the villages with the construction of new houses. The second phase was marked by the "descent" of houses from the valley sides towards the communications, where new villages in rows were created. In the third phase, most old villages are supplied with utilities (electric lighting, local water supply, car access roads). In these villages, development cores with central functions (shops, cultural centers, ambulances, schools, sports fields, etc.) are planned for the surrounding villages within the new local communities. Mountain villages, on the other hand, were already in a phase of depopulation with a pronounced tendency to age.

During the Austro-Hungarian rule, numerous and large facilities were built throughout BiH: residential, administrative, educational, court, and military buildings, water supply systems in urban settlements, water supply facilities in karst areas, regulation of larger watercourses, about 120 industrial companies, 1517 km railways, roads, and facilities in agriculture. The technical level and organizational forms of the construction industry were at the level of technically advanced countries. Investors, designers, and professional staff were foreigners, while the unskilled workers were local people. This factor caused a stronger tradition in construction not to be established on a domestic basis. In the period between the two world wars, the construction industry almost wholly died out in BiH. In the period 1919–1925, housing construction intensified. There are no significant public works, except for a slight upgrade of the 150-km-long railway network. During the Second World War, public buildings were mostly destroyed, and the traffic was destroyed. As much as 40% of the total housing stock was destroyed.

After the Second World War, the ruralization of urban settlements in BiH created particular difficulties in accommodating rural migrants and meeting their needs. In areas where the highest quality housing stock has been built, there is the largest housing shortage and the worst living conditions (Sarajevo, Banja Luka, Tuzla, Zenica, Mostar) with significantly polluted watercourses and air, as well as heavy traffic jams. In 1961, urban settlements had already grown in population. A total of 11 urban settlements had over 10,000 inhabitants, and 57.1% of the total urban population lived in them. The following stood out in size: Sarajevo with 143,117, Banja Luka with 50,650, Tuzla with 37,760, Mostar 35,284, Zenica with 32,536, Brčko with 17,949, Bijeljina with 17,340, and Bihać with 15,763 inhabitants (Fig. 12.4).

Since 1957, after the introduction of contributions and the formation of funds for housing construction, there has been a significant increase in the volume of housing construction. Between 1946 and 1991, about 850,000 housing units were built, and investments in housing construction were significant and amounted to up to 10% of national income in specific periods. Significant results have been achieved in constructing infrastructure and communal economy, especially in supplying urban settlements with the necessary amounts of water and electricity, expanding the PTT (Post and Telegraph Organization) network, green and recreational areas, etc. In the modern period, mostly urban settlements in BiH have a sewerage system.

Accelerated industrialization and electrification; depopulation of rural areas and settlements; expansion of the transport network and development of modern production, exchange, and consumption; and the emergence of various types of migration caused the rapid expansion of cities or urbanization rural landscapes. The consequence is the emergence and transformation of urban settlements around which urbanized zones are created.

It is a feature of BiH that a large share of the deagrarized population is non-urbanized, maintaining this fairly high level in almost all inter-census periods and ultimately reaching a maximum of 46.5% in 1981, followed by a decrease to 37.2% in 1991. In 2013, it was 44.6%. This part of the deagrarized population that works but does not live in the urban settlements has a significant impact on daily migration and, at the same time, in a specific way, affects the settlement system. This population could not settle permanently in the urban areas, so they tried to get closer to the place of work by reproducing the dispersed system of construction outside the urbanized area.

In 1991, the largest urban settlements were Sarajevo with 416,497, Banja Luka with 143,079, Zenica with 96,027, Tuzla with 83,770, Mostar with 75,865, Bihać with 45,553, Brčko with 41,406, Bijeljina with 36,614, Prijedor with 34,635, Doboj with 27,498, Bugojno with 22,641, and Trebinje with 21,870 inhabitants (Fig. 12.5).

The aggressive war on BiH (1992–1995) systematically destroyed urban settlements (urbicide was carried out). Cultural and historical monuments, mosques, cemeteries, and old towns were particularly damaged. Various infrastructure was severely damaged, and agricultural and forest land was destroyed or damaged (Bublin 2000).

Applying the rules of order of magnitude, the persistence of large and largest urban settlements to keep their places in the overall order was observed. In addition to Sarajevo as the center of the state, the network of urban centers also includes regional centers, among which the cities of Banja Luka, Mostar, Tuzla, Zenica, Bihać, and Doboj stand out in terms of the development of functions, tradition, number of urban population, and size of the gravitational-functional area. Some regularity was expressed in those urban settlements that developed intensively after the Second World War, such as Sarajevo, Banja Luka, and Mostar, when development planning and coordination came to the fore. That part of the urban settlement, inherited from the past, shows that there was no planning, but it developed spontaneously and organically. This development is characterized by an irregular and fragmented traffic network, irregular and unregulated building block. At the same time, urban settlements on the terrain of great slopes have inherited steep streets that cannot adapt to modern traffic because they were inherited from the reign

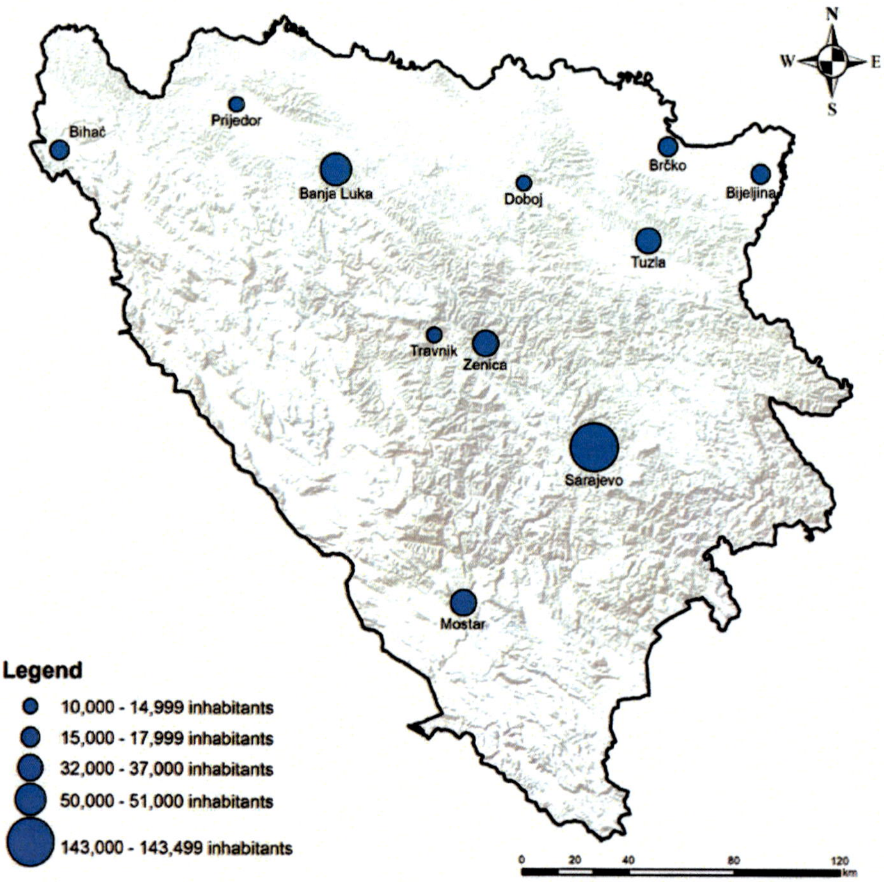

Fig. 12.4 Largest urban settlements in Bosnia and Herzegovina in 1961

of horses and pedestrians. This heritage of the cities of BiH is an aggravating circumstance in planning their future development because the central areas in question cannot be so easily abandoned.

Urban settlements in BiH have so far failed to develop entirely new urban cores. Historical centers from the Ottoman and Austro-Hungarian periods still do not lose their significance. Generally, urban settlements in river valleys have longitudinal development: Sarajevo, Mostar, Banja Luka, Doboj, Goražde, Tuzla, Travnik, etc. A smaller number of urban settlements that have developed at the estuaries of rivers have two or more development directions, while cities outside narrow river valleys such as Brčko, Bijeljina, Prijedor, Bugojno, and Livno are evenly distributed. Urban settlements are located mainly at altitudes up to 800 m, and their largest representation is up to 500 m. There is no correlation between the size of urban settlements and altitude in BiH. Urban expansion after the Second World War was largely monofunctional (residential and industrial zones) so that the multifunctional compactness of urban fabric was not ensured.

It is impossible to determine the correlation between the size of urban settlements and population density in BiH. There are significant differences between the population densities that they have. There are significant differences in population densities within urban settlements ranging from 100–150 inhabitants/ha in the central zones to 15–30 inhabitants/ha on the periphery. Some residential zones have a density of 250–350 inhabitants/ha. In the historic cores of BiH cities, the possibilities of increasing the density by new construction or reconstruction are mostly limited, and the dominant number of floors is G + 1 in the historic cores from the Ottoman period or G + 3 from the Austro-Hungarian period.

In 1991, BiH had a total of 1,315,756 housing units with 1,205,641 households. The average living space per capita was about 17 m^2, and the average apartment size was 55.5 m^2 versus 32 m^2 in 1951. If we consider the occurrence of a large number of apartments for leisure and recreation, there was a deficit of apartments primarily in urban areas (Bublin 2004). Almost three-quarters of the housing stock in 1991 was privately owned, while the remaining number of housing units—mostly apartments in apartment buildings—were socially owned. From 1992 to 1995, about 35% of housing units in BiH were partially or completely destroyed. Of this number, almost 80% of the housing units were either

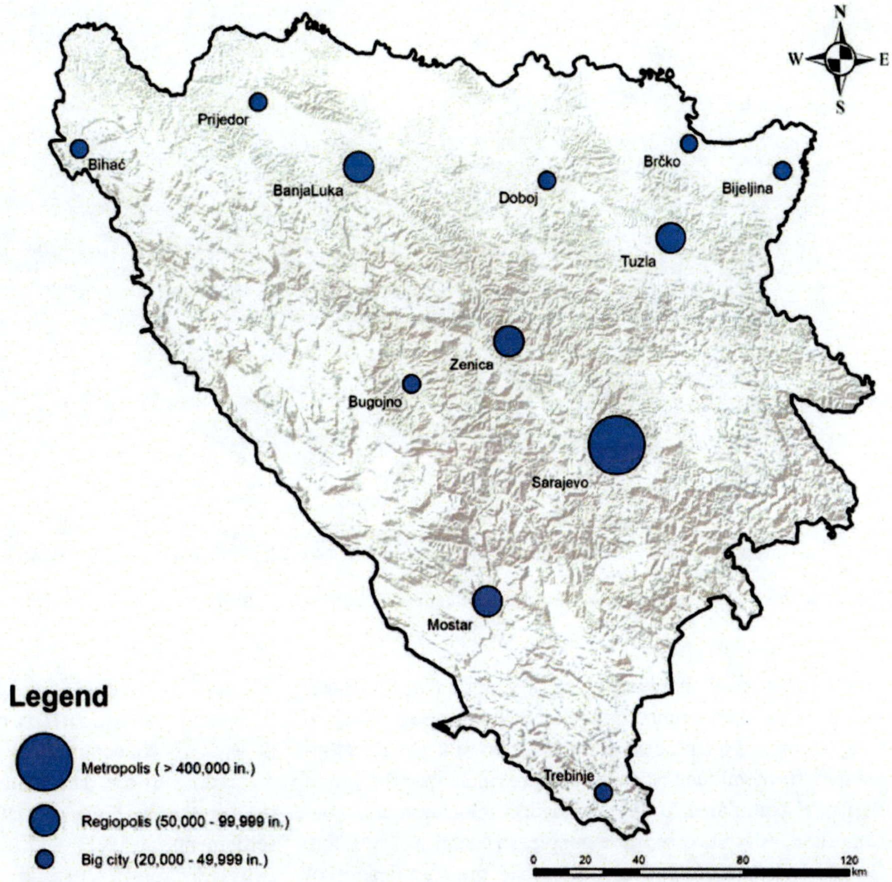

Fig. 12.5 Largest urban settlements in Bosnia and Herzegovina in 1991

destroyed or badly damaged—from that 66.6% in the entity of FBiH, 30% in the entity RS, and 3.4% in Brčko District. The reconstruction followed the return process, so since the Dayton Peace Accords Agreement by 2005, around 260,000 housing units have been reconstructed with over 170,000 with donor funds. In the years immediately following the war, the international community's participation in financing reconstruction in BiH was almost 100%, after which domestic sources gradually became involved, allocating very significant budget funds at all levels of government in BiH. By 2019, a total of about 345,500 housing units have been renovated in BiH (Ministry of human rights and refugees of BiH 2006). In 2013, BiH had a total of 1,607,998 housing units (Fig. 12.6) with 1,155,552 households. The average living space per capita was about 34 m^2 and was twice as large as in 1991, while the average apartment size increased to 74 m^2. There were a large number of vacant flats, namely, 385,659, which represents 24% of the total number of housing units and is a consequence mainly of forced migrations from the period 1992–1995 and economic migrations in the last two decades (BHAS 2016).

The increase in urban population was largely dependent on the magnitude of migration movements and less influenced by the natural increase in urban population. In the context of rapid economic development, industrial and urban centers were increasingly attracted to the population from closer and further surroundings, and the migration of population from villages to cities (related to industrialization and changing economic composition of the population) was a characteristic phenomenon in this period. Population migration towards cities has led to a much faster increase in urban population than was the increase in the total population of BiH.

The most dynamic changes in the period of the strongest industrialization up to 1981 were next to the cities and their settlements in the immediate vicinity as well as along the main roads. In this period a weak, still insufficient diversification of rural municipalities was recorded manifested by reducing the share of the active population in the primary and increasing the number of employees in the secondary and tertiary sectors (Gekić and Bidžan-Gekić 2016).

Polarization in the urban development of BiH is reflected in the very dynamic monocentralism of the metropolis, i.e., the growing dominance of Sarajevo and the macro-regional centers of Banja Luka, Tuzla, and Mostar, which was intensified by the war, during which most refugees and

Fig. 12.6 Holiday homes in Ravno Rostovo near Bugojno. (Photo by Aida Bidžan-Gekić)

displaced persons inhabited these cities, and now, for objective reasons, they cannot or are slowly returning. This, in turn, intensified the process of polarization of urban development of the mentioned cities, further provincialized the rest of BiH, and stimulated further population relocation and concentration, as is the case, for example, in Trebinje, Foča, and Goražde. Before 1992, these cities had the significance of regional centers, which in modern times, due to the Dayton political-territorial structure, they lost.

The problem that occurs in the cities of BiH is the process of ruralization of the cities of BiH that takes place in modern times, directly related to a more pronounced concentration of population. Namely, due to the war and post-war rural exodus, high unemployment (unemployment rate in 2020 was 33.7%), poverty (GDP per capita in 2020 in BiH was about 6100 USD), low level of education of the population inhabiting cities and suburbs in the last 15 years, and overpopulated and spontaneous construction, the periphery of Bosnian cities often adopt a semirural way of life, which rarely reaches the city center. It is pretty certain that the process of urbanization, i.e., the penetration of the urban way of life into the surroundings of BiH cities, is more substantial and more relevant. However, we should not lose sight of the problems arising from rapid urbanization, often more quantitative than qualitative. These are very complex sociological and socio-geographical processes of faster life transformation, which sometimes cause several psychosocial problems (according to Nurković 2006).

With 69 in./km^2 in 2013, BiH is one of the least populated countries in Europe. It has a total of 6141, mainly small settlements, with an average of 575 inhabitants. On every 100 km^2 of territory, there is an average of 12 settlements. According to that, BiH is one of the countries with a low degree of concentration of inhabitants and settlements. According to the data of the Agency for Statistics of BiH used in the 2013 census (BHAS 2013), there are 110 urban settlements in BiH and 6031 other settlements (mostly extremely rural). There are significant differences between settlements in the number of inhabitants, types of functions, population structure, and other characteristics. Since the 1981 census, the administrative and legal criteria have been used to determine and declare the type of settlement in BiH through legal regulations. The settlements are divided into two groups: urban and other. It was allowed for a local level, a municipality, to designate a settlement as urban if it has historical or tourist significance or constitutes an important protected urban heritage. In the 2013 census, the Agency for Statistics kept the types of settlements that were in 1991.

12.2 Number and Size of Nonurban Settlements

Out of the total number of settlements in BiH, 98.2% are settlements that are not urban and which the Agency for Statistics of BiH classifies in the group of others (nonurban or mostly rural). In 2013, there were 2,030,091 inhabitants, or 57.5% of the total population of BiH in these settlements. On average, 337 inhabitants live in each rural settlement. Due to its size, the structure of rural settlements is very fragmented. It arose as a direct reflection of the dependence on rural settlements' micro-location, genesis, and functions. As a

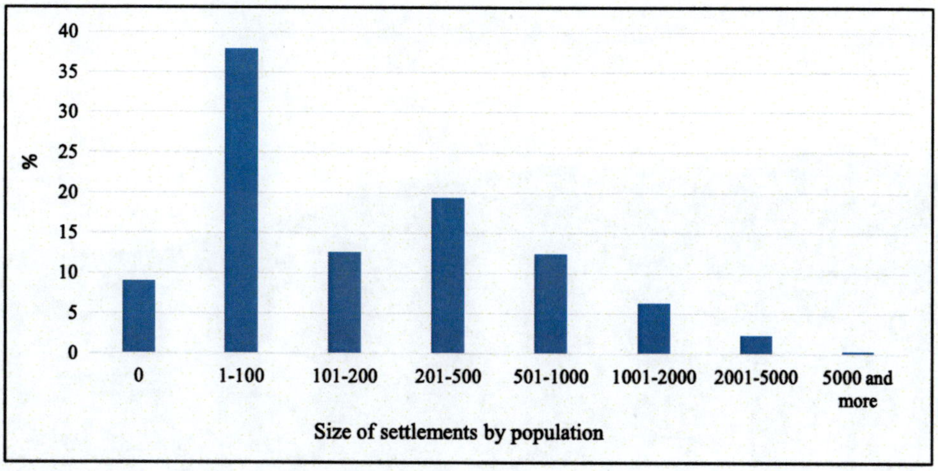

Fig. 12.7 Size of rural settlements by population in BiH in 2013

result of the war between 1992 and 1995, many settlements with no inhabitants remained. Their total number is as much as 542 or 9% of the total number of rural settlements, and their distribution is mainly related to geographically unfavorable mountainous and high-karst areas of eastern, central, and western BiH. There are 2286 or 37.9% of settlements with up to 100 inhabitants, and an average of 34 inhabitants live in them. There are 758 or 12.6% of rural settlements with 101–200 inhabitants. They have an average of 147 inhabitants. There are 1175 rural settlements with 201–500 inhabitants, and the average number of inhabitants is 331. Therefore, in the category of typically small settlements up to 500 inhabitants, there are 4756 or 78.9% rural settlements. They are inhabited by a total of 576,864 or 16.3% of the population of BiH. Settlements of 501–1000 inhabitants have 746 and 527,692 inhabitants. In the category of 1001–2000 inhabitants, there are 380 settlements in which a total of 508,622 inhabitants live. In the penultimate category are rural settlements with 2001–5000 inhabitants, of which there were 139 (Fig. 12.7). In 2013 they had a total of 380,862 inhabitants. In the category of rural settlements with over 5000 inhabitants, there were only 5 settlements with a total of 36,051 inhabitants. The largest settlement of this type in BiH is Janja, which is located in the northeast of BiH near Bijeljina.

With the number of inhabitants and the area of the territory, BiH has a very developed network of rural settlements. The fragmented and uneven distribution of rural settlements arose due to the action of various natural and socioeconomic, historical, and other factors. It is adequate to the fragmented individual agrarian-livestock production and the archaic social-property organization. Small settlements usually do not have even a minimum of utility services. They have difficulties with educating children, health care, and raising all forms of social standards. In this respect, there are significant differences between the northern Peripannonian, central mountain-valley and karst, and the southern karst regions of BiH (Figs. 12.8 and 12.9). Lukomir is a mountain village (rural settlement) located on the southern slopes of Bjelašnica at 1495 meters AMSL (above sea level), which makes it the highest inhabited place in BiH and the only one over 1300 m (Fig. 12.10).

Based on the number and socioeconomic structure of the population and other economic-geographical characteristics in BiH, we can single out the following groups of rural settlements: lonely settlements of the agrarian type represented in mountainous parts; classic rural settlements with different means of agrarian production (agricultural, agricultural-livestock, and livestock) which represent units outside the narrower zones of influence of urban settlements; rural settlements in suburban areas in which the most pronounced socioeconomic and social stratification has occurred, the emergence of a mixed population, and which are transforming quite rapidly under the influence of urban settlements; there are also rural settlements with a specific function such as mining, mining-agricultural, forestry, etc.; and then resort settlements—settlements that have become social and gravitational centers but that have not yet grown into urban settlements.

By classifying rural households into four main groups (nonagricultural, gardens, smallholdings, and farms) according to the amount of land cultivated and number of livestock kept, the Rural Household Survey established that just over half of rural households (51%) have no more involvement with agriculture than their urban or suburban counterparts, with over a third (36%) having no agricultural production at all. More than a third (36%) operate "smallholdings," producing a significant share of their own food requirements, but with a relatively small involvement in agricultural markets. The net cash income they generate from

Fig. 12.8 Characteristic old-type rural houses in mountainous part of western Bosnia near Blidinje Lake. (Photo by Aida Bidžan-Gekić)

Fig. 12.9 Rovna—contemporary mountainous village near Bugojno in Central Bosnia. (Photo by Haris Gekić)

agriculture represents just a few percent of the total household incomes, but the food they produce has a significant income-saving effect in addition to this. Around 13% of households can be considered as full- or part-time farms, producing significant quantities for sale. However, even this group obtains more than three-quarters of household income from nonagricultural sources (mainly regular employment and social benefits), and so is more dependent on the nonagricultural economy and on social policies than on the agricultural economy and its policies. In fact, less than 1% of households can be classified as "commercial farms." Livestock production is very important in BiH, with 43% of all

Fig. 12.10 Lukomir—the highest inhabited place in Bosnia and Herzegovina. (Photo by Aida Bidžan-Gekić)

households in the Rural Household Survey keeping livestock of some kind (UNDP 2013).

12.3 Number, Size, and Spatial Distribution of Urban Settlements

During the Austro-Hungarian rule in 1879, BiH had 47 cities and after the First World War in 1921, 59. In 1961, BiH had 66 urban settlements, and already in 1971, it had 96. In 1991, there were 105 urban settlements, while in 2013, there were 110. In 1991, the average size of an urban settlement was 17,377 inhabitants and in 2013, 13,646. Urban settlements in BiH covered about 1560 km^2 or 3% of the total area of the territory in 2013. They had 1,501,068 inhabitants or 42.5% of the total population. The average urban population density in BiH is 962 in./km^2.

The urban settlements (towns) with the highest population density in 2013 were (in in./km^2) Novi Travnik, 4415; Mostar, 3307; Višegrad, 3073; Zvornik, 2948; Široki Brijeg, 2627; Breza, 2430; Visoko, 2296; Jablanica, 2193; Velika Kladuša, 2082; and Konjic, 2064. The largest urban settlements in BiH had the following population densities (in in./km^2): Sarajevo, 1917; Banja Luka, 1354; Tuzla, 1538; Zenica, 1614; Ilidža, 698; Mostar, 3307; Bijeljina, 1499; Brčko, 1685; Bihać, 470; Prijedor, 1651; Doboj, 1721; and Trebinje, 1042. The least populated urban settlements are the following (in./km^2): Istočni Stari Grad, 25; Bosansko Grahovo, 54; Trnovo-FBiH, 69; Kupres-FBiH, 81; Neum, 89; Glamoč, 112; Grude, 155; Ljubinje, 169; and Bosanski Petrovac, 169 (calculated from BHAS 2016). One of the examples of urban settlements with above-average population density is Bugojno, located in central BiH at the crossroads of important roads connecting Sarajevo in the east with split in the Republic of Croatia in the west and Mostar in the south with Banja Luka in the north. Bugojno has a population density of 1360 in./km^2; it is home to 1% of the urban population of BiH and occupies 0.7% of the urban area of BiH (Fig. 12.11) (Gekić 2021).

The urbanization pattern in BiH is mainly oriented to larger cities (urban settlements). The BiH is going through intensive processes of depopulation and demographic aging, with pronounced regional disproportions of the population in the last two decades. BiH is insufficiently urbanized country in twenty-first century. In the period from the first official census during the Austro-Hungarian rule (1879) in BiH, the total population grew from 1.16 million to 1.9 million by the beginning of the First World War. The share of urban population was very low and ranged from 12.8% to 14.6%. The urban population growth rates were constantly higher than the total population growth rate, so in the period 1895–1910, the urban population grew at an annual average rate of 2.1% and the total population at an average annual rate of 1.4%. Then between the two world wars, due to the economic policy of the country's development and orientation towards agriculture, the share of urban population declined to 10.1% in 1931 and after the Second World War in 1948, to 10.8%. The total population increased from 2.32 million in 1931 to 2.56 million in 1948. At the same time, urban population growth averaged 1.1% annually, compared to 0.6% of the average annual population growth. In the period 1953–1961, the share of urban population increased from 12.7% to 20.6%, while the average annual growth rate of urban population was 10.8% compared to the average annual growth rate of the total population of 1.9%.

In the period 1961–1991, the total population increased from 3.28 million to 4.38 million at an average annual rate of 1.1%, while the share of urban population increased from 20.6% to 39.7%, from 657,275 to 1,737,797 at an average annual rate of as much as 5.2%. In this period, the urban population increased by 2.6 times in 30 years. According to the estimates of the Agency for Statistics of Bosnia and Herzegovina, after the 1992–1995 war, BiH had about 732,000 less of the total population (16.7% less) and about

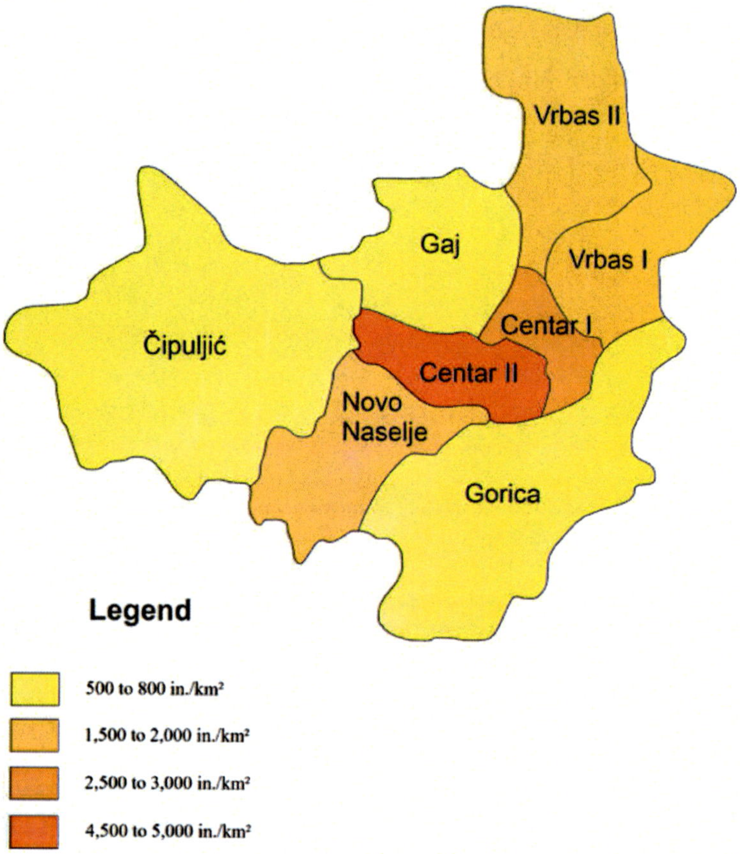

Fig. 12.11 Population density in urban settlement of Bugojno in 2013

190,000 less of the urban population (10.8% less) in 1996 compared to 1991. This is due to the suffering of the population, emigration from outside the country, and the relocation of the population from villages to cities. The total share of urban population is 42.5% in 2013, which is 2.8% more than in 1991 (Fig. 12.12). However, essentially the urban population is down by about 240,000. If we take into account the estimated number of inhabitants from 1996, we can conclude that the number of urban population in BiH is smaller by one large city of 50,000 inhabitants, and in comparison with 1991 by a city the size of Sarajevo from 2013. After 1996, there is a stagnation in the increase in urban population share. BiH in the last interval between census 1991 and census 2013 saw growth of share of urban population in total population by 2.8% mostly because decrease in total population was higher than decrease in urban population (−19.3% and −13.7%) (Gekić and Bidžan-Gekić 2020).

Some of the main characteristics of contemporary urban development in BiH after the period of war which lasted from 1992 to 1995 are strong regional disparity in the recent urban development, which is reflected in uneven urban population distribution within BiH; strong demographic growth of urban areas and strong depopulation of rural areas; and spatially differentiated concentration and polarization of the demographic development of BiH. In 2013, the largest urban settlements in BiH were Sarajevo with a population of 271,194, Banja Luka with 138,963, Tuzla with 74,457, and Zenica with 70,553. There is a total of 16 urban settlements with populations in excess of 15,000, more than half of which are situated in the north and only two in the south, and not a single one in the east or west part of our country (BHAS 2016). The level of urbanization achieved in 2013 shows that BiH is one of the countries with a moderate level of urbanization in the world. At the same time, it has the lowest level of urbanization in Europe (measured by the share of urban population), similar to Moldova.

To show the evolution of the distribution of urban settlements of different sizes over time, Gekić and Bidžan-Gekić (2020) analyzed the size and number of urban settlements by dividing them into six classes by population number for 1981, 1991, and 2013 with respect to the average urban settlement size of the following: I (100,000+), metropolis; II (50,000–99,999), regiopolis; III (20,000–49,999), big town; IV (10,000–19,999), midtown; V (5000–9999), small town; and VI (below 5000), micro town. The number of smallest urban settlements (VI class) decreased by 24% from 1981 to 1991, while the number of small urban settlements (V class) increased by 26%. At the same time,

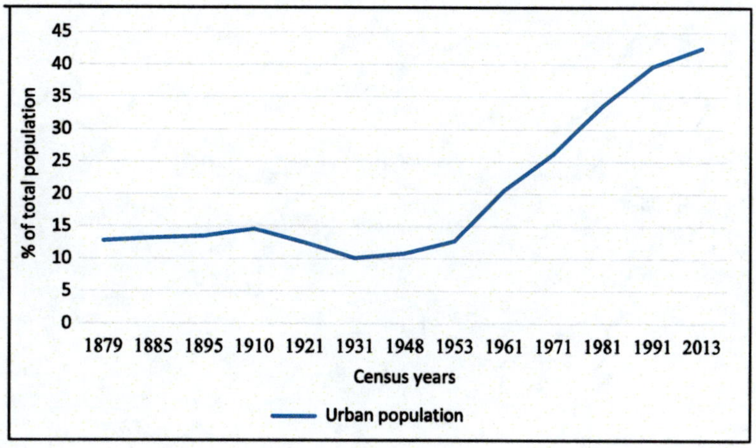

Fig. 12.12 Urban population of Bosnia and Herzegovina from 1879 to 2013

Table 12.1 Number of urban settlements by classes, 1981–2013

Class	1981	1991	2013
I (100,000 +)	2	2	2
II (50,000–99,999)	3	3	4
III (20,000–49,999)	5	7	6
IV (10,000–19,999)	21	25	22
V (5000–9999)	27	34	33
VI (below 5000)	45	34	43
Total	103	105	110

Gekić and Bidžan-Gekić (2020)

Table 12.2 Urban population in classes of urban settlements (%), 1981–2013

Class (population)	1981	1991	2013
I (100,000 +)	33.7	32.2	27.3
II (50,000–99,999)	13.8	14.7	17.9
III (20,000–49,999)	10.5	13.3	13.4
IV (10,000–19,999)	19.4	20.1	18.5
V (5000–9999)	13.1	13.5	14.7
VI (below 5000)	9.5	6.2	8.2

Gekić and Bidžan-Gekić (2020)

the share of urban population in the smallest group of urban settlements decreased from 9.5% to 6.2%. Due to the war and related emigration movements, the number of urban settlements in the VI class increased by 9, and the share of urban population in this group increased by 2.0% (Table 12.1).

The great dependence of urban settlements on economic development was demonstrated in the period 1981–1991. Apart from the top five largest settlements, urban settlements with the strongest industries and the largest investments in the industry have grown the fastest. Examples of such urban settlements were Brčko (45% growth) and Bugojno (38% growth). The average population size of urban settlements was 13,450 in 1981, 16,550 in 1991, and 13,646 inhabitants in 2013.

The largest changes in population number in the period 1991–2013 were in urban settlements with the majority Bosniak population. Sarajevo lost 23% of the population or Zenica 26.5%, compared to 1991. By contrast, urban settlements that were affected by far fewer war events (1992–1995), as in the entity of Republika Srpska, have more inhabitants in 2013 than 1991 (Bijeljina—an increase of 15.5%).

In 2013, as well as in 1981 and 1991, urban settlements up to 5000 inhabitants (VI class—micro towns; Table 12.2) had the largest share of urban settlements. The share of these settlements in the total number of urban settlements in BiH decreased from 43.7% in 1981 to 39.1% in 2013. This points to a significant feature of the urban network of BiH, which is mainly composed of small urban settlements. In contrast, in this urban settlement class, only 8.2% of the urban population lived in 2013. The largest share of the urban population of BiH of 27.3% lives in the two largest urban settlements (metropolises), Sarajevo and Banja Luka. In 2013, the

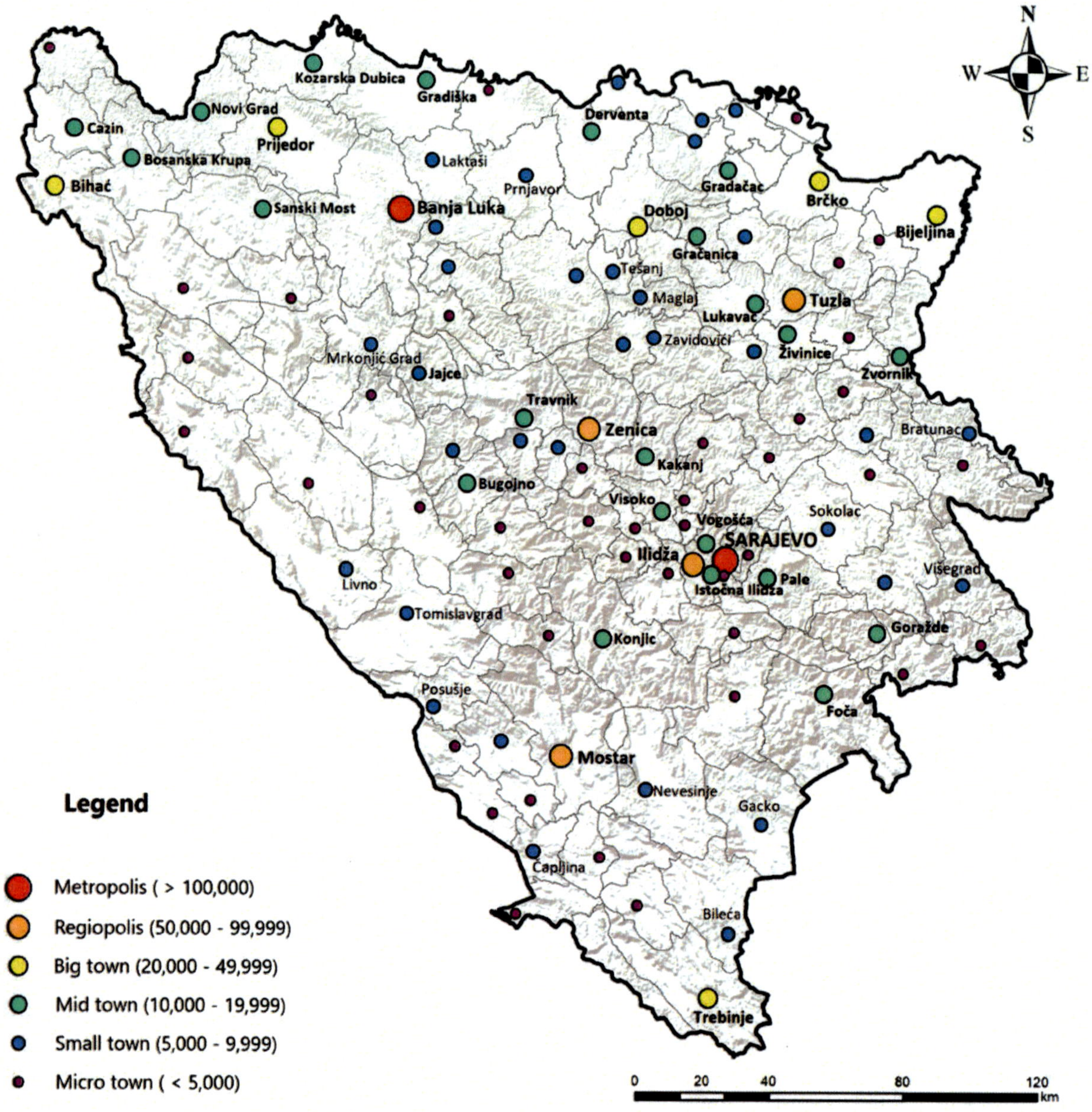

Fig. 12.13 Class categories of towns in Bosnia and Herzegovina in 2013

urban system of BiH had only 12 urban settlements with more than 20,000 inhabitants (10 in 1981 and 12 in 1991).

In 2013, the largest urban settlements were Sarajevo with 271,194, Banja Luka with 138,963, Tuzla with 74,457, Zenica with 70,553, Ilidža with 68,366, Mostar with 60,195, Bijeljina with 42,278, Brčko with 39,893, Bihać with 39,690, Prijedor with 29,555, Doboj with 25,132, and Trebinje with 23,770 inhabitants (Fig. 12.13). The current urbanization is most influenced by the emigration of young skilled population to larger urban areas, and especially to European Union countries such as Germany.

A large number of small urban settlements in 2013 stand out, as many as 43 or 39.1% with less than 5000 inhabitants. It should be noted that several settlements from this group had less than 2000 inhabitants. These are Istočni Stari Grad (37), Trnovo-FBiH (67), Istočno Novo Sarajevo (84) (separate parts of Sarajevo and Trnovo after the establishment of IEBL), Bosansko Grahovo (651), Kreševo (1025), Kalinovik (1040), Šekovići (1322), Rudo (1760), Han Pijesak (1820), and Glamoč (1885). In the territorial distribution of urban settlements, it is noticeable that small and sparsely located urban settlements are characteristic of western and eastern BiH. There are no urban settlements in these parts of the country that can be considered big (20,000–50,000 inhabitants). A comparative analysis of the size of urban settlements leads to the conclusion that Sarajevo is in great disproportion with the urban settlements of the east, south, and west of BiH. 42.5% of the total population of BiH lives in all 110 urban settlements, and a slightly higher share of the employed population is 46.5% in urban settlements in relation to the total number of employees in BiH (based on results from BHAS 2016).

12.4 Characteristics of Urban System

The genesis of almost half of the urban settlements with less than 5000 inhabitants shows that they are mostly road settlements and local centers of the rural environment. After the Austro-Hungarian occupation, between the two world wars, and in the socialist period, most of these settlements were not interesting for investments in the industry due to traffic isolation and other causes. Private capital was invested in numerous sawmills in Pale, Trnovo, Šipovo, and Čelinac (Mišković 1978). Wood processing has been extended to the modern period. Some urban settlements from this group have benefited enormously from the emigration of young labor to European Union countries, which send significant remittances to BiH every year. This primarily refers to Cazin, Velika Kladuša, and Ključ in the northwest of BiH, which are spatially very close to the European Union. Some urban settlements, such as Pale-RS, have significantly increased their population with the establishment of the IEBL and the establishment of administrative and governing and other functions of the entity of RS in this settlement.

The urban system of BiH is approximately following the rank size rule according to Zipf (1949) in the distribution of cities at national level. The second-ranked city of BiH does have the population as almost half the population of the largest city. Sarajevo is the largest urban settlement, and in 2013, its population was 271,194, and the second largest city, Banja Luka's population in 2013 was 138,963, just a little bit more population than one-half of Sarajevo. Even the fifth-ranked city Mostar had concentration of people very close to one-fifth of Sarajevo (Gekić and Bidžan-Gekić 2020).

The polarization of BiH's urban development is reflected in the very dynamic monocentrism of the capital, i.e., the growing domination of Sarajevo and macro-regional centers such as Banja Luka, Tuzla, and Mostar, which have become more intensive as a result of the war (migration of a large number of refugees and displaced persons). This has amplified the polarization of urban development in said cities, while further provincializing other parts of BiH and stimulating further migration and concentration such as in the case of Trebinje and Bugojno. Before 1992, these cities were considered regional centers, a role which they have lost under the current political and territorial system established by the Dayton Peace Agreement.

As the strongest work center, Sarajevo rests on the edge of the network, while some parts of the region lack stronger work centers. This is particularly evident in the eastern part of central BiH. In addition to the asymmetry in terms of the size of work centers, the density of the urban network is also uneven. As far as network density is concerned, central BiH stands out, particularly the road directions towards Zenica and Doboj. The poorest network, however, is observed in eastern BiH. This will certainly influence the individual directions, especially the intensity and level of urbanization thereof (Nurković and Gekić 2011).

Significant changes occurred in the top 20 urban settlements when comparing the situation in 2013 to 1981. The first four urban settlements maintained their rank, while Ilidža took over the fifth rank from Mostar by separation from Sarajevo urban settlement. Brčko, Bihać, Doboj, Prijedor, Sanski Most, Bugojno, Travnik, and Gradiška also lost their ranks for one to three positions. Derventa experienced the largest decline in what was once the top 20 ranked urban settlements from the 14th to the 25th position, mainly due to the large loss of Bosniak and Croat population and the collapse of industrial production (85% fewer Bosniaks and Croats combined). In 2013, Živinice occupied the 14th place, and they used to be on the 37th, thanks to the proximity of Tuzla, one of the largest urban settlements of BiH, and the development of industry. The urban settlement of Pale climbed to the 19th from the 92nd position, as it became one of the most important centers of development of the entity of Republika Srpska by the Dayton Peace Agreement.

The major five urban settlements of BiH remained the dominant in Bosnian urban system; they remained the top metropolitan cities after 1991: Sarajevo, Banja Luka, Tuzla, Zenica, and Mostar. The large urban centers like Sarajevo act as the major magnets for the migrants after the 1995. These cities are becoming more popular recently. It is happening because these cities are getting more attention for the urban development, whereas the small and medium cities are not getting much attention by the governments.

Generally speaking, the distribution of urban settlements in BiH according to the rank size rule is quite regular when compared to other developing countries. Since the post-Second World War period, the urban system has been increasingly moving towards a state adjusted to the rank size rule in line with the economic development of the country, which is unfortunately disrupted by recent war events, and many urban settlements that were ranked medium- and small-sized in 1991 today have a smaller population and an economy dominated by tertiary activities, and the industry has not been restructured. The analysis of the distribution of urban settlements according to the rank-size rule order from 1991 to 2013 shows no significant deviations from the new,

Dayton, political-geographical structure of BiH, with the exception of the appearance of Ilidža, Cazin, and Pale in the top 20 largest urban settlements. In the future, with the planned infrastructure projects, the growth of the top eight largest urban settlements is expected to be much higher than in the rest. A research of mobility in the ranking of urban settlements by size (1981–2013) reveals that the very top of the urban system hierarchy is far more stable than the part where small urban settlements are located. With greater investment especially in the industry, some urban settlements were moving from lower to higher hierarchical ranks.

In 2013, the urban system of BiH is characterized by desirable urban primacy with a value of 0.49. We can say that BiH has almost a primate city. Bosnia has more than one dominant city of economic and political importance, Sarajevo as its administrative and economic center and Banja Luka as second economic center and regional center. In 1981, the urban settlements of BiH were mostly occupied by industrial workers. In medium- and small-sized urban settlements, the share of employees in industrial activities exceeded 55% and in large urban settlements 35%. Already in 1991, significant changes in the development of urban settlements were noticeable, with larger urban settlements having an increasing share of employees in tertiary activities (on average 60–70%), while medium- and small-sized urban settlements slightly increased this share from 35% to 40% (Gekić and Bidžan-Gekić 2020). In the period 1992–1995, the industry of BiH has experienced a major collapse. Important markets have been lost; large numbers of capable and educated workforce have left the country; and manufacturing technology has been destroyed, devastated, obsolete, etc. The industry has not been able to return or successfully restructure, and tertiary activities still do not give much impetus to the development of urban settlements as it was done by industry until 1991.

In the eight largest urban settlements of BiH in 2013, 257,365 inhabitants were employed of which only 18% worked in industrial activities and over 80% in tertiary activities. Over 80% of employees in tertiary sectors were in Sarajevo, 82.6% (79,000); Banja Luka, 80.2% (39,370); and Mostar, 80.1% (15,526). The share of employees in tertiary activities is high in medium- and-sized small urban settlements, with an average of between 60% and 70%, and examples are Ljubuški, 81.5%; Neum, 84.3%; Vitez, 77%; Livno, 76%; Bugojno, 69% (Fig. 12.14); Visoko, 61%, etc. In the tertiary sector in urban areas of BiH, most of the population is employed in the wholesale and retail trade sector (16.8%) and as much as 12.1% in the public administration and defense and compulsory social security sectors (this last share is sufficient to indicate that the Dayton Peace Agreement made the administrative structure of BiH more complex; only 5.1% of employees were in 1991) (BHAS 2016; Gekić and Bidžan-Gekić 2020).

Given below is the calculation of the degree of urbanization for all municipalities/towns (hereinafter: municipalities), taking into account the share of urban in the total population in 2013. The first group, rural municipalities, includes 31 municipalities. These are mainly municipalities small in surface, created by establishing an inter-entity border line (IEBL) through the Dayton Peace Agreement, which were previously highly rural with a dominance of agricultural function.

In the second group of municipalities with a very low degree of urbanization (up to 20.00% share of urban population), there are 21 municipalities. Among them, the ones with the smallest share stand out: Istočni Stari Grad (3.3%), Trnovo-FBiH (4.5%), Kalesija (6.2%), Velika Kladuša (11.2%), and Tešanj (12.2%). Most municipalities were in the third group, with a low degree of urbanization (from 20.00% to 39.99%), a total of 49. Among them stood out those with the lowest share: Hadžići (20.9%), Cazin (21.0%), Široki Brijeg (21.3%), Breza (21.3%), and Busovača (21.7%) on one side, and those with the highest share: Kotor Varoš (38.6%), Bratunac (38.6%), Bijeljina (39.2%), Nevesinje (39.8%), and Bosanska Krupa (39.9%), on the other hand.

In the fourth group with a medium degree of urbanization (from 40.00% to 59.99%), there were 25 municipalities. Among them, those with the lowest share stood out: Vogošća (40.1%), Jablanica (40.1%), Sanski Most (40.8%), Derventa (42.4%), and Konjic (42.7%) on one side, and those with the highest share: Goražde (56.5%), Mostar (56.9%), Kupres-FBiH (57.0%), Vlasenica (58.6%), and Gacko (59.7%), on the other hand.

A total of ten municipalities had a high degree of urbanization (from 60.00% to 79.99%), namely, Rogatica (60.8%), Foča-RS (61.4%), Pale-RS (61.7%), Zenica (63.7%), Neum (64.7%), Tuzla (67.1%), Bileća (70.5%), Bihać (70.6%), Ljubinje (73.9%), and Banja Luka (75.1%).

The very high degree of urbanization (80.00% and more urban in the total population) had six municipalities, four of which are part of the capital city of Sarajevo: Centar Sarajevo (96.2%), Novi Grad Sarajevo (99.4%), Novo Sarajevo (98.7%), and Stari Grad Sarajevo (98.3%). Ilidža (95.2% share of the urban population) is an almost inseparable part of the city of Sarajevo, but it is administratively separated, while Istočna Ilidža (93.6%) is a former part of Ilidža which is rapidly developing under the agglomeration influence of Sarajevo. Trebinje, in the extreme south of BiH, has a share of the urban population of 81.4% and is an example of decades of development and population concentration predominantly within the urban border lines, mainly due to unfavorable geographical conditions in other parts of this municipality (Fig. 12.15).

Migrations from urban settlements abroad, in relation to smaller and economically less developed urban settlements to large and economically more developed urban settlements,

Fig. 12.14 Western part of the town of Bugojno. (Photo by Haris Gekić)

and still in the rural-urban relation, have the greatest influence on the movement of the number of inhabitants in urban settlements in the last two decades. The functional analysis of urban settlements in BiH is based on the data of the last census in 2013. The number of employed population by activities that the census recorded in 21 features was analyzed. Urban settlements are classified into types according to the predominant activity if it exceeds 25% of the total number of employees in the settlement. If there was no such activity, we determined the type according to the predominant group of activities (most often, the service type or industrial type). There are a total of seven functional types of urban settlements (towns): urban settlements of primary production, mining, industrial, commercial, tourist, health, and diversified service (Fig. 12.16).

In the first group of urban settlements of primary production (agriculture, forestry, and fishing), there were seven settlements. These are Bosansko Grahovo and Glamoč in the west, Čajniče and Han Pijesak in the east, Kalinovik and Ljubinje in the southeast and south, and Kneževo in central BiH. Their common feature is a small number of inhabitants (below 4000) and an unfavorable natural geographical environment (high mountains, karst fields, etc.). In Bosanski Grahovo, 34.5% of employees worked in primary activities. Only in Kneževo did the total number of employees exceed 1000 (1153).

In the second group were mining urban settlements, of which there were three. These are Banovići, Breza, and Kakanj, which for more than 100 years have based their development on the exploitation, and partly processing, of the coal ore. In Banovići, 30.1% of employees were in mining. The third group included industrial towns (manufacturing, construction industry, electricity production), of which there were 40. Some of these towns had over 40% of employees in the industry. These are Gacko (53.2%), Kotor Varoš (51.0%), Ugljevik (44.9%), Gradačac (43.1%), Gračanica (42.2%), Brod (41.4%), Goražde (41.0%), Maglaj (40.9%), and Derventa (40.3%). Only one town in this group, Trebinje, exceeds 20,000 inhabitants in size.

In the fourth group, there were urban settlements with a dominant trade function, of which there were ten. These are Vitez, Posušje, Široki Brijeg, Čitluk, Ljubuški, Srebrenik, Grude, Živinice, Bijeljina, and Brčko. As many as half of these settlements are located in southwestern BiH near the border with the Republic of Croatia, of which Posušje stands out with 28.7% of employees in trade activities. Vitez used the favorable geographical position in central BiH, through which the main route connecting Sarajevo with Banja Luka passes, by building a special trade zone, which is one of the largest in BiH. Vitez is an urban settlement with the largest share of employees in trade activities in BiH (32.6%). Brčko and Bijeljina, located in the extreme northeast on the border with the Republic of Croatia and the Republic of Serbia, have taken advantage of the favorable geographical position for developing trade activities.

In the fifth group, there was one urban settlement, Neum, where the tourist function dominates. It is the only urban settlement in BiH on the sea. Of the total number of employees, 29.1% work in tourism and catering. In the sixth group, there was one urban settlement, Fojnica in central BiH, with the largest share of people employed in health and social work (30.8%). Fojnica is a place known for its healing waters and spas.

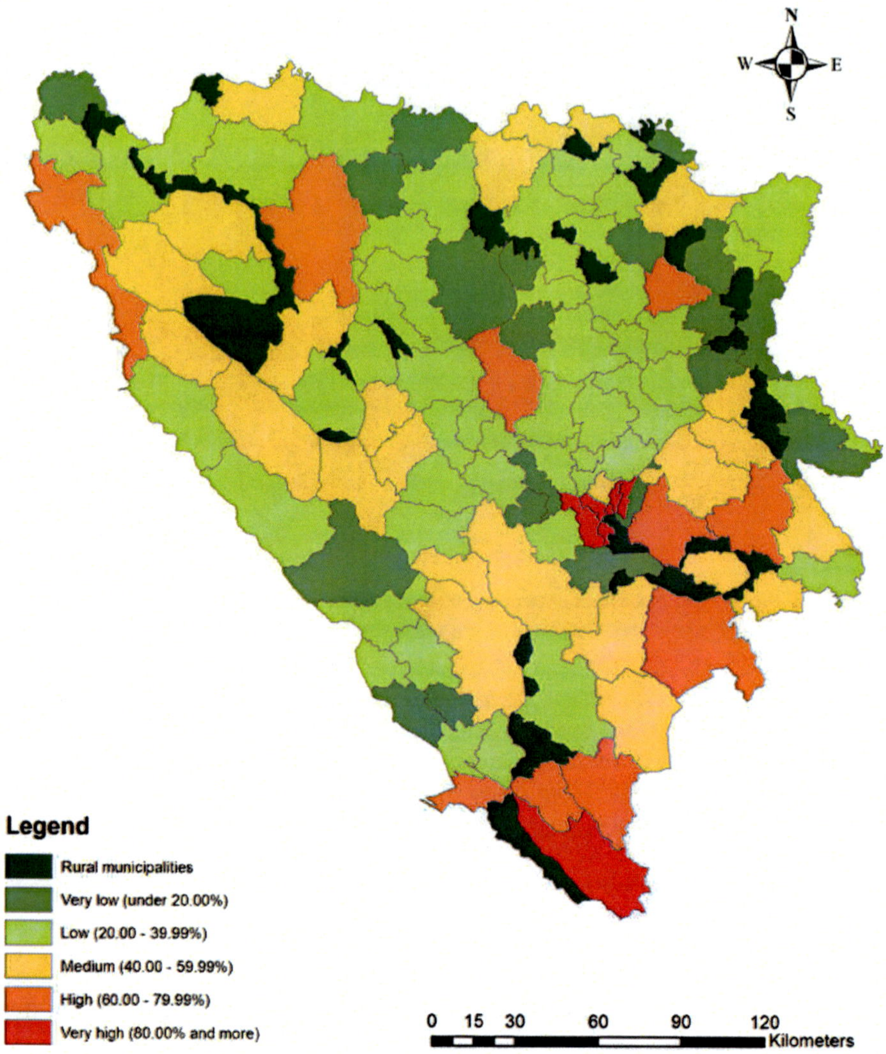

Fig. 12.15 Degree of urbanization in Bosnia and Herzegovina in 2013

In the last seventh group, there were the most urban settlements, 48, where there is no dominance of some particular activity, but the largest share is held by service activities together. In these settlements, over 65% of employees are in service industries. Settlements with over 80% of employees in service activities (Sarajevo, Istočna Ilidža, Foča, Banja Luka, Mostar, and Orašje) stand out (Fig. 12.17).

In the modern spatial structure of urban settlements in BiH, agricultural and forest areas still occupy a significant share, ranging on average from 25% to 35% individually (except for the smallest urban settlements in terms of area). Land use for residential purposes averages 15–20%, followed by traffic areas (5–10%), industrial areas (3–6%), communal areas (1–2%), business and trade zones (0.5–1.0%), and other. For example, in the town of Bugojno, industrial areas occupy 4% of the city territory, while business and trade zones occupy only 0.8% (Gekić 2011). Sarajevo has been the most important city in BiH for centuries. In the modern period, Sarajevo is ultimately the most important political, cultural, economic, and educational center of BiH. Despite everything that happened, especially during the aggressive war of 1992–1995, when Sarajevo was under complete siege and when a real urbicide was happening, Sarajevo today still lives a multinational, multireligious, and multicultural life (Bidžan 2009). The Army of the Republika Srpska burned down the City Hall among numerous destroyed and damaged buildings (Fig. 12.18).

It was the largest and most representative building from the Austro-Hungarian period in Sarajevo. Only in one night of August 25–26, 1992, 80% of the total library collection was destroyed by the aggressor grenades, and this beautiful and valuable cultural monument was destroyed. About two million books, articles, and magazines representing BiH's cultural treasures have been irretrievably destroyed. After being renovated, the City Hall reopened in 2014 (History of City Hall 2020).

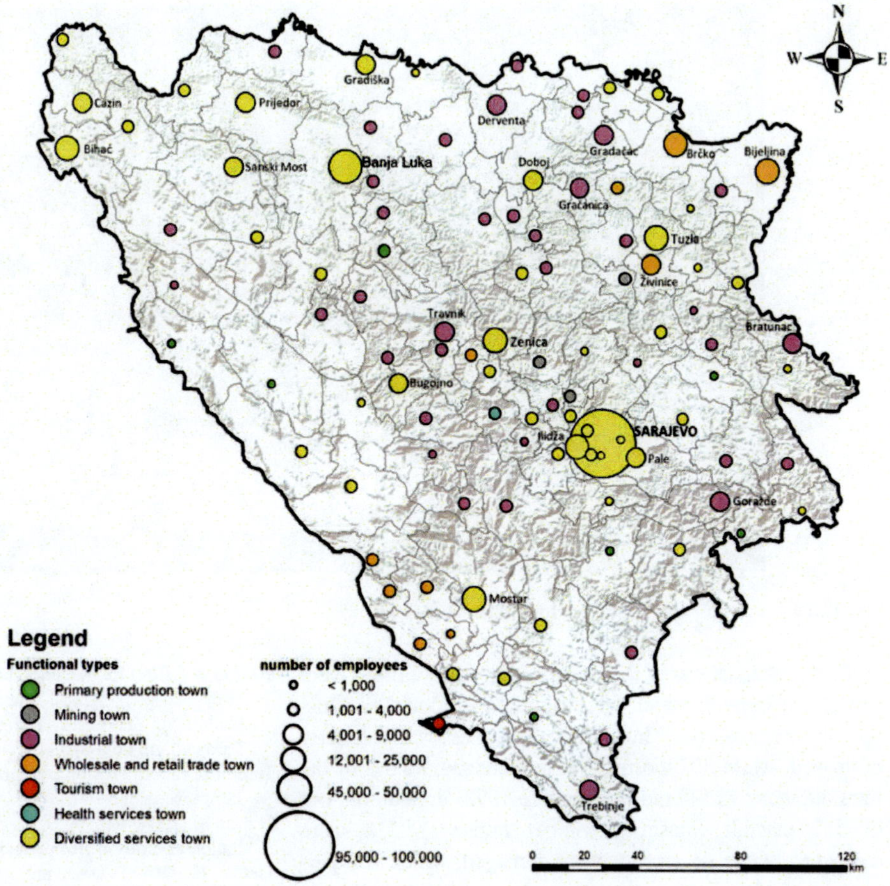

Fig. 12.16 Functional typology of towns in Bosnia and Herzegovina in 2013

Fig. 12.17 City of Sarajevo. (Photo by Aida Bidžan-Gekić)

Fig. 12.18 Sarajevo City Hall. (Photo by Aida Bidžan-Gekić)

During the last war in BiH, many religious buildings were destroyed, some of which were significant cultural and historical monuments. Thus, 614 mosques were utterly destroyed, while 307 mosques were damaged, 80.5% of the total number of 1144 mosques before 1992. In addition, out of 557 masjids (mosques without minarets), 218 were destroyed, while 41 masjids were damaged, which is a percentage of 46.40% of masjids (Islamic community in Bosnia and Herzegovina 2015). A significant number of Orthodox (297) and Catholic churches (353) and other buildings were also destroyed or damaged (Slobodna Evropa 2021).

References

Bakaršić S (1961) O nekaternih činiteljih urbanizacije v Bosni in Hercegovini v turškem obdoblju [On some factors of urbanization in Bosnia and Herzegovina in the Turkish period]. Geografski obzornik 8(1):10–19

Bakaršić S (1962) O nekim društvenim faktorima u izgledu i teritorijalnom razvitku bosansko-hercegovačkih gradskih naselja [On some social factors in the appearance and territorial development of BiH urban settlements]. Zbornik VI Kongresa geografa Jugoslavije, Ljubljana

Bakaršić S (1973) Još jednom o procesu urbanizacije za turskog doba u Bosni i Hercegovini [Once again about the process of urbanization in the Turkish period in Bosnia and Herzegovina]. Geografski pregled 14–15:214–230

BHAS 2013, Agency for Statistics of Bosnia and Herzegovina (2013) Sistematski/sustavni spisak općina i naseljenih mjesta Bosne i Hercegovine [Systematic list of municipalities and settlements of Bosnia and Herzegovina]. Sarajevo

BHAS 2016, Agency for Statistics of Bosnia and Herzegovina (2016) Census of population, households and dwellings in Bosnia and Herzegovina 2013. Sarajevo

Bidžan A (2009) Regionalno-geografski značaj Sarajeva kroz vijekove [Regional-geographical importance of Sarajevo through the centuries]. Zbornik 2. Kongresa geografa Bosne i Hercegovine, Sarajevo, pp 665–675

Bublin M (2000) Prostorno planiranje [Spatial planning]. Univerzitetska knjiga, Sarajevo

Bublin M (2004) Rehabilitacija gradova Bosne i Hercegovine [Rehabilitation of cities in Bosnia and Herzegovina]. Buybook, Sarajevo

Gekić H (2011) Društveno-geografske determinante kao osnova prostornog plana općine Bugojno [Socio-geographical determinants as the basis of the spatial plan of the municipality of Bugojno]. Master thesis, Department of Geography, Faculty of Science, University of Sarajevo, Sarajevo

Gekić H (2021) Demografska slika općine Bugojno [Demographic profile of the Municipality of Bugojno]. Dobra knjiga, Sarajevo

Gekić H, Bidžan-Gekić A (2016) Urbanization in Bosnia and Herzegovina: a case study of Uskopaljska Valley. Theor Appl Sci 38(6):147–154

Gekić H, Bidžan-Gekić A (2020) Urban hierarchy inside a divided urban system of Bosnia and Herzegovina. In: 1st Istanbul international geography congress proceedings book. Istanbul University Press, Istanbul, pp 492–510

Hadžibegović I (2004) Bosanskohercegovački gradovi na razmeđu 19. i 20. stoljeća [Bosnian cities at the turn of the 19th and 20th centuries]. Institut za istoriju, Sarajevo

Handžić A (1977) O značaju putova za razvitak gradskih naselja u Bosni u XVI i XVII stoljeću [On the importance of roads for the development of urban settlements in Bosnia in the 16th and 17th centuries]. Prilozi Instituta za istoriju XIII, Sarajevo, pp 73–78. Available at http://iis.unsa.ba/wp-content/uploads/2019/09/13-Prilozi-%C4%8Clanak-Adem-Hand%C5%BEi%C4%87.pdf. Accessed on 16 Aug 2020

History of City Hall (2020) Gradska vijećnica Sarajevo. Available at https://www.vijecnica.ba/en/home. Accessed on 17 Nov 2020

Islamic community in Bosnia and Herzegovina (2015) A day of remembrance for 614 demolished mosques. Available at https://www.islamskazajednica.ba/index.php?option=com_content&view=article&id=22240:nis. Accessed on 31 July 2021

Kamberović H (1998) Osnovna obilježja razvoja društva u Bosni i Hercegovini od 1945. do 1953.godine [Basic features of the development of society in Bosnia and Herzegovina from 1945 to 1953]. Časopis za suvremenu povijest 30(2):359–376

Kovačević-Kojić D (1978) Gradska naselja srednjovjekovne bosanske države [Urban settlements of the medieval Bosnian state]. Veselin Masleša, Sarajevo

Kreševljaković H (1949) Gradska privreda i esnafi u Bosni i Hercegovini od 1463 do 1831 [Urban economy and guilds in Bosnia and Herzegovina from 1463 to 1831]. Glasnik istorijskog društva BiH 1

Mastilo N (1959) Kasaba kao vrsta naselja [Kasaba as a type of settlement]. Geografski pregled 3:103–120

Ministry of human rights and refugees of Bosnia and Herzegovina (2006) Stambeni i urbani profil Bosne i Hercegovine [Residential and urban profile of Bosnia and Herzegovina]. Sarajevo

Mišković M (1978) Funkcije gradskih naselja i gradova u SR Bosni i Hercegovini [Functions of urban settlements and cities in FR Bosnia and Herzegovina]. Geografski pregled 23:75–91

Nurković S (2006) Suvremeni sociogeografski problemi regionalnog razvoja Bosne i Hercegovine [Contemporary sociogeographical problems of regional development of Bosnia and Herzegovina]. Annales, Series historia et sociologia 16(1):203–212

Nurković R, Gekić H (2011) New spatial development processes of urbanization of Sarajevo. J Settl Spat Plan 2:109–114

Pašalić E (1960) Antička naselja i komunikacije u Bosni i Hercegovini [Ancient settlements and communications in Bosnia and Herzegovina]. Zemaljski muzej Sarajevo, Sarajevo

Redžić H (2009) Srednjovjekovni gradovi u Bosni i Hercegovini [Medieval towns in Bosnia and Herzegovina]. Bemust, Sarajevo

Renier H (1931) Fizičko-geografske i kulturne osnove naselja u Bosni i Hercegovini [Physical-geographical and cultural bases of settlements in Bosnia and Herzegovina]. Glasnik Geografskog društva 17

Slobodna Evropa (2021) Hiljade srušenih bogomolja u BiH—zločini koji bez kazne padaju u zaborav [Thousands of destroyed places of worship in BiH—crimes that fall into oblivion without punishment]. Available at https://www.slobodnaevropa.org/a/dzamije-crkve-rusenje-rat-bih/31397067.html. Accessed on 17 Aug 2021

Sućeska A (1968) Neke specifičnosti istorije Bosne pod Turcima [Some specifics of the history of Bosnia under the Turks]. Prilozi 4:42–54

UNDP (2013) National human development report 2013, Rural development in Bosnia and Herzegovina. United Nations Development Programme in Bosnia and Herzegovina, Sarajevo

Zipf G (1949) Human behavior and the principle of least effort. Addison-Wesley, Cambridge, MA

Part IV

Economic Geography of Bosnia and Herzegovina

13 Natural Resources of Bosnia and Herzegovina

Abstract

This chapter presents the main characteristics of the Bosnia and Herzegovina natural resources important for economic development. The physical-geographical attributes of our country and related natural resources are one of the factors of economic development. They are a constant and necessary condition for economic development, and the success and speed of economic trends depend on them. All geographical components: geological composition and relief of the territory, climatic and hydrological conditions, pedogeographic characteristics, as well as part of space, solar, and geothermal energy, influenced that, under certain conditions, the economy developed faster or slower on the territory of Bosnia and Herzegovina. There are resources from each of these spheres, renewable and nonrenewable, which enable the development of certain economic branches and do not depend on the import of resources from other territories. Natural conditions have been and remain a significant factor in economic development and the formation of the economic structure in Bosnia and Herzegovina. In addition to land, as a primary economic resource, Bosnia and Herzegovina have significant forest, water, and ore resources.

Keywords

Economy · Agricultural land · Forests · Water resources · Mineral resources

13.1 Agricultural Land

Land is considered a primary natural and economic resource. According to their basic characteristics, land resources can be determined in two basic system divisions: automorphic and hydromorphic. Automorphic soils of Bosnia and Herzegovina (hereinafter: BiH) participate with 84.5%, while hydromorphic soils with about 11.1%, while the remaining 4.4% fall on other areas with watercourses, lakes, and other reservoirs. As we can see from the above, automorphic soils participate in a much higher percentage than hydromorphic soils. They are not affected by excess water. Among them is a large scale of different soil types that differ in depth and fertility. The use of these lands for nonagricultural purposes must be limited as they are highly suitable for intensive agricultural production. Hydromorphic soils participate in smaller percentages and are characterized by excess water, i.e., soil saturation with water, which can have different periods of duration in different years (winter-spring, which sometimes includes part of the autumn period). This state of humidity makes it impossible to perform agricultural work promptly, and plants often suffer from the presence of a large amount of moisture. The end effect of such influences is a meager yield and an uncertain harvest.

Agricultural land in BiH occupies about 2.2 million ha, 43% of its total area. Arable land occupies about 1.6 million ha, 72.7% of agricultural land, and 602 thousand ha are uncultivated surfaces, 27.3% of agricultural land (Table 13.1) (Strategic Plan for Rural Development of Bosnia and Herzegovina 2018–2021).

In terms of agriculture, the most valuable land (I and II credit rating class) in BiH is small (300 thousand ha). According to the data of the BH MAC for 2020, the total mine suspected areas cover 956 km^2 (1.9% of the total area in BiH), and the largest part of these areas is in the category of agricultural land around IEBL (Inter-Entity Boundary Line).

According to the share of ploughlands in the total area of the municipalities/cities (hereinafter: municipalities), we classified them into four groups in 2019 (Fig. 13.1). In the first group with more than 50% share of ploughlands in the total area of the territory, there were 16 of them, and they are all located in the northeastern and northern Peripannonian plain part of BiH. These are the areas of the most intensive agricultural production and are rightly called the granary of BiH. The following municipalities stand out in terms of

H. Gekić et al., *The Geography of Bosnia and Herzegovina*, World Regional Geography Book Series,
https://doi.org/10.1007/978-3-030-98523-3_13

Table 13.1 Agricultural land in Bosnia and Herzegovina in 2019

	Category of land used	Hectares (in 000)	%
Arable land	Ploughlands and gardens	1035	47.1
	Orchards	101	4.6
	Vineyards	5	0.2
	Meadows	457	20.8
	Total	**1598**	**72.7**
Uncultivated surfaces	Pastures	599	27.2
	Reeds and ponds	3	0.1
	Total	602	**27.3**
Total		2200	100

Strategic Plan for Rural Development of Bosnia and Herzegovina (2018–2021)

share: Pelagićevo (88.4%), Orašje (76.8%), Brčko (75.7%), Odžak (70.9%), Bijeljina (68.5%), and Usora (65.0%). In the second group with a share of ploughlands from 25.00% to 49.99%, there were 35 municipalities located mainly in the northern and northeastern Peripannonian plain part of BiH, with the exception of Dobretići (27.8%), Prozor (31.3%), and Kiseljak (31.4%) in the central mountain-valley part and Čapljina (41.6%) in the karst southern part of BiH. The highest shares in this group were in Tešanj (49.9%), Velika Kladuša (48.7%), Oštra Luka (47.3%), Cazin (46.8%), and Doboj-Istok (46.6%). The third group consists of municipalities with a share of ploughlands from 10.00% to 24.99%, and they are connected in the north to the areas of the previous group and are distributed mainly in the valleys of significant rivers (Vrbas, Bosnia, Neretva, Lašva). There were a total of 42 municipalities, and the highest shares in this group were in Teslić (22.2%), Maglaj (21.8%),

Fig. 13.1 Share of ploughlands in the total area of municipalities in Bosnia and Herzegovina in 2019

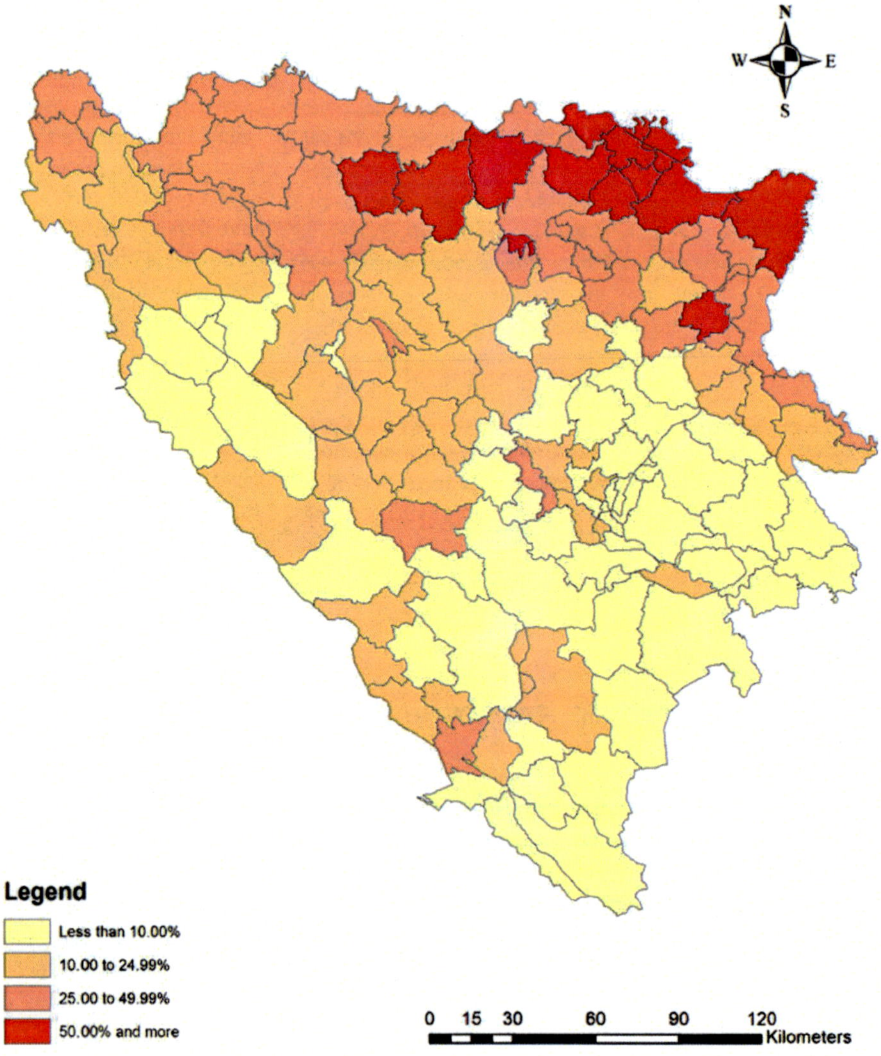

Srebrenica (20.9%), Tuzla (20.8%), and Breza (20.4%). In the last group with the share of ploughlands in the total area below 10.00%, there were 50 municipalities, mainly in unfavorable mountain-valley areas of central and eastern BiH, and mountain and karst areas of western, eastern, and southern BiH. With extremely low share in this group, below 3%, stood out municipalities in which unfavorable mountain areas dominate within the territory: Pale FBiH (2.5%), Jablanica (1.3%), Han Pijesak (0.8%), Petrovac (0.3%), and Istočni Drvar (0.1%).

Municipalities with the largest areas under ploughlands in BiH are located in the northern and northeastern Peripannonian plain. These are Bijeljina (50,250 ha), Banja Luka (43,850 ha) (Fig. 13.2), Prijedor (34,220 ha), Prnjavor

Fig. 13.2 Cultivated ploughlands near Banja Luka. (Photo by Aida Bidžan-Gekić)

(31,875 ha), Gradiška (31,250 ha), Derventa (31,240 ha), Brčko (30,445 ha), Doboj (24,900 ha), Sanski Most (23,400 ha), and Laktaši (20,200 ha) (Agency for Statistics of FBiH 2020; ISR 2020).

In 2019, 45.7% of the total area under ploughlands in BiH was unused and under fallow land. There are municipalities in all parts of BiH with high shares of uncultivated ploughlands. A total of 36 municipalities have a share of uncultivated ploughlands of 75.00% and more, and 42 from 50.00% to 74.99%. Municipalities with over 90% of unused ploughlands stand out: Istočni Drvar, Istočni Mostar, Višegrad, Čajniče, Olovo, Srebrenica, Foča RS, Teslić, Drvar, Prozor, Kotor Varoš, Istočni Stari Grad, and Kalinovik. Observing the total unused ploughlands, most of them are in the following municipalities: Prijedor (24,945 ha), Prnjavor (20,317 ha), Banja Luka (17,631 ha), Teslić (17,387 ha), Derventa (14,759 ha), Bihać (11,493 ha), Kozarska Dubica (10,416 ha) in the northern, and Srebrenica (10,406 ha) in the eastern part of BiH (Fig. 13.3) (Agency for Statistics of FBiH 2020; ISR 2020).

At an altitude above 500 m, there is 60% of arable land for growing vegetables. Quantitative indicators show that on average in BiH, there is 0.62 ha of agricultural land per capita or 0.45 ha of arable land per capita. About 14.2% of the land belongs to the first three classes of use-value. About 200,000 ha of land are endangered by floods, while about 70% of the area is exposed to erosion. According to the data obtained by the Census of Population, Households, and Dwellings, conducted in BiH in 2013, the total number of households engaged in agricultural activities is 363,394, of which 56,609 or 16% are commercial or market-oriented.

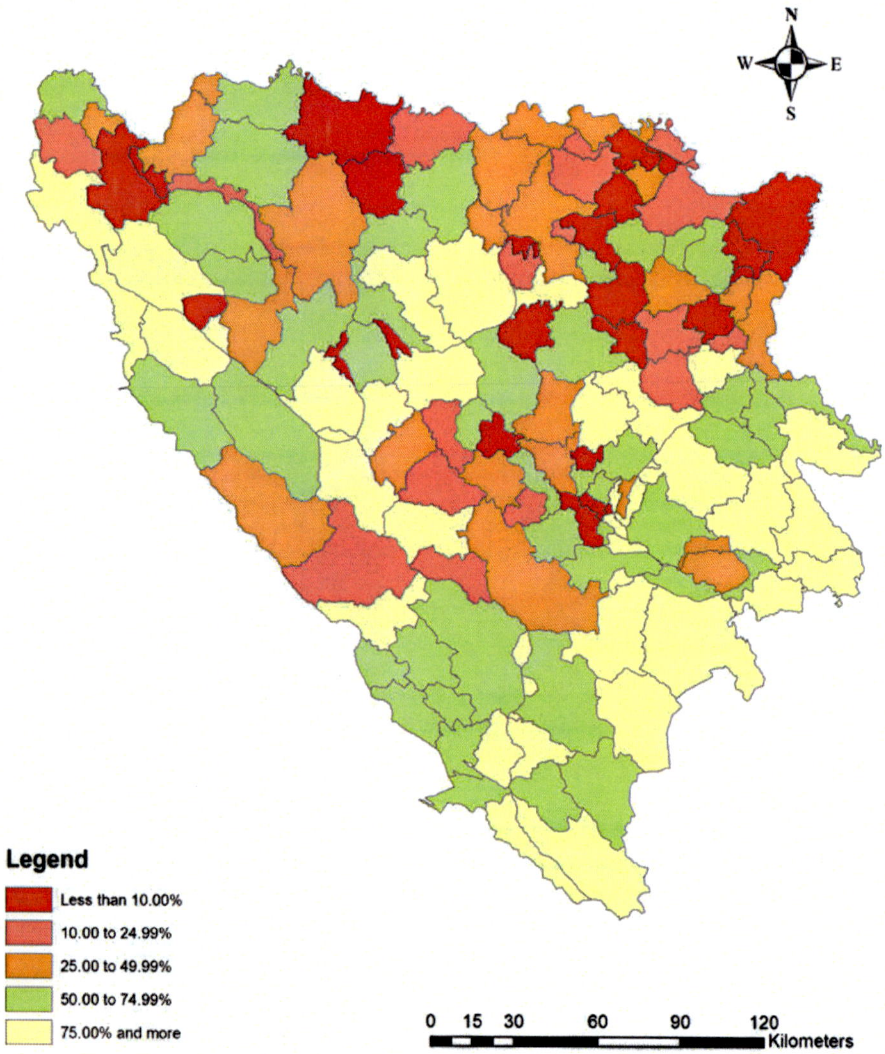

Fig. 13.3 Unused ploughlands and fallow lands in Bosnia and Herzegovina in 2019

13.2 Forests

Of the total land area of BiH, forests and wooded areas cover about 27,000 km^2 (53%); however, due to uncontrolled logging, blasting, forest fires, reservoir construction, etc. in the past 10 years, it is believed that the area under forests is reduced (Granić et al. 2008). Due to their natural and diverse structure and significant natural regeneration, they represent a crucial resource for the development of BiH.

The country is geographically well-positioned in terms of different climatic influences (the Mediterranean, sub-Mediterranean, and temperate continental climate zones) and owns more than a hundred tree species (Fig. 13.4). The main species are fir, spruce, white and black pine, beech, various species of oak, while less represented species of noble deciduous trees, such as maple, elm, ash, and fruit trees (cherry, apple, pear, plum). Forests occupy a significant place among the raw materials for economic development in our country.

In terms of forest ownership, about 72% of forests are state-owned and 19% privately owned. Unforested forest land occupies about 5000 km^2 or 18.5%, of which about 3900 km^2 is suitable for afforestation, and about 1100 km^2 is unsuitable for afforestation (FAOUN 2015). Significant devastation of forests, mostly around urban areas, also occurred in the previous war. The consequence of the war is still a large mined area of close to 1000 km^2, which has been lost for management for a long time, and given the damage, it is a potential hotbed for the development of plant diseases and insects (Fig. 13.5).

Biomass as a renewable energy source can be divided into wood, agricultural, non-wood, and waste, within which the following can be used: forest biomass, wood biomass, wood industry waste, agricultural mass (cereals—corn, wheat, barley, oilseeds, oats, rye), agricultural masses (fruits—plums, apples, vineyards, cherries, pears, etc.), non-wood growing biomass (fast-growing algae and grasses), and urban and industrial waste. For energy production, the most important source of biomass in BiH is forest wood mass (firewood and forest residue) and waste from the wood industry. Biomass obtained from agricultural production has significant energy potential; both biomasses have tremendous potential in northern and northeastern BiH (Đonlagić 2010).

The total wood stock in forests in BiH is 317.5 million m^3 or 203.6 m^3/ha (62% deciduous and 38% coniferous). The annual volume increment of forests in BiH is 9.5 million m^3 or 6.1 m^3/ha, and the possible annual volume of felling is 7.4 million m^3, although, in 2009, only 4.5 million m^3/ha were obtained (Doleček and Karabegović 2011).

Significant forest exploitation began during the Austro-Hungarian period. That is when the development of the wood industry, as a very important economic branch in BiH, started with the exploitation of wood mass. More than a thousand kilometers of the narrow-gauge railway for wood exploitation has been built in BiH. Nineteen large wood industry companies were established, while small sawmills were erected almost all over the country. Forests are characterized by various forest types, from coastal Mediterranean forests to mountain forests in central Bosnia (Rodić 1979).

13.3 Mineral and Energy Resources

Ore mining and its processing have been known in the area of today's BiH since ancient times. The migration of Germans and Slavs covered the continuous development of ancient mining in late antiquity and early medieval times in Bosnia. Mining continued to exist, with reduced capacity, in the wake of ancient mines. Greater momentum in the development of mining belongs to the newly developed Middle Ages. The reason for this improvement lies in the fact that, after the shortage of precious metals in Western and Central Europe, a general search for them started which was reflected in this area. For this purpose, the German miners, Sasi, were the

Fig. 13.4 Preserved natural forest around the Malo Plivsko lake. (Photo by Aida Bidžan-Gekić)

Fig. 13.5 Destruction of forest complexes for the development of tourism in mountainous areas; example of Vlašić mountain. (Photo by Aida Bidžan-Gekić)

most engaged, specializing in finding and organizing ore exploitation. Stronger mining development in Bosnia can be observed during the fourteenth and fifteenth centuries. Lead and copper mines are opening, and thanks to silver, it reached its culmination, becoming the basic economic branch in Bosnia. The most famous mines in Bosnia are located in the central Bosnian (Kamenica, Olovo, Dusina, Deževice) and eastern Bosnian basins (Srebrenica and its surroundings).

Silver mines attracted a larger number of foreigners. There was the formation of mining settlements, colonies, caravan stations, and squares, which had positive effects on certain areas and even on the overall prosperity of medieval Bosnia. Along with miners, and silver traders with capital, came artisans and traders of consumer goods so that, at that time, Bosnia got a stable economy with ore exploitation in the center. The ore was processed on the spot to a semi-finished product that could be more easily transported to the coast and then to Venice. Initially, the mine in Olovo (and Kamenica) had the greatest prosperity due to the high demand for lead; the same one Srebrenica had due to silver. The silver was melted near the mines, turned into furnaces, and then transported by caravans to the market. The lucrative earnings allowed foreigners to take full exploitation into their own hands. To control it completely, they bought individual "holes" (fosse) for a certain period from the Bosnian rulers and organized excavations and smelting. Silver was measured in liters (about 300 grams), and the price of silver in Dubrovnik reached the price of 7.5 Venetian ducats per liter. In addition to ore, silver was exported from Bosnia through Bosnian-style items (Kovačević 1961).

Mining is a very old industry and is of great importance for the economic development of BiH in general. Mining is divided into mining of metallic, nonmetallic, and energy raw materials. Metallic raw materials are further divided into ores that contain or do not contain iron. Examples of the latter are lead, zinc ores, and bauxite reserves in BiH. Nonmetallic reserves include clay, magnesite, quartz sand, ornamental stone, as well as primarily rock salt. The energy raw material is considered to be typical, low-calorie lignite (Study of mining and chemical industry in Bosnia and Herzegovina 2001).

After the Second World War, mining was a significant basis for industrialization, and after 1980, its importance diminished due to the depletion of the site. During the war (1992–1995), production was almost completely stopped, and for the needs of the electric power industry, production in coal mines was renewed the fastest. The most important deposits are black coal (Banovići, Ugljevik, Đurđevik, Zenica, Kakanj, Breza) and lignite (Lukavac-Dobrinja and Burkinje-Lipnica in the vicinity of Tuzla, Gacko); coal is mainly used in thermal power plants. There are iron ore mines near Ljubija and in Vareš. Also important are bauxite (Čitluk, Posušje, Lištica, Mostar), rock salt (Tuzla, Tušanj), and barite (Velika Kladuša, Kreševo). The mining sector contributes to the manufacturing industry, electricity generation, and exports.

Natural geographical conditions enabled speedy economic development of BiH; rich deposits of coal and salt conditioned the rapid growth of industry and settlements; the relief configuration and river flows also influenced the direction and expansion of the settlements in the valleys. The area of BiH is known for its numerous deposits of mineral resources that are important for today's industrial development (Andijašević 1981).

Especially important are the deposits of quality iron ores (magnetite, hematite, limonite siderite, etc.), which form the basis of the development of ferrous metallurgy; that is why this industry has a vibrant tradition. The richest iron ore deposits are located in central Bosnia (Vareš), Sana (Ljubija and Omarska), and in the western part of central Bosnia in the area of Raduša-Jajce-Mrkonjić Grad and Konjic-Jablanica-Prozor. Total iron ore reserves are estimated at more than two billion tons (Čičić 1979). The production of certain nonmetallic raw materials in BiH has constantly been increasing. Thus, the most significant increase was achieved in the production of asbestos, more than 100 times, and the increase in the production of rock salt was 617 index points, soda 503, and coke 108. Salt reserves are estimated at 350–450 million tons (Vujović and Stojković 1985).

Economically significant coal reserves refer to brown coal and lignite. Balance reserves are estimated at 2669 million tons, of which 1460 million tons or 54.7% lignite and 1209 million tons or 45.3% brown coal. The largest coal reserves are concentrated in the five most important coal basins: Central Bosnian, Banovići, Ugljevik, Kamengrad, and Herzegovina (Mostar and Gacko) (Table 13.2). BiH participates in world coal reserves with 0.5% and, in Europe, about 2% (Čičić 1979; Archive of Center for Investigative Reporting 2015).

In 2019, 13.3 million tons of coal were mined, of which 6,971,178 tons of lignite and 6,405,472 tons of brown coal were produced, which is at the level of production since the early 1970s (Agency for Statistics of Bosnia and Herzegovina 2020). Most of BiH's coal is used by thermal power plants, which are locally connected to its sources. Thus, the largest thermal power plant Tuzla (five blocks of 786 MW power) was built on the wealth of coal in the Tuzla valley, and the thermal power plant Kakanj (five blocks with 580 MW power) on the wealth of coal in the Central Bosnia Basin (Nurković 2003). Later, however, two additional thermal power plants were built: TPP Gacko in Herzegovina and TPP Ugljevik in the Ugljevik basin, so with the development of mining and energy, inevitable consequences arose, which refer to endangering and degrading the environment. Therefore, it is necessary to apply the most modern environmental protection measures and find reasonable solutions in the planning and landscaping of BiH.

Table 13.2 Balance reserves of coal in Bosnia and Herzegovina in 2018

Basin	Reserves of brown coal in millions of tons	Basin	Reserves of lignite in millions of tons
Central Bosnian	513	Kreka	744
Banovići	194	Stanari	107
Mostar	90	Gacko	278
Ugljevik	186	Bugojno	11
Kamengrad	112	Ramići	38
Miljevina	21	Livno	76
Lješljani	8	Tomislavgrad	206
Đurđevik	60	–	–
Drvar	7	–	–
Cazin	2	–	–
Kotor Varoš	16	–	–
Total	**1209**	**Total**	**1460**

Archive of the Center for Investigative Reporting (2015)

BiH is an important producer of nonmetallic ores, especially porphyry, asbestos, rock salt, ornamental stone, etc. In 2019, BiH produced 1.4 million tons of iron ore. On the territory of BiH, we also find significant reserves of nonferrous metal ores, among which bauxite reserves are significant. This ore is associated with limestones and dolomites of the Triassic, Cretaceous, and Tertiary ages. The first exploitation of bauxite began before the First World War near Bosanska Krupa. It is geographically distributed in three larger regions: Herzegovinian with localities in Tomislavgrad, Posušje, Čitluk, Žitomislići, and Stolac; Bosnian Krajina with localities in Bosanska Krupa, Bosanski Petrovac, Ključ, Mrkonjić Grad, and Jajce; and the eastern Bosnian region with sites near Vlasenica. Total reserves are estimated at about 41 million tons, and the quality of the ore is satisfactory (over 50% of aluminum oxide and over 5% of silicon oxide).

There are several smaller manganese ore deposits in BiH. The most important localities are Bužim, Čevljanovići (larger reserves), and near Banja Luka, Kozara, Kotor Varoš, and Rudo (smaller reserves are not yet exploited). Lead-zinc ore in BiH is distributed in three regions: around Srebrenica, Olovo, and Vareš and on the stretch Blagaj-Ljubija in the area of Sana river. Reserves have been insufficiently explored, and significant exploitation began in the 1960s. Occurrences of chromium ore (Dobroštica near Vareš), copper ore (Sinjakovo near Mrkonjić Grad, Borovica near Vareš, Kozara, and Srebrenica), antimony (Potkozarje), pyrite (Bakovići near Fojnica), mercury (Draževići near Srednji, Kreševo area), and gold (the upper course of the Vrbas river and its tributaries, Lašva river and Fojnica river, and Vareš) are known in the area of our country. However, the identified reserves of these metal raw materials do not represent sufficient quantities for their profitable exploitation, and more detailed research is needed. In 2019, BiH produced 1 million tons of bauxite, 41,000 tons of manganese ore, and 25,000 tons of lead-zinc ore. Reserves of nonmetallic raw materials in BiH are quite large, but they are still insufficiently geologically and technically examined. Particularly significant are the reserves of pyrophyllite (100% of the reserves of the former Yugoslavia were in BiH), asbestos (70%), and kaolin (47%). Pyrophyllite is only exploited in the Parsovići mine near Konjic. The richest asbestos deposits are in the mountain massif of Ozren. Kaolin deposits are numerous, and they are exploited on Motajica (Srbac) and near Bratunac. Reserves are also located near Kozarska Dubica, Zvornik, Gračanica, Tešanj, and Šipovo. The most important clay deposits are near Prijedor, Sanski Most, Busovača, and Kiseljak. The largest rock salt deposits are in Tuzla. Deposits of other nonmetals are known on the territory of BiH, such as barite (at the sites in Kreševo, Vareš, Velika Kladuša, Gornji Vakuf, Blagaj, Novi Travnik, and Prača), magnesite (at the sites around Banja Luka, Prijedor, Teslić, Maglaj, Zavidovići, Kladanj, and Rudo), gypsum (at the localities Volari near Šipovo, Donji Vakuf, Bosanski Novi, Sanski Most, Jablanica, Prozor, and Foča), quartz (at the localities near Kiseljak, Kreševo, Busovača), marlstone (at the localities near Kakanj, Livno, Bileća), gabbro, marble, dolomite, and limestone. In 2019, BiH produced 1 million tons of rock salt, 1800 tons of raw barite, 99,000 tons of raw gypsum, 680,000 tons of construction sand, 253,000 tons of kaolin, and 92,343 tons of bentonite (Marić 1991; Agency for Statistics of Bosnia and Herzegovina 2020).

Chrome ore can be found at several locations: near Duboštica, not far from Vareš, and in the Krivaja valley. Molybdenum ore used for the production of specialized steel is located near Jajce. Cobalt ore is located in the vicinity of Fojnica. Copper can be found in several places in Ljubija, Busovača, Sinjakovo, Fojnica, Bugojno, Jablanica, and Foča. There is Antimona near Čvrsnica, not far from Fojnica. Gold, either primary quartz gold, can be found in the central Bosnian ore-bearing basin around Baković and Fojnica and the mountains of Vranica and Vilenica. There is also gold in the deposits of the Vrbas and Lašva rivers and their tributaries. The gold content in 1 ton of pyrite in the central Bosnian basin near Fojnica is 20 grams. Silver is found in the vicinity of Srebrenica, Olovo, Kreševo, and Fojnica (Fig. 13.6) (Rodić 1979).

Oil and gas exploration in BiH began in 1889, when Austrian geologists Katzer and Hofer explored the Majevica area, in the villages of Zavid and Ražanj near Priboj, on the Tuzla-Zvornik road. Austrian geologists drilled several wells up to 50 m deep, from which they obtained small amounts of oil. The works were soon suspended due to a lack of funds. Oil was found in several wells in Požarnica and Simin Han near Tuzla. In 1939, about 300 tons of oil were extracted and transported to the Bosanski Brod Oil Refinery, but due to small production, expensive transport, lack of equipment, personnel, and the onset of war conditions, production ceased in 1943. From 1963 to 1973, the right to explore for oil in BiH belonged to the company INA-Naftaplin from Zagreb, which expanded the exploration from the northeast of Bosnia to the area of the central and outer Dinarides. Then, in addition to gravimetric and geomagnetic surveys, several deeper wells were drilled (Tuzla I at a depth 3539 m; Vareš 1 at 1202 m and Glamoč I at 4212 m) (Spatial Planning Strategy of the Federation of Bosnia and Herzegovina 1997).

In 1973, Energoinvest from Sarajevo received the right to explore oil and gas in BiH. In that period, the research was carried out by Energoinvest-Bosanski Brod Oil Refinery, where the Oil and Gas Exploration Sector was established in 1974. The refinery provided most of the funds for exploration work itself, and in 1983 it established the Working Organization for Oil and Gas Exploration in BiH. In the period from 1963 to 1991, extensive oil and geological research were performed. The exploration was carried out as part of two main projects: (1) the North Bosnia project, which was financed from its funds and a World Bank loan in the amount of US $ 2.5 million, and (2) the Dinarides project, which was fully funded by the American company AMOCO, which for this purpose formed the AMOCO-Yugoslavia Petroleum Company with headquarters in Zagreb, in the premises of the representative office of the Bosanski Brod Oil Refinery. The Northern Bosnia project covers the exploration area of northern Bosnia, which consists of the southern parts of the Pannonian Tertiary Basin. The exploration area is about 12,000 km^2. Based on the current level of exploration in northeastern Bosnia, crude oil reserves are considered to exist in four priority areas: the area south of Bosanski Šamac, the estimated reserves are 64.5 million barrels (about 9.2 million tons); area southwest of Orašje, reserves are estimated at 42.5–108.4 million barrels (about 6.1–15.5 million tons); Tuzla region in the Tinja river valley, estimated reserves are 99.8 million barrels (about 14.3 million tons); and the area around Lopare with an area of about 21 km^2 reserves is estimated at 83.2 million barrels (about 11.9 million tons). Estimated reserves at these four sites, which have been singled out as promising in terms of oil discovery, amount to about 355 million barrels or about 50 million tons of crude oil (Spatial Planning Strategy of the Federation of Bosnia and Herzegovina 1997). BiH imports about 50% of its oil from Islamic countries, mainly Iraq, Libya, and Iran (Nurković and Rewucki 2018).

Gračanica mine is a lignite mine located on the border of the municipalities of Gornji Vakuf-Uskoplje and Bugojno. It is located in the southeastern part of Bugojno coal basin, on the right bank of the Vrbas river, in the area of Rosulje. From 1938 to 1977, only pit exploitation was carried out, which, in 1977, was switched to surface exploitation. It has its own equipment for the production and preparation of coal with a capacity of approximately 750,000 t/year. In the postwar period, the largest buyers of coal are "Elektroprivreda BiH" and "Natron-Hayat" Maglaj, where the mine sells 95% of its total production. The remaining quantities of 5% are placed in mass consumption. In 2014, 349,093 tons of lignite were produced, the largest consumers of which are the Kakanj and Tuzla thermal power plants. The Gračanica mine employs about 200 employees and operates within the concern PC Elektroprivreda BiH JSC Sarajevo. Since the Bugojno Neogene basin has substantial coal reserves (reserves of 1 billion and 300 million tons), this mine has favorable conditions for long-term operation and increasing the current capacity (Archive of Gračanica Coal Mine).

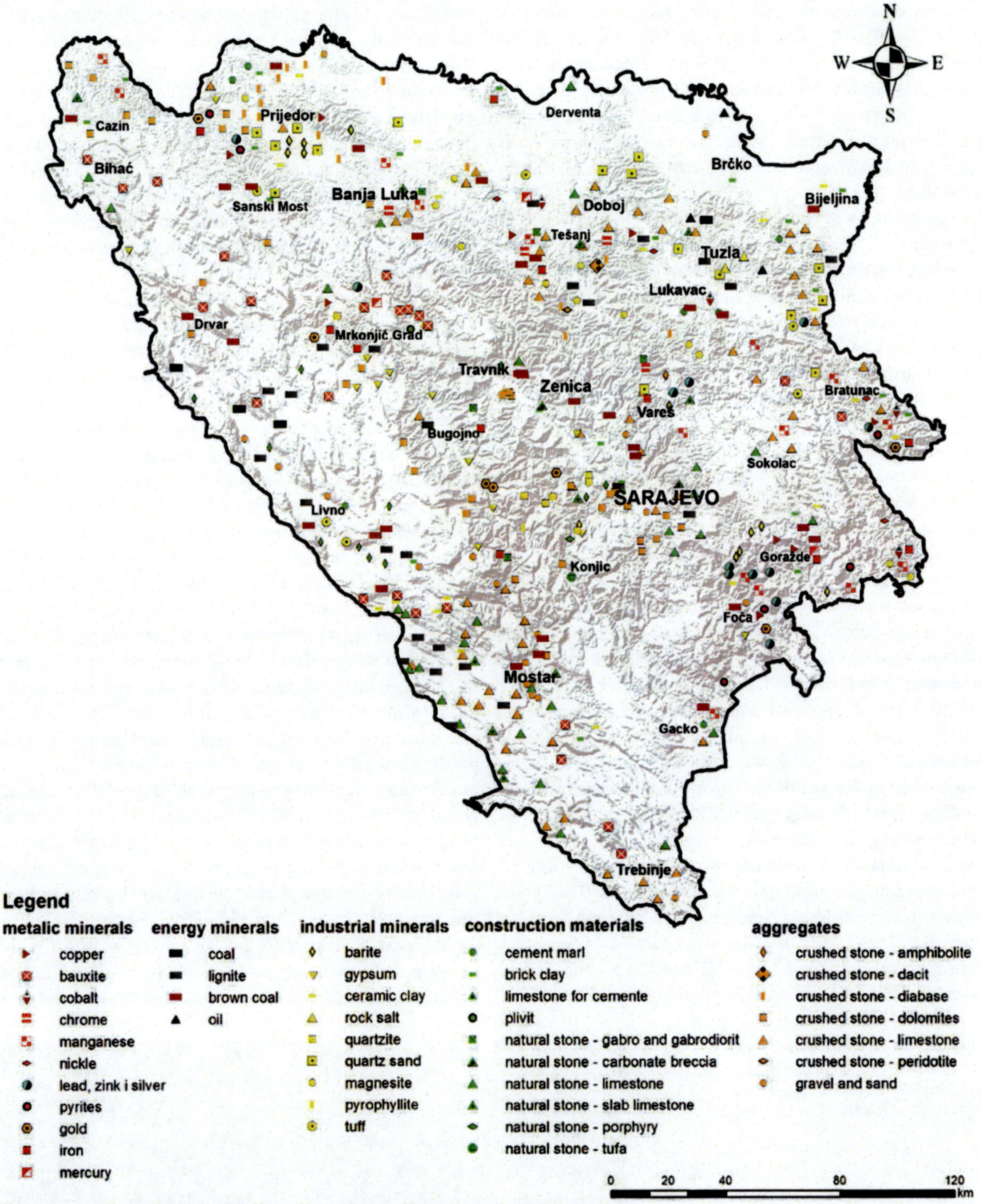

Fig. 13.6 Distribution of mineral resources in Bosnia and Herzegovina in 2019

Kakanj mine is a Bosnian brown coal mine located in the municipality of Kakanj. In the area of this municipality in 1898, Austro-Hungarian experts began geological research intending to start coal exploitation. Somewhat later, the first mine called Kakanj-Doboj was opened. Since 1902, it has been officially called Kakanj mine, with headquarters in Vienna at the time. Until 1979, when the open-pit mine Karaula was opened, only pit exploitation was performed. Since 2009, the mine has been operating within the concern PC Elektroprivreda BiH JSC Sarajevo. The Kakanj mine employs 2020 persons. The average annual production of coal is about 1,100,000 tons, and over 90% of the produced coal is used for the needs of the Kakanj thermal power plant. The Kakanj brown coal mine has a total geological reserve of 425.05 million tons of coal, of which 241.40 million tons are balance sheet reserves, while the exploitation reserves amount to 156.21 million tons of coal (Archive of Kakanj Coal Mine).

The first tons of coal in *Kreka mine* were excavated on September 22, 1884, at a daily mine on the outcrop of the main layer on the left bank of the Kreka stream for the needs of the Solana in Simin Han, opened in 1885, for which brown coal was extracted. Within the Subsidiary Company, the Kreka Mines in Tuzla have three mines: Mramor, underground exploitation; Šikulje, surface exploitation; and Dubrave, surface exploitation (Archive of Kreka Coal Mine). The coal deposit under the Dubrave lignite mine is the central part of the Kreka southern synclinorium. In a broader sense, it is an area bounded by the Spreča river and its right tributaries Rijeka and Gribaja. Two kilometers from the southern border, there is a road Tuzla to Zvornik and the railway Tuzla to Zvornik, so the mine is well connected. The mine had existed since 1980 when coal mining began. 90% of production is placed in TPP Tuzla, and the rest is placed for mass consumption. The total reserves of the narrow Kreka syncline are 449,622,000 tons. According to the mentioned reserves, the lignite mine "Dubrave" in Dubrave is designed for 30 years of operation with an average production of 3,000,000 tons of coal (Archive of Dubrava Coal Mine). Mramor coal mine is a lignite mine in Mramor near Tuzla. It is located in the part of the deposit of the northern Kreka syncline. This is the largest mine with underground exploitation, a mining tradition of over 50 years, and reserves and capacities that provide coal exploitation and employment of the surrounding population for the next 45 years. The lignite exploited in the "Mramor" mine is of the Pliocene age with high quality and calorific value. It is located in four coal seams: floor layer, main layer, II roof layer, and I roof layer. In 1998, the exploitation of the first roof layer was completed, and as a replacement capacity, the main coal seam was gradually prepared and opened. The second roof layer was still in operation at that time. In 2002, the exploitation of the second roof layer was completed, but the replacement capacity of the main layer—"Dobrnja"—took over production and labor and, at the same time, the preparation and opening of the main layer—"Marići" and "Mramor"—continued. The main layer pit, with its three districts "Dobrnja," "Mramor," and "Marići," has demarcated exploitation areas with 14,510,000 tons of exploitation reserves (Archive of Kreka 1 Coal Mine).

Zenica brown coal mine started operating on May 5, 1880. Today it operates within the Concern PC Elektroprivreda BiH JSC. This mine produces coal in three pit capacities: Stara jama, Raspotočje, and Stranjani, and it processes coal in the separation plant. It has the largest balance and potential reserves of brown coal in the Zenica-Sarajevo coal basin. It is the highest quality coal that has undergone a complete metamorphosis in converting to hard coal. The coal of the main coal seam of the pit "Raspotočje" can be treated as younger hard coal to the upper thermal energy value. The total reserves in the exploitation fields are today estimated at around 956 million tons. The production of coal from the company's pits in the last 5 years amounts to an average of 320,000 tons per year (Archive of Zenica Coal Mine).

The first written data on the existence of *Tuzla's natural salt* springs date back to the early Middle Ages, although it can be said with certainty that salt has been obtained in this area since time immemorial. This is indicated by the name previously used for this area (Salines in Latin means the city of salt pans). According to the model of colonial industrialization in BiH, the industrial exploitation of salt from the deposit in Tuzla began in 1885 (Figs. 13.7 and 13.8) (Archive of the saltworks JSC Tuzla). By drilling in 1978, the prospect was confirmed by discovering the rock salt deposit "Tetima." The reservoir's exploration and contouring was carried out for the next 6 years with 24 exploratory wells, of which 17 were drilled with a salt bases. In the plan, the deposit has an irregular elliptical shape, about 2.2 km long and up to

Fig. 13.7 Pump for saltwater from the Ottoman period kept in Salt Museum in Tuzla. (Photo by Aida Bidžan-Gekić)

Fig. 13.8 Reconstructed tower of a salt well in Tuzla. (Photo by Aida Bidžan-Gekić)

1 km wide. It was deposited in the eastern part of the Dokanj syncline and has a Dinaric route of direction with a maximum capacity of up to 200 m. After revitalizing mining facilities, plants, and devices that lasted from 1996 to 2000, production was restarted in early 2001. In the last 10 years, with the increase of capacity by saltwater consumers, the production of saltwater has intensified, and thus all the mining works that accompany it. Thus, between 2005 and 2014, 16 exploration and exploitation wells were drilled as capital facilities, making 21 out of 100 projected wells out of prewar 5. In this way, the demands of consumers were fully met. For 2015, 2 exploration and exploitation wells are planned to continue providing sufficient quantities of saltwater for the next period for many different products (Archive of Tuzla Salt Mine).

Based on the knowledge of the geological structure of the deposit and verified reserves of rock salt in the Tetima deposit, and the expressed needs for saltwater of approximately 3,000,000 m^3/year, we can conclude that the remaining service life of the deposit is around 50 years. By increasing the diameter of the chambers from 70 to 75 m and exploiting the ceiling part of the chambers (up to the roof breccias) in the final phase of exploitation of the reservoir, it is expected to increase the service life of the reservoir by about 15–20% (Archive of Tuzla Salt Mine).

Iron ore mines "Ljubija" JSC Prijedor was established to exploit ore deposits in the municipalities of Prijedor and Sanski Most. The ore-bearing area covers an area of about 1200 km^2. The research of the Ljubija metallogenetic area has determined about 347 million tons of geological reserves of iron ore and significant occurrences of mineral raw materials such as quartz sand, clay, limestone, dolomite, barite, fluorite, zinc, and lead. Industrial production in the Ljubija metallogenetic area began in 1916, and significant progress in iron ore production was achieved in the middle of the twentieth century, with the introduction of mass exploitation and the construction of plants for the preparation of iron ore. The mass mining production enabled the construction of Central mines, Eastern mines, and Omarska mines, which operated within the Iron ore mine "Ljubija" as three independent technical-technological units for exploitation and preparation of iron ore. The mine was revitalized in 2004, with the exploitation and production of iron ore concentrating under the auspices of a new company, ArcelorMittal l.t.d. Prijedor, whose founders are iron ore mine "Ljubija" JSC Prijedor (49%), and Mittal (51%), now ArcelorMittal. ArcelorMittal JSC Prijedor operates with a projected capacity of the Omarska Mine of 1.5 million tons per year of iron ore concentrate. During the 85 years of its operation, this mine produced about 74 million tons of commodity iron ores (Archive of Ljubija Iron Ore Mine).

13.4 Wind Energy

Since wind depends on weather conditions, its strength and speed are not always constant, making it difficult to establish and predict its characteristics. If the terrain's topography is complex, which is the usual case in BiH, the accuracy of wind potential prediction with the help of calculations at greater distances from the measuring point decreases. Hence, it is necessary to perform measurements at the potential location of wind farm construction. Measurements should last at least a year, and preferably longer, before the final decision to launch the investment (Tadić and Vujadinović 2011). If we look at the wind characteristics in BiH, it can be concluded that BiH has good wind potential. This does not mean that the entire area of BiH is highly suitable for constructing wind farms. Namely, BiH has many windy areas, but the problem is that the wind in these locations is not constant: it is either too weak or too strong. The bora in Neum is an example of a very irregular and often too strong wind. Such a wind is not suitable for energy use. Fortunately, BiH has many more locations that have satisfactory wind potential. Measurements of specific wind characteristics, via LIDAR, SODAR, and anemographs (speed, direction, frequency), showed that the area of Herzegovina is more favorable for the use of wind energy than the rest of BiH (Zlomušica 2013).

The assessment of potential locations for wind farms in BiH resulted in a list of 30 sites in the southern part of the country in the area of about 50 km along the border with Croatia, which by all observed characteristics represent the most significant wind potential in BiH (Đonlagić 2010). The total potential of the observed sites from the point of view of space availability is estimated at approximately 900 MW. The total technical potential for the use of wind energy is much higher and is estimated at approximately 2000 MW, taking into account that this amount was derived from considering the availability of suitable wind farms in our country without taking into account possible limitations (grid connection, environmental protection, etc.). It is estimated that the realistic goal of using wind energy in 2015 should be set between 400 and 600 MW (Tadić and Vujadinović 2011). BiH currently has three wind farms worth over 400 million BAM (204 million EUR), which produce more electricity than 150 small hydropower plants a year. Two wind power plants have been operating since 2018 in the area of Tomislavgrad. The third one, which is at the Podveležje plateau above Mostar, will soon start operating and has already produced the first test kilowatts these days. Of the three wind farms, two are state-owned and one is privately owned, built mainly with loans from the German Development Bank and will pay for itself in 13 years. According to available data, one wind farm is under construction, and three are planned (Archives of the Electric Power Industry of Bosnia and Herzegovina 2019).

The first wind farm in BiH, Mesihovina in the area of Tomislavgrad, started operating in March 2018 and is one of the largest wind farms in Southeast Europe. Mesihovina wind farm is located in the central part of the Tomislavgrad municipality and has 22 wind turbines with a total installed capacity of 50.6 MW and an annual production of about 165.17 GWh. With an average household consumption of 6000 kilowatt-hours per year, this wind farm can supply electricity to about 27,500 households. At the beginning of 2019, the second wind farm Jelovača, also in the area of Tomislavgrad, known as the largest private investment in the Federation of BiH, started operating. It is a wind farm with 18 turbines with a total installed capacity of 36 MW and produces 110 GWh per year. The third is the Podveležje wind farm, about 10 kilometers east of Mostar, which will produce about 130 GWh of electricity per year with 15 wind turbines, each with an individual capacity of 3.2 MW and a total capacity of 48 MW, an energy equivalent of 46 small hydropower plants (Federal Ministry of Energy, Mining and Industry 2018).

The first wind farm in the entity of Republika Srpska, Grebak, is being built for energy production in the Nevesinje area. Ten turbines will be installed, one of which will be able to produce enough electricity to supply Nevesinje. In 2021, the Hrgud wind farm in the Berkovići municipality, with 16 wind generators with 3 megawatts each, which will produce 126-gigawatt hours of electricity, should be built (Institute for Protection and Ecology of RS 2021). Elektroprivreda BiH (Public Enterprise Electric Utility of Bosnia and Herzegovina) is working to realize the Bitovnja wind farm project near Konjic with the planned 25 wind turbines installed capacity up to 60 MW and annual production of 110 GWh of electricity. They are also working on the realization of the project of the Vlašić wind farm project with the planned 18 turbines with a total capacity of 50 MW and the expected annual production of 115 GWh (Archives of the Electric Power Industry of Bosnia and Herzegovina 2021).

13.5 Solar Energy

The Sun is a free, almost inexhaustible source of energy available to us on an unlimited scale and without harmful climatic influences. The greatest importance of solar energy today, in our country, is its ability to be used in everyday life. The application of solar energy includes space heating and cooling, lighting, water heating, etc. BiH has an average of 1800 h of sunshine a year, the southern region even more, up to 2300 h of sunshine. Based on the analysis of numerous experts, it is evident that there is a significant potential for the application of solar energy in BiH, which amounts to 70.5 million GWh of irradiated energy of total radiation per year. When planning a photovoltaic power plant, we must make sure that the modules are not shaded, i.e., that they are shaded only for a shorter period. Continuous sunshine should be ensured throughout the year between 9.00 AM and 3.00 PM because, in that period, the Earth receives 80% of the sun's radiation energy. We determine the timing and duration of the shadowing by entering the position of the obstacles in the diagram of the Sun's orbit. In the solar orbit diagram, the orbit of the Sun across the horizon is represented by the angular height of the Sun and the azimuth. In BiH, modules are usually installed with an inclination of 30° to 35°. The area required to reach 1 kW of power depends on the type of module and the setting method (Bago 2012).

In BiH, the value of the daily amount of radiation on the horizontal surface ranges from 3.4 kWh/m^2 (Brod) to 4.22 kWh/m^2 (Trebinje), or between 1241 and 1540 kWh/m^2 per year. Due to the different daily amounts of radiation for different parts of BiH, the production and profit from electricity will be different. Comparing solar potential with other European countries (Germany, Poland, Sweden, etc.), which use solar energy significantly, BiH has a considerably higher annual capacity (10–30%) of solar radiation, which is unfortunately underutilized (according to research, solar energy potentials in our country amount to 70.5 million GWh/year) (Umihanić et al. 2009). The first solar power plant in BiH was opened in March 2012 in Kalesija, and it was built by the owners of the company "Eko Energija." It was installed on the roof of the City Hall in Kalesija, and its construction cost 800,000 BAM (408,000 EUR). With the help of solar modules, they convert the collected solar energy into electricity, which they send to the Electric Power Industry of BiH system. The solar power plant has a power of 120 kW, and the solar panels are located at 1200 square meters. The solar power plant annually produces between 130,000 and 150,000 kWh, and the annual salary is around 110,000 BAM (56,000 EUR) (FERK 2011).

Thanks to the team of the company "EMT Solar," a solar power plant was built in the Municipality of Goražde, where a unique adjustable steel structure was installed, the first of its kind in BiH. The construction is unique because it is seasonally adjustable, giving about 20% better yields than structures with a fixed angle and slope. The construction has three positions for four seasons, intended for geographical areas with extreme weather conditions (up to −300 C). This power plant will have an annual production of 200 MWh, i.e., it will be able to supply 100 rural households (Elektro test 2014). As far as private households are concerned, solar energy is currently used in two ways: solar thermal energy is used to prepare domestic hot water, and in solar cells, the photovoltaic effect generates electricity from the sunrays.

Many businesses and households have recognized the potential of solar energy, and more and more solar panels and photovoltaic models can be seen on the buildings. Thus, in BiH, there are 180 solar power plants in the incentive system (in the entity of Republika Srpska 28, while in the entity of Federation of BiH 152 solar power plants) (Business-magazine 2019).

13.6 Water Resources

The hydro potentials of BiH occur in all forms: surface watercourses, groundwater, Adriatic sea waters, and lake waters. BiH is one of the few countries, not only in the region but in the whole of Europe, which has the highest degree of diversity of its nature, including the hydrosphere, or hydrobiosphere or water sphere, making her a recognizable point on a global level. Given its overall geo-ecological environment and quality, BiH still has remarkably well-preserved reserves of quality water in geo-ecological and sanitary, microbiological terms, which can serve as an exceptional source of quality drinking water, the most sought-after resource in the world market.

The total technologically usable power of water resources is estimated at 17.7 billion kWh, but the level of utilization (about 20%) of the total hydropower potential is still low (Table 13.3). The use of water resources has not yet been fully completed in any of the basins, although a larger number of hydropower facilities were built in the postwar period. However, there is a growing need for electricity, primarily in industry and other economic activities; faster and more efficient construction of hydropower plants is required, which would significantly affect the level of investment in the future. The strength of the hydropower potential of BiH consists of the rivers of the Sava river basin: Drina, Bosna, Vrbas, and Una, as well as Neretva and Trebišnjica (Fig. 13.9) with their tributaries (Kozar and Livnjak 2015).

After the Second World War, several hydro facilities were built on the rivers, mainly on the Neretva river. Among them is HPP Jablanica, with 144 MW, built in 1955. In 1981, two more power plants were put into operation on Neretva: HPP Salakovac with 210 MW and HPP Grabovica with 114 MW, Rama HPP on the Rama river with 16 MW, while, on the Trebišnjica River, a 180 MW HPP was built in 1968. HPP Čapljina, one of the strongest hydropower plants with 420 MW, was also built in 1979. HPP Bočac, with a capacity of 110 MW, was constructed on the Vrbas river in 1981. HPP Višegrad, with a capacity of 43.2 MW, was built in 1988 on the Drina River. Thanks to numerous newly built energy facilities in the mentioned period of economic development of BiH, electricity production is growing steadily. The technically usable hydro potential in our country is 19,182 GWh and 5594 MW. In the period up to 1985, 5599 GWh (1587 MW) was built, and the unbuilt potential was 13,583 GWh and 4007 MW. Even today, some authors claim that the technically usable hydro potential is 24,000 GWh and 6800 MW (Bačinović et al. 2007).

Recently, a large number of hydropower facilities have been proposed for construction (Fig. 13.10). When the

Table 13.3 Hydro potential of rivers in Bosnia and Herzegovina (2009)

Rivers	Hydro potential		Used		
	MW	GWh/year	MW	GWh/year	%
Una	318	1400	88	40	2.8
Vrbas	839	1799	188	719	40.0
Bosna	356	1818	8	40	2.2
Drina	1948	4741	135	1100	23.2
Neretva	1846	4606	703	2768	60.1
Trebišnjica	953	3392	746	2515	74.1
Cetina	80	169	–	–	–
Total	**6340**	**17,925**	**1868**	**7182**	**39.7**

Federal Ministry of Energy, Mining and Industry (2010)

Fig. 13.9 Bileća artificial lake used for HPP Bileća on the river Trebišnjica. (Photo by Aida Bidžan-Gekić)

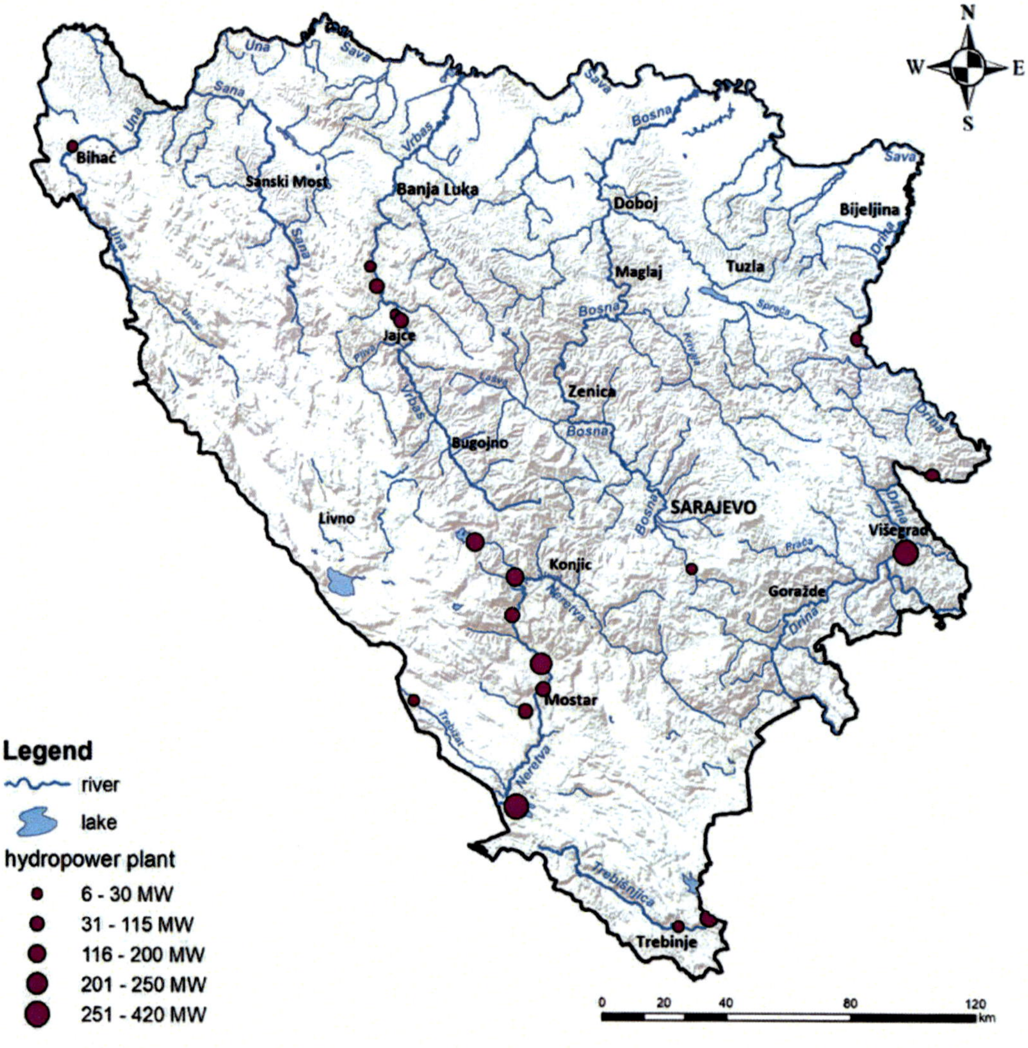

Fig. 13.10 Hydro potential in Bosnia and Herzegovina (2019)

characteristics of hydropower plants—candidates for possible construction are added up for all river basins, the following total hydro potential is obtained: 990 MW and 2913 GWh per year for the entity of Federation of BiH, and 1763 MW and 5824 GWh per year for the entity of Republika Srpska (Fig. 13.11). From the traffic aspect, the rivers and lakes of BiH are not of significant importance, except for the Sava river (Federal Ministry of Energy, Mining and Industry 2008).

BiH also has a considerable water potential, ideal for fattening cold-water and hot-water fish species. Along the shores of Neum bay, there are favorable conditions for the development of sea fishing. Unpolluted reservoirs, lakes, and rivers provide good natural conditions for fish farming. Artificial ponds and cage-type fish farms are located on the hydro-accumulations of the Neretva and Drina rivers. Fisheries are an up-and-coming agricultural sector in BiH. This is supported by the fact that fish is one of the products of animal origin that can be exported from BiH to the European Union market without any customs duties (within export quotas). About 6000 tons of fish are produced annually in BiH. Hatcheries and fish farms are numerous and have different capacities, areas, and annual production. According to FAO data for the Fisheries and Aquaculture Sector, 98 salmonid (trout), 28 carp, and 2 marine fish ponds can be found in BiH.

The most important fish species in the aquaculture sector of BiH are salmonid (California trout, brown trout, and brook trout), cyprinids (carp, grass carp, silver carp), marine (sea bass, gilthead sea bream, and dentex), and mollusks (mussels and oysters). To further develop, the fisheries sector needs the support of state institutions. Although the quota for the export of salmonid species has been increased, the question of the impossibility of exporting live fish, for which the producers of BiH have the capacity, remains open. Regular incentives and the participation of foreign investors are also crucial for the development of this production (FAOUN 1 2015).

13.7 Geothermal Energy

Geothermal energy represents the heat from the Earth. It is clean and self-sustainable. Geothermal energy sources range from surface sources to hot water and rocks located a few kilometers below the surface and, even deeper, all the way to magma. In nature, geothermal energy most often appears in the form of volcanoes, hot springs, and geysers. In our country, geothermal energy has been used for centuries in the form of spas or recreational and healing bathing.

The exploration of geothermal springs in BiH is very low. About 30 deep wells were drilled throughout the country; on a small number, the thermal parameters of well fluids were

Fig. 13.11 The Zvornik Lake on the river Drina, which belongs to BiH, is used by the Republic of Serbia for production of electric energy. (Photo by Aida Bidžan-Gekić)

examined, while the thermal characteristics of the rocks were not examined anywhere. The total possible installed capacity of geothermal sources with 44 locations where, according to the set criteria, heat exploitation is possible in the whole of BiH is 9.24 MWt if we consider only the possibility of space heating (temperature reduction up to 50 °C), or 90.2 MWt, if geothermal energy for space heating and recreational and balneological needs is observed (bathing, temperature reduction up to 20 °C). With the use of all these sources, with a utilization factor of 0.5, it is possible to produce 145.75 TJ of energy in 1 year only for space heating, or a total of 1421.75 TJ of energy if combined space heating and bathing (Simić and Ikić 2011; Di Pippo 2005). The geothermal potentials of BiH for space heating are estimated at approximately 33 MWth. The temperature at the locations in Šamac (85 °C), Kakanj (54 °C), and Sarajevo (58 °C) is too low to start electricity production, which is the main reason why these reserves are taken into account only when it comes to exploitation of thermal energy. The primary geothermal deposits are located in Triassic and Cretaceous limestones and are formed by accumulations of geothermal waters with a temperature of 35–150 °C (Simić and Ikić 2011). In BiH, there is a possibility of producing electricity from geothermal energy. Still, because it is specific investment, and BiH is rich in other energy sources, the production of such electricity will not begin soon.

References

Agency for Statistics of Bosnia and Herzegovina (2020) Bulletin of industrial production. Available at https://bhas.gov.ba/data/Publikacije/Bilteni/2020/IND_00_2019_TB_0_BS.pdf. Accessed on 26 Aug 2021

Agency for Statistics of FBiH (2020) Statistical Yearbook of FBiH. Available at http://fzs.ba/wp-content/uploads/2020/01/Godisnjak2019.pdf. Accessed on 26 Aug 2021

Andijašević J (1981) Prilog ekonomskoj ocjeni prirodnih resursa [Contribution to the economic assessment of natural resources]. Ekonomski glasnik, Sarajevo

Archive of Center for Investigative Reporting (2015) Everything about the energy potential of BiH in one place 2015. Available at https://cin.ba/sve-o-energopotencijalu-bih-na-jednom-mjestu/. Accessed on 17 June 2020

Archive of Dubrava Coal Mine. Available at https://www.kreka.ba/index.php/2016-02-12-10-52-36/dubrave. Accessed on 25 May 2020

Archive of Gračanica Coal Mine. Available at http://www.rugracanica.ba/Orudniku.php. Accessed on 25 June 2020

Archive of Kakanj Coal Mine. Available at https://www.rmukakanj.ba/v2/stranica/1. Accessed on 22 Apr 2020

Archive of Kreka 1 Coal Mine. Available at https://www.kreka.ba/index.php/2016-02-12-10-52-36/mramor. Accessed on 23 June 2020

Archive of Kreka Coal Mine. Available at https://www.kreka.ba/index.php/istorijat. Accessed on 22 Apr 2020

Archive of Ljubija Iron Ore Mine. Available at https://rzrljubija.com/. Accessed on 11 May 2021

Archive of the Saltworks JSC Tuzla. Available at https://solana.ba/historijat-muzej/. Accessed on 14 Apr 2020

Archive of Tuzla Salt Mine. Available at https://rudniksoli.ba/historijat. Accessed on 07 May 2020

Archive of Zenica Coal Mine. Available at https://rmuzenica.ba/home/. Accessed on 08 May 2020

Archives of the Electric Power Industry of Bosnia and Herzegovina (2019) Invests in the wind farm on Vlašić. Available at https://www.epbih.ba/novost/10586/elektroprivreda-bih-investira-u-vjetropark-na-vlasicu. Accessed on 25 May 2021

Archives of the Electric Power Industry of Bosnia and Herzegovina (2021) Development plans at the Energy Summit "Trebinje 2021". Available at https://www.epbih.ba/novost/32552/o-razvojnim-planovima-epbih-na-samitu-energetike-trebinje-2021. Accessed on 25 May 2021

Bačinović D, Džananović I, Jahić A (2007) Tehno-ekonomska analiza rada male hidroelektrane Modrac [Techno-economic analysis of the operation of the small hydropower plant Modrac]. In: Konjić T (ed) Zbornik radova VIII savjetovanje BH Komiteta Cigre. BH K CIGRE, Neum, pp 385–392

Bago D (2012) Solarna energija u Bosni i Hercegovini [Solar energy in Bosnia and Herzegovina]. Association of Businessmen Posušje, Posušje, pp 2–17

BH MAC (2020) Bosnia and Herzegovina Mine Action Centre. Available at http://www.bhmac.org/?page_id=747&lang=en. Accessed on 20 May 2020

Business-magazine (2019). Available at http://business-magazine.ba/2019/08/25/solarne-elektrane-isplativa-investicija/. Accessed on 06 Nov 2020

Census of Population, Households and Dwellings in Bosnia and Herzegovina (2013) Final results. Agency for Statistics of Bosnia and Herzegovina. Available at https://www.popis.gov.ba/popis2013/knjige.php?id=0. Accessed on 06 Apr 2020

Čičić S (1979) Energetske mineralne sirovine [Energy minerals]. Geoinžinjering, Sarajevo

Dairy industry, Foreign Trade Chamber of Bosnia and Herzegovina. Available at https://www.komorabih.ba/wp-content/uploads/2021/01/Mlijecna-industrija-BHS.pdf. Accessed on 18 Aug 2020

Di Pippo R (2005) Geothermal power plant: principles, applications, case studies and environmental impact, Elsevier Science, North Dartmouth

Doleček V, Karabegović I (2011) Biomasa kao energetski potencijal obnovljivih izvora energije u Bosni i Hercegovini [Biomass as an energy potential of renewable energy sources in Bosnia and Herzegovina]. Academy of Sciences and Arts of Bosnia and Herzegovina, Sarajevo

Đonlagić M (2010) Obnovljivi izvori energije [Renewable energy sources]. Center for Ecology and Energy, Tuzla

Electro test (2014) Available at http://www.elektro-test.ba/solarni-paneli-solarne-elektrane/u-gorazdu-pustena-u-promet-novoizgradena-solarna-elektrana-2/. Accessed on 26 May 2021

Energy Sector Study in Bosnia and Herzegovina, ESSBIH. Available at http://www.mvteo.gov.ba/data/docs/reports/Summary_Report_hr_sr.pdf. Accessed on 24 Aug 2021

FAOUN (2015) Analysis of the forestry sector in Bosnia and Herzegovina. Available at http://www.fao.org/3/au015o/au015o.pdf. Accessed on 25 Aug 2021

FAOUN 1 (2015) Fisheries and Aquaculture Sector in Bosnia and Herzegovina. Available at https://europa.ba/wp-content/uploads/2015/05/delegacijaEU_2015020309341807bos.pdf. Accessed on 28 June 2020

Federal Ministry of Energy, Mining and Industry (2008) Strategic plan and program for the development of the energy sector of the Federation of Bosnia and Herzegovina. Available at http://www.fbihvlada.gov.ba/bosanski/izdvajamo/SPP-GLAVNA%20KNJIGA/SPP-TEKST%20FINALNI.pdf. Accessed on 25 Sept 2020

Federal Ministry of Energy, Mining and Industry (2010) Construction of power facilities in FBiH. Available at http://www.fbihvlada.gov.ba/bosanski/izdvajamo/izgradnja.pdf. Accessed on 07 Sept 2020

Federal Ministry of Energy, Mining and Industry (2018) Archive Federal Ministry of Energy, Mining and Industry, News launched

Mesihovina wind farm. Available at https://fmeri.gov.ba/vijesti/posts/2018/march/pustena-u-rad-vjetroelektrana-mesihovina/. Accessed on 05 May 2020

FERK (2011) Electricity Regulatory Commission of FBiH. Available at http://www.ferk.ba/_ba/images/stories/2011/2212-dz-pdi-eko-energija-bs.pdf. Accessed on 26 May 2020

Granić G, Zeljko M, Moranjkić I, Martinez JA, Olano M, Jurić Ž (2008) Studija energetskog sektora u BiH [Energy sector research in BiH]. Final Report, Ministry of Foreign Trade and Economic Relations of Bosnia and Herzegovina, Sarajevo

Institute for Protection and Ecology of RS (2021) Environmental impact study, construction of energy project 2021—wind farm "Grebak", Nevesinje municipality. Available at https://www.vladars.net/sr-SP-Cyrl/Vlada/Ministarstva/mgr/Documents/Нацрт%20студије%20утицаја%20на%20жс%20изград?е%20вјетроелектране%20Гребак_428849853.pdf. Accessed on 26 May 2020

ISR (2020) Institute for Statistics of Republika Srpska, Cities and Municipalities of Republika Srpska. Available at https://www.rzs.rs.ba/static/uploads/bilteni/gradovi_i_opstine_republike_srpske/2020/Gradovi_I_Opstine_Republike_Srpske_2020_WEB.pdf. Accessed on 26 May 2021

Kovačević D (1961) Trgovina u srednjovjekovnoj Bosni [Trade in medieval Bosnia], Djela, vol 13. Department of Historical and Philological Sciences, ND NRBiH, Sarajevo, pp 15–105

Kozar A, Livnjak E (2015) Potencijal i aspekti ekonomičnosti malih hidroelektrana u Bosni i Hercegovini [Potential and aspects of economy of small hydro power plants in Bosnia and Herzegovina]. In: Milojković S (ed) Zbornik radova međunarodnog naučno-stručnog simpozijuma Infoteh-Jahorina 2015. Faculty of Electrical Engineering, Istočno Sarajevo, pp 854–858

Marić Đ (1991) Industrijalizacija Bosne i Hercegovine—geografska studija savremenog razvoja [Industrialization of Bosnia and Herzegovina—a geographical study of modern development]. Geographical Society of Bosnia and Herzegovina, Sarajevo

Nurković R (2003) Utjecaj industrije na urbanistički razvoj i infrastrukturu centralnih naselja Tuzlanske kotline [The impact of industry on urban development and infrastructure of the central settlements of the Tuzla valley]. In: Nurković R (ed) Zbornik radova međunarodnog naučnog seminara: Regionalno razvojna problematika BiH i susjednih zemalja u procesu približavanja EU. University of Tuzla, Tuzla, pp 191–199

Nurković R, Rewucki J (2018) Economic relations of Bosnia and Herzegovina and Islamic countries. Folia Geograph 60(2):Presov, 49–61

Rodić D (1979) Geografija Jugoslavije 2 [Geography of Yugoslavia 2]. Naučna knjiga, Beograd

Simić J, Ikić S (2011) Geotermalne elektrane i mogući razvoj u Bosni i Hercegovini, [Geothermal power plant and a possible development in Bosnia and Herzegovina]. In: Milojković S (ed) Zbornik radova međunarodnog naučno-stručnog simpozijuma Infoteh-Jahorina 2011. Faculty of Electrical Engineering, Istočno Sarajevo, pp 1039–1042

Spatial Planning Strategy of the Federation of Bosnia and Herzegovina (1997) I faza. Federal Ministry of Physical Planning and Environmental Protection, Sarajevo

Strategic Plan for Rural Development of Bosnia and Herzegovina (2018–2021) Ministry of Foreign Trade and Economic Relations. Available at http://www.mvteo.gov.ba/data/Home/Dokumenti/Poljoprivreda/BOS_SPRR_2018-2021_-_korekcija.pdf. Accessed on 05 Apr 2021

Study of mining and chemical industry in Bosnia and Herzegovina (2001). Available at https://www.esiweb.org/pdf/bridges/bosnia/GTZ_RudarstvoHemija.pdf. Accessed on 16 June 2020

Tadić A, Vujadinović M (2011) Vjetropotencijal i strategijarazvoja vjetroelektrana u Bosni i Hercegovini [Wind potential and wind farm development strategy in Bosnia and Herzegovina]. In: Milojković S (ed) Zbornik radova međunarodnog naučno-stručnog simpozijuma Infoteh-Jahorina 2011. Faculty of Electrical Engineering, Istočno Sarajevo, pp 1034–1038

Umihanić M, Ćehajić N, Salihović N (2009) Usporedna analiza proizvodnje električne energije fiksnih fotonaponskih sustava u različitim dijelovima Bosne i Hercegovine [Comparative analysis of electricity production of fixed photovoltaic systems in different parts of Bosnia and Herzegovina], Tehnički glasnik, vol 9. University of the North, Varaždin, pp 128–135

Vujović M, Stojković J (1985) Stanje sirovinske osnove i perspektive daljnjeg razvoja solarstva u Tuzli [The state of the raw material base and perspectives of further development of solar production in Tuzla]. Museum of Eastern Bosnia, Tuzla Salt Factory, Tuzla

Zlomušica E (2013) Particular review on SODAR and LIDAR measurements of Bora Wind in Mostar, Bosnia and Herzegovina. Int J Eng Technol (IJENS, Rawalpindi) 13:49–57

14 Economic Development of Bosnia and Herzegovina

Abstract

Bosnia and Herzegovina belongs to the group of European countries with the lowest GDP per capita. The agricultural activity is dominated by the cultivation of corn and fodder plants used in animal husbandry. Although there is a noticeable lack of modern technologies in the industrial sector, many smaller firms have become competitive in the European market. In the section on industry development, we have presented the most important industries with companies that we believe are the bearers of development. The development of the transport system is very slow. The quality of roads is weak and that is one of the limiting factors of economic development. There is no commercial domestic air transport of passengers and goods. Bosnia and Herzegovina has significant natural and anthropogenic potentials for tourism development, but they are very poorly used. Only a few major urban centers such as Sarajevo and Mostar can boast a significant contribution of tourism to overall economic development. The development of other service sector activities is heavily influenced by low investment such as education and health. In international trade, the coverage of exports by imports is only 60%. The most important export product is electricity with a share of 5% in total exports.

Keywords

Development · Agriculture · Industry · Services · Tourism

14.1 General Characteristics of Economy Development

The most dynamic economic development of Bosnia and Herzegovina (hereinafter: BiH) took place under the influence of industrialization after the Second World War. These critical processes of economic and social restructuring took place depending on the general Yugoslav policy and circumstances, and for Bosnia and Herzegovina, in scope and especially in structure, very unfavorable and far behind the developed European countries with Western democracies. The real socialist period took place in several distinct periods. From 1953 to 1962, centralist-administrative management prevailed, which had all the characteristics of copying the Soviet system. By introducing the so-called self-government (1950) and the abolition of centralized directorates (1952), relations in business, especially in agriculture and industry, were significantly liberalized in relation to other countries of Soviet Eastern Europe. Certain connections have been established with Western countries, until the so-called first economic reforms in 1965, when there was a period of intensive industrialization, with a general transfer of values from agriculture to industry.

In the fourth substage of the real-socialist period (1979–1990), significant shifts in the economic structure were not achieved because market orientation, which was the only way out, was impossible to include and adapt to the existing authoritarian and planning system. Therefore, in that period, the current economic and political crisis elements were deepened and increased. The processes of genuine restructuring began only with the first democratic, multiparty elections in 1990 and the creation of independent states on the territory of the former Yugoslavia in 1991.

In the period 1979–1990, 129 new factories were built in BiH, and 10,332 new jobs were created. If we look at the volume of total Yugoslav construction, then that number was very low (share of 5.9% of new factories built and only 7% of new jobs) (Marsenić 1995).

Before the war, Bosnia and Herzegovina was one of the middle-developed republics of the former Yugoslavia, and in terms of the level of development, it was behind Slovenia, Croatia, and Serbia. The achieved level of development resulted in a pronounced dynamics of postwar economic growth. The growth of the social product in the economy, for example, from 1956 to 1966 was 8.7% per year; from 1966 to 1967, it was 4.6%; while by it was 1991 3.8%. When

H. Gekić et al., *The Geography of Bosnia and Herzegovina*, World Regional Geography Book Series,
https://doi.org/10.1007/978-3-030-98523-3_14

analyzing the dynamics of this development, it should be borne in mind that the development of the economy of this republic after the Second World War began slowly and that such a pace of economic growth was a condition to get out of long-term backwardness (Statistical Yearbook of Bosnia and Herzegovina 1991).

Before 1991, the share of primary branches in the industry structure was 30% higher than in the former Yugoslavia. The focus on such economic development of the primary industry was strongly reflected in the development of other economic areas and branches. Nearly 50% of Bosnia and Herzegovina's economy was industry, while only 15% was trading, and all long-term economic activities had a much smaller share (Statistical Yearbook of Bosnia and Herzegovina for 1991). Their development was much slower in relation to industry, and in some economic activities, it was unfairly neglected, especially if we consider the available opportunities (agriculture, especially livestock, etc.). Given that the structure of industrial production was dominated by the production of raw materials, intermediate goods, and the production of lower stages of work, the industry did not have a greater incentive for the faster development of other economic activities. At the same time, this was reflected in BiH's uneven spatial distribution of the population and economic activities.

Regarding the conditions and factors of production, the most important changes in the economy of Bosnia and Herzegovina, in the period 1992–1995 (during the aggression on Bosnia and Herzegovina), occurred based on their permanent destruction from the face of the Earth or their permanent disabling for economical use. First of all, in this regard, it can be argued that during the war, the volume of foreign trade of the country was reduced many times and that there were radical changes in its internal structure and external regional orientation. An important economic characteristic is reflected in the radical change in exchanging economic goods in the internal market. That little money supply in the form of its transactional component represented only the function of money as a "unit of account," and cash circulated in a very heterogeneous monetary structure. Therefore, the value of direct economic war damage, i.e., the immediate negative economic consequence of the war, was about USD 90 billion in market value, representing between 25% and 30% of Bosnia and Herzegovina's total estimated economic potential at the beginning of the war (it is estimated that, at the end of 1991, the amount was about USD 300 billion) (Poverty Reduction and Economic Management Unit Europe and Central Asia Region 2005).

Indirect economic consequences of the 1992–1995 war are lost foreign markets, perhaps the most extensive single form of indirect war economic damage because a large number of domestic production capacities were built to place their products on the former ex-Yugoslav and/or on the world market. Therefore, since the GDP of Bosnia and Herzegovina in 1991 was about USD 15 billion, it can be concluded that the war damage caused in this way amounted to approximately USD 45–50 billion. Conducted research on the competitiveness of Bosnia and Herzegovina has shown that, because of the war in the information sphere only, BiH has further aggravated and slowed down the improvement of its position in the world ranking of competitiveness (Hasić 2006).

Many consider scientific backwardness to be the most consequential negative economic and overall war damage in the long run. Just the fact that, due to the 1992–1995 war, more than 40% of scientific and research staff left the country permanently is enough to understand what and how significant are the consequences for the society, state, and economy of Bosnia and Herzegovina. If we add that the share of allocations for science from GDP in 2004 (0.2%) was about ten times less than in 1991 (1.8%), we get an accurate picture of the negative consequences of the war on lagging in scientific development. The consequences of the war between 1992 and 1995 in the economic sphere are vast and complex. They are especially pronounced in the production component and transport. Due to the war, the physical volume of production was reduced to less than 10% of the prewar level, and almost the entire production was on the war track. Foreign trade has practically dried up, except for those types of cross-border trade that had an unhindered flow (aluminum, wood, cereals). The total direct and indirect war economic damages were estimated at around USD 115 billion in market value, with the proviso that they were more pronounced in the domain of the so-called invisible capital (lost markets, technological backwardness, privatization, telecommunications, informatics, management, marketing). The war has left Bosnia and Herzegovina in the transition process about 20 years behind the average of other countries in transition, significantly affecting its competitiveness on an international scale. In the scientific-technological and information sphere, this lag is even more pronounced. The long-term consequences in the domain of "brain drain" into the world are especially negative (about 40% of the scientific potential). Due to the enormous economic consequences of the war and the slow transition changes, Bosnia and Herzegovina is considerably delayed on the path to joining the relevant Euro-Atlantic integration, especially the European Union (Bašić 2006).

With the disintegration of the former Yugoslavia, the market was lost, and the business ties of many Bosnian companies with their clients were severed. From 1992 to 1995, due to the war in Bosnia and Herzegovina, on the example of a once sizeable industrial center of Bugojno, we can see a complete cessation of production in almost all companies. Former large companies, such as "Slavko Rodić," "Špedicija," "Tvornica kože, krzna i obuće," are long gone. What's left are parts of the plants that are nearly not as competitive as their predecessors (Gekić and Bidžan 2015).

Economic activities in Bosnia and Herzegovina are extremely unevenly distributed in space. On the one hand, we have an intense concentration of economic activities and jobs in five centers: Sarajevo, Mostar, Banja Luka, Tuzla, and Zenica, and on the other, much less developed areas. This is shown by the largest number of employees in the industry. In Bosnia and Herzegovina, in 1971, 53.7% of all employees worked in the secondary sector, in 1991 51%, and in 2018 only 32.1%. With the decrease in the share of the number of employees, the importance of the secondary sector in the acquisition of gross domestic product also decreased, and the importance of the tertiary sector and especially the primary sector grew in the last 10 years. In the branch structure of Bosnia and Herzegovina in 2018, most employees were in the tertiary sector −52.2% of the total number of employees, while as many as 15.7% of employees were in the primary sector. The total number of employees increased from 563,642 in 1971 to 976,384, but in 2018 was lower by 15.8% (822,000) (Statistical Yearbook of Bosnia and Herzegovina 1971, 1991; Agency for Statistics of BiH 2017).

Transition, i.e., the processes of restructuring the economy and society as a whole in Bosnia and Herzegovina, is taking place very intensively, but also in significantly more complex and unique conditions (Nurković 2005). The standard transition package, applied more or less in most post-communist countries, was completed by the World Bank and the International Monetary Fund in accordance with the principles of neoclassical economic ideology (Lorber 2009).

In 1991, large industrial labor centers predominated in Bosnia and Herzegovina. More substantial labor centers were of particular importance for developing continuous urbanized zones; there were five of them in 1991. Sarajevo is the largest labor center, with 184,674 or 46.09%; Banja Luka 65,026 or 16.23%; Zenica 54,991 or 13.72%; Tuzla 51,852 or 12.95%; and Mostar 44,124 or 11.01% of the total number of employees in Bosnia and Herzegovina. Indicators on the total number of employees in regional centers from 2011 indicate a profound crisis of economic activities. In 2011, the number of employees fell below the levels in 1991; Sarajevo remained the largest labor center, with 94,165 or 28%; Banja Luka 70,958 or 21.16%; Tuzla 60,994 or 18.18%; and Zenica 54,622 or 16.2% of the total employees in Bosnia and Herzegovina. In 2016, Sarajevo continued to have the largest number of employees (Agency for Statistics of BiH 2017).

Judging by the dynamics of job growth in the network of work centers, significant changes will not occur quickly. According to the dynamics of job growth, some other centers stand out, primarily municipal ones. These are, for example, Zenica, Tešanj, Travnik, Brčko, and Bijeljina. Regional development disparities are a global problem; European countries are also developed very differently, so there are differences between regions. Regional development differences in Bosnia and Herzegovina are characterized by regional polarization of population and functions (Nurković 2012). In this context, the first private accumulation of capital took place the fastest in trade, tourism and catering, finance, intellectual services, and the like, with a much slower restructuring of industrial production, where a powerful dominance of state ownership and influence occurred is still present.

In the period 1996–2000, high economic growth rates were achieved, which in some of those years exceeded 20%. However, due to a socioeconomic structure dominated by industry and mining and an underdeveloped primary sector, such rates have not been sustainable in the long run. Since 2000, the transition process in Bosnia and Herzegovina has accelerated, largely due to monetary stabilization (low inflation, stable currency, low external public debt), price liberalization, and financial system reform. In Bosnia and Herzegovina, economic development in 2017 in real terms increased by 6.2%, the highest GDP growth in the previous 5 years, while its average growth rate was 5.2%. By comparison, Bulgaria had GDP growth of 5.6%, Serbia 5.5%, Romania 5.2%, Albania 5.0%, Croatia 4.6%, Macedonia 4.0%, and Montenegro 4.0% (The World Factbook CIA 2018).

Bosnia and Herzegovina has a transitional economy with limited market reforms. The economy relies heavily on the export of metals, energy, textiles, and furniture as well as on remittances and foreign aid. A highly decentralized government hampers economic policy coordination and reform, while excessive bureaucracy and a segmented market discourage foreign investment. The economy is among the least competitive in the region. Foreign banks, primarily from Austria and Italy, control much of the banking sector, though the largest bank is a private domestic one. The konvertibilna marka (convertible mark)—the national currency introduced in 1998—is pegged to the euro through a currency board arrangement, which has maintained confidence in the currency and has facilitated reliable trade links with European partners. Bosnia and Herzegovina became a full member of the Central European Free Trade Agreement in September 2007. In 2016, Bosnia began a 3-year IMF loan program, but it has struggled to meet the economic reform benchmarks required to receive all funding installments.

Despite an absolutely and relatively slow increase in the number of employees, more complex economic development has generated a strong socioeconomic transformation. On the one hand, an intensive process of deagrarization, and on the other hand, a corresponding increase in the number and share of employees in the activities of the secondary and tertiary sectors was maintained. However, data on GDP measured by purchasing power show that Bosnia and Herzegovina's GDP per capita in 2017 was around USD 12,700, and only Albania was a country with a slightly lower GDP per capita in the region of Southeast Europe. Bosnia and Herzegovina's regional development is behind Montenegro, Serbia, and Macedonia (The World Factbook 2018). Bosnia and Herzegovina's GDP per capita is only 33% of the EU's GDP per capita. GDP growth is estimated at 3% per year. Foreign direct investment accounts for 2.2% of GDP. According to the Agency for Statistics in Bosnia and Herzegovina, there are about 2.4 million able-bodied inhabitants, of which about 822,000 are in active employment, which is 34.3%. The number of unemployed with higher and high education in 2018 is about 42,800.

To create a map of the development of municipalities in BiH in 2019, we used the development index that takes into account per capita tax revenues, employment rate, population trends, share of the elderly, and education level. We divided the municipalities into five groups (Fig. 14.1). The most

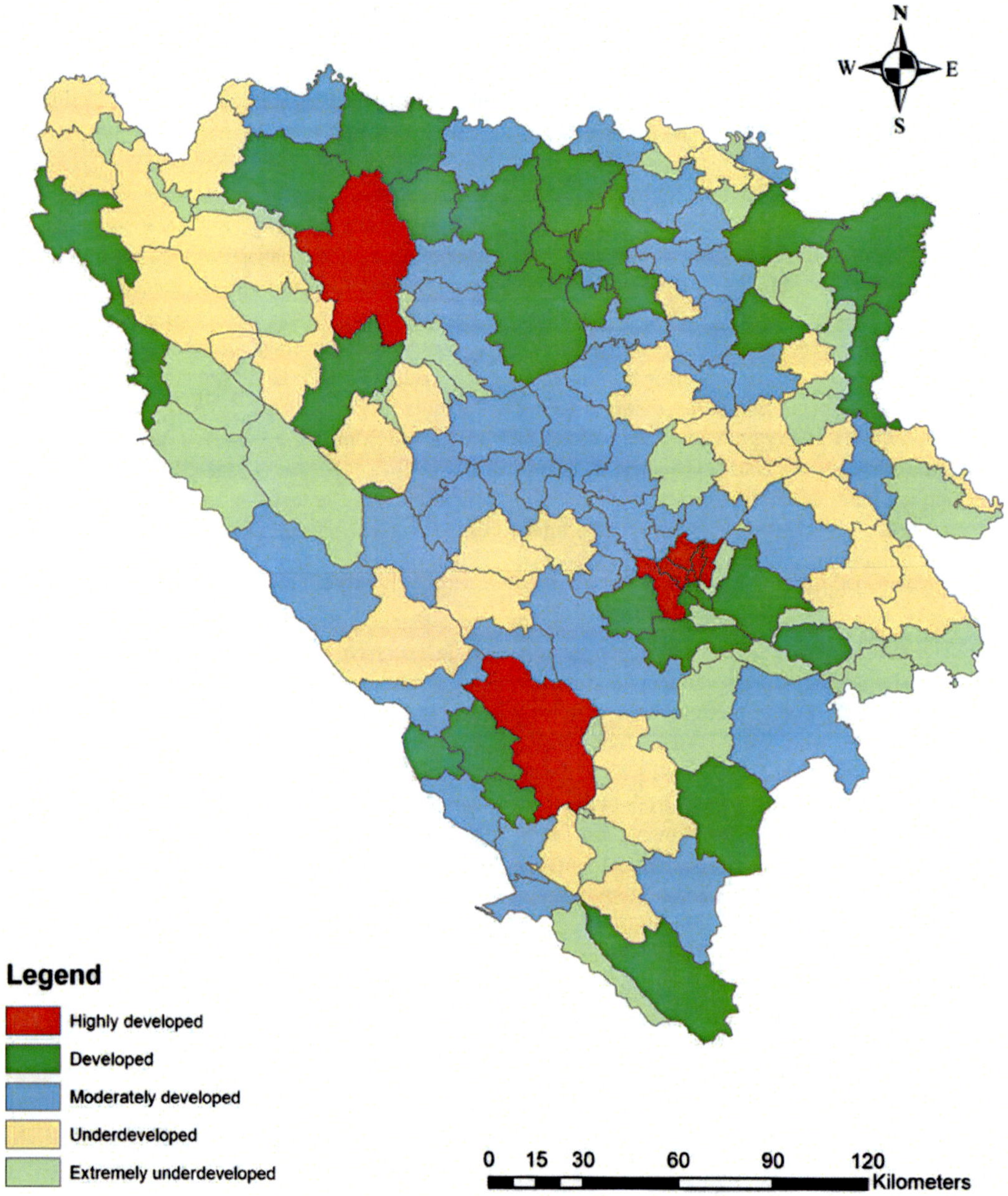

Fig. 14.1 The development of municipalities in BiH in 2019

developed municipalities in BiH were marked as highly developed, and there were only eight of them. These are mostly municipalities that are part of the City of Sarajevo (Centar Sarajevo, Novo Sarajevo, Stari Grad Sarajevo, Novi Grad Sarajevo) or are under its strong agglomeration influence such as Ilidža and Vogošća, but also Banja Luka as a center of development in the northern part of BiH, and Mostar in the southern part of BiH.

A total of 28 municipalities were marked as developed. Among them are those under the agglomeration influence of Sarajevo, such as Istočna Ilidža and Istočno Novo Sarajevo, Hadžići, Pale RS, and Trnovo, but also development centers in different parts of BiH such as Tuzla, Bihać, Široki Brijeg, Tešanj, Bijeljina, Brčko, Doboj, Trebinje, etc. A total of 42 municipalities are marked as moderately developed. Among them are Zenica, Banovići, Vitez, Visoko, Neum, Brod, Kozarska Dubica, Modriča, Travnik, etc. As underdeveloped local communities, we have marked 31 of them, and among others with weaker development are Kladanj, Olovo, Stolac, Bosanski Petrovac, Bratunac, Nevesinje, Ljubinje, Šipovo, and Vlasenica.

Similar to the next group, these are mostly highly depopulated areas. There were 34 of them in the group of municipalities marked as extremely underdeveloped. Most of them either have a small area of territory, were formed by the establishment of an Inter-Entity Boundary Line, and are rural local communities (municipalities), or they are located in the south, and in the mountainous areas of eastern BiH. In this group, the weakest developments are Istočni Drvar, Istočni Mostar, Dobretići, Bosansko Grahovo, Berkovići, Kalinovik, Novo Goražde, Rudo, Srebrenica, Pale FBiH, Sapna, and Vukosavlje (Agency for Statistics of FBiH 2020; ISR 2020).

14.2 Agriculture

The primary natural resource that determines the scope and structure of agricultural production is agricultural land, its area, topographic characteristics, and quality. The humid temperate continental climate of northern Bosnia and the distribution of fertile alluvial soil enable the cultivation of cereals, industrial plants, fruits, and vines. The Central Bosnian mountain-valley area with a changing continental climate offers excellent opportunities to produce Central European fruits, mountain cereals, vegetable crops, and livestock. The southern parts of BiH, with the characteristics of the Adriatic climate, are more favorable for cultivating numerous agricultural crops, especially vines, early vegetables and fruits, tobacco, etc.

The differences in the conditions of agricultural production within Bosnia and Herzegovina are significant. They are, as we have already said, a consequence of the action of natural geographical factors. These factors strongly influence the form and manner of organization of agricultural production and its productivity. Thus, for example, in the hilly and mountainous areas of Bosnia and Herzegovina, agricultural production is carried out mainly in the traditional way with less use of machinery and modern agrotechnical measures and means, resulting in lower productivity. Extensive livestock and agricultural production is represented here. In the plains of northern Bosnia and the valley-basin regions of central Bosnia and the Mediterranean area, agricultural production is carried out in a more modern way, with agrotechnical measures and means, which is why productivity is much higher. Also, one of the recent significant causes of insufficient production of primary agricultural products in Bosnia and Herzegovina so far has been insufficient and inadequate use of agricultural land. Still, with the emergence of the global pandemic COVID-19, more and more individuals are turning to their food production.

From Posavina to the coast of Bosnia and Herzegovina, given the physical-geographical and agricultural-livestock conditions, the territory of Bosnia and Herzegovina can be divided into six agricultural regions, which better reflect the natural conditions of agricultural production in BiH. Thus, going from north to south, the following alternate: plains 11.3%, hills 19.1%, hilly-mountainous area 10.2%, mountainous area 50.6%, Mediterranean mountainous area 6.5%, and Mediterranean region 2.3%. The largest part of arable land is located in the hilly agricultural region, namely, 44.4%, then in the plains 26%, mountain 25.6%, and Mediterranean 4%. Vineyards are most widespread in the Mediterranean region of 90.9%, and orchards are distributed in the same sizes and spatial distribution as arable land. There are the most natural meadows in the mountainous region, 71.9%, followed by 19.8% in hills, 5.9% in the plains, and 2.4% in the Mediterranean region. The distribution of pastures is somewhat different, considering the natural conditions and especially the morphological-biogeographical conditions. Thus, 46% of pastures are in the mountainous region, 30.4% in the Mediterranean region, 18.7% in the hilly area, and 4.9% in the lowland region.

Bosnia and Herzegovina spends more food than it produces, which certainly has a negative effect on the harmonization of the food balance, especially when it comes to overall export opportunities. The orientation towards the need for agricultural development in all settlements is therefore understandable. The development of agro-industry between 1995 and 2019 took place in Bosnia and Herzegovina under more favorable conditions than before; the basic preconditions and requirements for much faster growth of agricultural production have been created. Significant results have been achieved in the transformation of villages with more considerable investment of social funds and more organized efforts of the rural population in building infrastructure (electrification, construction of new and modernization of existing roads, health, water, and school facilities), the introduction of health, and pension insurance. The living standard of the rural population has increased, and working conditions in agriculture have improved.

Agricultural production possibilities in Bosnia and Herzegovina show that significantly larger quantities of food could be produced compared to current production; it is possible to increase production in agriculture by applying modern agro-technical measures with the education of the population. The contemporary transformation of our agricultural areas is primarily determined by the specific socioeconomic development of BiH, i.e., the increased industrialization of production methods and increasingly pronounced tertiary activities in rural settlements. Bosnia and Herzegovina does not use almost half of its arable land. The utilization of arable land is particularly low in western Bosnia, eastern Bosnia, and some municipalities in central Bosnia, primarily because displaced persons and refugees have not yet returned to their homes after deportation in the 1992–1995 war.

The area of used agricultural land in 2019 amounted to 1,016,682 hectares, which is 51,310 hectares or 5% more than in the previous year. In the structure of arable land, the most represented are sown areas with a share of 52%. The total sown areas amounted to 529,455 ha and are 1% less compared to the previous year. The decrease in sown areas was recorded in nurseries by 5%, cereals by 2%, vegetables by 3%, and industrial plants by 1% compared to the previous year. A significant increase in areas was recorded in nurseries by 59% (Table 14.1) (Ministry of Foreign Trade and Economic Relations of BiH 2019a).

Domestic seed production is insufficient, and large quantities of seeds are imported. Yields per unit area of all field and vegetable products produced in Bosnia and Herzegovina are low and represent one of the reasons for the weaker competitiveness of agricultural producers engaged in farming. The structure of sown areas with plant

Table 14.1 Ploughlands by way of use in Bosnia and Herzegovina in 2018 and 2019 in hectares

Way of use	2018	2019
Cereals	314,296	306,780
Industrial crops	15,324	15,225
Vegetables	71,758	69,735
Fodder crops	127,746	130,605
Nursery gardens	358	340
Rest on the ploughlands	4251	6770
Total sown	533,733	529,455
Fallows	64,218	77,040
Unused	367,421	410,187
Total ploughlands	**965,372**	**1,016,682**

Strategic Plan for Rural Development of Bosnia and Herzegovina (2018–2021)

crops in Bosnia and Herzegovina has not changed for many years. Cereals have the largest share in the total sown area of 58% (Fig. 14.2), followed by fodder crops 26%, vegetables about 15%, and industrial crops only 1%.

Cereal production is the foundation of the food security of the population, the most important segment of agricultural production. Cereals are the most represented in the overall structure of field production. Of the field crops, the most important is wheat production, as the most important bread grain, and corn, which is the main crop used in livestock nutrition.

The total realized production of grain in 2019 amounted to 1,666,184 tons and is lower by 4% compared to the total production from the previous year. Compared to the previous year, lower production was recorded in wheat by 9%, rye by 8%, barley by 6%, and corn for grain by 3%, while oats recorded an increase in production by 3%. Regarding the countries from which cereals are imported to Bosnia and Herzegovina, about 99% of imports are from four countries: Serbia (60%), Hungary (27%), Croatia (13%), and Italy (1%), while countries in which cereals are exported from BiH are Turkey (98%), Serbia (1%), and Croatia (1%) (Strategic Plan for Rural Development of Bosnia and Herzegovina 2018–2021).

In 2019, 51,251 ha were sown with wheat in BiH, while 50,936 hectares of ploughlands were harvested. The average yield was 4.6 tons per hectare. The extreme northeast is the real granary of BiH, and this lowland area between the Sava River and the extreme lower course of the Drina River is called Semberija. Bijeljina produced 20% of the total wheat production in BiH in 2019 and only then came Gradiška in the far north with 7.7%, Brčko with 6.5%, and Šamac with 4.9%. Of the other municipalities, the following stand out: Doboj (4.4%), Derventa (4.3%), and Modriča (3.3%) in the valley of the lower course of the river Bosna. Of the others, seven are characterized by only 2–3% of total wheat production in BiH: in fertile Posavina in the northeast Orašje, Pelagićevo, and Odžak, slightly eastern in Semberija, is Ugljevik, in the far northwest, Cazin, in the north Laktaši, and in the area of high karst and karst fields in the west, Glamoč. A total of 14 municipalities produce almost 70% of the total wheat production in BiH (Fig. 14.3) (Agency for Statistics of FBiH 2020; ISR 2020).

Vegetable production is one of the most represented and economically most profitable branches in agriculture. In Bosnia and Herzegovina, the conditions for growing various types of vegetables are very favorable. Due to significant climatic differences between individual regions, vegetables have almost continuous growth, from early vegetables in southern Herzegovina to late autumn in lowland and hilly and mountainous areas in the country's interior. Vegetables occupy about 12.5% of the total sown area. Almost about ½ the area under vegetables is sown with potatoes whose annual

Fig. 14.2 Crnoriječka plateau, fields prepared for planting buckwheat. (Photo by Aida Bidžan-Gekić)

production is about 350,000 tons. In 10 years, the average domestic production of potatoes is about 385,000 tons, with an average yield of about 10 tons per hectare. The rest of the area was sown with other vegetables: beans (22%), cabbage (14%), onions (14%), tomatoes (10%), peppers (10%), cucumbers (9%), carrots (5%), garlic (5%), peas (4%), lettuce (3%), watermelon (3%), and spinach (2%). Vegetable production varies over a period of 10 years in such a way that production increases in 1 year and decreases in the next (Table 14.2).

Growing vegetables is primarily conditioned by natural geographical factors, especially climate, height, and relief exposure (Fig. 14.4). Thus, potatoes, cabbage, and onions are grown mainly in hilly and mountainous areas. In the valley-valley and lowland areas, late vegetables are mostly grown, and in southern Herzegovina, in addition to early vegetables, vegetables are also produced during the winter (cabbage, lettuce, spinach).

Among the most important crops in the production of industrial plants, oilseeds stand out, especially soybeans and tobacco. During 2019, the realized soybean production of 23,753 tons was 2% higher than the previous year. The yield of soybeans was 2.6 tons per hectare and is 4% lower than last year. No less important is the fact that all soybeans produced in Bosnia and Herzegovina are not genetically modified (GMO-free), which gives it a unique quality and basis for use in the production of genetically unmodified livestock products: milk, meat, and eggs.

The production of industrial plants, too, is not negligible. In the structure of oilseeds production, about 76% of production is soybean production, and about 24% of production is oilseed rape production. Oilseeds were harvested on an area of almost 9000 hectares, from which production of 23,000 tons was achieved. The increase in the total harvested area of oilseeds was influenced by the increase in the harvested area of oilseed rape by 67% compared to last year. Sunflower is represented in the sowing structure sporadically, mainly as seed crop production, and in a small percentage. It is not included in the data on harvested areas, production, and yield of oilseeds.

Tobacco is an industrial crop grown in Bosnia and Herzegovina on very small areas, averaging about 1700 hectares, from which about 2300 tons of tobacco are produced annually. Tobacco yields range from 1.1 to 1.6 tons per hectare. The reasons for lower tobacco production and cultivation are reflected in the reduction of demand by the processing industry and its focus on raw material imports, lower purchase prices, and modest financial incentives for this type of production. Yet, these sown areas are important in ensuring farmers' incomes. Tobacco production in BiH takes place in the Posavina Canton and somewhat smaller areas in Herzegovina (Fig. 14.5).

In the structure of fodder production, about 55% of the total production is green corn (for silage), followed by legumes with a share of 16% and alfalfa with 15%. The total production of fodder plants in 2019 remained at the production level of 941,737 tons as in the previous year. Slightly higher yields increased the production of clover by 9% and alfalfa by 8%.

Although there are orchards with quality assortment and modern cultivation technology, generally speaking, fruit production in Bosnia and Herzegovina has the characteristics of

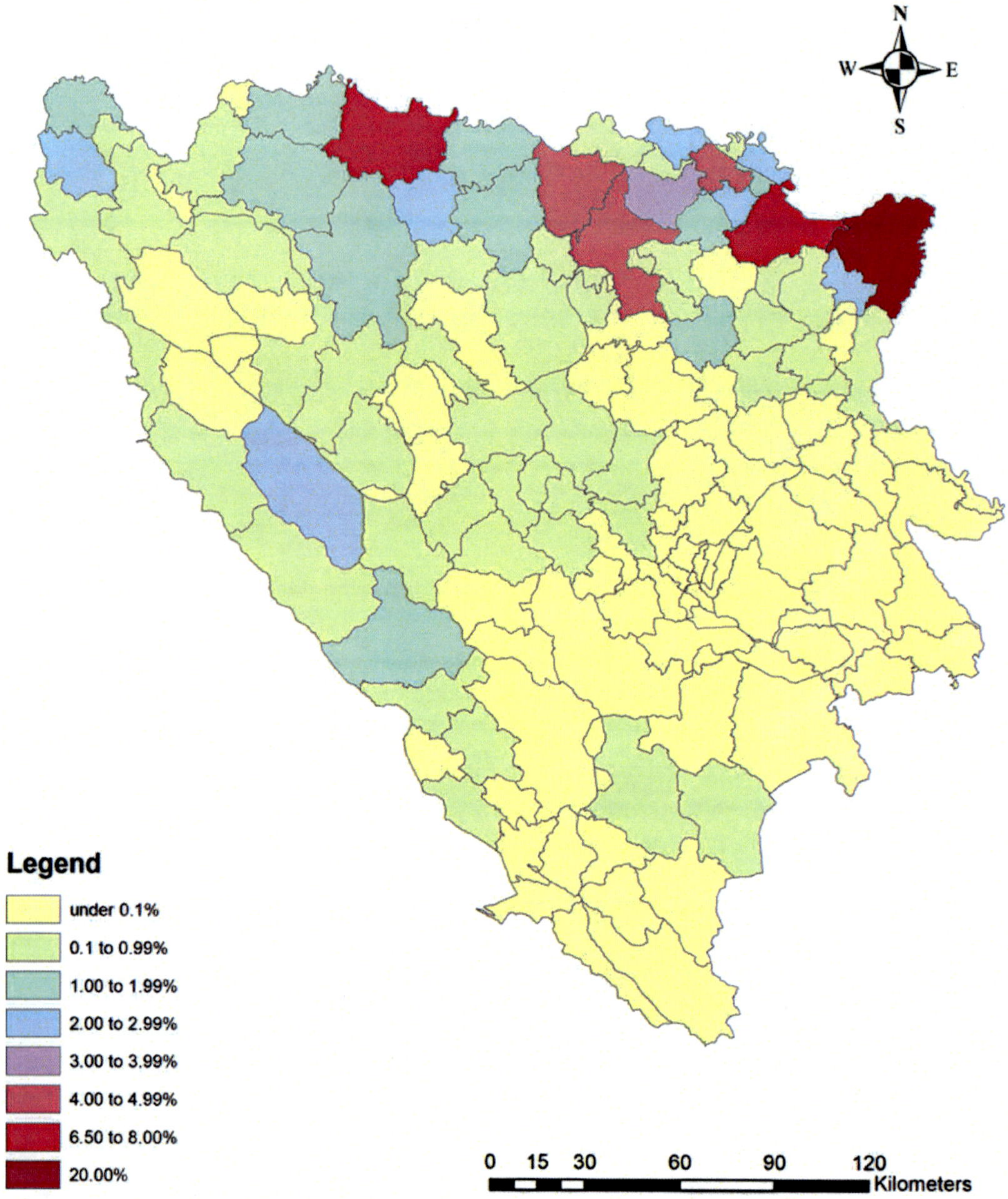

Fig. 14.3 The share of wheat production in Bosnia and Herzegovina in 2019

extensive production with a variety that is not interesting for the market, with outdated and uneconomical cultivation systems, unorganized and largely poorly controlled nursery production. The fruit market in Bosnia and Herzegovina is not organized and is mainly characterized by trade in goods outside the standard or of poorer quality. In addition to late frosts, other very unfavorable climate influences appear more and more often during the vegetation period, and, in the most sensitive phases of the fruit, they appear in the form of extreme drought, hail, or strong winds. Fruit production in Bosnia and Herzegovina takes place in areas estimated to occupy about 260,000 ha, on which an average of about 24 million native trees are planted. Plums (44%), apples (38%), and pears (12%) are the most represented in fruit production, although a large number of other fruit species are also grown (Table 14.3). According to the total production of raspberries (19,749 tons) in 2019, BiH ranks eighth in the world.

Bosnia and Herzegovina is known for its valuable indigenous varieties of fruits and vegetables. At the same time, there is an increasing demand for fruit that is organically produced and is not at all or, to a lesser extent, chemically treated. Awareness of the need to restore old grass orchards and the planting of old indigenous varieties, which are resistant to certain diseases, is slowly developing. Thus, there is a trend of planting indigenous varieties of apples. Old indigenous varieties of apples are in demand for consumption and the preparation of juice, jam, marmalade, jams, apple cider vinegar, brandy (rakija), and drying. The smell, taste, and aroma of home-grown varieties of apples, pears, plums (Fig. 14.6), cherries, peaches, apricots, medlars, mulberries, hazelnuts, etc. will hardly ever be forgotten by the one who ate them as a child. Fresh apples are also used to make typical dishes such as lijena pita (lazy pie), watered cider, tufahija (nut-filled apple), etc. Apple juice ("šira") is traditionally used fresh, although there are also options for canning. Apple jam is used as a sweet dessert in combination with sour cream or cream and for making cakes, primarily traditional (gurabija). Traditionally producing apple jam, with specific improvements, could become a recognizable product on a European scale. Some varieties of apples are used to make quality and expensive brandy (rakija).

Table 14.2 Production and yield of major crops in Bosnia and Herzegovina in 2019

Crop	Production in tons	Yield in tons/hectare
Wheat	264,769	3.8
Corn grain	1,235,596	6.4
Rye	10,449	2.9
Barley	74,462	3.6
Oats	27,231	2.7
Soy	23,753	2.6
Tobacco	1433	1.3
Potato	381,308	11.1
Carrots	10,906	8.1
Onions	40,398	8.5
Garlic	5641	3.3
Beans	12,718	1.7
Peas	4217	3.1
Cabbage	71,773	15.7
Tomato	43,700	12.6
Paprika	44,633	13.6
Cucumber	34,311	11.9
Clover	137,063	3.9
Lucerne	132,571	4.5
Corn silage	672,103	19.7
Strawberries	10,582	8.1
Raspberries	19,749	6.7

Strategic Plan for Rural Development of Bosnia and Herzegovina (2018–2021)

The production of berries in Bosnia and Herzegovina has experienced a substantial expansion in recent years, especially with raspberries and strawberries. The planted area of strawberries amounted to 1301 hectares and remained at the level of planted areas from the previous year. Realized strawberry production in 2019 amounted to 10,582 tons and is higher by 8%, while the average yield of strawberries per hectare was 8.1 tons. Raspberries were planted on an area of 2907 hectares, from which 19,749 tons were produced, which

Fig. 14.4 Arable farming in Vučipolje near Bugojno. (Photo by Aida Bidžan-Gekić)

Fig. 14.5 Scene of drying and chopping tobacco in the Rama region, Museum of the Franciscan Monastery Rama-Šćit. (Photo by Aida Bidžan-Gekić)

Table 14.3 Production and yield of most important fruits in Bosnia and Herzegovina in 2019

Fruit	Production (tons)	Yield (kg/tree)
Plums	114,415	9
Apples	98,265	13
Pears	30,345	11
Sweet cherries	9379	10
Sour cherries	4244	8
Apricots	1034	10
Mandarin	79	15
Lemons	15	7
Figs	980	13
Olives	776	12
Total	259,532	108

Ministry of Foreign Trade and Economic Relations of BiH (2019a)

is 28% less than last year's crop. The yield of harvested fresh raspberries per hectare was 6.7 tons and is lower by 41% than the previous year's crop. Reduction of planted areas by 15% and low yields resulted in lower raspberry production. The markets to which frozen raspberries and strawberries are placed are Serbia, Turkey, EU, and EFTA member countries.

The results of southern fruits and olives are very good, considering that the yields also increased compared to the previous year, lemons by 133%, olives by 31%, mandarins by 27%, and figs by 5%. The increase in fruit yields also affected the total fruit production, with the most significant increase recorded in lemons by 114%, mandarins by 41%, olives by 27%, and figs by 10%. The total area under vineyards is estimated to be about 3665 hectares (Fig. 14.7). In 2019, 39,289 tons of grapes were produced, and the total production is lower by 3% compared to last year. Although the number of fruit vines is lower by 858,000 pieces or 5%, the yield of grapes per vine was slightly higher by 4% than the previous year, but it still did not affect the production to exceed 40,000 tons achieved in the previous year. Bosnia and Herzegovina exports grapes to Croatia (78%), Lithuania (10), Serbia (6%), Montenegro (4%), and Russia (3%) (Ministry of Foreign Trade and Economic Relations of BiH 2019a).

Bosnia and Herzegovina has favorable conditions for the development of animal husbandry. Agrarian production cannot be economically justified on agricultural areas from 1 to 3 hectares, as it is not possible to base commercial livestock production on it. Livestock production, and especially meat production in Bosnia and Herzegovina, has encountered numerous problems in the past few years. Given the small production units and low level of milk production per dairy herd, the current situation in livestock breeding does not guarantee sufficient development in the coming period, and the competitiveness of the livestock sector in the open market cannot be ensured. Livestock production participates in the total value of agriculture with about 37%. This share is relatively low and indicates the structural problems facing the sector. In BiH, two basic types of cattle breeding can be differentiated: mountain-pasture and stable cattle breeding. Seminomadic mountain cattle breeding, which has a seasonal character, is represented in vast mountain pastures and meadows. In the mountainous and Mediterranean areas of Bosnia and Herzegovina, there are areas where animal husbandry has a long tradition and is the primary source of income. Such areas are Humine, Kupres plateau, Bjelašnica,

Fig. 14.6 Traditional plum orchard in Gradačac. (Photo by Aida Bidžan-Gekić)

Fig. 14.7 Traditional vineyard in Mostar. (Photo by Aida Bidžan-Gekić)

Vranica, and Vlašić. The situation in livestock in 2019 was as follows: cattle about 430,000 heads, pigs about 540,000 heads, sheep about 1 million heads, horses 14,000 heads, goats about 72,000 heads, and poultry about 22 million fowls (Ministry of Foreign Trade and Economic Relations of BiH 2019a).

The cattle sector is characterized by a limited number of large fattening farms and a large number of relatively small mixed farms producing milk and meat. In the hilly and mountainous areas of Bosnia and Herzegovina, autochthonous and primitive breeds of bush cattle and autochthonous breeds of sheep are bred: pramenka sheep, Kupres sheep, and Vlašić sheep (Figs. 14.8 and 14.9).

Following the 10-year trend of meat production, a decline in production of all categories, and an imbalance between the level of supply of domestic production and market needs, is

Fig. 14.8 Extensive sheep grazing in Gatačko polje. (Photo by Aida Bidžan-Gekić)

Fig. 14.9 Extensive cattle grazing in Gatačko polje. (Photo by Aida Bidžan-Gekić)

visible. The number of breeding cattle and fattening livestock was significantly reduced, which caused a significant decrease in the number of slaughtered heads in slaughterhouses. In addition, due to the lack of fodder from domestic sources, caused by unfavorable weather conditions, and the high prices of imported fodder, cattle production for fattening has become more expensive. In recent years, the prices of imported cattle for slaughter and fattening have also been slightly higher or almost at the same level as meat import prices. In the last 3 years, livestock imports have decreased, mainly from the European Union, at the expense of low-priced meat imports. This further reduced the number of slaughtered cattle, which brought the operations of the domestic slaughter industry to the brink of survival. The above resulted in a decline in domestic meat production and a shortage of meat on the domestic market, primarily fresh meat and raw materials for the meat processing industry, which was already based mainly on imported raw materials to meet its production capacity. Another fact that indicates the lack of meat from domestic production is the data on exports showing that mostly poultry meat is placed on the foreign market and only in the last few years has the export of beef increased, exclusively to the Turkish market.

Net production of all meat categories in 2019 amounted to 90,819 tons and is higher by 7% compared to the previous year. The higher level of total net meat production was influenced mainly by higher production of sheep meat by 11% and poultry meat by 8%, pork production by 6%, and beef by 4% compared to net production from the previous year. Meat production in 2019 was beef 11,939 tons (veal and beef up to 1 year 38.3%), sheep meat 1599 tons, pork meat 7749 tons, and poultry meat 67,165 tons.

When it comes to milk production, it is one of the essential livestock production branches and is still in a slightly more favorable position than meat production. The total production of raw milk in 2019 amounted to 643 million liters and is lower than the production from the previous year by 35 million liters or 5%. The total milk production is dominated by the production of cow's milk with a share of 97%, while the share of sheep's milk is 3% and goat's milk only 1%.

Thirty market-relevant dairies are operating in Bosnia and Herzegovina, while the number of dairies that have met all the conditions for exporting products outside BiH is 11, with a production capacity of 10,000–100,000 liters per day. In 2019, all dairies in Bosnia and Herzegovina purchased about 266 million liters of fresh raw milk for processing from milk producers. Drinking milk is the leading product in terms of production as well as export. The production of yogurt and other fermented beverages is a second-ranked product with an average production of 33 million liters (Dairy Industry 2020).

Of the other livestock products, the production of eggs, wool, and honey is significant. Wool production in 2019 amounted to 1381 tons, which is 4% less than last year, 722,877,000 eggs, and is higher by 10% compared to the previous year. The most significant decrease in production was recorded in honey production by 16%, compared to the production of the prior year when it amounted to 3992 tons (Ministry of Foreign Trade and Economic Relations of BiH 2019a).

Egg production is market-oriented and organized in large production units where modern technology is used in closed facilities. Additionally, part of the production is located on family farms where the obtained products (eggs and meat) are mainly used for their own needs. Such production takes place in extensive conditions, primarily in conditionally adapted facilities and a part of the free keeping of poultry.

As far as beekeeping in Bosnia and Herzegovina is concerned, it is a complementary occupation of farmers and a recreational occupation of non-farmers. Bee grazing in BiH is abundant and diverse. In the territory of Bosnia and Herzegovina, plants such as sage, heather, chestnut, almond, acacia, meadow flowers, etc. enable bees to produce high-quality honey. Climatic conditions allow the same plants to bloom in different areas at different times, so bee grazing is maintained throughout the seasons. Honey production is one of the most promising branches of agriculture in Bosnia and Herzegovina, which is increasingly expanding.

During 2018, in the agricultural sector in Bosnia and Herzegovina, the competent institutions in the field of agriculture carried out activities to implement key priorities, primarily the adoption and implementation of strategic documents, development of agricultural policy and rural development policy, improvement and promotion of incentives, implementation process harmonization of domestic regulations with European Union regulations, implementation of harmonized regulations, implementation of reforms in the context of European integration, and international cooperation.

Rural business development should be directed towards using raw materials already found in rural areas and the development of services needed by rural areas. Although a significant percentage of the rural population has confirmed that they have specific knowledge, experience, and skills to engage in certain occupations, additional training is needed on how to start activities such as wood processing, stone processing, processing of agricultural products, forest fruits, medicinal herbs, etc., and raw materials on the farm itself. Rural areas are still heavily burdened by the traditional division of labor into male and female, in which women have physical, heavier, and lower-paid jobs. Also, engaging in business activity in terms of ownership of business assets, management, and leadership is by definition destined for men. More and more women in rural areas have an education (even higher education) and the knowledge to successfully start and run a particular business activity on the farm itself or in the domain of performing a craft or some other entrepreneurial activity. In this, they should be given some support in holding specific training that will acquaint them with the conditions and ways of starting their own business and encourage them to take such steps.

One of the main weaknesses of the agricultural sector is characterized by the unorganized market of agricultural products, the system of purchase, low production volume, and the unprotected position of agricultural producers before the competition. The chance to overcome these weaknesses is reflected in connecting producers into cooperatives, associations, clusters, and other interesting forms of organization of producers. In this sense, groups of agricultural producers will be supported through professional services in terms of ways and forms of organization, legal support, and business plans development. The intensification of agricultural production requires and implies the introduction of new technologies and thus the acquisition of new knowledge and skills by agricultural producers to be able to apply new technologies on their own farms.

With better equipment and modern mechanization, it is possible to raise the level of production technology and manage production costs and product quality much better, which is directly dependent on the time required to perform certain work operations. The link between primary and secondary production and the creation of a supply chain represents a chance to improve the productivity and competitiveness of agriculture. In addition to providing the raw material base, many existing processing capacities require significant investments in the renewal and modernization of production and marketing technology. Several processing facilities must adapt to EU standards when it comes to food quality and safety. Processing capacities represent a significant potential for employment of workers in rural areas and the development of small and medium enterprises. For a systematic approach to the development of this area, it is necessary to build an operational investment program for agriculture and rural development (Development Strategy of Bosnia and Herzegovina 2010).

Fragmented and parceled farms are the main obstacle to improving the competitiveness of agricultural production. Since most agricultural land is privately owned, management authorities must find a way to stimulate farmers to exchange their plots, enlarge their holdings, and improve land quality and fertility by investing in land reclamation works. In addition, knowing the characteristics of land profiles and agroclimatic features of the territory would enable a more systematic approach to agricultural production planning. To achieve this, it is necessary to develop a project of digitalization of the land information system.

Linking primary and secondary production in the agriculture and food sector in terms of the planned organization of the production of those crops needed by the food industry would significantly facilitate the way for farmers to reach the market. A higher level of this cooperation would be achieved by uniting producers, purchasers, and processors of agricultural products in the supply chain (e.g., a cluster of healthy food). Many rural households in Bosnia and Herzegovina have a long tradition of processing various livestock and plant products directly on the farm. Products such as different meat products, cheese and clotted cream (kajmak), honey, and fruit and vegetable products represent an essential part of the rural economy and improve family farms' economic position.

14.3 Industry

Industry in Bosnia and Herzegovina began developing during the Austro-Hungarian rule in the second half of the nineteenth century. First to develop were the wood industry, ferrous metallurgy, and tobacco industry, while the processing industry developed very poorly and slowly. During the Kingdom of Yugoslavia, the development of the extractive and basic industries started, while the processing industry was underdeveloped. Throughout this period, Bosnia and Herzegovina was a significant raw material base for developing the processing industry in other areas of the Kingdom of Yugoslavia (Hrelja 1974).

The most dynamic economic development of Bosnia and Herzegovina was influenced by industrialization after the Second World War. These critical processes of economic and social restructuring took place depending on the general Yugoslav policy and circumstances. During this period, traditional industries advanced significantly: ferrous metallurgy, coal production, wood industry, and pulp and paper industry. In addition, many industries that did not exist before appeared in that period: the metal processing industry; the electrical industry; the machine, automobile, and aviation industries; and other types. However, even in this period in the industry structure in Bosnia and Herzegovina, the most important were the primary industries that produced raw materials and semi-finished products for the processing industry of other republics of the former SFRY. The main industrial areas were and remained to be Sarajevo-Zenica, Tuzla, and Banja Luka (Nurković 2007).

During the war (1992–1995), industrial production was almost completely stopped, and reconstruction was very difficult due to destroyed facilities and stolen devices. As a result, the remaining industry in 1997 worked with only 6–7% of prewar capacity, and only since 1998. It began to develop slowly but very poorly to this day. Iron ore is processed in the Zenica (Fig. 14.10), Vareš, and Ilijaš forges. The most important for nonferrous metallurgy was the alumina factory in Zvornik, which still operates today, and the aluminum factory in Mostar, which was closed in 2020. Before 1992, the most important were mechanical engineering and the metal industry (construction machinery, steel structures for construction, bearings, railway wagons, weapons, etc.). The main centers were still Sarajevo, Zenica, Tuzla, and Banja Luka.

A car factory (TAS) in Sarajevo produced various Volkswagen vehicles (Volkswagen Buba, Volkswagen Golf I, Volkswagen Golf II, Volkswagen Golf III, Jetta 1, Jetta 2, and others). Today, it produces only chassis components. The chemical industry is located in Tuzla, Lukavac, and Kreka (based on rock salt), Jajce, Sarajevo, Goražde, and Vitez, the paper industry in Prijedor, Maglaj, Kotor Varoš, Banja Luka, and Grude, wood in many places throughout the country, especially in Prijedor, Banja Luka, Maglaj, Travnik, Foča, Rogatica, Drvar, etc. The textile industry is developed in Visoko, Olovo, Kladanj, Travnik, Sarajevo, and Mostar. Other industries include leather (Visoko, Bugojno, Gračanica), food (Sarajevo, Gradačac,

Fig. 14.10 The plant of ArcelorMittal in Zenica, iron and steel production. (Photo by Aida Bidžan-Gekić)

Velika Kladuša, Mostar, Banja Luka, etc.), tobacco (Sarajevo, while Mostar and Banja Luka are closed), electrical (Banja Luka, Sarajevo, etc.), and many others.

14.3.1 Wood Industry

More than half of the total area of Bosnia and Herzegovina is covered with forests, which makes it one of the wealthiest countries in Southeast Europe, and its inhabitants are rightly proud of the long tradition of wood processing and exporting. The forests of Bosnia and Herzegovina are mostly natural forests, diverse by tree species, with a significant share of trees of larger average diameters, which makes a sound basis for the development of the wood industry. A wide range of products produced in Bosnia and Herzegovina are mainly based on raw materials: beech, fir, spruce, pine, and somewhat less walnut, oak, ash. The main strategic commitment of the wood industry is the highest possible level of finalization and the highest possible export of final wood products. Additionally, however, a significant share in production comprises products of a lower degree of processing, such as various materials, pellets, pallets, briquettes, and the like (Archives of the Foreign Trade Chamber of Bosnia and Herzegovina 2021a).

Also, there is a wide range of products for comfort, created in Bosnia and Herzegovina, ranging from solid wood furniture, through solid wood panels, all the way to construction joinery, parquet and flooring, and sawn timber. This industry is organized within many small- and medium-sized privately owned enterprises and is the only branch of the economy of Bosnia and Herzegovina that has had a surplus in international trade for many years. Given the quality raw material potential, mainly beech, and the pronounced trend of using wood as a geo-ecological and renewable material globally, it is realistic to expect an increase in production and exports of wood industry products in BiH.

The Association of Wood Industry and Forestry is a form of interest organization of members of the Foreign Trade Chamber of Bosnia and Herzegovina, whose main activity is producing and processing wood products. Members of the Association perform the following activities: production of wood and wood products (sawn timber and impregnation, wood veneer and boards, manufacturing of builders' carpentry and joinery, wooden packaging, etc.), other manufacturing industries (production of tables and chairs, production of other furniture for shops and business premises, production of kitchen furniture, production of other furniture), forestry, and forest exploitation.

In 2020, the wood industry of Bosnia and Herzegovina exported a total of wood and wood products worth 175 million EUR, wooden furniture worth 127.5 million EUR, and prefabricated houses worth 4.5 million EUR, which the total export in the value of 297.6 million EUR was realized. It imported wood products worth 85.7 million EUR (Archives of the Foreign Trade Chamber of Bosnia and Herzegovina 2021b).

Wood industry Malak Janj ltd Donji Vakuf is a recognizable brand in the wood industry in BiH, spread over the area of 35,000 m^2 (own sawmill, dryers with a capacity of 1400 m^3, machining department, furniture department, manual department, paint shop, glassmaker, and packaging plant). It was founded in 1901 during the Austro-Hungarian Empire as the Feldbauer sawmill. The factory is with more than 110 years of tradition. The factory has a technology that completes the entire production cycle, starting with the log and ending with the final product. The factory currently employs 200 workers. They produce a wide range of products from facade joinery, entrance and interior doors, beds, tables, other interior furniture, timber, facades, wooden shutters and shutters, and many other wooden products.

The environment rich in forest resources, a wide range of products, high level of installed capacity, experienced and skilled workforce, and the geostrategic position (located at the road junction and near seaports) are all factors that have led to Janj being a leading brand in the region when it comes to construction carpentry for decades now. We have millions of installed windows and doors in the homeland, in the countries of the region (Slovenia, Croatia, Serbia, Montenegro, and Macedonia), in the Middle East, Western Europe, North Africa, and in the countries of the former Soviet Union (Archive of the Malak Janj factory 2020).

14.3.2 Paper and Graphic Industry

The paper and graphic industries are branches of the economy with significant capacities and the need for significantly higher exports. For greater competitiveness in exports, they need substantial investments in the procurement of modern equipment, educated staff, and changes in business policy. The Association of Graphic and Paper Industry is a form of interest organization of members of the Foreign Trade Chamber of Bosnia and Herzegovina, whose main activity is producing and processing paper and printing. Members of the Association perform the following activities: manufacture of cellulose, paper, and paper products (production of pulp, paper, and cardboard, production of paper and cardboard products) and printing activity.

The total export of the graphic and paper industry of Bosnia and Herzegovina in 2020 amounted to 74.8 million EUR, and the total import of this industry amounted to 106.7 million EUR (Archives of the Foreign Trade Chamber of Bosnia and Herzegovina 2021c).

Violeta ltd Grude is the largest company engaged in the production of hygiene products in BiH (Fig. 14.11). In addition, Violeta distributes and imports hygiene and food products in Bosnia and Herzegovina. Violeta, the regional leader in the hygiene products market, was founded in 2001 and has developed into one of the leading brands of hygiene products in 20 years. Through dedicated work and investments, it quickly became a symbol of hygienic products. Today, Violeta employs 1125 workers and has a fleet of 341 vehicles in its distribution, while the range consists of 220 different products. In 2017, Violeta reached an agreement with the Austrian retail chain Spar to make its products available on the Austrian market (Archive of the Violeta factory 2020).

14.3.3 Metal and Electrical Industry

The metal sector in Bosnia and Herzegovina, along with primary metal production, is traditionally the strongest and most successful industrial and export sector, with an average of 34.2% of the country's total exports, which is a significant

Fig. 14.11 Violeta factory in Grude—the regional leader in the market of hygiene products. (Photo by Aida Bidžan-Gekić)

advantage over other export categories. The electrical industry achieved the most considerable positive change in exports (36.2%). The three most promising chains in Bosnia and Herzegovina are metal parts for the automotive industry, electrical industry, and machines and machine parts—all those segments in which the most significant added value in production is.

The beginnings of the automotive industry in Bosnia and Herzegovina were associated with manufacturing parts for Mercedes and then for Volkswagen. In BiH, in 1970, the TAS Sarajevo factory was established as a joint venture between Volkswagen and UNIS. Many companies have been involved in the supply chain for TAS, making the substantial development in the automotive sector of Bosnia and Herzegovina substantial.

In 2018, the automotive industry of Bosnia and Herzegovina employed over 11,000 workers and participated in exports with over 460 million EUR. It is essential to point out that the companies have ISO TS 16949 and IATF 2018 standard certificates and that they can compete with the world's best manufacturers.

Over 50% of exports are placed in four major markets: Germany (17%), Slovenia (15%), Austria (14%), and Croatia (10%), and having in mind the absorption power of these markets, export-oriented companies should continue the focus of their export promotional activities in these markets, leaving the possibility to open new markets (Archives of the FBiH Chamber of Commerce 2020).

Leading products are: automotive industry (seats, brake system parts, filters, transmissions, brake discs, gears, windshields, spark plugs, engines, pumps, springs, filter covers, wheels, sirens, rims, exhaust systems, shafts, compressors, bearings, car cables, bearings, batteries) electrical industry (transformer stations, elevators, crane drives, telecommunication cabinets, distribution cabinets, air-conditioning chambers, heat exchangers, electric motors, power cables, light bulbs), metal structures and components for construction, shipbuilding, energy, industry, etc.

The total export of the metal and electrical industry of Bosnia and Herzegovina in 2020 amounted to 950 million EUR, and the real import of this industry amounted to 1 billion and 358 million EUR (Archives of the Foreign Trade Chamber of Bosnia and Herzegovina 2021d).

Good representative of this industry is Prevent Group, an international concern of manufacturers of car upholstery, furniture, brake discs, protective equipment, automotive components, and ships. With about 14,000 employees, it generates an annual income of 458.3 million EUR. Prevent Group has companies in ten countries in Europe, South America, and Asia. In 1999, Prevent Group started the production of car coatings in Bosnia and Herzegovina, and in 2001 the first production plant was established in Zenica. To date, Prevent has spread to a total of 15 locations throughout Bosnia and Herzegovina, employing about 6500 workers. These factories are located in Visoko, Goražde, Bugojno, Tomislavgrad, Velika Kladuša, Sarajevo, Zenica, Srebrenica, Travnik, Tešanj, and Ključ. By entering the Bosnian market, Prevent first started producing car coatings, but with the opening of a production plant in Tešanj in 2003, it also started producing car parts. Later, Prevent started producing safety equipment and boats by opening factories in Goražde in 2008 and 2009. Only 3 years later, Prevent started producing car components in Bosnia and Herzegovina, for which a new factory was built in Srebrenica. In 2016, this company bought two footwear companies in Bugojno. In the same year, Prevent took over the forge Ilijaš, where the production of brake discs should begin (Archive of the Prevent Group company 2020).

14.3.4 Food Industry

The food industry of Bosnia and Herzegovina is one of the oldest industries. In the 1970s of the last century, this industry developed very quickly and comprehensively. Significant capacities for processing fruits, vegetables, cereals, meat, milk, oil, and confectionery raw materials were built during this time. The food industry has a special significance in the economic structure as an incentive factor for the faster development of primary agricultural production. At the same time, it affects the production orientation of agricultural producers, thus encouraging the production of market surpluses in agriculture as an essential basis for export orientation. In addition, the food industry is important from the aspect of employment (especially female labor) and the realization of a significant part of the total gross domestic product.

In the period 1992–1995, many facilities were destroyed or devastated. The war years disrupted many relations, reduced production in most plants, and lost markets. After the war, some plants were rebuilt, and privatization changed the ownership structure, resulting in large capacities never being able to regain the old image they had. Poor company management by new or old owners has led many plants to bankruptcy and almost cessation of operations. After the war, several new plants were built, but these are relatively small plants compared to prewar plants. For example, in 2010, the total production of fruit and vegetable products in all fruit and vegetable processing plants in Bosnia and Herzegovina, except for wine and fruit brandies, amounted to about 215,000 tons of finished products. The total installed capacity of refrigerators in fruit and vegetable processing plants was about 40,000 tons (Nerda 2011).

For years, even decades, the food industry has been declaratively considered as one of the strategic production branches on which Bosnia and Herzegovina should base its economic development. However, the reality is diametrically opposed; despite the continuous positive trend in production volume and foreign trade balance, BiH is still absolutely dependent on food imports. Large areas of fertile land have been uncultivated for years, and livestock is insufficient to meet the minimum needs of processing capacity and consumption. As one of the most successful export segments, the production of milk and dairy products is limited by the lag of competent institutions in creating the necessary regulatory and institutional frame.

Of course, we should also mention the positive examples of the fastest growing sector of the domestic food industry in recent years, namely, dairy, poultry, meat processing, fruit and vegetable production (especially gherkins, tomatoes, potatoes, raspberries, apples, pears, and plums), medicinal herbs and forest fruits, and aquaculture. The only branches of the food industry where exports are higher than imports are the dairy industry (milk and sour cream) and the mill industry (wheat flour). The largest production systems in this industry in Bosnia and Herzegovina are AS Holding (16 factories across BiH, Agrokomerc Velika Kladuša, Bimal Brčko (Fig. 14.12), Sava Semberija Bijeljina, Sarajevo Brewery, Studen-Agrana Brčko (Fig. 14.13), Barpeh Čitluk, Mak Zara Sarajevo, Coly-Company Lukavac, Jami Sarajevo, Norfish Blagaj, etc.

Milk production is considered a strategic agricultural sector in Bosnia and Herzegovina and relies on livestock, sheep, and goats. Milk production is the most common form of agricultural production, taking into account the fact that about 100,000 agricultural households keep cows. Of that number, about 33,000 households produce milk for commercial purposes. There are about 35 dairies in Bosnia and Herzegovina with an installed processing capacity of about 420 million liters per year, which represents a capacity utilization of 55%. In addition to registered dairies, there are several small family dairies, but they are exclusively focused on the immediate local market. The structure in the field of milk processing has not changed significantly for many years, where liquid products predominate, most of which are dominated by pasteurized and UHT milk with about 56%, yogurt, sour cream, fermented products with about 20%, cheese with 21%, and other dairy products with 3%. Today, several autochthonous kinds of cheese are produced in Bosnia and Herzegovina, such as Livno cheese, Vlašić cheese, bellows cheese, fatty, boiled, lean cheese, and other types of cheese. The largest production systems in this industry in Bosnia and Herzegovina are Meggle Bihać, Milkos Sarajevo, MI99 Gradačac, Eco cheese Puđa Livno, Kupres Milch, Mlijekoprodukt Kozarska Dubica, Mljekara Livno, Poljorad Travnik, ZIM Zenica, etc.

The meat processing sector is one of the most important and developed sectors in Bosnia and Herzegovina. Meat processing is performed according to EU standards, and exports to international markets evidence the quality. About 20 companies in Bosnia and Herzegovina process between 5 and 30 tons of meat per day into various meat products such as semipermanent and permanent meat products, such as boiled or dried sausages, pâtés, cold cuts, salami, etc. Annually, this production is over 18,000 tons. Fresh meat production in Bosnia and Herzegovina has decreased due to the destruction of livestock during the war and poor investment

Fig. 14.12 Bimal d.d.—factory for the production of edible oil. (Photo by Aida Bidžan-Gekić)

Fig. 14.13 Studen-Agrana sugar refinery in Brčko. (Photo by Aida Bidžan-Gekić)

in postwar development. Poultry production has experienced rapid growth in recent years, with products marketed locally and abroad. The largest production systems in this industry in Bosnia and Herzegovina are Akova Impex, Madi Tešanj, MI Trivas Prnjavor, Menprom Tuzla, Pavić Livno, Vizon Rogatica, ZP Komerc Bijeljina, Perutnina Ptuj-BH Breza, Bajra Travnik, Atlantic Argeta Sarajevo, Semić Visoko, Ovako Sarajevo, MI Brajlović, etc. (Archives of the Foreign Trade Chamber of Bosnia and Herzegovina 2021e).

Most manufacturers follow modern trends and the technologies in the production today and use the standards required by customers from EU member states, such as Global GAP and ISO standards. Also, today this sector can boast the use of highly sophisticated equipment, especially in fruit processing, where some companies have the most modern equipment available today.

When we talk about the sector of medicinal and aromatic plants, it should be noted that this sector includes herbs and essential oils, natural cosmetics, wild mushrooms, forest fruits, spices, and honey. Bosnia and Herzegovina has a long tradition of collecting and cultivating medicinal and aromatic plants, forest fruits, and mushrooms. The biodiversity of medicinal and aromatic herbs and mushrooms is excellent, making BiH very competitive in the international market. The quality is confirmed by the superb export results of both organic and conventional products. Export bestsellers are wild mushrooms (porcini, chanterelle, morel, and black trumpet) that are exported fresh, dried, or frozen, followed by herbs that make up about 20% of total exports of this sector, both wild (bear onion, sage, laurel, St. John's wort, hemp, pine, yarrow, marshmallow, bearberry, rosehip, elderberry, linden, nettle), and cultivated (chamomile, lemon balm, mint, calendula, lavender, basil, and immortelle), and forest fruits (cranberry, blueberry, strawberry, raspberry, blackberry, wild cherry, and chestnut). The production of essential oils in Bosnia and Herzegovina has increased dramatically over the last 10 years. Extremely popular oils are immortelle, sage, lemon balm, chamomile, laurel, spruce, lavender, heather, mint, thyme, spruce, and rosemary. Several companies specialize in producing conifer essential oils such as black pine, Scots pine, fir, and spruce. Also important is honey production, such as multi-flowered types of honey (Bosnian forest, Bosnian meadow, and Herzegovinian honey), chestnut, sage, and heather acacia honey. The largest production systems in this industry in Bosnia and Herzegovina are Faveda Sarajevo, Frost Ljubuški, Halilović Ilijaš, Imortelle Berkovići, Ljbilje Ljubinje, Novi dan Mostar, Agrofarm Žepče, Anđelić Trebinje, Heko Bugojno, Jukan eko hrana, Pharmamed Travnik, Seti Ilidža, etc.

Wine production in Bosnia and Herzegovina is divided into red wine with 45% and white wine production with 55% of total wine production. This division has been uniform in recent years. The autochthonous varieties Žilavka (white) and Blatina (red), as well as Vranac (red), which has already found its natural environment in Herzegovina, are the main trump cards of domestic wine producers. Larger producers also produce other varieties such as Cabernet Sauvignon, Merlot, Shiraz (red) and Chardonnay, and Sauvignon Blanc (white), which means that some well-known international varieties are represented in the form of a small but increasing share of used hectares, however, are the first three indigenous varieties most deserving of the recognition of Herzegovina as wine and viticultural region. Although wine is partly produced in northern Bosnia, we are primarily talking about wines from western and eastern Herzegovina, which make up about 85% of total wine production in Bosnia and Herzegovina. The quality of wine produced in Bosnia and Herzegovina from domestic grapes is unquestionable, and thanks to it, Herzegovina has already become a brand as a region. The largest production systems in this industry in Bosnia and Herzegovina are Hepok Mostar, AG Međugorje, Andrija Čitluk Cellars, Anđelić Trebinje Winery, Čitluk Winery, Herzegovina Product Čitluk, Jungić Čelinac Winery, etc.

Soft drinks and water production is a very important segment of Bosnia and Herzegovina's production, especially water. Although domestic bottling plants operate at 50% of their capacity, about 12 million EUR of water and soft drinks are exported from Bosnia and Herzegovina annually. Bottle bottling and juice factories are spread throughout Bosnia and Herzegovina, and the natural water potential of water quality and purity sets us apart from other countries in the region. Several years ago, a large number of capacities for the production of bottled and spring mineral waters and refreshing nonalcoholic beverages were built and modernized. The largest production systems in this industry in Bosnia and Herzegovina are Hepok Mostar, Sarajevski kiseljak Kiseljak, Bony Tuzla, Bosnaplod Brčko, Coca-cola HBC BH Hadžići, Hercegovačka pivovara Mostar, Milkos Sarajevo, Lukas TP Nakić Široki Brijeg, Pak centar Sarajevo, Pivara Tuzla, Hercegovinavino Mostar, Sarajevo Brewery, Vitaminka Banja Luka, Swity Gradačac, etc. (Archives of the Foreign Trade Chamber of Bosnia and Herzegovina 2021f).

Main representative of food industry in BiH is AS Holding company. The beginnings of this company are connected with the company AS Jelah, which started operating in 1988 as an independent trade store, while in 1995, the initial business grew into a private company. In 2016, AS Holding was established by developing an umbrella company with efficient corporate governance mechanisms that apply to 16 members. In 2018, this company recorded the greatest success in its business so far. Through additional investments and development, they retained existing customers and acquired new ones, expanded capacities, and made breakthroughs into new markets with the application of new technologies and an increase in the number of employees. There are 16 Bosnian companies within this company whose brands are highly positioned and well known to consumers in BiH and Europe, and the world. These are Zlatna džezva, Oaza, Tops, Zlatni puder, Tuzlanska so, Master strudel, Bosanski lokum, Kent napolitanke, Maza, Sprind jufke, Vispak tajna, Sprind zaliveni kolači, Klas kolači, etc., to name a few of over 2000 products in their portfolio (Fig. 14.14). For several years in a row, they have occupied the highest places in the overall ranking of the largest companies in Bosnia and Herzegovina. Together with its members, it employs over 4500 people, with a tendency to grow in all three business divisions: food, trade, and textiles (Archive of AS Holding 2020).

14.3.5 Textile and Leather Industry

The textile and leather processing industry in Bosnia and Herzegovina has a long tradition of clothing and footwear production, high-quality products and services, a good international reputation as a supplier and manufacturer, a skilled

Fig. 14.14 Vispak products from the AS Holding group—coffee blends. (Photo by Aida Bidžan-Gekić)

workforce with low production costs, and territorially well-positioned production facilities. These are some of the most significant potentials that attract international companies and investors to BiH. When added to the proximity of Western and Eastern European markets and short delivery times, all together result in enough space and opportunities for further expansion and improvement of this important industry.

Production is mainly based on "lohn" jobs for foreign partners (sewing, tailoring, finishing, assembly services) with a very small share of production of own products. Still, although the dominant place is occupied by internal finishing and processing, this branch of the economy in recent years has recorded outstanding export results (Sinanagić 2004).

The textile, leather, and footwear industry are import-dependent in terms of imports of raw materials that are mainly not produced in Bosnia and Herzegovina, and production is organized in small (up to 50 employees) and medium enterprises (up to 200 employees) with a wide spatial dispersion of production capacity. The structure of textile industry products ranges widely from men's and women's heavy clothing, light clothing, sportswear and equipment, car upholstery, underwear, protective clothing and equipment to medical and cooking programs, home and hotel textiles, carpets, rugs, and other products. The product structure of the leather and footwear industry includes all processes, starting from the processing of raw leather and wool (which requires special treatment); cutting leather men's, women's, and children's clothing; haberdashery accessories; and other leather products, to the production of men's, women's, and children's footwear; everyday footwear; special-purpose footwear for various ceremonies; work; sports; protective footwear; footwear for different seasons; etc.

In 2020, a total of products of this industry worth 5 billion and 528 million EUR were exported, while these products worth 8 billion and 790 million EUR were imported to Bosnia and Herzegovina (Archives of the Foreign Trade Chamber of Bosnia and Herzegovina 2021g).

The main representative of the industry is Confection factory Borac in Travnik. This factory was founded in 1952 by the decree of Josip Broz Tito. Production plants, which have become the backbone of the development of Travnik, and the wider region, employed 500 workers, among whom then, as today, were mostly women. The expansion began, and the confection factory Borac built production plants in Jajce, Gornji Vakuf, Prijedor, Luška Palanka, Zenica, and Banovići, where until 1992, close to 6000 employees worked. At the same time, finished products made in the halls of the Borac factory began to conquer the picky European market. In the mid-1960s, business cooperation was established with the company Kvelle. Among the significant foreign business partners, there was Hugo Boss, the world king of fashion. During the war (1992–1995), the plants of the Travnik confection were significantly devastated, but they never stopped working. With significant financial investments in production, Borac clothing, which has long become a recognizable brand on the world catwalks, thanks primarily to the quality and knowledge of 1723 employees, has renewed business cooperation with old but also gained new customers, including Hugo Boss, Pierre Cardin, Carl Gross, Otokern, Jupiter, Baldezarini, Burberry, etc. (Fig. 14.15) (Archive of the Borac Garment Factory 2020).

14.3.6 Construction Material Industry

Bosnian companies have gained a solid reputation in all aspects of construction projects, from design, construction, and construction to high-quality construction materials. The country is rich in natural resources such as wood, stone, gravel, sand, clay, cement (Fig. 14.16), and metal ores. The potential for growth of the construction market in the country is excellent, with the initial priority being to renovate and recover existing assets and infrastructure. Significant opportunities for the development of the construction industry of BiH rest on projects, so in addition to Corridor Vc, there are plans to build highways and railways and develop electricity generation. The development of construction lies in the construction of economic and tourist (spa, rural, mountain, sea tourism) facilities and the construction of facilities in energy (thermal power plants, pipelines, hydropower, etc.) These are large construction projects that require the establishment of a consortium for a wide range of works that would be competitive in international tenders.

One of the main representatives of industry is the construction company Širbegović which has been operating since 1990, first as the Architect, and then, since 1995, as a Construction Holding Company Širbegović. Since 2001, expansion has begun in neighboring countries, with the construction of the first facilities in Serbia, Croatia, and Slovenia. Constant investment in the development of production facilities and new technologies and investment in employee education has led to the company Širbegović becoming one of the leading construction companies in the region.

The company has built over 6000 facilities and over 11 million m^2 of space during its business so far. The company's main markets are Bosnia and Herzegovina, Croatia, Serbia, Montenegro, Austria, and Germany. It currently has over 400 employees, with this number increasing every year. This company specializes in performing construction works using modern technology of prefabricated reinforced concrete structures (Fig. 14.17). It strives to build facilities in the shortest possible time and achieve economic construction with high quality. The company is

Fig. 14.15 Confection factory Borac in Travnik. (Photo by Aida Bidžan-Gekić)

Fig. 14.16 Cement plant in Kakanj, HeidelbergCement Group. (Photo by Aida Bidžan-Gekić)

currently implementing several large projects—such as COVID hospital in Novi Sad, fruit processing facility in Nikšić, storage and production facility in Grude, a chicken farm in Jelah, Plodine supermarket in Makarska, etc. (Archive of the company Širbegović 2020).

14.3.7 Energy Sector

The energy sector in Bosnia and Herzegovina is one of the most important economic sectors that has a significant impact on the growth and development of BiH's economy, primarily through technological development and increasing the economy's competitiveness. Energy is a generator of development in technological, economic, scientific, and educational terms. Bosnia and Herzegovina is rich in natural resources necessary for the production of electricity—water and coal (Fig. 14.18). It is estimated that Bosnia and Herzegovina has over 5000 MW of hydro potential, of which only about 35% was used in terms of capacity and 38% in relation to the maximum possible production of electricity, which represents the lowest utilization of hydro potential in Europe. Of the about 5.8 billion tons of total geological coal reserves, the balance reserves amount to

Fig. 14.17 One of the factory halls of Širbegović inženjering in Gračanica. (Photo by Aida Bidžan-Gekić)

Fig. 14.18 Thermal power plant in Tuzla. (Photo by Aida Bidžan-Gekić)

2.540 billion tons, so significant funds are needed to explore potential reserves and translate them into balance (Archives of the Foreign Trade Chamber of Bosnia and Herzegovina 2021a).

Bosnia and Herzegovina has positive energy statistics, primarily in the context of electricity production and exports. It has a good connection with the power systems in the countries of the region, so the most significant volume of cross-border exchange is traditionally done with Croatia, then with Montenegro, and the least with Serbia. Croatia and Montenegro continue to have considerable balance sheet resource deficits and are leading importers on a regional scale.

Given the growth of electricity production from renewable sources, both in the world and in BiH, in addition to hydro potential, it is necessary to mention the prospects of wind and solar energy. Wind energy in Bosnia and Herzegovina is still insufficiently used for energy purposes, although there are projects whose implementation is expected in the coming period. According to preliminary analyses, Bosnia and

Herzegovina has significant potential for solar energy, primarily in the southern parts of its territory. The current use of solar energy in BiH is limited to the use of solar collectors in a small number of households and the use of smaller photovoltaic systems in some larger facilities. However, electricity production using solar energy requires solar systems, which are among the most expensive energy technologies.

There are solar power plants in Bosnia and Herzegovina, but they are mostly of smaller installed capacities and are connected to the systems of electric power companies that purchase the produced electricity. A large number of established solar power plants are privately owned.

Biomass energy is also interesting. The most important source of biomass for energy production in Bosnia and Herzegovina is wood pulp originating from forestry (firewood, forest residue) and wood waste from the wood industry. Significant energy potential is also represented by agricultural residues, biogas from farms, residues of cereals, waste from fruit growing, and residues of oilseeds. So far, only a small amount of biomass has been used for electricity production, mainly as firewood and briquettes for heating in households. Several plants in Bosnia and Herzegovina generate electricity using this method. Still, in the future, it is necessary to invest more funds in the development of new forms of energy generation (USAID—Energy Sector Assistance Project 2019).

14.4 Contemporary Aspects of Service Economies

14.4.1 Transport System

The infrastructure system is of great importance in each country's economic and social development; above all, it makes it accessible to all regions and thus affects regional development. According to the data on the density and length of roads and other networks, it is evident that there is enormous inequality between regions in this respect in Bosnia and Herzegovina (Čekić 2005).

In the transportation system of Bosnia and Herzegovina, the differentiated development of traffic branches is distinct. There are branches in expansion and those in stagnation or setback. This is clearly seen in the state of individual transport networks—in the tendencies of their development, arrangement, and equipment with means of transport and the size and performance of transport. Such differentiated development is a consequence of the unequal economic position of certain transportation branches, especially their position in transition conditions, then inherited relations and the influence of war events. Due to all the above, the role of certain transportation branches in the development of tourism and other activities of Bosnia and Herzegovina is different (Gekić 2011).

Road traffic in Bosnia and Herzegovina takes place on a network of roads with a total length of 24,800 km, of which the asphalt surface takes up slightly more than 63%. The density of this network is 48 km/100 km^2, which is below the density of the road network of developed countries in Europe. Local roads with 14,500 km dominate the structure, followed by the regional roads with 4496 km, main roads with 4039 km, and motorway with 198 km. The territory of Bosnia and Herzegovina is intersected by six main routes that are included in the European main road network:

- E-59: Croatian border—Izačić-Bihać-Ripač-Užljević.
- E-65: passage through Neum.
- E-73: Croatian border—Šamac-Doboj-Lašva-Sarajevo-Mostar-Doljani-Croatian border.
- E-661: Croatian border—Gradiška-Banja Luka-Jajce-Lašva.
- E-761: Bihać-Bosanski Petrovac-Jajce-Sarajevo-Višegrad-Vardište—border with Serbia.
- E-762: Sarajevo-Brod na Drini-Šćepan polje-border with Montenegro.

In 2019, 135 million passengers and 9.3 million tons of goods were transported in BiH by roads. Also, 257 people died in traffic accidents. In 2019, a total of 1,094,515 motor vehicles were registered in BiH, of which 86% were passenger vehicles. Most passenger vehicles were registered in the largest spatial units by population: Sarajevo 92,982, Banja Luka 64,252, Mostar 36,211, Bijeljina 33,626, Tuzla 32,574, Zenica 27,068, Ilidža 22,746, Prijedor 20,722, and Doboj 18,274 (Agency for Statistics of FBiH 2020; ISR 2020). In 2019, BiH had an average of 28 passenger motor vehicles per 100 inhabitants, which ranks it among the weaker motorized countries in Europe. At the local level, four groups stand out. In the group of municipalities with less than 20 passenger vehicles per 100 inhabitants, there were 27 municipalities, among which Dobretići (4), Donji Žabar (6), Kupres RS (12), Čelić (12), and Istočni Mostar (12) stand out. In the second group, with 20 to 29 passenger vehicles per 100 inhabitants, there were 73 municipalities, among which Breza, Gacko, Kiseljak, and Nevesinje stood out with 29. In the third group, with 30 to 39 passenger vehicles per 100 inhabitants, there were 39 municipalities, and Široki Brijeg (37), Laktaši (37), Čitluk (37), Petrovac (37), and Doboj Jug (39) stood out. In the group of municipalities with the most passenger vehicles per 100 inhabitants, 40 and more, there were only 4: Centar Sarajevo (42), Istočni Stari Grad (44), Trnovo FBiH (59), and Istočna Ilidža (70) which are under direct agglomeration influence of the capital Sarajevo (Fig. 14.19) (Agency for Statistics of FBiH 2020; ISR 2020).

Today's trends in the development of road traffic are aimed at the development of multimodal transport corridors in the directions of the largest transport flows, i.e., inclusion in the pan-European transport network. Pan-European

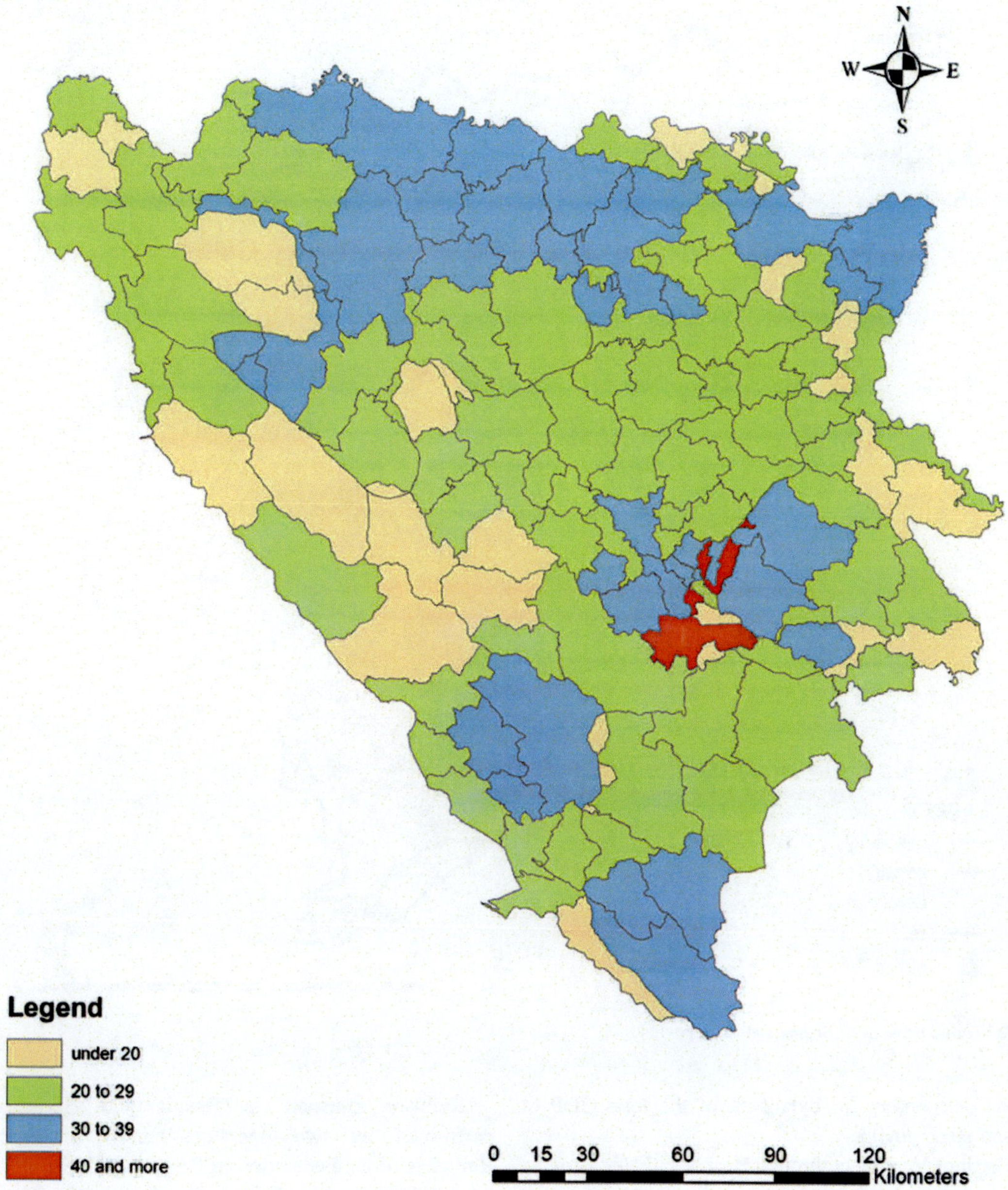

Fig. 14.19 Number of passenger motor vehicles per 100 inhabitants in BiH in 2019

Corridor V widened with branch C (Vc): Budapest, Osijek, Sarajevo, and Ploče. Branch Vc from the Croatian port of Ploče merged with the main branch of the pan-European transport corridor Vb, through the Neretva valley over the Ivan-sedlo bend, and the valley of Bosnia through Bosnia and Herzegovina in the southerly direction, to cross into Croatian territory again in Šamac, and in Budapest. So far, 113 km have been built out of a total of 335 km, which should be the part of the branch of this Corridor that goes through Bosnia and Herzegovina (Fig. 14.20) (Archives of Ministry of Communications and Transport 2021).

The construction of Corridor Vc is a significant factor in strengthening international processes in Bosnia and Herzegovina and its mutual spatial and economic connection. Future tourism development trends are related to this, which will undoubtedly contribute to faster and safer transportation of tourists to their destination. In the Corridor belt, 40 km wide (on less than 20% of the territory), live over 50% of the

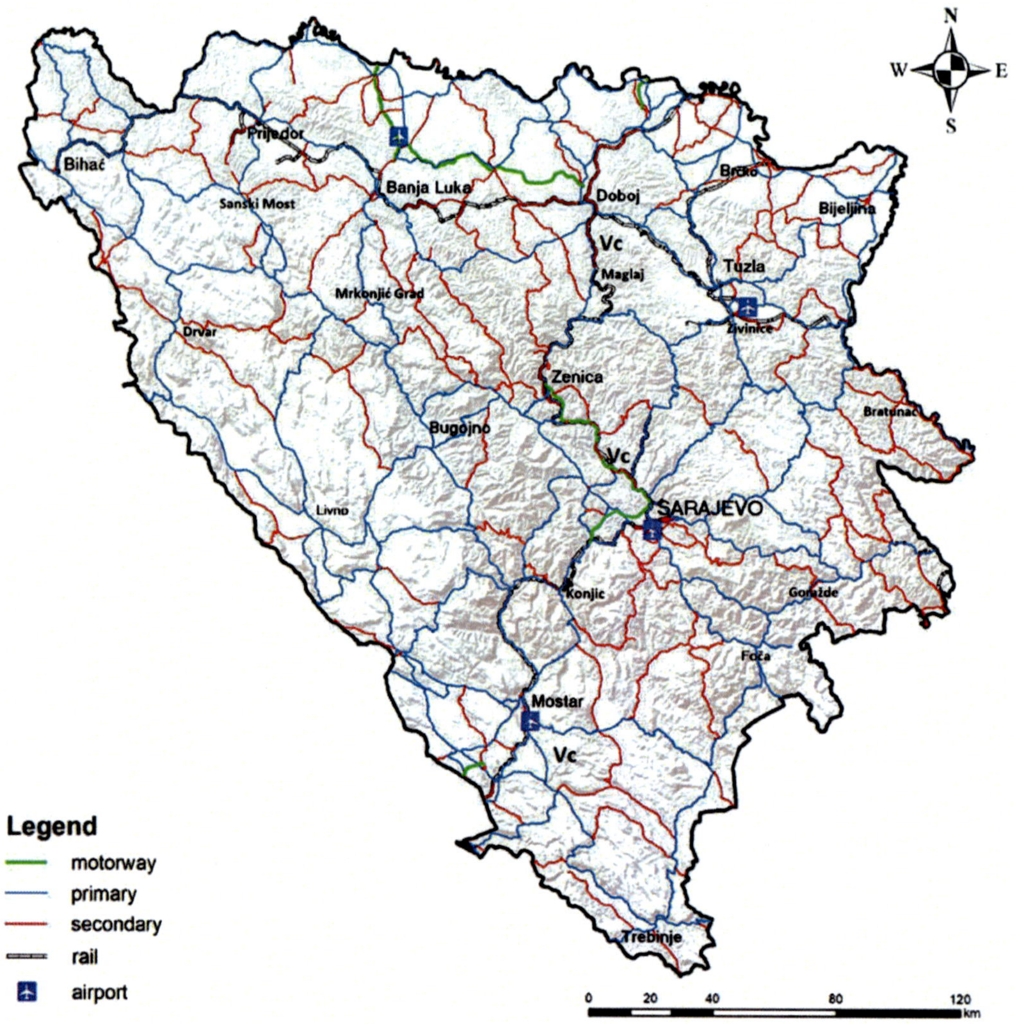

Fig. 14.20 Traffic network of Bosnia and Herzegovina

population, accounting for over 60% of the total GDP of Bosnia and Herzegovina.

The Corridor Vc project through Bosnia and Herzegovina is divided into four LOTs:

- LOT 1: Croatian border (Svilaj)—Doboj South (Karuše)
- LOT 2: Doboj South (Karuše)—Sarajevo South (Tarčin)
- LOT 3: Sarajevo South (Tarčin)—Mostar North
- LOT 4: Mostar North—Croatian Border (Bijača) (Archives of Ministry of Communications and Transport 2021).

Bosnia and Herzegovina's transportation infrastructure is still underdeveloped and cannot meet the needs of the discerning European visitor. At the same time, the same situation is followed by railway traffic, which is excluded in the area of targeted municipalities in BiH.

Railway transport in Bosnia and Herzegovina is performed by two operators—public companies: the Railways of the Federation of Bosnia and Herzegovina with its headquarters in Sarajevo and the Railways of the Republika Srpska with its headquarters in Banja Luka. The Railway Public Corporation of Bosnia and Herzegovina was established to manage the railway infrastructure. The total length of railways is 1031 km; industrial tracks of any significant production capacity should be added here. The length of the electrified railway is 795 km. This gives railway operators the possibility of direct access to collect and deliver all mass types of goods. There are 92 railway stations. The main routes that would bring the Bosnian railways into the pan-European railway network are certainly the pan-European Corridor Vc (Šamac-Doboj-Sarajevo-Čapljina-Ploče); Parallel 10 (transverse railway line Zvornik-Tuzla-Doboj-Banja Luka-Novi Grad-Republic of

Croatia); North-South direction (Dobrljin-Novi Grad-Bihać-Martin Brod-Knin, Republic of Croatia); and the railway Banovići-Tuzla-Brčko (Agency for Statistics of Bosnia and Herzegovina 2018). In 2019, 657,000 passengers and 13.5 million tons of goods were transported by rail.

The primary beneficiaries of railway transport services are the chemical industry in Tuzla, Arcelor Mittal Zenica, lignite mines in Zenica and Tuzla, Aluminum Plant in Mostar, an industrial complex in Sarajevo, distributors of petroleum products, etc. According to the railway network density, Bosnia and Herzegovina does not lag far behind compared to European countries. Still, the volume of transported goods and passengers (reduced to 1 km of a railway) is significantly lower than the European average. The reasons for this situation lie in the obsolescence of the railway infrastructure, which is why it cannot be used at standard capacity. There is no adequate number of passenger cars and locomotives for long-distance traffic. The dynamics of traffic on the railways of Bosnia and Herzegovina show a steady decline in passenger transportation because the competitiveness of road transport is higher. For the railway network in Bosnia and Herzegovina to meet European standards and become a significant part of the European railway network in the future, in addition to raising the level of regular maintenance of the existing condition, it is necessary to improve by investing in new vehicles and infrastructure capacities on lines where the volume of traffic is pronounced and on which an increase in traffic is expected.

Air traffic in Bosnia and Herzegovina developed in the 1960s with the opening of the airport in Sarajevo (Butmir). Even today, it is the main airport where most of the air traffic in the country takes place. In addition to Sarajevo, three more airports were built: Banja Luka (Mahovljani), Mostar, and Tuzla. There are no commercial internal air passenger and cargo traffic in BiH.

The most important airport in BiH is in Sarajevo. Air passenger traffic in this area began before the Second World War. In 1930, the line Belgrade-Sarajevo-Podgorica was opened. Smaller passenger planes were in use, and in that period, the military airport in Rajlovac was also used. After the end of the war, practically from 1947, regular civil air traffic to Sarajevo, along the lines between Sarajevo, Belgrade, and Zagreb, was reestablished. The new Yugoslav-Soviet company JUSTA performed the traffic, using the airport in Butmir for its activities. Due to the increased needs and the introduction of new types of jet aircraft, which required solid concrete bases for maneuvering surfaces. After many years of analysis and examination of the position of the future airport, it was decided to be located in Sarajevo field, near the existing grass airport of Butmir. The construction itself began in late 1965 and lasted just over 3 years. All the necessary airport infrastructure has been built: a runway, taxiways, a platform, a modest terminal, and an administrative building. The control tower was also part of the airport building. Radionavigation (R/NAV) equipment at the airport enabled the acceptance of all types of aircraft, including category C, and in the case of night flying and complex meteorological conditions (VFR and IFR flights). Finally, on June 2, 1969, Sarajevo-Ilidža Airport, as it was then officially called, was ceremoniously opened for civil air traffic. The opening of the new international airport also meant that new roads and opportunities for the development of civil aviation were opened.

The traffic of passengers and aircraft in this period increased gradually during each year, but Sarajevo Airport still had the role of "feeding airport," i.e., the role of the airport that provided passengers for airports in Belgrade and Zagreb. With the candidacy of Sarajevo for the 1984 Winter Olympics (ZOI '84), there is a need to expand the essential capacities of the Airport: extension of the existing runway, construction of a new parallel taxiway, a new terminal, and the Flight Control Service facility with a new control tower. At the same time, opportunities were being sought to install new R/NAV equipment to reduce the operating minima for aircraft operations and thus increase the frequency of landing and take-off procedures. This demanding idea was translated into reality before the very beginning of the '84 Winter Olympics. The runway has been extended by 150 m. A new taxiway has been made available, and the operating platform has been expanded. A complete new runway, approach, and taxiway lighting was installed, planned R/NAV assets (three VORs and a new ILS) were procured, and finally, a new, highly functional airport terminal building was built. The terminal building received all the facilities that are common at international airports. Thus, Sarajevo Airport has become one of the best-equipped airports in this part of Europe.

Until the beginning of the 1992, the traffic increased. This was a special contribution of the local charter airline AIR COMMERCE, transporting more than 40,000 passengers from October 1991 to March 1992. In 1992, all civilian flying was suspended entirely. In the period 1992–1995, the longest humanitarian air bridge for the supply of besieged Sarajevo was running. With close to 13,000 flights performed and a duration of more than 3 years, the scope and time of the air bridge established by the Western Allies after the Second World War to supply West Berlin were greatly exceeded. Since April 1996, part of the airport has been managed by the civilian authorities of Bosnia and Herzegovina. The struggle to return the Sarajevo Airport to its proper function—civil aviation—begins from that moment. Immediately after the opening of the Airport, Croatia Airlines established a connection between Sarajevo and Zagreb, and the Turkish company Top Air established a connection between Sarajevo and Istanbul. By the end of that year, Sarajevo Airport had 26,000 passengers and was the only civil aviation airport in Bosnia and Herzegovina. The general reconstruction of the airport terminal building, with entirely new technology and equipment of the highest quality, was successfully completed at the beginning of 2001. With the grand opening in March 2001, the airport meets all passenger needs in terms of facilities, which officially marks the beginning of a new era of quality work and top results. As the capital and seat of many embassies, Sarajevo is becoming an interesting destination for many well-known airlines. The professional attitude of Sarajevo Airport towards the activity it performs has been recognized by many: Lufthansa, Austrian Airlines, Turkish Airlines (Fig. 14.21), Croatia Airlines, Wizz Air, and other companies, with which we have been cooperating for years to mutual satisfaction (Sarajevo Airport Archive 2020).

The annual passenger traffic in 2019 at the airport in Sarajevo amounted to 1.1 million, at the airport in Tuzla 593,000, in Banja Luka 150,000, and the airport in Mostar about 33,000 passengers. All four airports are registered for international air traffic. In terms of cargo transport, the highest traffic in the amount of 2119 tons was realized at the Sarajevo airport, 110 tons at the Tuzla airport, 6 tons at the Banja Luka airport, and no cargo transport at the Mostar airport (Agency for Statistics of Bosnia and Herzegovina 2020).

However, the situation regarding transport capacities is not even close to satisfactory. The interest in Bosnia and Herzegovina for the use of air transport is apparent, so in the next period, a constant increase in the volume of transportation, the level and quality of transport services, and the level of safety are expected. In the long run, significant investments in these capacities can be expected to meet the needs in the coming period, creating space for private investment.

River and sea traffic currently takes place through the port of Ploče (in the Republic of Croatia) and along the Sava river along its entire length through Bosnia and Herzegovina. The capacity of the port of Ploče is approximately five million tons per year. This traffic is used exclusively to transport the base raw material, mining, energy, and chemical complex. Bosnia and Herzegovina realize the most considerable turnover in the port with 91.35%. In the Free Zone of the Port of Ploče are the Oil Terminals of the Federacija d.o.o. Ploče, occupying an area of about 15 hectares. The total current utilization of storage space is approx. 80,000 m3, and storage of goods from tanks on an annual basis of about 400,000 mT (Akta 2016). The Sava river has gained international importance, and as a waterway, it is directly connected with the Pan-European Corridor VII—the Danube river. There are also two important river ports on the Sava (Brčko and Šamac) with already built transshipment capacities. In the field of river traffic, especially in the port of Brčko (Fig. 14.22), there are significant opportunities to increase traffic. The ports of Ploče and Brčko, as well as the ports on the Sava River, have relatively good navigable, rail and road connections to the interior of the country, as well as abroad.

Postal and telecommunications traffic in Bosnia and Herzegovina provides a satisfactory range of services. In addition to the classic postal traffic and telecommunications, express mail, mobile telephony, satellite communications, and the Internet are used. In recent years, many PTT facilities and installations have been renovated and modernized in BiH, powerful digital transmission systems have been built, digital mobile telephony systems according to GSM standard, ISDN technology has been introduced, and roaming within the GSM network has been established.

Fig. 14.21 International Airport of Sarajevo. (Photo by Aida Bidžan-Gekić)

Fig. 14.22 Port of Brčko. (Photo by Aida Bidžan-Gekić)

The role of telecommunication networks in modern society is most important in the field where it should enable a fast, accurate, and timely exchange of information and preserve it in an original form. In the fixed telephony market of Bosnia and Herzegovina in 2019, there were 11 operators, of which 3 operators had a significant market power—Mtel a.d. Banja Luka, BH Telecom, and HT Mostar. The number of fixed telephony users of the dominant operators in Bosnia and Herzegovina in 2019 is estimated at around 582,000, with the rate of change compared to 2018 amounting to a negative 4%. The total number of users of alternative operators increased to about 144,000. The number of mobile telephony users was about 3.7 million, more than the population of Bosnia and Herzegovina. The number of internet operators was 65, and the number of internet connections was around 750,000 (Archives of Communications Regulatory Agency 2020).

14.4.2 Tourism Development

Tourism in Bosnia and Herzegovina has gone through several stages of development. Only since 1878, it is possible to talk of tourism as a socioeconomic phenomenon that has the characteristics of a tourist economy. The trips were individual and had military, political, economic, sports, and religious motives until then. A significant date in the history of tourism in Bosnia and Herzegovina is founding of the tourist club in 1884 by its first president Risto Samouković, a merchant from Sarajevo. In 1892, the first mountain house was opened in Trebević, Prenj, and Čvrsnica. Viktor Pojman, director of the Ilidža spa, was primarily responsible for promoting tourism and the increase of tourist traffic in Bosnia and Herzegovina. He organized trips to the mountains of Bosnia and Herzegovina, trained guides, organized visits to cultural and historical monuments, organized entertainment, and presented the folklore values of Bosnia and Herzegovina. Sarajevan Husein Pašalić provided the first professional guide service in 1879, and he engaged in this business, as one occupation, for several years (Beširević 1983).

The development of modern and mass tourism in the XX century in Bosnia and Herzegovina is a consequence of its favorable geographical position in relation to the SFRY and Western and Eastern Europe countries as the main areas emitting tourists. The central place of BiH in Yugoslavia has conditioned the development of transit tourism, which is today a significant form of tourist traffic.

Today's tourism of Bosnia and Herzegovina is based on numerous, diverse, and attractive natural and anthropogenic tourist attractions and is developing in almost all its parts. Some tourist values of Bosnia and Herzegovina are also of international importance; e.g., Sarajevo, with its Olympic mountains, is the greatest tourist value in Bosnia and Herzegovina and beyond due to its specific historical and geographical development, which can boast a very small number of cities in the world.

Geomorphological, hydrological, climatic, and biogeographical tourist attractions are essential for developing tourism in the diverse nature of Bosnia and Herzegovina. Of the geomorphological motifs, the mountains of Bosnia and Herzegovina (Bjelašnica, Igman, Jahorina, Vlašić, etc.) are of particular tourist value. The small coast of BiH (Neum), as part of the Mediterranean, i.e., European and world tourist values, together with thermo-mineral springs, around which spa tourist attractions were created (Terme Ilidža-Sarajevo, Reumal in Fojnica, Slatina in Banja Luka, etc.), and lakes (Boračko, Jablaničko, Prokoško, etc.), some rivers (Una, Neretva, Vrbas, Bosna, etc.) and tourist attractive waterfalls

(Šrbački buk, Martin Brod, Plivski, Kravice (Fig. 14.23), etc.), and river springs (Bosna, Buna, etc.), forms the basic hydrological motifs of Bosnia and Herzegovina. The natural tourist attractions of BiH are enriched by some climatic types and biogeographical tourist motives.

Protected cultural and historical heritage in BiH is very numerous. It is divided into movable and immovable monuments. The most important movable monuments include the following: Incunables, Charter of Kulin Ban, Miroslav Gospel, Altar partition of the early Christian basilica from Dabravine near Vareš, A monument of the Mithraic cult, Charter of King Stephen Tomašević, Charter of King Tvrtko I, the royal and judicial chair of Helen of Bosnia, The Sarajevo Haggadah, Hval's Codex, itd. Some of the most significant immovable monuments are: Old town Ljubuški, Blagaj fortress, Počitelj fortress, Gradačac Castle (Fig. 14.24), Doboj fortress, Srebrenik fortress, Kastel Fortress, Vranduk citadel, Travnik fortress, Jajce fortress, Tešanj fortress, Bobovac fortified city (Fig. 14.25), Baščaršija, Old bridge—Mostar, Old bridge—Konjic, Mehmed Paša Sokolović Bridge, City Halls (Sarajevo, Mostar, Brčko (Fig. 14.26)), Catacombs in Jajce, Stećci throughout BiH, Tito's Cave in Drvar, Tito's Cave near Kladanj, etc.

The culture of life (Fig. 14.27) and work of the inhabitants (Fig. 14.28) of BiH is important for tourism. The most important elements of this are folklore: *sevdalinka* (Bosnian

Fig. 14.23 Kravica Waterfall in Ljubuški. (Photo by Aida Bidžan-Gekić)

Fig. 14.24 Gradačac Castle (Kula Zmaja od Bosne) in Gradačac. (Photo by Aida Bidžan-Gekić)

Fig. 14.25 Bobovac, municipality of Vares—the best fortified town in medieval Bosnia. (Photo by Aida Bidžan-Gekić)

Fig. 14.26 City Hall in Brčko. (Photo by Aida Bidžan-Gekić)

folk song with a long and rich tradition and it marks longing, love zeal, love misery), Bosansko i Gluho Glamočko kolo are folk dances, also with a long tradition, etc. Then handicrafts in which we should single out the making of folk costumes (Rural, Dinaric, Central Bosnian, Herzegovinian, Posavinian, Urban), Bosnian Carpet Weaving, making copper engravings, woodcuts, etc. Also, traditional crafts should be singled out: carvers, potters (Fig. 14.29), goldsmiths, carpenters, arakijadžija (cap makers), *aščija* (traditional food cook), *mutabdžija* (traditional weaver), watchmaker, combers, slipper makers, *sarači* or saddlers, bakers, potters, etc. (Fig. 14.30). There is also a long tradition of wine making in BiH, where two indigenous grape varieties are included: žilavka and blatina. From the very diverse Bosnian cuisine, we will single out some dishes: Bey's soup, Bosnian pot, *ćevapi*, flatbread, pies, *klepe*, *sogan-dolma*, *pače*, okra, *baklava*, *tufahije*, *kadaif*, *Bosnian lokum*, *gurabije*, Herzegovinian grapefruit, Bosnian plum brandy, Bosnian coffee, etc.

Among the historical events important for BiH tourism, numerous battles and sessions from World War II stand out: Battles (Neretva, Igman March, Sutjeska, Kozara, etc.),

Fig. 14.27 Traditional Bosnian room settings provided relaxation and serenity in Etno avlija Mačkovac in Banovići. (Photo by Aida Bidžan-Gekić)

Fig. 14.28 Old coppersmith street in Sarajevo—kazandžija craft. (Photo by Aida Bidžan-Gekić)

Sessions of the Anti-Fascist Council for the National Liberation of Yugoslavia (Bihać and Jajce), Sessions of the State Anti-fascist Council for the National Liberation of Bosnia and Herzegovina (Mrkonjić Grad, Sanski Most and Sarajevo), and we should also point out the battles from the Defensive-Liberation War 1992–1995: The Battle of Žuč, Operation Autumn 94, Operation Summer 95, The Battle of Igman, etc.

Numerous events that take place in BiH are very attractive to tourists. From cultural events, we will single out the following: The Sarajevo film festival, Kastel Rock Fest, Kids Festival BiH, Kozara Ethno International Folklore Festival, Baščaršija Nights, The International Theater Festival MESS, Teatarfest, Jazz Fest Sarajevo, Sarajevo Winter Festival, International Festival of Friendship Goražde, Mostar Summer Fest, Semberija folk fest, West Hercegovina fest, Ilidža Folk Music Festival, BiH Theater Festival Fedra Bugojno, and Puppet Artists Biennial in Bugojno itd.

The following stand out in religious manifestations: Ajvatovica, Mawlid by the Buna river, Christmas concerts, and Bayram concerts, itd. Of the economic events, international fairs stand out as follows: ZEPS Intermetal Zenica, Mostar Fair, International Fair of Tourism, Catering and Trade "Trebinje," Interio Sarajevo, International Trade Fair Brcko, International Sarajevo Book Fair, Sarajevo Business Forum, Plum Days Fair in Gradačac, etc.

Among the attractive anthropogenic tourist motives, the most important cultural and religious institutions in BiH certainly stand out. From cultural institutions, we single out the following: museums (National Museum of BiH in Sarajevo, Museum of the Revolution in Sarajevo, Museum of the XIV Winter Olympic Games in Sarajevo, Jewish

Fig. 14.29 Pottery kept in Travnik castle. (Photo by Aida Bidžan-Gekić)

Fig. 14.30 The rural folk costume from the Kakanj area kept in Kraljeva Sutjeska Museum. (Photo by Aida Bidžan-Gekić)

Museum in Sarajevo, Homeland Museums throughout BiH, Salt Museum in Tuzla, Museum of Ivo Andrić's birth house in Travnik, AVNOJ Museum in Jajce, Sarajevo Tunnel Museum "Spas," Despić House in Sarajevo, Svrzo's House in Sarajevo, Sevdah Art House in Sarajevo, Tito's bunker in Konjic (Fig. 14.31), etc.), galleries (Art Gallery of Bosnia and Herzegovina in Sarajevo, Collegium Artisticum Gallery in Sarajevo, Roman Petrović Gallery in Sarajevo, RS Gallery of Fine Arts in Banja Luka, Franciscan Gallery in Široki Brijeg, International Gallery of Portrait Tuzla, etc.), and theaters (National Theater in Sarajevo, Chamber Theater 55 in Sarajevo, Youth Theater in Sarajevo, National Theater of RS in Banja Luka, National Theater Mostar, etc.). From religious institutions, we single out the following: places of worship (Bey's mosque in Sarajevo, Aladža mosque in Foča, Ferhadija mosque in Banja Luka, Church of the Holy Archangels Michael and Gabriel in Sarajevo, Cathedral of the Nativity of the Theotokos in Sarajevo, Church of Saint Anthony of Padua in Sarajevo, The Sacred Heart Cathedral in Sarajevo, Ashkenazi synagogue in Sarajevo, Sephardi synagogue in Sarajevo, Mosque of Sultan Ahmed Han II in Bugojno, etc.) and pilgrimages and shrines (Prusac near Donji Vakuf, Djevojačka cave near Kladnja, Dervish tekke Buna, Međugorje, Podmilačje, The Olovo church of Our Lady, monasteries: Moštanica, Gomionica, Liplje, Stuplje, Rmanj, Tvrdoš, Žitomislići, Papraća), etc.

As the visit of some anthropogenic tourist potentials and the organization of appropriate events is often not time-conditioned, tourist traffic can occur throughout the year and be used to extend the tourist season in designated tourist

Fig. 14.31 Tito's bunker in Konjic. (Photo by Aida Bidžan-Gekić)

centers and tourist-geographical regions. They manifest themselves as complex, independent, and complementary potentials, depending on several elements of general, historical, ethnographic, artistic, and cultural value and recognizability on the tourist market (Šećibović et al. 2006). According to the World Tourism Organization, Bosnia and Herzegovina is defined as one of the three tourist destinations in the world with a total growth potential of the tourist market of 10.5% (for the period until 2020). However, such a high growth rate also reflects the meager base value of the sector: currently, the tourism sector in Bosnia and Herzegovina represents only 0.2% of total European tourism. Among others, potential areas for the development of the tourism sector of Bosnia and Herzegovina are cultural and historical tourism, urban (city break), rural tourism, outdoor tourism, winter sports and mountain tourism, and spa and wellness (health) tourism, as well as business/conference tourism.

Bosnia and Herzegovina was under the influence of Roman, Illyrian, Slavic, Ottoman, Austro-Hungarian culture. Then, in the Second and First World Wars, BiH was the scene of many great battles of the national liberation struggle: Kozara, Sutjeska, Neretva, and Drvar, which went down in history, and today they are remembered through numerous monuments erected in honor of fallen fighters in these places. Due to their importance, they also represent a great tourist attraction. Also indispensable are anthropogenic tourist motifs from the period of the XIV Winter Olympic Games, which took place in 1984, when the capital of Bosnia and Herzegovina was the center of the sports world and fans of the Olympics. We certainly should not omit the period from 1992 to 1995, when the aggression was carried out against BiH. After that time, numerous monuments were erected to commemorate the suffering and war events that the people of Bosnia and Herzegovina went through. In BiH, a multitude of diverse and attractive natural and anthropogenic tourist motifs have been acquired, side by side, creating a mosaic. The functionality of the tourist-geographical position of Bosnia and Herzegovina as a contact area is reflected in the fact that it has a small part of the area along the Adriatic coast. Thus, as part of the Mediterranean, the Bosnian coast is one of the world's tourist values (Bidžan 2011).

The transit characteristic of the tourist-geographical position of Bosnia and Herzegovina is significant for the development of its tourism. Transit flows have a north-south direction, i.e., connecting two enormous and different spatial units, the Posavina-Sava area and the Adriatic. The tourist-geographical position of Bosnia and Herzegovina represents its spatial relationship to the main tourist destinations of Southeast Europe. By the way, tourist directions, as one of the three basic spatial categories of tourism development, most fully express the spatial aspect of tourist movements and their meaning.

14.4.3 Development of Tourist-Geographical Regions

There are 13 tourist-geographical regions in Bosnia and Herzegovina, namely, Sarajevo, Mostar, Bihać, Tuzla, Travnik, Banja Luka, Primorje, Doboj, Podrinje, Posavina,

West Herzegovina, Southwest Bosnia, and East Herzegovina (Bidžan 2012, 2016).

The number of tourist arrivals in 2019 in BiH was around 1.78 million, and according to tourist-geographical regions as follows: Podrinje (163,492), Mostar (317,311), Sarajevo (667,756), Banja Luka (127,033), Primorje (132,760), Doboj (107,490), Bihać (64,416), Tuzla (48,883), West Herzegovina (9740), Travnik (54,028), Southwestern Bosnia (12,949), East Herzegovina (64,169), and Posavina tourist-geographical region (10,975).

Table 14.4 shows the evaluation of tourist products of tourist-geographical regions of Bosnia and Herzegovina, which are ranked on a scale of 6, where rank 1 indicates underdeveloped tourist product, 2 bad, 3 good, 4 very good, 5 excellent, and 6 multifunctional tourist product (Bidžan 2011, 2012, 2016).

The following services (quality and quantity) were observed and evaluated in the products of tourist services: accommodation, catering facilities, entertainment facilities, activities offered, public transport, parking lots, accessibility, marking system, etc., where Bosnia and Herzegovina received an average grade of 3, which means that this product is well developed. The following services (quality and quantity) were observed and evaluated in the products of tourist informants: promotional materials, internet promotion, reservation system, destination information, communication with agencies and info bureaus, etc., where Bosnia and Herzegovina received an average rating of 1, which means that this product is underdeveloped. The following services (quality and quantity) were observed and evaluated in the products of tourist events: cultural events, sports events, and events, etc., where Bosnia and Herzegovina received an average grade of 2, which means that this product is poorly developed. In the products of tourist facilities, the quantity and quality of natural and anthropogenic tourist attractions were observed and evaluated, where Bosnia and Herzegovina received an average rating of 3, which means that this product is well developed. In the case of tourist experience products, services (quality and quantity) were observed and evaluated: city tours, getting to know parts of the region, etc., where Bosnia and Herzegovina received an average grade of 2, which means that this product is poorly developed. And with the products of tourist routes, where services (quality and quantity) were observed and evaluated: transitive, within regional and similar routes, BiH received an average grade of 2, which means that this tourist product is also poorly developed.

The maximum number of points in evaluating tourist products, which could get a functional tourist-geographical region in Bosnia and Herzegovina, was 36, and the minimum number of points is 6. None of the tourist-geographical regions have the maximum number of points due to their initial stages of tourism development and poor construction and improvement of tourist infrastructure, without which there is no development of tourism. The average value of the number of points of the tourist product was 13. Tourist-geographical regions whose sum of tourist product ratings are below 13 points are classified as potential, i.e., nonfunctional tourist-geographical regions and those with a value equal to or greater than 13 points are classified as functional tourist-geographical regions. Based on all the presented and processed data from official tourist boards and organizations, as well as the official websites of municipalities, we have made the following conclusions that in the group:

Table 14.4 Evaluation of tourist products of tourist-geographical regions of Bosnia and Herzegovina

Tourist-geographical region	Product of tourism service	Product of tourist informants	Product of tourist events	Product of tourist objects	Product of tourist facilities	Product of tourist routes	TOTAL points
Bihać	3	1	2	2	3	3	**14**
Tuzla	2	1	1	2	1	2	**9**
Travnik	4	1	2	3	2	2	**14**
Mostar	4	1	2	4	2	3	**16**
Sarajevo	6	2	4	6	5	6	**29**
Banja Luka	3	1	3	4	3	4	**18**
Posavina	1	1	1	1	1	1	**6**
Podrinje	1	1	1	3	1	1	**8**
Doboj	2	1	3	3	2	2	**13**
South-West Bosnia	1	1	1	1	1	1	**6**
East Herzegovina	1	1	1	1	1	1	**6**
West Herzegovina	3	1	2	3	2	2	**13**
Primorje	4	2	2	4	2	2	**16**
Average for BiH	***3***	***1***	***2***	***3***	***2***	***2***	***13***

According to Bidžan (2011, 2012, 2016)

nonfunctional tourist-geographical regions (Tuzla, Posavina, Podrinje, South-West Bosnia, and East Herzegovina) and functional tourist-geographical regions include the following: Sarajevo, Travnik, Mostar, Bihać, Banja Luka, Doboj, West Herzegovina, and Primorje.

According to Bidžan (2011), based on the analysis of tourist attractions and specific forms of tourism, which can be developed in Bosnia and Herzegovina's tourist-geographical regions, there is only one potential metropolitan tourist-geographical region in Bosnia and Herzegovina, and that is Sarajevo, followed by nine potential tourist-geographical regions for leisure and recreation: Bihać, Banja Luka, Tuzla, Posavina, Podrinje, Doboj, Southwestern Bosnia, East Herzegovina, and Primorje, as well as three potential tourist-geographical regions—cognitive, educational, and religious: Mostar, Travnik, and West Herzegovina. If we look at them through the seasonality of tourism, the Sarajevo metropolitan tourist-geographical region is visited throughout the year; tourist-geographical regions for leisure and recreation are visited mainly during the summer season, and tourist-geographical regions—cognitive, educational, and religious are visited most during two seasons. The metropolitan tourist-geographical region is the country's largest and central tourist-geographical region, with the best-developed tourist and other economic branches. It can only be the Sarajevo tourist-geographical region because it is the only one with the largest number of natural (21) and anthropogenic tourist attractions (125), the largest number of accommodation (4266) and catering facilities, the best traffic infrastructure and connections, the largest number of tourist organizations and agencies (47), as well as the largest number of developed specific forms of tourism in Bosnia and Herzegovina (Bidžan 2011, 2012, 2016; Bidžan-Gekić et al. 2017).

The following regions can be developed into true tourist-geographical regions for leisure and recreation: Bihać, East Herzegovina, Banja Luka, Tuzla, Posavina, Podrinje, Doboj, Southwest Bosnia, and Primorje, because these regions have the most tourist attractions where specific forms of tourism, related to excursions, sports, and recreation, e.g., the valleys of the Una, Vrbas, Sava, Drina, rivers, etc. are suitable for the development of the following specific forms of tourism: excursion, extreme, bathing, fishing, etc. Also, the mountains located in these regions can be associated with leisure and recreation, and we will only name some Grmeč, Konjuh, Zelengora, Maglić, etc. Numerous spas are also located in these areas and are suitable for leisure and recreation, such as Gata-Bihać, Ilidža-Gradačac, Dvorovi-Bijeljina, Laktaši-Laktaši, Mlječanica-Kozarska Dubica, Sanska Ilidža-Sanski Most, Slatina and Gornji Šeher-Banja Luka, etc. (Bidžan-Gekić and Gekić 2017). The following regions can be developed into true tourist-geographical regions—cognitive, educational, and religious: Mostar, Travnik, and West Herzegovina, and the reason are their tourist attractions where the specific forms of tourism can be developed, such as religious tourism (Medjugorje, Ajvatovica, Mevlud na Buni) and educational tourism (Jablanica, Jajce, Travnik), etc. (Bidžan-Gekić et al. 2019a, b).

According to the number of its functions, tourist-geographical regions are monofunctional (Tuzla, Posavina, Podrinje, Doboj, East Herzegovina, West Herzegovina, Southwest Bosnia, Primorje and Travnik) and polyfunctional (Sarajevo, Mostar, Banja Luka, and Bihać). The most important centers stand out in each tourist-geographical region (tourist centers) (Fig. 14.32). With that, in some tourist-geographical regions, two or more centers stood out, so we have a larger number of tourist centers in them. According to the number of tourist centers, tourist-geographical regions are divided into monocentric (Sarajevo, Tuzla, Primorje, Mostar, Bihać, Travnik, Banja Luka, East Herzegovina, Southwest Bosnia, and West Herzegovina) and polycentric (Posavina, Podrinje, and Doboj) (Bidžan 2012, 2016).

These tourist centers include several tourist sites with a complex structure, i.e., they have a large concentration of elements of the tourist offer. They are the carriers of tourist traffic and the poles of development of receptive tourist-geographical regions. The biggest obstacle in the balanced development of tourism in Bosnia and Herzegovina is the lack of investment in tourist-geographical regions, which would modernize and reconstruct roads and accommodation and catering facilities, differentiate the tourist offer, and create a general image of Bosnia and Herzegovina as a desirable tourist destination. Therefore, it is necessary to enrich the tourist offer of tourist-geographical regions in terms of constant innovation of tourist products and continuous improvement of the quality of all its components, and the primary goal is to create a recognizable and quality tourist product and developing unique identity tourism as the basis of economic prosperity. In the following text, we will briefly describe the four most important tourist-geographical regions of Bosnia and Herzegovina.

14.4.3.1 Sarajevo Tourist-Geographical Region

Sarajevo tourist-geographical region is located in the Sarajevo valley, which is connected to the Danube region and Central Europe in the north by the valley of the river Bosna, and to the Adriatic Sea and the Mediterranean in the south by the Neretva river valley. The tourist-geographical position of this region is very favorable because it is the capital of Bosnia and Herzegovina, well-liked by tourists. Also, the most important state roads pass through it. The largest international airport in Bosnia and Herzegovina is located there, which increases the importance of its tourist-geographical position, and, after the construction of Corridor Vc, its position will improve even more. It covers an area of 1406.5 km^2 with 415,643 inhabitants in 2013. Sarajevo's

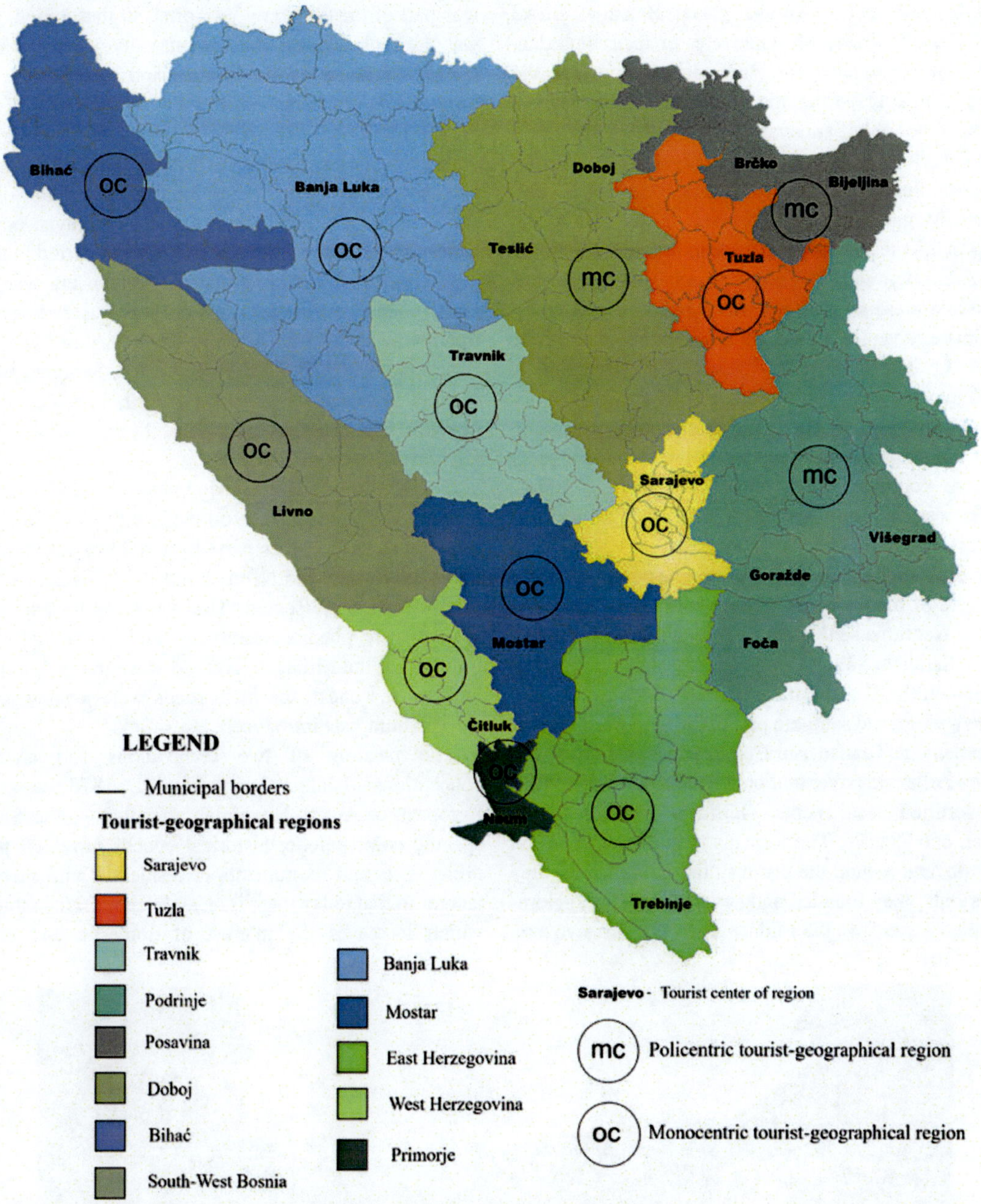

Fig. 14.32 Monocentric and polycentric tourist-geographical regions of Bosnia and Herzegovina

tourist-geographical region consists of four city municipalities: Stari Grad, Centar, Novo Sarajevo, and Novi Grad, and five other municipalities, namely, Ilidža, Trnovo-FBiH, Trnovo-RS, Hadžići, Vogošća, and Ilijaš (Bidžan 2016; Bidžan-Gekić 2020).

In the natural-geographical sense, it is located in the eastern part of the valley of the same name in the central part of the Dinaric mountain system of Bosnia and Herzegovina, in the inner mountain depression separated as the Sarajevo valley. The relief is mountainous mainly (above 700 m) and makes up about 78% of the area of this region, then hilly (from 550 to 700 m) and makes up 13%, and lowland (up to 550 m) with 9%. Sarajevo tourist-geographical region is located in the easternmost part of the Sarajevo plain, right in the widened valley of the Miljacka river and on the mountainsides that descend steeply towards it. The lowest altitude is in the area of Reljevo (460 m), while towards the southeast and southwest, the altitude increases and reaches a maximum at Bjelašnica mountain (2067 m). The average altitude of Sarajevo is 537 m. The valley is elongated from northeast to southwest and has an area of about 32 km^2. The rim of the Sarajevo valley is represented by the hills Žuč, Hum, Kobilja glava, Betanija, Breka, Sedrenik, Bistrik, Vraca, Mojmilo, etc. In their hinterland are the Olympic mountains: Bjelašnica, Igman, and Trebević, which are, primarily during the winter months, a favorite destination of not only Sarajevans but also their guests from all over the world. Also, we should not forget to mention that, in this region, there are several important speleological sites (Bijambare, Megara, Klokočevica, etc.). In the area of this region up to 750 m, a moderately warm and humid climate prevails. Areas between 750 and 1700 m have a forest-snow climate, while areas with an altitude above 1700 m have a tundra climate. Average annual air temperatures range from 6 to 10 °C. On average, it receives precipitation of 900–1250 mm per year. It has a dense river network, which belongs to the basin of the river Bosna, and its largest tributaries are Miljacka and Željeznica. In terms of biogeography, three ecosystems are represented in this region: mesophilic sessile and hornbeam forests, thermophilic beech forests, and dark coniferous forests.

Sarajevo's tourist-geographical region has been inhabited since the end of the Neolithic. From about 2400–2000 BC, there was a Neolithic settlement in Butmir, i.e., the western part of the Sarajevo valley. This area was inhabited in the Metal Age—in the II and I millennium BC. These are also sites in the western and southern part of the Sarajevo valley—Illyrian settlements Gradac near Butmir and Naklo on the left bank of the Željeznica river and on the slopes of Trebević—Illyrian fortified settlements, hillforts Debelo Brdo, Soukbunar, and Zlatište. The Romans built their fortress on Debelo Brdo, and around the warm sulfur springs in the area of today's Ilidž, they founded and developed a city agglomeration called Aqua S. In the Middle Ages, the Sarajevo area was part of the parish of Vrhbosna. In that period, Sarajevo was the ecclesiastical and administrative center of medieval Bosnia (Archive of the Commission to Preserve National Monuments of Bosnia and Herzegovina 2004).

The traditional date of the founding of Sarajevo is 1462, with Isa-beg Ishaković as its founder. He had the following profane and sacral buildings built: the Emperor's Mosque, the Emperor's Bridge, the hammam, a caravanserai with a bedestan, and shops. The most significant period in the urban development of Sarajevo in the Ottoman era was the sixteenth century during the reign of Gazi Husrev-beg, which is rightly called the golden age in the history of this city when Baščaršija was built. During this period, Sarajevo grew from a small town into a large sugar mill, in which numerous beautiful bridges were built (Figs. 14.33, 14.34, and 14.35) (Historical Archive 2007).

The Austro-Hungarian administration (1878–1918) also had a significant influence on the city (Fig. 14.36). Sarajevo was connected to other parts by a railway line through the valley of Bosnia and Neretva and the Miljacka canyon and further towards Višegrad. The water supply and sewerage system were planned, electricity and public lighting were introduced, the Miljacka riverbed was arranged, a quay was built on the coast, the main roads were asphalted, and an electric tram was introduced, etc.

The meeting of two civilizations and architectural conceptions—Islamic-Oriental until 1878 and Central European or Austro-Hungarian after that—gives Sarajevo a specific color palette; historical epochs have left traces on urban units and monuments of residential and sacral architecture so that today there is an ambient mosaic in this region, which increases the interest of domestic and especially

Fig. 14.33 Baščaršija—the heart of the Sarajevo tourist-geographical region. (Photo by Aida Bidžan-Gekić)

Fig. 14.34 Sarajevo's sebilj in Baščaršija. (Photo by Aida Bidžan-Gekić)

foreign tourists, towers, theaters, museums, galleries, etc. If we add to that the diversity of residential buildings (from small from the Ottoman period, through extensive and magnificent from the Austrian to the simple skyscrapers of today), as well as various urban communications (winding cobblestone alleys from the Ottoman, paved and a cube-covered wider street from the Austrian and wide boulevards in the newest parts of the city), then the mosaic is complete (Archives of the City of Sarajevo 2020).

In 1984, the XIV Winter Olympic Games were held in this region, which gives another peculiarity to this area (Fig. 14.37). Along with the traditional type of music, new music genres (pop, hit, chanson, jazz, rock'n'roll, heavy metal, hip hop, house, techno) have developed in Bosnia and Herzegovina in recent decades, which does not lag behind the other countries of the world. In this regard, numerous musical events are held in the Sarajevo Tourist and Geographical Region, such as the International Music Festival Jazz Fest Sarajevo, Baščaršija Nights, Novi grad Days, International Folklore Festival, etc.

"From its inception until today, Sarajevo has had a unique civilization and cultural development in which the charm of the past, spirituality, and sociability have become its recognizable code-multiethnicity-multiculturalism-multiconfessionalism-coexistence-tolerance. Everywhere and in every place, in the rich splendor of cultural-historical, religious and other monuments and buildings, it is recognizable in the past, which is one of the elements of Sarajevo's tourist identity" (Figs. 14.38 and 14.39) (Nurković et al. 2005).

Also, it is important to highlight the old town of Sarajevo, Baščaršija, which inherits the old Bosnian cuisine in its local eateries (aščinice), with the following savory dishes on their menu: sogan dolma, hadžijski ćevap, bey's soup, tarhana soup, Bosnian pot, okra, borek, ćevapi, etc. and sweet dishes such as baklava, hurmašica, kadaif, tufahija, tulumba, rahat and lokum, eaten with Bosnian coffee with sherbet made of rose, elderberry juniper, mint, etc.

It is also the center of road traffic in Bosnia and Herzegovina. Seven highways connect it with other parts of the country, to the north, M-5 in the direction of Travnik, Banja Luka, and Bihać region, M-17 (E73) towards Doboj and M-18 towards Tuzla, to the east water M-5 towards Podrinje, as well as M-19 (E761). To the south, the M-18 leads via Podrinje to Dubrovnik, and to the west, the M-17 (E73) in the direction of the Mostar region. Since 2003, it has been connected by the A1 highway with Ilijaš, Visoko, and Kakanj. The planned European highway, Corridor Vc, passes by Sarajevo, connecting it with Budapest in the north and Ploče in the south. This tourist-geographical region is located on a significant railway network—Ploče-Mostar-Sarajevo-Zenica-Doboj-Bosanski Šamac-Republic of Croatia and further to Central and Eastern Europe—which is one of the positive elements of tourist infrastructure and tourist product of this region. It also houses an international airport, which enables the connection of this area with emitting tourist destinations worldwide (Bidžan 2016; Bidžan-Gekić 2020).

Regarding tourist traffic, Sarajevo tourist-geographical region was visited by 667,756 tourists in 2019, of which 13.3% were domestic and 86.7% foreign. In the same year, they realized 1,309,113 overnight stays, domestic 11.3%, and foreign 88.7%. The number of tourist arrivals and overnight stays is increasing by about 18% by year. In 2017, this region was visited mainly by foreign tourists from Turkey, the United Arab Emirates, Croatia, Slovenia, Serbia, the USA, etc. (Sarajevo Canton in Figures 2020).

The base of the natural and cultural-historical heritage (Figs. 14.40 and 14.41) of this region provides an opportunity

Fig. 14.35 Gazi Husrev-beg Mosque. (Photo by Aida Bidžan-Gekić)

Fig. 14.36 Academy of Fine Arts (previously Evangelical Church) buildings from the Austro-Hungarian Period. (Photo by Aida Bidžan-Gekić)

for the development of the following specific forms of tourism cultural manifestation and congress (125 objects of cultural goods were recorded, as well as numerous cultural and artistic events of the world and European rank), winter-sports, hunting and adventure (Igman and Bjelašnica are the backbones of development with other mountains, Treskavica, Ozren, Zvijezda, Trebević, Crnorječka plateau), rural and eco-tourism (Crna Rijeka plateau, villages at the foot of Bjelašnica mountain: Dejčići, Šabići, etc., Tarčin, Pazarić, Trnovo, Lokve) and spa (Ilidža).

This tourist-geographical region is one of the metropolitan tourist regions and is of great importance for defining the tourist offer of Bosnia and Herzegovina; out of a total of 626 national monuments in Bosnia and Herzegovina, there are 95 national monuments in the Sarajevo region. All this characterizes it as the only metropolitan tourist-geographical region in Bosnia and Herzegovina and a unique identity in Southeast Europe and beyond. It is characterized by significant authenticity, representativeness, diversity, integrity, landscape-spatial value, with significant scientific,

Fig. 14.37 Welcome to the Olympic City. (Photo by Aida Bidžan-Gekić)

Fig. 14.38 The Church of Saint Anthony of Padua. (Photo by Aida Bidžan-Gekić)

educational, cultural-historical, social, economic, developmental, geo-ecological significance and function (Bidžan 2016; Bidžan-Gekić 2020). In 2006, the Lonely Planet tourist guide ranked Sarajevo as the 43rd most beautiful in the world, and in 2009, Sarajevo was chosen as one of the ten cities worth visiting (Archives of the City of Sarajevo 2006).

14.4.3.2 Mostar Tourist-Geographical Region

Mostar tourist-geographical region covers 3225 km^2 with 155,593 inhabitants in 2013, from Makljen and Ivan-sedlo in the north to Mostarsko polje in the south. The tourist-geographical position of this region is very favorable because the important Bosnian-Herzegovinian road M-17 passes through it, which connects the capital of Bosnia and Herzegovina and the tourist center of this region with the Adriatic coast. Its position will improve even more when the Vc highway is built through it, so tourists who come to the capital will find it easier to visit this area. It consists of five municipalities: Jablanica, Konjic, Mostar, Istočni Mostar, and Prozor (Bidžan 2016; Bidžan-Gekić 2020). The tourist center of this region is Mostar, which is also the political, economic, financial, and cultural center of Herzegovina.

The unique beauty of this region is contained in the fact that it is both mountainous and lowland and coastal and continental. In the areas around Jablanica, Konjic, and Prozor, there are rich biogeographical and hydrographic tourist potentials preserved from anthropogenic pollution.

It is located at the foot of the mountains of Velež, Hum, and Čabulja, in the valley of the Neretva river, at an altitude of 60–80 m. The northern area of this region consists of the municipalities of Prozor-Rama, Konjic, and Jablanica, which are located at altitudes of 200–1500 m. At higher altitudes, a snow forest climate with maritime influences prevails, and at lower altitudes, a Mediterranean climate. Average annual air

Fig. 14.39 Cathedral Church of the Nativity of the Theotokos. (Photo by Aida Bidžan-Gekić)

temperatures range in the northern parts from 10 to 12 °C and the southern from 14 to 16 °C. The average annual rainfall in the northern parts is around 1500–1750 mm, and in the southern parts from 1250 to 1500 mm. With such ideal climatic conditions, this tourist-geographical region has almost no dead tourist season, so it is accessible to tourists in every season. Also, this region, in addition to the beautiful green Neretva, has two significant reservoirs—Rama (Fig. 14.42) and Jablanica lakes—and numerous rivers such as Rama, Doljanka, Krupić, and other rivers, which is a vast water potential, which is currently only partially used for fish farming and could also be used for bathing, fishing, and adventure tourism. In the biogeographical sense, this region is located in seven ecosystems juniper pine, mountain ores on carbonates, beech-fir forests, thermophilic deciduous-deciduous forests with quercus, Bosnian pine forest, xero-thermic deciduous-deciduous forests, and floodplain forests.

Mostar tourist-geographical region has been at the crossroads of different civilizations for centuries, and each has woven a part of its uniqueness into it. According to archeological sites, this area was inhabited even in prehistoric times. Later, the site of today's city was a Roman colony—Martaritana or Musara. Today's Mostar is first mentioned in 1452 as a settlement by the bridge on the Neretva (Fig. 14.43) (Mihić 1968).

The famous Ottoman travel writer Evliya Çelebi wrote that Mostar was conquered by Sultan Fatih himself and assigned to the pasha in the Herzegovinian Sandjak. Mostar was named after the old chain bridge on the Neretva river, specifically after its guardian. Before the arrival of the Ottomans, the neighboring Blagaj was the capital of Herzegovina and the seat of the noble families Hrnić and Kosač, so the saying "šeher Blagaj, and the town of Mostar" has been preserved from that time. But when the Ottomans came to this area, they began to pay more attention to Mostar, and it soon became the capital of Herzegovina. In 1522, they moved to Mostar, the seat of the Herzegovina Sanjak, which had previously been in Pljevlja (Çelebi 1954).

During the Austro-Hungarian occupation, it was the seat of the Austro-Hungarian and French consuls. During this period, many European-style buildings were built, so Mostar began to look more and more like a Central European city (Fig. 14.44). Oriental-style buildings increasingly began to mix with Austro-Hungarian-style buildings, and this made for a very interesting combination. A large number of buildings at this time were built on the main street in Mostar. One of the most beautiful buildings of this time is the Mostar Gymnasium, reminiscent of the Sarajevo City Hall (Archives of the City of Mostar 2004).

Mostar is 60 km away from the sea. It hosts several famous cultural and historical monuments that are in its immediate vicinity: Kujundžiluk bazaar with the Old Bridge is on the UNESCO World Heritage List and many other monuments of Mostar's turbulent past. We must not neglect the beautiful old stone bridge and bazaar in Konjic (Figs. 14.45 and 14.46), the complex with the Museum of the Battle for the Wounded on the Neretva in Jablanica, over 42 sites with necropolises of stećak tombstones, and all this makes this tourist-geographical region very unique and attractive.

It is connected via the main roads M-17 (E73) with Sarajevo and Primorje tourist-geographical region, by the main road M-16.2 with Travnik and regional road R-419 (Jablanica-Posušje) with West Herzegovina tourist-geographical region, as well as the Republic Croatia, which is partially still a macadam road. The railway Sarajevo-Mostar-Ploče (Republic of Croatia) passes through this

Fig. 14.40 Inat kuća and Alifakovac mahala in Sarajevo. (Photo by Aida Bidžan-Gekić)

Fig. 14.41 Latin bridge built in the Ottoman period. (Photo by Aida Bidžan-Gekić)

region. There is also an international airport in the city of Mostar, which is of great importance for the future development of tourism in this area.

Regarding tourist traffic, the Mostar tourist-geographical region in 2019 had 317,311 tourist arrivals, of which 16% were domestic tourists and 84% foreign tourists. They realized 607,508 overnight stays, of which domestic tourists participated with 16%, and foreign tourists with 84%. Most tourists came from South Korea, China, Poland, Croatia, Italy, Germany, etc. (Herzegovina-Neretva Canton in figures 2020).

Mostar tourist-geographical region belongs to the cognitive, educational, and religious regions in terms of its functionality. It is of great importance for defining the tourist offer of Bosnia and Herzegovina; out of a total of 626 national monuments in Bosnia and Herzegovina, there are 64 national monuments in this area. Its unique natural and anthropogenic tourist potentials enable the development of many specific forms of tourism: bathing, sports-recreational, adventure, fishing, hunting, religious, cultural-event, etc. (Bidžan 2016; Bidžan-Gekić 2020).

Fig. 14.42 Rama Lake. (Photo by Aida Bidžan-Gekić)

Fig. 14.43 City of Mostar and Old Bridge (UNESCO World Heritage List). (Photo by Aida Bidžan-Gekić)

Fig. 14.44 The Mostar Gymnasium. (Photo by Aida Bidžan-Gekić)

Fig. 14.45 Konjic's čaršija—bazaar. (Photo by Aida Bidžan-Gekić)

Fig. 14.46 Konjic Old Bridge. (Photo by Aida Bidžan-Gekić)

14.4.3.3 Banja Luka Tourist-Geographical Region

Banja Luka tourist-geographical region is located in northwestern Bosnia, making the tourist-geographical position of this region very favorable because it connects central Bosnia with Croatia and Central Europe. The M-16 road (E-661) and the international airport give it particular importance. It has an area of 7977 km^2 and about 540,000 inhabitants. It includes the following municipalities: Banja Luka, Mrkonjić Grad, Prijedor, Ribnik, Kostajnica, Novi Grad, Kozarska Dubica, Oštra Luka, Gradiška, Šipovo, Jezero, Srbac, Laktaši, Čelinac, Kotor Varoš, and Kneževo (Bidžan 2016; Bidžan-Gekić 2020).

Banja Luka tourist-geographical region lies at an altitude of 90–900 m. It also hosts the Kozara National Park. Among the more significant geomorphological potentials are the mountains Manjača, Zmijanje, Uzlomac, and Lisina. Of the plain geomorphological potentials, the following are noteworthy Lijevče polje and Knešpolje. It is dominated by a moderately warm and humid climate and snow forest climate. Average annual air temperatures range from 6 to 12 °C. The average annual rainfall ranges from 800 to 1500 mm. In terms of biogeography, six ecosystems predominate, namely, floodplain forests, mesohygrophilous forests with pedunculate oak, montage beech forests, mesophilic sessile forests, and hornbeam forests, beech-fir forests, and dark coniferous forests.

A significant natural tourist potential is the Vrbas river (Fig. 14.47), which is the primary watercourse of this region. Its blue-green color and landscapes, created on its way from Zec mountain to the Sava river, make it one of the most beautiful rivers in Bosnia and Herzegovina. Many events and sports competitions are held on its banks and watercourses. It provides excellent conditions for water sports. Vrbas hosts many national and European competitions in kayak-canoe sports on wild waters, rafting, and fly-fishing. The specificity of this river is the boat dajak (Fig. 14.48), named after the pole used to push the boat.

There are also three spas (Gornji Šeher, Slatina, and Laktaši) with springs of thermo-mineral water, whose medicinal properties were used in the ancient period. It is still known for many green areas (parks and alleys), which is why it is called the "green area." The city of Banja Luka is the tourist center of this region and the second-largest city in Bosnia and Herzegovina and the largest in northwestern Bosnia. It is first mentioned in 1494 in the charter of the Hungarian king Vladislav II Jagelović, issued in Buda in Latin, although the city existed before. Prehistoric archeological sites and objects found on them provide evidence of the existence of the oldest human communities in this area. In the ancient period, the broader area of Banja Luka and western Bosnia was inhabited by the Illyrian tribes Maezaei and Oseriats, who left behind numerous hillfort settlements. After defeating the Illyrian tribes in Baton's war (first century AD), the Romans came to this area and included it in their Illyrian province. The city connected with the rest of the world when a Roman trade route passed through it, leading from Salona to the Servitium. Medieval life in Banja Luka and its surroundings flourished, as evidenced by many written documents and many fortified cities from the period from the twelfth to the fifteenth century (Fig. 14.49). With the fall of the medieval Bosnian state and the arrival of the Ottomans in 1528, Banja Luka became a stronghold of Hungarian and Ottoman domination. It gained special significance during the reign of Ferhat-pasha Sokolović (1574–1588), when it became the center of the Ottoman administrative unit—the Bosnian pashalik. With the arrival of Austro-Hungary, numerous private and public buildings were built in it, such as Royal grammar school, Railway station, and from private trade houses in Gospodska Street, which became the main trade point of the city (Archives of the Tourist organization Banja Luka 2019).

Banja Luka tourist-geographical region is located on the main road M-16 with international mark E661, an essential transit road in northwestern Bosnia, and connects this region with the Republic of Croatia, Doboj region, as well as the main roads M-14.1 and M-4. It is connected to the Bihać region by the main roads M-15 and M-5 (E761). Furthermore, the main road M-4 and M-16 connects it with the Travnik region. The Ploče-Sarajevo-Banja Luka-Zagreb railway, which passes through here, connects Central and Eastern Europe with the Adriatic coast. It also houses an international airport, which influences the arrival of foreign tourists.

Fig. 14.47 Vrbas river in Banja Luka. (Photo by Aida Bidžan-Gekić)

Fig. 14.48 Dayak boat on the river Vrbas. (Photo by Aida Bidžan-Gekić)

Fig. 14.49 Kastel Fortress is the oldest historical monument in the city of Banja Luka. (Photo by Aida Bidžan-Gekić)

Regarding tourist traffic, the Banja Luka tourist-geographical region in 2019 had 127,033 tourist arrivals, of which 43% were domestic tourists and 57% foreign tourists. They realized 217,798 overnight stays, of which domestic tourists participated with 46%, and foreign tourists with 54%. Most tourists came from Serbia, Croatia, Slovenia, Germany, Italy, Sweden, etc. (ISR 2020).

According to its functionality, the Banja Luka tourist-geographical region belongs to the region of leisure and recreation regions. It is of great importance for defining the tourist offer of Bosnia and Herzegovina; out of 626 national monuments in Bosnia and Herzegovina, 101 are in this area, which enables the development of cultural and event tourism (Fig. 14.50). Thanks to numerous natural tourist potentials, it is possible to develop bathing, spa, sports and recreation, adventure, fishing, hunting, and other specific forms of tourism (Bidžan 2016; Bidžan-Gekić and Gekić 2017; Bidžan-Gekić 2020).

Fig. 14.50 Monument to the Revolution in Kozara—a World War II memorial sculpture. (Photo by Aida Bidžan-Gekić)

14.4.3.4 Travnik Tourist-Geographical Region

Travnik tourist-geographical region is located in the central part of Bosnia and Herzegovina. Its tourist-geographical position is therefore very favorable. It is an important crossroads of Bosnia and Herzegovina's roads, and the international road E661 passes through it, enabling the development of transit tourism. The region has an area of 3189 km^2 with 254,686 inhabitants in 2013 and consists of the following municipalities: Travnik, Bugojno, Donji Vakuf, Gornji Vakuf-Uskoplje, Jajce, Dobretići, Novi Travnik, Vitez, Fojnica, Kreševo, Busovača, and Kiseljak (Bidžan 2016; Bidžan-Gekić 2020). The tourist center of the region is the municipality of Travnik.

It is located at an altitude of 470–2100 m. The most significant geomorphological tourist potentials among the mountains are the following: Vranica, Vlašić, Kalin, Bitovnja, Raduša, Rostovo, and some smaller mountains, as well as the canyon of Ljuta river, etc. This region's most attractive geomorphological tourist potential is the mountain Vlašić, which makes it an excellent place for skiing, hiking, sledding, and other sports and recreational activities. In the Travnik region, there are three climate types with their transitional subtypes: moderately warm and humid climate, forest-snow climate, and tundra climate. Average annual air temperatures range from 2 to 6 °C, and average annual rainfall ranges from 900 to 1250 mm. Hydrographic tourist potentials are very numerous and important for the development of tourism. We will mention only the most important, namely, Vrbas, Pliva, Ugar, Lašva, etc., numerous waterfalls and cascades, such as waterfalls on Kozica in Fojnica, Crindol waterfalls in Gornji Vakuf, Plivski waterfalls in Jajce (Fig. 14.51), and two waterfalls on Plačkovac in Travnik, followed by numerous natural and artificial lakes—Ždrimačko, Prokoško, Veliko, and Malo Plivsko, as well as multiple thermal and mineral springs—Reumal and Akvareumal in Fojnica, Toplica in Kiseljak, then Park, Kiseljačić, Podgaj, and Bio-han, also in Kiseljak. Biogeographic tourist potentials consist of the following ecosystems: mesophilic sessile and hornbeam forests, mountain ores on carbonates, dark coniferous forests, and beech-fir forests.

Also, there are many anthropogenic tourist potentials of this region (Fig. 14.52). Solely the data on the existence of over 1000 stećak tombstones from medieval Bosnia, with necropolises of 100 stećak tombstones (Fig. 14.53), hundreds of archeological excavations and finds collected in museums (Travnik, Jajce, etc.), and Franciscan monasteries in Fojnica and Kreševo and many other cultural and historical traces indicate not only valuable documents and objects but also about rich past and the creative work of the people of this area. The oldest written monument on Bosnian territory was found here (Umbro-Etruscan inscription on a vase from the sixth century BC, found at the Pod site, near Bugojno), and a bronze sword from Veliki Mošunj, with unique ornaments and aesthetic value (find about 1000 BC).

The tourist center of this region, Travnik, has historically been known as the seat of the Bosnian viziers (Fig. 14.54) and as the birthplace of Nobel laureate Ivo Andrić. It is rich in cultural and historical monuments and houses one of the best-preserved medieval towns. Many remains from the Roman

Fig. 14.51 The Pliva Waterfall in Jajce. (Photo by Aida Bidžan-Gekić)

period were found in the city itself: from pottery and tombstones to imperial money (Archive of the Travnik Homeland Museum 2020).

Also worth mentioning is the royal city of Jajce or the "open-air museum" that nurtures more than 22 national monuments. Many significant cultural and historical sites have been found in the area of Jajce, among which the temple of the god Mithras, which is the oldest monument in Jajce and dates from the fourth century, a fortress, medieval, built by the builder of Jajce, Bear Tower, mysterious underground catacombs associated with numerous legends, representing the temple and underground tomb of the last Bosnian king (Fig. 14.55) (Archives of the Tourist Board of Central Bosnia Canton 2019).

Fig. 14.52 Sultan Ahmed Khan II Mosque in Bugojno built in 1693

Fig. 14.53 Necropolis of Maculje with stećak tombstones in Novi Travnik. (Photo by Aida Bidžan-Gekić)

Also, it is important to point out that the tourist center of this region, Travnik, can offer famous Vlašić or Travnik cheese to gastronomy enthusiasts and ćevapi (grilled dish of minced meat) in one of the many local eateries on Žitarnica, trout on Plava Voda, and finally Lutvo's coffee (Fig. 14.56).

In terms of traffic, it is connected by the international road E661 (M-5) with the Doboj region; the main road M-16 is connected with the Southwest Bosnia region, the main road M-16.2 is connected with the Mostar region, and also with M-5 and M-16 is with the Banja Luka region. The most important communications stretch along the valleys of the rivers Lašva and Vrbas, where important urban settlements are located. The central geographical position of this tourist-geographical region connects the transit routes northwest-

Fig. 14.54 Travnik fortress. (Photo by Aida Bidžan-Gekić)

Fig. 14.55 Jajce—the old royal city. (Photo by Aida Bidžan-Gekić)

southeast (Bihać-Travnik-Sarajevo-Goražde) and northeast-southwest (Tuzla-Zenica-Travnik-Bugojno-Mostar-Ploče, i.e., Neum, and from Bugojno to the direction Livno-Split).

Regarding tourist traffic, the Travnik tourist-geographical region in 2019 had 54,028 tourist arrivals, of which 52.5% were domestic tourists and 47.5% foreign tourists. They realized 90,381 overnight stays, of which domestic participated with 48.8%, and foreign tourists with 51.2%. Most tourists came from Saudi Arabia, Croatia, Slovenia, Germany, the United Arab Emirates, Serbia, etc. (Central Bosnia Canton in figures 2020).

According to its functionality, the Travnik tourist-geographical region belongs to the group of cognitive, educational, and religious tourist-geographical regions. It is of great importance for defining the tourist offer of Bosnia and Herzegovina; out of a total of 626 national monuments in Bosnia and Herzegovina, 59 national monuments are situated in the Travnik region (Fig. 14.57). The inhabitants of this region have long realized that using the generosity of nature and cultural and historical wealth can attract tourists, which is why today, in addition to the excursion, business-transit and recreational tourism, sports-recreational, spa, hunting,

Fig. 14.56 Plava voda and Lutvo's coffee by the river Lašva. (Photo by Aida Bidžan-Gekić)

Fig. 14.57 The Watermills on the river Pliva—protected area of cultural heritage. (Photo by Aida Bidžan-Gekić)

fishing, religious, rural, and other specific forms of tourism are well developed (Bidžan 2016; Bidžan-Gekić 2020; Bidžan-Gekić et al. 2019a).

14.4.4 Education

The education sector in Bosnia and Herzegovina is a reflection of the state, defined by the Constitution of Bosnia and Herzegovina, the constitutions of the entities, cantons, and the Statute of the Brčko District of Bosnia and Herzegovina, which legally define competencies in education (Archives of the Ministry of Civil Affairs of Bosnia and Herzegovina 2018).

The education system in Bosnia and Herzegovina includes preschool education, 9-year primary education, compulsory and free of charge for all children from 6 to 15 years of age, secondary education optional, and higher education.

Preschool education and upbringing are intended for children from 6 months to entering primary school. A preschool preparatory program is implemented in the year before starting school, which is the first part of compulsory education and includes children between 5, 6, and 6 ½ years of age. Primary education is mandatory and lasts for 9 years. Public primary education is free of charge, and the only criterion for enrollment is the child's age—all children must be enrolled in primary school between the ages of 5 and 6.

Secondary education is available in general secondary education schools (grammar schools, gymnasiums (Fig. 14.58), vocational secondary schools, art schools, and other, less common types of schools. Secondary education lasts 4 years, and vocational secondary education lasts 3 or 4 years. Students are usually 15 years old when they enter high school, and they enroll based on their achievements from elementary school and the entrance exam. Higher education in Bosnia and Herzegovina is organized in three cycles (Bachelor, Master, Ph.D.). There are 10 public higher education institutions in Bosnia and Herzegovina (of which 8 are public universities) and 22 private higher education institutions, which offer a total of about 500 study programs. Bosnia and Herzegovina signed the Bologna Declaration in 2003 and in 2007 passed the framework law on higher education as the first and most important document governing this area. The Bologna Process has ten basic areas in which joint progress and development of the signatory countries are planned. Within this process, many bodies and institutions cooperate daily, exchange experiences, perform monitoring, etc., and in Bosnia and Herzegovina, some of them are students (Archives of Travel web portal VMC 2019).

The main problem of reducing the number of students in BiH is demographic in nature (reduction in the number of live births and emigration). In the school year 2015–2016, there were 111,354 students in BiH. In 2019–2020, that number was reduced to 85,907 or 22.8%. About 22% of students attend private higher education institutions. One third of all bh. students study in the Canton of Sarajevo. The largest number of students in the academic year 2019–2020 was at the following universities in BiH: University of Sarajevo 25,266 (29.4%), University of Banja Luka 10,432 (12.1%), University of Mostar 8158 (9.5%), University of East Sarajevo 7615 (8, 9%), University of Tuzla 7591 (8.8%), University of Zenica 3114 (3.6%), University of Mostar "Džemal Bijedić" 2781 (3.2%), and Pan-European University Apeiron 2615 (3.0%) (BHAS 2019).

14.4.5 Health Care

Bosnia and Herzegovina is one of the rare countries in the world that does not have a state ministry of health. According to the constitutional solutions, BiH's health care is in the full competence of the entity of Republika Srpska (RS) and the Brčko District of Bosnia and Herzegovina, and in the divided

Fig. 14.58 Gymnasium in Bugojno built in 1882. (Photo by Haris Gekić)

competence between the entity of the Federation of BiH (FBiH) and its ten cantons. Therefore, Bosnia and Herzegovina has 13 ministries of health: 1 in the entity of RS, 1 in the Brčko District of BiH, 1 in the FBiH, and 1 in each of the 10 cantons in the FBiH. At the state level, in addition to the Agency for Medicines and Medical Devices, there is the Ministry of Civil Affairs of Bosnia and Herzegovina with a special Department of Health, which deals mainly with statistics, i.e., information exchange with domestic and international institutions, such as the World Health Organization (Archives of the Center for Public Law Foundation 2019).

In some cantons in the FBiH and some parts in the entity of RS, i.e., in the Brčko District of BiH, healthcare institutions do not have the material or personnel conditions to provide the inhabitants of these areas with an appropriate level of health care. Therefore, the inhabitants of these areas are forced to use health services in the once very remote large medical centers, mainly in Sarajevo, Banja Luka, Tuzla, and Mostar. In Bosnia and Herzegovina, people under the age of 18 have the same rights as insured persons. Uninsured persons are provided with emergency medical care (free of charge) in life-threatening conditions, treatment of serious infectious diseases, with explicit quarantine, tuberculosis, HIV infection, SARS, etc. (Health Care Act 2013).

According to official data, the rate of allocations from GDP in Bosnia and Herzegovina was 9% in 2019. It is 215 EUR per capita. Almost all private consumption (99%) falls on direct household expenditures, while only a small part of private consumption (1%) goes to voluntary healthcare payments. More than half of the total health expenditure was spent on treatment and rehabilitation services, while ¼ funds were spent on medical resources. In BiH, healthcare spending is growing steadily. In 2015, the total healthcare spending amounted to 1 billion 364 million EUR, which is about 267.6 million less than in 2019 (Agency for Statistics of Bosnia and Herzegovina 2020).

The largest health centers in BiH are clinical institutions that have a tertiary level of health care, and they are university clinical centers located in large cities: Sarajevo, Banja Luka, Tuzla, Mostar, and Foča. Further in BiH, there are health centers that we call general and cantonal hospitals that are secondary level of health care, such as Cantonal Hospital "Dr Irfan Ljubijankić" in Bihać, General Hospital Sanski Most, Cantonal Hospital "Dr Safet Mujić" Mostar, General Hospital "Dr Mustafa Beganović" "Gračanica, Hospital" Sveti apostol Luka "Doboj", General Hospital Gradiška, etc.

According to the World Health Organization, in 2018, there were 19.1 family medicine doctors per 100,000 inhabitants in Bosnia and Herzegovina, which is important for achieving comprehensiveness and universality in terms of increasing access to healthcare services. There is a need for more family medicine doctors in the coming period in order to achieve coverage and to increase access to primary healthcare services, especially in the field of sexual and reproductive health. Their number should be higher in order to achieve coverage both in the European region of the World Health Organization (83.9 per 100,000 inhabitants) and among the members of the Health Network of Southeast Europe (77.1 per 100,000 inhabitants) (World Health Organization 2019).

14.4.6 Banking

Banking is one of the most vital sectors in Bosnia and Herzegovina. This sector is supervised by the Central Bank of Bosnia and Herzegovina, established on June 20, 1997, by the Law on the Central Bank adopted by the Parliamentary Assembly of Bosnia and Herzegovina. It monitors the financial system's stability and active participation in the development and implementation of the policy of stability and sustainable economic growth of Bosnia and Herzegovina by ensuring the stability of the domestic currency and overall financial and economic stability in the country. It maintains monetary stability in accordance with the currency board arrangement, which means that it issues domestic currency with full coverage in free convertible foreign currency at a fixed exchange rate of 1 BAM: 0.51129 EUR. The Central Bank defines and controls the implementation of the monetary policy of Bosnia and Herzegovina. It also assists and maintains appropriate payment and billing systems. It also coordinates the activities of the banking agencies of the entities, which are responsible for issuing licenses and supervising banks (Archives of the Central Bank of Bosnia and Herzegovina 2019).

There are 23 commercial banks in Bosnia and Herzegovina with approximately 10,000 employees. Only one bank is majority state-owned (Union Bank), while foreign banks own over 80% of the banking capital in Bosnia and Herzegovina. Some of these foreign banks are Addiko Bank AG, Komercijalna banka, NLB Group, Privredna banka Zagreb, ProCredit Holding, Raiffeisen Bank International, Sberbank, Steiermärkische Bank und Sparkassen, Ziraat Bankasi, etc. They collect deposits and savings from the broadest circle of clients from business and nonbusiness activities. Commercial banks grant loans to meet the needs of clients for funds. Bosna Bank International is the only bank in BiH that operates according to the principles of Islamic banking (Archive of the Agency for the Promotion of Foreign Investments in Bosnia and Herzegovina 2020). There are two investment development banks in Bosnia and Herzegovina, namely, the Development Bank of the Federation of Bosnia and Herzegovina and the Investment and Development Bank of the Republika Srpska.

The Development Bank of the Federation of Bosnia and Herzegovina is a financial institution tasked with implementing the economic policy of the Government of the Federation of Bosnia and Herzegovina for employment and economic development through incentive lending that includes development differentiated and stimulative interest rates, which are lower than market interest rates. This bank was established by the Law on the Development Bank of FBiH and started operating on July 1, 2008, as the successor of the Investment Bank of the Federation of BiH, which has successfully operated with its 109 employees since 1997. In addition to the existing commercial banking, its establishment has become an imperative on the way to faster and more rational economic recovery and development of the Federation of BiH. The Federation unconditionally guarantees all its obligations, and it is responsible with its property. The headquarters of this bank is in Sarajevo, and its branches are located in Mostar, Zenica, Orašje, and Bihać (Archives of the FBiH Development Bank 2019).

The Republika Srpska Investment and Development Bank (IRBRS) was established on December 6, 2006, in accordance with the Law on the Republika Srpska Investment and Development Bank. It is registered as a joint-stock company, in which RS has 100% ownership. Its strategic goals are to encourage investment and stimulate development in this entity. In addition, its role is the professional and efficient management of the assets of Republika Srpska, which is formally and legally registered in six funds. It does not have the character of a commercial bank, i.e., according to the Law, and it has no right to collect deposits or give guarantees and thus does not fall under the jurisdiction of the Banking Agency of Republika Srpska. It does not place funds intended for development projects directly but through microcredit companies and commercial banks. It is supported by some projects from international financial institutions, such as European Bank for Reconstruction and Development, World Bank, and European Investment Bank. IRBRS activities aim to finance projects that are important for the national economy and for which the private sector is not interested because they do not bring direct profit, require significant investments, or are too risky (Archives of the IRBRS 2019).

14.5 International Trade

The total trade of Bosnia and Herzegovina in 2019 amounted to 15.8 billion EUR. The value of exported goods was 5.9 billion EUR, and the value of imported goods amounted to 9.9 billion EUR, while a foreign trade deficit of 4 billion EUR was recorded. The trade deficit increased by 8.6% compared to the previous year. The main characteristics of foreign trade in 2019 were a decrease in the total foreign trade of Bosnia and Herzegovina (0.6%), a decrease in exports (3.4%), an increase in imports (1.2%), and a decrease in the rate of coverage of imports by exports for 2.8%.

The European Union is still the leading foreign trade partner of Bosnia and Herzegovina. In the total volume of trade, the EU participates with 66%, followed by the countries signatories to the "CEFTA 2006" agreement, which participate with 14%, while other countries participate with 20%. Exports to the EU account for 73% of total exports in BiH, and imports from EU countries account for 62% of total imports in BiH. In 2019, compared to the previous year, there was an increase in total trade with the EU (0.6%), a decrease in exports to the EU (3.4%), an increase in imports from the EU (3.6%), and a decrease in the foreign trade coverage ratio from the EU (5%), while the trade deficit increased (24%).

Exports to CEFTA countries participate with 17% of total exports in BiH, and imports from CEFTA countries participate with 12% of total imports in BiH. In 2019, compared to the previous year, there was an increase in the total trade of Bosnia and Herzegovina with CEFTA (1.1%), a decrease in exports in CEFTA (2.3%), an increase in imports from CEFTA (3.9%), and a decline in the rate of foreign trade coverage ratio with CEFTA (5%), while the trade deficit increased (35%) (Ministry of Foreign Trade and Economic Relations of BiH 2019b).

Exports to other countries (countries that do not belong to the EU and CEFTA) account for 10% of total export in BiH. In comparison, imports from other countries account for 26% of total imports in BiH. Of the main trading partners, we have the most extensive foreign trade coverage ratio with Austria of 151%, Slovenia 111%, Germany 72%, Croatia 69%, France 62%, Serbia 61%, and Italy 56%. In addition to these countries, significant trade partners of Bosnia and Herzegovina are China, the USA, Poland, and Russia. Still, the coverage of imports by exports with these countries is small, so with Russia, it is 29%, Poland 29%, the USA it is 8%, and China 2%.

In 2019, Bosnia and Herzegovina mainly exported to Germany (858 million EUR), Croatia (715 million EUR), Serbia (669 million EUR), Italy (665 million EUR), Austria (557 million EUR), Slovenia (516 million EUR), Montenegro (208 million EUR), and Turkey (149 million EUR). In the observed year, Bosnia and Herzegovina mainly imported from Germany (1.19 billion EUR), Italy (1.18 billion EUR), Serbia (1.09 billion EUR), Croatia (1.03 billion EUR), China (740 million EUR), Turkey (492.5 million EUR), Slovenia (463 million EUR), and Austria (368 million EUR).

The most exported products in 2019 were electricity, 293.6 million EUR; seats, 272.6 million EUR; other furniture and its parts, 172.6 million EUR; insulated wires, cables,

168.2 million EUR; and iron or steel structures, 165.6 million EUR. The most imported products in 2019 were petroleum oils, except crude, 806.8 million EUR; passenger cars, 453.3 million EUR; coal, briquettes 270.1 million EUR; medicines 253.1 million EUR; and electricity 138.3 million EUR.

In 2019, the share of industrial products in the total BiH exports was 93%, while agricultural products participated 7%. Industrial products account for 84% of total imports of BiH, while agricultural products account for 16% (BiH foreign trade in 2020).

In 2019, the export of agricultural products amounted to 420.1 million EUR, while the import of agricultural products amounted to 1.61 billion EUR. A deficit in the exchange of agricultural products of 1.19 billion EUR was recorded. Of the fruit crops for export, the most common are plum, apple, pear, grape, raspberry, strawberry, cherry, and forest fruits. The most important vegetable crops are potatoes, cabbage, cucumber, tomato, and paprika. Fresh plums, frozen and fresh raspberries, forest fruits, cherries, gherkins, tomatoes, potatoes, etc. are mainly exported. Four products for which Bosnia and Herzegovina has a significant trade surplus with other countries are plums, raspberries, cherries, and gherkins. Most were exported to EU and CEFTA countries.

References

Agency for Statistics of BiH (2017) The work report. Available at https://bhas.gov.ba/data/Dokumenti/Izvjesca/BHAS_Izvjestaj%20o%20radu_2017_BS.pdf. Accessed on 20 Mar 2020

Agency for Statistics of Bosnia and Herzegovina (2018) Bulletin BiH in issues. Available at https://bhas.gov.ba/data/Publikacije/Bilteni/2019/NUM_00_2018_TB_0_BS.pdf. Accessed on 21 May 2020

Agency for Statistics of Bosnia and Herzegovina (2020) Bulletin BiH in numbers. Available at https://bhas.gov.ba/data/Publikacije/Bilteni/2021/NUM_00_2020_TB_1_BS.pdf. Accessed on 21 Aug 2021

Agency for Statistics of FBiH (2020) Statistical yearbook of FBiH. Available at http://fzs.ba/wp-content/uploads/2020/01/Godisnjak2019.pdf. Accessed on 26 Aug 2020

Akta (2016) Business portal. Available at https://www.akta.ba/vijesti/znacaj-luke-ploce-za-bih-i-povezanost-luke-sa-koridorom-5c/71325. Accessed on 18 Aug 2020

Archive of AS Holding (2020). Available at https://www.asholding.ba/o-nama/. Accessed on 24 Apr 2021

Archive of Nerda (2011). Available at http://nerda.ba/pdf/Analiza_stanja_i_mogucnosti_poljprivredne_i_prehrambene_industrije_SI_BiH.pdf. Accessed on 29 Aug 2019

Archive of the Borac Garment Factory (2020). Available at https://lokal.ba/brands/borac-konfekcija-travnik. Accessed on 28 Apr 2020

Archive of the Commission to Preserve National Monuments of Bosnia and Herzegovina (2004). Available at http://old.kons.gov.ba/main.php?id_struct=6&lang=1&action=view&id=2500. Accessed on 15 May 2019

Archive of the company Širbegović (2020). Available at https://sirbegovic.com/bs/historija-kompanije-sirbegovic-inzenjering-prefabrikovana-konstrukcija-pps-ploce/. Accessed on 29 Apr 2021

Archive of the Malak Janj factory (2020). Available at http://www.malak.ba/ba/index.php?m=drvna_p. Accessed on 10 Mar 2020

Archive of the Prevent Group company (2020). Available at https://preventgroup.ba/kompanija/. Accessed on 25 Mar 2020

Archive of the Travnik Homeland Museum (2020). Available at https://muzejtravnik.ba/zavicajni-muzej-travnik/stalna-arheoloska-postavka/. Accessed on 15 Mar 2020

Archive of the Violeta factory (2020). Available at https://violeta.com/o-nama/nasa-prica/; https://violeta.com/o-nama/upravljanje-kvalitetom/. Accessed on 14 Mar 2020

Archives Ministry of Communications and Transport (2021). Available at https://www.jpautoceste.ba/o-autocesti/. Accessed on 18 Aug 2021

Archives of Communications Regulatory Agency (2020) Annual report. Available at https://docs.rak.ba/documents/0f76ec67-254c-4fb5-9be6-6e6e1d41a058.pdf. Accessed on 18 Apr 2021

Archives of Ministry of Communications and Transport 2021, available at https://www.jpautoceste.ba/o-autocesti/, accessed on 18/08/2021

Archives of the Agency for the Promotion of Foreign Investments in Bosnia and Herzegovina (2020). Available at http://www.fipa.gov.ba/atraktivni_sektori/bankarski_sektor/default.aspx?id=46&langTag=hr-HR. Accessed on 11 Aug 2020

Archives of the Center for Public Law Foundation (2019). Available at https://www.pfsa.unsa.ba/pf/wp-content/uploads/2015/01/Sistem-zdravstva-u-Bih.stanje-i-pravci-mogu%C3%A6e-reforme.pdf. Accessed on 25 June 2019

Archives of the Central Bank of Bosnia and Herzegovina (2019). Available at https://www.cbbh.ba/Content/Read/13. Accessed on 18 Aug 2019

Archives of the City of Mostar (2004). Available at https://www.mostar.ba/o-gradu-opcenito.html. Accessed on 26 Apr 2019

Archives of the City of Sarajevo (2006). Available at https://www.sarajevo.ba/srb/article/1310/sarajevo-43-najbolji-grad-na-svijetu. Accessed on 25 Aug 2019

Archives of the City of Sarajevo (2020). Available at https://www.sarajevo.ba/en/article/1214/sarajevo-u-austrougarskom-periodu. Accessed on 15 May 2019

Archives of the FBiH Chamber of Commerce (2020). Available at http://www.kfbih.com/mirsad-jasarspahic%2D%2Duciniti-sve-da-bosanskohercegovacke-kompanije-sacuvaju-poziciju. Accessed on 15 Mar 2020

Archives of the FBiH Development Bank (2019). Available at https://rbfbih.ba/o-nama/. Accessed on 20 Aug 2020

Archives of the Foreign Trade Chamber of Bosnia and Herzegovina (2021a). Available at https://www.komorabih.ba/o-privredi-bosne-i-hercegovine/. Accessed on 14 Mar 2021

Archives of the Foreign Trade Chamber of Bosnia and Herzegovina (2021b). Available at https://www.komorabih.ba/asocijacija-drvne-industrije-i-sumarstva/. Accessed on 07 Mar 2021

Archives of the Foreign Trade Chamber of Bosnia and Herzegovina (2021c). Available at https://www.komorabih.ba/asocijacija-graficke/. Accessed on 03 Mar 2021

Archives of the Foreign Trade Chamber of Bosnia and Herzegovina (2021d). Available at https://www.komorabih.ba/asocijacija-metalske-i-elektro-industrije/. Accessed on 02 Mar 2020

Archives of the Foreign Trade Chamber of Bosnia and Herzegovina (2021e). Available at https://www.komorabih.ba/wp-content/uploads/2019/02/Catalogue_milk_and_meat.pdf. Accessed on 01 Nov 2021

Archives of the Foreign Trade Chamber of Bosnia and Herzegovina (2021f). Available at https://www.komorabih.ba/wp-content/uploads/2019/03/Katalog-Proizvodnja-bezalkoholnih-pica-i-voda.pdf. Accessed on 25 Mar 2021

Archives of the Foreign Trade Chamber of Bosnia and Herzegovina (2021g). Available at https://www.komorabih.ba/asocijacija-tekstila-odjece-koze-i-obuce/. Accessed on 25 Jan 2021

Archives of the IRBRS (2019). Available at https://www.irbrs.org/azuro3/a3/?id=74. Accessed on 10 Aug 2020

Archives of the Ministry of Civil Affairs of Bosnia and Herzegovina (2018). Available at http://mcp.gov.ba/attachments/bs_Migrirani_dokumenti/Sektori/Obrazovanje/Obrazovanje-ostalo/Informacija_o_stat.pokazateljima_u_obraz._u_BiH_2018_(003)_bosanski_jezik.docx. Accessed on 24 June 2020

Archives of the Tourist Board of Central Bosnia Canton (2019). Available at http://www.tzsbk.com/turizam-sbkksb.html?showall=1. Accessed on 09 Apr 2019

Archives of the Tourist organization Banja Luka (2019). Available at https://www.banjaluka-tourism.com/index.php/sr/o-banjaluci/kolona-1/priroda-banjaluke/item/2115-rijeka-vrbas. Accessed on 14 Mar 2019

Archives of Travel web portal VMC (2019). Available at https://www.visitmycountry.net/bosnia_herzegovina/bh/index.php/o-bih/skolstvo. Accessed on 24 June 2019

Bašić M (2006) Osnove ratne ekonomije—s osvrtom na rat u Bosni i Hercegovini 1992–95. godine [Basics of war economy—with a review of the war in Bosnia and Herzegovina 1992–95. years], vol 57. Ekonomski pregled, Hrvatsko društvo ekonomista, Zagreb, pp 130–145

Beširević U (1983) Prilog istoriji planinarstva u Bosni i Hercegovini [A contribution to the history of mountaineering in Bosnia and Herzegovina]. In: Cecić I (ed) Enciklopedija Jugoslavije, separat SR Bosna i Hercegovina. Jugoslovenski leksikografski zavod, Zagreb, p 219

BHAS (2019) Statistics on education. Agency for Statistics of Bosnia and Herzegovina, Sarajevo

Bidžan A (2011) Turističko-geografsko rejoniranje kao osnova geografske regionalizacije Bosne i Hercegovine [Tourist-geographical zoning as the basis of geographical regionalization of Bosnia and Herzegovina]. Master's thesis, Faculty of Science, University of Sarajevo, Sarajevo

Bidžan A (2012) Turističko-geografski rejoni Bosne i Hercegovine [Tourist-geographical regions of Bosnia and Herzegovina]. In: Drešković N (ed) Zbornik III Kongresa geografa Bosne i Hercegovine. Geografsko društvo u FBiH, Sarajevo, pp 496–511

Bidžan A (2016) Regionalni aspekti vrednovanja turističkih potencijala Bosne i Hercegovine [Regional aspects of evaluating the tourism potentials of Bosnia and Herzegovina]. Doctoral thesis, Faculty of Science, University of Sarajevo, Sarajevo

Bidžan-Gekić A (2020) Regionalna turistička geografija Bosne i Hercegovine 1. Prirodno-matematički fakultet Univerzitet u Sarajevu, Sarajevo

Bidžan-Gekić A, Gekić H (2017) Regionalna povezanost i valorizacija turističkih potencijala Bihaćkog i Banjalučkog turističko-geografskog rejona [Regional connection and valorization of tourist potentials of Bihać and Banja Luka tourist-geographical region]. In: Drešković N (ed) Zbornik IV Kongresa geografa Bosne i Hercegovine. Geografsko društvo u FBiH, Sarajevo, pp 371–393

Bidžan-Gekić A, Gekić H, Nezirović S (2017) Cultural tourism impact on the development of Sarajevo tourist-geographical region. In: Drešković N (ed) Book of proceedings 2nd international tourism and hospitality management conference. Prirodno-matematički fakultet, Sarajevo, pp 223–234

Bidžan-Gekić A, Gekić H, Žunić L (2019a) Development of religious tourism in Mostar tourism-geographical region. In: Journal of tourism and hospitality management, vol 5. Faculty of Science, University of Sarajevo, Sarajevo, pp 143–168

Bidžan-Gekić A, Gekić H, Žunić L (2019b) Development of religious tourism in Travnik tourism-geographical region, Geografski pregled, vol 40. Geographical Society in Federation of Bosnia and Herzegovina, Sarajevo, pp 51–64

BiH foreign trade in 2020 (2020) Agency for Statistics of Bosnia and Herzegovina. Available at https://bhas.gov.ba/data/Publikacije/Bilteni/2021/ETR_00_2020_TB_1_BS.pdf. Accessed on 21 Aug 2021

Čekić Š (2005) Studija saobraćaja na području Kantona Sarajevo [Traffic study in the area of Sarajevo Canton]. In: Čekiš Š (ed) Separat studije: Robni tokovi na području Kantona Sarajevo. Fakultet za saobraćaj i komunikacije, Sarajevo, pp 150–172

Çelebi E (1954) Putopis, Odlomci o jugoslovenskim zemljama [Travelogue, Excerpts on Yugoslav Lands], vol I. Svjetlost, Sarajevo, p 274

Central Bosnia Canton in figures (2020) Federal Bureau of Statistics of BiH. Available at https://fzs.ba/wp-content/uploads/2020/06/SBK_2020.pdf. Accessed on 18 Apr 2021

Dairy Industry (2020) Foreign trade chamber of BiH. Available at https://www.komorabih.ba/wp-content/uploads/2021/01/Mlijecna-industrija-BHS.pdf. Accessed on 14 Feb 2021

Development Strategy of Bosnia and Herzegovina (2010). Available at http://mft.gov.ba/bos/images/stories/medjunarodna%20saradnja/BOS-A-Strategija%20razvoja%20BiH.pdf. Accessed on 04 May 2019

Gekić H (2011) Uticaj prometne infrastrukture na razvoj turizma u Bosni o Hercegovini [The impact of transport infrastructure on the development of tourism in Bosnia and Herzegovina]. In: Spahić M (ed) Zbornik radova međunarodnog naučnog skupa „Edukacija iz turizma i zaštite životne sredine kao pred uvjet turističkog regionalnog i prostornog planiranja". Prirodno-matematički fakultet Sarajevo, Konjic, pp 138–153

Gekić H, Bidžan A (2015) A geographical analysis of industry in Bugojno, Bosnia and Herzegovina, vol 6. Academic Research International, SAVAP International, Lodhran City, pp 157–167

Hasić D (2006) Ratne štete u Sarajevu 1992–1995 [War damage in Sarajevo 1992–1995]. Institut za istraživanje zločina protiv čovječnosti i međunarodnog prava, Sarajevo

Health Care Act (2013). Available at https://www.paragraf.ba/propisi/fbih/zakon-o-zdravstvenoj-zastiti.html. Accessed on 14 Aug 2020

Herzegovina-Neretva Canton in figures (2020) Federal Bureau of Statistics of BiH. Available at https://fzs.ba/wp-content/uploads/2020/06/HNK_2020.pdf. Accessed on 15 June 2021

Historical archive (2007). Available at http://historija.info/isa-beg-ishakovic-osnivac-sarajeva/. Accessed on 24 Apr 2019

Hrelja K (1974) Razvoj industrije u Bosni i Hercegovini do drugog svjetskog rata [Industrial development in Bosnia and Herzegovina until the Second World War]. In: Acta historico-economica Yugoslavia, vol I. Zagreb, pp 17–35

ISR (2020) Institute for Statistics of Republika Srpska, Cities and Municipalities of Republika Srpska. Available at https://www.rzs.rs.ba/static/uploads/bilteni/gradovi_i_opstine_republike_srpske/2020/Gradovi_I_Opstine_Republike_Srpske_2020_WEB.pdf. Accessed on 26 May 2021

Lorber L (2009) Transition in Slovenian rural areas, Revija za geografijo, vol 4. Oddelek za geografijo, Filozofska fakulteta Univerze v Mariboru Ljubljana, pp 103–116

Marsenić D (1995) Ekonomika Jugoslavije [Economy of Yugoslavia]. National Library of Serbia, Beograd

Mihić LJ (1968) Turistički motivi i objekti u Hercegovini [Tourist motives and objects in Herzegovina]. Turistička štampa, Beograd

Ministry of Foreign Trade and Economic Relations of BiH (2019a) Annual report in the field agriculture, food, forestry and rural development in Bosnia and Herzegovina, Sarajevo. Available at http://mvteo.gov.ba/Content/Read/poljoprivreda-prehrana-sumarstvo-ruralni-razvoj-izvjestaj. Accessed on 06 Apr 2020

Ministry of Foreign Trade and Economic Relations of BiH (2019b) Analysis of foreign trade of Bosnia and Herzegovina. Available at http://www.mvteo.gov.ba/attachments/hr_Home/Ostale_stranice/Me%C4%91unarodni_trgovinski_odnosi_i_sporazumi/Me%C4%91unarodni_odnosi_-_Statistika/Analiza_vanjskotrgovinske_razmjene_BiH_/Analiza_vanjskotrgovinske_razmjene_Bosne_i_Hercegovine_za_2019.godinu-hrvatski.pdf. Accessed on 25 Sept 2020

Nurković R (2005) Industrija kao faktor regionalnog razvoja Bosne i Hercegovine na primjeru Tuzlanske kotline [Industry as a factor of regional development of Bosnia and Herzegovina on the example of Tuzla valley]. In: Sundać D (ed) Zbornik radova problemi regionalnog razvoja Hrvatske i susjednih zemalja. Croatian Geographical Society, Zagreb, pp 173–182

Nurković R (2007) Distribution of industry in Bosnia and Herzegovina. J Geogr Inst "Jovan Cvijic", SASA 57:Beograd, 357–363

Nurković R (2012) Geographic views on regional planning and development of Bosnia and Herzegovina. In: Mironowicz I (ed) Proceedings 17th international conference on urban planning, regional development in the Information Society GeoMultimedia. AESOP, Schwechat, pp 315–319

Nurković S, Mirić R, Drešković N, Jahić H (2005) Regionalni aspekt turističkih potencijala Kantona Sarajevo—klasifikacija i valorizacija [Regional aspect of tourist potentials of Sarajevo Canton—classification and valorization]. In: Nurković R (ed) Zbornik radova Turizam kao faktor regionalnog razvoja. University of Tuzla Faculty of Science Department of Geography, Tuzla, pp 31–45

Poverty Reduction and Economic Management Unit Europe and Central Asia Region (2005). Available at http://web.worldbank.org/archive/website01012/WEB/IMAGES/BHCEMLOC.PDF. Accessed on 29 June 2019

Sarajevo Airport Archive. Available at https://www.sarajevo-airport.ba/Page/HIstorijski-razvoj-aerodroma. Accessed on 29 June 2021

Sarajevo Canton in Figures (2020) Federal Statistical Bureau of BiH. Available at https://fzs.ba/wp-content/uploads/2021/02/KS_2020.pdf.pdf. Accessed on 02 Apr 2019

Šećibović R et al (2006) Uvod u geografiju turizma sa osnovama prostornog planiranja [Introduction to the geography of tourism with the basics of spatial planning]. Publishing Center of the Faculty of Economics in Belgrade, Beograd

Sinanagić M (2004) Lohn poslovi-tranzicijska ili dugoročna strategija poslovne saradnje s inostranstvom [Lohn jobs-transitional or long-term strategy of business cooperation with foreign countries], Ekonomska revija, vol 4. University of Tuzla, Faculty of Economics, Tuzla, pp 145–151

Statistical Yearbook of Bosnia and Herzegovina (1971) Institute for Statistics of RBiH. Sarajevo, p 101

Statistical Yearbook of Bosnia and Herzegovina (1991) Institute for Statistics of RBiH. Sarajevo, p 103

Strategic Plan for Rural Development of Bosnia and Herzegovina (2018–2021) Ministry of Foreign Trade and Economic Relations. Available at http://www.mvteo.gov.ba/data/Home/Dokumenti/Poljoprivreda/BOS_SPRR_2018-2021_-_korekcija.pdf. Accessed on 05 Apr 2021

The World Factbook CIA (2018). Available at https://www.cia.gov/the-world-factbook/countries/bosnia-and-herzegovina/. Accessed on 25 Apr 2019

USAID—Energy Sector Assistance Project (2019). Available at https://usaideia.ba/wp-content/uploads/2016/05/Izvjes%CC%8Ctaj-o-trenutnom-stanju-i-potencijalu-u-BiH-za-izgradnju-kogeneracijskih-postrojenja-i-elektrana-na-biomasu-2.pdf. Accessed on 24 May 2020

World Health Organisation (2019). Available at https://gateway.euro.who.int/en/country-profiles/bosnia-and-herzegovina/#cah_1. Accessed on 15 Sept 2020

Part V

Geographic Regions of Bosnia and Herzegovina

15 Geographic Regionalization and Regions of Bosnia and Herzegovina

Abstract

The aim of this chapter is to present the issues of geographical regionalization and the modern regional geographical structure of Bosnia and Herzegovina. Research on the phenomenon of regional development and the very concept of regionalization of Bosnia and Herzegovina represents the first-class geographical interest, with the concept of regionalization as its theoretical and methodological procedure. The priority goal of Bosnia and Herzegovina on which a broad consensus has been reached is the process of integration into the European Union, and one of the conditions is the necessity of defining the statistical regionalization of the state, which creates preconditions for fulfilling obligations in this aspect. The chapter is conceived in two parts. The first part provides a methodological framework for the general concept of regionalization and a framework for conducting regional policy within the recent period frame from the aspect of their applicability in terms of general socioeconomic development. The second part serves the function of review and explanation of the current concepts of separation of regional units and a detailed presentation of the modern regional geographical structure of Bosnia and Herzegovina.

Keywords

Geographical regionalization · Regional policy · Geographical regions · Regional center · Municipalities

15.1 Geographical Regionalization and Regional Policy

Regionalization is one of the fundamental concepts of regional development. Its scientific and practical value must not be questioned. What is often controversial and insufficiently clear are the principles and criteria of regionalization that aren't unique, unchangeable, and generally accepted. These are the main reasons why the problems of regionalization are neither practically nor theoretically simple. Bosnia and Herzegovina is not unique in this regard, even though some of the country's specifics (historical, political, geo-transport, ethnic, etc.) give special weight to this problem. The principles and criteria of regionalization are sensitive to the socioeconomic system, space, and time (Osmanković 2004).

The area of Bosnia and Herzegovina has been the subject of the study of geographical sciences for a long time, making it one of the main interests for regional geography as well. During this lengthy period, a vast number of geographers, as well as others, contributed to the elucidation of the problem of regionalization of the area of Bosnia and Herzegovina, through attempts and concrete proposals for spatial differentiation. Reviewing these works, we encountered a large and wide range of approaches, principles, and methods that have been used in various attempts at regionalization. Through the analysis of the development of the political-territorial and administrative organization of the area of Bosnia and Herzegovina, the correlation between the formation of administrative units of different rank and the physiognomy of the space is noticeable. The fundamental construction of the political-territorial system in each period of development was based on the physiognomic understanding of space, no matter how much scientifically based knowledge about its differentiation was limited at a given historical moment. According to the basic, general principles of physiognomic differentiation, the foundations of the political-territorial division, which had its genesis in temporal and chorological development, were realistically stimulated. The correlation between the formation of administrative units of different rank and physiognomy of space is generally established in the primary phase of development of the Bosnian state, through all epochs of its historical-geographical development as an indication form of interactions of society at different levels of socioeconomic development and natural-

H. Gekić et al., *The Geography of Bosnia and Herzegovina*, World Regional Geography Book Series,
https://doi.org/10.1007/978-3-030-98523-3_15

geographical conditions in which it is organized through government and management mechanisms (Mirić 2011).

Speaking of complex geographical regionalization (as a procedure of differentiation of territories, i.e. space) according to established criteria, methods and principles that contain scientific elements of cognition, it is almost inevitable to incorporate elements related to political-administrative division (because it often conditions socioeconomic processes which interact in a particular space). Therefore, it is necessary to follow different approaches and methodological aspects of regionalization in order to determine the processes that interact to form the physiognomy of the same, giving an idea of real and cognitive reality, in its identity and diversity. Historically and geographically, the first attempts at geographical regionalization of the Balkan Peninsula, and within it Bosnia and Herzegovina, date back to the second half of the nineteenth (such as the attempts of NN, Kljajić etc.) and the first half of the twentieth century (such as the attempts of Cvijić, Krebs, Marek, Chataigneau-J. Sion etc.). All of them are basically characterized by simplicity, lack of theoretical-methodological construction, and vaguely articulated regional concept. The relatively weak theoretical basis, expressed through poor knowledge of regional-geographical features of a given area, and very narrow methodological basis of regionalization due to the use of only some physical-geographical or anthropogeographic "criteria," characterize, in our opinion, these attempts generally descriptive, but of historical-geographical value for research on the issue of regionalization of the Balkan Peninsula and Bosnia and Herzegovina (Nurković and Mirić 2005).

The first separate regionalization of the territory of Bosnia and Herzegovina after the Second World War was done by Kanaet (1954). He believed that geographical areas should not be viewed exclusively as natural geographical units, because they reflect socioeconomic and cultural-historical influences. Ilešič's geographical regionalization of Yugoslavia is particularly significant due to the pioneering use of dual regionalization, at least in the case of Bosnia and Herzegovina. According to this regionalization, in terms of physiognomy, three macro-regional zones stand out in this area—Pannonian, Dinaric, and Primorska. On the other hand, the author singled out 5 macro-regions (Sarajevo-Zenica macro-region, Northeast Bosnia, North-West Bosnia, Mostar macro-region, and the Group of West Bosnian mesoregions), which were then divided into 15 mesoregions (Ilešič 1961).

Under the influence of Petrović and Ilešič, most geographers adopted the regional differentiation of the Pannonian, Dinaric, and coastal parts of Bosnia and Herzegovina, with certain modifications. This group includes, among others, Marković, who in his works in 1967 and 1980 was the first to develop a three-level regionalization of Bosnia and Herzegovina, according to which the mesoregions Peripanonian Bosnia, Planinsko-Kotlinska (Mountain-Basin) Bosnia and Herzegovina, and the Middle Coastal Region fit as part of wider geographical areas, while they are simultaneously divided into nine regions or micro-regions (Marković 1980). The prewar evolution of the geographical (physiognomic) regions of Bosnia and Herzegovina is completed by Smlatić (1983) and Rogić (1990), who established the differentiation of Bosnia and Herzegovina into four macro-regions—Northern Bosnia, Central Bosnia, High Karst Region, and Low Herzegovina. After the last war, this approach was accepted with certain corrections by Nurković and Mirić in 1998 and 2005, and it gained popularity through the educational system of teaching geography in Bosnia and Herzegovina (Mirić et al. 2017).

Effective regional policy requires optimal management of resources and use of development potentials, with the aim of improving the overall economic and social situation and providing additional support to developing territorial units, so that all areas can contribute to a sustainable and competitive economy. The characteristics of regional policy in Bosnia and Herzegovina are primarily determined by inherited problems in the very concept of regional development, which are manifested through a deeply emphasized non-harmonized socioeconomic plan and strategy, and are a consequence of the political system of the country. Some of them are failure to adopt the concept of creating NUTS regions in Bosnia and Herzegovina; political resistance to the adoption and functioning of the European Commission's proposal to establish a nomenclature of territorial units for statistics; separate (entity) program documents in terms of drafting documents related to spatial and regional planning (development strategies, spatial plans); lack of the ministry for spatial planning at the state level; inconsistent concepts of regional policy that are created separately at the entity level; inconsistent criteria at the entity level that are necessary for rational planning of regional development policy (e.g., determining the level of development/underdevelopment of local self-governing units); very complex and politically conditioned coordination mechanisms for the application of positive European practices in the field of regional policy; non-fulfilment of obligations under the Stabilization and Association Agreement regarding the submission of statistical data to EUROSTAT (Law on Statistics of Bosnia and Herzegovina is not harmonized with the principles of the European Statistics Code of Practice), whereby the structure of the statistical system (entity level) makes it difficult to coordinate and efficiently produce statistical data; and the statistics provided by Bosnia and Herzegovina to EUROSTAT is only to a certain extent timely, relevant, and reliable.

The fact that Bosnia and Herzegovina is one of the few countries that does not have a spatial plan at the state level contributes to this, which makes it extremely difficult to plan and implement regional development policy and establish integrated planning, programming, management, and financing of regional development. The methodology for NUTS statistics must be met, because no country has become a member of the European Union without first performing the nomenclature of statistical regional units made in accordance with the prescribed standards that are clearly stated by EUROSTAT. In terms of achieving program objectives and aligning with EU regional policy, Bosnia and Herzegovina still does not have a system of financial management, control and revision for EU funds, except for cross-border cooperation programs.

15.2 Geographic Regions

Based on the principles of complex geographical regionalization, Nurković and Mirić (1998) differentiate Bosnia and Herzegovina into 4 macro-regions in the first, and 11 mesoregions in the second degree of regionalization, which for the first time points to one complex geographical regionalization of Bosnia and Herzegovina, which in hierarchical and qualitative-quantitative terms is not related to the area of the former state. In the theoretical-methodological sense, Rogić's drafts of physiognomic regionalization and regionalization of the Central Mountain Area of the former SFR Yugoslavia served as the basis for this regionalization. The same authors gave in 2005 in the paper "Review of the geographical regionalization of Bosnia and Herzegovina" a proposal for a new regional division of our country:

1. *Peripanonian Bosnia* with mesoregions:
   1a Una-Sana mesoregion
   1b Posavsko-donjovrbaska-donjobosanska mesoregion
   1c Posavsko-sprečansko-majevičko-donjodrinska mesoregion
2. *Planinsko-Kotlinska (Mountain-Valley) Bosnia* with mesoregions:
   2a Gornjovrbasko-plivska (Upper Vrbas-Pliva) mesoregion
   2b Gornjobosansko-lašvanska (Upper Bosnia-Lasva) mesoregion
   2c Upper Drina Mesoregion
3. *Bosnia and Herzegovina High Karst* with mesoregions.
   3a Mesoregion of karst fields and highlands of southwestern Bosnia
   3b Mountain Herzegovina
4. *Mediterranean Bosnia and Herzegovina Macroregion* with mesoregions.
   4a Mediterranean West Herzegovina Mesoregion
   4b Mediterranean East Herzegovinian Mesoregion
   4c Neretva Mesoregion
   4d Mesoregion of the Coastal Bosnia (Nurković and Mirić 2005) (Fig. 15.1).

15.2.1 Peripannonian Bosnia

The northern parts of the territory of Bosnia and Herzegovina—Peripannonian Bosnia—belong to the southern edge of the Pannonian Basin. From the geomorphological aspect, it is the area of a slightly undulating plain divided by numerous river flows. Peripannonian Bosnia is part of the macro-regional area that includes the edge of the Pannonian Basin in a broader transnational sense. The northern borders of this region are represented along the Sava River, which also forms the state border. This region reaches in the southern direction to the mountains Grmeč, Srnetica, Dimitor, Manjača, Čemernica, Konjuh, and Javor. It covers an area of 22,249 km^2, which is about 43.4% of the territory of Bosnia and Herzegovina.

It has a very favorable geographical location within Bosnia and Herzegovina and within the broader regional context as a transit area between higher hierarchical units in the European geospace—the Pannonian Basin and the Adriatic Sea. The lower and middle reaches of the larger rivers of this region (Vrbas, Una, Bosnia, and Drina) with their low watersheds in the southern direction are naturally predisposed to road communications. In the area of this macro-region, there are three of the four endpoints determining the astronomical and geographical position of Bosnia and Herzegovina: the far north (Kozarska Dubica municipality), the far west (Bihać), and the easternmost (Bratunac municipality) (Fig. 15.2).

The distance in the east-west direction is 309 km and speaks of a pronounced extension of this macro-regions area from the eastern to the western borders of Bosnia and Herzegovina. The morphostructural structure is dominated by lowland and hilly-plain relief and consists of low hills, vast alluvial plains, and valley extensions in the lower reaches of the Vrbas, Una (Fig. 15.3), Bosnia, and Drina whose physiognomic homogeneity is disturbed by the lone lower mountains Kozara, Motajica, Majevica, Prosara, Ozren, and Majdanska mountain, while at the far southern physiognomic boundaries, the altitude, geological structure, and geomorphological structure are gradually modified. The dominance of denudation-tectonic relief is represented by antiforms of mountainous morphostructures with foothills, which form the natural borders of the Peripannonian macro-region towards the southern parts of Bosnia and Herzegovina. Its southern physiognomic borders towards the (Planinsko-kotlinska) Mountain-Valley macro-region is closed by the mountain massifs of Dimitor, Manjača, Čemernica, Vlašić, and Ravan mountains, and in the far eastern part by the slopes of Konjuh and Javor (Drešković and Mirić 2017). In the synforms, the dominant valley

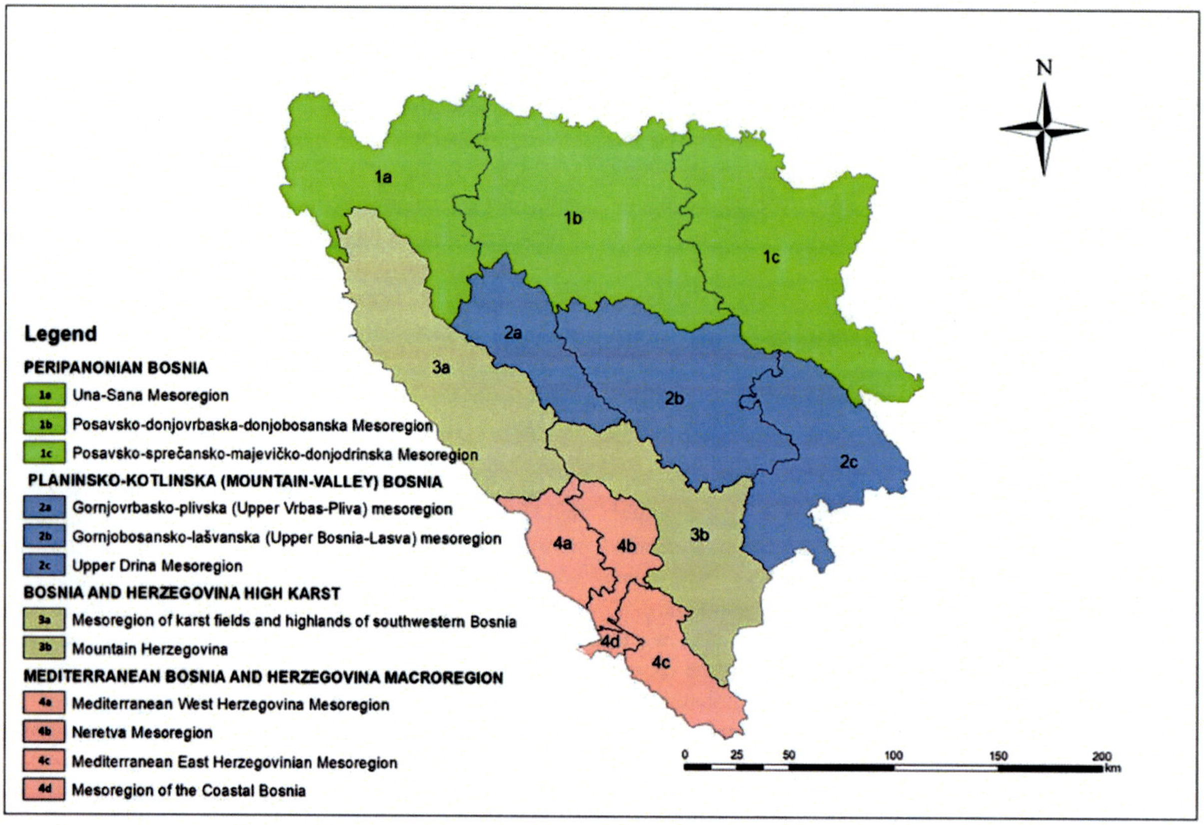

Fig. 15.1 Regions of Bosnia and Hercegovina

extensions predisposed to fluvial accumulation work, such as Prijedor, Banja Luka, Sprečko, Ivanjsko, etc., are also encountered. The lowest values of vertical dissection (0–5 m/km^2) and slope (0 °–5 °) morphologically characterize the subhorizontal surfaces of floodplains and terrace lowlands in the north (Bosanska Posavina, Semberija, Lijevče polje). The mountain morphostructures of the Inner Dinarides (Ozren, Ludmer, Majevica, Kozara, Motajica) and the hills are characterized by higher values of vertical dissection (100–300 m/km^2) and slopes (13 °–32 °).

In terms of geotectonic assembly, the block horst-trench structure formed by Neogene radial tectonics dominates. The neotectonic uplift of the antiforms of this macro-region in the northern parts (the area of the Sava trench) took place in parallel with the neotectonic descent in the valleys (Prijedor, Ljevča, Ivanjska, Sember, and Gornja Sprečanska). It is marked by fault-block mountain elevations—horst-anticlinoriums of Majevica, Trebovac, Kozara, Motajica, and Prosara and relief depressions of the Sava graben, Šamac-Brčko, Tuzla, Ljevča, Knežpolje, and Dubica ditch-synclinoriums. The gradual uplift of the terrain to the south and southeast is characterized by low foothills that are directly neotectonically associated with older fault-block mountain morphostructures with foothills. By neotectonic lowering of the terrain along fault lines, inner mountainous basins and valleys were formed in which layers of Neogene and Quaternary sediments were deposited (valley extensions of Bosnia, Una, Sana, Vrbas, Ukrina, Spreča, Drina, and their tributaries) with predominant fluvial-fluvioudenudation accumulation processes and forms. The most striking structures with a northwest-southeast direction are the Prijedor Basin (area 252 km^2) and the Upper Spreča basin (area 212 km^2) with lengths greater than 30 km.

Spatially smaller basin depressions: Tuzla, Banja Luka, Občko, Dubičko, Knežpoljsko, Prnjavorsko, Kotorsko, Dobojsko, Stanarsko, Blagajsko, Ugljevičko, etc. orographically frame the surrounding mountain elevations and low hills. The northeastern area of the macro-region is orographically expressed by the mountainous elevations of the central ophiolite belt. These are fault-block morphostructures: Ljubić 594 m above sea level, Čavke 549 m above sea level, Ozren 915 m; and wrinkled-onlay morphostructures: Uzlomac 1002 m, Borja 1078 m, Velika Manjača 1358 m, Tajan 1297 m, Velež 1128 m above sea level, Radovan Jelika 1290 m above sea level, Konjuh 1326 m, and Smolina 1274 m above sea level. They represent

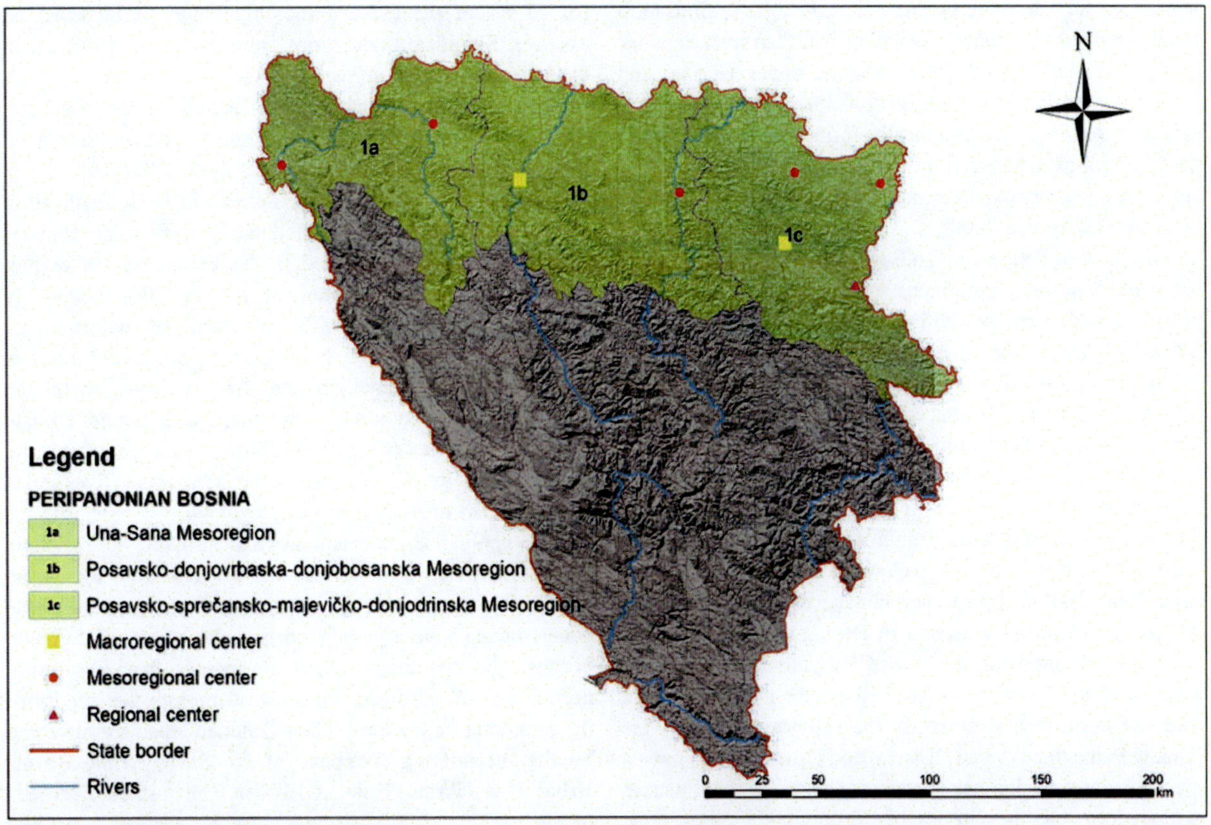

Fig. 15.2 Peripanonian Bosnia macro-region

Fig. 15.3 River Una in Bihać. (Photo by Aida Bidžan-Gekić)

neotectonically raised forest massifs and ridges with foothill steps dissected by a densely developed surface river network of the Krivaja, Rzav, Bosna, Vrbanja, upper Ukrina, and Usora river basins. The complex slope profiles of the central ophiolite zone are polygenetically formed on the heterogeneous lithological substrate of the volcanogenic-sedimentary complex (Jurassic magmatism, clastic, carbonate and silicon sediments) (Lepirica 2009).

The relief of Peripannonian Bosnia highlights micro-relief forms of elevations, depressions, and plateaus of technogenic origin formed along the ore-bearing basins of Kreka, Tuzla, Stanari, Ugljevik, Lopare, etc., embankments along the Sava in Bosanska Posavina, etc. Modern intensive denudation processes and forms, recent sinking of a part of the urban zone of Tuzla, and other processes and phenomena of mostly destructive character have been initiated by anthropogenic-technogenic interventions.

The increase in altitude from north to south and the wide openness to the Pannonian area have caused specific climatic conditions. The most important climate modifier is the orographic structure of the terrain of Bosnia and Herzegovina. As already pointed out, the area of Peripannonian Bosnia is wide open to the influence of cold continental air coming from the north and northeast. In the valleys of larger rivers (Una with the Sana, Vrbas, Bosnia, and Drina), which have a predominantly meridional direction of flow, these influences extend deep into the interior of the mountain-valley area. More specifically, the temperate-continental type of climate prevails in these areas, with quite pronounced thermal contrasts: harsh winters and warm summers (Drešković and Mirić 2017).

Areas of the Peripannonian macro-region along the Sava River with altitudes below 500 m have annual isotherms ranging from 10 °C to 15 °C, while in the higher parts to the south (including the Kozara massif), the annual isotherm ranges from 5 °C to 10 °C. Average annual temperature fluctuations up to hypsometric levels of about 1500 m above the sea level are above 20 °C, i.e., continental climatic influences can be determined as dominant in this region. The thermal regime at the level of average annual temperatures shows zonal characteristics: The warmest areas are in the northeast, from where the isotherm values decrease towards the mountain-valley zone of Bosnia and Herzegovina. Annual precipitation levels range from 700 mm to 1200 mm. The structure of precipitation is dominated by rain, with snow being a characteristic of all winter months. In this region, near the settlement of Domaljevac (area of the immediate Sava basin), the value of the lowest annual amount of precipitation in Bosnia and Herzegovina (706 mm) was registered. The average monthly precipitation is 90.0 mm, and the maximum precipitation is in the period May to July. The main precipitation minimum falls in the period February to March, during which about 69.0 mm (or 77.0% of the average monthly value) is recorded on average. Significant deviations from these indicators are in the higher areas that converge to the south and north of this region. The driest areas are connected to the northeast (Semberija area), while in the zone of the northernmost parts of the region, it reaches values up to 1200 mm.

According to the criteria of Köppen's (1936) classification of climate, climate class C (moderately warm rainy climate) is predominantly represented in the Peripanonnian region. The most common climate subtype is Cfb—moderately warm and humid climate with warm summers, which covers almost the entire area of this region. Climatic monotony is disturbed by the area in the far northeast, with the characteristics of a moderately warm and humid climate with hot summers (Cfa). This climatic subtype covers the lowland area east of the Tolisa River to the Janja River and is the driest part of Bosnia and Herzegovina.

The geological-geomorphological structure and climate have conditioned a very dense surface river network and an abundance of streams. In addition to hydrographic, these watercourses have a very important climatological function because through their valleys air masses from the north, northwest, and northeast are drawn deep into the interior of the mountain-valley area. Their flattened areas are dissected by the meandering riverbeds of the Sava, Drina, Bosna, Vrbas, Una, Ukrina, Tinja, and Tolisa. The unstable riverbeds of the mentioned plain watercourses are shallowly cut into younger Quaternary deposits. The riverbed of the Sava is the widest, averaging 250–450 m. Then the width of the lower course of the Drina follows (on average 170–300 m) then the Una, Bosna, and Vrbas, while the riverbeds of other smaller rivers are narrower. The largest river meanders are shaped by the lateral erosion of the Sava, the water-richest stream in BiH, whose average annual flows downstream from the mouth of Bosnia exceed 1000 m^3/sec. The average slope of the longitudinal profile of this large plain watercourse on the stretch from Jasenovac to Bosanska Rača is only 0.03‰. The meanders of the Sava are in some places over 10 km long, like the one in the mature phase of development upstream from Orašje with a meandering neck narrower than 400 m. Larger artificial reservoirs are characteristic for this area (Modračko Lake, Saničani fish pond, Bardača, Prnjavor fish ponds, Snježnica and Ugljevik Reservoirs, etc.). Along numerous fault and dislocation zones, thermal and mineral springs are very common (Gata near Bihać, Slatina, Kulaši near Prnjavor, Teslić, Tešanj, Guber near Srebrenica, Kiseljak near Tuzla, etc.).

In the area of Peripannonian Bosnia, the soils from the automorphic division are the most represented. It is dominated by district cambisol on acid silicate rocks, which covers most of the central area of the macro-region, and acrylic luvisol on limestone. Pseudogley is also very widespread, covering larger areas in the immediate Sava River

Basin (in the northeastern part from Bosnia to the Janja River), then in the Prijedor valley, and a larger area between the lower reaches of the Vrbas, Ukrina, and Bosnia. The southern border areas are dominated by calcomelanosols and chalcocambisols with luvisols, and luvisol on silicate rocks. In some places along the rivers Una, Vrabas, Ukrina, Bosna, Drina, and Spreča, the fluvial soils like fluvisols and semiglays are characteristic. The southern parts are also characteristic for the appearance of limited areas of cambisol and luvisol on serpentinite and peridotite.

The general biogeographical characteristics of this area are defined by the presence of chestnut (*Castanetum sativae*) forest communities in the northern part. The river flows are dominated by hygrophilous communities of willow and poplar (*Salicetumet Populetum*) and alder, in combination with the biome of moist forests of pedunculate oak and field ash. To the south, with the increase of altitudes, the vegetation consists of communities of mesophilic forests of sessile oak and hornbeam (*Querco-Carpinetum*). Communities of oak, white hornbeam, and cer appear in limited areas on the slopes of the hilly relief (the larger area is the far eastern part of the macro-region along the upper and middle course of the Drina). At higher altitudes, there is a phenomenon of mostly thermophilic community of beech forests, especially in the farsouth-eastern parts of the macro-region.

From the aspect of general economic development, the Peripannonian region has about 70% of arable land, with the dominance of growing cereals and industrial plants, fruits, vegetables, etc. Dominantly large potentials are located in the area of Lijevče field, Posavina, Semberija, Sprečko, and Prijedor fields. The primary vegetation of deciduous forests has been replaced in favor of creating agricultural arable land. The ore-mineral base is based on the exploitation of very diverse resources: from coal, bauxite, iron ore to salt, asbestos, marl, gypsum, kaolin, and a wide range of building stone (bihacite, marble, quartz sand, etc.).

According to the 2013 census, there were 1,991,723 inhabitants (or 56.4% of the total population of Bosnia and Herzegovina) within this macro-region, with an average population density of 89.9 inhabitants per km^2 (Bosnia and Herzegovina—68.9 inhabitants/km^2), and according to the above data represents the area of demographic focus within the national territory (BHAS 2016). The area of this mesoregion consists of 65 administrative units of local self-government, with different status (municipalities, cities). Within this macro-region is the Brčko District, as a separate administrative unit in Bosnia and Herzegovina. This macro-region is divided into three mesoregional units.

15.2.1.1 The Una-Sana Mesoregion

The Una-Sana Mesoregion is a northwestern, physiognomically homogeneous area within the middle and lower reaches of the Una River and its largest tributary, the Sana. Traditionally, this area belongs to the historical and geographical unit called "Krajina." It covers an area of 6107 km^2, which is about 12% of the territory of Bosnia and Herzegovina. According to the 2013 census, there were 420,377 inhabitants (or 12% of the total population of Bosnia and Herzegovina) within this mesoregion, with an average population density of 68.8 inhabitants per km^2, which is almost identical to the values for the state territory, but significantly lower from the indicators for the area of the Peripannonian macro-region (BHAS 2016).

In terms of landscape, the morphological structure includes hills and karst fields, alluvial plains and valleys, river valleys, and low mountains. A slight descent of the terrain from the south to the north is characteristic. The terrain is divided into several morphological units, namely, Una-Korana plateau, Majdanska mountain, Sana valley, Prijedor valley, Kozara, Lušci (Jelašinovačko) field, Gomjenica, Knešpolje, and in the far southwest the slopes of the Plješevica mountain. The left valley side of the Una is dominated by low hills intersected by smaller streams of geologically relatively younger origin. The vast Bihać valley (Bišće polje) is located between the slopes of Plješevica (in the west) and Grmeč (in the east) and represents the area of one of the two administrative and urban regional centers of this mesoregion—Bihać. The border of this mesoregion in the southeast is represented by the slopes of Grmeč, Srnetica, and Bobija, while in the far south there is the highest point—Gola Kosa (1651 m above sea level).

The area west of the Una valley is dominated by lowland-hilly relief, with relatively short streams belonging to the Glina and Korana basins (Kladušnica, Glinica, Mutnica, and Toplica). The relief of the Glina and Korana basins is dominated by lowland-hilly relief with spatially differentiated smaller mountain areas that have morphological features of the mid-mountain relief. Accordingly, it can be stated that the source zones of these watercourses are positioned at lower altitudes, in the zone from 500 to about 700 m above sea level. More specifically, within the catchment area, there are a number of peaks with altitudes above 500 m, such as Radač (626 m), Pećka brda (520 m), Skokovi (509 m), Veliki Bisovac (499 m), and others. Due to the predominant low mountain relief in this area, numerous settlements have been formed, mainly in the zones of morphological depressions and alluvial plateaus of watercourses. The highest morphostructure that extends in a north-westerly direction along the left side of the Una River canyon is the peak of Gomila (797 m). In the southwestern part, the terrain descends towards the rivers Una and Bihać (231 m), and on the north eastern part towards Bosanska Krupa (178 m) (Fig. 15.4) and Sanski Most (182 m) (Una-Sana Canton Environmental Protection Plan 2014).

The Una river springs in Lika and is 212 km long, and its lower course represents the border of Bosnia and

Fig. 15.4 Sojenica (Stilt house) on the river Una in Bosanska Krupa. (Photo by Aida Bidžan-Gekić)

Herzegovina with Croatia. According to its physiognomy, it belongs to the edge of the Pannonian plain, and the elbow turn to the northeast near Krupa indicates the original confluence with the Pannonian Sea, whose outflow extended the course further north. The current estuary is located near Jasenovac, and Knešpolje represents its macro floodplain. The Prijedor valley represents the largest tectonic extension, which was additionally shaped by fluvial erosion of Sana and Gomjenica, and due to the wetlands of these terrains, fishponds were built here. Southwest and south of this valley, over the low Majdanska Mountain, it passes into the area of the Una and Sana valleys, and the northern frame is represented by the slopes of Kozara, and in the far north Motajica with the river Sava.

The mesoregional center of the northern part of the Una-Sana area is the city of *Prijedor*, in which, according to the 2013 census, 89,397 inhabitants lived on an area of 834.06 km^2 (about 21.3% of the total population of this mesoregion). It is the fourth largest city in the Peripannonian macro-region. It is characterized by a low degree of urbanization (36.2%), and an unfavorable age structure in which the share of the old (16.1%) is higher than the contingent of the young population (12.7%) in the total population (BHAS 2016). Prijedor is traditionally an industrial environment that has based its economic development on the exploitation of natural resources (iron ore, gypsum, limestone, clay, quartz sand, and wood). It represents the center of the mining and processing area which gravitates to the northern part of the mesoregion. The modern economic development of Prijedor is significantly determined by three industrial branches: mining, agro-food industry, and tourism. On the slopes of the Majdan Mountains, there are surface iron ore mines (Ljubija and Omarska). These mines are part of the multinational company "Arcelor Mittal," one of the largest exporters in Bosnia and Herzegovina and represent the backbone of the industrial development of the mesoregion. The Faculty of Mining is located in the city itself and represents a scientific unit within the University of Banja Luka. The agro-processing industry is based on agricultural production, and in addition to farming (corn) and livestock production (milk), the carriers of economic development are food factories: biscuits and waffles factory "Mira" and fruit processing factory "Prijedorčanka." Prijedor has a regional hospital, a professional theater, a museum, and other cultural and educational institutions, as well as a sports airport (Urije). The tourist potential is primarily focused on the valorization of the Kozara Mountain and the Sana River.

The municipal areas that gravitate to it are *Novi Grad* (27,115 inhabitants), *Kozarska Dubica* (21,524 inhabitants), and *Kostajnica* (5997 inhabitants) (BHAS 2016). The morphological structure of the western and southern part of this mesoregion is represented by hilly-plain relief, alluvial plains, and basins that are divided by numerous river valleys. The terrain around Sanski Most and Ključ is part of the Sana-Una Paleozoic area, where in the far south (upper Sana) the relief passes to higher altitudes of mountain massifs on the border with the mountain-valley macro-region (Bobija, Dimitor, Grmeč in the northwest). Discoveries of deposits of brown coal and quartz sand, bauxite in Bosanska Krupa, manganese in Bužim, barite in Velika Kladuša, clay in Cazin, and bihacite in Bihać are significant. Of the metal resources, manganese (Bužim and Velika Kladuša) is significant. South of Bihać (area of the settlement Kulen Vakuf) gypsum reserves (about seven million tons) have been determined. The hydropower potentials of the Una and the upper course of the Sana and the thermo-mineral water in the settlement of Gata (near Bihać) are significant. In this part of the mesoregion, the area of the upper course of the Una was declared in 2008 the first National Park in BiH. The total area of NP Una is 19,800 ha, of which 13,500 ha is a zone of strict and targeted protection.

The municipality of *Sanski Most* with an area of 781 km^2 is one of the largest local government units in the Una-Sana Mesoregion, which according to the 2013 census had 41,475 inhabitants (53.1 inhabitants per km^2). A rarity is the existence of nine rivers that are located in the area of this municipality. The biggest potential is the Kamengrad brown coal mine (estimated reserves are around 300,000 tons) and the tourist valorization of the Sana River. South of Sanski most is the municipality of *Ključ* (16,744 inhabitants, 46.8 inhabitants per km^2) located at the mouth of the Sanica River. It is characterized by a low degree of urbanization (29.3%) and a negative natural increase (2017—4.4 per thousand).

Bosanska Krupa (25,545 inhabitants, 45.3 inhabitants per km^2) has a very negative natural increase (−7, 9 per thousand in 2015). Smaller municipalities with a very low population density are *Oštra Luka* (2786 inhabitants, 13.6 inhabitants per km^2), *Krupa na Uni* (1597 inhabitants 18.9 inhabitants per km^2), and *Ribnik* (6048 inhabitants, 11.8 inhabitants per km^2). In morphological sense, they represent a set of rural settlements without an urban center.

The mesoregional center of this area is the city of *Bihać*, located on the river Una. The area is 900 km^2 (which is 212 km^2 more than the prewar administrative borders). The area of the city makes up about 15% of the mesoregion, but the share in the population is much lower (about 13%). Bihać is one of the few mesoregional centers that has a smaller population than the municipalities that make up the mesoregion (Cazin has a larger population). According to the 2013 census, there were 56,261 inhabitants in the Bihać municipality (population density was 62.5 inhabitants per km^2). About 64% of the population inhabited the urban area. The City of Bihać has the largest gross domestic product per capita in the Una-Sana Mesoregion (7239 BAM in 2013) (BHAS 2016). The main reason is that the bodies of public institutions (Assembly and Government of Una-Sana Canton with line ministries, Cantonal Directorate for Roads, Public University, etc.) as well as institutions and companies of the service sector, the banks and BH Telecom, Elektro-distribution company, PTT, Railways, etc.) Today, Bihać is home to one of the largest breweries in Bosnia and Herzegovina, which produces two brewery brands, Preminger and Unski biser. Bihać dairy is owned by the German company Meggle. In addition to them, the carriers of economic development are the companies RIZ Krajina, Bor, Sani global, HUT Aduna, Opal group, Čavkunović BP, Čavkunović Trade, and Elumatec Maschinen. The development of the city is largely burdened by the failed privatization of large industrial plants (Krajinametal, Kombiteks, Polietilenka, etc.) and the huge influx of foreign migrants in recent years. The development opportunity is primarily in tourism and airport construction, and attracting foreign investment (the most important foreign investors are Meggle Dairy-Germany, Concern Agram-Croatia, and Austrotherm from Austria).

Of the other local self-government units, the town of *Cazin* stands out (66,149 inhabitants - after Prijedor, the most populous local government unit in the mesoregion). With the highest population density in the mesoregion (185.8 inhabitants per km^2, which is almost three times higher than the mesoregional average in 2013) and a very dynamic economy (over 3000 registered agricultural holdings, over 2400 registered small and medium enterprises), despite negative migration trends (in the period 2011–2015, a negative migration balance was recorded, which averaged −64 inhabitants). The municipality of *Velika Kladuša* (40,419 inhabitants, 122.1 inhabitants per km^2) covers the far north western part of the mesoregion and Bosnia and Herzegovina. The neighboring municipality of *Bužim*, with 19,340 inhabitants (2013), has the largest share of young people (population from 0 to 14 years) 21.8%, and the lowest share of old people (7.2%—population aged 65 and over) in the total population of all local communities in Bosnia and Herzegovina. The average age is also the lowest in the country (32.26 years) despite the high negative migration balance.

15.2.1.2 The Posavina-Lower Vrbas-Lower Bosnia Mesoregion

Posavsko-donjovrbaska-donjobosanska Mesoregion (The Posavina-Lower Vrbas-Lower Bosnia mesoregion) includes valley extensions of the lower reaches and mouths of the great rivers of the Black Sea basin, Bosnia and Vrbas, and their tributaries (Vrbanja, Usora, Spreča). The rivers from the immediate Sava basin located on low hills and alluvial lowlands, such as Jablanica, Jurkovica, and Ukrina also belong to this unit. The entire territory in the north overlooks the Sava River, while its southern physiognomic borders towards the Mountain-Valley macro-region are closed by the mountain massifs of Manjača, Čemernica, Ravan Mountains, and in the far eastern part the slopes of Konjuh. The eastern physiognomic border of this mesoregion was drawn over the low watershed of the rivers Krivaja and Spreča, and the mountain massifs of Ozren and Trebovac and the western rim of Šamačko field.

Traditionally, this area is part of a historical-geographical unit called "Krajina." It covers an area of 8565 km^2, which is about 16.7% of the territory of Bosnia and Herzegovina. According to the 2013 census, this mesoregion had 679,362 inhabitants (or 19.2% of the total population of Bosnia and Herzegovina), with an average population density of 79.3 inhabitants per km^2, which is significantly above the average for the state territory, but also lower than the indicators for the Peripannonian macro-region (89.9 inhabitants per km^2). In the relief structure, the most noticeable structures are the stepped fluvial-denudation surfaces dissected by a network of valleys. It is characterized by a permanent increase in altitude from the far north to the southern parts of the terrain, with the area of this mesoregion gaining an increasingly physiognomic character of the highlands, making a gradual transition to the mountain-valley macro-region. The terrain is divided into several morphological units.

The western part of the mesoregion consists of the Banja Luka valley and low mountains that frame it from the south, east, and west. This zone is intersected by the Vrbas canyon, whose composite valley has several shorter canyon parts and smaller extensions (Bočac, Krupa, Karanovac) (Fig. 15.5), after which it enters the Banja Luka valley. The average width of the valley is about 5 km, and the altitude is 163 m. In the northeast, there are the low hills of Crnivrh, from where the area descends into Lijevče field. Southwest is Manjača (1289 m above sea level), and the vast surface of Zmijanje. To the west, the terrain slopes slightly towards the Sana Valley and the Maidan Mountains. From the Banja Luka valley, the Vrbas has the characteristics of a typical lowland river, which built the macro-floodplain Lijevče field. The rest of the earlier estuary is today the Ramsar area of Bardača.

Banja Luka is a macro-regional center, and after Sarajevo, the largest city in Bosnia and Herzegovina. In 2013, it had a population of 185,042 inhabitants, with an average density of 149.36 inhabitants per km^2. With its area (1238.9 km^2), it is the largest local self-government unit in the country. The urban area covers 179 km^2, inhabited by as much as 87.7% of the total population of the city. According to the 2013 census, a positive rate of natural increase was recorded (birth rate 11.2, mortality 10.1 per thousand). Banja Luka recorded a decrease in the total population of 7.9% compared to 1991. In the age structure, the number of inhabitants over the age of 65 increased in 2013 by 53.3%, compared to 1991, the number of inhabitants under the age of 14 decreased by 35.8%. In the period from 2012 to 2016, a positive balance of population migration was observed, which was the highest in 2014 (925) and the lowest in 2016, when the lowest difference between immigrants and emigrants was recorded (BHAS 2016). The city also has the headquarters of various companies that, in addition to the local, also operate in the regional and international market, such as: Mtel Banja Luka, LANACO information technology, Molson Coors d.o.o. Banja Luka, Nestro petrol d.o.o. Banja Luka, BEMA d.o.o. Banja Luka, ELAS Metalexpertd.o.o. Banja Luka, SHP Celexa.d. Banja Luka, Drvoprodexd.o.o. Banja Luka, and others. In addition to the institutions of the entity (government center, national assembly, judiciary, etc.), and state level (Indirect Taxation Authority of BiH, Concessions Commission and Deposit Insurance Agency of BiH, etc.), the city is the home to financial institutions (Banja Luka Stock Exchange, Central Bank of BiH-Banja Luka branch). In economic terms, the largest number of employees (19.4%) and the largest revenues are generated by the trade sector (almost 2/5 of the total revenues of the City in 2016). The most important companies are Molson Coors d.o.o. Banja Luka, Nestro petrol d.o.o. Banja Luka, Elnos BL, DE-MI

Fig. 15.5 Bočac—River Vrbas. (Photo by Ranko Mirić)

promet, Centrum trade, Tropic, Dukat Banja Luka, etc. The manufacturing sector (which employs approximately the same number of workers) generates only 1/5 of total revenues. There is a noticeable permanent increase in the share of the information and communication technology sector (2016, about 10% of revenues, 8.5% of total employees) (Development Strategy of the City of Banja Luka 2018).

With 10,217 BAM of gross domestic product per capita (2014/15), Banja Luka ranks first in the Peripannonian macro-region. A special feature in relation to other regional centers is the institutions of education, culture, and health care, so that the gravitational influence of Banja Luka exceeds the borders of the region. The University of Banja Luka is the second largest higher education institution in Bosnia and Herzegovina, with 17 faculties. The University Clinical Centre is the largest and best-equipped health institution in this part of the macro-region, and in addition to it, the primary health care is provided by the health center in 35 facilities.

In terms of cultural content, the most important are: the city center with the Kastel Fortress, the Ferhadija Mosque, and museums (Museum of Contemporary Art, Museum of Republika Srpska) and numerous institutions (Archive, National and University Library, National and Children's Theatre) and events (Theatre Fest, Dukat fest, Kočić'ss Choir, Summer on the Vrbas, etc.). Attractive tourist sites are the monasteries Gomionica, Liplje, Stuplje, the towns Zvečaj Grad and Greben Grad, spas Laktaši, Slatina, and facilities in the valley of the Vrbas (hunting, fishing, sports—rafting, kayaking). In 2016, Banja Luka was visited by 82,171 tourists, who spent 130,790 nights. The location of the international airport is a very important infrastructural element in the economic development of Banja Luka.

Gradiška is the fourth largest city in this mesoregion (761.65 km^2) and the third most populous (51,727 in 2013). It has an extremely favorable traffic position (multinodality) and is located on the highway from Doboj to the state border. The most important resources are agricultural land Lijevče field (Agroimpex is one of the leading companies in Bosnia and Herzegovina in the production of apples and pears). Other municipal centers that gravitate to Banja Luka are in population, urban and functionally sense relatively rounded units. *Prnjavor* is a municipality in the northeastern part of the mesoregion, covering an area of 630 km^2 (57 inhabitants/km^2) with 35,956 inhabitants (it is recognizable by the name "Little Europe"—12 different nations).

Laktaši (35,000 inhabitants, 388.4 km^2) is a municipality north of Banja Luka. It is the headquarters of one of the largest construction companies (Integral Engineering). Srbac (17,587 inhabitants), *Čelinac* (15,548 inhabitants), *Kneževo* (9793 inhabitants), and *Kotor Varoš* (19,710 inhabitants) belong to the population of smaller municipalities with a population density below 45 inhabitants per km^2, and the degree of urbanization below 40%. *Dobretići* is the smallest municipality in this part of the mesoregion (59 km^2) and the smallest population (1629, or only 27.61 per km^2) without an urban center. The western part of the mesoregion consists of the area of the river Ukrina and the low hills of Krnjin (265), Ljubić (594) Vučjak (367), and Trebava (692), and slightly higher areas of the mountain border massifs of Ozren (918) and Borja (1007 m). In the relief of lowland terrains, the most pronounced synforms are the Doboj valley and Posavina, within which the Ivanjić field stands out. After the Doboj valley, Bosnia became a real lowland river.

The city of *Doboj* is a mesoregional center (Fig. 15.6). It was created at the intersection of important traffic routes, the one that goes through the valley of Bosnia and the cross-route Banja Luka-Tuzla (valleys of Usora and Spreča). It is

Fig. 15.6 City of Doboj. (Photo by Ranko Mirić)

characterized by a low degree of urbanization (36.2%) and an unfavorable age structure. The share of the old in relation to the contingent of the young population in the total population is higher by 3%. It has taken on a newer significance as the main railway junction of Bosnia and Herzegovina, where the railway line "45. Parallel" (Novi Grad-Banja Luka-Tuzla-Zvornik) intersects with the main route on the planned part of the corridor Vc. This importance has increased due to the transport of iron ore from Prijedor, coke from Lukavac, and limestone (quarry Ševarlije near Doboj) for the Zenica ironworks. With the completion of the highway from Banja Luka and the continuation through Modriča to Bijeljina, and corridor 5c (Ploče-Sarajevo-Doboj-Orašje), the transit importance of this city will gain even more in geo-traffic importance. The area of the territory of the city of Doboj in 2013 was 772 km^2. Without the newly established municipality of Stanari in 2014, it is 653 km^2. According to the 2013 census, the city had 71,441 inhabitants (excluding Stanari 69,343). Natural increase is negative (−331 in 2014).

It is the fourth largest city in the Peripannonian macroregion. The city has institutions of administrative and judiciary importance (Centre of Security Services, District Court, Prosecutor's Office and Commercial Court) and is recognizable by educational and cultural institutions (Faculty of Transport and Traffic Engineering, Museum of the Region, Main Library). The city also has very attractive tourist facilities (Doboj Fortress, Sports and Recreation Centre Preslica, Goransko Lake, etc.). The most famous sports event is the "International Handball TV Champions Tournament," which has supranational significance. Also, the international small business fair DOBOJEXPO brings together a large number of domestic and foreign businessmen, which contributes to the establishment of direct links between the local and international market. In 2014, the Stanari separated into an independent municipality, on whose territory the 300 MW thermal power plant "Stanari" started operating.

South of Doboj is *Tešanj*, the third largest local government unit in the mesoregion (43,063 inhabitants in 2013) (Fig. 15.7). It is, after the municipality of Doboj-Jug, the most densely populated area in this mesoregion (276 inhabitants per km^2). Among the 100 most successful companies from BiH in 2017, as many as 23 are from Tešanj (HIFA Group, AS Jelah-Tešanj, MADI, Prevent FAD d.d. Tešanj Mann Hummel BA). In the immediate vicinity of Doboj, there are three smallest municipalities in this mesoregion: *Usora* (49.8 km^2), *Doboj-Istok* (41 km^2), and *Doboj-Jug* (10.2 km^2). They are characterized by a high population density (over 130 inhabitants per km^2). The municipality of Doboj-Jug (405 inhabitants per km^2) is the most densely populated unit of local self-government in the Peripannonian region. *Maglaj* (23,146 inhabitants), *Zavidovići* (35,988 inhabitants), and *Žepče* (30,219 inhabitants) are municipalities in the valley of the river Bosna. Maglaj is one of the industry centers for the production of pulp, paper, paper products, and wood products (Natron-Hayat).

At 26 kilometers southeast of Doboj along the Usora valley, there is the municipality of *Teslić*, which is the second largest territory (838 km^2) in the Peripannonian macroregion. The backbone of the development is the health and tourist complex Banja Vrućica (one of the largest centers of congress tourism in BiH), which is among the most visited destinations in Bosnia and Herzegovina (average length of stay of tourists is 4.9 days). Derventa (517 km^2, 27,404

Fig. 15.7 Old Town Tešanj. (Photo by Aida Bidžan-Gekić)

inhabitants in 2013) was formed at the contact of the low hills of Motajica (west), Krnjina (south), Vučjak (east), and Ukrina. *Brod* (517 km^2, 16,619 inhabitants in 2013) is a city formed on the right bank of the Sava, immediately after the mouth of the Ukrina. (The phenomenon of "dual" cities on the Sava River: Gradiška-Stara Gradiška, Brod-Slavonski Brod, Šamac-Slavonski Šamac). The main economic entity of the city is the oil refinery.

15.2.1.3 The Posavina-Spreča-Majevica-Lower Drina Mesoregion

Posavsko-sprečansko-majevičko-donjodrinska mesoregion (The Posavina-Spreča-Majevica-Lower Drina mesoregion) represents the third mesoregional unit within Peripanonian Bosnia. In the east and north, its borders are hydrographic and at the same time state borders (Sava River, i.e., Drina), while the southern border towards the mountain-valley macro-region is extremely physiognomic—it stretches over the mountain massifs Konjuh and Javor. It covers an area of 7485.5 km^2, which is about 16.7% of the territory of Bosnia and Herzegovina. According to the 2013 census (BHAS 2016), this mesoregion had 891,984 inhabitants (or 25.2% of the total population of Bosnia and Herzegovina), with an average population density of 119.2 inhabitants per km^2, which is significantly above the average for the state territory, and for Peripannonian macro-region (89.9 inhabitants per km^2). The area of this mesoregion includes as many as 32 administrative units of local self-government, with different status (municipalities, cities, districts). Within this mesoregion is Brčko, as a separate administrative unit in Bosnia and Herzegovina with the status of a district.

In terms of physiognomy, this mesoregion, as well as the previous units of the Peripannonian macro-region, are characterized by various elements of relief. The southern higher areas are dominated by the massifs of Javor, Glogova Mountain, Javornik, and the hills and plateaus (Osat, Birač, Ludmer), which gradually turn north into lowland terrain. The border in the east is represented by the composite valley of the river Drina, the largest tributary of the Sava. The middle (from the Tijesno canyon to the Zvornik extension in the length of 133 km) and the lower part of the Drina course (from Zvornik to the confluence with the Sava 91 km long) belong to this mesoregion. In the lower course, the Drina flows through Semberija as a lowland-plain river with a small drop and an average flow at an estuary of 395 m^3/s. The area of the mesoregion north of the Spreča-Zvornik line has a geological soil structure (predominantly marl-clayey and sandy sediments) characterized by poor physical-mechanical properties of the soil. This, in combination with unfavorable hydrological conditions (the appearance of extremely high torrential water levels with long-term precipitation), is manifested in the form of soil instability and the occurrence of landslides with floods, which were most intense in 2014. Majevica (915 m above sea level) represents the natural border between Semberija in the northeast and the Sprečanska valley in the south. To the north is the area of Posavina, which completes this mesoregional area.

In addition to rivers, Mountain Rivers and rapids, several larger and smaller artificial lakes have been created and exist: Modrac, Bistarac, Rajsko Lake (Fig. 15.8), Hazna, Vidara, Snježnica, Mačkovac, Vijenac, and Pannonian Lake. Large valley areas are source of mineral and thermal springs, some

Fig. 15.8 Rajsko lake near Bašigovci. (Photo by Aida Bidžan-Gekić)

of which are used for spas—Salt Spa in Tuzla, Spa Ilidža in Gradačac, Terme Gračanica, and Kladanjska voda near Kladanj. These potentials include the area of the Protected Landscape "Konjuh," which is located in the municipality, Banovići, Kladanj municipality, and the town of Živinice (Tuzla Canton Development Strategy 2021).

For better connection of this mesoregion, the planned highway between Sarajevo and Belgrade, which leads through Žepče, Tuzla, Brčko, and Bijeljina to the border with Serbia, is of vital importance. The route of the highway from Tuzla to Žepče (connection to corridor Vc) is 55.4 kilometers long.

Semberija is a Quaternary macro-floodplain of the Drina River, which moved the course of the Sava several kilometers to the north with an area of about 300 km^2. It is a physiognomically homogeneous area in the northeastern parts of the mesoregion and Bosnia and Herzegovina.

The western border consists of a line starting from the Sava to the south and encompassing the settlements of Donji, Srednji, and Gornji Dragaljevac on the west side, turning to the southeast and encompassing the settlements of Suho Polje and Hase, and ending on the Drina near Batar, a few kilometers south of Janja. It is open in the north towards the Pannonian Basin, while in the west and south it is closed by the slopes of Majevica and other mountains. Watercourses flow mainly towards the Sava. In the northern part, near the Drina and Sava, there is also wetland (Gromiželj et al.). It is a fertile area with developed agriculture (cereals, fruits). Bijeljina (45,291 inhabitants, 2013) is an economic and cultural center (Croatian Encyclopedia 2021).

The city of *Bijeljina* is a mesoregional center, and with an area of 733.85 km^2, the largest unit of local self-government in this mesoregion. According to the 2013 census, the total population of Bijeljina was 107,715 inhabitants (after Banja Luka and Tuzla, the city with the largest population in Peripannonian Bosnia). It is one of the densely populated areas (147 inhabitants per km^2), which is significantly above the macro-region average (89.9 inhabitants per km^2). Unfavorable trends were recorded in terms of natural increase, which is negative (2013—956 births, 1303 deaths). Bijeljina is located in the center of the fertile Semberija plain and is one of the centers of agricultural production, primarily cereals (wheat and corn) and vegetables (cabbage, peppers, and tomatoes). The city is a university and cultural center. On the territory of the city, there are two border crossings towards Serbia over the rivers Sava and Drina, which are more frequent crossings for traffic from the west and south from Bosnia and Herzegovina to Serbia and Belgrade. In addition to the only urban city center (42% of the population of Bijeljina), the specifics are densely populated rural areas Janja (11,700 inhabitants, or 288 inhabitants per km^2), Dvorovi (4873 inhabitants, or 331 inhabitants per km^2), Velika Obarska (3942 inhabitants, or 104 inhabitants per km^2), and Patkovača (2619 inhabitants, with as many as 1018 inhabitants per km^2). Bijeljina is one of the few cities in Bosnia and Herzegovina that has a higher number of immigrants than emigrants (in 2010 the migration balance was +772 inhabitants). About 68% of the population of Bijeljina lives in rural settlements. Immigration is very intensive, as evidenced by the fact that out of the total population, only 33,855 were born in this city.

In addition to the economy, Bijeljina is also a city of agriculture; 6427 agricultural holdings were registered. In average high harvest years, about 55,000 tons of wheat, 160,000 tons of corn, 1200 of barley, and 65–70,000 tons of vegetables are produced. Two large companies are leading in the processing of agricultural products—Food Industry "Sava-Semberija" (processing and canning of fruits and vegetables) and "Žitopromet" (mill and bakery industry), as well as "Spektar-drink," "Tanasić," and "Balatunka" (processing and canning of fruits and vegetables) (Local Development Strategy of the City of Bijeljina 2018).

Of the industry, the most important are the plants for service and production of equipment of the Aviation Institute "Orao" and the production of spare parts for cars (S.E.P.). In the southern part of Bijeljina, there is the municipality of Ugljevik (15,710 inhabitants, 165.17 km^2 in 2013). The basis of the economy is significant coal reserves, on the basis of which the company Rudnik and the Ugljevik thermal power plant operate. The municipality had the second largest GDP in the Peripannonian micro-region (9407 BAM in 2013).

Brčko District is the second mesoregional center located 40 kilometers from Bijeljina and 80 km from Tuzla. It is located in the river valley that is part of the basin of the rivers Sava, Brka, and Velika Tinja. Altitude, depending on the terrace formed by the rivers, ranges from 85 m to 200 m (85%). The only exceptions are the northern slopes of Mount Majevica in the southern part of the District with an altitude of 200 to 400 m. According to the official results of the 2013 census, the Brčko District had 83,516 inhabitants on an area of 493.3 km^2 (with a very high population density of 169 inhabitants per km^2). The degree of urbanization was 54.50%. According to the 2013 census, the Brčko District is a multinational environment (Bosniaks make up 42.4%, Croats 20.7%, and Serbs 34.6%).

It is a multimodal transport hub (road, river, and railway traffic) where east-west and north-south traffic intersect (Fig. 15.9). Taxes in the Brčko District of BiH are more stimulating for new investments compared to the rest of BiH and the region. The most important companies in the food industry are: Bimal, Studen Agrana, Bosnaplod, Voćar, Briž, Biljana, Žitopromet, Elektromlin Stokić, Mlin Vujičići, and Mlin Bukvić (production of edible oil, sugar, jams, juices, teas, cereal processing). When it comes to the wood industry, the most important companies are: Malagić, Oglavina, and Duka Bosna.

Fig. 15.9 Port of Brčko on the Sava River. (Photo by Ranko Mirić)

The Brčko District of BiH has significant natural resources for primary agricultural production. Agricultural land accounts for 71.52% of arable land of high quality. It is dominated by corn, wheat, sunflower, oilseed turnip, soybeans, and other agricultural crops with the orientation of primary agricultural production towards processing capacities (Oil Factory "Bimal" d.d.). The land is also suitable for growing many types of fruit. The largest areas of fruit are under plums (450 ha in 2019) and apples (58.83 ha). Based on official data on the number of arrivals and overnight stays in hotels, 13,958 tourists came to the Brčko District of BiH in 2019, of which 57.2% were foreign tourists (Development Strategy of the Brčko District of BiH 2021).

The typical physiognomic area of Posavina is occupied by a large number of territorially smaller municipalities in this mesoregion. Modriča is the largest municipality in this area in terms of area (320 km^2) and population (25,720 in 2013), recognizable by its oil refinery and footwear industry. Orašje covers 121.8 km^2 (19,861 inhabitants in 2013) and is the most densely populated municipality in this part (163 inhabitants per km^2). It is the administrative center of Posavina Canton. Odžak covers 158.4 km^2 (18,821 inhabitants in 2013). Both municipalities record negative rates of natural increase (around −6 per thousand in 2015). Šamac has similar indicators (17,273 inhabitants in 2013, area 158.4 km^2) with a natural increase of −9 per thousand (the share of the old population in the total compared to the young is higher by 7.5%). The municipalities of *Domaljevac Šamac*, *Pelagićevo*, *Vukosavlje*, and *Donji Žabar* are rural areas without a city center, with a population not exceeding 5220 (Vukosavlje). They are characterized by a negative natural increase (Vukoslavlje as much as—11.3 per thousand in 2015) and a higher share of the elderly in relation to the young population in the total population.

Southwest of this unit, there is the Sprečko-Majevica area, with a macro-regional center (Tuzla) and municipalities that represent the most important mining and industrial basin in Bosnia and Herzegovina. Over a billion tons of exploitation reserves of lignite and brown coal, as well as significant amounts of rock salt, quartz sand, limestone, and magnesite, are located in this area.

With 110,979 inhabitants according to the results of the 2013 census, the City of *Tuzla* is the third largest city in Bosnia and Herzegovina (after Sarajevo and Banja Luka) (Fig. 15.10). It is the second macro-regional center in the Peripannonian area, and the administrative center of Tuzla Canton. The curiosity is that as many as 2151 residents of Tuzla did not comment (or left the column unanswered) on the issue of ethnicity. The rate of natural increase per thousand inhabitants in 2017 is −3.39, which is significantly higher than the negative natural increase in population compared to 2012, when it was −0.46. Unfavorable trends were also recorded in terms of age structure (higher share of the old than the contingent of young population in the total by as much as 2%). Railway traffic is represented exclusively when it comes to freight transport, because the current state of railway infrastructure does not meet the conditions for the development of passenger rail transport.

Unlike railway traffic, the development of air traffic in the area of TK is on an upward trajectory. The establishment of the PE "Tuzla International Airport" created the basic preconditions for the development and improvement of air

Fig. 15.10 Salines Soli square in Tuzla. (Photo by Aida Bidžan-Gekić)

traffic in the area of Tuzla Canton, and with additional investment in the development of air infrastructure and modernization of Tuzla International Airport services, passenger air traffic has experienced significant expansion. In 2019, the only airline operating from Tuzla International Airport, WIZZ AIR, carried a record 593,050 passengers to 12 destinations in Europe, which is four times more than in 2014 (Tuzla Canton Development Strategy 2021).

In 2013, GDP per capita amounted to 9252 BAM, which puts it in second place in the mesoregion (behind Ugljevik). The economy of Tuzla is export-oriented, and the most significant share in it have companies such as Solana d.d., Elsta Mosdorfer Bosnia d.o.o., Elatec d.o.o., Umel dalekovod montaža d.o.o., and many others. The mines "Kreka" Tuzla Dubrave, Šikulje, and Mramor have established balance reserves of coal amounting to about 664 million tons. In 2015, the production volume amounted to 1807,828 tons, and the number of employees was 2754 workers. Soli Mine d.d. Tuzla deals with the exploitation and distribution of salt and technological water. In the sector of processing industry, the food industry has its most significant capacities in the mill-bakery, meat, dairy, production of beer, soft drinks and mineral water, etc. The largest retail chain in Bosnia and Herzegovina, Bingo, is based in Tuzla. By using saltwater resources to build a unique complex of salt lakes Panonika with a salt waterfall, the city of Tuzla has made a certain turnaround in the development of the economy, placing a significant focus on the development of tourism. The number of registered tourists in the period 2012–2017 had a trend of continuous growth, and the number of tourists increased from 15,247 tourists in 2012 to 30,365 tourists in 2017, which is an increase of about two times. JU University in Tuzla has a total of 12 faculties and the Academy of Dramatic Arts. Due to the increased soil pollution for certain areas near the Tuzla Thermal Power Plant, the cultivation and consumption of agricultural products are not recommended. According to data from 2014, over 2000 landslides have been registered, most of which are those that directly endanger people and their material goods, or make it impossible or difficult for them to live and work in the endangered area (Tuzla City Development Strategy 2018).

According to the 2013 census, *Srebrenik* had 42,762 inhabitants and is one of the most densely populated areas (160 inhabitants per km^2). Only 20% of total population live in the city center. Of the total area of the municipality, 55.88% is agricultural land. "Corn Flips" (production of flips, chips, packaging of coffee and powder products) is a recognizable brand in the BH market, and "Zaharex," the largest exporter and largest employer in this municipality. Along the lower course of the river Spreča, there are the municipalities of Gračanica (2013 population 45,220 inhabitants, population density 210 inhabitants per km^2) and Petrovo (6474 inhabitants). *Gračanica* is one of the most economically vital areas in the mesoregion, thanks to a large number of very flexible business firms with a small number of employees (up to ten workers). Petrovo is a rural area with extremely unfavorable characteristics of the population (natural increase −13.1 in 2015, the share of the elderly in the total population as much as 22.1%). The municipality of *Lukavac* covers an area of 337 km^2 with a population of 44,520. It is located in the valley of the river Spreča (railway Tuzla-Doboj), and part of the slope of the mountains Konjuh, Ozren, and Majevica. Natural increase in the municipality of Lukavac is negative and averages −179 inhabitants, or 1.08% of the total population per year. The degree of urbanization is about 27%. According to the data, the number of live births for 2017 was 341, and the number of deaths was 473, that is, the difference in births/deaths is

−132 inhabitants, only on the basis of natural population movements, not taking into account migrations to surrounding municipalities or abroad (Integrated Development Strategy of Lukavac 2019). This municipality is one of the most industrially developed areas, with a concentration of ore-mineral resources and processing plants (lignite is exploited in the municipality of Lukavac on the surface mine Šikulje and Lukavačka Rijeka). The three largest industrial companies are Global Ispat Coke Industry (fourth largest exporter in Bosnia and Herzegovina in 2017), Sisecam Soda Lukavac d.o.o. (sixth largest exporter in Bosnia and Herzegovina in 2017), and Cement Factory d.o.o. Lukavac. These three companies employ more than a third of the workers and generate over 50% of the municipality's revenue.

According to the 2013 census, there are a total of 57,765 inhabitants in the municipality of *Živinice*. With almost twice the share of the young (17.1%) compared to the contingent of the old population (9.3%) in total, it has a favorable demographic picture. The most important resources are coal deposits (Dubrave and Đurđevik—total reserves of brown coal are 60 million tons and lignite per 100 million tons). In industrial production, the company Alfe MI doo Živinice stands out, engaged in the production of metal containers (over 50% of the total value of exports of the municipality).

According to the 2013 census, the municipality of *Banovići* had 22,827 inhabitants on an area of 183.24 km^2. The backbone of economic development is RMU "Banovići" (annual production of over 1500,000 tons of coal) with 2787 employees (2016). The municipality of Kalesija had 33,053 inhabitants on an area of 201 km^2 in 2013. With a population density of 164 inhabitants/km^2, it belongs to the group of densely populated municipalities, but it is also an area with one of the smallest shares of urban population in the country (below 10%). *Kladanj* is located in the valley of the river Drinjača, at the foot of the mountain Konjuh, 52 kilometers south of Tuzla. The area of the municipality is 331 km^2; in 2013, there were 12,348 inhabitants. On the territory of the municipality of Kladanj is about 83.16% of the area under forests. The most significant natural value in the municipality is the Protected Landscape "Konjuh." *Sapna* municipality is located in the northeast of Bosnia and Herzegovina. The area of the municipality is 118 km^2, and according to the results of the 2013 census in Bosnia and Herzegovina, it had 12,136 inhabitants, one of the least developed municipalities in the mesoregion (GDP in 2013 was 1714 BAM). 61.64% of the territory of the municipality is under forests.

According to the 2013 census, the city of *Gradačac* had 39,340 inhabitants on an area of 218 km^2. It is one of the densely populated areas (180, 46 inhabitants per km^2). It is recognizable for the development of fruit growing (cultivation and processing of plums—Gradačac Fair has international significance). Recently, there has been a continuous growth of entrepreneurial activity through partnerships between domestic and foreign investors. Parts of production facilities are located in the city (Cimos, Wagner Automotive) for the world's leading automotive industries (Volkswagen, PSA, Audi, BMW) and in the food processing industry (Lactalis and Zott) are located in the city. Gradačac is the leading local community in the production plums and among the leading local communities in terms of exports in Bosnia and Herzegovina. Leading the way are TMD Cimos (ninth largest exporter in Bosnia and Herzegovina in 2017), TMD Group, Wagner Automotive (metal processing industry), and Kula Gradačac (textile industry).

The municipality of *Teočak* covers an area of 30.85 km^2. It was founded in 1992 by separating from the municipality of Ugljevik. It is recognizable by the artificial lake "Sniježnica." *Čelić* (136.8 km^2, 10,502 inhabitants in 2013) is a municipality located on the slopes of Majevica and parts of Posavina, known for its berry fruit, which represents the brand and tradition of this region.

Zvornik is the regional center of the mesoregion, and one of its industrial centers. In 2013, there were 58,856 inhabitants living on an area of 279.13 km^2 (of which 20.7% in the city center). It is characterized by a high population density (156.47 inhabitants per km^2). Today, the city of Zvornik is among the local communities with the largest share of the returnee population (36.49%) which should be provided with socioeconomic conditions for sustainable return. Mineral waters in Kozluk represent the raw material basis for production in AD "Vitinka" in Kozluk. The Drina is the largest right tributary of the Sava, on which HPP "Zvornik" was built in 1955. The Zvornik Lake Reservoir is 25 km long, 200 m to 3 km wide, and up to 39 m long. The Drinjača River also has a significant hydropower potential (flow 6.5 m^3/s) (five concessions for the construction of small hydropower plants) (Strategy for Integrated Development of the City of Zvornik 2018).

At the northern exit from the city, there is an alumina factory Birač (bauxite processing from the Milići mine). There are significant mineral reserves of building stone and clay (Jošanica, Čelopek), mineral, and drinking water (Kozluk and Kiseljak). The Faculty of Technology operates in the city. In the far southeastern part of the mesoregion, there are the municipalities of *Bratunac* (20,340 inhabitants, 293.5 km^2 in 2013) and *Srebrenica* (13,409 inhabitants, 25.5 km^2), areas with a low level of development and extremely unfavorable social and economic structure. In Srebrenica, these are the direct consequences of the suffering of the population in the last war, which, as previously stated, was characterized by the International Court of Justice as genocide.

Vlasenica (11,447 inhabitants, 225.32 km^2 in 2013) is a municipality below Mount Javor, economically focused on the exploitation and processing of wood (about 2/3 of the area of the municipality) and raw aluminum (aluminum hardware). The Department of Forestry of the Faculty of Agriculture, University of East Sarajevo, operates in Vlasenica. Municipality of *Milići* was formed from a part of the territory of the municipality of Vlasenica. In 2013, 11,441 people lived on an area of 279.13 km^2 (only 20.7% in the city center). The economic development of the municipality is based on the exploitation of bauxite ore (geological reserves are estimated at 22.5 million tons). The carrier of the economic development of the municipality is the company "Boksit" a.d. Milići.

15.2.2 Mountain-Valley Macro-region

The spatial coverage and geographical position of the Planinsko-Kotlinska (Mountain-Valley) macro-region indicate its central position in the context of the national territory, which is often identified in the literature from the aspect of its belonging to the Central and Inland Dinarides. Observed from the aspect of historical-geographical evolution, this area represents the core of Bosnian statehood, which started from the period of mention of "horion Bosnia," and is also referred to in the literature as a region of the mountain core (Smlatić 1983). It covers the central parts of Bosnia and Herzegovina between Peripannonian Bosnia and the high karst region in the south. It is separated from Northern Bosnia by the mountain ranges Dimitor-Manjača-Čemernica-Vlašić-Konjuh-Javor, and from the high karst, it is separated by the mountain massifs of Zelengora, Treskavica, Bjelašnica, Bitovnja, Raduša, Vitoroga, and Lisina. In the west-east direction, this region runs from the upper course of the Sana to the republican border of BiH between BiH, Montenegro, and Serbia (Fig. 15.11).

It covers an area of 12,961.8 km^2, which is about 25.3% of the territory of Bosnia and Herzegovina. According to the 2013 census, there were 1,060,469 inhabitants (or 30.03% of the total population of Bosnia and Herzegovina) within this macro-region, with an average population density of 81.8 inhabitants per km^2 (Bosnia and Herzegovina-68.9 inhabitants/km^2), and according to the above data, it represents the demographic focus of the national territory (BHAS 2016). Favorable geographical location within Bosnia and Herzegovina is manifested by the position of the upper reaches of large rivers of this macro-region (Vrbas, Bosnia, and Drina), which are naturally predisposed to road communications, and their physiognomy and purpose formed a large number of valley extensions within a very complex mountain morphostructure of this geographical area. This is a distinctly mountainous area dominated by a sharias structure with wide thrust fault. The mountains are of the type of tectonic antiforms (horst anticlines and anticlinorium) with heights often over 2000 m. The lower, intermountain area is intersected by a large number of river

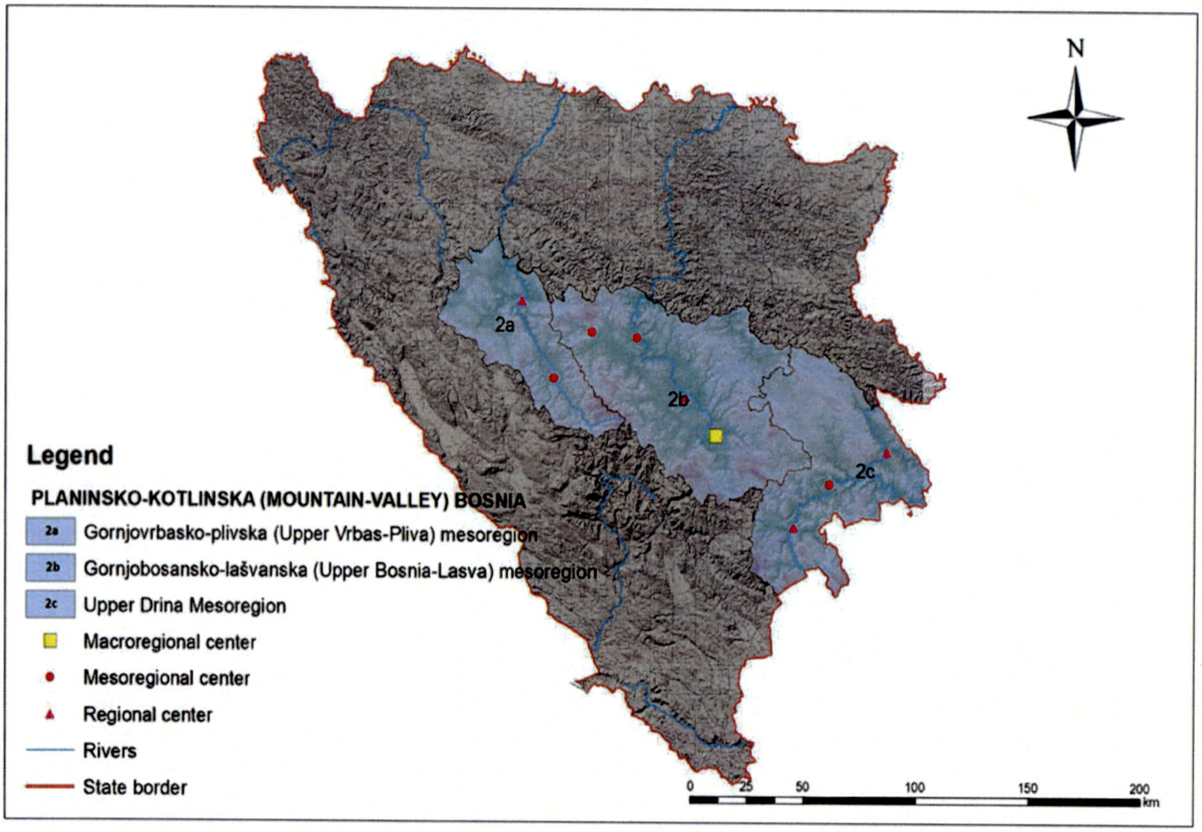

Fig. 15.11 Planinsko-Kotlinska (mountain-valley) macro-region

valleys, which are, most often, polyphase, polygenetic, polymorphic, and often composite and of canyon type, and as a whole they manifest as antecedent and epigenetic forms. They have a very mismatched and stepped longitudinal profile. The direction of their provision, depending on the angle they have in relation to the directors of the basic morphostructures, is transversal, diagonal or transverse. This type of valley has, in the upper part of the flow, mostly all right tributaries of the Sava: Bosnia with Krivaja and Lašva, Drina with Prača and Drinjača, and Vrbas with Pliva. More extensive valley areas with low terrain are found in the central part of the area, and among them, the most famous is the Sarajevo-Zenica field, as well as some terrains in the composite extensions of the Drina (Višegrad, Foča, and Goražde). They are tectonically predisposed, filled with Neogene coal deposits and subsequently formed mainly by a fluvial process. The overall morphosculpture is dominated by fluvial forms represented, on the one hand, by the river bed and valley sides, and on the other hand, by proluvial alluvium, alluvial accumulative terraces, alluvial field plains, and river islands. On the sloping sides, there is a sloping morphosculpture formed by the processes of washing, dredging, flowing, and sliding. In addition to them, glacial, paleoabrasion, and paleoeol forms are found in fragments, most often as relics. Within this area, several tectonic units can be distinguished: the central ophiolite zone, the anticlines of the Drina and Prača Paleozoic, the Jurassic and Cretaceous flysch zones, the Central Bosnian Shale Mountains, the Manjača-Grmeč zone, the Sana-Una Paleozoic, and a number of freshwater Neogene basins. Climatic characteristics in these areas show more pronounced thermo-pluviometric contrasts compared to the previous region.

More specifically, the basic characteristics of this climate are harsh winters with absolute temperature minimums up to −35 °C), while summers are warm and with absolute maximum temperatures up to 36 °C). The average annual rainfall is from 1000 mm to 1200 mm. In the structure of precipitation in the lower hypsometric levels, rainfall predominates, while in the highest hypsometric levels, snowfall is very abundant. A very important feature of winter anticyclones is temperature inversions, which are characterized by inverse thermal air stratification, which causes vegetation turns in altitude vegetation zonation. A typical area for such synoptic situations is the mountain-valley area of Bosnia and Herzegovina, where winter anticyclones are the most represented in terms of intensity and duration (Drešković and Mirić 2017).

Most of the macro-region is characterized by the Cf climate type, which includes the lower parts of the upper reaches of the Drina, Bosnia, and Vrbas with their tributaries. Regularly with the increase in altitude, the subtypes (Cfa to Cfc) take turns. Climatic type Df is represented by an average of 900 m above sea level to the zone enclosed by an isohypsus of 2000 m above sea level. It is continuously spread over mountain massifs in the east (Sljemenska, Devetak, Konjuh, Javor) in the central part (Treskavica, Jahorina, Romanija, Zvijezda), in the north and west (Vlašić, Vranica, Bitovnja), and northwest (border mountains Dimitor and Čemernica). The highest peaks of the mountain massifs of Vranica, Bjelašnica, and Treskavica have limited tundra climate areas (ET).

Hydrographic Basin Vrbas in its central and upper parts of the flow belongs to the mountain-valley macro-region of Bosnia and Herzegovina. The total area of the basin is about 6288.59 km^2, and about 50% belongs to this macro-region. The upper and source part of the river Bosna, which belongs to this macro-region, is extremely rich in surface flows and with branched hydrological network. In the upper course, from the source to Zenica (about 120 km), it flows through the Sarajevo, Visočko, Kakanjsko, and Zeničko fields, which separate the gorges and breakthroughs. In the middle course, it breaks through gorges cut into geologically resistant rocks (Vranduk-Nemila). The main tributaries of the Bosna River in the mountain-valley macro-region are Željeznica (26.9 km), Miljacka (35.9 km), Stavnja (30.4 km) on the right, and Fojnička river (46 km), Lašva (49.4 km) on the left. The only hydropower plant (Bogatić) is on the river Željeznica with an installed capacity of 6.5 MW. The Drina basin in this macro-region covers its upper part up to Peručko Lake, and due to the shape of the basin, about 60% of its surface belongs to the mountain-valley macro-region.

The pedological structure is very complex and is a reflection of a complex geological-geomorphological terrain, a developed hydrographic network of surface flows, and the presence of a large number of climatic types and variants in this area. The most widespread soils are from the automorphic division, primarily district cambisol, rendzina-calcocambisol-luvisol complex on limestone or dolomite, rendzina, and lithosol-rendzina complex on dolomite. Calcocambisol and calcomelanosol-calcocambisol mosaic on limestone along the flows of the Vrbas, Bosnia, and Drina fluvial soils are formed to a significant extent.

In addition to the ecosystem of mesophilic deciduous oak and hornbeam forests (predominantly represented in the foothills of the central and continental Bosnian Dinarides), the macro-region abounds in thermophilic beech communities, followed by mesophilic communities of beech forests. At higher altitudes (over 1300 m), communities of mixed beech-spruce forests have been formed, i.e., mixed beech-spruce-fir forests.

The specificity of the vegetation of the mountain-valley macro-region is the ecosystem of mountain Rudine on silicates represented only on the highest areas of high silicate mountain massifs, most often above the ecosystem of juniper pine and mountain heaths, i.e., in Bosnia and Herzegovina, it occurs only in high habitats of Vranica (where the *Pinetum mugi silicicolum* ecosystem was primarily developed) (Drešković and Mirić 2017).

The settlement network consists of a system of 2491 settlements, which are located in 45 local self-government units, 4 of which have the status of the city (Sarajevo, Istočno Sarajevo, Goražde, and Zenica). There are very few underdeveloped municipal centers without urban population (these are mostly newly formed municipalities after 1995). Due to the complexity of the space of this macro-region in terms of settlement structure, population, and resource base, the emphasis in the development of these contents is given in the consideration at the mesoregional level.

15.2.2.1 The Upper Vrbas-Pliva Mesoregion

The Upper Vrbas-Pliva mesoregion represents the smallest mesoreginal unit within mountain-valley macro-region. In the east, its borders are represented by a watershed between Vrbas and Bosnia (Vranica, Komar), while in the north and northwest, it borders the area of the Peripannonian macro-region (Čemernica, Manjača, and Dimitor). The southwestern border (towards the macro-region of High Karst) is represented by the mountain massifs of Raduša, Vitorog to the top of Gola Kosa. It covers an area of 2708.4 km^2, which is about 5.3% of the territory of Bosnia and Herzegovina. According to the 2013 census, this mesoregion had 121,754 inhabitants (or 3.4% of the total population of Bosnia and Herzegovina), with an average population density of 44.9 inhabitants per km^2, which is significantly below the average for the mountain-valley macro-region (81.8 inhabitants per km^2).

Physiognomically, it represents the catchment area of the upper and part of the middle course of the Vrbas (only the far western parts belong to the Una and Cetina basins to a small extent). The oldest geological formations are connected to the area of the Central Bosnian Palaeozoic, whose backbone is located in the group of Vranica Mountains. The oldest Lower Palaeozoic formations in the area are represented along the southern slopes of Mount Komar (phyllite, amphibolite, quartzite, marbles, etc.). Silurian-Devonian deposits spread in the far southeastern parts. Young Palaeozoic deposits are present in the wider area around Šćit and the southeastern slopes of Dobruška Vranica and the catchment area of the Desna River (in the far southeastern part of the Vrbas Basin). Carbonates from the Carboniferous form a surface continuity with Early Palaeozoic deposits in the part of the source ridge of the Vrbas River. The deposits belonging to the transitional Permian-Triassic formations spread in a wide zone of the Radalj Mountain. Mesozoic formations are widespread mainly in the central and southern part. Lower Triassic deposits extend along the left valley side of the Vrbas River from Gornji Vakuf to the settlement of Drežev Dol, and in the wider area of the Stožer Mountain, i.e., along the right valley side of the Vrbas River from Plivsko Lake to Mrkonjić Grad. Middle Triassic formations are only limited in a few small isolated areas in the southern part of the Mesoregion. The middle and upper Triassic in the form of a wide and continuous belt extends from the Veliki Sjever (1561 m) to the top of Lisina (1469 m) in the municipality of Mrkonjić Grad.

Geomorphological characteristics of the terrain are primarily determined by mountain-valley morphology and fluviodenudation, fluvial, fluvio-karst, and karst type of morphosculpture. The relief has extremely dynamic features and represents one of the most morphologically striking parts of Bosnia and Herzegovina. Fluviodenudation relief dominates the slope sides of mountain morphostructures that form clastic rock masses. On them, the slope processes of washing, dredging, sliding, and flowing are very pronounced, which form various accumulative forms of relief. High floods are frequent in this mesoregion. Soils from the automorphic division are dominant from the aspect of surface distribution: district cambisol on acid silicate rocks, rendzina-calcocambisol-luvisol complex on limestone or dolomite, rendzina and lithosol-rendzina complex on dolomite. In the downstream segment of the Vrbas flow, fluvial soils fluvisol were formed. The least surface distribution belongs to the soils of eutric cambisols on bases rich in silicate rocks. The vegetation characteristics of this mesoregion are presented with several communities. The ecosystem of mesophilic deciduous forests of sessile oak and hornbeam is predominantly represented in the foothills of the central and continental Bosnian Dinarides.

More specifically, the vegetation communities of this belt were developed in the hypsometric belt of 300 and 800 m above sea level in which the Cfb climate type dominates. Further in higher altitudes (from about 700 m above sea level), mainly thermophilic communities of beech develop, and above them mesophilic communities of beech forests (*Fagetum montanum*). In the belt above 1300 m formed communities of mixed beech-spruce forests (*Abieti-Fagetum*) or mixed beech-spruce-fir forests (*Piceo-Abieti-Fagetum*) dominate. In the zones of mountain peaks above 1800 m above sea level, pine juniper and a floor of mountain pastures are represented, which further turn into mountain barren lands (Drešković and Mirić 2017).

The resource base of this mesoregion is coal reserves (Bugojno, Jajce basin), then deposits of ferrous metals (iron, pyrite, and manganese), and deposits of nonmetals (barite, gypsum, dolomite, quartz, etc.). There are also lignite deposits in the area of the Vrbas valley ("Gračanica" mine near Bugojno and Gornji Vakuf), gypsum (Gornji Vakuf-Uskoplje), quartzite, mercury, copper, lead, zinc (Gornji Vakuf-Uskoplje), and building materials (technical and building stone-Bugojno, Jajce). Large hydro potentials are relatively well valorized in energy (hydropower plants) and

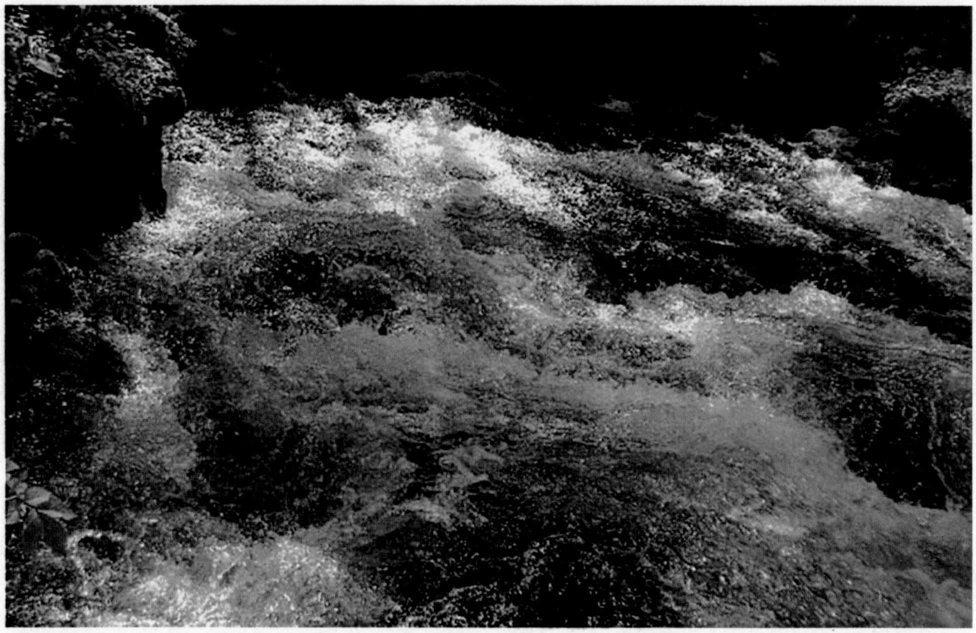

Fig. 15.12 Source of Pliva. (Photo by Ranko Mirić)

tourism (Vrbas, Janj, Pliva (Fig. 15.12)) sector. The mesoregion has a very favorable traffic position, because it connects the northern Peripanonnian areas (Banja Luka) with the Neretva valley and the Adriatic Sea by the main road through the Vrbas valley (Makljen pass).

Seven units of local self-government represent the centers of the settlement structure. Except for the municipality of Jezero, all others have a developed city center with an urban population, but the degree of urbanization does not exceed 50%. (About 37% of the total population lived in the cities of this mesoregion.) The municipality of Bugojno is a mesoregional center, with the highest concentration of population and functions (it makes up 36% of the urban and 26% of the total population of the mesoregion), and is the only local unit with a population density above the macro-regional average. The regional center is the municipality of Jajce with a low degree of urbanization (27%). Compared to the 1991 census, all municipalities in this mesoregion recorded a population loss in 2013 (Bugojno and Jajce approximately one third, Jezero one half). The total number of settlements in the area of the mesoregion in 2013 was 351, of which only 6 were urban.

The municipality of *Bugojno* is a mesoregional center in the southwestern part of this region. The area is 361 km^2. In 2013, the total population was 31,470, which is 32.9% less than in 1991. The population density is above the average for Bosnia and Herzegovina (87.2 in./km^2). The area of this municipality basically consists of the valley of the upper course of the river Vrbas, which is called Uskopaljska valley (in the narrower sense in the area of Bugojno it is also called Bugojanska valley), and the mountains surrounding this valley reach 1700 m in height. The average annual air temperature is about 9 °C, and the average annual rainfall is about 830 mm. The Vrbas River flows through the middle part of the municipality from north to south, and the eastern half of the municipality has a much more developed river network than the western one. The forest cover of the municipality is as high as 61%, and hunting grounds in the mountainous east and west (especially Koprivnica) are known throughout Europe.

The municipality of Bugojno records extremely negative demographic trends, and out of a total of 78 settlements, 25 have no permanent residents. The urban center of Bugojno with 15,555 inhabitants, which is one of the largest urban settlements in BiH, has lost 31.3% of its population compared to 1991. The unemployment rate in the municipality is very high (around 32%). Significant reserves of lignite are present, while the exploration of the present phenomena of iron ore, pyrite, copper, gold, zinc, and silver is very low. Of the nonmetallic mineral raw materials, the reserves of abrasives, dolomite, and architectural stone are significant. The basis of economic development is the manufacturing industry and trade. Bugojno was one of the favorite vacation spots of the president of the former SFRY, Josip Broz Tito, where he also received numerous world officials. The most important tourist attractions in addition to extremely beautiful and preserved nature (mountains Kalin and Rudina, the area of Ravno Rostovo, wooded areas of Koprivnica and Duboka with numerous springs, etc.) are: Sultan Ahmed Mosque, Church of Saint Anthony of Padua (the second largest

Catholic church in BiH), then numerous archeological sites and sites from prehistoric (Pod site where the oldest written trace was found on BiH soil), ancient (Pod, Grudine) until the Middle Ages (remains of medieval towns and fortifications Vesela Straža, Susid, Kaštel), Sulejmanpašić and Rustempašić towers, Türbe of Malkoč and Skenderpašić, there are numerous stećak tombstones at 14 sites, the Memorial of the People's Liberation War Pršljani, the Gymnasium building, and many others (Gekić 2021) (Fig. 15.13).

The southernmost area of this mesoregion belongs to the municipality of *Gornji Vakuf-Uskoplje*. According to the results of the 2013 census, there were 20,933 inhabitants (52.86 inhabitants/km^2) in the municipality. It is located mainly in the Uskopaljska valley which is surrounded by the Raduša and Vranica massifs. The valley is agriculturally valorized (vegetable growing is developed, especially cabbage growing). The backbone of economic development is a significant mineral wealth (Gračanica lignite mine, magnetite, gypsum in the area of Bistrica and Voljica, quartzite) and forest resources (covering about 60% of the territory of municipalities). The degree of urbanization is low (23.1%), but the share of young people (17.5%) in the total population is much higher compared to the old population (10.7%). Along with Donji Vakuf, it has the lowest average age in the mesoregion (36.7 years).

The municipality of *Donji Vakuf* is located on the important main route M-5 towards the Lašva valley. The valley of the river Vrbas divides the territory of the municipality into two almost equal parts, with the largest part being a distinctly hilly and mountainous area (up to 550 m above sea level accounts for only 7% of the territory of the municipality). According to the results of the 2013 census, there were 13,985 inhabitants in the municipality (43.7 inhabitants per km^2). Significant ore-mineral resources are the basis for the development of the nonmetallic industry (gypsum, quartz, marble deposits). About 65% of the territory is under forests (wood industry). The production of prefabricated buildings (Promo Company) is significant. About half of the population lived in the only urban settlement (67 others are rural settlements). Donji Vakuf has the lowest average age in the mesoregion (36.4 years), the highest share of young people (17.9%), and the lowest share of old people (10%) in the total population of the municipality (BHAS 2016).

The regional center, municipality of *Jajce*, is located at the confluence of the Pliva andVrbas rivers. It covers an area of 363 km^2, on which, according to the results of the 2013 census, there were 13,985 inhabitants (80.41 inhabitants per km^2). Trends in negative natural increase were recorded (2015–3.6%). The municipality is well connected with other parts of Bosnia and Herzegovina by the main roads M-5 (Travnik) and M-16 (in the direction of Banja Luka). Hydrographic conditions are extremely favorable. The rivers Pliva and Vrbas with several tributaries flow through this municipality, and the Great and Small Pliva Lakes are significant, as well as the attractive Pliva waterfalls in the very center of the city. Forests cover 54% of the total territory of the municipality. The most important economic resources and the basis of development are hydroelectric power plants—Jajce I and Jajce II (total capacity 60 MW). There are several mineral and ore deposits in the municipality, including bauxite deposits (Poljane and Bešpelj). There are also deposits of nonmetallic raw materials (building stone in Komotinski potok, clay in Divičani, and gypsum anhydrite on the shores of Lake Pliva). The decorative stone Plivit (made of travertine materials) is widely known. Jajce Alloy Wheels (metal processing industry) is the largest exporter in the municipality. Jajce is one of the

Fig. 15.13 Sultan Ahmed street in the city of Bugojno. (Photo by Haris Gekić)

most important tourist centers in the country (exceptional natural environment with 30 national, very important cultural and historical monuments from all periods).

Upstream along the river Pliva to the southwest is the municipality of *Šipovo* with an area of 553.41 km^2, and the lowest population density in the mesoregion (18.6 inhabitants per km^2). It is located in the valleys of the rivers Pliva and Janj, surrounded by the massifs of Vitorog, Lisina, and Čardak. According to the results of the 2013 census, there were 10,293 inhabitants in the municipality with pronounced negative demographic characteristics (in 2015, the natural increase was −9.5‰). The backbone of development consists mainly of nonmetallic mineral resources (high-quality reserves of gypsum, bentonite, dolomite, and brown coal), and forest resources (about 50% of the municipality's territory). The degree of urbanization is about 40%, a special tourist and the canyon and valley of the river Janj with attractive "Janjski islands" (Fig. 15.14).

The municipality of *Jezero* is located on the shores of Lake Pliva, between Šipovo, Mrkonjić Grad, and Jajce, from which it is equally distant (about 11 km). With an area of 55.6 km^2 and 1144 inhabitants (2013), it is the smallest municipality in this mesoregion (population density is only 20.6 inhabitants per km^2, excluding urban population) and a very poorly developed economy. It has one of the lowest GDPs in the country (2013/2014 only 1102 BAM).

The largest (677.43 km^2) municipality of this mesoregion by area, *Mrkonjić Grad*, is located in the far northwest. It is located at a favorable intersection of the main roads M-5 (Jajce-Bihać) and M-16 (Banja Luka-Jajce-Split). According to the results of the 2013 census, there were 16,667 inhabitants in the municipality (24.6 inhabitants per km^2). The population lives in 38 settlements (44.2% of the population lives in the city center) (BHAS 2016). It has a very unfavorable age structure, in which the old population makes up as much as 20% of the total (young only 13.8%). The larger companies are the Bema footwear factory and the RS silicon factory (production of silicon metal). The highest revenues of municipalities are generated from the trade sector, which employs about 15% of workers. From the tourist aspect, the most interesting is the mountain Lisina and the Balkana Lake at its foot, and the historical complex.

15.2.2.2 The Upper Bosna-Lašva Mesoregion

The Gornjobosansko-Lašvanska (The Upper Bosna-Lašva) mesoregion is spatially determined by its central geographical position within the territory of Bosnia and Herzegovina, and along with the Gornjovrbasko-plivska mesoregion, it is the only geographical unit that does not have access to the international border. It consists of as many as 28 administrative-political units of local self-government, with a dominant concentration of a larger number of urban centers and population. On an area of 5849.83 km^2 (about 11.4% of the territory of Bosnia and Herzegovina) according to the 2013 census, this mesoregion gathered 844,680 inhabitants, which represented about 24% of the total population in Bosnia and Herzegovina (which is immediately behind Posavsko-sprečansko-majevičko-donjodrinska mesoregion).

This mesoregion is the area with the highest concentration of population in Bosnia and Herzegovina with an average population density of 144.4 inhabitants per km^2, which is approximately twice the value of the national territory (68.9 inhabitants/km^2), but also significantly above the average in the mountain-valley macro-regions (81.8 inhabitants per

Fig. 15.14 Janjski waterfalls near Šipovo. (Photo by Ranko Mirić)

km^2), occupying 44.55% of the total area of the mountain-valley macro-region.

The area has a very complex geological composition and belongs to different geotectonic and facial units. In the extension structures of the Central Bosnian Međugorje basins: Sarajevo Field, Visočka, and Lašvanska Basins, thick deposits of Neogene sediments were deposited, which were covered with younger Quaternary sediments. The mentioned relief depressions dominantly marked by accumulation processes and increased lateral fluvial erosion express fragments of the former spacious pull-apart structure of the Sarajevo-Zenica basin. The flattened surface of the Sarajevo field with an area of 44 km^2 is the largest basin in the macro-region of the Bosnian highlands (Lepirica 2009).

Neogene complexes cover about ¼ the surface of the mesoregion. Quaternary strata (the youngest geological formations) are mostly found next to larger rivers where accumulative planes are formed. They have the largest distribution in the area of Sarajevo field. The Miocene complex (12.5% of the total area of the mesoregion) is most widespread in the Sarajevo-Zenica basin (it contains the main coal series). Mesozoic deposits cover 2/3 of the surface of the mesoregion. They are represented on the northeastern edge of the Sarajevo-Zenica basin. These layers cover the area from the valley of the river Ljubina in the east, all the way to the far northwestern parts of the basin (just below the mountain Vlašić). Cretaceous sediments also occur in the northeastern part of the mesoregion, in the valley of the rivers Bioštica and Stupčanica. The southeastern and eastern part of the perimeter of the basin is mostly built of Triassic sediments. These deposits are present along the eastern edge of the mesoregion all the way to the Zvijezda mountain, i.e., in the valleys of the rivers Lepenica, Fojnička River, and Kreševka, in the far northwestern part of the mesoregion. These formations also occur on the far southeastern edge of the mesoregion (southwestern parts of the Romanian plateau).

Palaeozoic deposits (about 17% of the territory of the mesoregion) are spread on parts of the Bosnian shale mountains (Krušćica, Radovan Mountain, parts of the Vranica mountain) and the valley of the river Bistrica. In the form of smaller fragments, they include parts of Rostovo, Novi Travnik, Fojnica, and Kreševo. The Upper Bosnian-Lašva mesoregion is the central area of the macro-region dominated by the succession of depressions (Sarajevo-Zenica valley together with a smaller composite valley of the river Lašva in the northwest of the mesoregion) and the mid-mountain-highland relief on the edges with higher hypsometric levels. Altitudes of 260–500 m have a share of almost 7% and refer to the lowest terrains of the mesoregion or river alluvium. Areas from 500 to 1000 m make up 56.43%, in the total area of the mesoregion (low mountains around the valley of the river Bosna and its tributaries and the rivers Bioštica, Stupčanica, Krivaja, and Duboštica in the far northeast). The hypsometric level from 1000 to 1500 m has a share of 31.7% and includes the mountain morphostructures of Zvijezda in the northeast, Čemerske Mountains and Sarajevo's Ozren in the east, Trebević and Igman in the south, and the slopes of Vlašić (far northwest). The hypsometric level from 1500 to 2000 m accounts for about 5% of the surface. These include the mountains of the Inner Dinarides, which with their northwestern, western, and southern parts enter the area of the Upper Bosnian-Lašva Mesoregion. This hypsometric level covers a part of the high mountain massif Vlašić in the northwest, parts of Vranica in the west, and in the south the northern parts of Igman, Trebević, Bjelašnica, and Treskavica, and in the east part of Jahorina. Only 0.5% of the surface area falls to the highest hypsometric level (from 2000 to 2110 meters). These are the highest peaks of the Upper Bosnia Lašva mesoregion: the Nadkrstac (2010 m) on Vranica, Velika ćaba (2088 m) on Treskavica, and Krvavac (2062 m) on Bjelašnica. In terms of geomorphological structure, this area is dominated by fluvio-denudation relief forms that make up about ¾ the surface of the mesoregion. Fluvio-accumulative forms have a proportion of approximately 12%. The fluvio-karst form is present only in the share of 0.34% in the area of the mountain Vlašić, which represents the farnorth-western border of the mesoregion.

Underground karst morphosculpture also developed on the dominant limestone-dolomite substrate. One such in the eastern part of the territory is the Bijambare Cave, which is a karst form of relief. Various karst accumulative forms of cave jewellery, stalactites and stalagmites, developed there (Drešković and Mirić 2017).

The nontectonic uplift of the Inner and Central Dinarides was reflected in the narrowing and closure of the Sarajevo-Zenica Neogene basin, which resulted in the formation of the Central Bosnian hills formed on Miocene clasts between the Busovača-Ilidža fault in the southwest and the Sarajevo-Banja Luka flysch zone in the northeast. Due to dynamic geotectonic (fault) processes, secondary valleys (Lašva, Fojnička rivers, Stavnja, Prača, etc.) were formed which are directed towards the main river valleys.

When it comes to the spatial definition of climate types according to Köppen's classification of climate in the Upper Bosnian-Lašva Mesoregion, three climate classes (C, D, E) and ET are represented. The dominance of the Cfb climate in the valley of the Bosna River and its larger tributaries can be observed, while the Cfc climate type occupies a wider area of the mesoregion. Dfb climate type is specific for the southern, southwestern, and northern edges of the mesoregion (i.e., hypsometric levels from 1000 to 1500 m above sea level). Dfc climate type covers the areas of the mesoregion from 1500 to 2000 m above sea level. ET climate type is represented at the highest altitude of the mesoregion which is 2110 m (Vranica) in the far western part of the mesoregion.

The average annual temperature in the lower altitudes of the mesoregion is about 11 °C (area along the basin of the river Bosna and its tributaries). The annual rainfall is about 1100 mm. It gradually decreases towards the interior where in the northern and northeastern part, it averages about 850 mm. The Bosna River is the main watercourse and hydrographic backbone in the Upper Bosnia-Lašva Mesoregion. It belongs to the Sava River Basin with an area of 10,759 km^2 or 21% of the total area of Bosnia and Herzegovina.

The hydrographic source of the river Bosna is represented by over 35 water-producing springs that were formed at the contact of the water-permeable and water-impermeable rocks of Igman and the Sarajevo field. Downstream with the confluence of river Željeznica, the river Zujevina flows into the river Bosna from the left. At about 2.9 km downstream of the river Bosna, the river Dobrinja flows from the right, which is mainly fed by the periodic flows of the river Sušica. The next river that flows into Bosnia from the right is the Miljacka (it originates at the foot of Romanija from Paljanska and Mokranjska Miljacka) (Fig. 15.15). The rivers Vogošća and Ljubina are the next right tributaries of the river Bosna. The right tributary of Bosnia at a distance of 9 km downstream is the river Misoča in the Ilijaška valley, followed by the rivers Stavinja, Fojnica, Lašva, and Ribnica.

Fig. 15.15 Paljanska Miljacka source. (Photo by Ranko Mirić)

The structural-lithological contact of limestone with the watertight deposits of the Inner Dinarides is morphologically emphasized in the relief plastic by sudden changes in the slope and the appearance of strong karst springs. Such examples are the vertical limestone section in the western slopes of Bukovik (1531 m above sea level), down which the 98 m high Skakavac waterfall collapses vertically. Hydrogeological active zones of permanent springs: Paljanska and Mokranjska Miljacka, Bioštica, Orlje, Žepa, Plave Vode, Dugovještice, Banjice, Krupa, etc. formed along faults at the contact of the steep slopes of the limestone massifs of the Inner Dinarides with permeable deposits of valley depressions.

Numerous occurrences of mineral and thermo-mineral waters are found in the area of the Upper Bosnia-Lašva Mesoregion. Occurrences of mineral, thermal and thermo-mineral waters occur in the direction of two tectonic faults, namely, Sarajevo-Kiseljak-Busovača and the direction Knežina-Olovo-Orlje.

A complex pedological substrate was formed in the area of the Upper Bosnian-Lašva Mesoregion. More specifically, a number of different soil types were formed from the automorphic and hydromorphic divisions. The most common are district cambisols on acid silicate rocks (49% share in the mesoregion area) which dominate in the northeastern, northern and northwestern areas of the mesoregion all the way to the southeastern borders (mountain areas or terrains with an altitude of 700 to 1500 m). The second most dominant soil is calcocamisol and calcomelanosol-calcocambisol mosaic on limestone, which are present not only in the southeastern and eastern part but also in the far western part of the mesoregion (occupying about 22% of the mesoregion territory). Fluviosols or alluvial soils have developed on river sediments in the zones of large watercourses of the Bosna, Željeznica, Miljacka, Fojnička Rivers (tributaries Kreševka and Lepenica), and the Lašva valley. The fluviosoles are followed by soils from the rendzina-district cambisol-pseudogley complex on flysch-like sediments and are the third most represented soil in this area (share of 8.1% of the mesoregion area). The more widespread soils of the mesoregion also include the eutric cambisol and the ranker-cambisol and cambisol-luvisol series on serpentinite and peridotite, which are especially present in the central part of the Bosna river basin (4.6% of the area). Fluvisols account for about 3.4%.

About 65% of the territory of the mesoregion is covered with forest vegetation, of which the most common are deciduous (about 35%), followed by mixed forests (15.13%), while coniferous forests account for about 12.6%. The largest area of the mesoregion belongs to the Euro-Siberian-

Boreamerican biogeographical region, which is in contact with the far northern parts of the Adriatic province. The Euro-Siberian-Boreo-American biogeographical region is represented by two biogeographical provinces (Illyrian and Moesian). In the Illyrian province, in the area of the mesoregion, belts of beech forests (*Fagion Illyricum*) with subbands of thermophilic beech forests, mountain mesophilic beech forests, and mountain beech-fir forests stand out. In the zone of the hypsometric belt from 1000 to 1800 m above sea level, a belt of dark coniferous spruce and fir forests stands out.

The ecosystem of mesophilic deciduous forests of sessile oak and hornbeam (*Carpinion betuli Illyricum Horvat*) is mostly represented in the mesoregion with altitudes of 300–800 m above sea level, in which the Cfb climate type dominates. The sub-belt of mesophilic beech forests covers a large area of the central and continental Bosnian Dinarides, in a wider morphological zone of transition from the upper parts of the mountain to the lower parts of the mountain belt (Drešković and Mirić 2017).

In terms of ore-mineral resources, the modern industrial development of this area relies on a rich raw material and energy base and reserves of ore and mineral raw materials. In the Sarajevo-Zenica valley there are large reserves of brown coal (580 million tons or 29% of the reserves of Bosnia and Herzegovina). In the Vareš area, there are deposits (already on the threshold of commercial exploitation) of iron ore of magnesite, lead, zinc, copper, barite, refractory clay, marl, limestone, and quartz sand. In the area of Kreševo, there are fortified deposits of pyrite, copper, mercury, and antimony. There are numerous quarries of limestone, dolomite, quartz keratophyre, granite, and quartzite. The southwestern area of the mesoregion is especially rich in various types of nonmetals (barite, quartz sand, calcite reserves, and various other minerals). Silver was found in the area of the central Bosnian shale mountains in the vicinity of Busovača, and pyrite in Bakovići near Fojnica.

The Upper Bosnian-Lašva Mesoregion is the most economically developed part of Bosnia and Herzegovina. The concentration of industrial-energy and processing capacities is in connection to the area of the Sarajevo-Zenica valley (the main producer of brown coal, iron, and steel). The Sarajevo-Zenica coal basin with quality brown coal covers an area of about 900 km^2. In the postwar period, all these mines were equipped with modern mechanization. The secondary sector is reflected in the development of industry, construction, and energy (Zenica, Kakanj, Sarajevo), and the tertiary sector, which is dominated by trade, transport, tourism, health, and culture. The primary sector is reflected in the development of forestry, and in a much smaller percentage through agricultural activities. According to the 1991 census, the Upper Bosnia Lašva Mesoregion had 1,026,641 inhabitants, while in 2013, that number amounted to 844,680, which represents a decrease in population by 17.8%. (Only 5 of the 28 municipalities of the mesoregion recorded an increase in population in this period.) There is also a significant polarization in population distribution. More than half of the municipalities (15 of them) have a lower population density than the average. The remaining 13 municipalities have indicators that are higher than average and gather about 72% of the population of the mesoregion to only 26% of the total territory (Sarajevo city municipalities, Zenica, smaller municipalities within Istočno Sarajevo, Breza, Visoko, and Vitez). Of the total population according to the 2013 census, males account for 48.4%, while females account for 51.6%. The status of the city has four municipalities within the mesoregion, namely: Istočno Sarajevo, Sarajevo, Visoko, and Zenica.

The capital *Sarajevo* is the area of the most intensive economic, political, cultural, tourist, and general socioeconomic development of Bosnia and Herzegovina. In administrative terms, the city consists of four municipalities that from east to west represent an uninterrupted urban structure along the valley of the river Miljacka (Stari Grad, Centar, Novo Sarajevo, and Novi Grad). Together with Vogošća, Ilidža, Trnovo (FBiH), Hadžići, and Ilijaš, they form the Sarajevo Canton. According to the 2013 census, there were 275,524 inhabitants (high 2470 inhabitants per km^2) on the territory of 4 city municipalities. The young population makes up 16% of the total population of the city (the contingent of the old has a smaller share −13.8%).

Development and its functions have conditioned it to be a traffic center in BiH. Seven highways connect it with other parts of the country. In the north are the routes M-5 (Travnik - Banja Luka-Bihać), M-17 (Zenica-Doboj), and the main road M-18 (towards Tuzla). To the east are the main routes M-5 (towards Goražde and Višegrad) and M-19 (towards Zvornik). To the south, in the direction of M-18, there is a connection with Dubrovnik (via Foča), and the main road M-17 (Mostar). The Šamac-Ploče railway line connects Sarajevo with the far northern and southern parts of Bosnia and Herzegovina. Sarajevo Airport is the best infrastructurally equipped for this type of traffic in Bosnia and Herzegovina (about 1.4 million passengers in 2019). City traffic, in addition to bus traffic, consists of trams (since 1885) and trolleybuses (since 1984). The city itself is located in the valley of the same name, which spreads to the west into the Sarajevo field (area of the municipality of Novi grad). It is surrounded in the immediate vicinity on the south side of the Trebević slope with smaller elevations to the west (Vraca pass, Hrasno, and Mojmilo hill) (Fig. 15.16), while the northern rim is represented by Crepoljsko, Bukovik, Lopata, Saračevac, Poljine, Hum, Orlić, and Žuč hills. The City of Sarajevo is the center of the most important state, entity, and cantonal institutions (Parliament of Bosnia and Herzegovina and Parliament of the Federation of Bosnia and Herzegovina,

Fig. 15.16 The Vraca Memorial Park dedicated to the World War II victims in Sarajevo. (Photo by Ranko Mirić)

Council of Ministers of Bosnia and Herzegovina and the Government of the Federation of Bosnia and Herzegovina, courts, and other institutions of state entity and cantonal importance). There are numerous diplomatic and consular missions (currently 44 embassies are located in Sarajevo, of which a smaller number are not in city municipalities), economic missions, and missions of international institutions and organizations operating in BiH (Delegation of the European Union, OSCE Mission NATO Headquarters Office UN Office of the High Representative UNHCR, World Bank missions, etc.).

In terms of education, the City of Sarajevo has 26 primary and 37 secondary (vocational and gymnasium) schools recognizable by quality in Bosnia and Herzegovina. In 2014–2015, 22,374 students attended primary school and 14,154 secondary school students (about 14% of the total population of the city). The University of Sarajevo was founded in 1950 (the first members were the Faculty of Philosophy and Veterinary Medicine). Today, as a respectable higher education teaching institution in the international framework, it gathers 30 institutions, namely, 22 faculties, 3 academies, and 5 scientific research institutes, with over 550 study programs and about 30,000 students, which make this university the largest in Bosnia and Herzegovina.

In economic terms, Sarajevo is the most attractive area for business investment and business due to its position as the capital. In 2019, 24,275 business entities were registered in the city (the highest concentration is in the municipality of Centar 9747). For comparison, the city of Banja Luka had 9182 registered businesses in the same year. In terms of structure, over 80% of business entities in the city are from the tertiary sector (over 80%, approximately the same number of employees), dominated by trade (shopping centers), finance and banking (headquarters of 14 large banks), health (Clinical University Center, General Hospital), and education, information, and communication activities (46 businesses), hospitality and hotels industry (over 130 facilities with about 7000 beds) and related tourist activity. In 2018, the City of Sarajevo was visited by 323,554 tourists, who realized 654,232 overnight stays.

This represents 22% of the total number of tourists in BiH in 2018, and 21.5% of overnight stays at the state level. Different historical periods have left a deep mark on today's Sarajevo. From the late Neolithic (locality Debelo brdo), the settlement of Illyrian tribes (Desitiati) from the Bronze and Iron Ages, Roman ancient culture (up to the fifth century AD), many remains of material culture have been found. With the arrival of Slavic peoples in this area, it represents the beginning of permanent and continuous settlement but also the creation of embryos of the Bosnian medieval state (the first written trace of the Bosnian state was left by the Byzantine emperor Constantine VII Porphyrogenitus in his work "De administrando imperio"). The origin of the first settlements is connected to the end of the thirteenth century (Vrhbosna).

The most representative remains from that period are stone tombstones-stećak (written in the medieval Bosnian alphabet—Bosančica). The urban foundations of Sarajevo, the capital of Bosnia and Herzegovina, were laid in 1462 by Isa-beg Ishaković, thus beginning the period of Ottoman rule, and the influence of Oriental-Ottoman architecture in the expansion of the city. In the seventeenth century, Sarajevo had about 5550 houses, with 93 neighborhoods ("mahala"). With the weakening of the Ottoman Empire, the rule of the Austro-Hungarian monarchy was established (1878 to 1918). The presence of Central European culture and architecture is still visible today (several Western European architectural styles—eclectic, renaissance, seecession, Bosnian style). The period from 1918 to 1945 was marked by stagnation in the general urban and economic development of Sarajevo, which lost its main work. During the Second World War, there were huge sufferings from the fascist regime of the independent state of Croatia and the German occupation forces (about 10.6% of the population of Sarajevo). In the mid-1960s (the adoption of the general plan for the development of the city), its rapid expansion in a westerly direction along the Miljacka valley began. Several representative business facilities were built (Energoinvest, Skenderija Cultural and Sports Center, State Hospital, Maternity Hospital, Hepok Shopping Center, Elektroprivreda) administrative (building of the former government of SRBiH and the Republic Assembly), cultural (radio-television building), educational (Faculty of Science, Student Dormitory Nedžarići), and residential areas (Alipašino polje, Otoka, Čengić Vila). The last stage was marked by infrastructural works due to the IVX Winter Olympic Games (1984), when many business, tourist, and residential buildings were reconstructed (Zetra sports hall, Ciglane, Mojmilo and Dobrinja settlements, Oslobođenje newspaper house, Holiday hotel).

The most important cultural resource of the Sarajevo Canton and Bosnia and Herzegovina is the diverse cultural heritage that has been preserved for centuries in Europe, thanks to the geographical, political, and religious specifics of medieval Bosnia and the unique cultural and religious tolerance of the Ottoman Empire. Eastern and Western Christianity, Islam, and Judaism left a great cultural heritage on the territory of Sarajevo Canton, but also Bosnia and Herzegovina, and created a unique Bosnian identity (Bibanović 2015).

It is almost impossible to objectively single out certain cultural and historical attractions in Sarajevo, due to their number and concentration. This is best illustrated by the decision of the State Commission for the Preservation of National Monuments of Bosnia and Herzegovina, which declared the area of Sarajevo's core in the part of Stari Grad and Centar to be a national monument as a historic urban landscape with 87 individual national monuments and properties. Recognizable in the world are the City Hall, the Latin Bridge (the site of the assassination of Crown Prince Franz Ferdinand by Gavrilo Princip), Baščaršija, the Bey's Mosque with a madrasa, the old Orthodox Church, the Catholic Cathedral, the Old Jewish Temple, Sebilj, the Eternal Fire, the National Museum, and many more. A special factor in the tourist offer of Sarajevo is the Sarajevo Film Festival, which has become a recognizable event in the world. The city of Sarajevo with its municipalities is a unique urban area along the valley of the river Miljacka. The Municipality of *Novi Grad Sarajevo* (47 km^2 in 2013) had 118,553 inhabitants (2511.72 inhabitants per km^2). It is the only city municipality that had a positive rate of natural increase (1.2%) with a low average age (39 years—2015). The urban population has a share of 99.43%. It covers the western part of the Sarajevo field (the area of the lowest elevation in the city of Sarajevo—482 m above sea level in the settlement of Reljevo). It is characterized by a modern urban structure with collective housing facilities (Alipašino, Mojmilo, Dobrinja, etc.). The northern part of the municipality is primarily built of individual settlements (Buća potok, Boljakov potok, Rajlovac, etc.).

Novo Sarajevo, with an area of 9 km^2, is the smallest municipality in Bosnia and Herzegovina. In 2013, it had 64,814 inhabitants and the highest population density in the state (6546.87 inhabitants per km^2). It is located in the Miljacka valley between the Hum hill and the Vraca pass. A negative rate of natural increase (−1%) was recorded. The average age is 41.1. The urban population makes up 98.67% of the total.

The Centar Sarajevo municipality (33 km^2) is part of the city area, located in the valley of Miljacka and on the slopes of the surrounding hills in the north (Nahorevo, Mrkovići, Poljine −64.6% of the area is above 700 m above sea level). In 2013, it had 55,181 inhabitants (1672.15 per km^2—the third most densely populated municipality in BiH). It has a negative rate of natural increase (−2%). Average age is 42.2. The share of urban population was 96.19% (Fig. 15.17).

The municipality of *Stari Grad Sarajevo* (50 km^2) is located in the eastern part of the Sarajevo valley, which is surrounded by Trebević on the south and the mountains Crepoljsko and Bukovik on the north. It is recognizable for the Baščaršija (the old city center from which the city of Sarajevo developed). Stari Grad is the largest municipality in the city of Sarajevo (covering 37% of its territory). In 2013, it had 36,976 inhabitants (719.38 inhabitants per km^2), a negative rate of natural increase (−3%). The average of the years was 41.4. The urban population had a share of 98.28% in the total.

The municipality of *Ilidža* is located in the extreme southwest of the Sarajevo field, which administratively does not belong to the City of Sarajevo, but is physiognomically and functionally part of the urban city space (tram city traffic). It covers an area of 136 km^2, where in 2013, there were 66,730 inhabitants with a high population density (465.34 inhabitants per km^2), and a positive rate of natural increase (2%). The average age was 37.5 years. The urban population has a very high share (95.2% in total). The city area is located in the area of Sarajevo field and on the southwest side is the

Fig. 15.17 The Centar municipality. (Photo by Ranko Mirić)

mountain Igman. The area has a very rich hydrographic network, the backbone of which is Bosnia (springs at the foot of Mount Igman) and the rivers Željeznica, Zujevina, Miljacka, Dobrinja, Tilava, and Večerica. It is globally recognizable for The Big Alley (Velika aleja), Vrelo Bosne (protected area), and spa tourism. The city of Sarajevo supplies drinking water from the Bačevo spring.

The sectoral structure indicates that the municipality is dominated by the tertiary sector in which 57.4% of registered entities employ 48.9% of employees. The secondary sector employs 24.6% of employees who, according to the number of registered business entities, participate with 14.1% in the total number of business entities in the municipality (Development Strategy of the Municipality of Ilidža 2014). The process of building residential, business, and hotel capacities (Malak, Hollywood, Hills) in the city center is very intensive (In 2018, Ilidža registered 417,947 overnight stays).

As part of the once unique metropolitan area of Sarajevo (a city community that consisted of 10 local self-government units until 1991), local self-government units were formed from parts of Sarajevo municipalities in the eastern part of the Sarajevo valley and fields (except Pale RS municipality, which had the status of a municipality before 1991).

The city of *Istočno Sarajevo* is a complex unit of local self-government. In the area of this mesoregion, only the municipality of Sokolac does not belong to it. The area within this region is 742 km^2, in which 49,495 inhabitants lived in five local self-government units in 2013, with an average population density of 66.7 inhabitants per km^2. The degree of urbanization in 2015 was 56.2%. The city of Istočno Sarajevo in this mesoregion consists of the municipalities of Istočna Ilidža, Istočno Novo Sarajevo, Pale, Istočni Stari Grad, and Trnovo (RS).

The administrative center of the city of Istočno Sarajevo (City Administration) is located in the municipality of *Istočno Novo Sarajevo*. It consists of two physiognomic units: a hilly-mountainous area in the northeast (Trebević, Jahorina, Romanija, and Ozren) and the southeastern part of the Sarajevo field (Fig. 15.18). The Dayton Agreement established the newly formed municipalities from the former rural settlements (Lukavica, Kasindo, Vojkovići, etc.), most of which in modern times have experienced a very dynamic spatial and functional transformation into modern urban settlements. Municipal, city, entity, and some state institutions are located there, as well as the seat of the University of Istočno Sarajevo (16 faculties and 2 academies). In recent years, with the construction of new apartment blocks, new buildings of municipal, city, republic, and state institutions, construction of new cultural, school, and economic facilities, parks, sports, and other facilities, Istočno Sarajevo is taking on physiognomic and functional features of the city (Strategic development plan of the city of Istočno Sarajevo 2010).

Trnovo (RS) with an area of 110 km^2 in 2013 had 2050 inhabitants (17.64 inhabitants per km^2). It is characterized by a very unfavorable demographic trend. The share of the

Fig. 15.18 Jahorina mountain. (Photo by Ranko Mirić)

elderly (21%) is twice as high as the share of the young population in the total population, with a high negative rate of natural increase (−10‰) and an average age of 45.7 years (2015). The urban population makes up 46.63%. The city headquarters are located on the river Željeznica and are surrounded by mountain ranges of Treskavica and Jahorina (Trnovo owns 30% of the territory of the sports and recreation center Jahorina).

Pale (RS) is the largest municipality in eastern Sarajevo (492.82 km^2). In 2013, it had 20,909 inhabitants (43 inhabitants per km^2). It is dominated by mountain relief (Jahorina, Gosina, Romanija, and Ozren). The municipality is located only 15 km from Sarajevo (main road M-5). About 70% of the area is covered by forests. It has a negative rate of natural increase (−4.1‰). The urban population accounted for 61.72% of the total. Carriers of economic development are forestry, wood, and construction industry and tourist and catering industry (tourist center Jahorina as one of the most developed in BiH, ski resort Ravna planina, sports center Dvorišta Jahorina, and Orlovac and Novakova cave). The highest revenues are generated by the tourism, trade, and catering sector, especially in the winter part of the year (accommodation capacities in the municipality are around 3500 beds).

Smaller but more densely populated municipalities are located in Sarajevsko field. *Istočna Ilidža* (29 km^2) in 2013 had 14,763 inhabitants (529.14 inhabitants per km^2), which is an increase of 71% compared to 1991. The rivers Željeznica, Kasindol, and Dobrinja flow through the territory of the municipality. From a rural area, this municipality has become a highly urbanized area (93.65% of the urban population). It has a negative rate of natural increase (−6‰, 2015). In economic development, trade, catering, and other service activities dominate.

Istočno Novo Sarajevo (38 km^2) in 2013 had 10,642 inhabitants (306.8 inhabitants per km^2) and a zero rate of natural increase. Since 1991, there has been a 161% increase in population (far ahead of all other local government units in BiH). The average age of the population is 40.1 years. There are several companies in the metal processing industry (Famos, TAT Thermal Appliances, LASTA Istočno Sarajevo).

The municipality of *Istočni Stari Grad* (88 km^2) in 2013 had 1131 inhabitants (16.19 inhabitants per km^2). It is characterized by an extremely low rate of natural increase (−18‰ in 2015), with an average age of 48.1 years. The urban population makes up only 3.27% of the total. The old population has a share of as much as 27% in the total population (the young account for only 10%). It occupies the area of the gorge valley of the river Miljacka and parts of the mountain massifs of Trebević and Ozren.

Vogošća (72 km^2) in 2013 had 26,343 inhabitants (367.41 inhabitants per km^2). It stretches from the very edge of the Sarajevo field to the confluence of the Misoča and the Bosna rivers in Ilijaš. One of the few municipalities that had a positive rate of natural increase (3‰, 2015) and a significantly higher share of young people compared to the old population (by 6.7%). The average age is only 37. The urban population makes up 40.12% of the total. The area of the protected landscape of the Skakavac waterfall extends partly to the municipality of Vogošća. Prior to the Dayton Accords, Vogošća was a city municipality and was connected by trolleybus transport to the center itself (6 km away). It is a modern urban settlement where service activities dominate.

Ilijaš (309 km^2) is a municipality that had 19,603 inhabitants in 2013 (63.52 inhabitants per km^2). The area occupied by the municipality and the town itself is located in the valleys of the rivers Bosna and Misoča, which to the north passes into the mountainous area of Crnoriječka plateau, and the mountains Čemernica, and the valleys of the rivers Rača and Ljubina. In 2015, the municipality had the highest birth rate in BiH (15‰ with a positive natural increase of 4.3‰). The average age is 37.08. The urban population makes up 25.1% of the total. The area of the protected landscape of Bijambare belongs to this municipality. Significant investments are made in the cultivation of ecologically healthy agricultural crops (Nišići). The largest companies are in the wood processing industry (Standard furniture factory with 700 workers). According to the 2013 census (BHAS 2016), the municipality of *Hadžići* (273 km^2) had 23,891 inhabitants (87.42 inhabitants per km^2), and a positive rate of natural increase (2‰). It is a suburban area of Sarajevo with a low share of urban population (20.9%). It is located on the highway and railway line Sarajevo-Mostar. It extends to Ivan pass in the south, and it also includes the southern parts of the mountains Bjelašnica and Igman-Hadžička valley open to the Sarajevo field. The average age was 37.13. The age structure is favorable (as in the municipalities of Ilijaš and Vogošća) with a higher share of young compared to old (by 7%). Due to the proximity of Sarajevo, six industrial zones have been formed here, which bring together many domestic and foreign companies (Coca-Cola Beverages BiH, Akova group, Argeta, Teloptic-dairy, Sinalco, Würth, and many others).

The municipality of *Trnovo* (F BiH) has an area of 341 km^2. In 2013, it had 1502 inhabitants (only 4.44 inhabitants per km^2). It is located 30 km southeast of Sarajevo in the area consisting of the mountains Igman, Bjelašnica, Treskavica, and Jahorina. It has a pronounced negative natural increase (as much as −21‰, which is also the lowest rate of natural increase in the Upper Bosnia-Lašva Mesoregion), and the highest mortality rate (27‰). The average age is 49.14 years, and it has only 4.46% of urban population. The average annual air temperature in Trnovo is 7.3 °C, and with plenty of snowfall, it is the backbone of the economic development of the municipality. Apart from the Olympic Mountains, the tourist attractions are the Rakitnica Canyon and the original ambience of the Bjelašnica village of Umoljani.

The municipality of *Pale (FBiH)* covers 87 km^2 with a population of 904 in 2013 (1.83 inhabitants per km^2). It had a negative natural growth rate (−10‰, 2015). The average age is 42.2. It has no urban settlements. It is located on the banks of the river Prača in a valley bordered by the mountains Jahorina and Romanija, 36 km east of Sarajevo.

On the development axis Sarajevo-Zenica, there are several municipalities bordering the local self-government units of the Sarajevo urban area. Northeast of Sarajevo is the municipality of *Olovo*, about 60 kilometers from the BiH capital. With an area of 408 km^2 (in 2013), it had 10,175 inhabitants (24.95 inhabitants per km^2). Unfavorable trends are manifested through a negative rate of natural increase (−6‰), but there is still a higher share of young people compared to the old population (by 0.7%). The urban population makes up only 24.23% of the total. It has significant natural resources (68.1% of the municipality's area is covered by forests). Thermal waters are the basis for the development of health tourism (Aquaterm). The city center itself is the place of origin of the river Krivaja (confluence of Stupčanica and Bioštica).

At 23 km from Sarajevo, the river Stavnja flows into Bosnia from the north. Along the course of this river to the north, there are two municipalities of this mesoregion. *Breza* (73 km^2) in 2013 had 14,168 inhabitants (194.35 inhabitants per km^2), a negative rate of natural increase (−4‰), and an average age of 39 years. The urban population makes up only 21.27% of the total. Breza is 25 km away from the capital of BiH, Sarajevo. The most important natural resource is coal, whose exploitation is carried out within the economic entity Breza Brown Coal Mine. A significant resource is geothermal mineral water in the valley of the river Stavnja (mineralization is 1.2 g/l and belongs to the chlorine-sulphate-calcium-sodium-magnesium group of medicinal waters). Breza mine employs over 60% of the workforce. Of the tourist attractions, the most important is the late antique Basilica, which is relatively well preserved.

The municipality of *Vareš* (390 km^2) in 2013 had 8892 inhabitants (22.79 inhabitants per km^2), and a negative rate of natural increase (−12‰). The average age is relatively high (44.6). The urban population accounts for 32.80% of the total population. Located at 829 meters above sea level (the valley of river Stavnja), it is surrounded by mountains Zvijezda and Perun. The specificity of the area of the municipality of Vareš is the remains of once very developed mining-artificial lakes created by ore exploitation and ore disposal (surface mine Spruce and disposal Veovača). Mining once employed about 5000 workers (all plants were shut down long ago). It is one of the most forested municipalities in BiH (77% of the area under forests). Comparing the census from 2013 with the one from 1991, the number of inhabitants decreased by as much as 60%. The rate of natural increase is extremely negative (−11.2‰ in 2015). It is 8% older than the contingent of young people. It is one of the most underdeveloped municipalities with a ubiquitous emigration trend.

The municipality of *Visoko* (233 km^2) is the regional center of this part of the mesoregion. In 2013, it had 39,938 inhabitants (173.04 inhabitants per km^2) and a positive natural increase (0.1‰). The average age is only 36.6. The urban center (28.1% of the population) is located at the confluence of the Fojnica and Bosnia rivers. Visoko is a town with a long tradition (the historical center of the medieval state of Bosnia, where the Charter of the Bosnian Kulin Ban was written in 1189, the oldest preserved Bosnian state document), and it had developed trade. The area of the municipality has relatively low altitudes in the valleys of the rivers Bosna and Fojnica. Stocks of limestone and brick clay have conditioned the development of production of construction materials (the company IGM is the leading producer of brick and styrofoam products in Bosnia and Herzegovina). In the food industry, the leading producers in the state are Vispak, Akova Group, Semić, and a number of other smaller companies. The leading company in the processing industry is Ugarak (production and installation of high-quality PVC joinery and aluminum hardware, with a large export orientation). In the last decade, the Ravne Archaeological Park with the Bosnian Pyramids complex has paid special attention to Visoko's tourist offer. Without entering into discussions about the scientific basis of their existence, Visoko has become a globally recognizable tourist destination, thanks to this locality, with a sharp increase in the influx of (especially foreign) tourists.

The municipality of *Kakanj* (377 km^2) in 2013 had 37,441 inhabitants (99.31 inhabitants per km^2), and a negative rate of natural increase (−1‰). 31.51% of the population lives in urban settlement. The municipality is located in the alluvial plain (Dobojsko field) through which the river Bosna flows (larger tributaries in this municipality are Zgošća, Ribnica, Trstionica). The most important natural resource is a coal basin about 22 km long (brown coal geological reserves are estimated at about 400 million tons). Forests cover about 60% of the municipality. Mining, cement, and electricity production are the basis for the development of the municipality of Kakanj. The share of mining and electricity production in the social product is over 45%. According to the data submitted by JP Elektroprivreda BiH d.d. Sarajevo, Subsidiary Brown Coal Mine "Kakanj" d.o.o. Kakanj, in 2016, employed a total of 1730 workers, which is 520 workers less than in 2006 (Revised Development Strategy of the Municipality of Kakanj 2018). Very important economic entities are Cementara Kakanj (Heildeberg Group) and Termoelektrana Kakanj (thermoelectric Poweplant—450 MW).

The city of *Zenica* with an area of 558.5 km^2 and 110,663 inhabitants (198.14 inhabitants per km^2) is the second largest urban and regional center of the Upper Bosnian-Lašva Mesoregion (after the Sarajevo municipality of Novi Grad, it has the largest population in macro-regional context). In 2013, a zero rate of natural increase was recorded. The average age is 38.4 years. Around 64% of the total population lived in the city center. The municipality has an excellent position, because in addition to being located on the route of the Vc corridor, the Lašva loop also connects traffic to the western part of the country.

Zenica is located on a relatively large erosive extension of the river Bosna, which belongs to the valleys of the central Bosnian low mountains. It has a favorable geo-traffic position and is the economic center of the geographical region of central Bosnia. The territory of the municipality, practically through the city, is crossed by the railway and the main road M17, i.e., the future highway within the traffic Corridor V-c. The distance of Zenica from the capital is 74 km. Of particular importance is the administrative and political function of the center of Zenica-Doboj Canton (brings together 12 municipalities) and the concentration of public institutions (ministries, health, administrative services). Zenica has a university (it has eight member faculties). In 2015, a natural growth rate of 0.5‰ was recorded (average age is 38.4). 63.8% of the total population lived in the city center.

In the area of the city of Zenica, a significant wealth of natural resources has been determined (agricultural land represented by approximately 23%, forest land with approximately 58%, and developed hydrographic network), but still it is underutilized. Extensive coal deposits have been the basis of economic development since the end of the nineteenthth century. The Zenica brown coal mine started operating in 1880. It currently operates within the Concern of the Public Company Elektroprivreda BiH. The total reserves in the exploitation fields are today estimated at about 956 million tons, of which the balance reserves are approximately 177 million tons, potential 720 million tons, and currently off-balance sheet approximately 58 million tons of coal. The production of coal from the Company's pits in the last 5 years averages 320,000 tons per year (Spatial plan of the city of Zenica 2016). It employs about 1450 workers. Zenica was a production and processing center of ferrous metallurgy in the former Yugoslavia. With the failed privatization and the loss of the world market, the once largest mining and metallurgical plant now operates within ArcelorMittal (with several other companies). This company has been the largest exporter in BiH for many years and employs about 2400 workers. Other companies include Almy (400 employees, production of concrete and concrete accessories), ITC (trade of agricultural machinery and agricultural equipment), Metalno Zenica, Zenica Dairy Industry (800 subcontractors), and many others.

The municipality of *Kiseljak* (160 km^2) covers the eastern edge of the central Bosnian shale mountains, in contact with the Neogene sediments of the Sarajevo-Zenica valley. Due to the proximity of the capital (25 km) and the main road in the country, it has a favorable geographical position. Kiseljak is widely known as a trade center, and a large number of companies engaged in this activity are registered here. In 2013, the municipality had 20,722 inhabitants (125.59 inhabitants per km^2), and a negative rate of natural increase (−3‰). The degree of urbanization is low (only 17.15%). Kiseljak has forest and mineral resources (primarily kaolin clays). Natural mineral water is the most important resource of the municipality of Kiseljak, given that the largest company in the municipality operates precisely on the basis of its exploitation. The product Sarajevski Kiseljak is known as a brand of the municipality and is exported all over the world.

Sarajevo kiseljak is one of the best alkaline mineral waters with its quality.

According to the 2013 census, the municipality of *Kreševo* (146 km^2) had 5273 inhabitants (35.39 inhabitants per km^2). It covers the area of the Kreševo valley surrounded by branches of the mountains Bitovnja, Zec, and Pogorelica. It is 48 km away from Sarajevo. Natural resources are forests (about 75% of the area) and barite deposits. It has a negative rate of natural increase (−6‰), and an equal share of young and old population (about 15% in total). The urban population accounts for only 19.44% of the total population of the municipality. Kreševo was a metal processing region, which in the modern age has restructured its economy with flexible smaller economic entities (e.g., MIS Kreševo—a leading producer in the meat industry). The company with the highest revenues in the municipality is Boreas (distributor and representative of world-famous brands for Bosnia and Herzegovina such as Heineken and Philip Morris).

Fojnica (301 km^2) hadin 2013—12,356 inhabitants (40.38 inhabitants per km^2), and a negative rate of natural increase (−3‰). The urban population (28.89%) lives in the eponymous city center. Forests occupy 75.2% of the total area of the municipality. Formerly, traditional mining has been reduced to two dolomite mining sites. On the basis of thermo-mineral waters of different temperatures (21 to 30.5 °C) and chemistry, the Fojnica spa complex was developed, which is the basic carrier of the development of the entire municipality (basis for spa and bathing tourism). The complex has a number of facilities that make up the infrastructure. There is a modern equipped medical rehabilitation center with the Reumal Hotel (complete medical facilities for rehabilitation), a hospital with all necessary medical treatment, and the Aquareumal apartment complex (for accommodation, rest, and recreation, with 74 apartments). The center also includes the Aquapark (1000 m^2 of water surfaces of various contents). Among the tourist attractions are the Kozice waterfall, the Franciscan Monastery, and Prokoško Lake (protected area—a natural monument).

In 2013, *Busovača* (157 km^2) had a population of 17,910 (113.35 inhabitants per km^2), a positive rate of natural increase (1‰), and a low average age of 36.5 years (2015). The vitality of the population is reflected in a significantly higher share of young people (17.7%) compared to the old population (10%). It has a small percentage of urban population (21.74%). It is located on the slope of the Krušćica Mountain, right next to the confluence of the Lašva and Bosnia rivers (one of the largest fault lines in BiH is named after this town). The municipality of Busovača lies on four rivers (Ivančica, Klokotnica, Lašva, and Kozica). It is 63 km away from Sarajevo. It has larger quantities of limestone and clay used for the production of brick products (production of construction materials in the companies Tamex, Grand, Paleta plus). About 62% of the area of the municipality is under forests.

Along the valley of the river Lašva (left tributaries of Bosnia) are located the densely populated municipalities of Vitez and Novi Travnik, and the mesoregional center, the Travnik city. In 2013, the municipality of *Vitez* (158 km^2) had 25,836 inhabitants (162.49 inhabitants per km^2), with a share of urban population of 24.5%. It is located in the valley of the Lašvansko field and on the slopes of the Kruščica Mountain. Location between the larger centers of this mesoregion is especially important (Vitez is 70 km away from Sarajevo, 30 km from Zenica and 14 km from Travnik). It has a positive rate of natural increase (0.5‰) and a low average age of 36, 45 years (the share of young people in the total population is the largest in the macro-region—18.1%). The northern area of the municipality is an area of population concentration, where all 34 settlements around the river Lašva were formed. The municipality is recognized for its very flexible and adaptable economic and trade economy, with an extremely large number of companies from many manufacturing and service industries. Among the leaders in Bosnia and Herzegovina are FIS Vitez (among the leading trading companies with 3000 employees, and its own production of furniture and textiles), TIOIL (trade in oil and petroleum products) KIM-TEC (one of the leading distributors of IT products and services in BiH), and CM-cosmetic market (chain of cosmetics stores).

The municipality of *Novi Travnik* (242 km^2) in 2013 had 23,832 inhabitants (98.48 inhabitants per km^2). A positive rate of natural increase (1‰) was recorded, and the average age (36.3 years) was the lowest in the macro-region. The urban population had a share of 37.8% in the total (2015). The municipality is located mostly in the valley of the river Grlovnica, between the mountains Komar, Radovan, and Vranica. Of the total area of the municipality, 70% is covered with forest. The city was literally created in the 1950s of the last century, thanks to the construction of a factory for the purpose-built industry "Bratstvo" (it employed about 10,000 workers). By radically reducing the capacity of the purpose-built industry factory, the municipality turned the former economic giant into an industrial zone and attracted several investors in the metal processing industry (which participates in the total economy of the municipality with about 70%). "Bratstvo" Novi Travnik is now a state-owned company with reduced capacity and employs about 350 workers (production of military equipment and weapons).

The city of *Travnik* (529 km^2) in 2013 had 53,482 inhabitants (101.1 inhabitants per km^2). It has a favorable geographical position in the valley of Lašva (which springs on the territory of Travnik), which surrounds Vlašić, and the massifs of Komar and Vilenica (forests cover 52.5% of the city area). The upper course of the river Ugar (Vrbas basin) also belongs to this town. The advantage of the geographical position is reflected in the vicinity of other centers in Bosnia and Herzegovina (Zenica 37 km, Banja Luka 98 km, Bugojno 44 km, and Sarajevo 92 km). Travnik is a mesoregional center and the administrative-political seat of the Central Bosnia Canton (gathers 12 local self-government units) with a large number of public institutions. The University of Travnik (a private higher education institution with five faculties) has its headquarters in this city. In 2015, a negative rate of natural increase was recorded (−0.7‰). The average age is quite low (37.5), and the age structure is favorable (the share of young people in the total is higher by 5.5% compared to the old population). Urbanization is low (urban population makes up only 28.7% in total). It has a brown coal mine (Abid Lolić—production of about 150,000 tons). In the area of the municipality of Travnik, there is a significant number of economic entities (recognizable in the state) that deal with the textile and footwear (Borac, Fortitudo, Nobil, Prevent, Olip) and pharmaceutical industry (Pharmamed). The food industry is especially developed (companies Bajra and Smajić-meat industry, Best-juice production). The world-famous product is Vlašić sheep cheese. In the context of tourism, the city contains attractions from a cultural and historical aspect (the birthplace of Nobel laureate Ivo Andrić, Old Town, Colourful Mosque, and the most important natural motif is Mount Vlašić, which is a popular destination for mountain and winter tourism contents) (Development Strategy of Central Bosnian Canton 2021).

15.2.2.3 The Upper Drina Mesoregion

The Upper Drina Mesoregion covers the territory in the far east of Bosnia and Herzegovina, with a wide access to the international border. It consists of ten administrative-political units of local self-government, with a dominant concentration of a larger number of urban centers and population. According to the 2013 census, on an area of 4403.5 km^2 (about 8.6% of the territory of Bosnia and Herzegovina), this mesoregion gathered 94,035 inhabitants, which represented about 2.7% of the total population in Bosnia and Herzegovina. With an average population density of 21.35 inhabitants per km^2, which is three times lower than the value for the national territory (68.9 inhabitants / km^2), and makes up only 11.1% of the population in the mountain-valley macro-region (BHAS 2016).

It spatially covers the area east of Romanija, Ozren, and the eastern slopes of Jahorina, Treskavica, and Zelengora. The northern border is the mountain massif of Javor and the valley of the river Žepa. In the far south, Maglić is the highest mountain in Bosnia and Herzegovina. The mesoregion is characterized by a diverse geological structure of the terrain. Geological formations from the oldest Palaeozoic, through Mesozoic to the youngest, Cenozoic deposits are represented in this region. The oldest geological formations are related to the Eastern Bosnian Palaeozoic, whose backbone is located in the Drina and Prača Palaeozoic cores (Lower and Middle Devonian deposits). It belongs to the upper part of the Drina river basin and the tributary system in the part between Foča and Goražde. Mesozoic geological formations formed during the Triassic period are widely represented in the southern and central parts of the Drina basin. They were developed in the part of the Drina and Sutjeska rivers as well as the Govza River. Jurassic deposits have a limited surface distribution in the eastern and southern part of the basin. The morphostructural backbone of the basin consists of a series of high mountains of the southeastern Bosnian highlands, Maglić (2386 m—the highest peak in BiH), and lower massifs such as Crni vrh (1405 m), Devetak (1424 m), Veliki Žep (1537 m), Zvijezda (1675 m), and the spacious Glasinačka plateau.

The hydrographic backbone of the basin is the river Drina. It is formed by the confluence of the rivers Piva and Tara in the zone of Šćepan field. In the downstream segment of the flow, about 5 km from the formations, the first significant left tributary flows into the Drina river —the river Sutjeska. The river Velika Bjelava flows 11 km downstream from the mouth of the Sutjeska. The next left tributary is the river Bistrica. It has a very developed source head formed on the slopes of Treskavica on one side and Jahorina on the other. At about 4.2 km downstream, in the vicinity of the town of Foča, a water-producing right tributary of the Ćehotina flows into the Drina River. In the downstream segment of the flow to the next Višegrad valley extension, a large number of rivers flow into the Drina River (Jošanica, Kolunska River, Osanica, Ocka). Practically the entire part of the natural bed of the Drina River upstream from Višegrad to Ustiprača was flooded by the construction of the Višegrad reservoir (the regime is fully adapted to the operating regime of HPP "Višegrad"). Among the most significant tributaries of the Drina in this segment are Lim and Prača. Lim springs from Lake Plav (at about 900 m above sea level) only in the downstream segment belong to Bosnia and Herzegovina. In the zone of the settlement of Borač, the river Rakitnica flows into it, on which the settlement of Rogatica was formed. It flows into the river Drina in the zone of the settlement of Ustiprača.

The most common soil types belonging to the mosaic are district cambisol on acid silicate rocks and calcocambisol and calcomelanosol-calcocambisol on limestone. In the far north of the mesoregion in the compact area, there is a mosaic calcocambisol-luvisol, and a complex of rendzina-calcocambisol-luvisol on limestone or dolomiteand peridotite and eutric cambisol on bases rich in silicate rocks. Fluvisol occurs along the Drina valley (from Foča to Goražde). Along the narrow valley of the Drina, communities of sub-Mediterranean xerothermic deciduous forests of cer and malt oak dominate. In the basin of Ćehotina and Bistrica with Govza, ecosystems of the sub-belt of mountain beech-fir forests are represented, which continue in the strip area along the lower parts of Treskavica and Jahorina to the mountain Devetak. Larger areas at higher altitudes of mountain massifs of this mesoregion are the dominant communities of barley and spruce from the ecosystem of dark coniferous forests.

The ecosystem of relict and endemic forests of Pančić spruce developed under the influence of the orographic structure of the terrain and other ecological conditions within the hypsometric zone from 800 to 1600 m above sea level, in the area of the upstream part of the catchment area of the river Drina, and the downstream part of the catchment area of the river Lim. The predominant locations where Pančić spruce develops are the south-exposed slopes of the said belt (Drešković and Mirić 2017).

The resource base of this mesoregion has significant amounts of gypsum ore, limestone, travertine, and mineral and thermo-mineral water, brown coal reserves (Miljevina), and large forest reserves. In the 1990s, this mesoregion was a highly developed industrial area with large state-owned companies employing over 1000 workers. With the failure of privatization and the loss of the market, most of these production capacities experienced an economic collapse. Restructuring the economy in conditions of lack of investment has left its mark in all segments of socioeconomic development.

Ten units of local self-government represent the centers of the settlement structure. Except for Foča (FBiH) and Novo Goražde, all other municipalities have a developed urban center with an urban population, where the degree of urbanization varies from 22 (Rudo) to 61% (Foča RS). Approximately half of the total population lived in urban settlements. A characteristic of this mesoregion is the extremely large number of settlements in some municipalities (Višegrad 163, Goražde 144, Rogatica 118). A total of 873 settlements form a settlement network, in which there are only 8 urban ones. The city of Goražde is the mesoregional center with the highest concentration of population and functions (it makes 1/4 of the urban and 22% of the total population of the mesoregion) and is the only local unit that has a population density above the macroregional average. In addition to Goražde (Fig. 15.19), the regional centers are Foča and Višegrad. Compared to the 1991 census, all municipalities in the Upper Drina Mesoregion recorded a population loss in 2013 (the total population deficit is as high as 42.2%), which is a consequence of the war and the low level of socioeconomic development that caused depopulation processes (in 2015, according to the data of the entity statistical institutes, the number of deaths was twice as high as the number of births).

The municipality of ***Rudo*** is located in the eastern part, near the border between Bosnia and Herzegovina and Serbia. It stretches between the mountain peaks of Vučevica, Varda, and Tmor, which intersect the river valleys of Lim and Uvac (400–500 m above sea level). The municipality covers an area of 347.63 km^2, with a total of 7963 inhabitants according to the 2013 census (low degree of urbanization of 22%). There are unfavorable demographic indicators such as negative natural increase (−5.5‰, 2015) and the ratio of the share of young (12.9%) and old (20%) in the total population. In the absence of larger economic entities in the secondary sector, with agriculture that is poorly developed, Rudo has the lowest GDP in the mesoregion (only 1895 BAM in 2013/2014).

Between the mountains Cicelja and Čivči Brda, on the river Janjina (right tributary of the Drina), lies the municipality of ***Čajniče***. It is located on the border of BiH, Serbia, and Montenegro. It covers an area of 274.6 km^2, which, according to the results of the 2013 census, is inhabited by

Fig. 15.19 City of Goražde. (Photo by Aida Bidžan-Gekić)

slightly more than 4895 inhabitants (17.3 inhabitants per km^2). About 45% of the population of the municipality lived in the city center. The basis of development is agriculture and animal husbandry. The municipality does not have significant ore-mineral resources.

Foča (RS) is one of the regional centers of the mesoregion, recognizable by its large area (1134 km^2) in which (census from 2013) lived 18,228 inhabitants. This municipality occupies 1/4 of the territory of the mesoregion but makes up only 19% of the total population (population density is only 17 inhabitants per km^2). Compared to the 1991 census, it has lost more than half of its population (forced migration during the war1992–1995), and due to the collapse of the most important economic complexes (wood industry "Maglić" based on forest resources covering about 70% of the municipality) and the poor economic situation, this trend has continued in modern times. It is located on the main roads M-20 (Gacko-Foča-Ustiprača) and M-18 (Sarajevo-Foča-Nikšić, Montenegro). It represents a transitional zone in terms of climate (average annual temperature is 10 °C—temperate climate).

The municipality has a very dense network of watercourses (17 rivers and streams), and the Drina, Tara, Ćehotina, Sutjeska, and Bistrica are very attractive. They cut into high mountain massifs (Maglić, Volujak, Vučevo, Ljubišnja, and the eastern part of Zelengora). About 70% of the territory of the municipality has a height of over 750 meters. It has a high degree of urbanization (61%). There are unfavorable demographic indicators such as negative natural increase (−5.3‰, 2015) and the ratio of the share of young (12.2%) and old (19.5%) in the total population. The absence of larger economic entities caused a high unemployment rate (around 41% in 2015). The development chance of the municipality is in tourism. The most attractive is the national park Sutjeska (in 2015, there were 6905 registered overnight stays), rafting on the Tara (a large number of camps), sand pyramids in Miljevina (Fig. 15.20), and rich cultural and historical heritage (Aladža mosque, obituary of stećak tombstones in Crkvina).

The city of *Goražde* is a mesoregional center, which is also the administrative center of the Bosnian-Podrinje Canton. It is located in the central upper course of the river Drina (important tributaries are Prača, Kolina, and Osanica). It is characterized by hilly and mountainous relief (slopes of Jahorina). Goražde covers an area of 248.8 km^2, where today live 31,582 inhabitants. The share of urban population was 56.5%, and it is the only local self-government with a population density higher than the macro-regional average (84 inhabitants per km^2). Negative demographic trends are present, but to a much lesser extent compared to mesoregional ones. The share of the old population is higher by 1.5% compared to the young population, and the natural increase is also negative (−3.2‰ in 2015). Although it has a newly opened higher education institution (International University in Goražde), most residents continue their education in Sarajevo (99 km away), which with its attractive functions influences the outflow of young and able-bodied population from Goražde. The concentration of administrative and functional contents and favorable conditions for the development of investments has conditioned Goražde to become the center of various economic activities in the entire mesoregion. Metal processing, chemical, special purpose industry, and plastic processing industry are the most developed processing branches. Goražde is one of the few cities that has the dominance of the secondary sector in terms of employment and income. The largest companies are Ginex (dedicated industry majority owned by the FBiH Government), Emka Bosnia (metal and plastic industry), Pobjeda Technology (ammunition), and Prevent group (car covers, protective equipment, about 1000 employees). The basis of the tourist potential is

Fig. 15.20 Sand pyramids in Miljevina near Foča. (Photo by Ranko Mirić)

the Drina River, on which the city itself lies (rafting, fishing) and forest wealth (hunting grounds). The proximity of Sarajevo as an emitting market is a positive factor in the field of tourism development.

The municipality of *Novo Goražde* covers an area of 123 km^2 with 3117 inhabitants according to the 2013 census. Negative natural increase (−5.5‰, 2015), unfavorable ratio of young (10%) and old (17.5%) in the total population, and average age of 44.3 years (highest in the mesoregion) with permanent emigration (in 2015 alone, 3.3% of the population emigrated) speak in favor of the unfavorable demographic picture. In 2019, only ten students enrolled in the first grade of primary school (Development Strategy of the Municipality of Novo Goražde 2020). The municipality has no urban population. In economic terms, trade is the most important sector alongside public administration. It is one of the underdeveloped municipalities.

Foča (FBiH) is located on the banks of the river Drina (Cvilinsko field), on an area of 224.65 km^2 (only 11 inhabitants per km^2). The administrative seat of the municipality is in Ustikolina (often added to the name of the municipality). This area is characterized by a large number of foggy days (194). There are unfavorable trends in demographic processes (negative natural increase of as much as −11.9‰, 2015). The share of the old population is 10% higher than the share of the young in the total population. About 55.5% of the territory is under forests. Millbakery Company "MLIN" Ustikolina is the only significant processing company.

Rogatica (644 km^2) is located in the northern part of the mesoregion. It is located in the valley of the river Rakitnica (525 m above sea level). The highest mountains are Devetak (1417 m) and Sjemeć (1497 m). About 80% of the territory of the municipality is at an altitude of 800 to 1900 meters. According to the 2013 census, the municipality of Rogatica has 11,603 inhabitants (only 49% of the population compared to the 1991 census). 60% of the municipality's population lives in the city center. Population trends were marked by negative indicators (natural increase in 2015 was—5.7 per thousand). The wood processing industry is the most developed (the company Ikonić exports 90% of the production of doors and windows). The treatment plant (in the 1990s formed the backbone of the municipality's economic development with 750 employees) today employs less than 20 workers.

Sokolac with an area of 693.45 km^2 is after Foča (RS), the largest municipality in this mesoregion. It covers the area of the karst plateau Glasinac (between the mountains of Romanija and Kopita). In the northwest, Bioštica, Kaljina, Kruševica, and Bereg cut their gorges. It is administratively part of the city of Istočno Sarajevo. It is located on the important highway M-19 (Sarajevo-Zvornik). According to the 2013 census, this municipality had 12.21 inhabitants (17.4 inhabitants per km^2). About half of the municipality's population lived in the city center (47.1% of the total). Negative natural increase (−7.9 per thousand in 2015) and unfavorable age structure are the main characteristics of the population of Sokolac. In economic terms, on the basis of forest resources, the wood industry (export-oriented) has developed, which in total, makes up one-third of the economic entities of the municipality. The municipality has great potential in the field of animal husbandry (milk production). Most of the revenue is generated in the area of trade and construction.

Han Pijesak covers an area of 322.9 km^2 (the lowest population density in the mesoregion of 10.9 inhabitants per km^2). According to the 2013 census, the municipality had 3530 inhabitants. The town of Han Pijesak (51% of the total population) is located at 1100 m above sea level. It has a typical mountain climate (average annual air temperature is 7.4 °C). The most important natural resource is forests, which cover 2/3 of the municipality (one of the richest forest areas in BiH). There are unfavorable demographic trends (higher % of older than the young population, and the average age of the population, which is the highest in the mesoregion is 46 years, 2015). The sector of agriculture and forestry, in the total turnover at the level of the economy, generates over half of the municipality's revenues.

In the far northeast along the river Drina, on the slopes of Varda and Zvijezda is the second regional center of this area, the municipality of *Višegrad*. It covers an area of 448.14 km^2 with 10,668 inhabitants (according to the 2013 census), which is less by as much as 52.3% compared to the 1991 census. The M-5 highway is the backbone of the traffic. It has a negative natural increase (−10‰, 2015) and an unfavorable ratio of the share of young (12%) and old (21%) in the total population (average age is 44.6 years). The basis of development is natural resources, on the basis of which the economy has developed. The carrier of development is the company Hidroelektrane na Drini (about 300 employees), which generates the highest revenues of the municipality. It is important to mention the thermo-mineral potentials, which conditioned the development of the spa-rehabilitation center Vilina vlas with accompanying accommodation capacities. Tourist potentials are very promising, thanks to the global recognition of the Mehmed-pasha Sokolović Bridge (on the UNESCO list), the Drina River with lakes, the Dobrun Monastery, etc.

15.2.3 Macro-region of the Bosnian High Karst

Southwest of the mountain-valley macro-region, the macroregion of the Bosnian high karst stretches along the Dinaric direction. Its very name clearly emphasizes the geographical elements that distinguish it, namely, hypsometry and geological material. This region, with its territorial form, reveals its zonal character. Like other macro-regions of Bosnia and Herzegovina, it is part of a wider spatial unit, which refers to the high karst of the Dinaric Mountains, from Slovenia in the northwest to Albania in the southeast. The very pronounced dominance of the widespread Mesozoic-Paleogene carbonate platform with a dominant change of antiforms (mountain ridges and massifs) and synforms (karst fields) give a physiognomic stamp to this macro-region. In Bosnia and Herzegovina, it consists of two mesoregions (Mirić et al. 2017) (Fig. 15.21).

The high karst of Bosnia and Herzegovina, in addition to its very pronounced physiognomic elements of homogeneity, is the only area that has a border with all other macro-regions in the national framework. In the far northwest and north (due to the specific shape of the territory of the City of Bihać), the border with the Peripannonian macro-region extends over the mountains Ujilica, Bobara, Grmeč, and the north eastern edge of Bravsko field. To the south, it crosses the massifs of Srnetica and Bobija and the peaks of Gola Kosa and Ovčara. It is separated from the mountain-valley macro-region by Vitorog, the northern edge of the Kupres field, and the mountain massifs Stožer, Raduša, Bitovnja, Ivan, Bjelašnica, Treskavica, Zelengora, and Volujak. The border with the Mediterranean macro-region stretches along the southern slopes of Zavelim, and over the top of Oštrec, it connects to the northeastern slopes of Čvrsnica, and through the massifs of Prenj, Velež, Snježnica and Trusina, and Sitnica and Viduša, the macro-region extends to the state border with Montenegro.

It is an area of very powerful karst deposits, whose depth is often over 3000 m. The mountain systems have a predominantly Dinaric direction of extension (northwest-southeast), dominated by mountains with altitudes of over 2000 m: Prenj, Čvrsnica (Fig. 15.22) and Cincar. In their internal structure, there are mainly thrust faults and ridges that are subsequently severely fractured and broken by radial processes. Between the mountains or over them, vast plateaus, karst-corrosion surfaces, and fields have developed, which gradually descend towards the Adriatic basin.

A small number of karst fields have been cut into Paleogene flysch deposits (Dabar and Fatnica), while most of them have been developed on limestones and dolomites. Among the most important are Glamočko, Livanjsko, Duvanjsko, Kupreško, Nevesinjsko, Popovo, Gatačko, and other smaller fields. Most of these fields have fissure channels that have a

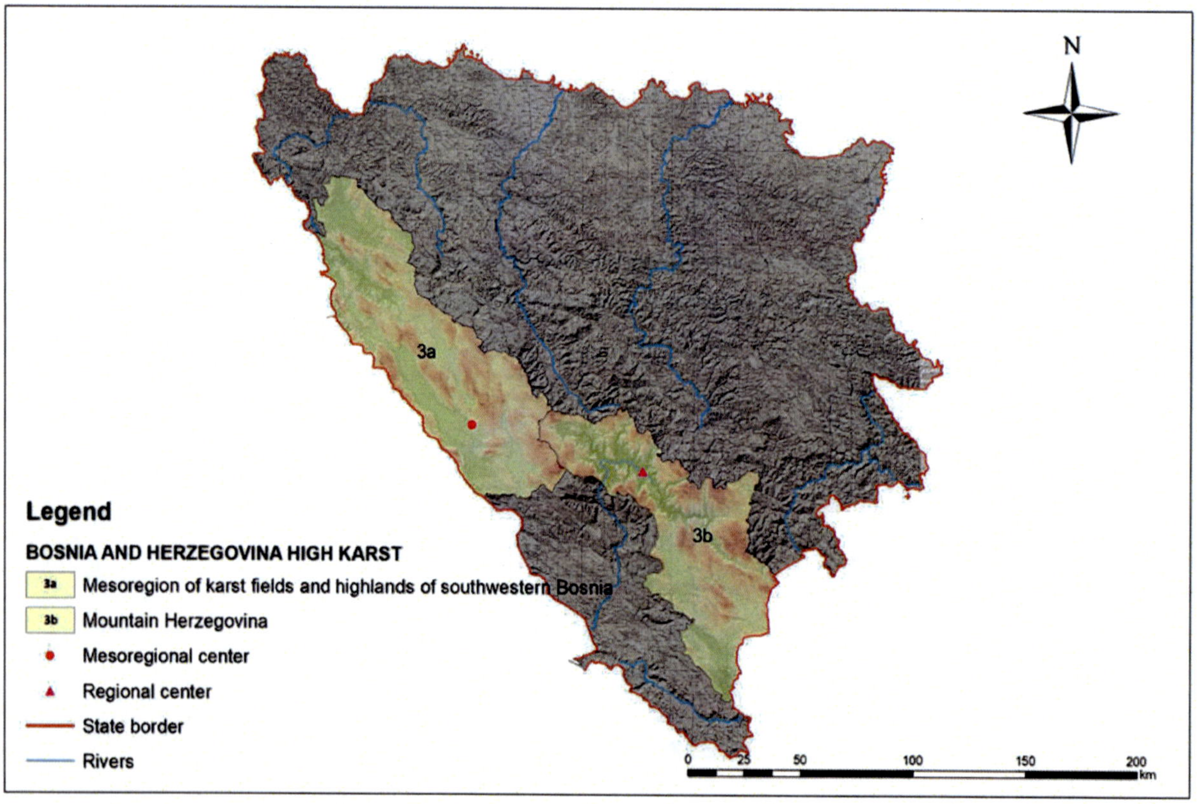

Fig. 15.21 Macro-region of the high karst

Fig. 15.22 The southern slopes of the Čvrsnica mountain. (Photo by Ranko Mirić)

dual hydrographic function: during floods, they become springs, so the bottoms of the fields become flooded and become lakes, while during drought, lake water flows into the interior of the limestone section. The dominant morphosculpture was developed on a limestone-dolomite substrate and is represented by karst-corrosion relief of the Holo-karst type, with all possible forms: karst fields, karst bays, various sinkholes, scrapes, shells, muzzles, bogs, etc. Macro and most mesoforms are tectonic predisposed and then shaped by karst-corrosion, and partly fluvial and gravitational process (polyphase). The Holo-karst is also typical by numerous karst underground forms—pits and caves—which are also characterized by unique morphosculpture (Drešković and Mirić 2017).

As a part of the central mountain area, this macro-region has a transit character, due to the spatial contact with the Mediterranean (in the south) mountain-valley (in the north or northeast) and Peripanonian (in the northwest). A significant number of passes (Oštrelj, Mliništa, Kupreška vrata, Makljen, Ivan, Čemerno) represent connections between the northern and southern areas of Bosnia and Herzegovina).

According to the 2013 census, 176,521 inhabitants in 17 local self-government units lived on an area of 10,794.5 km^2 (approximately 21.1% of the territory of Bosnia and Herzegovina), which ranks this macro-region among the least populated areas in the country. The average population density is far below the national average—only 16.35 inhabitants per km^2. The reason for this is the traditionally pronounced emigration trends, the long-term trend of low development, and the extremely negative socio-economic consequences of the war in the period between 1992 and 1995 (forced migrations, destruction of economic capacities, etc.) The specificity of this area is reflected in the size of the area of local self-government units, of which the six largest (Bosansko Grahovo, 780; Glamoč, 1033.6; Konjic, 1169; Livno, 994; Nevesinje, 877; and Tomislavgrad, 967.4 km^2) cover as much as 54% of the territory of the macro-region (the size of the territory of these units far exceeds the national average of 360.6 km^2).

In the relief of the region, the area of the western and northwestern parts (from the upper course of the Una to the middle course of the Neretva) lies in the higher zone of high karst. It has the appearance of a well-dissected arch with rows of mountains separated by vast fields. Among the many mountains, whose height increases to the southeast, stand out: Grmeč (1605 m), Srnetica (1305 m), Klekovača (1962 m), Jadovnik (1656 m), Šator (1875 m), Vitorog (1906 m), Cincar (2006 m), Golija (1890 m), Raduša (1807 m), Ljubuša (1797 m), and Dinara (1913 m) on the border of Western Bosnia and Dalmatia (Smlatić 1983).

Dominant synformal forms were formed by polygenetic activity of tectonic, fluvial, and karst erosion processes. This primarily refers to the most extensive karst fields, whose genesis is linked to the processes of descent along fault lines, which also represented the border with the orogenic elements of the relief. In further evolution, these synforms were filled with water during the Neogene, and after the Limno-Lacustrian phase, they were remodelled by exogenous processes (fluvial erosion and to a lesser extent corrosion), and from fossil depressions to recent forms of recognizable karst fields. Long-term biomass sedimentation processes in these depressions have led to the characteristic appearance of caustic biolites, primarily coals of different calorific value.

This macro-region consists of two regional units of lower hierarchical rank, whose names indicate a high degree of determination from the aspect of physiognomic and geological-geotectonic characteristics. There are relatively pronounced climatic differences between the northwestern and southeastern areas of this macro-region. In the area from Grmeč to Raduša, the region has a moderate thermal regime with values of annual isotherms from 9.0 to 3.0 °C. The pluviometric regime shows altered Mediterranean features with annual isohyetic values ranging from 1150 mm to 1750 mm in the areas of the surrounding contact peaks of the Dinara Mountain.

To the east, the areas of high mountain massifs cross to the southeast into the area of high karst fields of Eastern Herzegovina and Rudina, and the value of isotherms changes. In the area of the lower parts of the terrain of the upper part of the Neretva river valley, they amount to about 12 ° C (Jablanica 12.2 ° C, Konjic 10.9 ° C, Kalinovik 7.9 ° C) and decrease to about 3 ° C in the zones of the highest peaks. In the southeastern part, the values of annual isotherms are from 10 to 14 ° C and decrease with increasing altitude. In this part of the macro-region, due to the present influences of the Adriatic Sea, the Mediterranean pluviometric regime is pronounced. Higher mountain massifs in the north (Zelengora, Lelija, 1500 mm), Gatačko field (around 1600 mm), are characterized with lower average annual precipitation, while towards the southeast, they increase over 2000 mm. The predominant type of climate in the karst fields of this macro-region is Cf (moderately warm and humid climate). This has conditioned the dominance of three climate types in this area, which alternate from south to north due to altitude and continentality. The area of Livanjsko field and Buško blato covers a compact area of Csb climate. The areas in the upper course of the narrow valley of the Neretva and its tributaries (Trešanica, Rakitnica, etc.) and the river Rama with the lake of the same name have the same climatic characteristics. In the far southeast of the macro-region, this climatic type is present in the Fatničko field all the way to Lake Bileća, where it acquires the characteristics of the Csa climate. With increasing altitude and continentality, subtypes shift from Cfa (hot summers) to Cfb (warm summers) and Cfc (fresh summers).

Areas of humid boreal climate (Df) continue on this climate type, which includes higher hypsometric areas of mountain massifs of this macro-region that frame karst fields (starting from Grmeč and Osječenica in the northwest, all the way to Lelija, Crvanj, and Herzegovinian Bjelašnica in the southeast). The occurrence of isolated, spatially smaller areas of ET climate type (tundra climate) is pronounced in areas with over 2000 m above sea level, in the zones of the highest mountain morphostructures in the macro-region (peaks of Vran, Čvrsnica, Zelengora, and Lelija).

The area of the macro-region belongs to the basins of the Una (tributary Unac in the northwest), Cetina (in the western parts—the river Šuica), Neretva (upper course with tributaries Rakitnica, Rama, Doljanka, etc.), and Trebišnjica (Zalomka). The appearance of natural (mountain) lakes (there are eight of them on Zelengora only) as well as artificial hydro-accumulations with large areas of water area (Buško—57 km^2, Jablanica, Ramsko, Klinje, Bilečko) is also characteristic.

The pedological features are related to the dominant limestone-dolomite substrate. A particularly characteristic soil of this area is calcomelanosol, which occurs in Karst Mountains, most often in a mosaic combination with chalcocambisol and luvisol. Calcocambisol and calcomelanosol-calcocambisol mosaic (brown soils on limestone and dolomite) and calcomelanosol-luvisol mosaic dominate. Within these areas, rendzina and lithosol-rendzina complex on dolomite (in the form of an isolated area in the Neretva valley from Glavatičevo to Konjic), district cambisol on acid silicate rocks (in the area of the northern part of the macro-region from Jablaničko lake to Bitovnja, and in the form of a strip area in the source and upper course of the Neretva) are characteristic. In addition to them, fluvisol is present in a larger area in parts of Livanjsko, Glamočko, and Fatničko fields, and eugley in the northwestern and southeastern part of Livanjsko field (Fig. 15.23) and Glamočko field. The rendzina-calcocambisol mosaic on marl and limestone forms a wider distribution on the Vučevo-Lebršnik line (occurs in a smaller area and in Kupreško field).

The area of this macro-region belongs to the area of permeation of vegetation communities belonging to the Euro-Siberian-Boreo-American and Alpine-High-Nordic biogeographical region, with a large number of climatogenic ecosystems. The sessile and pedunculate oak forests in the larger compact area are connected to the Livanjsko field area. At lower altitudes, there are communities of xerothermic deciduous forests of cer and malt (peripheral Livanjsko and Duvanjsko fields, Ramsko lake, Nevesinjsko, and Gatačko fields). The ecosystems with the highest distribution belong to the communities of beech and beech-fir forests. In the northwestern part of the macro-region, the mountain massifs are under the dark coniferous forests of the fir and spruce community (*Abieti-Piceion Ilyricum*). The ecosystem of the subbands of thermophilic Mediterranean-mounted beech forests (*Seslerio-Fagetum moesiacae*) occurs in an isolated area on Tušnica and Zavelim. Juniper pine (*Pinus mugo*) is

Fig. 15.23 Livanjsko field with the mountain Kamešnica. (Photo by Ranko Mirić)

preserved on the highest mountain peaks of Lunjevača, Ljubuša, Šator, and Crvanj. The entire area of the macro-region is rich in endemic species and communities that are extremely endangered due to the great anthropogenic impact on them.

As already mentioned, this is the least populated macro-region, which is characterized by a low degree of urbanization, population density far below the national average, and pronounced depopulation processes with unfavorable trends in the age structure of the population. According to the 2013 census, the settlement network of this area consisted of 851 settlements. Of these, as many as 691 are rural settlements with less than 200 inhabitants (or 82% of all settlements in this macro-region). In the area of Bosnia and Herzegovina's high karst, 187 settlements with less than 16 inhabitants were registered (which is an approximate population density of the macroregion). A very negative phenomenon is the extinguishing of settlements; according to census data in 2013, as many as 168 were without registered residents (or one in five settlements in the macro-region) (BHAS 2016).

15.2.3.1 The Mesoregion of the Bila, Karst Fields, and Highlands of Southwestern Bosnia

The mesoregion of the bila, karst fields, and highlands of southwestern Bosnia is generally identified with the structural-facial zone in the outer Dinarides, which is marked as the "Higher zone of the High Karst" (Vidović 1978). It consists of ten administrative-political units of local self-government, with the absence of larger urban centers and a pronounced dispersion of settlements. According to the 2013 census, on an area of 5921.1 km^2 (about 11.5% of the territory of Bosnia and Herzegovina), 92,195 inhabitants were registered, which represented about 2.6% of the total population in Bosnia and Herzegovina. With an average population density of 15.6 inhabitants per km^2, which is over four times lower than the value for the national territory, with Livno as a regional center with a low degree of centrality, and population-depleted municipalities in the northwest of this mesoregion, the whole area is stagnant or declining from the aspect of socioeconomic development. The karst fields of this mesoregion are the dominant synformal structures. In the far northwestern part, which is drained by the river Una, there are smaller forms (Petrovac—47.5 km^2, Drvarsko, and Bravsko), to which the fields belonging to the Cetina basin continue. It is also the most extensive field complex in the Dinaric karst, and it consists of four large fields: Livanjsko (the largest field in the Dinaric karst with an area of 405 km^2), Duvanjsko (121.6 km^2), Glamočko (129 km^2), and Kuprеško (94, 8 km^2) (Fig. 15.24).

They are characterized by a large area, closed by higher mountain terrain, and hydrological communication that is possible only by underground flows. These fields are arranged in stages at different heights: at the highest altitude is Kupreško, and at the lowest Livanjsko field (or polje). Each field has its own specifics. Kupreško field is at the highest level (over 1100 m); it is characterized by a dissected relief with three large and many smaller depressions, but also mounds. Glamočko field stands out with its elongation and almost completely flat surface and is divided into upper and lower field, of which only the latter belongs to the Cetina river basin. The main feature of the Duvanjsko field is its

Fig. 15.24 Sinkholes in Kupreško field. (Photo by Ranko Mirić)

hydrological specificity because it is the only field in which a unique hydrological system has been developed. The hydrological network gathers in the course of the Šujica, and then all the waters of the Duvanjsko field flow away through the Kovači pit. Due to its extensiveness and different hydromorphological characteristics, the Livanjsko field is divided into three parts: the Livanjsko field in the narrow sense, Srđevićko field, and Buško blato (Jagušt 2019).

The characteristic morphological form of mountain massifs is relatively wide and slightly wavy ridges with rounded peaks called bila and represent a typical physiognomic seal of antiform relief elements of this mesoregion (the most representative are Grmeč—1604 m, Jadovnik—1656 m, Raduša 1956 m, and Ljubuša 1797 m). Wrinkle-onlay mountain ridges with foothill steps in the shape of a backdrop extend to the typical Dinaric direction NW-SE (Osječenica 1798 m above sea level, Klekovača 1962 m above sea level, Lunjevača 1707 m above sea level, Staretina 1633 m above sea level, Golija 1891 m above sea level, Tušnica 1541 m above sea level, Vitorog 1907 m above sea level). The western border (at the same time the state border with Croatia) consists of the mountain ridges of Ujilica (1654 m above sea level), Dinara (1913 m above sea level) and Kamešnica (1856 m above sea level). In the area of this mesoregion, mountain massifs are a relief frame with large fields in karst. Šator, Staretina, Golija, and Krug separate Livanjsko and Glamočko fields. Lib, Midena, and Ljubuša frame Duvanjsko, while Kupreško field is located between the mountain complexes of Cincar, Vitorog, Plazenica, and Raduša.

The hydrographic feature of this mesoregion is the occurrence of occasional and constant flows, with high water levels in the winter when the surrounding areas of karst fields are flooded. In Livanjsko field, all watercourses outside the winter period (except Bistrica, Sturba, and Žabljak in the central part) dry up. In terms of vegetation, the largest areas are occupied by forest communities of beech and fir with spruce (*Piceo-Abieti-Fagetum*). Larger in the Grahovo field are the forests of pedunculate oak and hornbeam. Pine juniper (*Pinetum mugi*) communities are found at higher altitudes in the Šator and Lunjevača mountain massifs. On Klekovača, mountain maple populations have been preserved. Automorphic soils characteristic of the limestone-dolomite substrate, such as distric and eutric cambisol, calcocambisol, calcomelanosol, luvisol, rendzine, regosol, and hydromorphic soil of fluvisol, and smaller areas of peat soils, predominate.

The most important mineral raw materials are brown coal and lignite (Tušnica and Prolog in Livanjski, Kongora in Duvanjski basin), and peat (Crni Lug in Kupres and Ždralovac in Bosansko Grahovo). Thanks to natural values (forest-preserved complexes with rich fauna, attractive mountain landscapes, speleological objects, rich bio and zoo diversity in the whole area), artificial values (Buško Lake), and rich cultural and historical heritage, there are opportunities for various types of tourism.

From the aspect of population characteristics, negative trends are expressed. Compared to 1991, the census in 2013 showed that in most municipalities, there was a significant decline in population (Bosansko Grahovo by 70.53%, Drvar by 54.14%, and Glamoč with 69.35%, Kupres by 42.77%, Bosanski Petrovac by 52%). Only the municipality of Tomislavgrad recorded an increase of about 5%.

The city of *Livno* is the mesoregional center of this part of the Bosnian karst. The area is 994 km^2 (which makes Livno one of the largest local governments in Bosnia and Herzegovina). The number of inhabitants from the last official census from 2013 was 34,133 inhabitants (about 25% less than in the 1991 census). The population density is the highest in the entire macro-region (34.4 inhabitants per km^2). It is located on extremely karst terrain at an altitude of 730 to 2008 meters. The average annual temperature in this area is 9.82 °C (average annual rainfall around 1200 mm/m^2). It abounds in submerged rivers (Bistrica, Žabljak, and Sturba), which drain into the Cetina basin.

At the Livanjsko field itself, there is a complex hydropower system that collects water and takes it to HPP Orlovac in the Republic of Croatia. Part of this system is the Buško Lake, Lake Mandek, Drinovački Channel (which collects the waters of Bistrica, Sturbe, and Žabljak) and finally Lake Lipa, where all the water is collected and drained through a tunnel to HPP Orlovac (Drešković and Mirić 2017).

In the sociogeographical sense, the area of the City of Livno records extremely negative demographic trends (high mortality rates and negative natural increase), and emigration (2013—59), especially of the young and able-bodied population. All 59 settlements recorded a decrease in population (the city center lost about 3000 inhabitants compared to 1991). The age structure is unfavorable (the old is 2.2% higher than the young in the total population) (BHAS 2016). In the area of the City of Livno, the percentage of unemployed is high and amounts to 47.6%, which means that almost half of the population is not employed and does not participate in the gross domestic product.

Mineral raw materials in the area of the City of Livno are associated with the appearance of lignite (mainly to meet the needs of the local population—Tušnica and Podgledala mines). This area is extremely suitable for the development of agriculture, especially livestock. Livno has created a recognizable brand of dairy products (Livno cheese). There are a large number of monuments that can be used for the development of tourism (Old Upper Town, mosques, churches, monasteries, clock towers). Livno is recognizable by the Livanjsko field as the largest karst field in the world and the largest wetland habitat in BiH (it is included in the Ramsar list of habitats). Mount Cincar and the Krug plateau offer a rare tourist attraction - wild horses. The basis of economic development is the food processing dairy industry (dairies Puđ, Livno, Šestić, etc.). The position of Livno is defined by the great distance from the macro-regional centers in Bosnia and Herzegovina (Mostar-127 km, Sarajevo 190 km, Banja Luka 189 km), which significantly affects the modern socioeconomic phenomena of traditional emigration to neighboring Croatia (The distance to Split as a larger center is only 70 km).

The area of the municipality of *Tomislavgrad* is 967.4 km^2 (fifth in area in Bosnia and Herzegovina). According to the 2013 census, there were 31,592 people living in this area (urbanization only 17.7%, population density 32.7 inhabitants per km^2). It includes the area of Duvanjsko field and Šujička plateau, which are closed by the massifs of Cincar, Ljubuša, and Kamešnica with the accumulation of Buško Lake. The specificity is the large number of windy days a year (about 280—used by the construction of the Mesihovina wind farm), over 600 km^2 of natural pastures, arable land and meadows, and significant water potentials and runways for smaller tourist aircraft. It has mineral deposits (lignite in the Kongora deposit, bauxite). It is export-oriented (power cables, plastic products, pellets, timber, etc.). It is the only municipality in the mesoregion with an increase in the number of inhabitants compared to the 1991 census (by about 5%), and a higher share of young people (16.4%) compared to the old population (14.5%) followed by the average age of 39 years (the youngest population in the mesoregion). Tourist potentials are relatively well used, and the most important are Buško Lake, winter center Risovac, Blidinje Lake and Eco-village Grabovica. In 2017, the municipality also had 11 registered capacities intended for the accommodation of guests (about 400 beds). Traditionally, a large number of Tomislavgrad residents work abroad.

Through the Šujica valley and across the Malovan pass, we reach the municipality of *Kupres*. This unit of local government covers an area of 569.8 km^2 on which, according to the 2013 census, there were 5057 inhabitants (with a population density of low 8.9 inhabitants per km^2).

However, the municipality has other data, i.e., the number is much lower, even for about 2000 inhabitants. The prevailing opinion is that this difference lies in the fact that about 2000 inhabitants are temporarily in the Republic of Croatia or Western European countries due to economic migration (Kupres Municipality Development Strategy 2016).

Kupres is equally distant from larger centers in Bosnia and Herzegovina (about 130 km from Mostar and Banja Luka) and approximately the same to Split in the Republic of Croatia. It includes Kuprečko field (with smaller karst plateaus—Vukovsko field and Ravno). The lower parts are surrounded on all sides by mountains (Stražbenica, Stožer, Lupoglav, Malovan, Vitorog). The most important watercourses are Mrtvica and Milač, and from the lakes Kukavičko and Turjača. The average annual temperature is

5.7 °C with an average of 55 days of snowfall (the Čajuša ski winter center was developed on the basis of this resource). The age structure is unfavorable (higher share of the older population compared to the younger population by 8%), with a pronounced negative natural increase (2015: −5.9 per thousand).

Between the Kupreško and Glamočko fields lies the Hrbine plateau. The municipality of *Glamoč* covers an area of 1033.6 km^2 (fourth in area in the country) where, according to the 2013 census, there were 3860 inhabitants (only 3.7 inhabitants per km^2, of which about 48% is urban). The vast Glamočko field is surrounded by the mountains Cincar, Staretina, Šator, and Vitorog. The average annual temperature in Glamoč is 7.2 °C (fresh summers and cold winters). There is not a single processing plant in Glamoč. The largest number of inhabitants is employed in the primary sector (agriculture and livestock). Every fourth inhabitant of the municipality is over 65 years of age (the share of young people is only 13% in the total population). Emigration and negative natural increase (about −10‰) further complete the negative demographic picture. The most significant cultural event in the history of this municipality was in 1982 when Glamočko kolo was included in the UNESCO World Heritage List.

To the northwest is the Unac river basin and the area of the municipality of *Drvar*, which covers an area of 589.3 km^2. According to the 2013 census report, there were 7506 inhabitants in the municipality of Drvar (only 12.7 inhabitants per km^2, of which 53% were urban). It is located in the valley of the eponymous karst field surrounded by mountain massifs (Osječenica, Klekovača, Vijenac, Šator). The river Unac flows through the municipality (up to 600 m above sea level), while the surrounding mountain massifs reach close to 2000 meters. Annual precipitation is 1108 mm (the height of the snow cover is up to 90 cm). The area of the municipality has the characteristics of Cfc climate which turns into Df (maximum absolute annual temperatures are 38.4 °C, and minimum −27.2 °C). In terms of traffic, it is poorly connected with other parts of the mesoregion (Livno 97 km). The nearest centers are Bihać (80 km) and Banja Luka (125 km). Natural resources include forest resources (over 60% of the total territory of the municipality) and pastures (developed sheep breeding). The natural increase of the population is negative (about 15 per thousand), with three times higher share of the old (29.1%) compared to the young (10.3%) in the total population. Tito's Cave is the most important tourist attraction of the municipality (it is visited annually by about 10,000 tourists from the region).

The municipality of *Bosansko Grahovo* covers an area of 780 km^2. It includes Grahovsko and the southern part of Livanjsko field, which surround the mountains Vjenac and Šator, and Dinara and Ujilica. The average annual rainfall is 1990 mm, and the average temperature is 5.6 °C. According to the 2013 census, the population was 2449 (urbanization only 26.6%, with a low population density of 3.1 inhabitants per km^2). In 2015, an extremely negative natural increase (−15.1‰) was recorded. The age structure of the population is unfavorable (the young population makes up only 7.6%, while the old population makes up one-third of the total). The average age of the population of this municipality is the second highest in Bosnia and Herzegovina (52.1 years). Among the tourist resources are the Ledenica cave and the potential for hunting tourism. It is one of the few municipalities where most employees are in the field of agriculture, forestry, and fishing (about 50%) and in public administration (about a fifth of all employees) and is one of the most underdeveloped local governments.

Bosanski Petrovac is located in the far northwestern part of this mesoregion. In the area of 709 km^2 in 2013, 7328 inhabitants were registered (10.3 inhabitants per km^2). Urbanization is 48% (municipal center). It has an extremely favorable traffic position on the main road M-5 (Bihać-Sarajevo). It occupies the area of Petrovačko field, which is surrounded by the mountains Grmeč, Risovac, and Osječenica. The specificity is the absence of larger surface flows. Negative demographic trends are also present in other municipalities (the number of inhabitants decreased by over 50% compared to 1991). With a negative natural increase (about 13 per thousand), the young population has the lowest share (11.8%), while the share of the elderly (25%) is twice as high. Emigration is pronounced (in 2017, 97 inhabitants). Forests cover 63.61% of the municipality's territory and represent the basis of economic development (the seven largest exporters in the municipality are from the wood industry, but with a small degree of product finalization). Livestock is also developed, while the largest number of employees is in the field of trade. According to economic indicators, Bosanski Petrovac belongs to the extremely underdeveloped local communities. The tourist zone Oštrelj (air spa and ski resort) and Medeno field (historical area) represent recognizable tourist contents of the municipality. In the southwest, there is the municipality of *Petrovac* with an area of 145 km^2 (before the Dayton Agreement, part of the municipality of Bosanski Petrovac). According to the 2013 census, it had only 361 inhabitants (3 inhabitants per km^2). It is a rural area without an urban settlement, with negative demographic characteristics (the average age of the population is 46 years, and the share of the population over 65 is about 30%). In this municipality, there is a rainforest reserve "Lom" with a core size of 55.8 hectares, which is a strict wildlife reserve (category Ia according to IUCN classification).

Istočni Drvar with a total of 79 inhabitants (according to the 2013 census) is the smallest municipality in Bosnia and Herzegovina, specific according to many indicators. Within the state, it has the lowest share of young people (only 3.8% in the total), and the largest share of men (67.1%) in the total population. Almost the entire area of the municipality is covered with forests, mostly conifers. There is no running water or municipal settlement in the municipality.

Kupres (RS) is the smallest municipality in this macro-region (47.8 km^2), which according to the 2013 census numbered only 300 people. Looking at the state level, it has the largest share of the elderly population in the total (as much as 45.3%) and the highest average age (53.7 years) and represents an extremely underdeveloped rural area with very low economic development (BHAS 2016).

15.2.3.2 Mountainous Herzegovina

Mountainous Herzegovina in the southern part of the territory covers the structural-facial zone in the outer Dinarides, which is marked as the "Lower Zone of the High Karst," while in the north and northeast, there is the shift between the areas of Mesozoic limestone and dolomites and the central flysch zone (Vidović 1978). It consists of seven administrative-political units of local self-government, with a pronounced polarization in the distribution of population—2 municipalities in the northwestern part (Konjic and Prozor Rama) gathered almost 50% of the population of this mesoregion. According to the 2013 census, there were 84,326 inhabitants on an area of 4873.44 km^2 (about 9.5% of the state territory), which represented only 2.4% of the total population within the state. The average population density was 17.3 inhabitants per km^2, which is almost four times lower than the value for the national territory (68.9 inhabitants per km^2).

In terms of physiognomy, the mesoregion is an area characterized by the succession of valley-basin synforms represented by the Rama and Jablanica valleys in the north, a narrow valley of the upper Neretva with a source in the east, and vast karst fields that gradually descend to the southeast (Nevesinje, Gatačko, western parts of Dabarsko field, Cerničko and Fatničko) into the smaller Bilečko field (with the artificial lake of the same name) and plateaus (Morine). The mesoregion from the northwest to the southeast is framed by wrinkled ridges and anticlines of the mountain massifs of Ljubuša, Raduša, Ivan, Bjelašnica, Visočica, Treskavica, Lelija, Zelengora, Gatačka Bjelašnica, and Viduša in the far south. The western border of the mesoregion towards the Mediterranean macro-region was defined in the previous considerations. In the area of Mountain Herzegovina, rocks and sediments from the Palaeozoic to the Quaternary are represented. Palaeozoic deposits are represented mainly by clastic, and less by carbonate formations (in some locations they are metamorphosed). They are mainly found on the northern edges (around Ivan Sedlo, Konjic, and in some places Jablanica). Mesozoic deposits are widespread in the central and outer Dinarides, to which the area of the mesoregion (clastic carbonates, volcanic-sedimentary, and flysch deposits) also belongs. Cretaceous deposits occupy about 50% of the total area (limestones and dolomites). The Paleogene alternates with Cretaceous deposits in wider or thinner banded areas. Neogene deposits have been developed in the areas of rivers and lakes (coal deposits). Areas in the far south (Nevesinje, Fatnica field) have the characteristics of Csb climate, and Bilečko field of Csa. In the far north, the narrow valley of the Neretva is penetrated by a maritime influence (climate type Csb). Other areas to the north mostly have the characteristics of Cf climate, which at higher altitudes turns into Df (humid boreal climate). The backbone of the hydrographic network is Nerteva (with tributaries Rama, Doljanka, Rakitnica, Ljuta, Trešanica, etc.)

The dominant pedological substrate is represented by an automorphic division, and it consists of calcomelanosol (most often in a mosaic combination with calcocambisol and luvisol). Smaller areas of rendzina and district cambisol on acid silicate rocks are found in the area of the northern part in the source and upper course of the Neretva, and in the area of Kalinovik. From Vučevo to Lebršnik, there is a wider distribution of rendzina with calcocambisol. Vegetation characteristics are almost identical in terms of the distribution of climatogenic ecosystems. Cer and malt are distributed around Ramsko Lake, Nevesinje and Gatačko fields, and in the wider area south of Fatničko field. On them, with the increase of altitude (towards the north and northwest of the mesoregion), communities of beech and beech-fir forests continue. The ecosystem of mountain ores and juniper pines is connected to the highest zones of the mountain Crvanj.

The mesoregion has significant deposits of caustic biolite (coal), reserves and deposits of construction materials (limestone, ornamental stone), hydropower potential (Neretva, Rama, mini hydropower plants), and extraordinary potential in terms of natural (primarily mountain) but also cultural and historical tourist resources. Like the entire macro-region, it is characterized by a low degree of socioeconomic development, depopulation, and a low degree of urbanization. According to the 2013 census, all municipalities in the mesoregion recorded a decline in population compared to the 1991 census (Nevesinje at least −7.2%, Kalinovik at most—52.2%).

In the far northwest is the municipality of *Prozor* Rama. The city center and Ramsko Lake are located in the valley area, which is surrounded by the Vran, Ljubuša, Raduša, and Makljen massifs. In the area of 477 km^2 in 2013, 14,280 inhabitants were registered (30 inhabitants per km2). Urbanization is only 24% (municipal center). It has a favorable traffic position, considering that it is connected to the road communications in the Vrbas valley by the main road through the Makljen pass. The Rama River (33 km long) has a significant hydropower potential that is well used. Natural increase is negative (−5.3 per thousand, 2015), but

Fig. 15.25 Ramsko lake. (Photo by Ranko Mirić)

the share of the young population is 4.5% higher than the share of the elderly in the total population (the most favorable ratio in the entire macro-region), followed by the lowest average age of the population (38 years). The basis of economic development is the Hydroelectric Power Plant "Rama" (172 MW), and a number of smaller plants of the food industry (cheeses and dairy products). In terms of tourism, Ramsko Lake (Fig. 15.25) is an exceptional destination in terms of natural and cultural-historical resources (the Franciscan monastery on Šćit and Ramsko Lake and the archaeological site of Varvara from the Bronze Age).

In the southern parts of the municipality, there is a hydro-accumulation of Lake Jablanica, which in addition to Prozor (11% of the area) also belongs to the municipalities of Konjic and Jablanica.

The smallest municipality (301 km^2) in this mesoregion, *Jablanica*, is located in the valley of the river Neretva, between the mountains Prenj and Čvrsnica. According to the 2013 census, it had 10,111 inhabitants (population density was 32.08 inhabitants per km^2). The only urban settlement (40.1% of the population of the municipality) is located on several river terraces of the Neretva. It is located at the intersection of the main roads Sarajevo-Mostar and Mostar-Banja Luka, as well as on the railway line Sarajevo-Ploče. The area belongs to the changed Mediterranean climate (Csb), with the exception of high mountains (Df). The most important mineral raw materials belong to the group of architectural and construction stone gabbro (known as Jablanica granite). Abundant water resources (Jablanica Lake, Doljanka, and Sanica rivers) are a great treasure for this region. The negative natural increase in 2013 was −1.8 (2015—5.1) with a negative migration balance (43 more emigrants). In the structure of the population, the number of inhabitants over the age of 65 is growing (16.55% in 2019), which is an increase of 3.48% compared to 2015 (13%). The basis of economic development are Hidroelektrana Jablanica (180 MW) and the companies Granit and Voćepromet. Jablanica has significant tourist resources, and global recognition is a monument from World War II, a railway bridge over the Neretva with the museum "Battle of the Neretva," and the necropolis Dugo polje (150 stećak tombstones).

The regional center of this mesoregion is *Konjic*, the third largest unit of local self-government in Bosnia and Herzegovina (1169 km^2). According to the results of the 2013 census, the total population in the municipality of Konjic was 25,148 (a decrease compared to the 1991 census by 43.2%). In the period between the last two censuses, the municipality of Konjic lost more inhabitants (19,114) than all other municipalities in the mesoregion combined (15,569). The degree of urbanization was 42.7% (population density around 22 inhabitants per km^2). There are as many as 168 settlements, of which as many as 151 have less than 200 persons (according to the 2013 census, 21 settlements without inhabitants were registered). The main route M-17 (Sarajevo-Mostar) passes through Konjic, and the railway route Sarajevo-Ploče in the length of 38.71 km. The alternating Mediterranean climate along the Neretva valley to the north reaches all the way to the village of Šunji (600 m above sea level). There are very large reserves of extremely high-quality dolomite and quartzite, barite, and abrasive stone.

The municipality of Konjic is the richest municipality in Bosnia and Herzegovina in terms of hydro potential: 145 km of river flow and 14 km^2 of lake areas. The water area of the municipality of Konjic consists of the following watercourses and water surfaces: the river Neretva with its tributaries,

Boračko lake—glacial lake and Jablanica lake—artificial lake (Integral development strategy of the municipality of Konjic 2018).

In 2016, the share of young and old population equalized (15% in total). The negative natural increase in 2015 was −3 per thousand with a negative migration balance (134 more emigrants, or a loss of 0.45% of the population). The emigration is a consequence of the Konjic location in the immediate vicinity of the macro-regional centers of Sarajevo and Mostar, about 60 km away. In 2016, tourism employed about 500 people (share in the municipality's GDP of only about 2%). In agriculture, the most important are the cultivation of potatoes and, more recently, goats (milk). About 50% of employees work in industry. Holders of municipal economic development, primarily from the metal and wood sectors, are the company Igman (employs 1149 workers), Graewe-TADIV (about 350 employees), Sur Tec Eurosjaj (260 employees), and the Konjic-karton cardboard packaging factory.

Towards the east and the source of the Neretva, the municipality of *Kalinovik* is located (681.11 km^2), which includes a typical mountain area of high and rocky intertwining mountains Treskavica, Visočica, Crvanj, Lelija, and Zelengora. As much as 83.5% of the municipality has an altitude of over 1000 meters, with only 1.75% of flat terrain. The canyon-shaped valley of the upper Neretva in the length of 40 km in this municipality (with the tributary Bistrica) is the backbone of the hydrographic network, which is completed by mountain lakes (Kotlaničko, Štirinsko, Borilovačko, Crvanjsko, and Kladopoljsko). Gvozno polje is a karst area with the location of stećak tombstones and a small lake of the same name. According to the 2013 census, it had 2029 inhabitants (3 inhabitants per km^2). With the lowest natural increase in BiH (−20.7 per thousand) and a very small share of the young population in the total population (only 9.8%), it has a very unfavorable demographic picture. The basis of economic development is forestry with a low degree of finalization.

The municipality of *Nevesinje* has an area of 923.4 km^2. According to the 2013 census, it had 21,961 inhabitants (15 inhabitants per km^2). It belongs to the hilly and mountainous area (average altitude is 860 m) which is dominated by the Nevesinje field surrounded by the massifs of Velež, Crvnj, Crna Gora, and Snježnica. Zalomka is a sinkhole that has an average surface flow of about 200 days (planned construction of HPP Gornji horizonti). It has unfavorable demographic trends (negative natural increase of −8.1 per thousand, about 1/5 of the population is older than 65). The backbone of economic development is agriculture (potatoes about 50% of sown areas, oats, onions, carrots) and the dairy industry (Perfeto Dairy).

The municipality of *Gacko* covers an area of 736 km^2. According to the 2013 census, Gacko had 8990 inhabitants (12 inhabitants per km^2). Hydrographic features are sinkholes (Mušnica), but a surface hydrographic network has been developed (Neretva springs in Gacko, Izgorka and Jabučnica form Sutjeska). It shares the area of the Sutjeska National Park with the municipalities of Foča and Kalinovik (Perućica rainforest and a historical site from World War II). The most important of the lakes is Klinje (the first artificial lake in BiH was built in 1896 for the needs of Gatačko polje irrigation). Gatačko field is located between the mountain massifs of Zelengora, Maglić, Baba, and Gatačka Bjelašnica. The Čemerno mountain pass is important (the main road from Sarajevo to Trebinje). The lignite coal deposits in Gatačko field are of the greatest importance (concrete and marl reserves are also significant). The Gacko thermal power plant (300 MW) was built on the basis of coal reserves, which is the backbone of economic development and contributes to the municipality having one of the highest GDP in the country (10.547 BAM in 2013/14). The present negative demographic trends are less pronounced compared to other municipalities in the mesoregion.

In the territory of *Bileća* (632.7 km^2) of the southernmost municipality of the macro-region, in 2013, there were 10,807 inhabitants (17 inhabitants per km^2). The mountains Baba, Viduša, and Sitnica frame parts of four karst fields in this municipality (Bilečko, Fatničko, Plansko, and part of Dabarski). The source of Trebišnjica is in Bileća. With a negative natural increase (−8.9 per thousand), the share of the elderly in relation to the young population by 4.4% in the total population has a very unfavorable demographic picture. The most important economic entities are the dairy Pađeni (120 employees) and the textile company Trendtekst. The necropolis of the Bunčići stećak tombstones is a tourist attraction.

15.2.4 The Mediterranean Macro-region of Bosnia and Herzegovina

The Mediterranean macro-region of Bosnia and Herzegovina is a physiognomic unit with a dominant influence of geographical factors of the Adriatic Sea. In the literature, this area is often determined as "Low Herzegovina" (Marković 1980). However, having in mind the larger hierarchical unit within the regionalization of the entire Eastern Adriatic area to which it belongs and its physiognomic identification within the area of this macro-region, and the dominant influence of Mediterranean natural geographic factors, we consider justified the use of names that more precisely determine this area. In 2013, 302,446 people lived on the surface of this smallest macro-region of Bosnia and Herzegovina (5324.7 km^2 or 10.4% of the state territory), which is a share of 8.5% of the total population of the country. With a population density of 56.8 inhabitants per km^2, it is significantly below the national average (68.9 inhabitants/km^2). This macro-region has 14 administrative-political units of local self-government, with a pronounced polarization in the distribution of the population. The city of Mostar (macro-regional center) gathers 35% of the total population, and approximately half of the total urban population of the region (49.4%) (BHAS 2016). The area of this macro-region is a submediterranean-Mediterranean area between the high karst of Bosnia and Herzegovina and central and southern Dalmatia in the Republic of Croatia. In the direction northwest-southeast, it stretches from the Duvanjsko polje to the mountain Orjen, i.e., the border of Montenegro in the southeast. The Mediterranean macro-region includes several smaller spatial units: Bijelo and Mostarsko polje (Bišće), Bekija (southeastern continuation of Imotsko polje), Humine (the area between Ljubuški and Ljubinje), Popovo polje, and parts around the middle course of the Neretva. It extends between 42° 34′ and 43° 40′ north latitude and 17° 05′ and 18° 34′ east longitude (Fig. 15.26).

The northernmost and westernmost points of the Mediterranean macro-region are located on the borders of the municipality of Posušje, while the easternmost and southernmost points are in the municipality of Trebinje. The southernmost point of the macro-region is also the southernmost point of Bosnia and Herzegovina and is located near Podštirovnik on the border between Bosnia and Herzegovina, Croatia, and Montenegro.

The Mediterranean macro-region of Bosnia and Herzegovina borders the Republic of Croatia in the west, while in the south, it reaches the Adriatic coast in the sector Neum-Klek and towards the mountain Orjen to the southeast it again has land border with Croatia. The sea border is 13.4 km long. This is the oldest part of the Bosnian state border. The northern and eastern borders of this region coincide with the border of the Bosnian karst region. This region is separated from the region of Bosnian high karst by high mountain structures: Zavelim, Lib, Čvrsnica, Prenj, Crna Gora, Velež, Sniježnica, and Viduša. The southeastern border is the state border with Montenegro, in the sector from Lake Bileća it extends to the interstate border.

The geological structure of the Mediterranean macro-region is dominated by Mesozoic limestones, Paleogene flysch sediments, and Quaternary deposits at the bottom of karst fields and valleys, while older Palaeozoic formations are not present. Mesozoic sediments have absolute dominance over most of the terrain of the Mediterranean macro-region. They are represented by limestones and dolomites, mostly of Cretaceous age. Slightly older Jurassic sediments

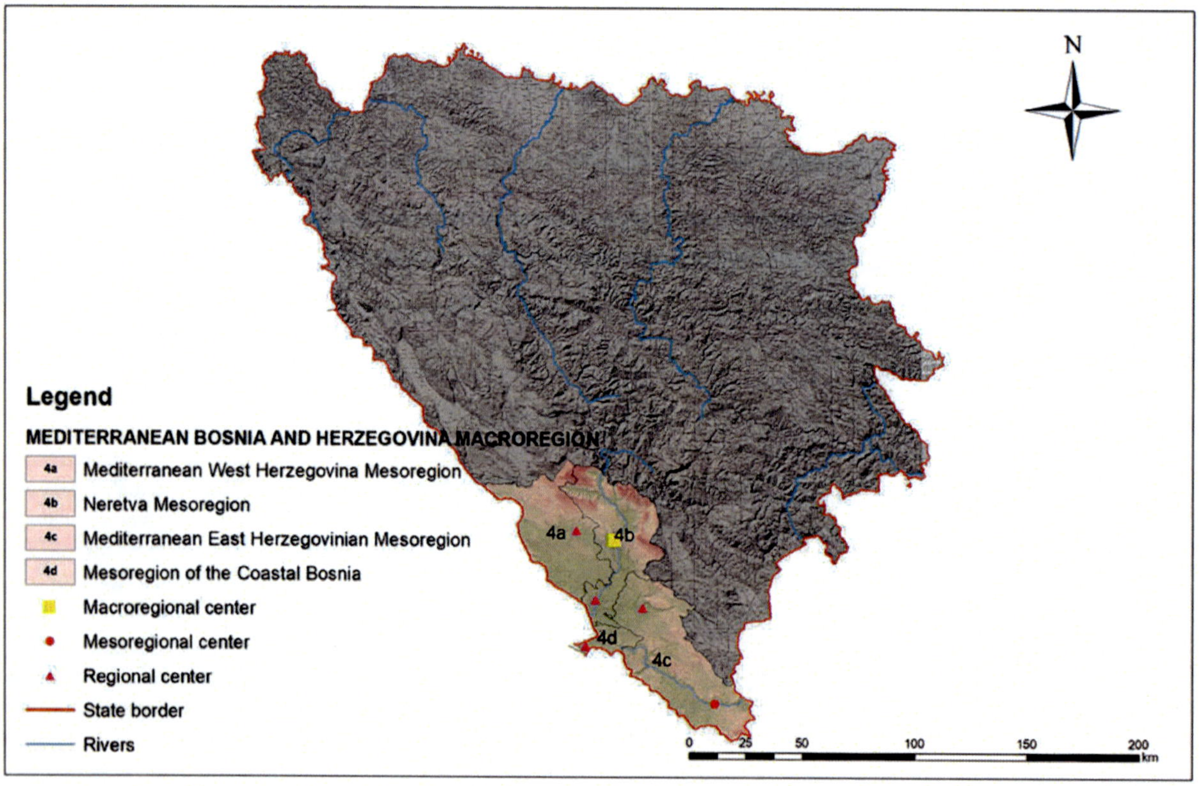

Fig. 15.26 The Mediterranean macro-region

can be found only on higher mountain morphostructures, such as Prenj and Čvrsnica in the north, and Ljubomir and Leotar north of Popovo polje. The oldest Mesozoic limestones and dolomites are from the Triassic period; they are very rare in this area (mostly only in the highest parts of Prenj). Paleogene flysch is represented by marine sediments of the carbonate platform characteristic of these areas. It is widespread in the eastern and western areas of the macro-region where it meets synonymous structures (karst fields and other depressions in the karst). It is also characteristic of the Klek peninsula.

Alluvial valleys are the most common in the Neretva and Trebižat valleys. The bottom of the Mostar valley is covered with 30 m thick layers of fluvio-glacial sediment in which the Neretva has cut its bed and three terraces. The Neretva valley between Bočević and Počitelj has a canyon-like appearance; it is cut in the karst areas of Brotnjo and Dubrava. The youngest Quaternary sediments can also be found in the karst fields of this macro-region. Over 95% of the territory of the Mediterranean macro-region covers the geocomplex of the outer Dinarides. Within this unit, two structural-facial zones stand out: the lower zone of high karst and the higher zone of high karst. These two zones are approximately separated by the mountains Zavela and Čabulja. These mountains form that higher part of this complex, which in the north is separated by a passive continental rim from the central Dinarides. Since it is also a relatively narrow belt, most of the region belongs to the lower zone of high karst, which consists of Humine and Rudine, and karst fields.

The only small part of the area of the Dinaric foothills extends through this macro-region as the inner Adriatic-Ionian zone. The very narrow belt of this zone covers only a few kilometers wide coastal part. The far northern rim of this macro-region belongs to the zone of Mesozoic limestones and dolomites, which is part of the geotectonic complex of the inner or central Dinarides. The Mediterranean macro-region is characterized by a relatively lower relief in most of the territory and mountainous in the transitional parts towards mountainous Herzegovina. The Neretva Valley, which stretches in a north-south direction, divides this area into two parts in the traditional aspect—eastern and western Herzegovina. The Neretva Valley south of Mostar is cut into karst areas, Brotnjo on the right and Dubrava on the left. Great composition is shown by the Neretva valley and north of Mostar, where it breaks between large mountains. In the lower course, numerous floodplains that follow the Neretva itself, and its tributaries Trebižat, Bregava, and Buna, are also characteristic. Among the mountain structures, the following stand out: Bjelašnica (1867 m), Ljubomir and Leotar in the area of Trebinje, Viduša (1328 m), Radimlja (1135 m), Hrgud (1103 m) in the east, and Čabulja (1661 m) in the western part of this region. The highest mountains are located in the far northern part, namely, Prenj (Zelena Glava 2103 m), Velež (1967 m), and Čvrsnica (2228 m). In the south, at the very exit to the Adriatic coast, you will find the mountain Žaba (954 m).

Exogenous processes also have a great influence on the relief of this area. The relief of this macro-region is characterized by many fields and bays, low plateaus and humps, pits, caves, and dry blind valleys, and the fluvial relief forms are dominated by floodplains, river terraces, narrow valleys, and gorges. Among the karst fields of this region, Popovo polje (59.5 km^2), Mostarsko polje (35 km^2), Mostarsko blato (35.7 km^2), and Bijelo polje (13 km^2) stand out. The climate of the Mediterranean macro-region is influenced by several factors, such as latitude, altitude, proximity to the sea, the direction of the mountains, air masses, and winds. The Mediterranean macro-region is located in the area around the meridian of 43° N, which means that it is located in the northern part of the northern subtropical zone. Generally speaking, the altitude increases from south to north, i.e., from the coast to the mountains in the interior, whose highest peaks exceed 2000 meters above sea level. Going from west to east, you can see the lower terrain in the central part, in the valley of the river Neretva, while the east and west are slightly more surface.

The proximity of the sea, specifically the Adriatic, which is part of the larger Mediterranean, is one of the key climatic factors in this region. Marine impacts are felt almost in the entire territory of the region. The direction of the Dinarides is almost parallel to the Adriatic coast and as such prevents the penetration of cold air masses from the north and retains warm masses moving from south to north. The impact of large air masses must be viewed in the context of the wider region. The most significant is the action of warm air masses from the Sahara, then the Genoa cyclone, and the Azores anticyclone. The main winds on the coast are bora and jugo, zmorac, and burin. Bura and jugo are mostly winter winds. The first is cold and dry, and the second is warm and humid. Zmorac and burin blow only in summer when the weather is nice, and zmorac has a cooling effect. Farther from the shore, the influence of zmorac and burin is not felt, but the danik and nocnik can blow. The Mediterranean Cs climate is widespread in almost the entire region, with its hot Csa type dominating for the most part, and the warm or modified Csb type in the smaller northern parts. In addition to the Mediterranean climate, in the hypsometrically highest area of the macro-region, in the north, there is a widespread mountain Df climate.

Csa climate, as said, dominates most of the territory of the Mediterranean macro-region, and this applies to the coast of Bosnia and Herzegovina, the Neretva basin downstream of Mostar, and the area of Popovopolje. The characteristics of this climate are hot and dry summers, and mild and rainy winters. The limits of the Csa climate here almost coincide with the annual isotherm of 12 °C, while the average July temperatures climb up to 25 °C (such as for Mostar), and in January they are not lower than 5 °C. Mostar is the warmest city in Bosnia and Herzegovina, and one of the warmest in Europe with a record measured temperature of 46.2 °C. Most of the area with this climate receives up to 1750 mm of precipitation, with the exception of Popovopolje where this amount exceeds 2000 mm. The driest is the Neretva valley with a little more than 1000 mm of precipitation. Most of this precipitation is excreted in the winter, mostly as rain, rarely in the form of snow.

Csb, or altered Mediterranean climate, occurs in the north of the macro-region, at the transition to the mountainous area, then in the sector of Dabar field, and in the far southeast. It is characterized by slightly lower temperatures, so the average annual temperature drops to 10 °C, and in hypsometrically higher areas even lower. This climate receives more precipitation than Csb one. Thus, the wettest area in Bosnia and Herzegovina is located right here, in the farsoutheast of the region, where Mount Orjen is. It receives more than 3000 mm per year. Other areas receive between 1750 and 2000 mm. Representatives are the towns of Posušje and Široki Brijeg.

Df climate, i.e., mountain type, is represented by a very small part in this region, in the highest parts of the mountains Prenj and Čabulja. It is characterized by low temperatures (average annual to 6 °C), and a large amount of precipitation (over 2000 mm).

The Mediterranean macro-region has the most pronounced homogeneity both in terms of pedology and vegetation in relation to other macro-regions in Bosnia and Herzegovina, although there is also diversification and change of several pedological and vegetation types and subtypes. Since the soils in this area were mainly formed on carbonate rocks, the dominant soils are characteristic of carbonate parent rocks such as lithosol, regosol, calcocambisol, calcomelanosol, rendzina, red soils, etc. With the exception of red soil, other soils are mostly poorly developed and have low fertility, and bare karst is present on a significant area. Thus, calcocambisols and calcomelanosols, which are based on limestone rocks, occur in the largest distribution here. These soils are permeable and have a weakly acidic reaction. They spread over almost the entire territory, with the fact that they alternate somewhere and combine with other types of soil. However, with watercourses and generally in areas where there is a greater influence of water on soil formation, their occurrence is less frequent. In the narrow valley of the Neretva, and on the surrounding mountains Čabulja, Čvrsnica, and Velež, the combination of these soils with very scarce lithosol soils is pronounced. Lithosol is also found in smaller segments in other parts of the region, such as the Popovo polje area. On the other hand, calcocambisols and calcomelanosols in combination with terra rossa (red soil) give much more fertile and usable soils.

The main distribution of terra rossa is in western Herzegovina, and in the area of Popovo polje, and the peripheral sectors of these areas form the transition from mainly chalcocambisol to terra rossa. These soils are 30 to 70 cm deep and are the most fertile in this region, occurring on both limestone and dolomite. They are most often used for growing vines. These are skeletoid, skeletal black, and black-brown humus-carbonate soils on a hard limestone and dolomite base. It has the largest distribution in Prenj, and these soils are also present on the mountain Orjen, Podveležje, and as smaller isolated areas in western Herzegovina. In the Mostar valley and around Blidinje Lake, rendzina is formed on moraine material. Regosol also stands out from the carbonate soils, but its distribution is smaller and refers to the far northwest of the region. There are still some hydromorphic soils in the Mediterranean macro-region. These are soils formed under the influence of stronger soil moisture from precipitation, catchment, flood, and groundwater. In the first place, there is fluvisol. It occurs in river valleys, so it is most pronounced in the lower course of the Neretva, Trebišnjica, Trebižat, and in the karst fields of Imotski and Dabar. In addition to fluvisol from hydromorphic soils, there are also eugley and semigley. They are formed under the influence of water, which is present at a depth of 80 cm, with the difference that in semigleys this influence is somewhat smaller than in eugley, but in both cases they are wetlands. They occur in Hutovo and Mostarsko blato and the eastern part of Dabar field. In Hutovo blato, along with eugley, peat soil or histosol also occurs. On the steep slopes of the mountains in the north, colluvialdeluvial soils also occur, especially in the Drežanka canyon.

In terms of vegetation, the dominance of sub-Mediterranean forests is absolutely emphasized. Only in the mountainous north, the more obvious plant diversification is expressed, according to the hypsometric principle, i.e., with altitude shift. In most of the so-called Low Herzegovina, the dominant forest ecosystem is the Macedonian oak (querciontrojanae). At the transition to the hypsometrically higher north, it alternates with malt oak forests, with the belt of these forests being narrower in the northwest, due to the sudden rise of mountains such as Čabulja, Čvrsnica, Prenj, and Velež, while in the northeast and east, it is much wider, and in the area of eastern Herzegovina, it represents the dominant vegetation, which spreads even in parts of Popovo polje. The coastal sector of this region has a very different ecosystem, because this area is under the direct influence of the Mediterranean. Here are the evergreen forests of Česvina. Along the lower course of the Neretva, there is a strip area of hygrophilous forests and willow and poplar shrubs, which require a larger amount of moisture. Wetland and marsh vegetation is located primarily in Hutovo, but also in Mostarsko blato. The Imotsko polje also has specific vegetation, represented by cer forests. When it comes to the northern, mountainous part of the macro-region, then the general conclusion is that although this part covers a smaller area, it has a much more complex arrangement of vegetation. Thus, on the left side of the Neretva on the mountains Velež and Prenj, the Macedonian oak and malt oak forests are replaced by the forests of Moesian beech and fir, then forests of munika, then mountain pine, while part of Prenj above 2000 meters represents very sparse vegetation. The situation is similar on the right side of the Neretva, with the belts of pre-mountain and mountain Moesian beech being more emphasized here. It should be mentioned that similar vegetation dominates in the far southeast in the area of the mountain Orjen, with thermophilic species of Moesian beech being pronounced at its foot.

It consists of four regional units of lower hierarchical rank, in the differentiation of which the leading attributes are represented by a combination of natural geographical homogeneity and traditional cultural-historical conception of this area. Although poorer in terms of economic resources when compared to some other macro-regions, the Mediterranean macro-region cannot be seen as passive, as it has different natural resources. Despite the fact that most of its terrains are rocky and barren, the lowlands situated along the river valleys and in the fields are, in combination with the Mediterranean climate, favorable for the cultivation of such crops (e.g., tobacco, grapes, early fruits, vegetables, and flowers) which are placed on the domestic and foreign markets. As far as ore deposits are concerned, there are significant deposits of bauxite (not far from Mostar) and nonmetallic barite ores. There are significant and interesting attractions in the Mediterranean macro-region that are favorable for the development of tourism. In addition to the climate, which is a factor of great importance for the summertime and bathing tourism, both on Bosnia and Herzegovina's coast and in the inland bathing areas, there are also natural beauties and extremely rich artifacts and sites of millennial cultural and historical heritage throughout the macroregion that represent a great advantage in the field of tourism.

The total arable agricultural land covers about 1/5 of the territory of the macro-region. (downstream the Neretva River and in the Popovo Polje). Traditionally, there is a well-developed production of tobacco, grapes, early fruits, vegetables, and flowers, but also a developed viticulture that is based on a great range of domestic resources (grapevines called Žilavka and Blatina, both of which are used to produce the best Bosnian wines). In the areas of Mostar, Čitluk, Ljubuški, and Čapljina, there are 3800 ha planted with vines and covered with vineyards. The area around the lower course of the Neretva River is known for the production of early fruits (cherries, peaches, apricots) and then for figs and almonds. The Mostar region is famous for the production and export of cherries and peaches. In terms of fruit growing, three municipalities stand out: Mostar (Fig. 15.27), Čapljina, and Stolac. Aromatic and medicinal plants thrive in this region, with an ever-increasing development recorded in terms of ecologically clean cultivation, with animal husbandry, which has secondary importance. Opportunities for the development of beekeeping are not used to a sufficient extent, but it is rapidly developing, thanks to the private sector.

The energy baseline of this region includes the Neretva River and considerable ore reserves of brown coal available in the Mostar field. The hydropower of the Trebišnjica River

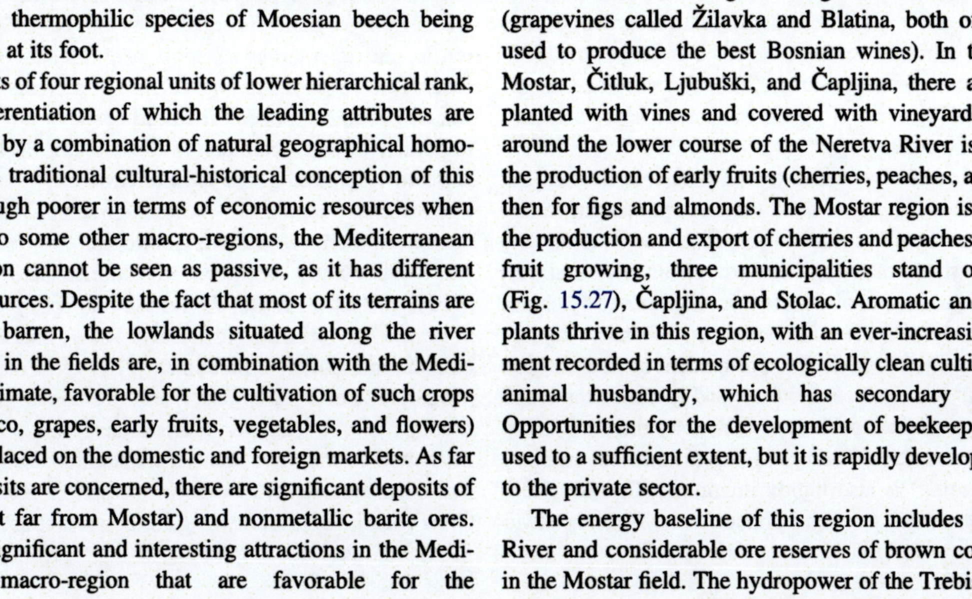

Fig. 15.27 City of Mostar. (Photo by Aida Bidžan-Gekić)

is used by the hydroelectric power plants of Trebinje. A large accumulation lake with a volume of 1.1 billion m^3 was created as part of the hydropower system of the Trebišnjica River, and it is supported with the Grančarevo dam that is 123 m high. The reversible HPP Čapljina was constructed on the waters of the Trebišnjica River, and its power amounts to 420 MW. The middle course of the Neretva River has an enormous hydropower potential. The HPP Mostar (69.5 MW), the HPP Grabovica (133 MW), and the HPP Salakovac (210 MW) are also situated in this area. Recently, more and more is being invested in renewable energy infrastructure (Podveležje wind power plant, in Mostar, solar energy). Bauxite deposits with the main deposits and ores situated around the towns of Posušje, Široki Brijeg, Stolac, Mostar, and Ljubuški are important as a basis for the development of nonferrous metallurgy. There is a site of an important nonmetallic ore of barite situated near Mostar's settlement of Blagaj.

The textile industry in Čapljina, Stolac, and Mostar, but also the tobacco industry (FDM), the metal processing industry in Mostar (Aluminij Mostar), Čapljina, and Trebinje, the wood-processing industry, the production of transport systems, the production of electrical appliances and machines in Mostar have also been developed in this macro-region, while in Posušje, there is the production of construction materials and the chemical industry, and in Grude the wood pulp, celluloses, and paper industry (Violeta) and the printing and graphic industry (Grafotisak), while the food industry (meat processing) is also highly important. The main economic centers have been developed in the Neretva River Valley and on the crossroads of local and regional importance.

In the area of the Mediterranean macro-region, a transversal road has been developed that connects the north and south of Bosnia and Herzegovina. The Mostar-Metković road (M-17) is categorized as an international road, while a number of main roads on the stretch from Ljubuški-Čapljina-Stolac to the town of Trebinje (M-4) are also important. The construction of the future corridor Vc is of great importance for this macro-region, which would run through the Neretva valley, towards the port of Ploče in Croatia. Railway traffic is also well-developed, with the most important Metković-Sarajevo railway, which is 177.5 km long, and it was built in the period between 1885 and 1891. Regarding air traffic, the international airport near Mostar (Ortiješ) has recorded an ever-increasing volume of services. PTT traffic has also been developed. The head office of one of the most important phone operators in Bosnia and Herzegovina—Eronet—is situated in the town of Grude.

This macro-region has excellent conditions for the development of tourism, which is developing ever more successfully. Particularly important for tourism are the natural reserve of Hutovo Blato, the Buna River spring, the Kravica Waterfall on Trebižat, the Neretva River, and a number of cultural and historical monuments: Mostar, Počitelj, Stolac, Mogorjelo near Čapljina, and religious tourism in Međugorje and summer tourism in Neum. The Hutovo Blato Nature Park is a Ramsar habitat and the largest ornithological reserve of its kind in this part of Europe, both in terms of size and diversity (Fig. 15.28). It is situated just 5 km

Fig. 15.28 The Hutovo Blato Nature Park. (Photo by Aida Bidžan-Gekić)

away from Čapljina and 35 km from Mostar. The Buna River Spring with a tekke from the sixteenth century is one of the largest and most beautiful in Europe. Međugorje is also known as one of the largest Catholic sanctuaries in the world. The apparitions of the Virgin Mary began around the 1980s, when pilgrimages from all over the world began. It is estimated that 15 million people have visited this small town to date. Not far from the small human settlement called Ravno, there is the largest grotto and cave system discovered in Bosnia and Herzegovina—which is known by the name of Vjetrenica. The length of the cave corridors amounts to a total of 6 kilometers. Structures worth mentioning in this macro-region are the historical buildings that belong to the Roman basilica, such as, e.g., Mogorjelo near Čapljina, whereas the world-famous symbol of Mostar is the Old Bridge. This unique architectural achievement connects the banks of the Neretva River with one stone arch situated in the very center of the city (which is declared as the protect cultural monument on the UNESCO list of protected cultural assets). Počitelj is a medieval town that was developed during the era of the Turkish Empire, and it is the site of the longest-living art colony in Southeast Europe. A thread of tourist activities (kayaking, rafting) has been developed on the Neretva River, which flows through deep canyons and then reaches with its calm currents the estuary in the Adriatic Sea. The town of Trebinje stands out clearly with its appearance and construction, and it is known for its unique Mediterranean-oriental architecture and great expansion in the development of tourist infrastructure (accommodation facilities in Ivanjica).

15.2.4.1 The Mediterranean West Herzegovina Mesoregion

The Mediterranean West Herzegovina mesoregion occupies the territory in the far west of the macro-region with a wide access to the international border. It consists of five administrative-political units of local self-government, and in addition to the municipality of Čitluk, it is spatially correlated with the West Herzegovina Canton. According to the 2013 census, covering an area of 1543.3 km^2 (about 3% of the territory of Bosnia and Herzegovina), this mesoregion gathered 113,038 inhabitants, which represented 37.4% of the population of the macro-region and only 3.2% of the total population in Bosnia and Herzegovina. The average population density of 73.2 inhabitants per km^2 was slightly higher than the value for the national territory (68.9 inhabitants/km^2). The urban population accounted for only 22% of the total population (Posušje had the largest share of 31.2%). Široki Brijeg is a regional center (at the same time a cantonal center) and gathers the largest number of inhabitants, but the primacy in relation to another city (Ljubuški) is only 1.45.

The Mediterranean Mesoregion of West Herzegovina is situated in the southwestern part of Bosnia and Herzegovina, which borders the hinterland of the central Dalmatian coast (Republic of Croatia). The northern border is surrounded by the mountains Vran and Čvrsnica, the eastern by Čabulja and Mostarsko Blato, the western by the slopes of the mountains Zavelim and Vran all the way up to Blidinje Lake. Along the peripheral mountain ranges, the relief is characterized by the dominance of the Brotnjo Plateau, a number of karst fields (Rakitno, Posuško, Imotsko, Ljubuško, Veljačko, Vitinsko, etc.) and the emergence of caves (Ravlića, Vrbine, Hajdučka, Bubalova, etc.). Due to the large difference in altitude (from 22 m to over 2200 m), the climate is highly diverse, but most of the area is under a strong Mediterranean influence (Fig. 15.29).

A blanket of snow has been recorded, and it covers up to 1 cm; however, its duration is symbolic and it lasts 3–9 days. Thanks to mild winters, vegetation of certain crops (grass, cereals, winter vegetables, etc.) is possible throughout the year. Summer droughts require the crop irrigation (Waste Management Plan of the West Herzegovina Canton 2021).

The largest river Trebižat is an underground river, which changes its name three times with the change of its course (Tihaljina, Mlade, Trebižat). It is a tufa-bearing river contributing to the creation of waterfalls (the largest and most famous is the Kravica Waterfall, 26–28 m high). Its most common forms of flora include the macchias of hornbeam, ash, dogwood, field maple, maple, pedunculate oak and pubescent oak, pomegranate (in the south), and a number of coniferous species such as, e.g., pine, fir, juniper, and the relict Heldreich's pine, whose largest complex in Europe grows in a forest reserve called Masna Luka within the Blidinje Nature Park (in the north) (Galić 2015).

Traditional agriculture (tobacco, fruits and vegetables, vines) is located in karst fields where municipal centers have been developed (Imotsko-Bekijsko, Ljubuško, Posuško Polje fields). Over the past decade, trade has developed there, and Western Herzegovina has become an important transit area. In addition to a considerable number of trading companies, among the most important business entities are companies operating in the field of metal-processing, meat-processing, construction, and mineral industries. There are well-developed industries that manufacture beer, candles, footwear, graphic products, construction products, electrical cables, interior equipment, sanitary ware, hygienic products, and take part in processing of plastics-polymeric materials. The fact is that the current structure of the economy is the result of market development of the economy (Waste Management Plan of the West Herzegovina Canton 2021).

Four local self-governments have recorded an increase in the size of population by comparing the results of the last two official Population Censuses (2013 and 1991): Čitluk by 20.03%, Posušje by 19.51%, Široki Brijeg by 7.32%, and Grude by 5.81%. Ljubuški has recorded a decrease of 0.55%. The town of *Široki Brijeg* is a regional center in the

Fig. 15.29 Ljubuško field and city of Ljubuški. (Photo by Ranko Mirić)

Mediterranean macro-region of Bosnia and Herzegovina. It is situated at the point of transition from the sub-Mediterranean to the continental area at the slopes of Čabulja and Čvrsnica. It is connected to Mostar by the main road M-6.1 (22 km). It occupies an area of 387.60 km^2, which according to the 2013 census had a population of 28,929 (74.64 inhabitants/km^2). Trends in the demographic structure are positive (more young population compared to older population by 3%, recording a positive rate of natural population increase). Široki Brijeg has the highest GDP in the mesoregion and the highest average wage. The economy is focused on a number of flexible smaller companies in the manufacturing, trade, and construction industries. The main factors of economic development are the companies such as FEAL (aluminum profiles), Hering (construction), TT Kabeli (energy cables) MEPAS, Nemeks, and MCI (trade). The Academy of Fine Arts in Široki Brijeg is part of the Mostar University.

The municipality of *Ljubuški* occupies an area of 289 km^2 where, according to the most recent 2013 population census, there were 28,184 inhabitants (96.29 inhabitants/km^2). It is characterized by a low share of urban population (only 14.3%). It is located in the lowlands of the Trebižat Valley, at the crossroads towards Mostar (36 km), Čapljina (18 km), and Croatia (Makarska 55 km). The Kravice Waterfalls are a recognizable part of the image and an important factor in the development of this municipality. The number of visitors, measured by ticket and voucher sales at the entrance to the site, is constantly increasing, and according to the data received from the Public Company Parkovi, in 2019, it amounted to 306,509 visits (Development Strategy for the Town of Ljubuški 2021).

The municipality of *Posušje* occupies an area of 461.1 km^2, and according to the most recent 2013 population census, it had 20,447 inhabitants, recording the lowest population density in the mesoregion (44.4 inhabitants per km^2). It has the most favorable demographic characteristics in this area—a positive rate of natural population increase, a higher share of younger compared to older population in the total number of population (by as much as 7.2%), and the lowest average age (36.4 years). Trade is the most important sector (Vokel, Binvest, Planet), in addition to which there are significant production capacities (Veltplast, Lager and Miviko).

The municipality of *Grude* occupies an area of about 220.8 km^2. According to the 2013 population census, the municipality of Grude had 17,308 inhabitants. It is situated in the area of Imotsko-Bekijsko Polje on the border with Croatia (Imotski—19 km). Grude is the only urban settlement (about 20% of the population of the municipality). There is a domination of the private sector, and the main factor of economic development is the Violeta Company (regional leader in the production of hygiene products, with more than 1000 employees) and Grafotisak (printing company).

According to the 2013 population census, the municipality of *Čitluk* had 18,140 inhabitants. It is the second smallest municipality in the macro-region (181 km^2), but also the second most densely populated local community (100.2 inhabitants/km^2). It is characterized by a low degree of urbanization (18.3%). It is one of the most economically developed municipalities in the country, thanks to Međugorje (the most visited religious pilgrimage site and tourist destination in Bosnia and Herzegovina), and viticulture (the largest vineyards in the country with 21 registered wineries). The largest producer is the Čitluk Winery. According to the generated revenues, Premier World Sport (a chain of sports betting house) stands out in the private sector.

15.2.4.2 The Mediterranean East Herzegovina Mesoregion

The Mediterranean East Herzegovina Mesoregion is located in the far south and southeast of the macro-region and at the same time Bosnia and Herzegovina, with a wide access to the international border with Montenegro and Croatia. It covers the area between Snježnica and Dubrava in the north, Dabarsko polje, Sitnica, Viduša, and Bilećko Lake in the east, while the extreme southern and southeastern parts represent the state border. The relief structures of Popovo polje and Humina, the massifs of Hrgud, Herzegovinian Bjelašnica, Leotar, and a number of smaller karst fields (Stolačko, Ljubinjsko, Strujičko, Mokro, etc.) give a stamp to the physiognomy. The most famous speleological objects are Vjetrenica and Pavlova caves. This unit consists of five administrative-political units of local self-government, of which Trebinje has the status of a city and represents a mesoregional center. It is the largest unit in the Mediterranean macro-region (2040.3 km^2 or about 40%), which is approximately 4% of the territory of Bosnia and Herzegovina. According to the 2013 census, this mesoregion had 52,544 inhabitants (only 17.4% of the population of the macro-region and 3.2% of the total population of Bosnia and Herzegovina). The average population density of 25.8 inhabitants per km^2 was over 2.5 times lower than the value for the national territory (68.9 inhabitants/km^2). About 61.5% of the total population in the mesoregion lived in urban settlements. Trebinje had the largest share of the city population (87.6%), which gathers the largest number of inhabitants, and the urban primacy in relation to the other city (Stolac) is extremely high (6.46).

The main geological characteristic of this area is conditioned by the fact that the entire area is part of a vast Mesozoic-Tertiary complex of the Outer Dinarides. It is dominated by Jurassic-Cretaceous Mesozoic limestones and dolomites, and Quaternary deposits that have a significant presence in the lower parts (Popovo, Stolačko, Ljubinjsko fields). As a consequence of the development and action of geological processes in carbonate substrates, karst relief prevails in the shape of specific relief forms with unique characteristics (Vjetrenica Cave). Fluvioaccumulative forms are associated with the Trebišnjica River Valley. The direction of the Dinaric morphostructures in the south partly prevents the penetration of marine influences towards the northern parts, while in the lower parts (Popovo Polje and the Trebišnjica valley), there is a presence of the Csa climate, which turns into Cfb in the hinterland. In higher hypsometric areas, Cfc climate type (Leotar, Hercegovačka Bjelašnica) occurs in the smaller range areas. At the extreme southeast, there is the Mountain Orjen (1895 m above sea level, with Dfb climate type), which receives the highest amount of precipitation in Bosnia and Herzegovina (3347 mm).

The mesoregional center is the city of *Trebinje* (Fig. 15.30). An exceptional geographical position, a very favorable climate, and abundance of water surfaces have created conditions for the intensive development of the tourist sector (Dubrovnik 30 km, Mostar 115 km). According to the 2013 population census, there were 29,198 inhabitants occupying the territory of 854 km^2, which makes more than one-half of the population of this mesoregion. Over the past few years, Trebinje has recorded a positive migration balance but also a negative rate of natural population increase. It

Fig. 15.30 Trebinje city. (Photo by Aida Bidžan-Gekić)

represents a highly urbanized area (with 81.4% of the urban population share in the total population) with a low population density (only 34 inhabitants per km^2). Very rich and diverse resources have created conditions for the economically very progressive development of a wide range of economic activities (such as energy—hydroelectric power plants on the Trebišnjica River, food industry—Swisslion, agriculture, construction, and tourism). Lagging behind Mostar, Trebinje also records the highest GDP per capita within the macro-regional borders (BAM 8235 in 2013). The most intensive growth is recorded in tourism. The total number of tourist arrivals in 2015 was by 86% higher than in 2011, thanks to foreign tourists, which in 2015 were more than three times as much compared to 2011 (Development Strategy of the Town of Trebinje 2017). This sector has created conditions for the development of other activities, primarily construction (an enormous rate of increase in the housing prices).

The Municipality of *Stolac* is the regional center in this area. It is situated in the karst field of the same name in the valley of the Bregava River. Highly important are the highway M17.3 (Buna-Neum) and the vicinity of Mostar (about 25 km). With 14,502 inhabitants (2013), it is the most densely populated area in the mesoregion (43.81 inhabitants per km^2), which is characterized by a low degree of urbanization (26.3%), and negative trends in the rate of natural population increase. The age structure is very markedly unfavorable. The municipality of Stolac has the largest number of population in the age group of 15–64 years, which amounts to 9078 or 69.89%, and the smallest number of population in the age group of 0–14 years, which amounts to 1352, or 10.40%, which is why the number of population in the 65+ age group amounts to 2558 or 19.69% (Development Strategy of the Municipality of Stolac 2018).

There are strong unfavorable trends in the economy, with a high share of unemployment. Stolac is a town with 30 sites that are on the list of National Monuments of Bosnia and Herzegovina, the most famous of which are the necropoli of *stećak* tombstones at Radimlja and Boljuni. According to the 2013 population census, the three remaining municipalities of this mesoregion combined had less than 10,000 inhabitants (*Berkovići* 2114, *Ljubinje* 3511, and *Ravno* 3219). They are characterized by low population density (Berkovići 8.5, Ljubinje and Ravno 11 inhabitants per km^2). Apart from Ljubinje (79% in the town center), there is no urban population in the other two municipalities.

15.2.4.3 The Neretva Mesoregion

The Neretva Mesoregion, with its central position within the Mediterranean macro-region, represents a meridionally elongated area from Čvrsnica, Diva Grabovica, and Prenj in the north, to Hutovo Blato and the state border with Croatia in the south. The composite valley of the Neretva River and its tributaries give a stamp to the physiognomy. It covers the area of three administrative-political units of local self-government, of which Mostar has the status of a city and represents a macro-regional center. It covers an area of 1516.2 km^2 or about 3% of the territory of Bosnia and Herzegovina. According to the 2013 census, this mesoregion had 132,211 inhabitants (about 43.7% of the population of the macro-region and 3.75% of the total population of Bosnia and Herzegovina). The average population density of 87.2 inhabitants per km^2 was significantly higher than the value for the national territory (68.9 inhabitants/km^2). The urban population of the mesoregion accounted for 54.2% of the total. Mostar, as the primary city, had ten times more urban population than Čapljina, while Istočni Mostar is a municipality with an exclusively rural population.

Physiognomically, this space is highly complex, because it consists of spaces with large differences in altitude (the highest elevation reaching 2102 m and the lowest 2.5 m). The areas with the highest altitudes are situated in the western, northwestern, and eastern parts of the mesoregion (Prenj, Velež, Čvrsnica, and Čabulja mounts). In the far north, they partly belong to the geotectonic structural-facial zone of the higher high karst zone. Mesozoic deposits prevail in the geological structure.

Upper Cretaceous formations prevail in the wider area of the middle and lower Neretva basin. These are dominant Turon-Senonian deposits, which are petrographically represented by limestones, limestones with hornblende, dolomites, and marls (Drešković and Mirić 2017). Paleocene-Eocene deposits run within these areas in the form of elongated areals having a Dinaric direction of stretching (the so-called Liburnian limestones). The lower parts of the area (fields and valleys) are characterized by Quaternary deposits that occur in several isolated areas. The bottom of the Mostar valley is covered with powerful accumulation deposits of fluvio-glacial deposits (30 meters thick) in which the Neretva riverbed is formed with three well-defined terraces. The relief gradually shifts to the lower parts towards the south (Bijelo polje, Mostarska valley, and Bišće polje fields). In the area around the lower course of the Neretva River, we find a well-defined Brotnjo-Dubrave plateau, the best-preserved parts with the height of about 270 m. The Brotnjo-Dubrave plateau is best developed in pure Cretaceous limestones. Towards the mountains around the middle Neretva, the plateau abruptly stops or gradually turns into higher, less pronounced and uncertain levels, e.g., Raška Gora and Podvelež (Roglić 1957).

Towards the south lies the area of Humina, with the fertile alluvial valley of the Neretva delta and the depression of Hutovo Blato, which is the lowest part of the mesoregion. In the tectonic depression of the Čapljina-Hutovo cryptodepression, there is a lithological domination of tertiary clastic deposits represented by predominantly impermeable,

Neogene and Paleogene sandstones, clays, marls, and conglomerates. Hutovo Blato is a cryptodepression caused by the sideway sinking of the Adriatic coast, and it is both tectonically and hydrographically in the closest connection with the Neretva delta (Hrelja 2017).

Karst and fluvio-karst types of relief are characterized by the emergence of numerous ravines, sinkholes, bays, and fields along the Neretva River. Fluvio-denudation relief is present on the slopes of the Prenj and Čvrsnica mountains. The fluvio-accumulation type of relief is represented by alluvial plains and river terraces in the Neretva river valley. These forms are present in the valley of the Neretva River (with a small interruption on the stretch from the Drežanka River estuary to the Neretva River, all the way to Salakovac, and in the area of the middle and lower course of the Buna River and in the area of Mostarsko Blato) (Fig. 15.31). In the extreme south, the climatic conditions are under maritime influences, and they have the characteristics of the Csa climate type, which gradually passes into the Csb climate type along the Neretva Valley (from Bijelo polje). Towards the north and northeast, with the increase in altitude, the climate assumes the characteristics of the D climate type (Prenj, Velež).

In line with the geological substrate, the highest level of presence is associated with automorphic soils from the mosaic of calcomelanosol, lithosol, and cambisol on limestone, and of terra rossa scattered in smaller areals. Hydromorphic soils are associated with narrow river valleys (fluvisols) and wider areas of Mostar and Hutovo Blato (semigley, eugley). In terms of vegetation, due to the relief and the specific features of climate, the mesoregion is an area with a high degree of biodiversity. It is characterized by the emergence of climatogenic vegetation in the south (swamp vegetation, pine, laurel) to the typical maquis and garrigue, to the communities of beech-fir and spruce-beech forests in the north and at the higher altitudes.

The City of *Mostar* represents one of the four macro-regional centers, and it is the center of the Mediterranean macro-region of Bosnia and Herzegovina. It has an extremely favorable position situated on six rivers (Neretva, Radobolja, Buna, Bunica, Jasenica, and Drežanka) and on one of the most important highways in Bosnia and Herzegovina (M-17, Šamac-Doljani), where it intersects with the M-61 (Mostar-Nevesinje) and R-424 (Mostar-Čitluk). The core urban area is situated in the area of three ravines, and it constitutes one-fifth of the territory of the City of Mostar (the degree of urbanization is about 57%), while the remaining four-fifths are made of hilly and mountainous relief (over 60% of altitude above 1000 meters). The Neretva River runs through the center of the city, and it forms the ravine-valley area in which it is situated (altitude 60–80 m). The surrounding areas are surmounted by mountains with a specific climatic and vegetation environment, which varies from the mild mountainous to Mediterranean climates. There are two bus stations and one railway station (Šamac-Ploče) in the City itself. In the southern suburbs of Mostar, called Ortiješ, there is an international airport. Mostar is the administrative and governmental seat of the cantonal (Herzegovina-Neretva

Fig. 15.31 Mostarsko blato. (Photo by Ranko Mirić)

Canton), entity (Federation ministries) and State-level institutions (the Food Safety Agency, the State Intellectual Property Institute, and the Institution of the Ombudsman for Consumer Protection) and cultural (14 institutions; the Archives and the Museum of Herzegovina, the Pavarotti Music Centre, the National Library), and the educational seat of the macro-regioin (with two higher education institutions: the University of Mostar and the "Džemal Bijedić" University). The City territory covers an area of 1175 km^2 (it is the second largest city after Banja Luka, excluding Eastern Sarajevo) and the fifth largest city by the criteria of population size according to the 2013 population census (105,797, or 90 inhabitants per km^2). The general trend of demographic characteristics is associated with the lower negative rates of natural population increase (−1.9 per thousand in 2015) and a very low level of positive immigration (7 inhabitants, 2013), while the share of younger population compared with the older population is lower by only 0.5% in total figures. Thanks to a higher concentration of public offices and economic development, the City of Mostar has one of the largest gross domestic products per capita in Bosnia and Herzegovina, with a value amounting to BAM 10,916 (which is, after Sarajevo, the largest of all macro-regional centers).

Thanks to its pool of natural resources, Mostar has all the prerequisites for greater economic development, especially from the point of view of a developed infrastructure, which includes the proximity of the airport and railway, access to the Croatian port of Ploče, proximity to the Adriatic Highway, and the future corridor Vc. There are great potentials for further development that include minerals and ores (bauxite, gypsum, brown coal, architectural stone), wind energy potentials (wind farm of Podveležje), and sun energy potentials, a wealth of groundwater and surface waters of the Neretva River and its tributaries, and numerous tourist attractions. Mostar is best known for its cherry fruit yields (about 500 tons, making it number one producer in Bosnia and Herzegovina). There are nine fish farms in the Neretva River Basin in the territory of the City of Mostar. The success of tourism in the City of Mostar is mainly based on the tourist offer related to the sites of cultural and historical heritage. The Old Town is an area under UNESCO protection since 2005, and it is the basis for the development of the cultural and historical area of Mostar. A tourist attraction is Blagaj (dervish tekke, at the springs of the Buna River). Rujište Nature Park (on the edges of the Prenj and Velež mountains) is an air spa with the largest amount of oxygen in this part of Europe (Development Strategy of Herzegovina-Neretva Canton 2021).

According to the data provided by the Federation Bureau of Statistics, in 2019, there were 108,915 tourist arrivals and 140,408 overnight stays, with an average length of stay of 1.3 days (Herzegovina-Neretva Canton in figures 2020). The area of the Old Bridge and the Old Town of Mostar are on the UNESCO list of protected sites of cultural heritage. The largest economic entities are the Public Company of Autoceste FBIH, Aluminij (in the phase of restructuring), the Public Company of Elektroprivreda HZ HB, ATACO. There are also some fast-growing, modern companies situated in Mostar, such as, e.g., Sik, NSoft, TD Hercegovina promet, Sunita-Commerc, Amitea, and More Screens.

The municipality of *Čapljina* is situated south of Mostar, in the lower alluvial basin of the Neretva River, only 30 kilometers away from the Adriatic Sea. The position of Čapljina itself is determined by the position of some extremely important and frequent routes that connect the Adriatic coast with the Neretva River valley towards the hinterland. It belongs to the Csa climate type (Mediterranean climate with hot summers and mild winters). The municipality of Čapljina occupies an area of 256 km^2, with a population of 26,157 in 2013 (which is the most densely populated area in the Mediterranean macro-region, with 102 inhabitants per km^2). There are negative trends in the rate of natural population increase and migration (in 2014, 570 inhabitants emigrated, or 2.2% of the total). Important roads (M-17, route of corridor Vc and railway line to Ploče in Croatia) run through the municipality, which serve as a link between the Adriatic Sea and the Central Europe. The Doljani border crossing is one of the busiest in the country. The area of the municipality is characterized by a typical karst relief (the surrounding hills rise above 500 m of altitude above seal level), in combination with the fertile alluvial valley of the Neretva delta (altitude is 2.5 m above sea level). The area of the municipality is intersected by four rivers: Neretva, Bregava, Trebižat, and Krupa, which make a fertile alluvial plain that continues towards the territory of the Republic of Croatia. Its tourist resources include Mogorjelo (one of the most important Roman monuments in the country), a rich archaeological site on the right bank of the Neretva River called Gabela (as a National Cultural Monument of Bosnia and Herzegovina), Počitelj (an example of an oriental city), and the Hutovo Blato Nature Park.

Istočni Mostar is the smallest municipality in this macro-region in terms of its size (85 km^2) and population (257 inhabitants in 2013). Without the urban center and urban population, it is a rural hilly and mountainous area with three settlements.

15.2.4.4 The Mesoregion of the Coastal Bosnia

The mesoregion of the coastal Bosnia physiognomically represents the area of the bay Neum-Klek (Fig. 15.32) with a narrow hinterland to the mountain massif Žaba. Administratively, it covers the territory of the municipality of Neum with an area of 225 km^2 (about 0.4% of the territory of Bosnia and Herzegovina) in which in 2013, there were 4653 inhabitants or 20.68 inhabitants/km^2. The urban

Fig. 15.32 The bay Neum-Klek. (Photo by Ranko Mirić)

population of this mesoregion accounted for 70% of the total. As the only city in Bosnia and Herzegovina on the Adriatic coast, it is a center of coastal tourism, thanks to the typical Mediterranean influences.

The town of Neum is situated on the hilly coast of the Neum-Klek bay. The town itself has developed on the steep slopes of the bay during the 1970s of the last century (with the construction of the Adriatic Highway in 1965, and the establishment of Neum Municipality in 1977). The area of Neum Municipality belongs to the Inner Adriatic-Ionian structural-facial zone in the geotectonic complex of the Dinaric foothills. In the northern, northeastern, northwestern, and central part of the municipality, there is the prevalence of permeable rocks (stratified limestones and dolomites). The extreme eastern part is originating from the age of the Lower and Upper Cretaceous with predominantly limestone rocks. The Klek peninsula belongs to the Upper Cretaceous formations with stratified limestones and dolomites. Geomorphologically, several units stand out in the area of Neum. The waters of the Neum-Klek Bay is with an average depth of up to 27 m, a coastal belt with a rocky shore (with an altitude of about 160 m), and a hilly-mountainous belt with smaller plateaus (maximum height up to 953 m—Mount Žaba). Deeper in the hinterland, there are small fields such as Gradac and Moševići (Quaternary deposits). The marine aquatorium situated in the area of the Neum-Klek Bay is the only saltwater area that belongs to Bosnia and Herzegovina. This area occupies parts of the Neum Bay and half of the Mali Ston Channel, with about 12.2 km^2 representing the territorial waters of Bosnia and Herzegovina. The marine waters of this bay, as part of the Adriatic Sea, play a crucial role in the climatic conditions of the southern part of Bosnia and Herzegovina (Drešković and Mirić 2017).

The average annual temperature in the zone of the coastal belt of Neum is around 15.9 °C. Pronounced Adriatic maritime influences are felt mostly in the coastal zone and the immediate hinterland. The warmest months are July and August, with values averaging around 24.5 °C. On average, the coldest month is January (7.1 °C). Fishing is an activity that the inhabitants of the municipality of Neum have always been engaged in, and it has significant potential for further development and investments. There are two fish farms in Neum Bay used for breeding high-quality fish (mainly sea bass and gilthead sea bream). The most important tourist accommodation capacities include the hotels Zenit, Neum, and Sunce, and over the past decade, the construction of other accommodation capacities has been extremely intensified. In 2019, Neum recorded a visit of 132,760 tourists, who spent 292,773 overnight stays (average length of stay was relatively low 2.2 days). This accounted for 8% of the number of registered tourists in Bosnia and Herzegovina that year, and about 8.7% of overnight stays at the state level. Neum is surrounded by two border crossings and faces some major problems in everyday life. Due to the procedures that are required to be followed in order to enter the European Union, the possibilities for free movement of persons and goods are quite limited, which is especially evident during the tourist seasons. Neum has 7000 beds in high category hotels, as well as resorts and private accommodation, which are accompanied with appropriate infrastructure (the Neum

Hospital, post office, banks, and a huge network of shops). In addition to its hotel capacities, Neum organizes (or is involved in) a network of tourist (excursion and sightseeing) travels by bus and boat to neighboring tourist places (e.g., Dubrovnik, Mostar, and Međugorje are about 70 km away).

References

BHAS (2016) Agency for Statistics of Bosnia and Herzegovina (2016) Census of population, households and dwellings in Bosnia and Herzegovina 2013, Sarajevo

Bibanović E (2015) Prirodno i kulturno naslijeđe Sarajeva [Natural and Cultural Heritage of Sarajevo]. Savart, Sarajevo

Croatian Encyclopaedia (2021). Available at https://www.enciklopedija.hr/natuknica.aspx?id=55333. Accessed on 22 Nov 2020

Development Strategy for the Town of Ljubuški (2021). Available at https://ljubuski.ba/wp-content/uploads/Nacrt_Strategija_razvoja_2021_2027.pdf. Accessed on 18 Aug 2021

Development Strategy of Central Bosnian Canton (2021) Available at https://sbk-ksb.gov.ba/images/Nacrt_strategije.pdf. Accessed on 11 July 2021

Development Strategy of Herzegovina-Neretva Canton (2021). Available at https://www.vlada-hnz-k.ba/sites/default/files/17.11.strateska_platfroma_hnk_2021_2027_nacrt_mm_1.pdf. Accessed on 10 Sept 2021

Development Strategy of the Brčko District of BiH (2021). Available at https://skupstinabd.ba/2-registar/ba/Odluke/2021/2290.%20Strategija%20razvoja%20BD%20BiH___bos.pdf. Accessed on 17 Sept 2021

Development Strategy of the City of Banja Luka (2018). Available at https://www.banjaluka.rs.ba/wp-content/uploads/2018/03/situaciona_analiza.pdf. Accessed on 11 June 2020

Development Strategy of the Municipality of Ilidža (2014). Available at https://www.opcinailidza.ba/uploads/files/shares/Strategija%20razvoja%20/strategija2014-2020.pdf. Accessed on 4 Oct 2020

Development Strategy of the Municipality of Novo Goražde (2020). Available at https://novoGoražde.rs.ba/strategije/?pismo=lat. Accessed on 23 Mar 2021

Development Strategy of the Municipality of Stolac (2018). Available at http://stolac.gov.ba/index.php/opcinska-uprava/opcinski-dokumenti/category/11-strategija-razvoja-opcine-stolac. Accessed on 11 Nov 2020

Development Strategy of the Town of Trebinje (2017) Available at https://www.investintrebinje.com/wp-content/uploads/2019/03/Strategija-razvoja-grada-Trebinja-2018-2027.pdf. Accessed on 17 Mar 2020

Drešković N, Mirić R (2017) Regionalna geografija Bosne i Hercegovine I [Regional geography of Bosnia and Herzegovina I]. Prirodno-matematički fakultet Univerziteta u Sarajevu, Sarajevo

Galić J (2015) Demografski problemi Zapadnohercegovačke Županije i njihov utjecaj na morfološko-funkcionalne promjene naselja [Demographic problems of West Herzegovina County and their impact on morphological and functional changes of settlements]. Prirodno-matematički fakultet, Departman za geografiju, turizam i hotelijerstvo, Univerzitet u Novom Sadu, Novi Sad

Gekić H (2021) Demografska slika općine Bugojno [Demographic profile of the municipality of Bugojno]. Dobra knjiga, Sarajevo

Herzegovina-Neretva Canton in figures (2020) Federal Bureau of Statistics of BiH. Available at https://fzs.ba/wp-content/uploads/2020/06/HNK_2020.pdf. Accessed on 15 Aug 2021

Hrelja E (2017) Modeli održivog upravljanja zaštićenim prirodnim područjima Bosne i Hercegovine [Models of sustainable management of protected natural areas of Bosnia and Herzegovina]. Doctoral thesis, Sveučilište u Zagrebu, Prirodoslovno-matematički fakultet, Zagreb

Ilešič S (1961) Geografska regionalizacija Jugoslavije [Geographic regionalization of Yugoslavia]. Geografski vestnik 33:3–23

Integral development strategy of the municipality of Konjic (2018). Available at https://www.konjic.ba/ba/component/phocadownload/category/31-strategije.html. Accessed on 12 July 2020

Integrated Development Strategy of Lukavac (2019). Available at https://opcina.lukavac.ba/podaci/2019/01/Strategija-integriranog-razvoja-Op%C4%87ine-Lukavac-2019-2027.pdf. Accessed on 17 Nov 2020

Jagušt I (2019) Hidrogeografska obilježja rijeke Cetine [Hydrogeographical features of the river Cetina]. Prvostupnički rad, Prirodoslovno-matematički fakultet, Zagreb

Köppen W (1936) Das geographische System der Klimate [The geographic system of climates]. In: Köppen W, Geiger R (eds) Handbuch der Klimatologie, vol 1. Borntraeger, Berlin. http://koeppen-geiger.vu-wien.ac.at/pdf/Koppen_1936.pdf. Accessed on 20 Dec 2021

Kupres Municipality Development Strategy (2016). Available at http://www.opcinakupres.ba/attachments/article/216/Strategija_razvoja.pdf. Accessed on 7 Sept 2020

Lepirica A (2009) Reljef geomorfoloških makroregija Bosne i Hercegovine [Relief of geomorphological macroregions of Bosnia and Herzegovina]. Zbornik radova Prirodno-matematičkog fakulteta Univerziteta u Tuzli 6(6):7–53

Local Development Strategy of the City of Bijeljina (2018). Available at https://www.gradbijeljina.org/UcitaneSlike/admin/Dokumenti/strategija-lokalnog-razvoja-grada-bijeljina-2014-2023,-revidovana-za-period-2019-2023.pdf. Accessed on 11 Sept 2020

Marković Đ (1980) Regionalna geografija SFR Jugoslavije [Regional Geography of SFR Yugoslavia]. Građevinska knjiga, Beograd

Mirić R (2011) Kompleksnost geografskih implikacija političko-teritorijalne organizacije Bosne i Hercegovine u svjetlu savremenih evropskih integracijskih procesa [Complexity of geographical implications of the political-territorial organization of Bosnia and Herzegovina in the light of modern European integrationprocesses]. phD thesis, Faculty of Science, University of Sarajevo, Sarajevo

Mirić R, Avdić B et al (2017) Geografska regionalizacija Bosne i Hercegovine u kontekstu promjena u njenoj političko-teritorijalnoj strukturi [Geographical regionalization of Bosnia and Herzegovina in the context of changes in its political-territorial structure]. Zbornik radova Četvrtog kongresa geografa Bosne i Hercegovine, Sarajevo

Nurković S, Mirić R (1998) Bosna i Hercegovina – geografske regije (karta) [Bosnia and Herzegovina - geographic regions (map)]. In: Zupčević O (ed): Atlas svijeta – za osnovne i srednje Škole. IP "Sejtarija", Sarajevo

Nurković S, Mirić R (2005) Osvrt na geografsku regionalizaciju Bosne i Hercegovine [Review of the geographical regionalization of Bosnia and Herzegovina]. Geografski radovi, Tuzla

Osmanković J (2004) Regionalizacija i regionalni razvoj Bosne i Hercegovine u poslije- ratnom razdoblju –prilog. Zbornik Ekonomskog fakulteta u Zagrebu 2(1):33–44

Revised Development Strategy of the Municipality of Kakanj (2018). Available at http://kakanj.com.ba/v4/wp-content/uploads/2018/07/Prijedlog-revidirane-strategije-razvoja-op%C4%87ine-Kakanj-2007-2020.pdf. Accessed on 20 June 2021

Rogić V (1990) Enciklopedija Jugoslavije [Encyclopaedia of Yugoslavia], 7. Jugoslavenski leksikografski zavod, Zagreb

Roglić J (1957) Zaravni u kršu [Plateaus in the karst]. Geografski glasnik 19:103–134

Smlatić S (1983) Geografske regije Bosne i Hercegovine [Geographical regions of Bosnia and Herzegovina]. In: Cecić I (ed) Enciklopedija Jugoslavije Separat SR Bosna i Hercegovina. Jugoslovenski leksikografski zavod, Zagreb, pp 61–76

Spatial plan of the city of Zenica (2016). Available at http://www.urbzdk.ba/upload/pdf/PPZE-Nacrt-Prostorni_razvoj_okoline.pdf. Accessed on 11 Apr 2021

Strategic development plan of the city of Istočno Sarajevo (2010). Available at https://www.opstinasokolac.net/dokumenti/strategije/strateskiplan-istocnosarajevo.pdf. Accessed on 12 Nov 2020

Strategy for Integrated Development of the City of Zvornik (2018). Available at https://gradzvornik.org/wp-content/uploads/2020/08/Strategija-integrisanog-razvoja-grada-Zvornik-za-period-2018-2027.-godine.pdf. Accessed on 12 Nov 2020

Tuzla Canton Development Strategy (2021). Available at http://www.vladatk.kim.ba/Ministarstva/MP/2021/SR/STRATEGIJA_RAZVOJA_TUZLANSKOG_KANTONA_ZA_PERIOD_2021.-2027.pdf. Accessed on 12 June 2021

Tuzla City Development Strategy (2018). Available at https://grad.tuzla.ba/wp-content/uploads/2019/03/Revidirana-Strategija-razvoja-Tuzle.pdf. Accessed on 8 Aug 2020

Una-Sana Canton Environmental Protection Plan (2014). Available at https://vladausk.ba/v4/files/media/pdf/59c4b9bcf3c679.65823756_Plan%20zastite%20okolisa%20USK.pdf. Accessed on 10 July 2020

Vidović M (1978) Geotektonsko poznavanje terena Bosne i Hercegovine [Geotectonic knowledge of the terrain of Bosnia and Herzegovina]. IX Kongres geologa Jugoslavije, Sarajevo

Waste Management Plan of the West Herzegovina Canton (2021). Available at http://www.skupstina-zzh.ba/Documents/2021/2702/Prijedlog%20Odluke%20o%20dono%C5%A1enju%20plana%20upravljanja%20otpadom%20ZZH%202021-2031.pdf. Accessed on 17 Sept 2021

Index

H. Gekić et al., *The Geography of Bosnia and Herzegovina*, World Regional Geography Book Series,
https://doi.org/10.1007/978-3-030-98523-3

D

E